Lecture Notes in Computer Science 4773

Commenced Publication in 1973
Founding and Former Series Editors:
Gerhard Goos, Juris Hartmanis, and Jan van Leeuwen

Shingo Ata   Choong Seon Hong (Eds.)

Managing Next Generation Networks and Services

10th Asia-Pacific Network Operations
and Management Symposium, APNOMS 2007
Sapporo, Japan, October 10-12, 2007
Proceedings

Volume Editors

Shingo Ata
Osaka City University
Graduate School of Engineering
3–3–138, Sugimoto, Sumiyoshi-ku, Osaka 558–8585, Japan
E-mail: ata@info.eng.osaka-cu.ac.jp

Choong Seon Hong
Kyung Hee University
School of Electronics and Information
1 Seocheon, Giheung, Yongin, Gyeonggi 449-701, South Korea
E-mail: cshong@khu.ac.kr

Library of Congress Control Number: 2007936094

CR Subject Classification (1998): C.2, D.2, D.4.4, K.6, H.3-4

LNCS Sublibrary: SL 5 – Computer Communication Networks
and Telecommunications

ISSN 0302-9743
ISBN-10 3-540-75475-X Springer Berlin Heidelberg New York
ISBN-13 978-3-540-75475-6 Springer Berlin Heidelberg New York

Springer is a part of Springer Science+Business Media

springer.com

Printed in Germany

Typesetting: Camera-ready by author, data conversion by Scientific Publishing Services, Chennai, India
Printed on acid-free paper SPIN: 12169795 06/3180 5 4 3 2 1 0

Preface

We are delighted to present the proceedings of the 10th Asia-Pacific Network Operations and Management Symposium (APNOMS 2007), which was held in Sapporo, Japan, on October 10–12, 2007.

The Organizing Committee (OC) selected the theme of this year's symposium to be *"Managing Next Generation Networks and Services."* Recently, various convergences in wired and wireless networks, as well as the convergence of telecommunications and broadcastings, have taken place for ubiquitous multimedia service provisioning. For example, broadband IP/MPLS wired networks have been actively converged with IEEE 802.11e wireless LAN, IEEE 802.16 Wireless MAN, 3G/4G wireless cellular networks, and direct multimedia broadcast (DMB) networks. For efficient support of service provisioning for ubiquitous multimedia services on broadband convergence networks, well-designed and implemented network operations and management functions with QoS-guaranteed traffic engineering are essential.

This year, the APNOMS call for papers resulted in 161 paper submissions from 24 different countries, which included countries outside the Asia-Pacific region (from Europe, the Middle East, and North and South America). Each paper was carefully reviewed by at least three international experts. Based on review scores and the discussions that followed, APNOMS 2007 Technical Program Committee reached the consensus that 48 high-quality papers (29.8% of the submissions) should be selected as full papers and 30 as short papers. Accepted papers were arranged into 10 technical sessions and 2 short paper sessions (poster presentations). These sessions focused on the management of distributed networks, network configuration and planning, network monitoring, and especially the area of management of wireless networks and security.

The Technical Program Committee (TPC) Co-chairs would like to thank all the authors who contributed to the outstanding APNOMS 2007 technical program. We also thank the TPC, OC members and reviewers for their support throughout the paper review and program organization process. Also, we are grateful to IEICE TM, Japan, and KICS KNOM, Korea, for their sponsorship, as well as IEEE CNOM, IEEE APB, TMF, and IFIP WG 6.6 for their support of APNOMS 2007.

October 2007

Shingo Ata
Choong Seon Hong

Preface

Organizing Committee

General Chair

Hiroshi Kuriyama, NEC, Japan

Vice Co-chairs

Kyung-Hyu Lee, ETRI, Korea
G.S. Kuo, National Chengchi University, Taiwan

TPC Co-chairs

Shingo Ata, Osaka City University, Japan
Choong Seon Hong, Kyung Hee University, Korea

Tutorial Co-chairs

Hajime Nakamura, KDDI R&D Laboratories, Japan
Kwang-Hui Lee, Changwon University, Korea

Special Session Co-chairs

Kazumitsu Maki, Fujitsu, Japan
Taesang Choi, ETRI, Korea
Yan Ma BUPT, China

DEP Chair

Nobuo Fujii, NTT-AT, Japan

Exhibition Co-chairs

Seiichi Morikawa, Cisco, Japan
Dongsik Yun, KT, Korea

Poster Co-chairs

Naoto Miyauchi, Mitsubishi Electric, Japan
Young-Seok Lee, CNU, Korea

Publicity Co-chairs

Hiroshi Uno, NTT, Japan
Young-Myoung Kim, KT, Korea
Gilhaeng Lee, ETRI, Korea
Qinzheng Kong, HP APJ, Australia

Finance Co-chairs

Toshio Tonouchi, NEC, Japan
Hong-Taek Ju, Keimyung University, Korea

Publication Chair

Jun Kitawaki, Hitachi, Japan

Local Arrangements Co-chairs

Kouhei Iseda, Fujitsu Laboratories, Japan
Mitsutomo Imazaki, NTT Comware, Japan
Yoshiaki Yamabayashi, CIST, Japan

Secretaries

Hikaru Seshake, NTT, Japan
Young-Woo Lee, KT, Korea

Advisory Board

Graham Chen, EPAC Tech., Australia
Makoto Yoshida, University of Tokyo, Japan
Masayoshi Ejiri, Japan
Doug Zuckerman, Doug Zuckerman Associates, USA
Seong-Beom Kim, KT, Korea

Steering Committee

Nobuo Fujii, NTT, Japan
Hiroshi Kuriyama, NEC, Japan
James Won-ki Hong, POSTECH, Korea
Kyung-Hyu Lee, ETRI, Korea
Young-Tak Kim, Yeungnam University, Korea
Yoshiaki Tanaka, Waseda University, Japan

International Liaisons

Ed Pinnes, Elanti Systems, USA
Raouf Boutaba, University of Waterloo, Canada
Carlos Westphall, SCFU, Brazil
Marcus Brunner, NEC Europe, Germany
Rajan Shankaran, Macquarie University, Australia
Alpna J. Doshi, Satyam Computer Services, India
Teerapat Sanguankotchakorn, AIT, Thailand
Borhanuddin Hohd Ali, University Putra, Malaysia
Victor W.J. Chiu, Chunghwa Telecom, Taiwan
Luoming Meng, BUPT, China

Technical Program Committee Members

Aiko Pras, University of Twente, Netherlands
Antonio Liotta, University of Essex, UK
Carlos Becker Westphall, UFSC, Brazil
Chi-Shih Chao, Feng Chia University, Taiwan
Eiji Takahashi, NEC, Japan
G.S. Kuo, NCCU, Taiwan
Gabriel Jakobson, Altusys, USA
Graham Chen, EPAC Technologies, Australia
Haci Ali Mantar, Gebze Institute of Technology, Turkey
Iwona Pozniak-Koszalka, Wroclaw University of Technology, Poland
Jae-Hyoung Yoo, KT, Korea
Jianqiu Zeng, BUPT, China
Jose-Marcos Nogueira, UFMG, Brazil
Joseph Betser, Aerospace, USA
Kenichi Fukuda, Fujitsu, Japan
Kwang-Hui Lee, Changwon National University, Korea
Lin Zhang, BUPT, China
Lisandro Zambenedetti Granville, UFRGS, Brazil
Marcus Brunner, NEC Europe, Germany
Mehmet Ulema, Manhattan College, USA
Nazim Agoulmine, University of Evry, France
Prosper Chemouil, France Telecom, France
Qinzheng Kong, HP APJ, Australia
Radu State, LORIA - INRIA Lorraine, France
Rocky K.C. Chang, Hong Kong Polytechnic University, Hong Kong
Seongjin Ahn, Sungkyunkwan University, Korea
Shuang-Mei Wang, Chunghwa Telecom, Taiwan
Tadafumi Oke, NTT Comware, Japan
Taesang Choi, ETRI, Korea
Teerapat Sanguankotchakorn, AIT, Thailand

Yan Ma, BUPT, China
Yoshihiro Nakamura, Nihon University, Japan
Young Choi, James Madison University, USA
Yuka Kato, Advanced Institute of Industrial Technology, Japan

Additional Paper Reviewers

Adetola Oredope, University of Essex, UK
Alexandre Menezes, UFSC, Brazil
Aujor Andrade, UFSC, Brazil
Carla Merkle Westphall, UFSC, Brazil
Chiara Mingardi, NEC Europe, Germany
Clarissa Marquezan, UFRGS, Brazil
Cristiano Both, UNISC, Brazil
Cristina Melchiors, UFRGS, Brazil
Daniel W. Hong, KT, Korea
Denis Collange, France Telecom, France
Deok-Jae Choi, Chonnam University, Korea
Dong Hoon Lee, Korea University, Korea
Dong-Sik Yun, KT, Korea
Fabrice Clerot, France Telecom, France
Fernando Koch, UFSC, Brazil
Georgios Karagiannis, University of Twente, Netherlands
Gil-Haeng Lee, ETRI, Korea
Hajime Nakamura, KDDI R & D Labs. Inc., Japan
Hassnaa Moustafa, France Telecom, France
Hideo Imanaka, NTT, Japan
Hikaru Seshake, NTT, Japan
Hiroomi Isozaki, Osaka City University, Japan
Hiroshi Uno, NTT, Japan
Hisoshi Kuriyama, NEC, Japan
Hong-Taek Ju, Keimyung University, Korea
Hoon Lee, Changwon National University, Korea
Jae-Oh Lee, University of Technology and Education, Korea
James Hong, POSTECH, Korea
Jitae Shin, Sungkyunkwan University, Korea
Jong-Tae Park, Kyungpook National University, Korea
Kamel Haddadou, LIP6, France
Katsushi Iwashita, Kochi University of Technology, Japan
Kazuhide Takahashi, NTT DoCoMo, Japan
Kazumitsu Maki, Fujitsu, Japan
Ken Hashimoto, Osaka City University, Japan
Ki-Hyung Kim, Ajou University, Korea
Kohei Iseda, Fujitsu Laboratories, Japan
Kyung-Hyu Lee, ETRI, Korea
Ling Lin, University of Essex, UK
Luciana Fujii Pontello, UFMG, Brazil
Luiz Henrique Correia, UFLA, Brazil
Makoto Takano, NTT West, Japan
Marat Zhanikeev, Waseda University, Japan
Mi-Jung Choi, POSTECH, Korea
Myung Kim, Korea University, Korea
Naoto Miyauchi, Mitsubishi Electric, Japan
Nobuo Fujii, NTT-AT, Japan
Paulo Silva, UFSC, Brazil
Quoc Thinh Nguyen Vuong, University of Evry, France
Ramin Sadre, University of Twente, Netherlands
Remco van de Meent, University of Twente, Netherlands

[illegible]

Yi Zhu, University of [illegible]

[illegible]

Korea

[illegible]

Table of Contents

Session 1: Management of Distributed Networks

Session 2: Network Configuration and Planning

Session 3: Network Security Management I

Session 4: Sensor and Ad-hoc Networks

Session 5: Network Monitoring I

Session 6: Routing and Traffic Engineering

Session 7: Management of Wireless Networks

Session 8: Network Security Management II

Session 9: Network Monitoring II

Session 10: Security on Wireless Networks

Sessions S1, S2: Short Papers

Design of a Digital Home Service Delivery and Management System for OSGi Framework*

Taein Hwang[1,2], Hojin Park[2], and Jin-Wook Chung[1]

[1] School of Information and Communication Engineering, Sungkyunkwan University, 300 Chunchun-dong, Jangan-gu, Suwon-si, Gyeonggi-do, 440-746, Korea
tihwang73@hotmail.com, jwchung@songgang.skku.ac.kr

[2] Digital Home Division, Electronics and Telecommunications Research Institute, 161 Gajeong-dong, Yuseong-gu, Daejeon, 305-700, Korea
hjpark@etri.re.kr

Abstract. Digital home services have been provided separately by each service provider who has a closed service delivery infrastructure for delivering home services. Because of this, 3rd party service providers who do not have the infrastructure to deploy home services have difficulties in participating in the home network market. Also, it is difficult for 3rd party service providers to provide a control-device-specific graphic user interface to service users due to the limitation of home gateway resources. In order to solve these problems, we propose a digital home service delivery and management system for small business companies that need to provide various home services to users. Also, this system supports functions for providing a user friendly graphic user interfaces dedicated to mobile device. By using the proposed system, the service aggregator can lead the competition of 3rd party service providers in the home network service market, and help it grow rapidly.

1 Introduction

Traditional service providers such as the telephone and the cable TV companies all have dedicated wires into the home. However, this configuration will be unmanageably complex in the networked homes of the future. Furthermore, diversified portfolio of services such as home security, health monitoring, telephony, and audio/video media may employ different communication technology protocol, which will complicate the configuration. To mitigate this potentially chaotic situation, developers have proposed the home gateway; a centralized device that interfaces between the external Internets and internal home device and appliance networks. Some common hardware components of a home gateway include a processor, persistent storage, networking, and device interfaces, which are typically powered by an operating system or real-time operating system [1,2]. Home devices can be connected to the Internet via the home gateway that supports many network interfaces such as Power Line Communication (PLC), Ultra Wide Band (UWB), Wireless Local Area Network (WLAN), Fast

* This work was supported by the IT R&D program of MIC/IITA[2006-S068-01, Development of Virtual Home Platform based on Peer-to-Peer Networking].

S. Ata and C.S. Hong (Eds.): APNOMS 2007, LNCS 4773, pp. 1–10, 2007.

Ethernet, and Bluetooth. Through these interfaces, the outdoor home users can monitor and control indoor home devices.

In the well established home infrastructure, a service provider can provide a health monitoring service that can send health data measured by an indoor healthcare device to a remote doctor, and deploy crime prevention and disaster prevention services by using a home viewer device. To offer these services to service users, the service providers need an outdoor infrastructure such as a security system, accounting system, and portal system as well as an indoor home infrastructure to provide services to the users. But, it is difficult for 3rd party service providers, who do not have a service delivery infrastructure, to deploy home services because of the expensive overhead cost of building the infrastructure. Thus, 3rd party service providers need an open Digital home Service delivery and Management (DSM) System to provide the basic functions for service provisioning.

Fig. 1 illustrates the traditional service delivery architecture. This architecture is closed because the digital home services are provided separately by each service provider who has a closed service delivery infrastructure for delivering home services. Thus, the 3rd party service providers have difficulties in participating in the home services market. Furthermore, it is difficult for the service users to find necessary services among many available ones.

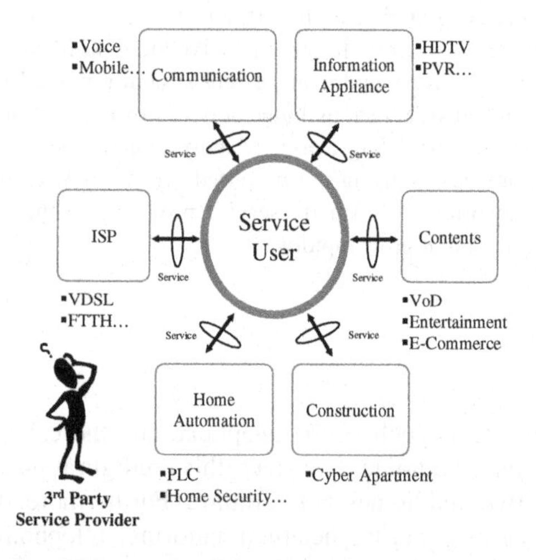

Fig. 1. Conventional Service Delivery Architecture

To solve this problem, we propose the DSM system illustrated in Fig. 2. This system provides the functionalities of a service gateway to deliver various services to a service user through a single path so that the 3rd party service providers can participate in the market. Furthermore, service users can easily search necessary home services among many available services. With this configuration the device manufacturers do not need to develop a service dependent device. Service providers can easily provide convenient service to service users. As a result, the device manufacturers and the 3rd party service providers can easily participate in the home service market through the DSM system that performs function of service delivery between a home gateway and a service provider server.

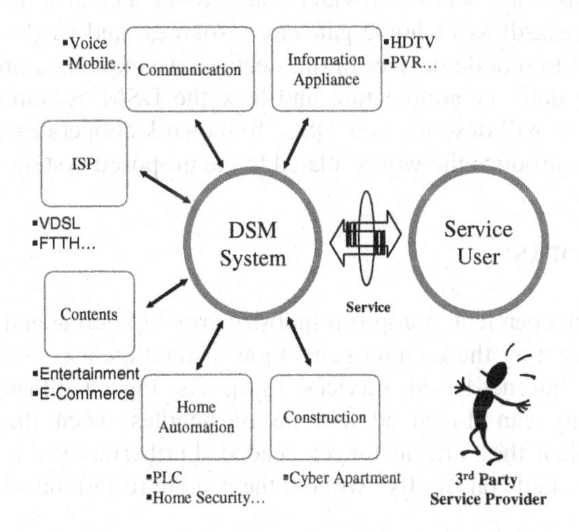

Fig. 2. The Proposed Service Delivery Architecture

Another issue on home service delivery is how to deploy a Graphic User Interface (GUI) for each remote control device. Conventionally, home service management systems deliver to a home gateway service bundles that include the GUI module for controlling a home device. Thus, each home gateway should have all of these GUI modules for controlling devices such as a cell phone, a Personal Digital Assistant (PDA), and a personal computer.

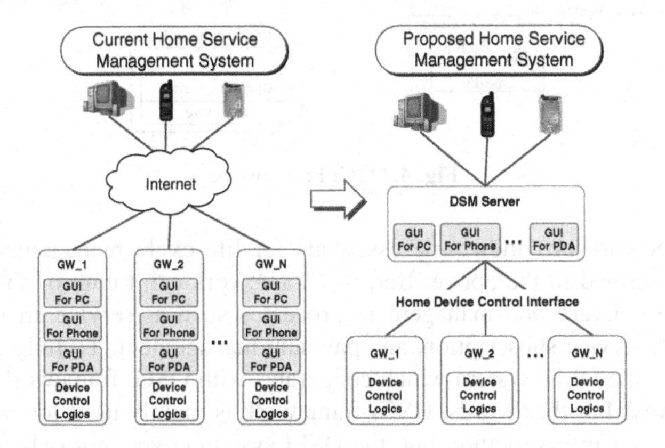

Fig. 3. Deployment of GUI Module dedicated to Control Device

But, the home gateway is a physical device limited in such hardware resources as CPU, memory, and storage, thus this approach is not a good choice for this type of system. Also, service providers are difficult to change a GUI module for each mobile device in case of applying this solution to the system. To solve these problems, the proposed DSM system installs only device control module on a home gateway, and the DSM system manages the GUI module for control devices as illustrated in Fig. 3. For this reason, 3rd party service providers can provide a control-device-specific GUI to service users regardless of home gateway resources, and easily upgrade the GUI module dedicated to mobile device. In this section, we explained problems with conventional service delivery architecture and how the DSM system can handle these problems. Next, we will describe how OSGi framework cooperates with the proposed system, and then introduce the works related to the proposed system.

2 Related Works

The OSGi is an independent, non-profit industry group to define and promote an open standard for connecting the coming generation of intelligent consumer and business appliances with Internet-based services [3,6]. As Fig. 4 illustrates, the OSGi-compliant gateway can download and install bundles when they are needed or uninstall them when they are no longer needed. Furthermore, the modules may be installed or uninstalled "on-the-fly" without the need to restart the whole system.

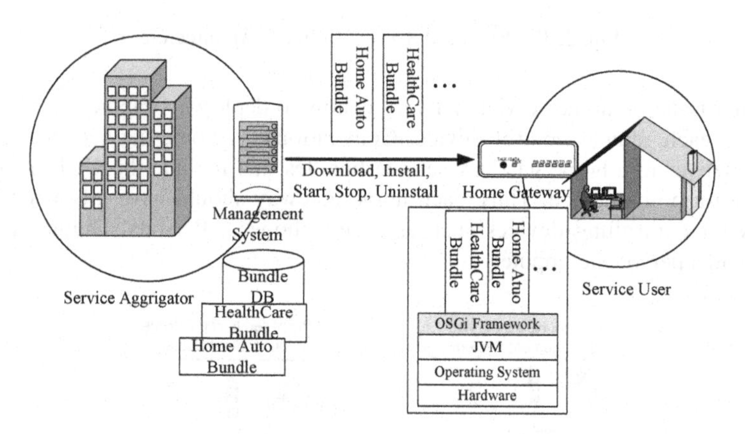

Fig. 4. OSGi Framework

OSGi framework provides basic essentials for life cycle management of service bundles as described in the above. But, this framework can't control and manage all of the service delivery and management processes such as service creation, service advertisement, service subscription, and payment management. To fully control all of them, we need the DSM system which cooperates with OSGi framework installed on a home gateway. In other words, OSGi framework is mainly used for end user's devices in an indoor infrastructure, but, the DSM system covers not only an indoor infrastructure but also an outdoor one.

ProSyst developed mPower Remote Manager (mPRM) using OSGi framework [4], which enables the seamless delivery of services, remote administration and management as well as the exchange of data between devices and manufacturers, network operators and external service providers. mPRM can be easily integrated with existing enterprise systems, such as application servers, databases, billing systems or messaging middleware. To be sure mPRM has powerful service management functions, but indeed it is not an end user friendly system because it is mainly focused on a service provider or a gateway manager. mPRM offers to service users a home portal which is installed on home gateway. This portal enables a remote user to control home devices or monitor the state of home security. But, this approach is not suitable for applying to a variety of remote control devices. That is because a new GUI module has to be installed on a home gateway whenever new remote devices are added. Thus, this approach may cause the shortage of a home gateway resource in a case where remote control devices are very diverse. To alleviate these problems, we divide each service bundle to a heavy GUI and a device control module, and deliver only the device control module to a home gateway with leaving the GUI module in the DSM system. So, we can easily add and update the GUI modules dedicated to remote control devices.

Telia unveiled home gateway and management system [5], solution for service transferring network. With a motto of "No New Boxes", the system stripped itself of the existing method to put home gateway in each home so as to place service gateways on network. This new method makes distinctive difference from other conventional types of products. The service gateway solution installed on network is proposed to work as cost-effective and flexible solution to provide distribution services as well as better security and scalability while helping customers get new methods to deal with new service in their home. In addition, service gateways can be managed by the management system in the network as it is required to make concentrated management of services by external network or service network. A drawback of this approach is that applicable area is limited to All-IP environment. This is because service gateways can't directly control non-IP home devices as they are installed on access network not home network. But, the DSM system targets the service gateway on the home network and dynamically delivers device control bundles into it. Thus, The DSM system can directly control non-IP home devices connected by various home network interfaces. In the following section, we design the DSM system for the Open Service Gateway initiative (OSGi) framework.

3 Design of the DSM System

Before we show the detail architecture of a DSM system, we explain how services such as the healthcare and cyber education are delivered by the DSM system, which is illustrated in Fig. 5. First, a service provider requests the service development to a service and contents manufacturer, by whom the developed services are then registered on the DSM system through the service registration interface for the service provider. The registered service is now advertised to service users by the DSM system, and then the service users can subscribe to the registered service, which the DSM system will install and start on the OSGi-compliant home gateway which is in the users' home. Based on the duration and frequency of usage of each service, the user will receive a single bill that summarizes all user activities.

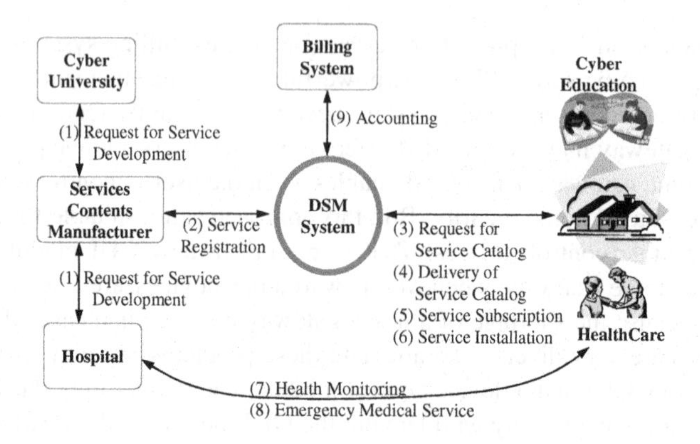

Fig. 5. Service Delivery Scenario

In this paper, we propose a DSM system illustrated in Fig. 6. The DSM System consists of a Control Module (CM), Management Module (MM), Service Provider Module (SPM), and Management Agent (MA).

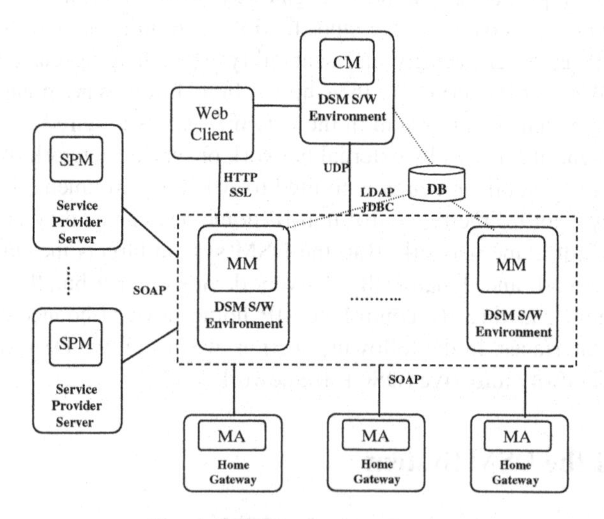

Fig. 6. DSM System Architecture

The CM and the MM are logically detached modules for system scalability. They can be located in a single physical server or be separated for load balancing of the MM. The CM monitors the MM status and balances the load of the MM. When a home gateway requests an IP address of a MM to the CM, and the CM checks the state of the MM, and then assigns the home gateway to the lowest load MM. The MM is the main engine for the service management, the home gateway management, the home device control management, the user authentication, and the account management. Also, the MM has portals for the service provider, service user, and system manager. The SPM performs function to connect a SP server and a MM. Thus, this module has service interfaces that can invoke services provided by a MM. By using them, service providers can register service packages as well as service bundles, and authenticate service users. Lastly, the MA acts as the agent to handle the messages received from the MM. It also bootstraps to initialize a home gateway.

The DSM system supports three types of service interfaces. One is for invoking services to initialize home gateways, handling fault events occurred by home gateways, and locating family [7]. Another is for handling event messages occurred by home devices, locating device drivers for controlling home devices [8]. The other is invoked by SP servers and it is used for the service registration, the user accounting, and the user authentication [9]. We define these service interfaces using WSDL (Web Service Description Language). Also, SOAP (Simple Object Access Protocol) is used for exchanging messages defined by WSDL between the MM and the MA or between the MM and the SPM. Though remarkable other protocols such as SNMP (Simple Network Management Protocol) exist, the DSM is implemented based on SOAP which has lightweight and platform independent feature. That is why the DSM system can communicate with a variety of home gateways and SP servers. All of the interfaces described in the above were standardized in TTA (Telecommunications Technology Association), which is the IT standards organization in Korea that develops new standards and provides one-stop services for the establishment of IT standards as well as providing testing and certification for IT products.

4 Home Service Delivery Process

In this section, we describe the service delivering process from service registration by a service provider to service usage by a service user. First, a service provider registers service bundles via the service provider portal of the MM, and then he/she creates a new service after binding the registered bundles and inputs detailed service information such as the service accounting policy, service description, and bundle list.

Upon end at the service registration process, a service user can subscribe to the service. After the service user subscribes to a service in the service categories, the resource manager checks the available disk space and the processing power of the service user's home gateway. If his/her home gateway has sufficient resource, the service bundle is installed on the home gateway, and then the service user can use the subscribed service via the DSM system. Fig. 7 shows each of the above procedures in detail. Next, we explain the registration process of the home gateway illustrated in Fig. 8. When a home gateway is booted, of which the bootstrap management agent requests an IP address of the MM to the CM. Upon reception of the IP address, the bootstrap management agent sends the identification number of the home gateway to the MM with the received IP address. After receiving the URL of the management agent from the MM, the bootstrap management agent downloads and starts a home gateway management agent from the received URL. After successfully started, the home gateway management agent sends the platform profile of the home gateway to the MM, and then receives the URLs of system bundles such as the resource management agent, framework management agent, accounting agent, and permission management agent bundle. Finally, the home gateway management agent downloads the system bundles from the received URLs, and then starts the bundles.

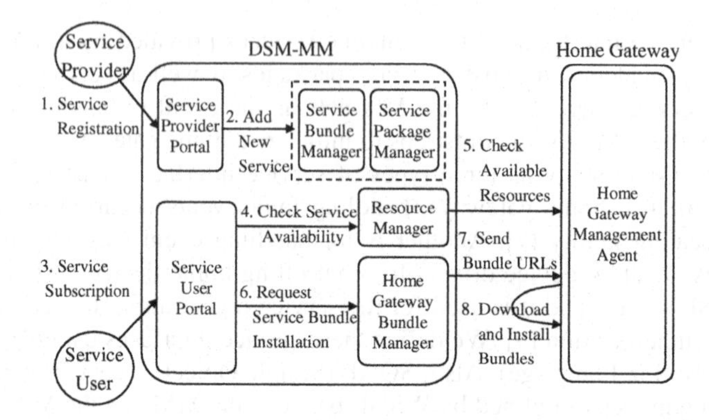

Fig. 7. Service Delivery Process

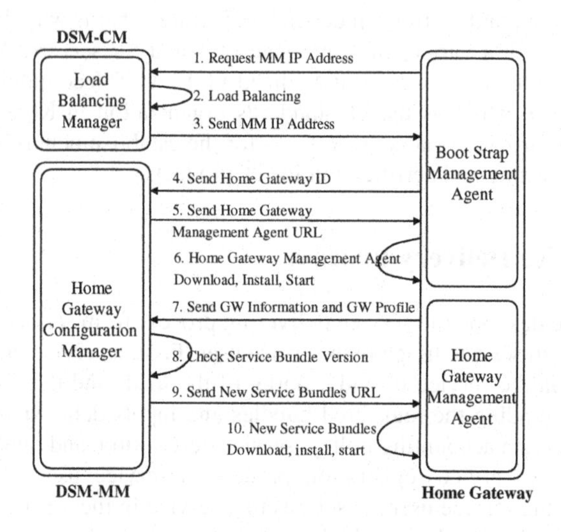

Fig. 8. Home Gateway Registration Process

Fig. 9 shows how new home devices are detected and registered by the DSM system. First, upon connection in home network, a home device is detected by a base driver. The base driver requests a device driver for handling the detected device to the home device control agent, which successively requests the URL of the device driver to the home device control manager, and receives it, and then downloads the device driver from the received URL.

After device detection, a service user downloads a client program on his/her mobile phone. With the client program, the user establishes the connection to the control connection manager of the DSM. Through the established connection, the control connection manager installs on the connected mobile phone the GUI module that enable the home devices to remotely control, and finally the service user can control home devices via the GUI module dedicated to the mobile phone. Fig 10 shows more detailed control procedures of home devices.

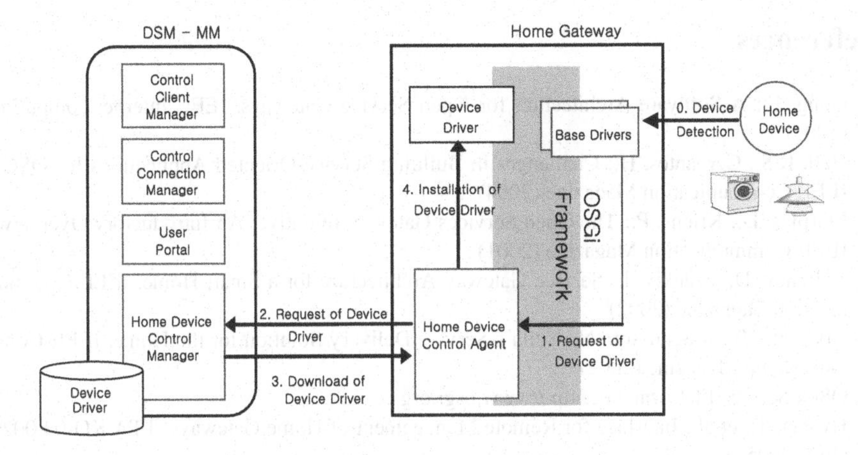

Fig. 9. Home Device Detection

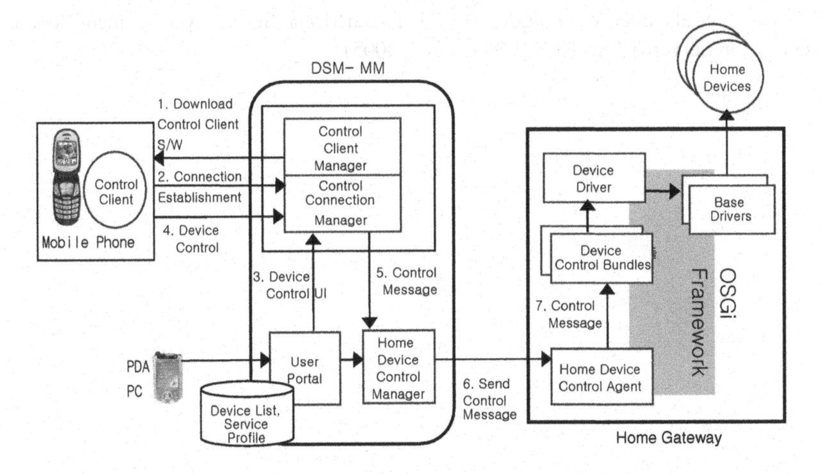

Fig. 10. Home Device Control

5 Conclusions

In this paper, we proposed the DSM system to deliver and manage 3rd party service provider's services to home service users. We described the functional modules of the DSM, and showed how home services of the service providers can be delivered to the service user via the DSM. The proposed DSM can lead competition of the 3rd party service provider in home services market. The service user can be provided with various services through reliable service aggregator, and receive a single bill for the subscribed services. Also service providers can dynamically provide a user friendly GUI module dedicated to remote control device.

References

1. Gong, L.: A Software Architecture for Open Service Gateways. IEEE Internet computing (2001)
2. Hall, R.S., Cervantes, H.: Challenges in Building Service-Oriented Applications for OSGi. IEEE Communication Magazine (2004)
3. Marples, D., Kriens, P.: The Open Services Gateway Initiative: An Introductory Overview. IEEE Communication Magazine (2001)
4. Valtchev, D., Frankiv, I.: Service Gateway Architecture for a Smart Home. IEEE Communication Magazine (2002)
5. Horowitz, B., Magnusson, N.: Telia's Service Delivery Solution for the Home. IEEE Communication Magazine (2002)
6. OSGi Service Platform 4.0, http://www.osgi.org
7. Hwang, T., et al.: Interface for Remote Management of Home Gateways. TTA.KO-04.0029 (June 2005)
8. Hwang, T., et al.: Interface for Remote Control of Home Devices. TTAS.KO-04.0033 (June 2005)
9. Hwang, T., et al.: Inter-Working Interface between Home Service Management Server and Application Server. TTAS.KO-04.0031 (June 2005)

A Self-managing SIP-Based IP Telephony System Based on a P2P Approach Using Kademlia

Felipe de Castro Louback Rocha and Linnyer Beatriz

Programa de Pós Graduação em Engenharia Elétrica,
Universidade Federal de Minas Gerais, Belo Horizonte, Brazil
louback@gmail.com, linnyer@gmail.com

Abstract. Most of Voice over IP(VoIP) systems are deployed using SIP protocol and based in a client-server architecture. A lot of people and companies could be benefited if the system did not require the configuration and maintenance of central servers. Some proposals are found in the literature. In order to improve the efficiency of such approaches and to provide a complete self-configuring service of IP telephony, we suggest a new solution based on the P2P algorithm Kademlia. According with the results of our experiments, our approach has a better performance when network experiences different kind of delays due to the fact that Kademlia supports concurrent searches, improving the latency in user location, and providing a faster call establishment.

1 Introduction

Voice over IP(VoIP) systems are becoming very popular nowadays, with many services being offered in the market. Deployment of services based on IP telephony have good advantages, such as low cost calls and mobility to the users. The configuration and maintenance of such systems requires the configurations of server and specialized employees to do so. Knowledge of the signalling protocols are need, for example. However, a lot of small-size to middle-size companies can't make use of an IP telephony system due to lack of specialized employees or lack of enough money to hire a third-party company to install and keep its telephony IP system.

Day by day, distributed applications based on a Peer-to-Peer architecture are becoming popular and with this DHTs (Distributed Hash Tables) are being more utilized. DHTs are known by its decentralization, fault tolerance and scalability characteristics.

In many scenarios, deployment of DHTs instead of centralized servers is a good deal. Telephony IP systems can make use of such benefits. Using these DHTs characteristics, we propose in this article another approach to replace VoIP centralized servers by an approach using DHTs, envisioning a complete self-organizing and self-configuring IP telephony system using Kademlia P2P algorithm.

S. Ata and C.S. Hong (Eds.): APNOMS 2007, LNCS 4773, pp. 11–20, 2007.

1.1 P2P Networks

There are a great number of definitions to P2P(Peer-to-Peer) system. In the most basic level, a P2P system is a system where multiple software applications interact with each other to realize a task. The node group, as a whole, is some times referred to as an *Overlay Network*. This contrasts with the traditional client-server model, where a piece of centralized software(the server) process tasks from different clients. The P2P or the client-server model choice is a matter of architecture decision of where the data-processing of the information will happen.

A P2P architecture does not necessarily implies that each node must offer all kind of services or store all available data. All nodes on the overlay network must provide all services collectively, but any node can provide just a fraction of the services. For example, a group of nodes may provide a database service and each one may store a small fraction of the database itself. If a big number of nodes share the task, the chances that some entry will be stored in one chosen node is small, but at least one node on the overlay network will store the entry, making sure that, as a group, the nodes offer a complete database system.

In a structured P2P architecture, nodes are connected to each other in a logical and well defined structure, such as a ring, a tree or a grid. Many arrangements are possible, but in general an unique identifier is given to a node when it is inserted into the overlay network. This identifier may be specified in a randomly way or determined by the hash of some property of the node, such as his IP address. The ID determines with which other nodes the node must establish contact. For example, a new node must connect to other nodes "near" him, to some notion of proximity.

A very special type of structured P2P system is widely deployed, called DHT (Distributed Hash Table). Some of the most discussed DHTs algorithms are Chord[1], Kademlia[2] and Bamboo[3]. In a DHT, not only the connectivity structure among the nodes is controlled in a mathematical way, but also the way the resources are shared. An ID is given to each resource in the same ID space that the ID given to the nodes. In other words, the range that the values that a resource ID can have is the same as the node ID. The resource identifier is the hash of some property of the resource, such as the filename or another keyword that identifies the resource. The hash of this keywork is done, yielding to the resource ID, and the node with ID "closer" stores the resource. The definition of *proximity* and redundancy are very dependent of the algorithm deployed.

1.2 SIP - *Session Initiation Protocol*

SIP [4] is a protocol based and derived from HTTP(*Hiper Text Transfer Protocol*), so its types of messages and traffic are very similar to the http traffic. SIP is a general protocol to establish and control multimedia sessions, but is has been mostly used in the VoIP field.

Devices implementing SIP may be configured to communicate directly with each other, but usually in systems with more than two UAs(User Agents) a central server or a proxy is used. In SIP's specification [4] the functions of a server are divided, including the proxy server and the registrar server function. This functions are often implemented together, and for the sake of clarity, in this article we will refer to the server realizing the previous cited functions as proxy server. A SIP proxy keeps information about user location and it is also responsible for all the routing and establishment of calls.

1.3 OpenDHT

With the goal to serve applications that would make use of a public DHT, it was created OpenDHT[5]. OpenDHT is a public shared DHT that runs in about 200 machines of the PlanetLab[1] globally distributed. OpenDHT implements the BambooDHT[3] algorithm. Nodes that run OpenDHT code are considered infrastructure nodes. Clients may run codes that uses OpenDHT through RPCs(Remote Procedure Calls), but they do not have OpenDHT code implemented. In this way, applications do not need to worry about DHT implementation. But on the other hand, nodes can not execute specific codes at the infrastructure nodes. In most of the applications today, the DHT is executed through a local library that implements it. The adoption of a library that implements the DHT allows the use of specific code and code optimized to the application, but each application must implement its own DHT. OpenDHT offers an opposite approach, with a smaller flexibility of utilization, but the application programmer does not need to worry about DHT implementation.

1.4 Kademlia

Kademlia[2] is a P2P Distributed Hash Table. Kademlia uses the same general approach as others DHTs. An 160-bit ID is associated to each node and it is given and algorithm that searches and locates successively closer nodes to some given ID, converging the search in a logarithmic way. Kademlia treat nodes as leaves in a binary tree, where the position of each node is determined by the smalled unique prefix of its ID.

To some given node, the tree is divided in a series of successively smaller subtrees that do not contain the node. The biggest subtree consists of half the subtree that doens't have the node. The next smaller subtree consist of half the remaining tree not having the node and so on.

Kademlia protocol assures that each node knows the location of at least one node in each of its subtree, if that subtree has a node. With this warranty, each node can locate another node through its ID by successively querying the best node the knows, trying to locate it in smaller subtrees and finally the search converges to the target node. Usually searches are made in a concurrent way, that is, more than one query is sent at the same time. The parameter α controls this behavior.

Each Kademlia node has an ID of 160 bits. Node IDs are usually randomly numbers of 160 bits. Key IDs are also 160 bit longer. To give a pair <key,

[1] www.planet-lab.org/

value> to a node, Kademlia uses the notion of distance between identifiers. Given two IDs of 160 bits, x and y, Kademlia defines the distance between them as their bitwise exclusive or(XOR) interpreted as an integer. Kademlia nodes keep contact information about each other to make routing of messages possible. For each $0 < i < 160$, each node keeps a list of tripes <IP, UDP Port, node ID> to nodes with distance between 2^i and 2^{i+1} of itself. These lists are called k-buckets.

More details about the algorithm can be obtained at [2]. Kademlia is the algorithm implemented in one of the most popular file share software, Emule.

2 Related Work

In [6] it is proposed the use of OpenDHT as an alternative to resource location servers of SIP architecture. SOSIMPLE [7] is a project that aims to implement a system without servers, based on a SIP P2P communications. In this project all P2P traffic is managed using SIP messages and the DHT implemented is based on the Chord algorithm. Skype [8] allows users to make free calls to other Skype users as all as paid phone calls to regular phone numbers. Skype is parcially P2P, once that it users centralized servers to authentication and authorization. Skype is based on a proprietary protocol and it is not compatible with other protocols such as SIP and H323. Kundan Singh and Henning Schulzrinne from Columbia University have a project [9] with a similar approach to SOSIMPLE, using Chord as the DHT. The main differences are related to the transport and routing of data inside the P2P overlay network.

All works cited above requires Internet connectivity to work, otherwise communication will fail or not even be established.

3 Our Proposal

We want our proposal to promote cost reduction, because there will be no need to buy physical servers, and management simplification, once it will not be needed any personnel to keep the system running. To achieve such goals the main point is to remove the need of a centralized resource location server in SIP based systems. The centralized server will be replaced for one adapter that will run locally at the machine and will be responsible for all the server functions.

It is a requirement that it will not be necessary to adapt existent UAs and that every UA that is SIP compliant will be able to use the adapter.

The adapter was written in C++ using the library Resiprocate[2] to treat SIP messages. The implementation using Kademlia used the code from Anulus project[3]. Anulus project implements Kademlia algorithm in C++ and one of the authors of this article is part of the Anulus developer team.

[2] http://www.resiprocate.org/

[3] http://code.google.com/p/anulus/

SIP has been the definitive standard adopted by many companies and most of the VoIP providers today have their system based on this protocol. This is why SIP was chosen as the signaling choice.

3.1 Adapter Behavior

Call establishment flow in a centralized approach can be made in one of two ways: either using a redirect server or a proxy server. When a server behaves like a redirection server, the caller sends a `SIP INVITE` to the server, showing the intention of initialize a session with a callee. The redirect server then locates the IP address of the callee. After the IP is located, the server sends back a `302 Moved Temporarily` message to the caller. The caller then sends out a new `SIP INVITE` to the callee using the IP address returned by the redirection message. All the later message flow is done directly between the caller and callee. When the server behaves like a proxy, basically it handles all the signalling between the two parties. It was adopted the redirection server behavior in our adapter. The redirection server does not need to keep track of the sessions established, making its implementation simpler.

3.2 The Adapter

The adapter is responsible for all the resource location(user location) process as well as user maintenance in the DHT. Basically the adapter can be divided into two main modules: the SIP module, who implements the redirection server functions and SIP message treatment, and the P2P module, responsible for locating and keeping the users in the P2P overlay as well as have the algorithm Kademlia running. Communication between these modules are done through an API(Application Programming Interface) exported by the P2P module, allowing registering and searching of users to be done.

3.3 Initialization Process

To enter the P2P overlay network, a node needs to find another node that is already part of the overlay. OpenDHT was used to store a reference to a node in the overlay, so that another node can find a participant node. The idea to store a reference in OpenDHT is based on the idea of Dynamic DNS, but using a public shared DHT to do such. In Dynamic DNS the entries are updated in real time.

The reference is store using as **key** `bootstrap.node.br` and as **value** related to the key *IP:Port* of a node that is already in the P2P network. In this way, when a node searches for that key, it will be returned the information need to contact a node in the P2P overlay.

When the adapter is initialized, it searches for the key `bootstrap.node.br` at OpenDHT. In case of a successful search, the adapter uses that information and starts the bootstrap process.

In case of an unsuccessful search, initially the node tries to register at OpenDHT the key `bootstrap.node.br` with its own IP, so that he can become a reference for future nodes that try to join the overlay. After the register attempt at OpenDHT, the node try to locate any other node at the P2P overlay sending a broadcast. *Broadcast* is used because the unsuccessful search for the key at OpenDHT may indicate, for example, that connectivity with the Internet was lost, or that this is an attempt to establish a telephony IP system in an environment that does not have connectivity to the Internet or even that this is the first node at the overlay.

If any node at the overlay notice the search broadcast, it answers it with its IP. The node who generated the request then starts the bootstrap process with that node. Only the first reply to the broadcast is considered.

3.4 User Registration

During the register process, the UA send a `SIP REGISTER` to the adapter. The SIP module notices that the packet is a register message. DHTs store pairs <key, value>. After a user is located in a telephony IP system, the UA need the information regarding the port and IP address of where the UA of the callee is running, so the call can be established. So, when using a DHT to store information regarding users location, pairs are stored using the following format: <User, IP:Port>. In this way, when a search is performed, it will be returned the IP and port where the callee is located.

3.5 User Maintenance

The main use of Kademlia today is file sharing. When a search is made for some key, nodes that participate in the search, that is, that are on the search path, cache a copy of the search. In this way, future searches to this key may be faster due to the probability that a node on the search path may have a copy in cache. In the case of telephony IP, the register of users can not be persistent because users may enter and leave the network at any time, and they may have a different IP address each time they come back. The behavior of Kademlia algorithm had to be changed in order to support the needs of telephony IP systems. An expiration time of 3600 seconds was added to the key, and after that the entry is deleted.

Softphones must update its registry with the servers at regular intervals. This is done by sending `SIP REGISTER` messages at regular intervals to the server. We took advantage of this feature to update the user registry expiration time at the DHT. When the adapter receives a `SIP REGISTER`, a new key is inserted in the DHT and the expiration time is initialized to 3600 seconds. 3600 seconds is the default time for most of the softphones available today. When it receiver another `SIP REGISTER` message, the node responsible for that user updates the registry expiration to 3600 seconds again, unless other value is specified at the message. Cached copies of user registries are not updated. But the expiration of cached copies is not a problem because, as a characteristic of Kademlia algorithm, all the searches converge to the node responsible for the key. So, in our case, even after all the cached copies expire, during a new search, new cached copies will be formed.

3.6 User Location and Call Establishment

As said before, our adapter behaves like a redirection server. When a UA wants to establish a call with another UA, a `SIP INVITE` message is sent to the adapter, who in turn searches for that user into the DHT. In case where the user is found, a `302 Moved Temporarily` is built according to SIP specifications[4], with all the information necessary to redirect the call. This message is sent back to the caller UA and then the caller sends a new `SIP INVITE` to the callee. When the user is not found inside the DHT, a `404 Not Found` message is returned.

3.7 Performance

The system can be self-configured and self-mainteined but the effects of this decentralization must be measured. In a centralized approach, user location is made in $O(1)$ while most of the DHTs, including Kademlia, is designed to locate a resource with $O(log(n))$ messages[2], where n is the number of nodes in the overlay network. This affects the call establishment time, once it will take more time to locate a specific user before the call can be established.

In order to measure the efficiency of our approach regarding call establishment time, we used three machines connect to each other by a 100Mbps full duplex link, using a switch to do so. In each node we started 20 adapters, in a total of 60, and after they were started, seven different users were registered.

Concurrent searches improve latency in resource location when timeouts for a request is experienced, since more than one request is sent at a time. In order to measure how the concurrency parameter(α) would affect user location time, we divided our experiments in three different scenarios. In our adapter, the request to search for a value has a timeout of 3 seconds before re-transmission. In the first scenario, all 60 nodes did not experienced delay. In the second scenario, half the nodes were programmed to hold the search response for 2 seconds before sending it to the node who made the search. In the third scenario half of the nodes were configured to hold the search responses for 4 seconds. So, in the first scenario no delay at all, in the second a delay smaller than the retransmission timout and on the third one a delay that would cause a new transmission due to timeout. 40 searches were made in each scenario and we repeated the experience for $\alpha = 1$ and $\alpha = 5$. The value 1 for α was chosen so we could see how DHTs that does not support concurrency would behave; the value 5 was chosen based on the studies presented at[10]. This way we could check what impact does the concurrency has when the network introduces delay(simulated by the induced delay on half of the nodes).

The results can be seen in figure 1 for $\alpha = 1$ and in figure 2 for $\alpha = 5$.

We could see on figure 1 that as delay increases on the network, the average time to locate a resource also increases. When a 2 second delay were introduced in half of the nodes, the requests that reached those node had a higher response

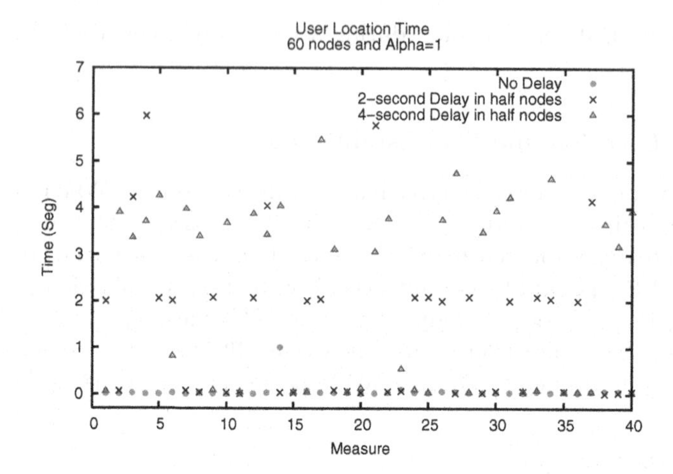

Fig. 1. Time to locate a user at Kademlia with $\alpha = 1$

time. When a 4 second delay were introduced, the requests that reached those nodes were re-transmitted to other nodes after the 3-second timeout. When the experiment were run with $\alpha = 5$(figure 2) we can see that the resource location time were not that much affected. That happens because five requests are sent concurrently, and if some requests reach nodes pre-programed to introduce a delay, some would still reach nodes with no programmed delay.

With these experiments, we can conclude that when multiple concurrent requests are sent we have a better chance to get a faster response even when some nodes on the overlay network are experiencing network congestion. When $\alpha = 1$, Kademlia is similar to Chord in message cost [2]. Chord does not support concurrent searches. We could see from our experiments that an approach that supports concurrency can improve the mean response time for user location. All other approaches found on literature, but Skype, use Chord as the underlying DHT.

4 Future Work

4.1 NAT *Traversal*

Users behind NAT is still an open issue. At the present time, we assume that all users can be found on reachable IPs. One alternative would be the implementation of some solution like STUN(Simple Traversal of UDP through NAT) on the adapters with valid IP address.

4.2 Authentication

At the present time we do not provide any sort of authentication and assume that all users can be trusted. But this can not be a valid assumption on various

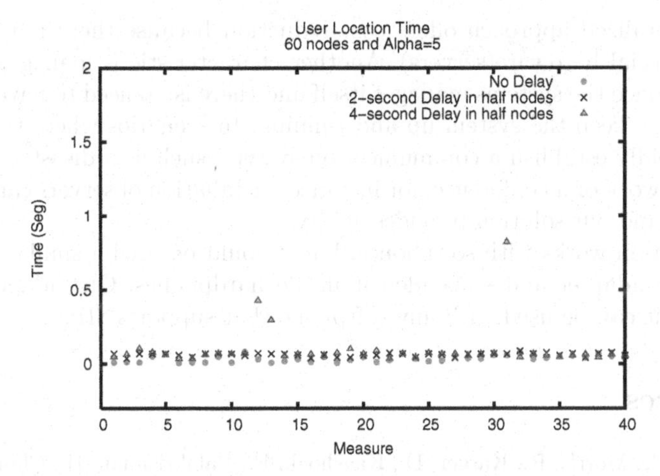

Fig. 2. Time to locate a user at Kademlia with $\alpha = 5$

scenarios. The solutions listed on section *Related Work* do not treat authentication in a very effective way. One alternative would be the PGP web of trust model. Another one would be to store centralized certificates in a central point in the network, but this would require a centralized authority to be consulted. In the case of ad-hoc or transiente networks, no verification may be required or desirable. An authentication system is still an open issue and is under analysis.

5 Conclusion

Our approach is effective in building a decentralized, self-configuring and self-managing telephony IP system. Self-configuring is provided by the fact that it is not necessary to install and configure a centralized server. Once the adaptor is initialized, not further configuration is required. Self-managing is provided by the fact that the system does not require any person to manage it. Users joining and leaving the system are handled by the P2P network without the need of a centralized server to store user information.

Our solution works well on environments with Internet connectivity or without it. For example, if the connectivity to the Internet of a company is lost, the IP telephony system would not stop in our solution. In scenarios where there may not be an Internet connection available, our approach still works.

Our solution is the first one that uses the algorithm Kademlia for such purpose. Concurrent searches can reduce significantly the mean delay time in resource location when nodes are experiencing packet loss or high traffic as we could see from our experiments. Other approaches found on the literature do not use concurrent searches because of the DHT deployed. The fact that our approach is the first one to use Kademlia also leads our work to be the first one to analyse how the concurrency parameter on Kademlia can benefit user location time on decentralized IP telephony systems under different network circumstances.

A decentralized approach offers cost reduction because there is no need to buy any special hardware(servers). Another characteristic is management simplification, once the systems manages itself and there is no need to have someone specialized to keep the system up and running. In scenarios where there is the need to rapidly establish a communication system, such as a disaster area with wireless network or a conference, for instance, installation of servers can be time-consuming and our solution provides agility.

Our solution works with softphones, but it could be used a similar approach to code the adapter and embedded it inside hardphones. Our adapter is SIP compliant, it can be used with any softphone that supports SIP.

References

[1] Stoica, I., Morris, R., Karger, D., Kaashoek, F., Balakrishnan, H.: Chord: A scalable Peer-To-Peer lookup service for internet applications. In: Proceedings of the 2001 ACM SIGCOMM Conference, pp. 149–160. ACM Press, New York (2001)

[2] Maymounkov, P., Mazires, D.: Kademlia: A peer-to-peer information system based on the xor metric. In: Druschel, P., Kaashoek, M.F., Rowstron, A. (eds.) IPTPS 2002. LNCS, vol. 2429, pp. 53–65. Springer, Heidelberg (2002)

[3] Rhea, S., Dennis Geels, T.R., Kubiatowicz, J.: Handling churn in a dht. Technical Report UCB/CSD-03-1299, EECS Department, University of California, Berkeley (2003)

[4] Rosenberg, J., Schulzrinne, H., Camarillo, G., Johnston, A., Peterson, J., Sparks, R., Handley, M., Schooler, E.: Sip: Session initiation protocol (2002)

[5] Rhea, S., Godfrey, B., Karp, B., Kubiatowicz, J., Ratnasamy, S., Shenker, S., Stoica, I., Yu, H.: Opendht: a public DHT service and its uses. In: Proceedings of the ACM Conference on Applications, Technologies, Architectures, and Protocols for Computer Communications, pp. 73–84. ACM Press, New York (2005)

[6] Singh, K., Schulzrinne, H.: Using an external dht as a sip location service. Technical report, Department of Computer Science, Columbia University, CUCS-007-06, New York (February 2006)

[7] Bryan, D.A., Lowekamp, B.B., Jennings, C.: SOSIMPLE: A serverless, standards-based, P2P SIP communication system. In: AAA-IDEA 2005 (June 2005)

[8] Baset, S.A., Schulzrinne, H.: An analysis of the Skype peer-to-peer Internet telephony protocol. Technical Report CUCS-039-04, Columbia University (September 2004)

[9] Singh, K., Schulzrinne, H.: Peer-to-peer internet telephony using sip. In: NOSS-DAV 2005, pp. 63–68. ACM Press, New York (2005)

[10] Li, J., Stribling, J., Gil, T., Morris, R., Kaashoek, F.: Comparing the performance of distributed hash tables under churn (2004)

A Collective User Preference Management System for U-Commerce

Seunghwa Lee and Eunseok Lee

School of Information and Communication Engineering, Sungkyunkwan University
300 Chunchun Jangahn Suwon, 440-746, Korea
{shlee,eslee}@ece.skku.ac.kr

Abstract. In the area of electronic commerce, the personalized goods recommendation system is a very important research issue that raises user satisfaction, and increases loyalty towards the content provider. For this, the correct analysis of user preferences is essential, and most existing researches use a purchase history or a wish list. However, due to the rapid development of information technologies, commerce has progressed from e-commerce to *U(Ubiquitous)-commerce*. In the ubiquitous environment, computing devices of various types, including the mobile device itself, exist in user space; in addition, a broad range of information related to user preferences is generated while using these devices. Hence, if the information is efficiently managed, a more effective recommendation strategy will be established. This paper proposes a multi-agent based U-commerce system to efficiently collect and manage diverse context information that can occur in the ubiquitous environment. Therefore, a more personalized recommendation, which is reflected by various user preferences, is possible. A prototype was implemented in order to evaluate the proposed system, then, through results, the existing recommendation method is compared and the effectiveness of the system is confirmed.

1 Introduction

In electronic commerce, personalized content and goods recommendation are necessary, in order to raise user satisfaction and promote commerce. Hence, the correct analysis of user preferences is very important.

To obtain user preferences, existing researches in the area of e-commerce use methods that either explicitly represents interest information such as response, purchase history, wish list, or inferred information by implicit user action, such as browsing time for each web page, and terms extracted from their favorite site list stored in the browser [1]. The obtained information is managed is a distributed manner in each shopping location, or in the latter case, although the preference information stored in the user's device, these different context data containing preferences are not unified.

Internet access is available anytime, anywhere due to the rapid development of network technology and the spread of wireless devices. Also, with the advent of ubiquitous computing, E(Electronic)-commerce is evolving to U(Ubiquitous)-comm-erce, and

S. Ata and C.S. Hong (Eds.): APNOMS 2007, LNCS 4773, pp. 21–30, 2007.

surpassing M(Mobile)-commerce. Hence, various researches are progressing to support the new commerce form employed in U-commerce [2]-[5].

The greatest difference between U-commerce and M-commerce is existing computing devices of various types as well as traditional mobile device, such as PDA and cellular phone. Hence, diverse information related to user preferences is generated while the user interacts with these devices, and new methods to collect and manage the information are required.

This paper proposes a multi-agent based U-commerce system to efficiently manage diverse context information that can be generated in the ubiquitous environment. Therefore, more personalized services are possible, also obtaining satisfactory results such as raising user satisfaction and increasing loyalty toward the content provider.

A prototype was implemented to evaluate and compare the proposed system, with existing recommendation methods, in terms of recommendation suitability. Through the results, the effectiveness of the system is confirmed.

The paper is organized as follows: Section 2 introduces related works. Section 3 describes the overall structure and behaviors of the proposed system. Section 4 confirms the scalability via a sample scenario, and evaluates the accuracy for recommendation results. Section 5 concludes this paper, and identifies future work.

2 Related Work

In the past few years, the existing E-commerce domain has been greatly enhanced by the development of various methods that recommend goods based on the user purchase history or goods of interest. Most recommendation systems use a method that combines *content-based recommendation* and *collaborative recommendation* in order to offer more accurate personalized information [6][7]. Content-based recommendation is a method recommending information by analyzing the similarity between user profile and content, and collaborative recommendation is a method recommending the information between users with similar profile. For this, generating the user's profile, which reflects accurate preferences, is very important.

While E-commerce is shifting to M-commerce, the greatest difference is that users can use commerce regardless of time and space [8]. By using this feature, user location based researches are progressing [9], gradually moving towards U-commerce.

U-commerce differs from E-commerce or M-commerce in the following ways. First, U-commerce uses various type devices surrounding the user, as well as traditional mobile devices such as PDA, and cellular phone using a wireless network. Second, everything (i.e., goods, refrigerator, shelf, counter, etc) related to commerce is interconnected via a network [2]. Hence, autonomous commerce is possible, in most devices, through interaction between devices. Therefore, diverse context information should be collected and integrated from heterogeneous devices, based on more correct user preference analysis.

Recently, various researches are progressing to support the new commerce. Representative studies are Metro Group's Future store [2] and Fujitsu's NextMart [3]. They are developing various recommendation services via built-in computing ability in different peripheral objects such as shopping cart, shelf, and mirror, which connect with each other using a network.

However, most existing research is insufficient to infer the user's preference by analyzing the relevance between the context information generated from using the different devices.

This paper proposes a novel U-commerce system that can more correctly analyze user preferences through embedding agents in the various computing devices and user circumstance and cooperation between these devices. In the next section, the proposed system is introduced in detail.

3 Proposed System

The proposed system is aimed at developing a U-commerce system which collects information related to preferences, from the heterogeneous user peripheral devices, these are based on recommending the goods, in order to support the commerce occurring in the ubiquitous environment.

3.1 The Proposed System Architecture

As presented in Figure 1, the proposed system is composed of three parts: *Client Module (CM)* is embedded in the different devices surrounding the user, and the client's information collected by the CM is integrated in a *User-data Management Server Module (USM)*. Then, a *Recommendation Server Module (RSM)* decides on the suitable goods taking into account the user preferences, via interaction with the USM, and then, recommends the goods. The USM is located at the default server of the user's home location or a portal server managing user accounts, and the RSM is located at the commercial off-line store. The role of each principal component is as follows.

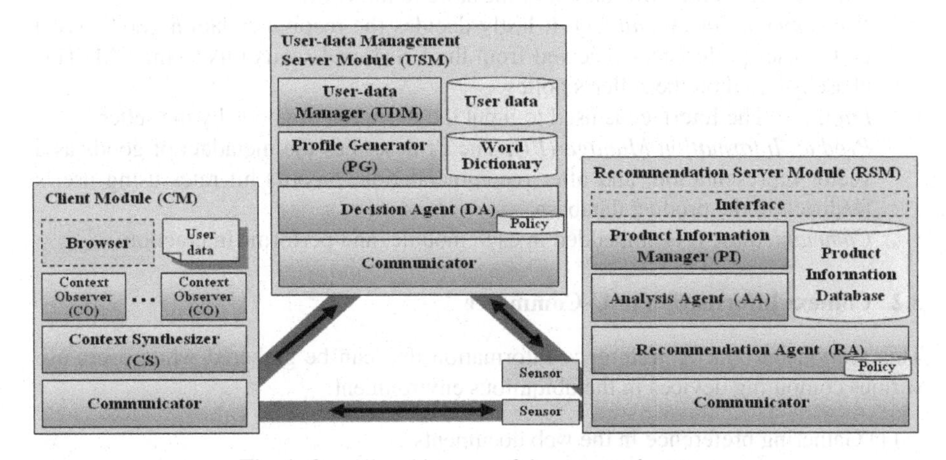

Fig. 1. Overall architecture of the proposed system

- *Context Observer(CO)*: The CO is embedded in the client device, and collects the data that can be used to infer the user preferences. It is developed as various forms, and extracts keywords implicated in documents built-in the browser, to monitor the user's web browsing pattern, analyze logs generated from applications, and so on.

- *Context Synthesizer(CS)*: The CS integrates the data collected from the several COs.
- *Browser*: The browser shows the received recommendation content.
- *Profile Generator(PG)*: The PG processes the information transmitted from the CM based on the Word Dictionary, and generates a user profile. Hence, *stop words* missed by the CO, are removed (the stop word means that is a commonly used word such as 'the'). The extracted keywords and appearance frequency are represented in the vector space.
- *User-data-Manager(UDM)*: The UDM renews the stored user data using the profile received from the PG, and clusters the users to the group with similar propensity.
- *Decision Agent(DA)*: When the user preferences are requested from the RSM, the DA measures similarity and decides scope of user profile, which will be informed to the RSM, included from the product's metadata, including the request message.
- *Word Dictionary*: The stop list and terms to recognize the synonyms of keywords are stored. It is designed and located at the server, and considers the resource constraints of mobile client devices. In future work, it should be provided to the *ontology*, including commerce domain, to improve the performance of the system.
- *Sensor*: The sensor recognizes the client device's identity.
- *Analysis Agent(AA)*: The AA finds the location of the USM, storing user information based on the information received from the sensor. Also, it selects some metadata of the principal goods in the Product Information database, in order to inform the information of the store to the USM.
- *Recommendation Agent(RA)*: It lastly decides the recommendation goods based on the user preferences received from the USM, and sends this to the CM. This phase is based on the seller's policy.
- *Interface*: The Interface is used to input the metadata of goods by the seller.
- *Product Information Manager(PI)*: The PI describes the metadata of goods as a vector representation, and after recommendation, records hit rates using user's feedback in the product database.
- *Communicator*: It is embedded in each module, and performs interaction.

3.2 Context Information in U-Commerce

This section introduces preference information that can be gathered while users use various computing devices in the ubiquitous environment.

(1) Gathering preference in the web documents

The user searches information using a computing device such as desktop, PDA. Hence analyzing of user preferences is possible by analyzing the characteristics of web documents viewed by the user. For this, TF-IDF methods [10], represent user interest in vector space, using the appearance frequency of terms appearing in the web document. This method is commonly used in the information retrieval domain.

However, in general, web pages contain several items that cannot be classified as 'primary content' (e.g., navigation sidebars, advertisements, copyright notices, etc.) [11]. Hence, this paper extracts the keywords containing user interests, by analyzing the hyperlink phrase or terms inputted in the search field, rather than by analyzing the entire page. These phases are presented in Figure 2.

```
Step 1. Finds the anchor tag and form tag by analyzing the html source
of a web page
  Step 1.1. Collects the URL and hyperlink phrase linked with anchor tag
  Step 1.2. Generates a mapping table
  Step 1.3. Analyzes the URL linked with form tag, and find the URL pro
cessing the query
Step 2. Analyze the next page's URL viewed by the user
  Step 2.1. Extracts the phrase linked with the URL from the mapping table
  Step 2.2. If next page is the URL linked with form tag, and extracts t
he query from the entire URL
Step 3. Removes the stop word(including articles and prepositions) defi
ned in advance
Step 4. Records the extracted keywords and the appearance frequency, an
d periodically sends it to the UMS
```

Fig. 2. The extraction phase of keywords from web pages

The collected keywords are presented in formula (1). Where w means the weight value for each term, and uses the appearance frequency of the term.

$$UserID_{context\_type} = \{(term_1, w_1), (term_2, w_2)..., (term_n, w_n)\} \quad (1)$$

An example that extracts the principal keywords from a hyperlink phase and query of a web page, is presented as Figure 3.

This method can apply to analysis using patterns of the web store. The user searches the required goods via keywords and category searching via the web store. The searching keywords inputted by the user, contain information such as brand name, product type, and specific good name. Hence, this data explicitly represents the user's interest even if it does not specify processing. Category searching can be used to extract the type of goods of interest to the user.

(2) Obtaining preference from the interaction TV

TV programs selected by users contain a lot of information that reflects the preferences of each user. For example, background music, actor's name, clothing, and place are information that can affect commerce.

Moreover, the TV is no longer a static system, evolving towards an interactive system, where users can select programs as desired. Hence, this paper used the VOD service on the Web as an early model of interactive TV. This service offers a web page that includes a summary or various explanations of content, such as a movie clip. Hence, important keywords can extract methods as introduced in the previous section.

In future work, if the *mpeg7* [12] is used, more efficient information collection is made possible.

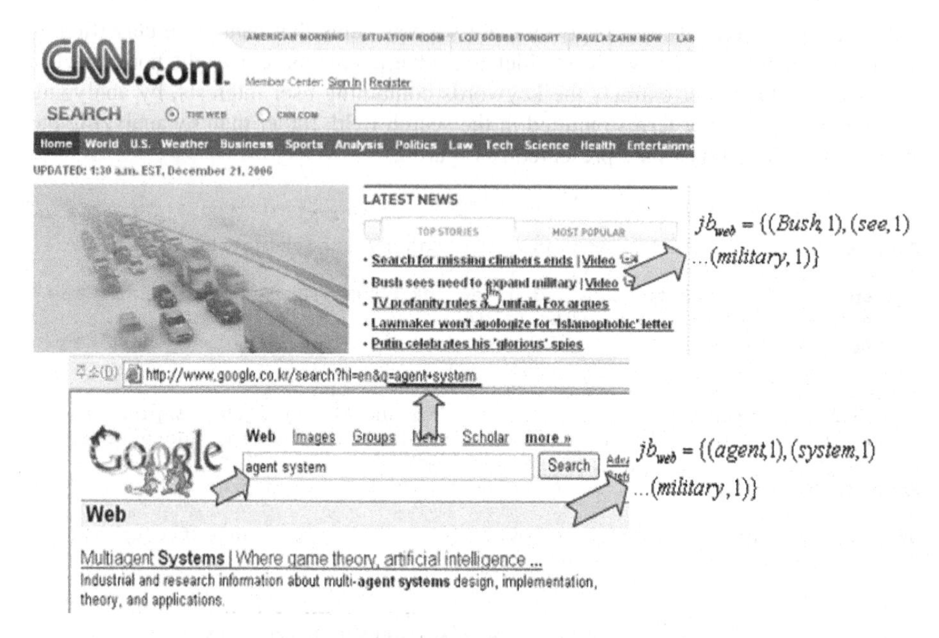

Fig. 3. Example of extracting the keywords from a web page

(3) Acquisition preference from the Navigation
The information for the user's destination is very important, because it can identify the user location and moving route. The current navigation executed on the PDA mostly records the destination information in log form. Hence, this paper collects user preferences by analyzing the log.

3.3 User Clustering Reflecting the Multi Preference

As introduced in the previous section, the information collected from the various devices is represented in the vector space. This section introduces the behavior of the User-data Manager (UDM), which clusters the users having similar preferences, using collected information.

First, USM compares the profile for product of a user randomly which is selected as the criterion, with the other user's profiles. The nodes, is representing users, are dispersed according to this result.

Second, the user profile for the TV program makes of a new axis, and the users are disseminated to the two dimension space. Then, it adds a new dimension using the user profile for the place collected from the navigation, and clusters the users. According to these phases, the users are grouped by similar preferences. The clustering of multi-dimension data like this is performed using the K-means algorithm, which groups proximity nodes using distance concepts. The clustering phases are presented in Figure 4.

After the group is constructed, the UDM extracts keywords that are not duplicated from the profile of other nearby users, and these keywords are mapped with the

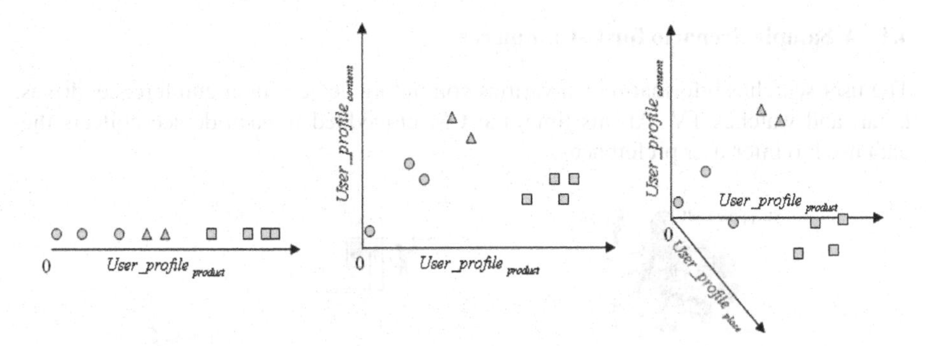

Fig. 4. Clustering phase by K-means algorithm

existing user profile. Therefore, a user can take recommended information from other users having similar preferences.

3.4 Recommendation Algorithm

The RSM requests the user preferences from the USM, in order to recommend the goods to the user visiting the store. At this time, the RSM sends the metadata for the main products selling at the store. This is vector set composed of keywords in the user profile.

The USM compares user profile with product information, and extracts principal keywords. The size of product information is restricted to prevent obtaining the entire user information by the RSM.

The similarity measurement between the user profile and the product information is performed using the *cosine similarity measure* commonly utilized in the information retrieval domain. Then, the result is transmitted to the RSM.

$$sim(u_i, p_j) = \frac{\sum u_i p_j}{\sqrt{\sum u_i^2} \cdot \sqrt{\sum p_j^2}} \tag{2}$$

* where, $sim(u_s, p_i)$: measurement of the similarity between user profile and product info.

The RSM decides recommendation goods using the result, and sends the recommendation information to the user.

Through these methods, item-to-item recommendation is issued not only via similarity between user profile and product information, but also by interactive recommendation between users in the same group.

4 System Evaluation

This section confirms the applicability of the proposed system using a sample scenario, and compares this system with existing recommendation method in order to evaluate the system.

4.1 A Sample Scenario for U-Commerce

The user searches information via various computing devices in circumference, drives a car, and watches TV. At this time, the CM embedded in each device collects the data used to infer user preferences.

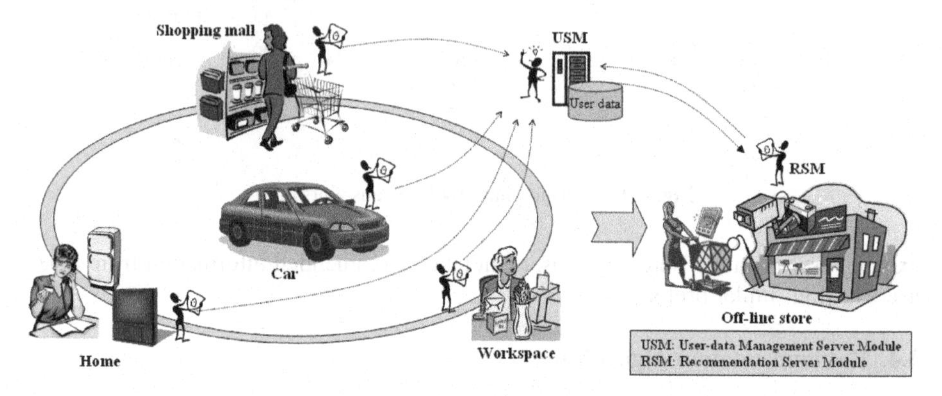

Fig. 5. Sample scenario for U-Commerce

A seller of an off-line store selling a type of fashion accessory has described the information of goods in detail, such as 'Actor Alice wore an accessory in a TV series called Friends', in order to recommend more suitable goods to the user. The RSM extracts the keywords from the information, generating product metadata.

Then, when users have been in the store, the RSM identifies the user data including a USM location, via user's portable device, and sends the product information to the USM storing the user's information.

The USM informs the some keywords related to the store's goods, among the user preference information to the RSM based on product type delivered from the RSM. Then, the RSM decides final recommendation goods based on this information. Therefore, users can receive recommendation goods related on TV programs that the user prefers. In this way, through the proposed system, applying a more diverse recommendation strategy is possible as an existing recommendation method, it considering only the purchase history or goods of interest.

In another way, this can extend services such as recommending TV programs preferred between users, like the goods of specific brands or those having similar life style, or recommending famous places, as well as goods recommendation.

The information, transmitted to the user, would contain useful information such as product location, hereafter, can adapt to appropriate type based on characteristics of user devices.

4.2 System Evaluation

We implemented a prototype to evaluate the system, compared existing recommendation methods, based on the purchase history and the interest goods in the web shopping mall, with the recommendation results of the proposed system based on the multi preference information.

In the experiment, we collected the purchase history and the interest goods list from 20 users. We implemented a web browser collecting the hyperlink information, and a log analyzer on the PDA, embedding the navigation, then took usage patterns about the web shopping mall, web TV, and recent destination information.

We also described the product information of 100 items, and users select the goods of interest among these lists, and then measured the correctness of the recommendation results for each method. A measure is used for the *precision* and the *recall,* which are broadly used for performance evaluation of information retrieval system. The precision means the ratio of the number of relevant information corresponding to user's intent among recommended information, and recall means the ratio of the number of relevant information recommended to the total number of relevant information in the database.

$$precision = \frac{|\{relevant\} \cap \{recommended\}|}{|\{relevant\}|} \qquad recall = \frac{|\{relevant\} \cap \{recommended\}|}{|\{relevant\}|} \tag{3}$$

As a result, the existing method exhibited low accuracy because most users haven't been buying the fashion accessories. However, the proposed system demonstrated relatively high accuracy due to being based on different user preferences.

Although clustering is not satisfactorily accomplished because the number of users that participated in the experiment was few; yet the proposed system demonstrated sufficient evidence only using the item-to-item recommendation method. The evaluation result is presented in Figure 6.

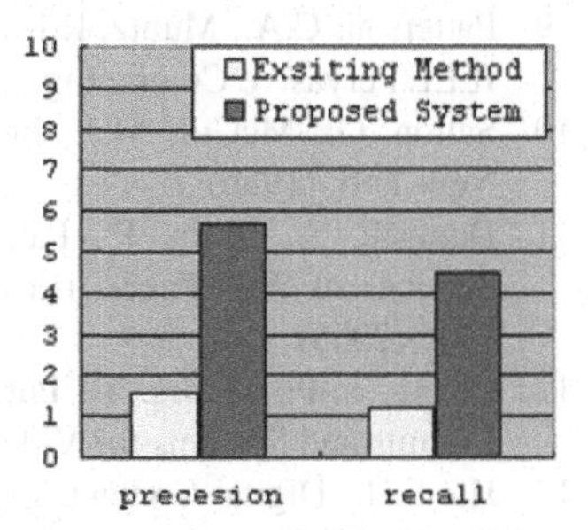

Fig. 6. Comparison of recommendation result

5 Conclusion

In this paper, we proposed a multi-agent based system that builds more flexible recommendation strategies using coordination management about the various user preference information generated in the ubiquitous environment.

We implemented the system prototype, and performed the comparison with the existing recommendation method, and as a result, confirm that the proposed system makes possible more accurate recommendation. This feature of the system is expected to make possible more effective commerce in the ubiquitous environment. As future work, the following topics will be investigated.

- A method to use the personal information of the specific user, which does not violate privacy when user sends a present to another user.
- A method that selects the most effective device to represent the recommendation content, and adapts the content according to the characteristics of the device or the user preference.
- A client agent that blocks unnecessary recommendation information.

References

1. Rucker, J., Polanco, M.J.: Siteseer: Personalized Navigation for the Web. Communication of ACM 40(3), 73–76 (1997)
2. http://www.future-store.org
3. http://forum.fujitsu.com
4. http://www.accenture.com
5. Wan, D.: Personalized Ubiquitous Commerce: An Application Perspective, Designing Personalized User Experiences in e-Commerce, pp. 141–160. Kluwer Academic Publishers, Dordrecht (2004)
6. Linden, G., Smith, B., York, J.: Amazon.com Recommendations: Item-to-Item Collaborative Filtering. IEEE Internet Computing 7, 76–80 (2003)
7. Balabanovic, M., Shoham, Y.: Fab: Content-Based, Collaborative Recommendation. Communications of ACM 40(3) (1997)
8. Lee, E., Jin, J.: A Next Generation Intelligent Mobile Commerce System. In: Ramamoorthy, C.V., Lee, R., Lee, K.W. (eds.) SERA 2003. LNCS, vol. 3026, pp. 320–331. Springer, Heidelberg (2004)
9. Patterson, C.A., Muntz, R.R., Pancake, C.M.: Challenges in Location-Aware Computing. IEEE Pervasive Computing 2, 80–89 (2003)
10. Salton, G., McGill, M.J.: Introduction to Modern Information Retrieval. McGraw-Hill, New York (1986)
11. Debnath, S., Mitra, P., Pal, N., Giles, C.L.: Automatic Identification of Informative Sections of Web Pages. IEEE Trans. on Knowledge and Data Engineering 17(9), 1233–1246 (2005)
12. Chang, S.-F., Sikora, T., Puri, A.: Overview of the MPEG-7 Standard. IEEE Trans. on Circuits and Systems for Video Technology 11(6) (2001)
13. Ha, S.H.: Digital Content Recommender on the Internet. IEEE Intelligent Systems 21(2), 70–77 (2006)

Distributed Cache Management for Context-Aware Services in Large-Scale Networks

Masaaki Takase, Takeshi Sano, Kenichi Fukuda, and Akira Chugo

Fujitsu Limited
4-1-1 Kamikodanaka, Nakahara-ku, Kawasaki, Kanagawa, Japan
{masaaki.takase,sano.takeshi,fukuda.kenichi, chugo}@jp.fujitsu.com

Abstract. In recent years, the number of messages transferred through networks has skyrocketed with the rising number of network nodes. For example, servers collect variant sensor information and store it. Although related works exist that present cache servers intended to reduce network costs, those cache functions do not work well for short lifetime content because most cached contents with short lifetimes are expired before referral. In this paper, we propose a method to resolve the problem for the coming ubiquitous network society, which is an asynchronous cache management method according to the application requirement. This method enables the CPU load of the servers to be reduced through service and application management for short-lifetime content, too. Furthermore, we propose a load-balancing method using autonomous message exchange instead of a management system. This method enables the CPU load to be balanced over multiple servers.

Keywords: Service management, Server management, Sensor information.

1 Introduction

In recent years, the number of nodes connecting networks has increased. For example, a sensor node, such as an RFID tag reader, is connected to a network. The information collected by its reader is often called the context. The provision of context-aware services has commenced [1], and the trend looks set to continue with the announcement by the Japanese Ministry of Internal Affairs and Communications of the u-Japan policy [2]. Generally, there are context generators, context providers and service providers, and they are located in different administrative domains to provide context-aware services. Context providers collect context such as sensor information and store it, and provide it to various service providers. This means there will be various node administrators in the future, meaning that future networks should incorporate a management function over different administrative domains in addition to the function for large-scale networks.

As such networks proliferate, the number of messages will also increase, for example, when servers collect sensor information, such as the IDs of RFID tags. In

S. Ata and C.S. Hong (Eds.): APNOMS 2007, LNCS 4773, pp. 31–40, 2007.

these cases, although the message size is small, the frequency of transmission is very high. Hence, one of the requirements for the servers is the capability to process large numbers of messages, which is also an inevitable requirement for future networks.

In this paper, we propose an asynchronous means of cache management and load balancing to solve this problem, and evaluate its advantages via a prototype system.

2 Background to the Research

2.1 Need for a Context Transfer Platform

We illustrate the example of providing multiple context-aware services in Fig. 2-1.

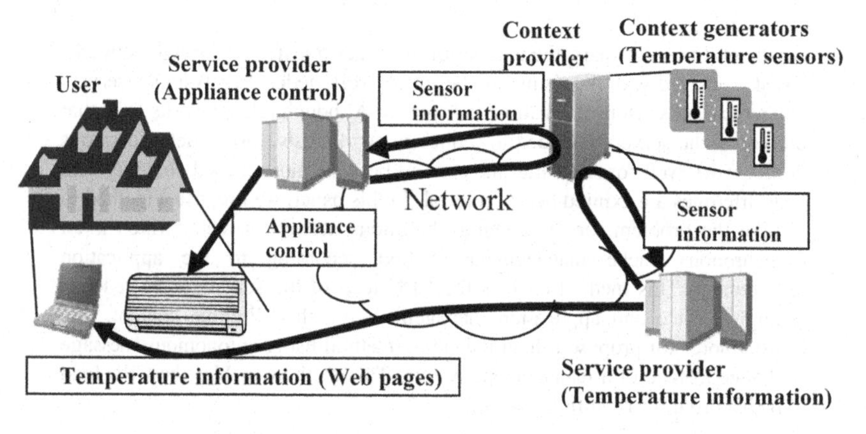

Fig. 2-1. The provision of context-aware services

In this case, there are a context provider and two service providers. The context provider publishes the information for many service providers and each service provider offers a context-aware service using such context information. Two services are the temperature information service and the appliance control service, and they are operated by different companies.

In recent years, it has been proposed that the service implement an environment to enable context information to be easily used by service providers, thereby enabling service providers to offer context-aware services easily. Moreover, many users can use the services, too. However, as many context providers and service providers will appear in future, we believe that a context transfer platform is desirable.

We also believe that a method of reducing the load on the context server using caching technologies is needed because there is a massive volume of contexts, because each context has a short lifetime and because not all contexts are used by service providers. Moreover, we think that cache servers should incorporate a management function over different administrative domains.

3 Requirements for Distributed Cache Management

3.1 Related Works and Problems

The number of services has increased and this caused an increase in the number of attributes and requirements, as well. There are various context-aware services. For example, some are a real-time services based on a context that reflects the latest sensor information, while others are non-real-time services based on a context at some point in time. The latter type of service allows a time range for context accuracy. In this paper, we focus on the non-real-time services, illustrate the problem that occurs if existing cache technologies [3][4] apply to such services. For example, it is assumed that a temperature information service provider requires a context every fifteen minutes and an appliance control service requires a context every minute. If a temperature sensor pushes the context to the context server every second, the context server sets one second as the context's lifetime. In this case, an introduced cache server also sets one second as the expiry time and each context expires after one second. In this case, most cached contexts with short lifetimes expire before referral, thereby preventing cache functions from working well. This occurs because the context referral interval is longer than the context's lifetime.

Moreover, each service is operated by different companies and each cache server is also operated by different companies. If an existing cache technology including hierarchical cache model [5][6] is applied to cache servers, the operator of the cache server has to publish the server's information and has to maintain the published information. Therefore there is the problem that it is impractical from viewpoint of ease of maintenance in the coming ubiquitous network society if the existing technology is applied.

3.2 Requirements for the Cache Server

We have identified the following two requirements for a cache server if the cache server is introduced to reduce the messages processed by a context server in the situation described above:

(a) The cache needs to be used effectively, even though the context is changed frequently and has a short lifetime, as in the case of sensor information.
(b) Load balancing needs to be conducted over different administrative domains autonomously without a central management system.

4 Asynchronous Cache Servers and Their Load Balancing

To meet the requirements described above, we propose an asynchronous cache system and a load-balancing function for the cache servers. As the non-real-time service described in section 3.1 is assumed, the cache server does not have to set the cache expiry time to correspond to the context lifetime based on the renewal interval of the sensor information because the service does not require the newest context. The feature of an asynchronous cache system involves optimally setting the cache refresh time instead of the cache expiry time based on the clients' requests, and look-ahead contexts based on cache refresh time. This system works well because the cache does not expire based on the setting made by the client, including the requirements of the application. Moreover, the load-balancing function does not involve managing the cache servers collectively, but simply switching the connection of servers via negotiation among the cache servers when one cache server's load is high.

4.1 Asynchronous Cache System

4.1.1 Overview of the Asynchronous Cache System

We illustrate the overview of the asynchronous cache system in Fig. 4-1. In Fig. 4-1-(a), the cache server updates its cache periodically according to the attribute that the client has indicated, and maintains the connection with the context server, which enables the cache server to have the optimal cache. In this case, the client is the application that provides services. The client can then indicate the application's attribute, such as the context referral interval. In this paper, we use the context referral interval as the application's attribute. Therefore the cache server can set the refresh time of the cached context. If there are several context referral intervals, the shortest one is adopted. In Fig. 4-1-(b), the cache server can return the previously cached context to the client when the cache server receives a request of the context in Fig. 4-1-(b). In this system, the update of the cache between the cache and context servers in Fig. 4-1-(a) and the requirement of context between the client and cache server are operated asynchronously. As above, the cache server updates the context according to the refresh time determined by the attribute from the client's request and maintains the cache refresh time. Therefore the context server can avoid excessive requests from the applications' requirements.

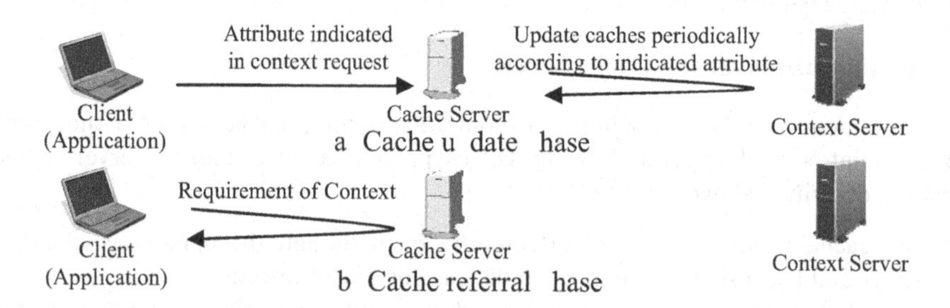

Fig. 4-1. Overview of the Asynchronous Cache System

4.1.2 Operation of the Asynchronous Cache System

Fig. 4-2 shows a sequence where the client requires context with the referral interval being indicated to the cache server.

The sequence works as follows:

1. A cache server receives a context referral request from a client. The context referral request contains the context address, referral interval and client ID.
2. The cache server checks whether it has a cache of the required context or not.
3. If the cache server has no caches related to the required context, it sends the context referral request to a context server.

4. The context server registers the cache server and load information. The latter is then used in the load-balancing function, described later.
5. The cache server registers the requirement from the client on a DB (database). The requirement contains the context address, referral interval and client ID.
6. The cache server registers the client information on the DB. The client information includes the client's address, the number of the client and total context size to which the cache server is referred. This information is used in the load-balancing function.
7. The cache server periodically updates the context with decided refresh time, A.

As described above, the cache server receives indications about the referral interval within the context referral request from clients. In other cases, if the cache server has a cache and the referral interval related to the context required from the client exceeds that of the cache, steps 3 to 5 are not executed. If the referral interval is shorter than that of the cache, steps 3 to 7 are executed in order to update the refresh time.

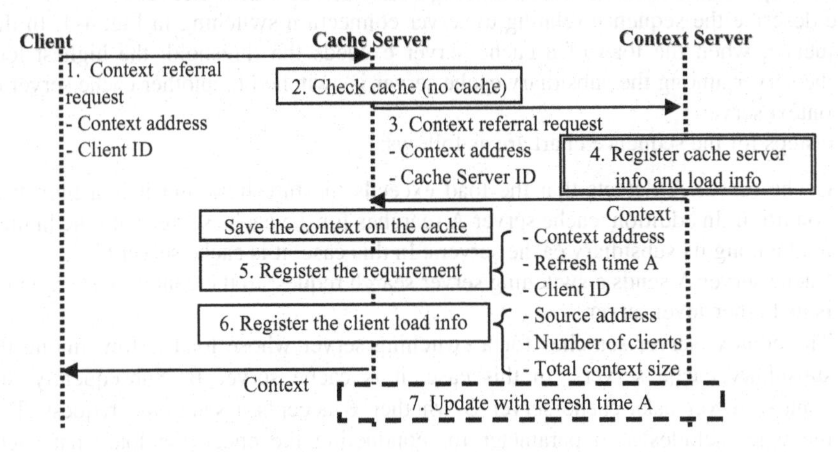

Fig. 4-2. Sequence of the Context Referral Request

4.2 The Load-Balancing Function for Cache Servers

4.2.1 Overview of the Load-Balancing Function for Cache Servers

In this function, a cache server whose load becomes high switches the subsidiary cache server(s) from itself to another server with a relatively low load, by negotiating with other cache servers (Fig. 4-3). The objective is to balancing the load among cache servers. From the viewpoint of ease of maintenance, the function proposed here has an advantage over other systems that spreads the load using a server list of other domains that is published by the administrator of the domains. This is because the load balancing proposed here can be applied to cases where cache servers are managed by different administrative domains without having a server list of other domains. The trigger for the switching of cache servers is the number of accesses per second and the total size of the context referred from clients and subsidiary cache servers. Fig. 4-3 represents an example. When the loan on cache server A becomes high, the cache and context servers search the switching server by negotiating with each other, and switch cache server C from cache server A to cache server B, whose load is low.

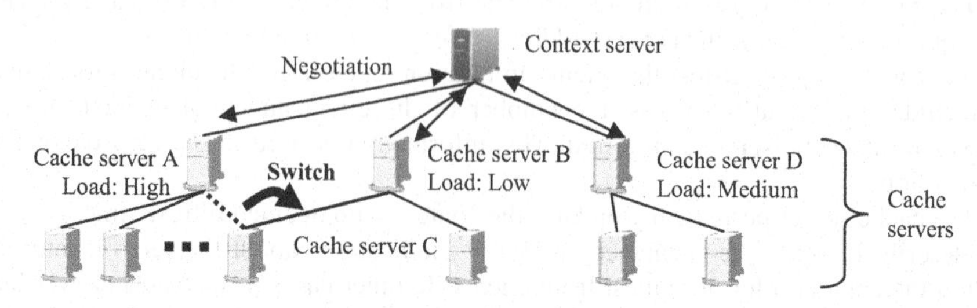

Fig. 4-3. Overview of the Load-Balancing Function for Cache Servers

4.2.2 Operation of the Load-Balancing Function for Cache Servers

We describe the sequence relating to server connection switching in Fig. 4-4. In this sequence, when the load of a cache server exceeds the threshold, the highest load cache server among the subsidiary cache server is switched to another cache server or a context server.

The steps for the sequence chart are as follows:

1. Cache server A detects that the load exceeds the threshold and it is a high load condition. In addition, cache server A searches for the cache server with the highest load among its subsidiary cache servers. In this case, it is cache server C.
2. Cache server A sends a switching server search request to the context server, which is its higher-level server.
3. The context server searches for a switching server whose load is low among the subsidiary cache servers. In this case, it is cache server B. Subsequently, the context server asks cache server B whether B accepts a switching request. This message includes as a parameter for conducting the check the load that cache server C has in relation to cache server A.
4. Cache server B checks its load using the received parameter to ascertain whether cache server C can be connected or not. In this case, cache server B determines the load to be low; meaning that cache server C can be connected, and thus returns a response to the context server.
5. The context server evaluates the response from cache server B. In this case, the response is OK, so it stops searching for switching servers. If the response is NG (meaning "no good"), which means cache server C cannot be connected to cache server B, steps 4 to 5 are executed. Subsequently, the context server returns the address of cache server B as the switching server to cache server A.
6. A switching request is sent from cache server A to cache server C. This message indicates cache server C to connect to cache server B.
7. Cache server C switches the connecting server from cache server A to cache server B, so cache server C sends a connection request to cache server B. The connection request consists of the set of context referral requests that cache server C referred to cache server A.

8. Cache server B registers the load information relating to cache server C, and sends the context referral requests that cache server B has not yet referred.
9. When cache server A receives the response from cache server C, it deletes the load information relating to cache server C and sends the context referral termination requests that are received from cache server C to the context server.

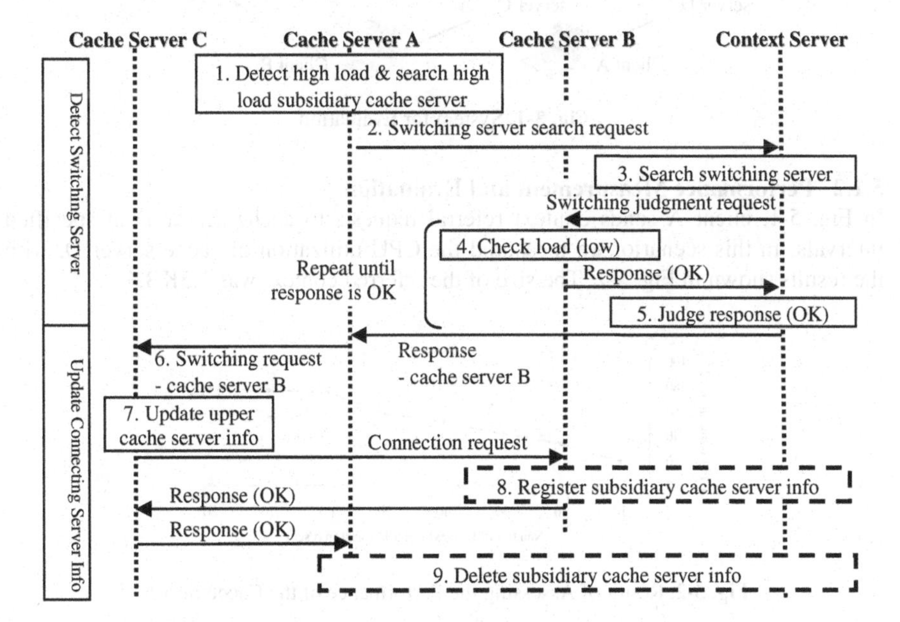

Fig. 4-4. Sequence of Server Connection Switching

5 Evaluation of Development Technologies

In this section we present the results of evaluation relating to the technologies described in section 4. As a result of evaluation, the operation and performance of the development technologies are shown to be confirmed and it is possible that our technologies are applicable in the assumed environment.

5.1 Performance of the Cache Server

5.1.1 Prototype System

This prototype system is implemented on the IA server using free software. The implementation environment is as follows:

- Hardware architecture, CPU and memory: Intel x86, Xeon 3.06GHz and 2GB, respectively.
- Operating system, JavaVM and application server: CentOS 4.4 (Kernel 2.6.9), Jrockit 1.5.0_08 and Jboss 3.2.7, respectively.

Fig. 5-1 shows our prototype system configuration used in the evaluation.

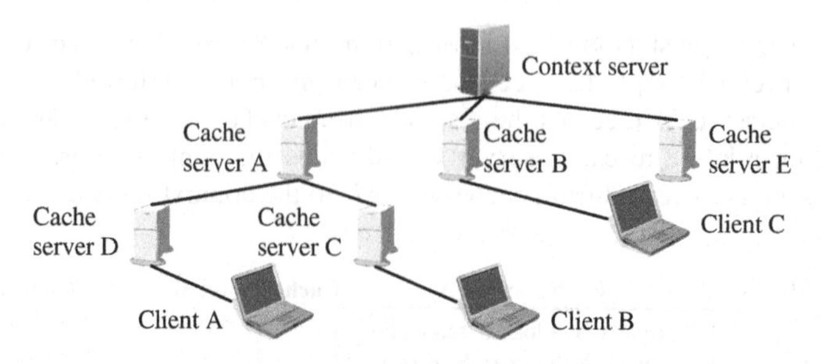

Fig. 5-1. System for Evaluation

5.1.2 Performance Measurement and Evaluation

In Fig. 5-1, client A sends context referral requests to cache server D at specified intervals. In this scenario, we measured the CPU utilization of cache server D, with the results shown in Fig. 5-2. The size of the referred context was 1.5KB.

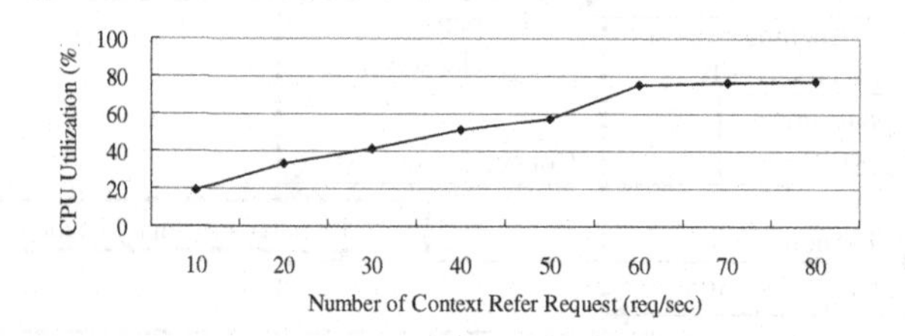

Fig. 5-2. Result of Assessing the Performance of the Cache Server

Fig. 5-3 shows the change in the CPU utilization of cache servers A and B when the load of cache server A becomes high, and cache server C is switched to cache server B.

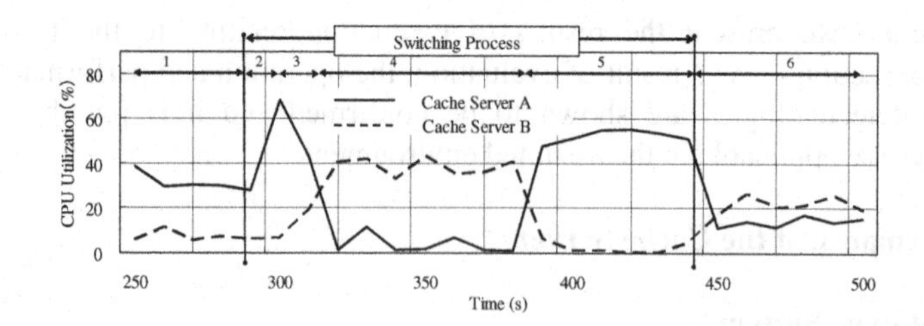

Fig. 5-3. Change in CPU Utilization at Connection Switching

During period 1, connection switching is yet to take place. The load of cache server A exceeded that of cache server B. Subsequently, during period 2, cache server A detects that the load is high, and during period 3, the load of cache server A is temporarily higher because cache server A searched the switched cache server. In period 4, cache server B registered the information that was referred, meaning its load increased, while during period 5, the load of cache server A rose by deleting information relating to cache server C. Period 6 came after connection switching, meaning that the load of cache server A declined.

5.2 Results of Simulating the Cache Server

5.2.1 Simulation Model

We set a load to each cache server per unit time, following which subsidiary cache servers, connecting to cache servers whose loads exceed the threshold ahead, are switched to other cache servers. There are two ways for searching for a switching server. One involves searching from neighboring servers, and the other involves searching from the route server in a tree structure. We adopted the former in order to reduce the number of searches. In addition, to avoid switching subsidiary cache servers, i.e. deepening the tree structure, we limited the depth.

We set the number of cache servers at 250,000, and simulated a one-year period assuming the unit time to be a single day.

5.2.2 Simulation Results

To evaluate the overhead via server connection switching, we counted the number of searches, i.e. the number of times switching servers were searched, in each process (Fig. 5-4).

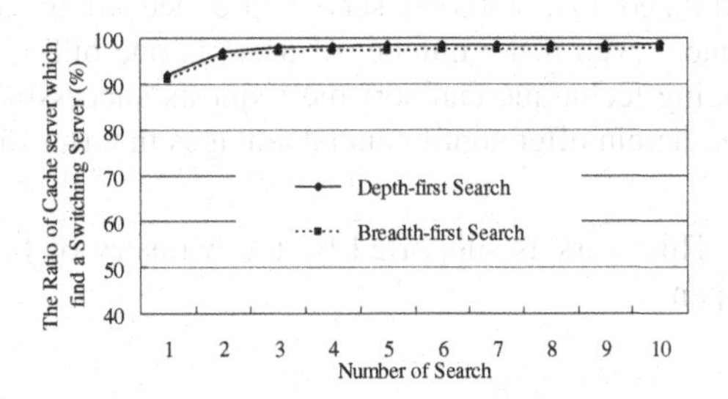

Fig. 5-4. Simulation Results

5.3 Discussion

As shown in the measurement results in Fig. 5-2, it was confirmed that our cache server processes about 80 context referral requests per unit time and is capable of caching contexts effectively according to clients' requirements. In addition, the CPU utilization of the cache server increases almost linearly with the increase in the number of context referral requests per unit time. In order to operate the cache server with CPU utilization of 80% or less, the switching process should be generated when CPU utilization exceeds 80%. Based on the result of Fig. 5-3, it was confirmed that our cache servers negotiate with each other and can operate switching processes when one of the cache servers has to deal with a high load.

Here, it is assumed that the number of clients sending context referral requests is one hundred million (which is assumed to be the number of cellular phone users), and each client sends a context referral request once every ten seconds. Under these circumstances, we need 125,000 cache servers to which the clients are directly connected. At this time, if we use an existing load balancer, it is unrealistic to expect it to sort the requests to 12,500 servers, and copy the context to the servers. We also assume that when cache servers are connected in a tree structure, and the cache hit rate in each cache server is assumed to be 50%, the tree structure becomes a binary tree. Based on these assumptions, the total number of cache servers can be provisionally calculated at about 250,000. Fig. 5-4 shows that almost 100% of the switching process is completed after double the number of searches. Therefore, if there are 250,000 cache servers constructed in a tree structure, the overhead generated by the server connection switching process has no effect on other processes, and connection switching among servers can be implemented. As described above, the possibility of load balancing over different administrative domains is confirmed.

6 Conclusion

In this paper, we proposed asynchronous cache management and a load-balancing technique in preparation for the coming ubiquitous network society, in which servers should handle short lifetime contexts over multiple administrative domains. We created a prototype system and evaluated its advantages in terms of performance. As a result, the cache server we proposed can process 80 context referral requests per second, and we estimated that 250,000 servers are needed based on the server's processing performance, even if the number of users is one billion. Furthermore, the proposed load-balancing technique can sort the requests successfully. Consequently, the methods proposed herein offer significant advantages in terms of scalability.

Acknowledgments. This work is supported by the Ministry of Internal Affairs and Communications, Japan.

References

1. Itao, T., Nakamura, T., Matsuo, M., Aoyama, T.: Context-Aware Construction of Ubiquitous Services. IEICE Transactions on Communications E84-B(12), 3133–3140 (2001)
2. Ministry of Internal Affairs and Communications: Information and Communications in Japan (2005), http://www.johotsusintokei.soumu.go.jp/whitepaper/eng/WP2005/chapter-1.pdf
3. Wessels, D.: Web Caching, O'reilly Media, Inc. (2001)
4. e.g. Squid Core Team: Squid Web Proxy Cache, http://www.squid-cache.org/
5. Rabinovich, M., Spatscheck, O.: Web Caching and Replication, Pearson Education (2001)
6. Wessels, D., Claffy, K.: Internet Cache Protocol (ICP), version 2, RFC2186 (1997)

Towards Low-Latency Model-Oriented Distributed Systems Management*

Iván Díaz, Juan Touriño, and Ramón Doallo

Computer Architecture Group
Department of Electronics and Systems,
University of A Coruña, Spain
{idiaz,juan,doallo}@udc.es

Abstract. Windows and Unix systems have been traditionally very different with regard to configuration storage and management. In this paper we have adapted our CIM-based model-driven management framework, AdCIM, to extract, integrate and modify management and configuration information from both types of OS in a multiplatform and seamless way. We have achieved very low latencies and client footprints without sacrificing the model-driven approach. To enable this functionality for a wide range of system administration applications, we have implemented both an efficient CIM XML dialect and a distributed object infrastructure, and we have assessed its performance using two different approaches: CORBA and Web Services.

Keywords: Management, Distributed Systems, WMI, CORBA, Web Services, CIM.

1 Introduction

System administrators have to take into account the great diversity of hardware and software existing nowadays in organizations. Homogeneity can be achievable in some instances, but also has risks; i.e., a monoculture is more vulnerable against viruses and trojans. The combination of Windows and Unix-like machines is usual, whether mixed or in either side of the client/server divide. System administration tasks in both systems are different due to a great variety of interfaces, configuration storage, commands and abstractions.

To close this gap, there have been many attempts to emulate or port the time-proven Unix toolset. Windows Services for Unix [1] are Microsoft's solution, enabling the use of NIS and NFS, Perl, and the Korn shell in Windows, but it is not really integrated with Windows as it is a migration-oriented toolset. Cygwin [2] supports more tools, such as Bash and the GNU Autotools, but it is designed to port POSIX compliant code to Windows. Outwit [3] is a very clever

* This work was funded by the Ministry of Education and Science of Spain under Project TIN2004-07797-C02 and by the Galician Government (Xunta de Galicia) under Project PGIDIT06PXIB105228PR.

S. Ata and C.S. Hong (Eds.): APNOMS 2007, LNCS 4773, pp. 41–50, 2007.

port of the Unix tools that integrates Unix pipelines in Windows and allows accessing the registry, ODBC drivers, and the clipboard from a Unix shell, but its scripts are not directly usable in Unix.

The subject of performance and low-latency issues in system administration is discussed in several works. Pras et al [4] find Web Services more efficient than SNMP for bulk administration data retrieval, but not for single object retrieval (i.e. monitoring), and conclude that data interfaces are more important for performance than encoding (BER vs XML). Nikolaidis et al [5] show great benefits by compressing messages in a Web Service-based protocol for residential equipment management, but only use Lempel-Ziv compression. Yoo et al [6] also implement compression and other mechanisms to optimize the NETCONF configuration protocol which uses SOAP (Web Services) messaging.

Also, there are several works that use XML to represent machine configurations, like the ones by Strauss and Klie [7], and Yoon et al. [8], both mapping SNMP to XML. The drawback is that SNMP has a very flat structure that does not represent aspects like associations as flexibly as CIM [9].

The framework used in this paper, AdCIM [10] (`http://adcim.des.udc.es`), can extract configuration and management data from both Windows or Unix machines, represent and integrate these data in a custom space-efficient CIM XML dialect, and manipulate them using standard XML tools such as XSLT. One objective of this framework is to support monitoring applications with low-latency gathering of structured data and small footprint. To achieve multiplatform interoperability and low-latency network messaging, two different distributed object technologies will be used: CORBA [11] and Web Services [12].

The paper is organized following the structure depicted in Fig. 1. Thus, Section 2 presents the XML Schema transformation and the advantages of the miniCIM XML format. Section 3 details the different processes used in Windows and Unix to extract miniCIM configuration data. Section 4 discusses the use and implementation details of both CORBA and Web Services middleware. Section 5 presents experimental results to compare the performance of both approaches. Finally, conclusions and future work are discussed in Section 6.

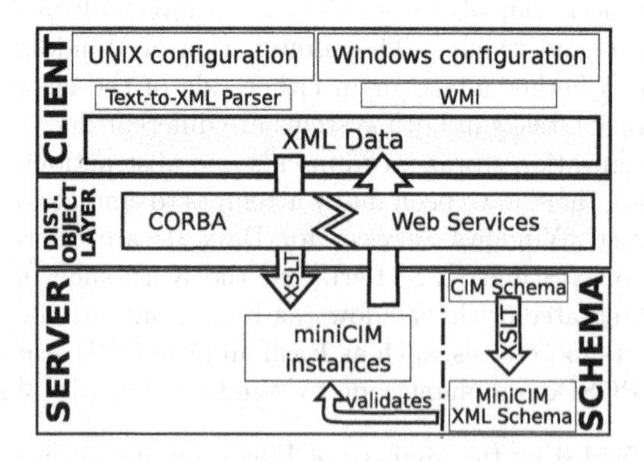

Fig. 1. Overview of the integrated management framework

2 Schema Transformation

This section details how the CIM schema is translated into an XML Schema, and then derived into an abbreviated XML syntax of CIM (which we denoted as miniCIM) that keeps all semantic constraints and helps to reduce latency and transfer times in our framework.

```
<CIM_InetdService namespace="dc=udc">
   <SystemCreationClassName>CIM_ComputerSystem</SystemCreationClassName>
   <SystemName>shalmaneser</SystemName>
   <CreationClassName>CIM_InetdService</CreationClassName>
   <Name>ftp</Name>
   <SocketType>stream</SocketType>
   <Protocol>tcp</Protocol>
   <Wait>nowait</Wait>
   <User>root</User>
   <Command>root/usr/sbin/tcpd</Command>
</CIM_InetdService>
```

Fig. 2. Example of miniCIM inetd service instance

cimXML [13] is the official DMTF representation of CIM in XML, which represents both the CIM schema and CIM instances. Since these two aspects are not separated it is very cumbersome to represent instances. Schema information can be merged from an external file, but there is still overhead; e.g, key properties must be present both in the name of the instance and as properties. Thus, a declaration of services of a machine in cimXML is 305Kb long, which is reduced to 3Kb in our approach by removing schema information from instances. This greatly reduces message size, which will be shown later as a main factor in the speed of the framework.

In order to simplify matters, we decided to translate the schema information to XML Schema [14], which supports type inheritance, abstract classes and keys, with XSLT. The resulting XML Schema defines a much terser instance syntax, miniCIM. Such an instance for inetd services can be seen in Fig. 2. Its format is property-value pair based, but semantic information is not lost, only moved to the XML Schema. Invalid instances, or dangling references, are reported by the XML validator.

An XSLT stylesheet processes the cimXML schema and its CIM classes depending on their abstractness status, superclass, and association type if applicable. These will be mapped to attributes of a new type in the final schema. For example, association-related properties are represented as application-defined attributes, and others like abstractness and superclass, supported using XML Schema inheritance constructs.

Association references are also mapped as properties. Constraints are mapped to XML Schema constructs, such as cardinality, addressed with `minOccurs` and `maxOccurs`. The reference properties `Antecedent` and `Dependent` in each association can have their names changed by child classes. To account for that, the `Override` qualifier is also supported.

3 Configuration Data Extraction

This section describes the methods used both in Unix and Windows systems to extract configuration data from system files. Configuration information is collected from two sources: flat text sources, such as files and internal commands, and the WMI (Windows Management Instrumentation) subsystem present in all post-2000 Windows systems.

Usually, Unix-based OS codify almost all configuration data in text files and directory structures that are seldom available in XML format, so our framework parses and transforms them to XML to facilitate further processing. To do so, data are serialized to plain text and processed with grammar rules.

```
  import Martel; from xml.sax import saxutils
  def Item(name): return Martel.Group(name,Martel.Re("\S+\s+"))
  fields =Item("name")+Item("socktype")+Item("proto")+Item("flags")+Item("user")+Martel.ToEol("args")
4 offline =Martel.Re("#<off>#\s*")+Martel.Group("off",fields)
  commentary=Martel.Re("#")+Martel.Group("com",Martel.ToEol())
  serviceline =Martel.Group("service",fields)
  blank=Martel.Str("\n")
  format=Martel.Group("inetd",Martel.Rep(Martel.Alt(offline, blank, commentary, serviceline)))
9 parser = format.make_parser()
  parser.setContentHandler(saxutils.XMLGenerator())
  parser.parseFile(open("inetd.conf"))
```

Fig. 3. Martel program used for parsing `inetd.conf` to XML

```
#echo  stream tcp nowait root    internal
ftp    stream tcp nowait root    /usr/sbin/tcpd /usr/sbin/proftpd
#<off>sgi_fam/1-2 stream rpc/tcp wait   root   /usr/sbin/famd fam
```

(a) Sample lines with original `inetd.conf` format

```
<doc>
  <commentary> <com>#</com>echo stream tcp nowait root internal </commentary>
  <line> <id>ftp</id> <ws> </ws> <id>stream</id> <ws> </ws> <id>tcp</id> <ws> </ws>
       <id>nowait</id> <ws> </ws> <id>root</id> <ws> </ws> <id>/usr/sbin/tcpd</id>
  <ws> </ws> <id>/usr/sbin/proftpd</id> </line>
   .........
  <off> <com>#</com>&lt;off&gt;#<ws> </ws>
    <line> <id>sgi_fam/1-2</id> <ws> </ws> <id>stream</id> <ws> </ws> <id>rpc/tcp</id>
    <ws> </ws> <id>wait</id> <ws> </ws> <id>root</id> <ws> </ws>
    <id>/usr/sbin/famd</id> <ws> </ws> <id>fam</id> </line>
  </off>
  </doc>
```

(b) Result of parsing `inetd.conf` to XML

Fig. 4. Input and output of parsing `inetd.conf` to XML

Grammar rules are described using Martel [15], a Python module to parse text files into SAX events, then directly transcribable as XML data. Fig. 3 shows an example of a Martel program that produces a structured XML file from the inetd services configuration file `/etc/inetd.conf` (shown in Fig. 4(a)). In this code, Martel operators `Re` and `Alt` represent the "*" and "|" regular expression operators, respectively, and operator `Group` aggregates its second argument into a single XML element. Finally, `ToEol` matches any text before the next end of line.

This Martel code defines the `inetd.conf` file as composed of three types of lines: off lines (line 4), commentaries (line 5), and normal lines (line 6). Every normal line maps to an enabled service, and off lines to temporarily disabled services. The program also has to discriminate between commentaries and the `#<off>#` sequence that begins an off line. Each line is partitioned as a list of items

```
import Martel; from xml.sax import saxutils
def Group(x,y): return Martel.Group(x,y)
def Re(x): return Martel.Re(x)
def Item(name): return Martel.Group(name,Martel.Re("\S+"))
def Date(name): return Martel.Group(name,Martel.Re("\S+\s+\d+\s+[0-9:]*"))
def Space(): return Martel.Re("\s*")
def Colon(): return Martel.Re(":\s*")
def Origin(): return G("origin",Re("\w+"))+Col()
def OriginPid(): return (Group("origin", Group("name", Re("[\w()_-]+")) +Re("\[")+
    Group("pid", Re("[0-9]+")) +Re("\]")+ Colon()))
fields =(Date("date") +Space()+ Item("host") +Space()+ Martel.Alt(OriginPid(),Origin(),Space()) +
    Martel.UntilEol("message") + Martel.ToEol())
format=Group("file",Martel.Rep(fields))
parser = format.make_parser()
parser.setContentHandler(saxutils.XMLGenerator())
parser.parseFile(open("m"))
```

Fig. 5. Martel program used for parsing `/var/log/messages` to XML

```
   import sys, win32com.client, pythoncom, time; from cStringIO import StringIO
   locator = win32com.client.Dispatch("WbemScripting.SWbemLocator")
   wmiService = locator.ConnectServer(".","root\cimv2")
 4 refresher = win32com.client.Dispatch("WbemScripting.SWbemRefresher")
   services = refresher.AddEnum(wmiService, "Win32_Service").objectSet
   refresher.refresh()
   pythoncom.CoInitialize()
   string = StringIO()
 9 for i in services:
     (string.write((
     "<SystemCreationClassName>"+unicode(i.SystemCreationClassName)+"</SystemCreationClassName>"+
     "<CreationClassName>"+unicode(i.CreationClassName)+"</CreationClassName>"+
     "<Name>"+unicode(i.Name)+"</Name>"+"<State>"+unicode(i.State)+"</State>"+
14   "<StartMode>"+unicode(i.StartMode)+"</StartMode>"+"</CIM_Service>").encode("utf8")))
   print string.getvalue()
```

Fig. 6. Python script to extract service information from WMI

that are mapped to fixed properties in the resulting instances. The abridged output in Fig. 4(b) is still a direct representation of the original data in Fig. 4(a), now structured.

Grammar rules can document the configuration format formally, following the original file format very closely, or simply describe the high-level format of the document (e.g. line-oriented with space separators). The latter approach makes it easier to process many formats without doing much work specifying rules, but the former has the benefit of early-on error checking and validation of configuration formats. Since Martel supports backtracking, multiple versions of the same file can be identified using different trees aggregated by an alternation (or) operator at their top.

The output of Fig. 4(b) is processed by an XSLT stylesheet, which can be executed server-side or in the client, but the former is preferred because servers usually have more processing power, and also to reduce footprint and latency in the client. Nevertheless, there are very efficient C XSLT processors [16] that can be used in some client nodes to reduce load on the server.

Figure 5 shows a more complex example of grammar rules that parses the `/var/log/messages` log file, composed of messages, warnings and errors from various system processes and the kernel. The format is line based, each line consisting of date, host name, optional originator and pid (process id), and a free-form message.

Windows discarded files in favor of the registry as system configuration repository as of the Windows 95 release. Thus, to extract configuration data it would seem necessary to manipulate registry data. Instead, we have used the Windows

WMI subsystem, which provides comprehensive data of hardware devices and software abstractions in CIM format, exposed using COM (Component Object Model), the native Windows component framework. WMI is built-in since Windows 2000, but it is also available for previous versions. Queries can also be remote using DCOM (Distributed COM). Its coverage varies with Windows version, but it can be extended by users.

WMI data can be uniformly retrieved using simple code, such as the one shown in Fig. 6, which uses the COM API and directly writes XML data of miniCIM instances. The code uses a locator to create a WMI COM interface named `SWbemRefresher` (see line 4) which makes possible to update WMI instance data without creating additional objects. In the next lines, instances contained in the refresher interface are queried and their data written as miniCIM instances. A `StringIO` Python object (line 8) is used to avoid string object creation overheads.

4 Distributed Object Technologies

This section describes the use of both CORBA and Web Services distributed object middleware in our framework. As can be seen in Fig. 1 the purpose of this middleware is to transfer efficiently miniCIM instances or raw XML data between clients and servers.

CORBA achieves interoperability between different platforms and languages by using abstract interface definitions written in IDL, from which glue code for both clients and servers is generated. This interface is a "contract" to be strictly honored. This enforces strict type checking, but clients become "brittle": any change or addition in the interface breaks them and needs their recompilation and/or readaptation.

Using an XML Schema validated dialect has two benefits: first, it promotes flexibility, since changes in format can be safely ignored by older clients and, second, preserves strict validation. Both aspects are important due to the extensibility of the CIM model, but in very time-critical instances a direct mapping of a CIM class to IDL is still possible.

To pass XML data via CORBA they are flattened to a string. This solution is not optimal, since time is lost in serialization and de-serialization. A more efficient solution would be to pass the data as a CORBA DOM Valuetype [17], which is passed by value with local methods. Then, the parsed XML structure would not be flattened, so clients would manipulate the XML data without remote invocations. Unfortunately, this is a feature not yet well supported in most production-grade ORBs. Our implementation of choice is omniORB [18], a high-speed CORBA 2.1 compliant ORB with bindings for both C++ and Python.

In contrast with CORBA, Web Services (WS) solutions provide an interoperation layer than can be both tightly coupled (using XML-RPC messaging) or loosely coupled (XML document-centric). Gradually, WS are being more oriented to support web-based service queries than to offer distributed object middleware, but there is a significant overlapping between the two approaches.

WS use XML dialects for both interface definition (WSDL) and transport (SOAP). It may seem that using XML dialects would promote synergy, but the use of XML as "envelope" of the message and representation of it is orthogonal at best. It is worse, in practice, since the message must be either sent as an attachment (which implies Base64 transformation), or with its XML special characters encoded as character entities to avoid being parsed along with the XML elements of the envelope. Additionally, two XML parsings (and the corresponding encoding) must be done, being added to message-passing latency. In the foreseeable future, there is not any support in view for platform independent parsed XML representation in WS.

As implementation we have chosen the Zolera SOAP Infrastructure [19], the most active and advanced WS library for Python.

5 Experimental Results

We have proceeded to evaluate the performance of our framework for various representative tasks and the impact of the integration technology (Web Services vs CORBA). The tests have been performed using Athlon64 3200+ machines connected by Gigabit Ethernet cards.

We have tested three different cases of use of our framework, each both in Windows and Unix (Linux). The parameters measured for these cases have been total time, latency and message size, with and without compression. Total time is defined as the round-trip time elapsed between a request is sent and the response is completely received. Latency is the round-trip time when the response is a 0-byte message. Two different algorithms have been used for compression: zlib and bzip2. The three cases tested are:

- CPU load retrieval, shown in Fig. 7. This case is representative of monitoring applications, usually invoked several times per second, which need fast response times and low load on the client.
- Service information discovery, shown in Fig. 8. This case represents discovery applications, invoked with a frequency ranging from minutes to hours.
- Log file information retrieval and parsing (data analysis), shown in Fig. 9. This case represents bulk data requests invoked manually or as part of higher-level diagnostic processes. These requests have unspecified total time and data size, so they are invoked ad-hoc, with little or no regularity.

The code examples of Section 3 have been used for the second and third cases. The code of the first case was omitted for brevity.

The first case in Fig. 7 shows lower latency and total time for CORBA vs WS in both platforms. Base latency for CORBA is roughly 0.2 ms, whereas it is 20-30 ms using WS, in great part due to the overhead of parsing the envelope and codifying the message. Compression benefits WS but slows down CORBA performance. Figure 7(b) shows the cause: WS messages are large enough to be slightly benefitted from compression, but CORBA messages (less than 50 bytes long) are actually doubled in size. In the second case (Fig. 8), web service times

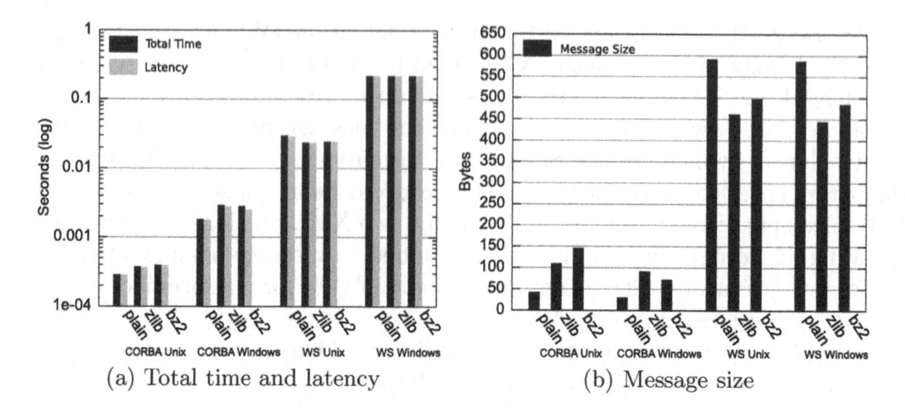

(a) Total time and latency (b) Message size

Fig. 7. Performance measurements for test one: CPU Load

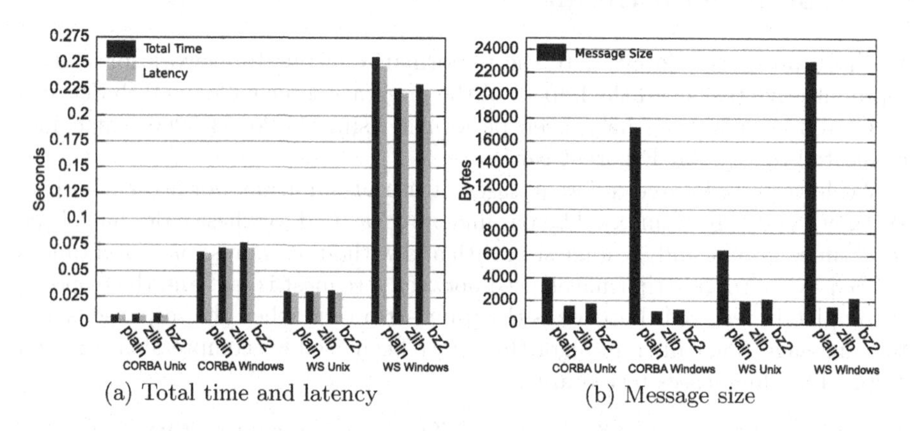

(a) Total time and latency (b) Message size

Fig. 8. Performance measurements for test two: Service Discovery

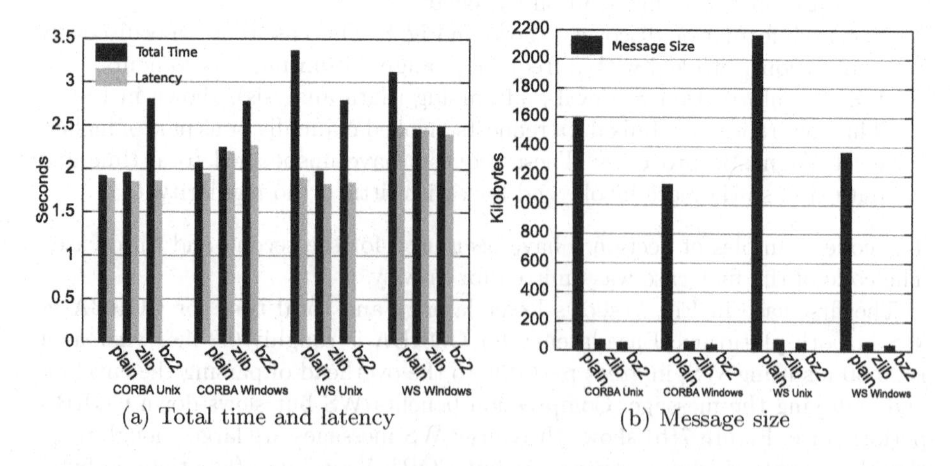

(a) Total time and latency (b) Message size

Fig. 9. Performance measurements for test three: Log Parsing

are very similar to those obtained in the previous case, since parsing overhead dominates total time. CORBA times are longer than in the first case, but shorter than those of WS.

In these two test cases, Windows times are higher than those of Unix, because of the overhead of operating with COM objects. Nevertheless, these overheads are smoothed over in the third test case. From figures 7 and 8, it is clear that message size determines total time, affecting WS much more, both due to their XML envelope and codification. The envelope size (almost fixed) clearly affects only the first case, but codification introduces a 20% message size overhead on average.

The third case (Fig. 9) shows a much narrower spread of values due to message size (1Mb+); thus, total time is dominated by transfer time, instead of by protocol overheads. In this test compression in WS achieves times comparable to those of uncompressed CORBA. This would be more noticeable with less bandwidth, as WS compression ratios of 40:1 are reported in Fig. 9(b).

In general, all times are very acceptable, although CORBA has a clear advantage. The benefits of compression are dubious, except in the third test for WS. bzip2 is slower and is oriented to larger data sets than zlib, which is a better choice for the tested cases. Although both WS and CORBA are acceptable solutions for information exchange in our framework, monitoring and low-latency applications strongly favor CORBA over WS due to its message compactness and better processing time.

6 Conclusions

We have explored the adaptation of our AdCIM framework to various system administration applications, focusing on the following relevant features:

- Definition of an XML Schema and a new XML mapping of CIM, named miniCIM, that simplifies the representation and validation of CIM data and allows the use of standard XML Schema tools to manage CIM seamlessly.
- Extraction of monitoring, service and log data from Windows and Unix into miniCIM instances. This is achieved using different techniques (text-to-XML parsing grammars, WMI scripts) due to the different management approaches supported by each OS.

We have also discussed methods and alternatives to implement multiplatform and low-latency transport methods using two different approaches: CORBA and Web Services technologies. Lastly, we have validated and tested the implementations by defining a testing framework for measuring total time, latency and message size.

The chosen domain of processor statistics, network services and log data analysis has illustrated the use of these methods for administration domains particularly dissimilar between operating systems. But the scope of our approach is not limited to such domains, as the model and implementation technologies are truly general and extensible.

As future work, we plan to implement CORBA interfaces based on Valuetypes, design real-time support agents that diagnose and aggregate global network issues, and extend WMI coverage (i.e. for text-based Windows configurations).

References

1. Microsoft Windows Services for Unix [Visited July 2007], http://www.microsoft.com/windows/sfu/
2. Noer, G.J.: Cygwin32 - A Free Win32 Porting Layer for UNIX Applications. In: 2nd USENIX Windows NT Symposium, pp. 31–38 (1998)
3. Spinellis, D.: Outwit - Unix Tool-Based Programming Meets the Windows World. USENIX 2000. In: Technical Conference pp. 149–158 (2000)
4. Pras, A., Drevers, T., van de Meent, R., Quartel, D.: Comparing the Performance of SNMP and Web Services-Based Management. IEEE Electronic Transactions on Network and Service Management 1(2), 1–11 (2004)
5. Nikolaidis, A.E., Doumenis, G.A., Stassinopoulos, G.I., Drakos, M., Anastasopoulos, M.P.: Management Traffic in Emerging Remote Configuration Mechanisms for Residential Gateways and Home Devices. IEEE Communications Magazine 43(5), 154–162 (2005)
6. Yoo, S.M., Ju, H.T., Hong, J.W.: Performance Improvement Methods for NETCONF-Based Configuration Management. In: Kim, Y.-T., Takano, M. (eds.) APNOMS 2006. LNCS, vol. 4238, pp. 242–252. Springer, Heidelberg (2006)
7. Strauss, F., Klie, T.: Towards XML Oriented Internet Management. In: 8th IFIP/IEEE Int'l Symposium on Integrated Network Management, IM 2003, pp. 505–518 (2003)
8. Yoon, J.H., Ju, H.T., Hong, J.W.: Development of SNMP-XML Translator and Gateway for XML-based Integrated Network Management. International Journal of Network Management 13(4), 259–276 (2003)
9. DMTF. Common Information Model (CIM) Standards [Visited July 2007], http://www.dmtf.org/standards/cim
10. Diaz, I., Touriño, J., Salceda, J., Doallo, R.: A Framework Focus on Configuration Modeling and Integration with Transparent Persistence. In: 19th IEEE International Parallel and Distributed Processing Symposium (IPDPS 2005). Workshop on System Management Tools for Large-Scale Parallel Systems, p. 297a (2005)
11. CORBA/IIOP Specifications [Visited July 2007], http://www.omg.org/technology/documents/corba_spec_catalog.htm
12. Web Services Architecture [Visited July 2007], http://www.w3.org/TR/ws-arch/
13. DMTF. Specification for the Representation of CIM in XML [Visited July 2007], http://www.dmtf.org/standards/documents/WBEM/DSP201.html
14. Lee, D., Chu, W.W.: Comparative Analysis of Six XML Schema Languages. ACM SIGMOD Record 29(3), 76–87 (2000)
15. Dalke, A.: Martel [Visited July 2007], http://www.dalkescientific.com/Martel/
16. XSLTC [Visited July 2007] http://xml.apache.org/-xalan-j/-xsltc/-index.html
17. Object Management Group. XMLDOM - DOM/Value Mapping Specification [Visited July 2007], http://www.omg.org/cgi-bin/doc?ptc/2001-04-04
18. Grisby, D.: omniORB - Free High Performance ORB [Visited July 2007], http://omniorb.sourceforge.net/
19. Salz, R.: ZSI - The Zolera SOAP Infrastructure Developer's Guide [Visited July 2007], http://pywebsvcs.sourceforge.net/zsi.html

OMA DM Based Remote Software Debugging of Mobile Devices*

Joon-Myung Kang[1], Hong-Taek Ju[2], Mi-Jung Choi[1], and James Won-Ki Hong[1]

[1] Dept. of Computer Science and Engineering, POSTECH, Korea
[2] Dept. of Computer Engineering, Keimyung University, Korea
{eliot,mjchoi,jwkhong}@postech.ac.kr,juht@kmu.ac.kr

Abstract. The mobile devices have gained much functionality and intelligence with the growth of network technologies, the abundance of network resources, and the increase of various services. At the same time they are also becoming complicated and related problems to services and resources of mobile devices frequently occur. Since it is not easy for the manufacturers to completely remove the software errors of the mobile devices before they sell them, the users face inconvenience caused by them and the credibility of the manufacturers also decreases. So far, no definitive method has been given to debug software errors of the sold mobile devices. In this paper, we propose a debugging method to find and correct the software errors of the sold mobile devices based on the Open Mobile Alliance (OMA) Device Management (DM) standard. We define the managed objects (MOs) for composing the execution image and design the management operations for collecting MOs at the central server. We present a system that we have developed based on the MOs and the management operations. We also present how to debug software errors with the execution image and JTAG debugger.

Keywords: Device Management, Software Debugging, Mobile Device Diagnostics, OMA DM, OMA DM DiagMon.

1 Introduction

Recently, the growth in ubiquitous and mobile computing systems has led to an early introduction of a wide variety of intelligent wireless and mobile network capable devices [1]. They have gained much functionality and intelligence as the hardware and software technologies are getting advanced. The mobile devices are becoming more complex continuously. The higher the complexity of a device becomes, the higher the possibility of errors in it [15].

* This research was supported by the MIC (Ministry of Information and Communication), Korea, under the ITRC (Information Technology Research Center) support program supervised by the IITA (Institute of Information Technology Assessment)" (IITA-2006-C1090-0603-0045).

S. Ata and C.S. Hong (Eds.): APNOMS 2007, LNCS 4773, pp. 51–61, 2007.

It is very difficult for manufacturers of mobile devices to completely remove software errors before they sell them. Therefore, the sold mobile devices usually still have software bugs, which cause the device to self-reset, freeze, or have system malfunctions. The most common method in solving these problems is to reset the system, which is only a temporary solution. Hence, the users still have to face these on-going problems. Furthermore, the manufacturers cannot find the root cause of these software errors even when they are reported to the service center, since it is difficult to reproduce the same errors in experiments. Therefore, it is difficult to locate the source code's exact error of the software. This leads to a decrease in the credibility of the current mobile devices, and there will not be a solution to fix the software errors in the products that are to be manufactured in future.

In this paper, we propose a debugging method to find and correct the software errors of the sold mobile devices. Through this method, the software errors can be corrected by the manufacturers and updated software versions can be provided to the users via service centers, and the corrected software can be applied to the future mobile devices as well. This will alleviate the users' inconvenience and increase the mobile device's credibility.

One of the methods for debugging the software error is to dump the execution image when an error occurs and to make use of it. The execution image includes registers, stack, key events, and so on. This method is used in common software debugging. For example, we debug using 'core dump' in the UNIX system, and report the gathered errors and logs to the Microsoft server in the Windows system. We also use this debugging method to correct the software errors with the execution image. That is, the software errors of the mobile device can be fixed by using the dumped execution image when the device is automatically or manually reset.

It is difficult to collect the execution image of the sold mobile devices due to the large size of data, and low bandwidth and high error rate of the wireless network environment. Moreover, the process of collecting the images in the central server is very complex. That is, when a reset occurs, the execution image must be produced, and the reset must be reported to the central server, where the image needs to be gathered. When debugging the software error, the system information of the mobile device like CPU type, memory size, etc. is also required along with the execution image.

To satisfy these constraints, the Open Mobile Alliance (OMA) Device Management (DM) framework [2, 3], which is the international standard for the mobile device management, can be used. The DM protocol proposed by the OMA is appropriate for collecting large-scale data in a wireless network environment. In addition, the DM protocol includes the management operations needed for collecting the execution image from the mobile device. In the OMA DM framework, the system information of the mobile device has already been defined as the standardized managed objects (MOs) [4]. We only need to add new MOs to define all of the information for debugging in the OMA DM framework. Therefore, we defined the MOs for creating the execution image and designed the process for collecting it.

In this paper, we present the remote software debugging system for user mobile devices. This system can collect MOs and create the execution image to debug the software errors. The developed system consists of a client and server. The client collects the system data related to the defined MOs, while the server collects such data and uses it to correct the software errors with the JTAG debugger.

The remainder of this paper is organized as follows. Section 2 describes the OMA DM and the OMA DM Diagnostics and Monitoring standard [11, 12]. Section 3 describes the management architecture, management information and management operations. Section 4 presents the system development for validating our proposed solution. Finally, conclusions are drawn and future work is discussed in Section 5.

2 OMA DM and OMA DM DiagMon

In this section, we describe the specification of the OMA DM and the OMA DM DiagMon standard. We present bootstrapping, device description framework, and OMA DM protocol as well as the functions defined by the OMA DM DiagMon Working Group (WG).

2.1 OMA DM

OMA has been established by mobile operators, information technology companies, wireless equipment vendors, and content providers. It has defined the standard for wireless mobile terminals. The OMA DM WG is one of the major WGs in OMA. It has proposed how to define the management information for the mobile devices in the form of DM tree and how to manage the mobile devices remotely using DM protocol [5], which is an SyncML [6] based protocol aimed at providing remote synchronization of mobile devices. The OMA DM standard includes three logical components such as device description framework (DDF) [7], bootstrapping [8], and OMA DM Protocol [5]. DDF provides necessary information about MOs in device for the server. Bootstrapping configures initiative setting of devices. The OMA DM protocol defines the order of communicated packages by the server and client. Each device that supports OMA DM must contain a management tree [9], which organizes all available MOs in the device as a hierarchical tree structure where all nodes can be uniquely addressed with a uniform resource identifier (URI) [10]. The management tree is not completely standardized yet. OMA allows each device manufacturer to

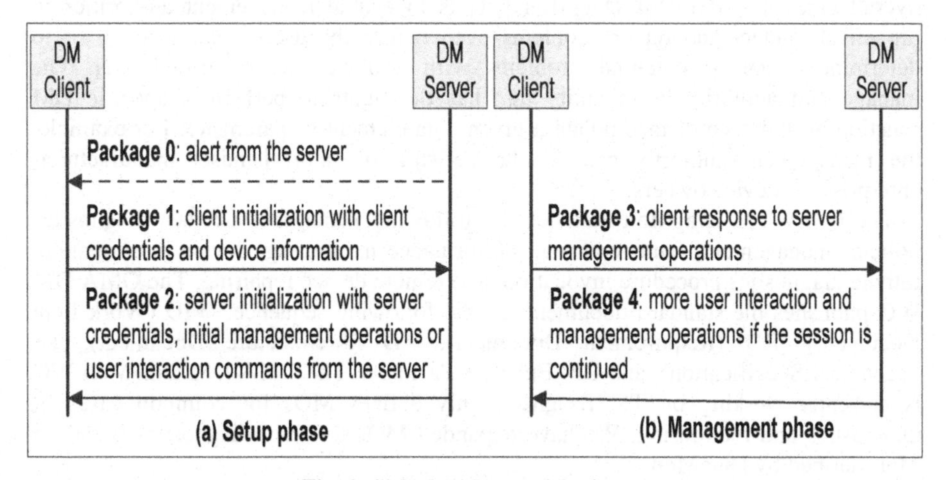

Fig. 1. OMA DM Protocol Packages

easily extend the management tree by adding new management functions in their devices by defining and adding the management nodes to the existing management tree. We show how this can be done in Section 3.

The OMA DM protocol consists of two parts as shown in Fig. 1: setup phase (authentication and device information exchange) and management phase [5]. The management phase can be repeated as many times as the server wishes, and the management session may start with package 0 (the trigger).

Table 1. OMA DM Commands

Feature	Description	OMA DM Command
Reading a MO content or MO list	The server retrieves the content from the DM Client or the list of MOs residing in a management tree.	***Get***
Adding a MO or MO content	A new dynamic MO is inserted	***Add***
Updating MO content	Existing content of an MO is replaced with new content	***Replace***
Removing MO(s)	One or more MOs are removed from a management tree	***Delete***
Management session start	Convey notification of device management session	***Alert***
Executing a process	New process is invoked and return a status code or result	***Exec***

Table 1shows the OMA DM commands, which are similar to SNMP operations [16, 17]. A management session is composed of several commands. The server retrieves the MO content or MO list from the DM client by the '*Get*' command. The server can add a new MO by the '*Add*' command. Moreover, the server can replace or delete by '*Replace*' or '*Delete*' command. The client can notify the management session by '*Alert*' command, while the server can execute a new process to the client by '*Exec*' command. We can design the diagnostic process by a composition of these commands.

2.2 OMA DM DiagMon

The OMA DM WG has introduced device management diagnostics and device monitoring functionality to remotely solve the problems of mobile devices. The overall goal of OMA DM DiagMon [11] is to enable management authorities to proactively detect and repair problems even before the users are impacted, or to determine actual or potential problems with a device when needed [12]. The management authority is an entity that has the rights to perform a specific DM function on a device or manipulate a given data element or parameter. For example, the management authority can be the network operator, handset manufacturer, enterprise, or device owners.

The OMA DM DiagMon includes the following management areas: diagnostics policies management, fault reporting, performance monitoring, device interrogation, remote diagnostics procedure invocation, and remote device repairing. The OMA DM WG publishes the standard documents as the following sequence: WID (Work Item Document), RD (Requirement Document), AD (Architecture Document), TS (Technical Specification), and EP (Enablers Package). The OMA DM DiagMon WG is currently working on TS. DiagMon only defines MOs for common cases of diagnostics and monitoring. We have expanded MOs for reset diagnostics based on MOs defined by DiagMon.

3 Management Architecture

Our goal is to provide an efficient method in order to debug software errors by collecting the reset data from the sold mobile devices. Fig. 2 shows the overall management architecture of our proposed solution, which is composed of the DM Server and DM Client. The DM Server sends the initialization and execution request of the reset diagnostic function to the DM Client. The DM Client, which is equipped in various mobile devices such as PDA, cell phone, lap top etc., replies the result of the request by the DM Server. The analysis server obtains the data related to reset and debugs the software errors by JTAG debugger. The presenter shows the data and the result of the debugging.

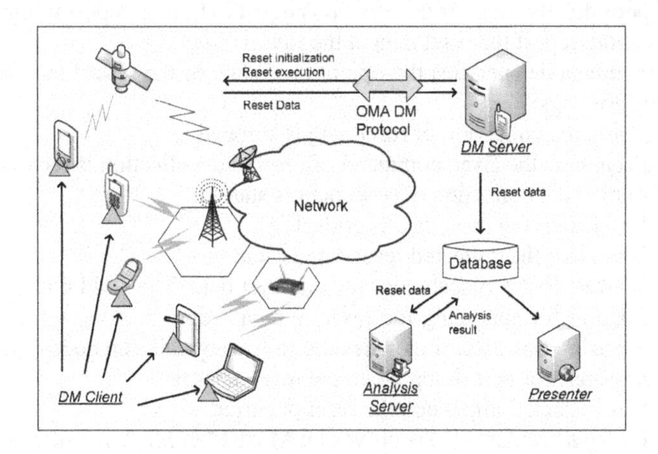

Fig. 2. Overall management architecture

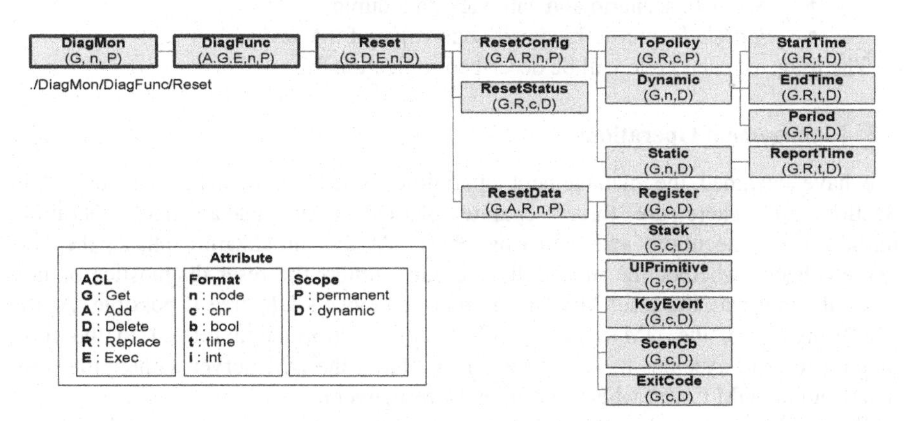

Fig. 3. DM tree for reset diagnostics

3.1 Management Information

We have defined the DM tree for the reset diagnostics (shown in Fig. 3) by expanding the *DiagMon* node and *DiagFunc* node defined by the OMA DM DiagMon WG.

These are not the standard nodes but under consideration of the standards. We have created the *Reset* node for the reset diagnostics. Each node has its own *access control list* (*ACL*), *format*, and *scope* attributes, which are denoted in the parenthesis in each node of Fig. 3. For example, the DM server can request *GET*, *DELETE*, and *EXEC* command on *Reset* node because its *ACL* is (G, D, E). The *format* defines the type of the node. The *scope* specifies whether the node is permanent or dynamic.

We have defined three children nodes under the *Reset*: *ResetConfig*, *ResetStatus*, and *ResetData*. The *ResetConfig* node is a placeholder for the reset configuration information. This interior node has following three children nodes:

- *ToPolicy*: the type of reporting schedule (value: *Dynamic*, *Static*)
- *Dynamic*: collect reset data from *StartTime* to *EndTime* and report it periodically. (e.g., if the *Period* is equal to 0, then report it immediately.)
- *Static*: report the reset data at the *ReportTime*.

The *ResetStatus* node specifies the operational state of the reset function. Its value is one of the followings:

- *None:* the collection of reset data is stopped
- *Prepared:* the Exec command of reset data collection is received
- *Active:* the collection of reset data is started
- *Processed:* the reset data is collected
- *Reported:* the collected reset data is sent

The *ResetData* node is a placeholder for the reset data. The child nodes contain the relevant information for analyzing the reset. We can reproduce the reset case on the JTAG debugger using this data. It includes the following children nodes.

- *Register*: register dump when the reset occurred.
- *Stack*: stack dump when the reset occurred
- *UIPrimitive*: last UI Primitives on MMI(Man Machine Interface)
- *KeyEvent*: event value of KEY_EVENT and data on MMI
- *ScenCb*: scenario and call back data dump
- *ExitCode*: pre-defined exit code value of the device

The usage of each node will be described in Section 3.2

3.2 Management Operations

We have designed the management operations based on the DM tree defined in Section 3.1. There are three separate phases in the management operation: initialization, execution, and gathering phase. At the initialization phase, the DM Server checks whether the mobile device can support the reset diagnostics or not. Also, it can create the reset MOs in the mobile device's DM tree if possible. At the execution phase, the DM Server sets the policy information related to the reset diagnostics and executes it. At the gathering phase, the DM Server gathers the reset data from the DM Client when it notifies the reset event.

By dividing the management operation into three phases as shown in Fig. 4, an efficient management operation can be achieved. First, each management phase consists of the same management commands. Hence, a single management command can process an operation of many MO addresses (Target LocURIs), which decreases the size of management package. Second, each phase can be independently used for its purposes. That is, to diagnose a reset, all three phases do not need to be repeated.

Once the initialization is processed, it does not need to be repeated. Also, after the execution phase, there is no need to repeat it to process gathering, as long as the policy for collecting the data remains unchanged. Therefore, it is more efficient than processing all three phases to diagnose a reset.

Fig. 4 (a) shows the initialization phase of the reset diagnostics. When the DM Server wants to initialize the reset diagnostics function, it needs to send the *NOTIFICATION* message [13] to the DM Client. When the DM Client receives the notification message, it sends the server-initiated *ALERT* command to the DM Server with the device name. Next, the DM Server sends the *ADD* command to initialize the reset diagnostics function and the *REPLACE* command to set *ACL* for *Reset as GET, DELETE,* and *EXEC*. For efficiency, as shown in the package *#4* of the sequence 5 and 9 in Fig. 4 (a), we have added many MOs by using one *ADD* command.

If the mobile device supports the reset diagnostics, it can add *Reset* to its DM tree and send a successful *STATUS* command (200) to the DM Server. If the addition of the *Reset* is successful, then the DM Server adds *ResetConfig*, *ResetStatus*, and *ResetData* step by step. Finally, the DM Server sends a completion message to the DM Client to finish this management session. After the initialization phase, the mobile device is ready to execute the reset diagnostics function.

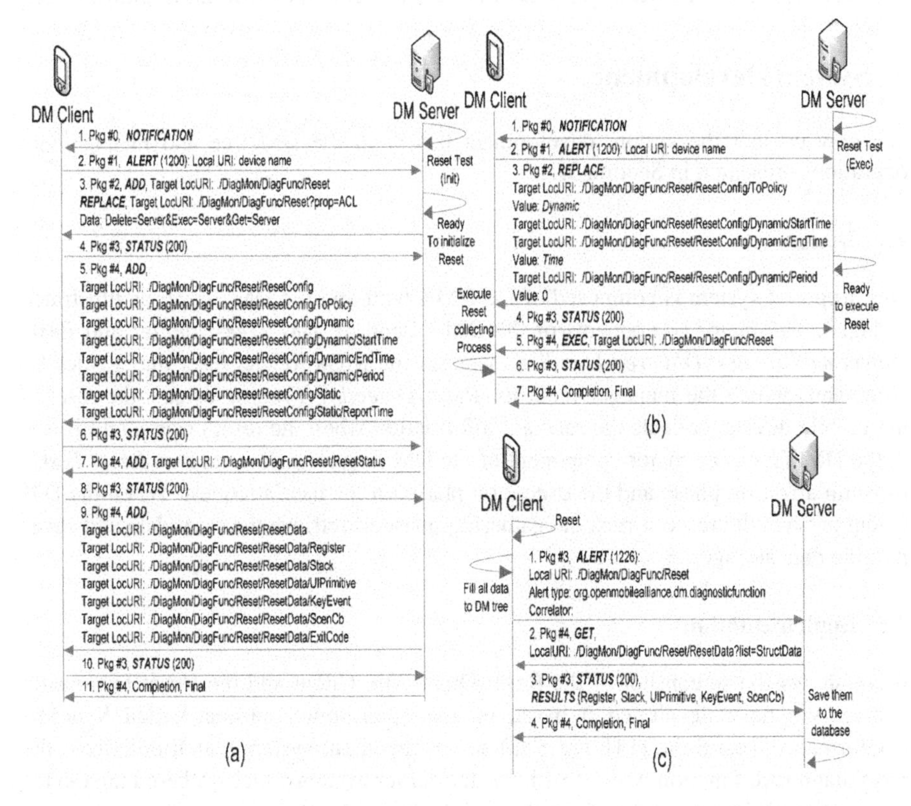

Fig. 4. Three phases of management operation: (a) Initialization phase (b) Execution phase (c) Gathering phase

Fig. 4 (b) shows the execution phase of the reset diagnostics. When the DM Server wants to execute the reset diagnostics function, it sends the *NOTIFICATION* message like the initialization phase. Then the DM Client sends the server-initiated *ALERT* command to the DM Server. The DM Server sends the *REPLACE* command to set the policy information. Since we need the real time data for analyzing the reset, the policy is dynamic in order to get the data from *StartTime* to *EndTime* periodically. The *Period* is 0 in order to receive the reset event immediately. If the mobile device has initialized the reset diagnostics function, then it sends the status code as 200 (success). Otherwise, it sends the status code as error code. Then, the DM Server sends the *EXEC* command to execute the reset diagnostics function. The DM Client executes the reset collecting process. Finally, the DM Server sends the completion message to the DM Client and the management session is finished.

Fig. 4 (c) shows the gathering phase of the reset diagnostics. When the reset occurs, the DM Client stores all relevant information to its DM tree. When it is ready to report, it sends the generic *ALERT* command to the DM Server. The DM Server sends the *GET* command for the *ResetData* node to retrieve all information related to the reset. It saves data to the database. As mentioned earlier, we can reproduce the reset case on the JTAG debugger using this data and find the error in the source code.

4 System Development

We now present the system development based on the DM tree and management operations presented in Section 3.

4.1 Design

Our proposed system is composed of the DM Client and the DM Server as illustrated in Fig. 5. The major components of the DM Client are *DM Tree Handler* and *Reset Detecting Process. DM tree Handler* manages the MOs for the reset diagnostics by commands which the manager requests. *Reset Detecting Process* detects the reset in the mobile device, collects the relevant information when the reset occurs, and fills it in the DM Tree. The major component of the DM Server is *Reset Tester. Reset Tester* runs initialization phase and the execution phase on the user's request. When the DM Client notifies the reset, it runs the gathering phase to retrieve the reset data and saves it to the data storage.

4.2 Implementation

Fig. 6 shows the screenshot of the Reset Diagnostic Client and the Reset Diagnostic Server. We have developed it based on the open source project called SyncML Conformance Test Suite [14]. Fig. 6 (a) shows the client system which initialized the reset diagnostic function. Fig. 6 (b) shows the server system which gathered the device information and the reset data from the mobile device 1.

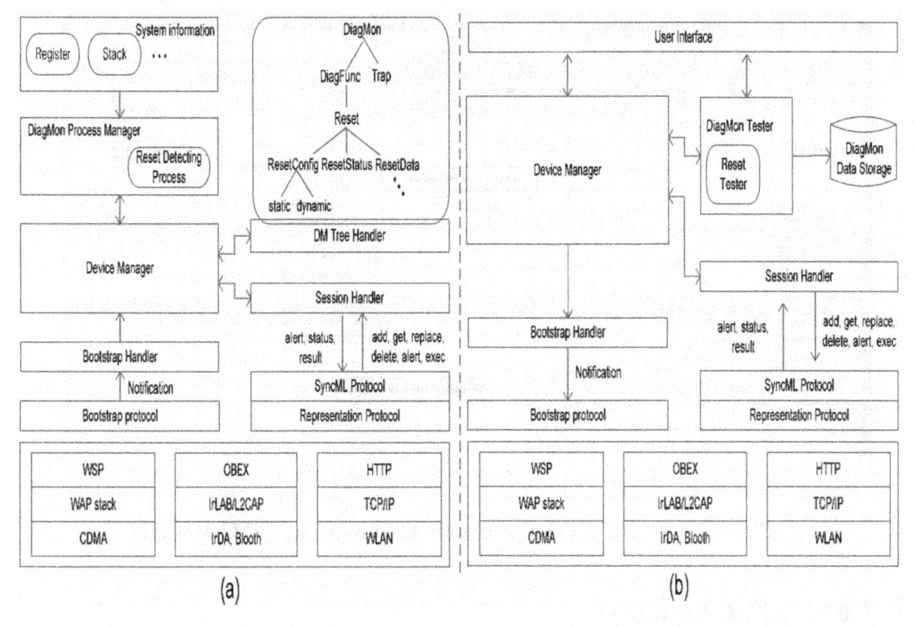

Fig. 5. System Architecture Design: (a) DM Client and (b) DM Server

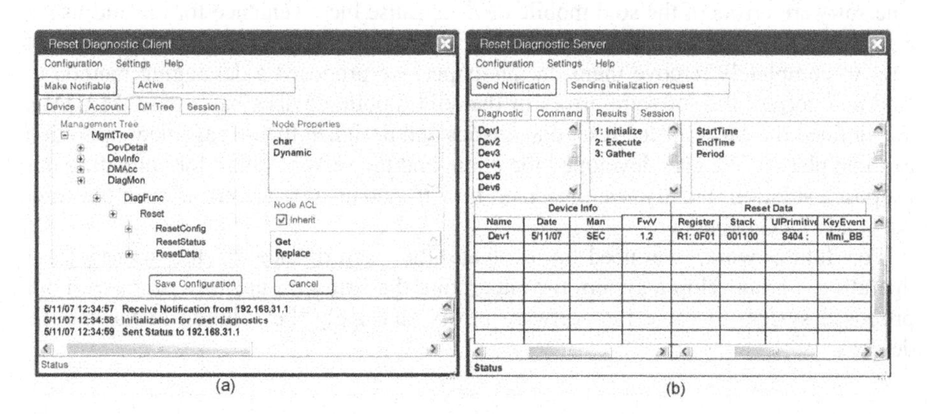

Fig. 6. (a) Reset Diagnostic Client system (b) Reset Diagnostic Server system

The server gathers the reset data from the client and the data is used to debug the reset error using JTAG debugger. Fig. 7 shows the screenshot of debugging using Trace32 debugger [18] as JTAG debugger. First, it loads the CMM file in accordance to the mobile device model and configures the debugging environment. The mobile device model information is recorded in the standard object *DevInfo*. This debugger can set the LR address in Register14 and we can find the source location of Register 14 address in the Data.list window. The name of source file and function can be found in the symbol.info window.

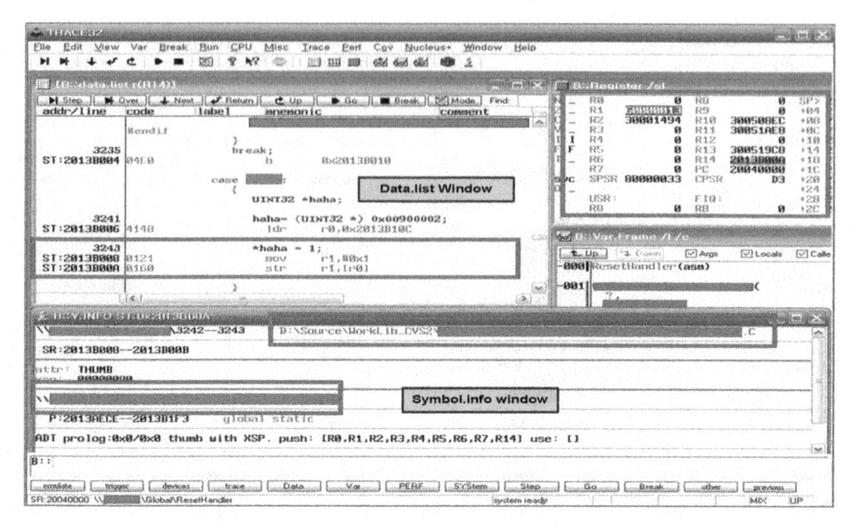

Fig. 7. Screenshot of debugging software error using JTAG debugger

5 Concluding Remarks

The software errors of the sold mobile devices cause inconvenience for the end-users. So far, we have used temporary solutions such as the system reset because it is not easy to completely remove them. In this paper, we proposed a debugging method to find and correct the software errors of the sold mobile devices based on OMA DM. We defined the MOs for the reset diagnostics and designed the management operation as three phases. We also developed the client and the server system for validating our proposed solution. Finally, we presented how to debug software errors with the reset data and JTAG debugger.

For future work, we need to evaluate the performance of the management operations and develop a system for integrating the firmware update system with our proposed system to solve the software errors and apply the solution to the mobile devices.

References

1. Chakravorty, R., Ottevanger, H.: Architecture and Implementation of a Remote Management Framework for Dynamically Reconfigurable Devices. In: ICON 2002, Singapore, pp. 375–381 (August 2002)
2. OMA (Open Mobile Alliance), http://www.openmobilealliance.org/
3. OMA DM (Device Management) Working Group, http://www.openmobilealliance.org/tech/wg_committees/dm.html
4. OMA, OMA Device Management Standardized Objects (2007)
5. OMA, OMA Device Management Protocol (2007)
6. SyncML Forum, SyncML Device Management Protocol, http://www.syncml.org/
7. OMA, OMA DM Device Description Framework, Version 1.2 (2007)
8. OMA, OMA Device Management Bootstrap (2007)

9. OMA, OMA Device Management Tree and Description (2007)
10. IETF, Uniform Resource Identifiers (URI) RFC 2396 (1998)
11. OMA, DiagMon (Diagnostics and Monitoring) Working Group.
12. OMA, DiagMon Requirement draft version 1.0 (June 2006)
13. OMA, OMA Device Management Notification Initiated Session (June 2006)
14. SyncML Forum, SyncML Conformation Test Suite, http://sourceforge.net/projects/oma-scts/
15. Adwankar, S., Mohan, S., Vasudevan, V.: Universal Manager: Seamless Management of Enterprise Mobile and Non-Mobile Devices. In: MDM 2004, Berkeley, CA, USA, pp. 320–331 (January 2004)
16. Stallings, W.: SNMP, SNMPv2, SNMPv3 and RMON 1 and 2, 3rd edn. Addison-Wesley, Reading, MA, USA (1999)
17. A Simple Network Management Protocol (SNMP), RFC 1157 http://www.ietf.org/rfc/rfc1157.txt
18. MDS Technology, Trace32 Debugger, http://www.mdstec.com/

Process Control Technique Using Planning and Constraint Satisfaction

Haruhisa Nozue, Hajime Nakajima, Haruo Oishi, Takeshi Masuda, and Tetsuya Yamamura

Access Network Service Systems Laboratories, NTT Corp.,
NTT Makuhari Bldg. 1-6 Nakase Mihama-ku Chiba-shi, Chiba 261-0023 Japan
{nozue,nakajima.hajime,oishi,tmasuda,yamamura}@ansl.ntt.co.jp

Abstract. We propose a new technique that realizes flexible process control by combining "the planning paradigm" and "the constraint satisfaction paradigm." Two factors should be kept consistent in business processes: business objects and activity sequences. We adopted the planning paradigm to control the sequence of processes and the constraint satisfaction paradigm to keep the objects consistent. We then developed a new process control technique that allows operation support systems to suggest executable activities and consistent values to users dynamically. We implemented a prototype system and confirmed its feasibility.

Keywords: constraint satisfaction, planning, process control, workflow.

1 Introduction

The need for flexibility of workflow management systems (WfMSs) has been recognized while they help to improve the efficiency of business processes. There are certain types of business processes where a WfMS may cause a decline in efficiency because of its insufficient flexibility. People tend to define only well-known flow patterns because the number of flows that they can define in advance is finite and the form for representing flow patterns leads people into modeling in such fashion. If someone has to proceed with an undefined flow, the system is no longer in operation and needs to be modified. In such cases, the efficiency of business processes may deteriorate greatly contrary to the purpose of WfMSs. As a result, the over all loss of efficiency may become too large to ignore even if the WfMS contributes to efficiency in regular cases. This does not matter if we can define all the flow patterns that can occur in advance or easily modify workflow schemas at any time. However, that is very difficult when various flow patterns are required or possible. First, it is often difficult to enumerate all the possible sequences because it is hard to anticipate all cases while constructing the system, and in that case the number of flow patterns might be too huge to enumerate. Second, it is very difficult to modify flow patterns because relations between activities in a WfMS tend to be interdependent and complicated.

The above issues occur frequently in some business processes such as access network resource allocation, because of the huge number of flow patterns that

S. Ata and C.S. Hong (Eds.): APNOMS 2007, LNCS 4773, pp. 62–71, 2007.

must be considered. Various flow patterns may be possible since dependency among activities may be weak in such processes. At the same time, with such processes, the aim is mainly to move the business objects into a particular state. In other words, there are certain business objects that are dealt with in a business process and certain conditions that the business objects must obey. Moreover, efficiency depends on whether the system can work with operations proper to the state of business objects at each moment. People tend to implement a WfMS in an ad-hoc manner to manage such business object conditions. This leads to difficulties when modifying workflow schemas. Hence, it is necessary to get the state of business objects and the sequence of activities to interact.

Our purpose is to provide a new flexible technique for managing business processes that are unsuitable for WfMSs. There are two kinds of flexibility required for such business processes, an adequate number of alternative flows and the capacity for modification. WfMSs lack both kinds of flexibility in a business process capable of (or requiring) a large number of flow patterns.

We set the following basic functions as requirements for our technique:

1. manage sequence of activities in the business process
2. ensure consistency of the business objects
3. make the process sequence and object state interact

Needless to say, the first is already met by WfMSs. However, the way to represent relations between activities embraced in general WfMSs causes the lack of flexibility. Therefore, we adopt a different paradigm to represent the relations between activities. Moreover, it is greatly preferable to consider the interaction between the process sequence and the business object state as mentioned above.

In this paper, we describe a new technique that achieves flexible process management by controlling the sequence of activities and consistency of objects (e.g. a suite of equipment in an optical access network service) and getting them to interact. Specifically, we control the consistency of objects with the *constraint satisfaction* paradigm and the sequence of activities with the *planning* paradigm, while causing each of their states to influence each other through a predefined mapping. Our technique helps to improve the process management method as regards both types of flexibility mentioned above, namely an adequate number of alternatives at runtime and the easy modification.

2 Preliminary

Now we briefly introduce the two paradigms that we adopt. There are certain solver libraries that can easily derive the solutions of either paradigm [1], [2].

2.1 Constraint Satisfaction Problem

A *constraint satisfaction problem (CSP)* concerns finding values that meet all the constraints [8]. *Constraint satisfaction* is a paradigm for finding solutions while formalizing a problem to a CSP.

Definition 1 (CSP). *A CSP is defined as a triple* $\mathcal{P}_C = (X, \mathcal{D}, \mathcal{C})$ *where:*

- $X = \{x_1, x_2, \ldots, x_n\}$*: a set of* n *variables.*
- $\mathcal{D} = \{D_1, D_2, \ldots, D_n\}$*: a set of finite domains of* X*, that is, each* $D_i \in \mathcal{D}$ *is a set of values assignable to* $x_i \in X$*.*
- $\mathcal{C} = \{c_1, c_2, \ldots, c_m\}$*: a set of constraints. A constraint is a subset of a direct sum of certain domains, i.e.* $c_j \subseteq D_{j_1} \times \cdots \times D_{j_k}$*. Each element in a constraint is a tuple of values simultaneously assignable to respective variables.*

Definition 2 (Solution of CSP). *We call* $(v_1, \ldots, v_n)$ *a* solution *of the CSP* $\mathcal{P}_C$ *if it is an element of* $D_1 \times D_2 \times \cdots \times D_n$ *and satisfies all the constraints in* $\mathcal{C}$*, that is,* $(v_{j_1}, v_{j_2}, \ldots, v_{j_k}) \in c_j$ *for all* c_j *in* $\mathcal{C}$*.*

2.2 Planning Problem

A *planning problem* is a problem related to finding a sequence of activities to execute under given conditions [4].

Definition 3 (Planning Domain). *A* planning domain Σ *is defined as a triple* (S, A, γ) *where:*

- $L = \{p_1, p_2, \ldots, p_n\}$ *is a set of propositional symbols.* S *is equal to (or a subset of)* 2^L*. Each element of* S *is a subset of* L *and called* state*.*
- *Each* $a \in A$ *is called* action *and characterized by a tuple of four subsets of* L*,* $(p^+(a), p^-(a), e^+(a), e^-(a))$*.* $p^+(a)$ *and* $p^-(a)$ *are called* preconditions *of* a*.* $e^+(a)$ *and* $e^-(a)$ *are called* effects *of* a*. A positive set and a negative set have no common elements, i.e.* $p^+(a) \cap p^-(a) = e^+(a) \cap e^-(a) = \emptyset$*.*
  If a state $s \in S$ *satisfies* $(s \subseteq p^+(a)) \wedge (s \cap p^-(a) = \emptyset)$ *for an action* a*, then* a *is* applicable *to* s*.*
- γ *that maps some of* $S \times A$ *to an element of* S *is defined as follows:*

$$\gamma(s, a) = \begin{cases} (s \setminus e^-(a)) \cup e^+(a) & a \text{ is applicable to } s, \\ \text{undefined} & a \text{ is not applicable to } s. \end{cases}$$

Definition 4 (Planning Problem and Solutions). *A planning problem is a triple* $\mathcal{P}_P = (\Sigma, s_0, g)$*, where* $s_0 \in S$ *is the initial state and* $g \subseteq L$ *is goal propositions. A* solution *of a planning problem* $\mathcal{P}_P$ *is a finite sequence of actions* $\langle a_1, \ldots, a_k \rangle$ *such that* $g \subseteq \gamma(\cdots \gamma(\gamma(s_0, a_1), a_2) \cdots, a_k)$*.*

We use the term "*activity*" to refer to atomic tasks in business processes rather than "action" that corresponds to that of planning problems.

3 Details of Our Method

Our technique achieves business process flexibility and business object consistency simultaneously while causing two paradigms, namely the constraint satisfaction and planning paradigms, to interact. The constraint satisfaction paradigm copes with the static aspects of the business process, that is, the consistency of constraints and various data. The planning paradigm copes concurrently with the dynamic aspect of the business process, that is to say, activities executed successively and the state that varies along with those activities. Furthermore, the state of one aspect influences the other through the mapping between them. As a result of the interaction between these two paradigms, the system with our technique accomplishes flexible processes management.

Until the goal conditions are satisfied, the following procedure repeats:

1. System → User: show executable activities.
2. User → System: select an activity to execute.
3. System → User: show its target variable and consistent values.
4. User → System: select a value for the variable.
5. If the state of planning does not meet the goal propositions, return to Step 1.

The system regards the addition of a constraint to fix a value to a variable as an activity. In Step 1, the system presents alternative activities that are executable. The system uses the planning paradigm to derive such activities. In Step 3, the system presents values that are consistent for all constraints. To derive such values, the system applies the constraint satisfaction paradigm.

3.1 Difference Between Workflow and the New Method

Although planning and workflow function similarly in that they suggest executable activities for each moment, they differ as regards the form of relations between activities. The form of the planning paradigm is superior to that of workflow in terms of both types of flexibility.

In workflow schemas, the definitions of relations between activities depend directly on the existence of particular activities. For example, the relation between two activities `A` and `B` is generally represented like "`A` is followed by `B`." Besides, all flow patterns must be described in advance. Therefore, people tend to model their business processes from the viewpoint of how to proceed with the processes and thus define only regular processes (over-specification). Hence, it is often necessary to modify the definition of the workflow schema when the business has changed or an exception has occurred. But then, people with workflow schemas are likely to define (and implement) detailed jobs in each activity considering their sequence. Moreover, in order to make sequences dependent on the state of the business objects, people often define and implement complicated ad-hoc rules. As a result, it is often very difficult to modify workflow schemas.

In planning description, the preconditions and effects of each activity are represented based on the state. Hence, we can define each activity independently of others. For example, activities `A` and `B` are handled as follows:

1. As a result of `A`, the state has changed into a certain state.
2. `B` is one of the activities that can be executed in this state.

This means that there is less interdependence that might cause over-specification and difficulty of modifying in WfMSs. Moreover, people can define very simply,

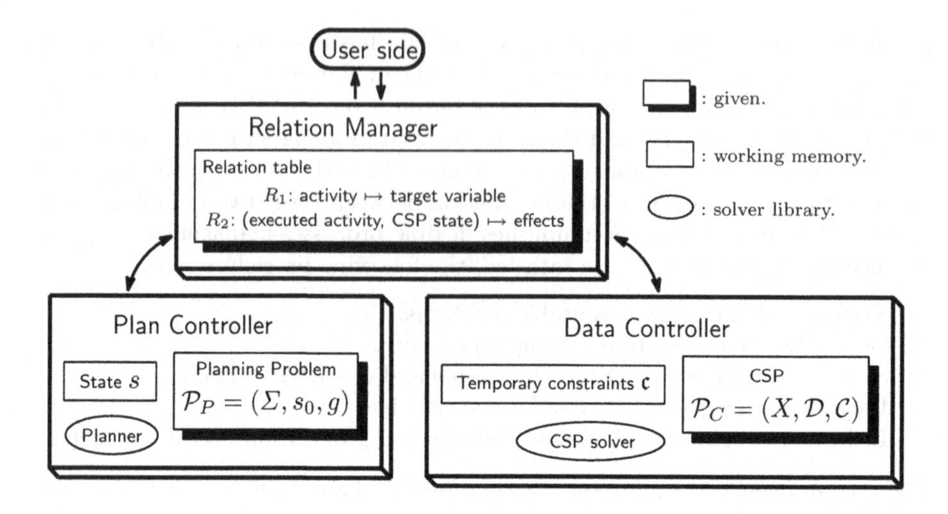

Fig. 1. System architecture

unlike with workflow, a process that has a large number of flow patterns because people have to describe not flow patterns but conditions for activities. Planning description has such a nature, and thus is suitable for flexible business processes.

Besides, with our technique, rules can be implemented more independently than the ad-hoc manner in WfMSs. Because our technique leads users to separate rules for activities and rules for business objects, and makes it possible to separate them from its implementation.

3.2 Components

The consistency is controlled by three components (cf. Figure 1), namely Plan Controller (PC), Data Controller (DC) and Relation Manager (RM).

Plan Controller. PC manages an activity sequence in a business process. As a result of the activity executed by the user, PC receives effects from RM that should operate on the state and updates the state with these effects. PC derives alternatives of activities that are executable for the state of that moment using the planning paradigm and returns these alternatives to RM. PC has a planning problem $\mathcal{P}_P = (\Sigma, s_0, g)$ defined in advance and a current state $s \in S$. Σ represents the activities in the business process. At the beginning of a business process, s is set to s_0. If s satisfies g, namely $g \subseteq s$, PC will notify RM that the process has finished. "Executable" means applicable and reachable to the goal, that is, an activity said executable if it is the first step of a solution of (Σ, s, g). We can regard PC as a function that maps a pair of positive and negative effects $e^+, e^- \subseteq L$ to executable activities considering the current state, i.e. $f^s_\Sigma : 2^L \times 2^L \to 2^A$. s will be replaced with $(s \cup e^+ \setminus e^-)$ as a side effect of the function. We show the details in Figure 2.

$f_{\Sigma}^{s}(e^{+}, e^{-})$

1: $s := (s \setminus e^{-}) \cup e^{+}$;
2: $act := \emptyset$;
3: **for each** $a \in A$
4: **if** (a is applicable to s) **then**
5: $s' := (s \setminus e^{-}(a)) \cup e^{+}(a)$;
6: **if** the planning problem (Σ, s', g) has solutions **then**
7: $act := act \cup a$;
8: **end if**
9: **end if**
10: **end for each**
11: **return** act;

Fig. 2. Algorithm for deriving executable activities and update the state

update the contents of each δ_i

1: $\mathfrak{c} := \mathfrak{c} \cup \{($ `x == v;` $)\}$;
2: $\varsigma :=$ *all solutions of the CSP* $(X, \mathcal{D}, \mathcal{C} \cup \mathfrak{c})$;
3: **for each** $i \in \{1, 2, \cdots n\}$
4: $\delta_i := \emptyset$;
5: **for each** $(v_1, v_2, \ldots, v_n) \in \varsigma$
6: $\delta_i := \delta_i \cup \{v_i\}$;
7: **end for each**
8: **end for each**

Fig. 3. Algorithm for deriving consistent values when assigning a value v to a variable x of $(\mathcal{P}_C, \mathfrak{c})$

Data Controller. DC maintains the consistency of business objects. DC has a CSP $\mathcal{P}_C = (X, \mathcal{D}, \mathcal{C})$ defined in advance and a set of constraints $\mathfrak{c}$. Each of X corresponds to business objects (e.g. fibers, connectors). $\mathcal{D}$ consists of sets of alternatives for each of X. $\mathcal{C}$ is the business object constraints (e.g. the fiber and the connector should be connected). $\mathfrak{c}$ varies with progress of the business process by the addition or removal of constraints. For example, a constraint will be added to $\mathfrak{c}$ when the user inputted a value to a variable (e.g. select a connector from the available connectors). Normally, there are two kinds of interactions between DC and RM. One is the addition (or removal) of constraints to fix a value to a variable, and the other is to pass valid domains (subsets of D_i that consist of consistent values) to RM. There are several methods for calculating valid domains. Which method is suitable depends on the scale of its CSP, that is, the number of variables, size of the domains and complexity of the constraints. In Figure 3, We illustrate one of the procedures for calculating valid domains. In this procedure, DC calculates and memorizes all valid domains $\delta_i \subseteq D_i$ for each variable x_i when a constraint is added to $\mathfrak{c}$. In reply to a request for the valid domain of x_i from RM, DC will return δ_i calculated in advance.

Relation Manager. RM serves as an interface to the user side and an intermediary between PC and DC. As relations between activities defined in the planning problem and the CSP, RM has the following two mappings:

1. $R_1 : A \to X$. This maps each activity to the variable that should be inputted.
2. $R_2 : A \times 2^{D_1} \times 2^{D_2} \times \cdots \times 2^{D_n} \to S \times S$. This maps an executed activity and a state of CSP domains to positive and negative effects. The intention of this mapping is for the state of DC to affect s. We regard $(\delta_1, \delta_2, \ldots, \delta_n)$ as a state of the CSP. Results of this mapping will be fed to f_{Σ}^{s}.

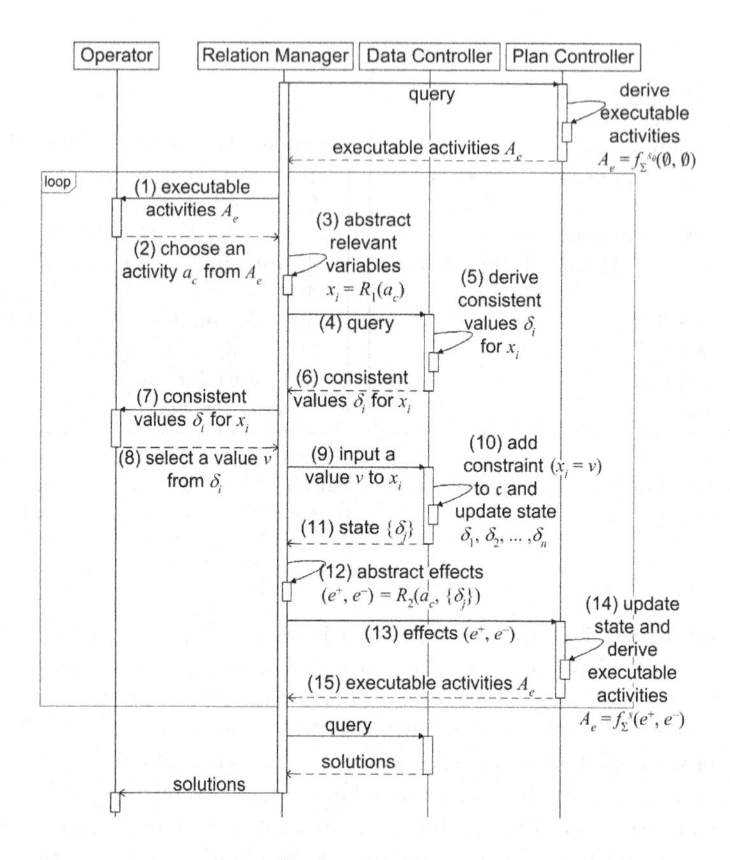

Fig. 4. Sequence diagram

According to R_1 and the selected activity, RM perceives that a variable x_i is the target to input in the activity. RM then retrieves δ_i (the set of consistent values of x_i) from DC and shows the contents of δ_i to the user. Next, RM receives a value inputted to the variable by the user, and passes it to DC. Thereafter, RM receives the state of the CSP from DC and abstracts effects to affect s according to R_2 and the received state. Finally, RM reports the effects to PC.

3.3 Procedure in Regular Cases

Figure 4 shows a sequence diagram that represents the procedure in regular cases. At the beginning of the process, the state of PC is set to s_0, which is the initial state of the predefined planning problem, and the set of temporary constraints $\mathfrak{c}$ is initialized to an empty set. First, PC derives executable activities in the initial state, and hands them to RM. After that, the loop section in Figure 4 is repeated continually until the state of PC satisfies the goal propositions.

planning domain
```
( define (domain travel)
 (:requirements :strips)
 (:predicates
  (selected ?data)
  (reserved ?resource))
 (:action input-max-cost
  :effect (selected trn))
 (:action select-start
  :effect (selected trn))
 (:action select-destination
  :effect (selected trn))
 (:action select-transport
  :effect (selected trn))
 (:action reserve-transport
  :precondition (selected trn)
  :effect (reserved trn)) )
```

planning problem
```
( define
  (problem reserve-transport)
  (:domain travel)
  (:init
    (not (selected trn))
    (not (reserved trn)))
  (:goal (reserved trn)) )
```

CSP
```
start == transport.from;
destination == transport.to;
max_cost >= transport.cost;
```

Fig. 5. Problem descriptions in the example

3.4 Other Features

In addition to the procedures mentioned in Section 3.3, our system can release values that were once fixed and accept inputs of inconsistent values. Users can release values assigned to variables whenever they like. When a "release activity" is performed, some effects may be given to PC. If a user inputs an inconsistent value, DC calculates the variables that it is appropriate to release to regain the consistency of the CSP at little cost [9]. Then, RM shows the result to the user, and finally some variables are released. That is to say, even if inconsistent values are inputted, the CSP in DC will quickly become consistent. As a result, the CSP in DC is kept consistent throughout.

4 Example

We provide a simple example that is an application to support travel planning. Specifically, a user makes a reservation after determining the mode of transport while inputting the starting point, destination and budget. Figure 5 shows descriptions of the problems. The description of the planning problem is in PDDL [3]. PDDL is a standardized language for describing planning problems. The CSP definition is written in independent forms designed for our system. Table 1 shows the mapping from activities to variables. Each activity except "reserve-transport" has one target variable. Table 2 shows the mapping from activities and CSP state to effects. After an arbitrary activity except "reserve-transport," if there are multiple consistent values for the variable "**transport**," no effect will affect. If the variable "**transport**" is settled on one value by the propagation caused by an input, the effect "`(selected trn)`" will affect the state of PC. As a result, the precondition of the activity "reserve-transport" is satisfied, therefore, the reservation will be executable. If the user executes it, "`(reserved trn)`" will affect the state of PC. Then, the state satisfies the goal

Table 1. Mapping from activities to variables in the example

activity name	target variable
input-max-cost	**max_cost**
select-start	**start**
select-destination	**destination**
select-transport	**transport**
reserve-transport	*no target*

Table 2. Mapping from activities and CSP state to effects in the example

activity name	state of CSP	effects
reserve-transport	any state	(reserved trn)
other activities	$\lvert\delta_{\mathbf{transport}}\rvert = 1$	(selected trn)
	otherwise	*no effects*

propositions and the procedure finishes. Input to each variable is executable at any time because the preconditions except that of "reserve-transport" are empty.

5 Implementation

We implemented a prototype of a middleware that employs our technique and confirmed the feasibility of our technique. We developed three applications, the network resource allocation, travel planning and clinical processes, on the middleware. Although the response time of our prototype system was not very long, the achievement of a practical response time for real (large) problems is a subject for study because the applications are so-called toy-problems.

6 Related Work

A method has been proposed that applies constraint satisfaction to network resource allocation [6]. Its purpose is to extend the degree of freedom greatly in activity sequences. Constraint satisfaction achieves the persistent consistency of business objects. However, a user must judge which activity to execute because the sequence of activities is completely free. Therefore, if there are rules about the activity sequences, it may be difficult to ensure that the rules are obeyed.

ConDec is a declarative language for modeling business processes [7]. This language succeeded in concisely representing various flow patterns by adopting a declarative notation instead of the conventional procedural notation. There is a technique that realizes a large variation in flow patterns by leaving room to define flows in the runtime [5]. However, it is difficult to change workflow schemas because people have to define relations between activities directly with these methods. Furthermore, they are not designed to deal with the consistency of business objects. Hence, the problem presented in this paper is not completely resolved by these methods.

7 Conclusions and Future Work

This paper has provided a new technique for realizing flexible process management by the interaction between two paradigms, namely constraint satisfaction and planning. With our technique, we can easily realize business process models capable of various alternative flows because relations between activities are defined as a planning problem suitable for describing business processes that have weak restrictions of sequences. Moreover, constraint satisfaction allows business processes to proceed while business object constraints are kept satisfied. In addition, we can enjoy flexibility as regards modification because each activity can be defined independently of other activities. Our technique includes the concept of workflow. That is, we can easily define and execute a business process equivalent to a process represented in a conventional workflow style. Therefore, system with our technique can deal with business processes suitable for WfMSs and business processes that require a large number of alternative flows. Moreover, these processes can coexist and interact without loss of flexibility.

Modeling processes and problems are considered as future work with a view to easing its difficulty. In addition, we plan to study scalability and the functions required for practical systems, for example, a reduction in computational complexity, connection to a database and the capacity for multi-user.

References

1. blackbox, `http://www.cs.rochester.edu/u/kautz/satplan/blackbox/`
2. Cream: Class Library for Constraint Programming in Java, `http://kurt.scitec.kobe-u.ac.jp/ shuji/cream/`
3. Ghallab, M., Howe, A., Knoblock, C., McDermott, D., Ram, A., Veloso, M., Weld, D., Wilkins, D.: PDDL—the planning domain definition language (1998)
4. Ghallab, M., Nau, D., Traverso, P.: Automated Planning Theory and Practice. Morgan Kaufmann, San Francisco (2004)
5. Mangan, P., Sadiq, S.: On building workflow models for flexible processes. Aust. Comput. Sci. Commun. 24(2), 103–109 (2002)
6. Oishi, H., Nakatani, T., Tayama, K., Ogasawara, S., Yamamura, T.: OSS architecture for flexible and efficient process control. In: NOMS, pp. 1–13 (2005)
7. Pesic, M., van der Aalst, W.M.P.: A declarative approach for flexible business processes management. In: Eder, J., Dustdar, S. (eds.) BPM 2006. LNCS, vol. 4103, pp. 169–180. Springer, Heidelberg (2006)
8. Russell, S., Norvig, P.: Artificial Intelligence: A Modern Approach, 2nd edn. Prentice-Hall, Englewood Cliffs, NJ (2003)
9. Tayama, K., Ogasawara, S., Horiuchi, S., Yamamura, T.: A solution search method that corresponds to changes in optical access network allocation results using the constraint satisfaction problem. In: Proceedings of the IEICE General Conference, pp. SE.1–SE.2 (2004)

A Mechanism of KEDB-Centric Fault Management to Optimize the Realization of ITIL Based ITSM

Bom Soo Kim, Young Dae Kim, Chan Kyou Hwang, and Jae Hyoung Yoo

KT Network Technology Laboratory,
463-1 Junmin-dong, Yusung-gu, Daejeon, Korea
{kbsjmj,webman,ckhwang,styoo}@kt.co.kr

Abstract. On the process of converting existing IT service management structure into ITIL based ITSM structure, many of operation management systems that introduced the guideline of ITIL are able to confront several side effects at operation of fault management process. The typical instances of the side effects are business disorder caused by refinement of fault management process, accuracy decrease of RCA (Root cause analysis) caused by reliability shortage of KEDB information, and delay of fault recovery time caused by the collaboration work to handle a fault between management steps, etc. In this paper, we proposed several mechanisms for KEDB-centric fault management in order to minimize the side effects of fault management process of the ITIL based ITSM. The main objects of proposed fault management mechanisms are to support effective collaboration system for handling a fault between management steps, to provide essential information for analyzing faults, and to provide accurate fault recovery path for prompt handling of various types of faults. The proposed mechanisms were applied to an implementation of Internet service resource management system in KT. As the result, the proposed mechanisms contributed to achieve business goals that are expected by introduction of ITIL based ITSM paradigm, such as TCO reduction through effective resource management and systematic management of service quality based on SLA with customer.

Keywords: ITIL, ITSM, KEDB, Service Support, Fault Management, Incident Management, Problem Management, Change Management.

1 Introduction

1.1 ITSM and ITIL

Common business goal that IT service providers have been pursuing continually is first, enhancement of customer satisfaction by guarantee of service quality that was contracted with customers and second, TCO (Total Cost Ownership) reduction by efficient IT resource management.

ITSM (IT Service Management) can be defined as a methodology to integrate all service components for service delivery and to manage these services effectively [5].

S. Ata and C.S. Hong (Eds.): APNOMS 2007, LNCS 4773, pp. 72–81, 2007.

ITIL (IT Infrastructure Library) is the most widely accepted approach to ITSM in the world. ITIL provides a comprehensive and consistent set of the best practices for ITSM, promoting a quality approach to achieving business effectiveness and efficiency in the use of information systems [1], [2], [3], [4].

In order to realize effective ITSM, ITIL provides two frameworks such as Service Delivery and Service Support. For the purpose of improving service quality, Service Delivery framework offers operational guidelines for Service Level Management, Availability Management, Capacity Management, IT Continuity Management, and Financial Management. And for the purpose of TCO reduction, Service Support offers operational guidelines for Service Desk, Incident Management, Problem Management, Change Management, and Configuration Management [5], [6], [7].

1.2 Realization of ITIL Based ITSM

Compared with existing IT management structure, ITIL has some peculiar management structure and operational guidelines in Service Support framework. For example, ITIL recommends service management process to be separated into several steps in order to deal with customers' VOC professionally and systematically. ITIL recommends relevant organizations to cooperate in handling customers' VOC based on their role and responsibility. Also ITIL recommends fault management process to utilize KEDB (Know Error DB) in order to handle repeatedly occurred faults accurately and rapidly. The expected effects according to conform these guidelines of ITIL can be summarized as below.

- Conversion of technical centric organization into process centric organization
- Enhancement of flexibility for various business requirement and environment changes
- Enhancement of customer satisfaction through accurate and rapid VOC handling
- Enhancement of operators' professionalism through process centric organization
- Enhancement of efficiency for IT resource management through professionalism

But, the common problems that come out when ITIL based ITSM paradigms are applied to existing IT management system in real field can be summarized as below.

- Operator's refusal feeling to newly introduced management paradigm
- Business blank by absence of organization which takes charge of new defined processes
- Business disorder by which existing single process is separated into multiple processes
- Business complexity by cooperation work for handling a job between several processes
- Difficulty of effective KEDB construction and management for various types of faults
- Reliability shortage of KEDB information for handling various types of faults
- Insufficiency of management tools suitable for each management processes

Most typical cases including these problems are the fault management that puts the operational guidelines of ITIL Service Support framework to practical use. We estimated that the problems can be occurred when unnecessary processes and activities of ITIL applied without optimization for real business environment, when there is no effective information delivery between management steps, and when there is no clear information structure for handling various types of faults in KEDB.

In order to overcome the problems, we proposed a mechanism of KEDB-centric fault management to optimize the realization of ITIL based ITSM. Main objects of the proposed fault management mechanisms are to provide operators with essential information for accurate fault management, to support an organic information delivery structure between management steps for effective cooperation, and to support correct and prompt fault process of repeatedly occurred faults through reliable KEDB information.

To achieve the objects, we suggested mechanisms such as KEDB-centric fault management process, ERP (Error Recovery Path) based KEDB construction mechanism, AERP (Assembling of ERP) mechanism, DRFM (Dashboard like Real-time Fault Monitoring) mechanism, and SIAM (Service Impact Analysis Map) mechanism.

The rest of the paper is organized as follows. Section 2 describes related study regarding motivation of proposed mechanism. Section 3 describes the architecture and operation principle of proposed mechanisms to optimize realization of ITIL based ITSM. Finally, Section 4 describes conclusions and future work.

2 Related Study

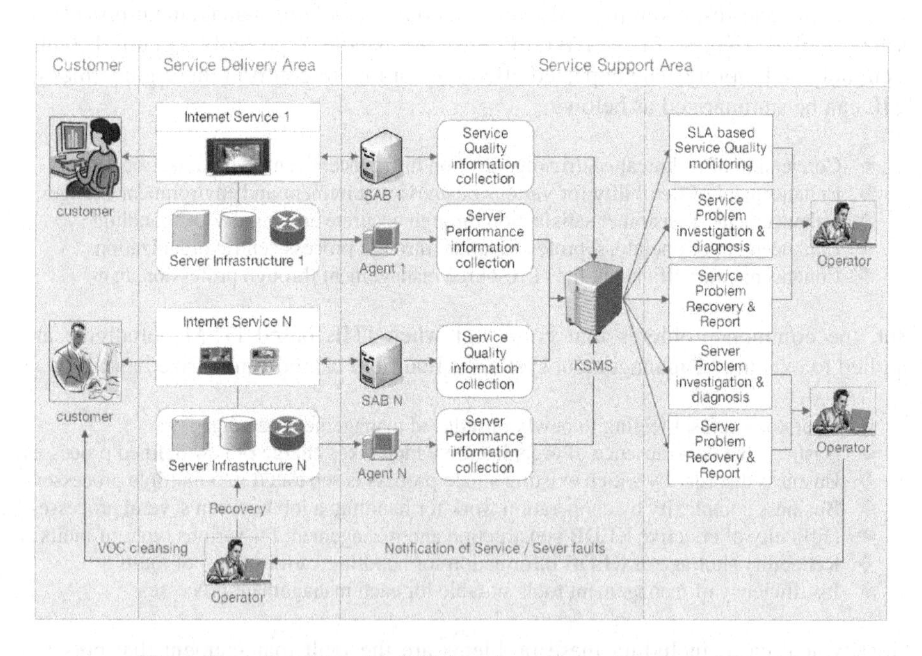

Fig. 1. Operational structure of Internet service resource management

KT introduced ITIL based ITSM into Internet service resource management system in order to change the operation paradigm from server centric management to service centric management. The object of Internet service resource management system is to manage various types of Internet services and server resources in integrated and systematic mode. For achieving this object, Internet service resource management system executes two main roles to monitor service connection quality with viewpoint of customer and to monitor service resource utilization with viewpoint of service provider. Fig. 1 shows operational structure of Internet service resource management.

In the initial version of Internet service resource management system applied ITIL based ITSM paradigms, the focus of design and implementation was fitted to faithful realization of ITIL standard. The critical problems that came out in the earlier version of Internet service resource management system are as below.

- Unnecessary procedure and the activity more than practical business use
- Unclear cooperation system between management steps for fault management process
- Inefficient information delivery structure for cooperation between management steps
- Most of fault processes are terminated at Incident Management abnormally
- Different results of fault analysis and recovery for same fault
- Difficulty of KEDB information construction
- Difficulty of KEDB information reuse for fault analysis and recovery
- Unclear configuration information change without clear RFC (Request for Change)

In order to overcome these problems, we need to fine new ideas to optimize ITIL based ITSM. So in later version of Internet service resource management system, the focus of optimization was concentrated to design concepts as below and proposed fault management mechanism was developed to support the design concepts.

- Offering of simple fault management procedure and activity
- Offering of simple cooperation system between fault management steps
- Offering of clear information delivery system between fault management steps
- Offering of simple methods to construct KEDB information more effectively
- Offering of simple methods to keep KEDB information more reliable
- Offering of easy methods to analyze and recover various types of faults

3 Proposed Fault Management Mechanisms

3.1 Design Objects and Considerations

In order to optimize the realization of ITIL based ITSM, proposed mechanism of KEDB-centric fault management has goal of three view point such as process, people, and tools. Design object of process view is to simplify ITIL based fault management process considering actual business environment. Design object of people view is to improve operators' skill through supporting accuracy information and tools regarding fault management. Finally design object of tool view is to support convenient solutions to monitor and analyze various types of faults.

We proposed KEDB-centric fault management process for the object of process view, ERP (Error Recovery Path) based KEDB construction mechanism and AERP (Assembling of ERP) mechanism for the object of people view, and DRFM (Dashboard like Real-time Fault Monitoring) mechanism and SIAM (Service Impact Analysis Map) mechanism for the object of tool view.

3.2 KEDB-Centric Fault Management Process

The main object of proposed KEDB-centric fault management process is to make fault management flows and cooperation flows simple and clear. To achieve the object, we suggested the core management concepts such as responsibility and delegation based operation principle, ERP based fault analysis and recovery, KEDB based information delivery structure, and traceability of fault management through information.

The proposed fault management process is operated based on responsibility and delegation of each management steps. The responsibility and delegation of each management steps are as below.

The main responsibility of Incident Manager is to carry out business to deal quickly with happened incidents through ERP registered to KEDB and to delegate the incident that can not solve to the Problem Manager.

The main responsibility of Problem Manager is to carry out businesses to identify solutions for the problems delegated from Incident Manager, to create new ERP that can solve delegated problems, and to delegate RFC (Request for Change) to Change Manager to request validation of new ERP.

The main responsibility of Change Manage is to carry out business to classify RFC delegated from Incident and Problem Manager, to identify ERP correspond to RFC through KEDB, to validate adequacy of ERP, to evaluate the influence of ERP, to register ERP to KEDB, and delegate Change Plan to Configuration Manager to request for configuration change based on ERP.

The main responsibility of Configuration Manager is to carry out business to identify the Change Plans delegate from Change Manager, to identify ERP correspond to Change Plan through KEDB, and to change the configuration based on ERP.

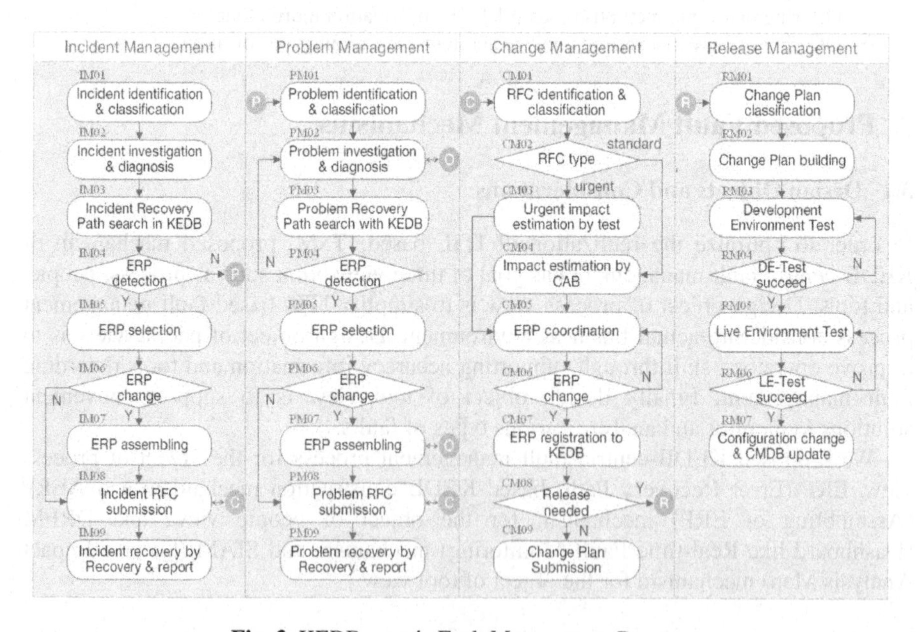

Fig. 2. KEDB-centric Fault Management Process

Fig. 2 shows basic scenarios of proposed fault management process. Detail business flows of proposed fault management mechanism are described as below.

- Incident Manager identifies and classifies the Event (IM01).
- Incident Manager analyzes a fault happened through DRFM and SIAM (IM02).
- Incident Manager refers to ERP regarding a happened fault at KEDB (IM03).
- If IERP regarding a relevant fault does not exist (IM04).
- Incident Manager transfers a happened fault to Problem Managers (PM01).
- If ERP for a relevant the fault exists (IM04).
- Incident Manager confirms information of ERP to solve a relevant fault.
- Incident Manager selects ERP to solve a relevant fault (IM05).
- Incident Manager delivers RFC to solve a relevant fault to Change Managers (IM08).
- If change of ERP regarding a relevant fault is necessary (IM06).
- Incident Manager changes ERP through Incident Resolution Assembler (IM07).
- Incident Manager delivers RFC to solve a relevant fault to Change Managers (IM08).
- Problem Manager classifies problem delegated from Incident Manager (PM01).
- Problem Manager analyzes a fault happened through DRFM and SIAM (PM02).
- Problem Manager refers to ERP regarding a happened fault at KEDB (IM03).
- If ERP regarding a relevant fault does not exist (PM04),
- Problem Manager analyzes the fault happened (IM04).
- Problem Manager confirms information of ERP to solve a relevant fault.
- Problem Manager corresponds, and selects ERP to solve a fault.
- Problem Manager delivers RFC to solve a relevant fault to Change Managers (PM08).
- If change of ERP regarding a relevant fault is necessary (IM05),
- Problem Manager changes ERP through Incident Resolution Assembler (IM07).
- Problem Manager delegates RFC to Change Managers (PM08).
- Change Manager classifies RFC delegated from Incident Manager (CM01).
- If type of RFC regarding a relevant fault is Urgent (CM02),
- Change Manager confirms ERP responded to RFC requested from KEDB (CM03).
- Change Manager evaluates influence of ERP through urgent test (CM03).
- If type of RFC regarding a relevant fault is Standard (CM02),
- Change Manager delivers RFC information requested to CAB.
- CAB confirms ERP information responded to RFC requested from KEDB (CM04).
- CAB evaluates an influence degree ERP apply through standard test (CM04).
- Change Manager carries out verification and coordination of ERP requested (CM05).
- If verification of ERP is succeed (CM06),.
- Change Manager registers ERP information to KEDB (CM07).
- Change Manager delegates Change Plan to Configuration Managers (CM09).
- If Release Management of ERP is necessary (CM06),
- Change Manager delegates ERP information to Release Manager (RM01).

3.3 ERP Based KEDB Construction Mechanism

The main object of proposed ERP (Error Recovery Path) based KEDB construction mechanism is to make KEDB information accurately constructed and easily utilized for analysis and recover of various types of faults. To achieve the object, we suggested ERP information structure. ERP information structure is composed with set of information such as service, server, error, error recovery procedure, etc.

The main features of ERP information structure are to offer essential information for handling repeatedly occurred faults easily and rapidly, to offer clear methods for fault detection and recovery through error recovery path information, to offer detail recovery step and operation in order to prevent operator's mistake, and to offer heuristic statistics information for operator to select error recovery path exactly.

Service	Server			Error		Error Recovery Path							
Service Type	Server Type	Resource Type	Server Maker	Error Event	Root Cause	ERP Path	ERP Rank	Try Count	Hit Ratio	Test result	Assess Results	Recovery Step	Recovery Operation
DHCP	DB	DBMS	MYSQL	Agent No Response	DB Locking	Erp1	1	133	98%	OK	OK	Action 01 Action 02 Action 03	Operation 01 Operation 02 Operation 03 Operation 04
DHCP	DB	DBMS	MYSQL	Agent No Response	DB Locking	Erp2	2	34	89%	OK	OK	Action 01 Action 02	Operation 01 Operation 02
DHCP	Server	S/W	IBM	Agent No Response	Process Down	Erp3	3	13	75%	OK	OK	Action 01 Action 02 Action 03 Action 04	Operation 01 Operation 02
IPTV	L3	H/W	Cisco	L3 Down	Port Fail	Erp1	1	56	68%	OK	OK	Action 01 Action 02 Action 03 Action 04	Operation 01 Operation 02 Operation 03 Operation 04
IPTV	L3	H/W	Juniper	L3 Down	Port Fail	Erp2	1	66	68%	OK	NOK	Action 01 Action 02 Action 03 Action 04 Action 05	Operation 01 Operation 02 Operation 03 Operation 04 Operation 05 Operation 06
IPTV	L3	O/S	Cisco	L3 Down	Process Down	Erp3	3	7	12%	OK	NOK	Action 01 Action 02	Operation 01 Operation 02

Fig. 3. ERP information structure of KEDB

Fig. 3 shows basic case of ERP information structure of KEDB. The basic flow of utilize ERP information to handle an occurred fault is as follows.

- Select service pattern corresponded to an occurred fault.
- Select server pattern corresponded to an occurred fault.
- Select the error pattern and root cause corresponded to an occurred fault
- Select ERP to recover an occurred fault.
- Confirm recovery steps and operation guidelines
- Execute recovery operations, and confirm execution results.

3.4 AERP Mechanism

The main Object of a proposed AERP (Assembling of ERP) management mechanism is to give operator ERP assembling method to manage ERP information conveniently. To achieve the object, we suggested the ERP assembling process and information structures as Fig. 4. The proposed AERP mechanism provides several of the information pools for convenient ERP assembling. ERPP (Error Recovery Path Pool) keeps the ERP that was registered before. ERSP (Error Recovery Step Pool) keeps the recovery steps that are components of ERP. EROP (Error Recovery Operation Pool) keeps practical process command language and alternative information registered to recovery procedure.

Incident Manager carries out role to assemble the ERP already registered for happened fault process. Problem Manager carries out role to generate necessary ERP in order to deal with a fault delegated to itself. Change Manager carries out role of verification regarding ERP requested from Incident Manager and Problem Manager, and testing and coordination of ERP.

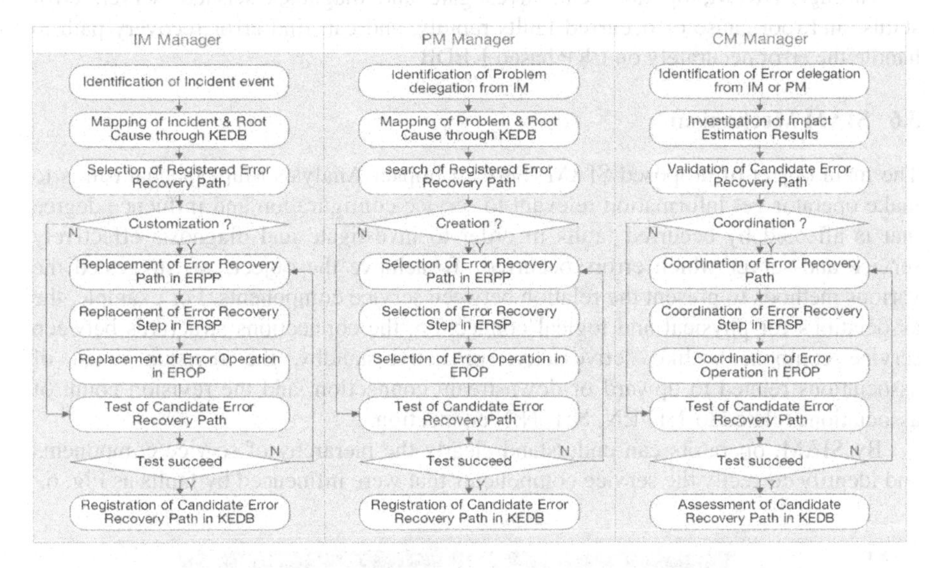

Fig. 4. Assembling Flows of ERP

3.5 DRFM Mechanism

The main object of proposed DRFM (Dashboard like Real-time Fault Management) mechanism is to provide operator with total fault information in real-time manner in order to identify and classify effectively service and server which errors occurred. So,

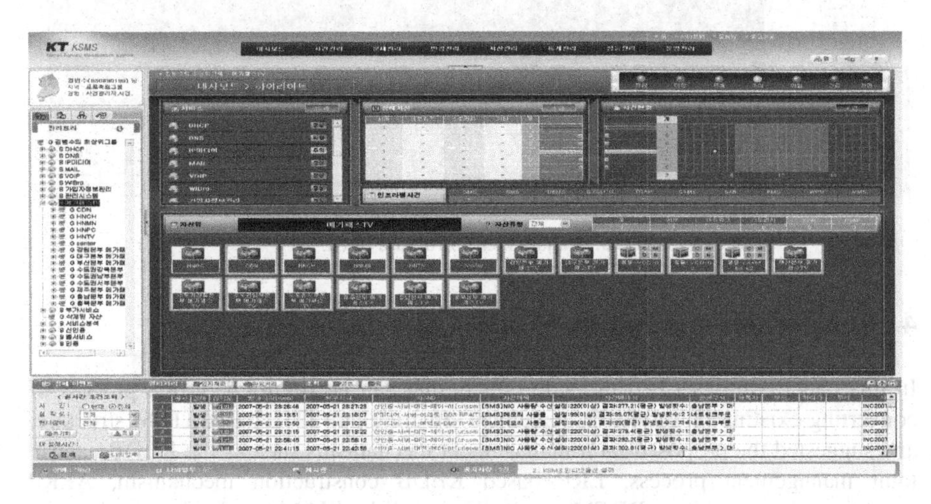

Fig. 5. Management Structure of DRFM

we suggested the method that enables service configuration information structuralized according to the relation of service configuration and method that display status of error occurring in dashboard manner as Fig. 5.

Through DRFM, operator can investigate and diagnosis service, server, error status, and root cause of occurred faults rapidly, and can find error recovery path to handle the error accurately on ERP based KEDB

3.6 SIAM Mechanism

The main object of proposed SIAM (Service Impact Analysis Map) mechanism is to make operator get information relevant to service configuration and influence degree that is affected by occurred faults in order to investigate and diagnosis effectively service and server which errors occurred. To achieve the object, we suggested the various methods to present the relation between service components. For example, the associations for physical and logical connection, the connections structures between service components like active-active or active-standby, the revision count of associations related to upward or downstream connection, and the revision count of associations related to 1:1, 1:N, N:1, N:N connection.

By SIAM, operators can understand clearly the hierarchy of service components and identify correctly the service components that were influenced by faults as Fig. 6.

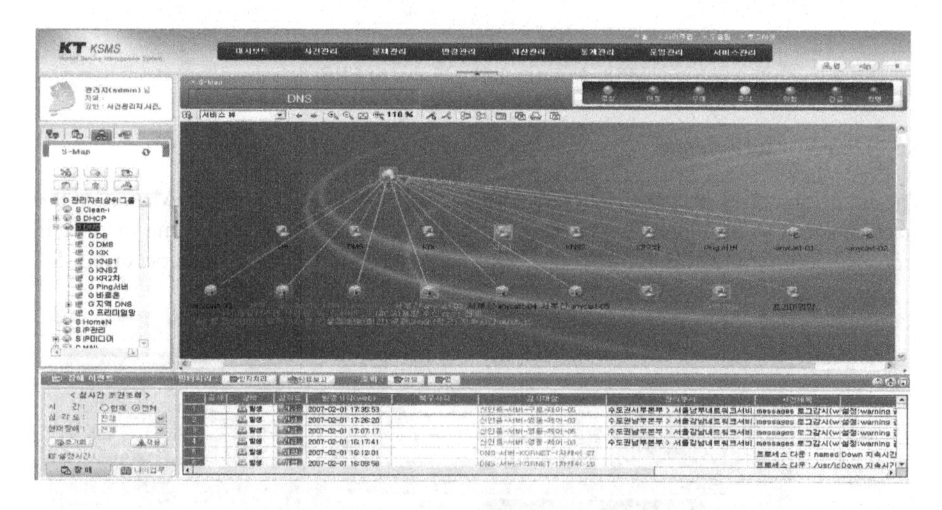

Fig. 6. Management Structure of SIAM

4 Conclusions and Future Work

In this paper, we described the side effects that can be occurred on the process of converting existing IT service management structure into ITIL based ITSM structure, and suggested the proposed fault management mechanisms such as the KEDB-centric fault management process, ERP based KEDB construction mechanism, AERP management mechanism, DRFM mechanism, and the SIAM mechanism in order to minimize these side effects. Through the proposed mechanisms, we were able to achieve almost of our management objects such as to provide operators with essential information for accurate fault management, to support an organic information delivery structure between management steps for effective cooperation, and to support correct and prompt fault process of repeatedly occurred faults through reliable KEDB information.

The direction of future work is focusing to improving of service operation degree in customer's view point and to advancing of service impact management according to various types of faults. So we are concentrating to researches such as the way of forecasting various faults relevant to SLA (Service Level Agreement), analyzing service impacts according to occurred faults, recovering faults relevant to SLA systematically and automatically, etc.

References

1. ITIL BOOKS, [Online], Available: http://www.itil.org.uk/
2. ITIL ORG, [Online], Available: http://www.itil.org/
3. ITIL OCG, [Online], Available: http://www.itil.com/
4. ITIL and ITSM WORLD, [Online], Available: http://www.itil-itsm-world.com/
5. IBM, ITSM Construction Methodology, IBM Technical Report (2005)
6. HP Korea, SLM, ITIL and HP Service Management Solution, HP technical Report (2003)
7. HP Korea, HP ITSM Assessment Introduction, HP technical Report (2003)
8. Hanemann, A., Sailer, M., Schmitz, D.: Service delivery: Assured service quality by improved fault management'. In: ICSOC 2004 (November 2004)
9. Hochstein, A., Zarnekow, R., Brenner, W.: ITIL as common practice reference model for IT service management: formal assessment and implications for practice. In: The 2005 IEEE International Conference on 29 March-1 April 2005, pp. 704–710 (2005)
10. Hochstein, A., Zarnekow, R., Brenner, W.: 'Evaluation of service-oriented IT management in practice Services Systems and Services Management. In: Proceedings of ICSSSM 2005 International Conference, June 13-15, 2005, vol. 1, pp. 80–84.
11. Bailey, J., Kandogan, E., Haber, E., Paul, P., Maglio: Designing for IT management: Activity-based management of IT service delivery. In: CMIIT 2007 (2007)

Automatic NE-adapter Generation by Interface Blending/Diagnosis Methods

Yu Miyoshi, Atsushi Yoshida, Tatsuyuki Kimura, and Yoshihiro Otsuka

NTT Network Service Systems Laboratories, NTT Corporation,
9-11, Midori-Cho 3-Chome Musashino-Shi,
Tokyo 180-8585, Japan
{miyoshi.yu,yoshida.atsushi,kimura.tatsuyuki,
otsuka.yoshihiro}@lab.ntt.co.jp

Abstract. Operating a next generation network (NGN) while new services and network elements (NE) are being intermittently introduced is a complex and difficult task. We are trying to establish flexible network operation technology. This article introduces a method of automatically generating NE-adapters in element/network management systems (EMS/NMS) to interact with NEs. This method alleviates the effort of NE configuration tasks by automating the development of functions that operators use to set up NEs. We call this interface blending/diagnosis method. Our approach consists of 1) a blending method that automatically generates NE-adapters for products from different vendors from commands and responses between an EMS and an NE and 2) a diagnosis method that verifies blended adapters with actual NEs. Moreover, this article introduces a configuration support tool that we developed. The tool complements the interface blending/diagnosis method even if there is no adapter to be possible to use as knowledge.

1 Introduction

In the near future, many telecommunication companies in the world will be migrating to a next generation network (NGN) based on IP (Internet protocol) technology. The IP-based technology is changing especially quickly, with new services appearing all the time while the life cycles of individual services tend to become shorter and shorter. And IP network consists of various types of network elements (NEs) such as routers, switches, and servers. With the recent growth of networks and services, network operators and OSS/BSS (operations support systems/business support systems) developers have had to handle an increasing number and variety of NEs, and their work has become very complicated. An OSS/BSS that supports the operation of these network services monitors the entire network including multiple types of equipment and performs configuration management which means inputting service orders or parameter settings to NEs. Without any doubt, the importance of the roles played by the OSS/BSS in providing network services will continue to increase in the future.

S. Ata and C.S. Hong (Eds.): APNOMS 2007, LNCS 4773, pp. 82–91, 2007.

In the NGN, interactions between various systems are performed than ever before by opened application programming interfaces (API). The New Generation Operations Systems and Software (NGOSS)[**1**] studied in TeleManagement Forum (TMF)[**2**] is establishing system interconnection technologies in the field of telecommunication management, and an OSS/BSS which is defined in upper layers such as service management layer (SML) or business management layer (BML) can connect flexibly other systems. However, in the case of connection in lower layers such as between element/network management layer (EML/NML) and NE layer (NEL), we must develop still a primitive interface because an EMS/NMS has to treat interfaces of various NEs of multi-vendor products. Moreover, frequent changes of NE specifications by recent service life cycle shortening hasten the progression of obsolescence of developed functions to absorb differences of interfaces and cost-effectiveness is hard to be provided. A general connection method with NEs is to add an "NE-adapter" which is written by light weight script languages. Many systems have adapters which described a different syntax for each NE.

We are trying to establish an NE-adapter automatic generation method to overcome a problem of differences of various NEs easily. By the approach called interface blending/diagnosis methods, we achieved generating a part of target adapters from existing adapter and vendor characteristics.

The paper is organized as follows. **Section 2** discusses issues in telecommunication management area and introduces related works. **Section 3** proposes interface blending/diagnosis methods, **Section 4** introduces detailed design of interface blending/diagnosis system, and **Section 5** describes our experimental study. **Section 6** describes the configuration support tool which supports to complement interface blending/diagnosis methods, and **Section 7** concludes this paper.

2 Issues in Telecommunication Management Area

2.1 Network Operations Issues

The network operator's job includes various tasks such as setting diverse service parameters, booting up equipment, updating operating software of NEs, and collecting error logs. Although these might seem to be simple tasks at first glance, they require different knowledge and techniques about the interface specifications and complex operating procedures of each NE, and there is no excuse for making mistakes while services are being provided. In particular, in a large-scale network, even a parameter modification needs a lot of work. As a result, operators are being required to make ever-increasing efforts. In NGN operations and also in the NGOSS, customer experience is regarded as one of the most important studies. Operators must respond quickly to a customer's service ordering. Because any delay of introduction of new service gives big influence to our business.

2.2 NE-adapter Development Issues

SNMP (simple network management protocol) is a well-known protocol for gathering information from NEs. Since it is a standard protocol in many types of equipment, it is widely used. However, it is weak at monitoring new technology because it is not possible to acquire information in the case of late implementation of management information base (MIB). Therefore, the use of a command line interface (CLI) to set parameters and acquire detailed information is currently the most popular approach in spite of being a legacy method.

An EMS/NMS typically uses a CLI, too. The EMS/NMS uses processing functions written by a script language and transmits its commands through CLI. Well-known script languages include Perl, JavaScript, Python, and Expect. This is because it is necessary to use interactive processing whereby the response to a transmitted command is used to select the subsequent processing and because processing statements are expected to be frequently updated to keep up with ongoing changes to NE specifications. These processing functions are called "NE-adapters".

However, a rapid introduction of developed NE-adapters has a problem. An NE-adapter needs different processing descriptions for different vendor NEs even if a command is the same meaning. For example, in situations where a new NE is released, it is preferable to match up the specifications by developing an NE-adapter at the same time as the EMS/NMS. But matching up is a difficult task when the NE and the NE-adapter have been developed in parallel. Even if it is assumed that the specifications can be matched up, it is still necessary to perform verification trials on the actual equipment. If a connection failure occurs during these trials, even more time must be spent on the specification design and NE-adapter coding.

With regard to the methods used to develop system functions, such as NE-adapters where results are needed quickly, attempts are being made to automate their creation by applying MDA (model driven architecture) [3], which is being studied by the OMG (object management group). However, since this method requires that the UML (unified modeling language) specifications of the functions are declared as models, it is necessary to support the creation of the models. Furthermore, we think it is important to consider enhancing efficiency by re-using existing adapters instead of creating functions from scratch with models. We must still address the problems of how to adapt quickly to the introduction of new NEs or modified interface specifications and how to continue operation smoothly.

3 Interface Blending/Diagnosis Method

In this section, we describe interface blending/diagnosis method that solve the above problems. Our proposal automates part of the work involved in operator settings input and NE-adapter development with the aim of reducing the workload (especially in the complexity of dealing with settings for different vendors and with frequent specification changes). Specifically, we are considering providing a system with "blending" and "diagnosis" methods that automatically bridge the differences between the interfaces of different vendors when performing similar processes (**Fig. 1**).

At first, our system extracts command strings from NE-adapters that have already been installed. After that, extracted commands are combined to create commands for a new NE or process. This method is called "blending". We think it is possible to operate in this way because many commands include words with similar meanings even though the NEs come from different vendors and have different specifications. Our method utilizes this characteristic. Moreover, using specifications defined by each vendor makes it more likely that commands and NE-adapters can be created automatically.

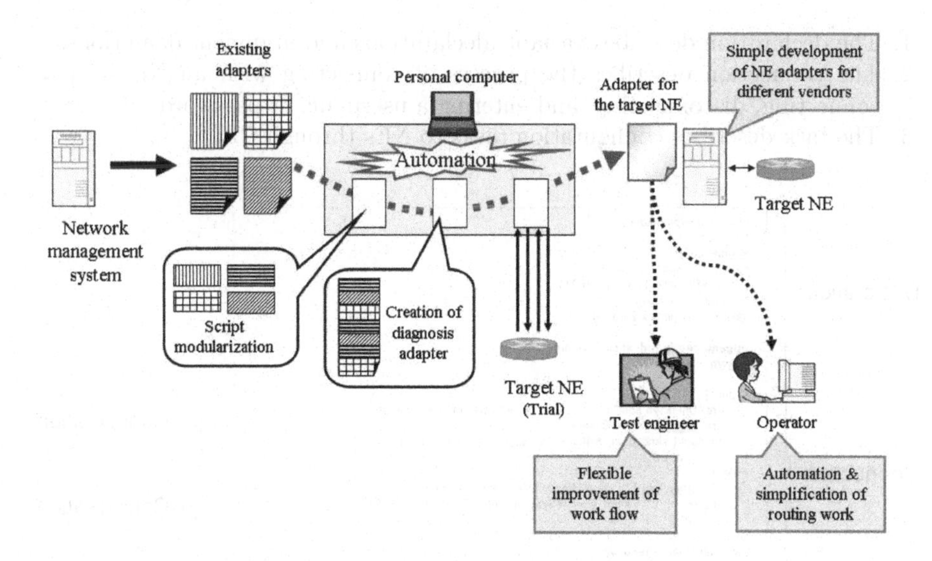

Fig. 1. Interface blending and diagnosis system

However, the NE-adapters generated by blending cannot be used without alteration. Therefore, our system is equipped with methods that use the adapter generated by the blending method to perform connection tests on the NE in question. We call this method "diagnosis" because it discovers the correct solution by connecting the system to the actual equipment. The adapters for the target NE will be generated automatically as long as our system acquires the answers it expects. This method will be able to check problems such as misconfigurations caused by human error. We chose to generate the adapters in script form, so they can be easily modified by operators and an adapter can be run as part of an EMS/NMS or as a tiny standalone application. If the methods become available, then the adapter development task can be directly performed within the work of NE and NMS connection trials, thereby contributing to the rapid introduction of a new service and it should be possible to greatly reduce the burden on operators and developer.

4 Detailed Design

This section presents details of the design that we implemented in a system for an interface blending and diagnosis method. There are four components, which are described in sections **4.1.** to **4.4.**

4.1 Analysis Component

First, we describe the component for analyzing input existing adapters. We found that all adapters have a similar structure for connecting to NEs. They also have a structure that can be divided into the following three parts (**Fig. 2**).

1. The declaration describes variable declarations and argument definitions.
2. The connection describes the process of connecting with an NE such as connecting, disconnecting, and entering a username or password.
3. The task describes configuration tasks to NEs through CLI.

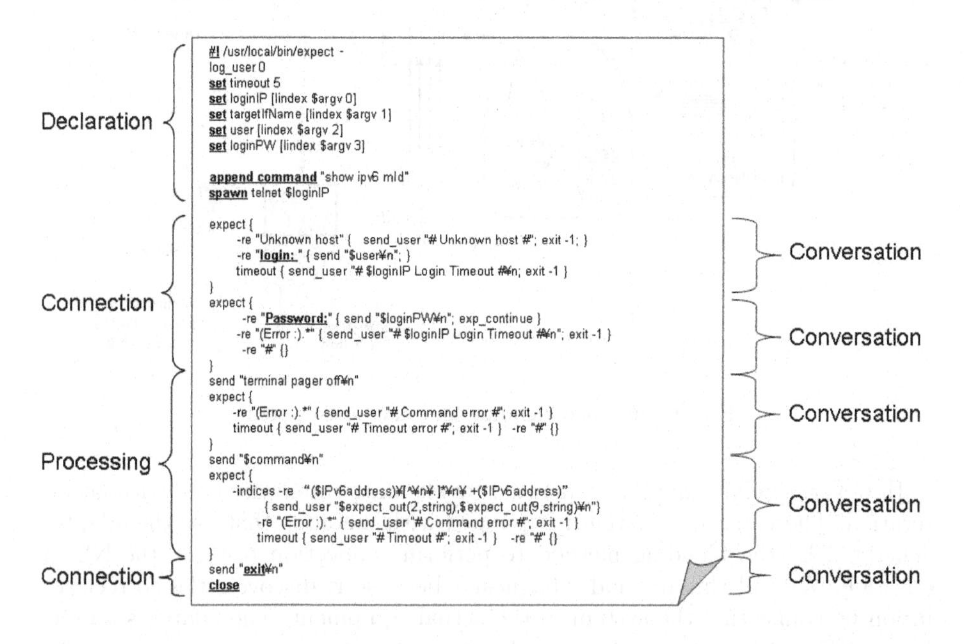

Fig. 2. Example of adapter description

Moreover we found that the connection and task parts can be divided more minutely and that the description of a command transmission can be paired up with the response. We call the pair of descriptions (a command and the response) a “conversation” and define it as the smallest unit that expresses adapters in our system. The analysis component divides an existing adapter into three parts, which are subdivided into conversations.

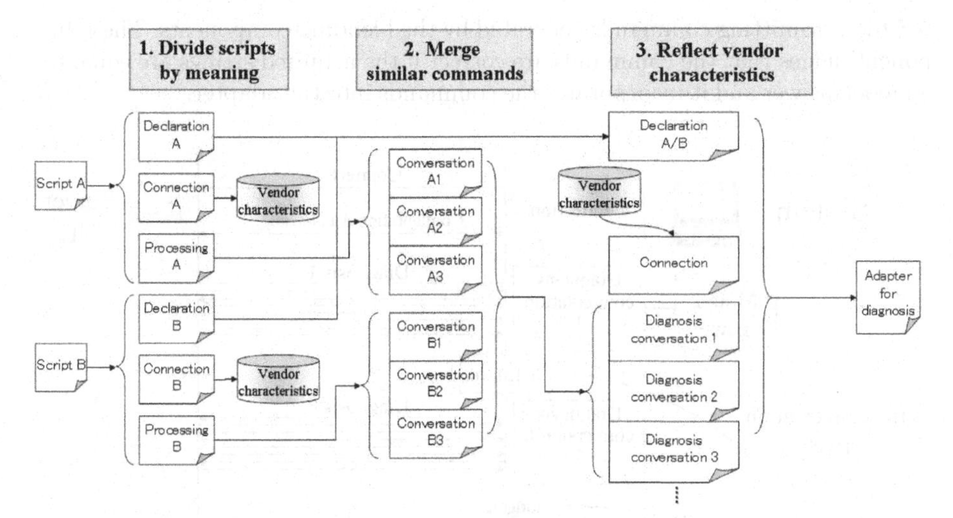

Fig. 3. Behavior of an analysis component and a blending component

4.2 Blending Component

The blending component generates an adapter for testing a target NE (**Fig. 3**).

In a survey of NE-adapters, we found similar patterns in many command strings. The interface specifications are not inherently compatible between different NEs, but NE command syntax is implemented considering usability. Therefore, we assume a system can obtain responses that enable it to generate correct adapters for a target NE if it queries the NE with similar commands of other vendors' NEs. The blending component generates trial adapters by recalibrating the existing commands.

On the other hand, an adapter has description parts that do not depend on meanings of processes. For example, a connection part for log-in is the same as long as the vendor is the same. Therefore, we thought that the efficiency of generating an adapter automatically would increase if a system recycles patterns pertinent to the vendor NE. And we implemented functions to store the vendor's characteristics in this component. If the target NE is supplied by a known vendor for our system, the vendor patterns are used prior to other blending methods in this component.

4.3 Diagnosis Component

The diagnosis component decides that an EMS/NMS can use the commands if the responses from the target NE match the model answers. In that case, it incorporates the character string in an adapter for a target NE. The behavior of the diagnosis component is shown in **Fig. 4**.

The diagnosis component must search for effective information in the responses of a target NE to transmitted commands. The component acquires model answers from an existing NE. Next, the component searches for correct answers for a target

NE by transmitting commands generated by the blending components. The component judges that the commands are correct if the acquired strings are equal to a model answer and it incorporates the commands into the adapter.

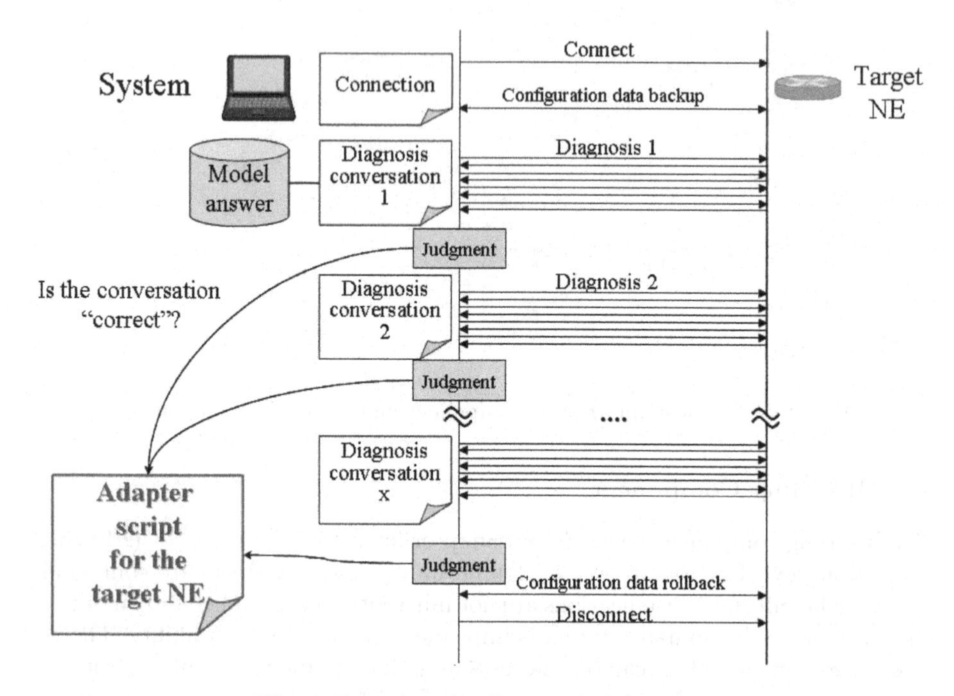

Fig. 4. Behavior of a diagnosis component

4.4 Complement Component

The interface blending/diagnosis system can handle simple patterns, but it is difficult to fill complicated patterns now. In addition, it is necessary not only to establish technologies for increasing the success rate of diagnosis, but also to strive to implement practical functions such as a system that can easily fill in missing parts of adapters through the intervention of developers. Therefore, the system should have a graphical user interface (GUI) for an operator to add commands if the diagnosis component could not be discovered. We developed a prototype configuration support tool which generated an adapter from NE setting of an operator as a method to complement semiautomatically. We describes it in **Section 6**.

5 Verification Trials of Blending Methods

We performed verification trials of blending methods.

In this trial, we assumed that an EMS/NMS developer implements an adapter for a new target NE and that the EMS/NMS already had adapters for other NEs.

We investigated how many correct answers our system could discover. For this investigation, we built a prototype of the blending/diagnosis components. As existing adapters, we used those of an NMS that we had developed previously [**4**]. All adapters were described in script language (Expect). They had commands for connecting to a target NE, getting data, and disconnecting from the NE. We prepared three target NEs from different vendors and tested whether our system could generate three new adapters for each target NE. In total, we tested the automatic generation of 9 adapters, 44 conversations. We deleted adapters for the target NEs from our system before testing, so the system did not know the correct commands of the adapters of the target NEs. The system had to search for effective commands from other adapters. We prepared active NEs in a test network and the system queried them.

First, we tested the diagnosis method using only three calibration rules implemented in the blending component.

Table 1. Results

		First try		Second try	
Target NE	Functions	Diagnosis success	Conversations generated automatically	Diagnosis success	Conversations generated automatically
Vendor A	Function 1	1/2	4/5	2/2	5/5
	Function 2	1/2	3/4	2/2	4/4
	Function 3	1/2	2/4	2/2	4/4
	Subtotal A	3/6	10/13	6/6	13/13
Vendor B	Function 1	1/2	5/6	2/2	6/6
	Function 2	1/2	6/7	2/2	7/7
	Function 3	1/2	5/6	2/2	6/6
	Subtotal B	3/6	16/19	6/6	19/19
Vendor C	Function 1	1/2	3/4	2/2	4/4
	Function 2	1/2	4/5	2/2	5/5
	Function 3	1/1	3/3	1/1	3/3
	Subtotal C	3/5	10/12	5/5	12/12
Total		9/17 (53%)	36/44 (82%)	17/17 (100%)	44/44 (100%)

As a result, our system found 36 effective replies in 44 conversations. 27 replies were found from vendor characteristics and 9 replies were found from a blending method. The correct rates are 53%. We think this result is good because more than half of the answers could be found in spite of using only simple three calibration rules. A developer may halve the effort for implementing a new adapter if our system can get 53% of the answers correct. And this result shows that there are many similar commands between different vendor NE interfaces. Afterward, we tried implementing new rules and vendor characteristics to automatically generate the commands that we were not able to confirm. As a result of adding new rules and having tested the diagnosis method again, our blending method was able to generate commands including those used for all 9 adapters and 44 conversations. New rules have general versatility, and they are useful in some other scenarios, too. These results show that our system can help developers of this EMS/NMS sufficiently during the introduction of a new NE.

We could get good results in this trial, but they depended on the number of stored patterns and defined calibration rules. Therefore, we should enrich the calibration rules and vendor patterns to enable the system to discover responses in many situations.

6 Configuration Support Tool for Complement

We noticed the success rate of diagnosis methods depends on quantity of vendor characteristics and existing adapters greatly. In early stage of operation environment with a few references, it is difficult to achieve an effect only by a blending/diagnosis method. Therefore we developed a "configuration supporting tool" which supports to generate an adapter even if there was no similar reference to raise a practical utility of interface blending/diagnosis methods (**Fig. 5**). When an operator performs only configuration tasks through CLI, and the adapter is made semi-automatically. The tool have a method to convert it from configuration of the operator immediately.

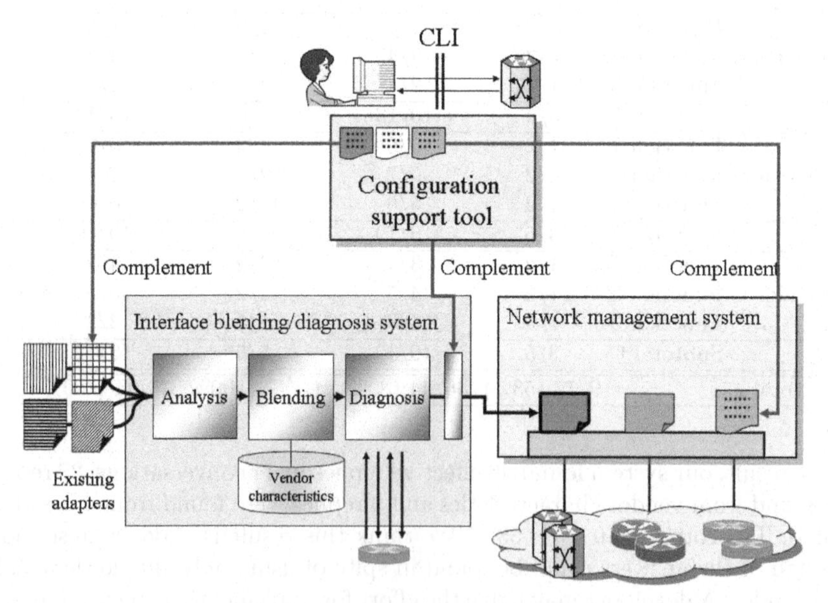

Fig. 5. Configuration support tool

At first an operator is connected to the target NE and performs configuration through CLI on the tool. The tool records a username, password, and command-line strings (commands and responses). Then, the tool captures communication history between the operation terminal and NE and extracts a transmitted message and the response one by one. By fitting in these strings in a skeleton model of a general-purpose script language, the tool generates a script file which the interface blending/diagnosis system can use as an adapter. Operators can modify easily a new processing pattern only by changing parameters. By these procedures, the configuration support tool generates an adapter without coding skills of programming/script languages. The current prototype tool can generate only description of sequential processing. By a function which adds advanced description such as a conditional branching, we are studying construction of a more convenient adapter development environment in future.

7 Conclusion

This article described interface blending/diagnosis method which alleviates the effort of EMS/NMS development by automating NE-adapters implementation task. We tested a prototype system to confirm its feasibility. In this trial, we were able to obtain effective responses for all commands that we wanted to generate. Moreover, we introduced the configuration support tool that generate NE-adapter from an operator's configuration history through CLI. The tool complements the interface blending/diagnosis method even if there is no adapter to be possible to use as knowledge. Therefore, we think our approach have proven the possibility of developing NE-adapters automatically.

References

1. Reilly, J.P., Creaner, M.J.: NGOSS distilled: The Essential Guide to Next Generation Telecoms Management. TeleManagement Forum (2005)
2. TM Forum: TM Forum Technical Programs - NGOSS 5.0 Overview and Release Notes (2005) TeleManagement Forum Web site, http://www.tmforum.org/browse.asp?catID=1913&sNode=913&Exp=Y&linkID=30811&docID=3650
3. Strassner, J.: A Model Driven Architecture For Telecommunications System Using DEN-ng. In: Technical Proceedings ICETE 2004, vol. 1 (2004)
4. Miyoshi, Y., Kimura, T., Otsuka, Y., Fujita, Y., Majima, S., Suda, K.: An Implementation of Service Resource Management Architecture. In: Technical Proceedings WTC/ISS2002 (2002)

Server Support Approach to Zero Configuration of Power Line Communication Modems and Coaxial Cable Modems

Daisuke Arai, Kiyohito Yoshihara, Akira Idoue, and Hiroki Horiuchi

KDDI R&D Laboratories Inc., 2-1-15 Ohara Fujimino City
Saitama Prefecture 356-8502, Japan

Abstract. In this paper, we propose a new approach to auto configure a power line communication modem and a coaxial cable modem with minimum user intervention. The proposed approach is based on the following strategies: (1) An ISP establishes servers, (2) An ISP operator inputs information to configure modems to the server before configuration and (3) The software for the zero configuration of modems is built into a residential gateway. We implement a system based on the proposed approach, and measure the configuration time. The result shows that the configuration time is about 280 seconds. The system allows us to zero configure modems with minimum user intervention and moreover, the configuration time is faster than manual settings.

1 Introduction

In recent years, always-on broadband Internet connections have become available at home, thanks to the diffusion of Fiber To The Home (FTTH) and x Digital Subscriber Lines (xDSL). Additionally, a wide variety of Networked Appliances (NAs) such as PC, printers, or IP set top boxes (STBs) are connected to home networks.

A number of methods have been developed for constructing a home network and in particular, Ethernet and 802.11 wireless LANs are generally used. Meanwhile, Ethernet and 802.11 wireless LANs have the following two problems, (1) Ethernet may often be eyesores in the room, due to the open wiring and (2) when an 802.11 wireless LAN is used on different floors of a house, not only the transmission rate decreases, but also the signal does not reach. With this in mind, the construction of home networks with power line communication modems (PLC) and coaxial cable modems, has attracted considerable attention, for it may represent potential solutions to the two problems mentioned above. Since PLC and coax cable modems use existing power lines and coaxial cables in the home, there is no risk of them spoiling the appearance of the room. Moreover, they have an insignificant effect on the transmission rate as compared with an 802.11 wireless LAN.

A PLC modem or a coaxial cable modem enables NAs to network access through power lines or coax cables by connecting to a power plug or a coaxial

S. Ata and C.S. Hong (Eds.): APNOMS 2007, LNCS 4773, pp. 92–101, 2007.

plug and connecting to a NAs. While, in order to provide secure communication, a modem must obtain a Network Membership Key (NMK)[1,2]. Additionally, configuration of the Quality of Service (QoS) is needed when connected to NAs such as STBs for delay sensitive multimedia applications. Moreover, modem firmware must be updated to run applications as reliably as possible without modems being prone to abnormal terminations due to bugs.

While, the configuration and updating of modem firmware require highly skilled and experienced users with technical knowledge of the Internet. This poses a barrier to Internet novices and raises technical issues. In order to break down this barrier, the zero configuration protocol by which configuration is completed only by connecting modems or with minimum user intervention (e.g., pushing a button) is needed. Recently, certain protocols and techniques [3,4,5,6,7,8,9,10] for the zero configuration have been developed, but virtually none of the existing protocols and techniques can fully address this issue: some require user intervention (e.g. input password) or devices such as Universal Serial Bus (USB) flash drives, while others are unable to accept firmware updates.

This paper proposes a new server support approach to configuration of a PLC modem and a coaxial cable modem to solve this issue. The proposed approach allows us to configure modem settings and update modem firmware with only pushing a button. The proposed approach consists of three stages: the first is an acquisition stage, in which a residential gateway (RGW) acquires information from a server for configuration, and update of modem firmware. The second stage is a detection stage, in which a RGW detects modems connected in a home network, while the final stage is a configuration stage, in which a RGW configures modems, based on information from the server and the result of the detection stage.

To show the effectiveness, we implemented a system based on the proposed approach, and measured the time of the NMK configuration in a network where three PLC and three coaxial cable modems were connected respectively.

This paper is organized as follows: In Section 2, we present an overview of a typical home network with PLC and coaxial cable modems, and address technical issues for zero configuration. We review recent related work in Section 3. In Section 4, we propose a new server support approach to zero configuration for PLC and coaxial cable modems. In Section 5, we implement a system based on the proposed approach, while in Section 6, we evaluate the time of NMK configuration in a network.

2 Typical Home Network and Technical Issues for Zero Configuration

2.1 Typical Home Network with PLC and Coaxial Cable Modems

We focus on the HomePlugAV (HPAV) modem [1] of the PLC modem, standardized by the HomePlug Powerline Alliance, and HomePNA (HPNA) modem [2] of the coaxial cable modem standardized by the International Telecommunication

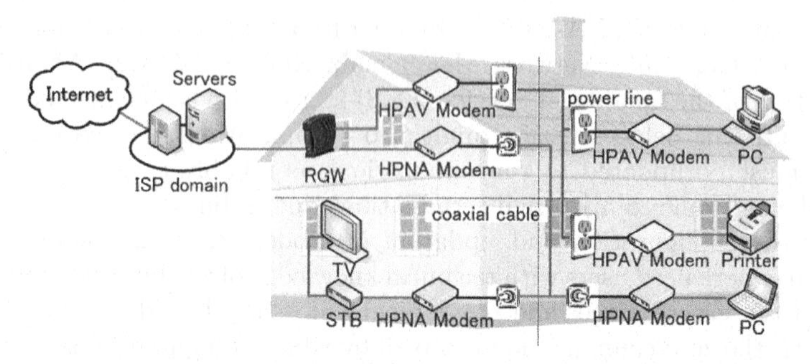

Fig. 1. Typical home network with HPAV and HPNA modems

Union. Figure 1 shows a typical home network with HPAV and HPNA modems. A RGW and two or more HPAV and/or HPNA modems, are shipped to the user from an Internet service provider (ISP) by contract with the ISP. Moreover, the ISP ships a configuration tool in removable media such as compact disk. A HPAV modem has one or more cat5 interfaces and a power line interface, while an HPNA modem has one or more cat5 interfaces and a coaxial cable interface. The user connects HPAV and/or HPNA modems to arbitrary places and NAs such as PC or STBs are connected to the RGW through HPAV/HPNA modems. The HPAV allows for network up to 200 Mbit/s and the HPNA does up to 240 Mbit/s.

When the user configures the NMK to HPAV/HPNA modems, the user typically connects their PC to an HPAV/HPNA modem, and runs the configuration tool. The configuration tool requests for input of the NMK to the user, and the user inputs the NMK. Thereby, the configuration tool configures the NMK to HPAV/HPNA modems through cat5 interface.

2.2 Technical Issues for Zero Configuration

The following technical issues should be addressed to achieve zero configuration of HPAV and HPNA modems by studying typical home networks in Section 2.1:

1. HPAV and HPNA modems should be auto configured with minimum user intervention (e.g. only connecting or pushing a button), to avoid users having to face the complicated and labor-intensive configuration task (Issue 1).
2. NMK Configuration should be performed for only intended modems to join the network, in order to ensure secure communication (Issue 2).
3. The updating of HPAV and HPNA modem's firmware should be performed fully within the auto configuration whenever applicable, in order to run applications as reliably as possible without any abnormal modem due to bugs (Issue 3).
4. The configuration time of a zero configuration scheme is faster than manual configuration for usefulness (Issue 4).

3 Related Work

Research and Development work on the zero configuration have been conducted. We summarize the recent related work commonly used in the home network. In addition, we show that the recent related work do not solely address all the issues in Section 2.2 simultaneously.

Dynamic Host Configuration Protocol (DHCP)[3][4] is a well-known practical protocol. It addresses Issue 1, in that a DHCP server centrally configures an IP address for NAs only the user connects the NAs to the network. However, it cannot address Issue 2. While, the configuration is automatically begun if A NAs is connected. Furthermore, it cannot address Issue 3 as they are restricted to link-local settings and a firmware update is outside the scope.

Universal Plug and Play Consortium (UPnP)[5] headed by Microsoft, design a set of auto-configuration protocols to enable plug-and-play, easy-to-use IP networking in IP networks lacking administrated network service. It address Issue 1, while, the configuration is automatically begun if A NAs is connected. However, it cannot address Issues 2 and 3 for the same reason DHCP.

The Zero Configuration Networking[6][7] by Internet Engineering Task Force, design a set of protocols for zero configuration. Nevertheless, they are still the same as DHCP for the Issues 1, 2 and 3.

Microsoft Windows Connect Now (WCN)[8] technologies support a Universal Serial Bus (USB) flash drive (UFD) configuration for the wireless setting. Users create network settings (e.g. Network Name (SSID)), store them in UFD, and propagate them to the wireless station and NAs via the use of UFD. The UFD configuration then addresses Issue 2. While, neither Issues 1 nor 3 can be addressed with the UFD configuration alone, since user intervention is required, and it is restricted to link-local settings (e.g. the sharing of SSID between a wireless station and wireless client devices) and a firmware update is outside the scope. In addition, the wireless station and wireless client devices require a USB interface.

A Wi-Fi Protected Setup (WPS)[9], as standardized by the Wi-Fi Alliance, supports a Personal Identification Number/numeric code (PIN) configuration for wireless setting. In the PIN configuration, all devices supporting the PIN configuration have a PIN. A PIN is provided by the following methods, (1) a fixed label or sticker may be placed on a device and (2) a dynamic PIN can be generated by a device and shown via a device's graphical user interface (GUI). Users input the PIN of the wireless client devices through their PC to a wireless station or they input the PIN of a wireless station through their PC to wireless client devices. The wireless client devices will then join the network that has the PIN of the wireless station or that has registered the PIN to the wireless station. The PIN configuration addresses Issue 2, although it cannot address Issues 1 and 3, because it is the same as the UFD configuration.

WPS and the One Touch Secure System (AOSS)[10], by Buffalo Technology Inc., supports a Push-Button Configuration (PBC) for wireless setting. In the PBC configuration, all devices supporting the PBC configuration have a button. Users push the button of the wireless station and those of the wireless client devices which join the network within a specific period of time. The wireless client devices can then join the network of which the button was pushed. Although the PBC configuration addresses Issue 1, it cannot address Issues 2 and 3 with PBC configuration alone. Moreover, users using the PBC configuration should be aware that there is a very brief setup period between pushing the buttons of the wireless station and those of the wireless client device during which unintended devices within range can join the network.

4 Server Support Approach to Zero Configuration

The findings in Section 3 motivated us to propose a new server support approach to zero configuration to meet all issues in Section 2.2, which will be presented in the following sections. The new approach target configuration of HPAV/HPNA modems in typical home network in Section 2.1.

4.1 Assumptions and Design Principles

Assumptions
We have the following assumptions.

1. An ISP establishes servers (a customer information system), which manage the user information.
2. An ISP operator inputs the following information before configuration, repeated the same times as the number of modems to the customer information system, (1) A Media Access Control (MAC) address of a modem, (2) NMK, (3) QoS parameter (e.g. combination of the following values, service type, priority, latency and jitter), (4) Firmware version, and (5) A user identifier (UID).
3. An ISP installs software for configuration of modems to the RGW. In addition, a UID and user password (e.g. a user password is automatically generated by the customer information system from a UID) are saved on the RGW.
4. The RGW has a mechanism to execute the software easily, such as via a push-button.

An ISP ships the RGW and two or more HPAV and/or HPNA modems to the user. The user connects HPAV and/or HPNA modems to arbitrary places and NAs such as PC or STBs are connected to the RGW through HPAV/HPNA modems. When the user configures the HPAV/HPNA modems, the user runs the software (e.g. pushing the button).

Design Principles
The following three stages are executed.

Acquisition stage
: In the acquisition stage, the RGW acquires the following information from the customer information system, (1) A MAC address of a modem, (2) NMK,

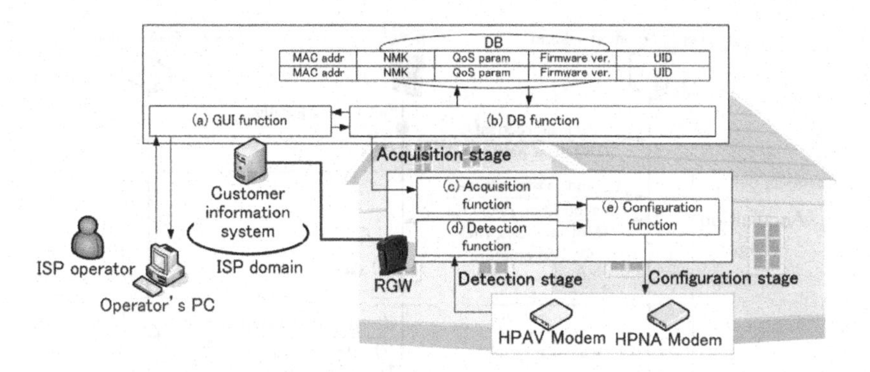

Fig. 2. Functional Architecture

(3) QoS parameter and (4) The firmware version. The RGW acquires the information for each modem, and then stores it.

Detection stage

In the detection stage, the RGW detects the following information from modems; (1) The MAC addresses of the HPAV/HPNA modems in a home network and (2) The NAs MAC addresses bridged by the HPAV/HPNA modem.

Configuration stage

In the configuration stage, the RGW compares the information acquired during the acquisition stage with that obtained during the detection stage. Consequently, the RGW configures NMK and QoS parameters to existing modems in a home network. In addition, the RGW compares the firmware version acquired during the acquisition stage with that stored in the previous acquisition stage. If the modem firmware has been updated, the RGW requests the modem to execute a firmware update. Two or more modems are configured from the remote one to the local one connected directly to the RGW.

The proposed approach configures the NMK and QoS parameters, and updates the modem's firmware only by the pushing the button, thus the proposed approach address Issue 1 and 3. Furthermore, the proposed approach acts only on the modem registered in the customer information system, thus the proposed approach address Issue 2.

4.2 Functional Architecture

Figure 2 shows the functional architecture of the proposed approach. The customer information system has the following functions: (1) GUI function (Fig.2 (a)) and (2) DB function (Fig.2 (b)). The GUI function provides a user interface for the input of the information to configure modems. An ISP operator uses a PC to input the information through the GUI function. The inputted information is stored to the DB of the customer information system by the DB

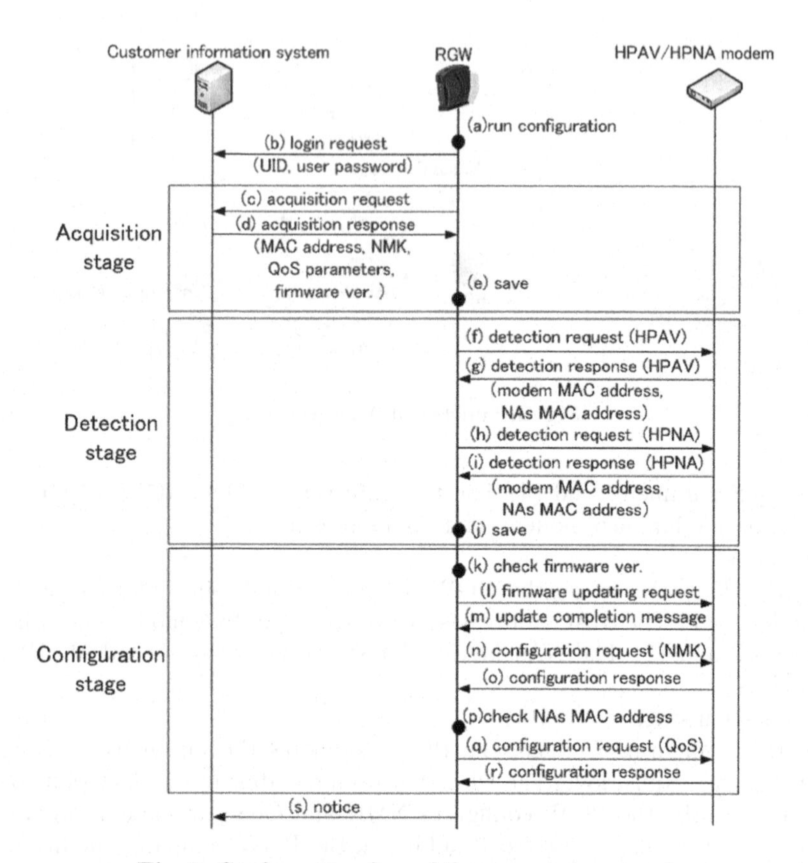

Fig. 3. Configuration flow of the proposed approach

function. Meanwhile, the RGW have the following functions, (1) Acquisition function (Fig.2 (c)), (2) Detection function (Fig.2 (d)), and (3) Configuration function (Fig.2 (e)). Each function of the RGW corresponds to the each stage shown in Section 4.1.

4.3 Configuration Flow

Figure 3 shows the configuration flow of the proposed approach. When users connect HPAV and/or HPNA modems to a RGW, they also run configuration program presented by a RGW (e.g. pushing a button of a RGW) (Fig.3 (a)). The RGW then sends a login request to the customer information system, including a UID and user password, whereupon the customer information system uses the UID and user password to authenticate the RGW (Fig.3 (b)). The UID and user password were presented to a RGW, when the RGW was shipped. If the authentication is successful, the RGW sends an acquisition request to the customer information system (Fig.3 (c)). The customer information system transmits the information to the RGW, including a MAC, NMK, QoS parameters

Table 1. System specifications

<table>
<tr><td rowspan="3">Customer information system</td><td rowspan="3">Linux server</td><td>CPU</td><td>Intel Core 2 T5500 1.66GHz</td></tr>
<tr><td>Memory</td><td>512MBytes</td></tr>
<tr><td>OS</td><td>FedoraCore 6</td></tr>
<tr><td rowspan="6">RGW</td><td rowspan="3">PC</td><td>CPU</td><td>Broadcom 4704 266MHz</td></tr>
<tr><td>Flash / RAM</td><td>8MBytes / 64MBytes</td></tr>
<tr><td>OS</td><td>FedoraCore 6</td></tr>
<tr><td rowspan="3">Windows server</td><td>CPU</td><td>Intel Pentium 4 1.4GHz</td></tr>
<tr><td>Memory</td><td>768MBytes</td></tr>
<tr><td>OS</td><td>Windows XP</td></tr>
</table>

and the firmware version for each modem (Fig.3 (d)). The RGW then saves that information to the local disk (Fig.3 (e)).

The RGW detects the following information: (1) The MAC addresses of HPAV and HPNA modems connected to a home network and (2) The MAC address of a NAs connected to a home network via HPAV and HPNA modems (Fig.3 (f), (g), (h) and (i)). The RGW saves that information to the local disk similarly to the acquisition stage (Fig.3 (j)).

The RGW compares the firmware version (Fig.3 (k)). If the modem firmware has been updated, the RGW requests a firmware update (Fig.3 (l)). When the modem has finished the firmware update, the modem sends an update completion message to the RGW (Fig.3 (m)). A firmware update (Fig.3 (k), (l) and (m)) is repeated the same times as the number of modems, while upon completion, the RGW configures the NMK of the HPAV and HPNA modems connecting to a home network (Fig.3 (n), (o)). The RGW also configures the QoS parameters after the NMK configuration and has information concerning the MAC addresses of NAs connected through HPAV and HPNA modems. Thereby, the RGW can configure QoS parameters for the modems selectively, such as the modems connected to NAs for a delay sensitive multimedia application (e.g. STBs) (Fig.3 (p), (q) and (r)). Finally, the RGW notices the customer information system of the configuration completion (Fig.3 (s)).

5 Implementation

We implement a system based on the proposed approach and describe a brief overview below, with certain implementation restrictions. Table 1 shows the system specifications.

We implement the customer information system, using a Linux server. The Linux server implements the functions described in Section 4.2 (Fig.2 (a), (b)). Meanwhile, the RGW is implemented using a poor PC and a Windows server. We deploy the Windows server to configure the HPAV modems, since program libraries to generate Ethernet frames for HPAV modems work only on the Windows systems. The poor PC implements the following functions, (1) Acquisition function (Fig.2 (c)), (2) Detection function for HPNA modems (Fig.2 (d)), (3)

Configuration function for HPNA modems (Fig.2 (e)), and (4) VPN function. In addition, the Windows server implements the following functions, (1) Detection function for HPAV modems, (2) Configuration function for HPAV modems, and (3) VPN function. The Windows server is established with the Linux server, which are connected with 100Base-TX. Moreover, the Ethernet frames to configure HPAV modems are generated by the Windows server, and the Ethernet frames arrive to the poor PC through the VPN function.

6 Evaluations

We measured the time of the NMK configuration, using an implemented system based on the proposed approach to show that the proposed approach addressed Issue 4. Figure 4 shows the measurement environment. We measured the time ten times. Table 2 shows the average time of each stage, and the total time.

The results show that the total time is 284.7 seconds. When users configure modems, they should also be familiar with the configuration steps. Since users have multiple steps involved for checking configuration steps, let us suppose that they check the configuration details with the help of an ISP call center. The average time spent on the phone to the call center is about 6 minutes [11], meaning the configuration time with the proposed approach is faster than the manual configuration. Thus, the proposed approach allows us to achieve zero configuration with minimum user intervention in an acceptable configuration time. The inquiry as to configuration from users may be reduced, leading to the decrease in the opportunity of the call center operations.

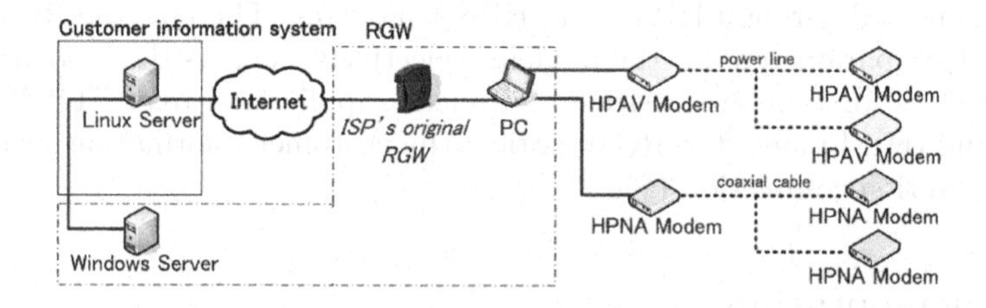

Fig. 4. Measurement environment

Table 2. NMK configuration time of the proposed approach

	Time(s)	Time(s)
Acquisition stage	22.1	284.7
Detection stage (HPAV)	20.3	
Detection stage (HPNA)	68.2	
Configuration stage (HPAV)	74.4	
Configuration stage (HPNA)	99.7	

7 Conclusions

This paper proposed a new approach to configure the HomePlugAV (HPAV) and the HomePNA (HPNA) modems. We show four issues for the zero configuration in a home network. The lack of a protocol addressing all issues simultaneously motivated us to design a new approach based on three stages, (1) Acquisition stage, (2) Detection stage and (3) The configuration stage. An elaborated description of the configuration flow of the proposed approach was shown. It allowed us to configure HPAV and HPNA modem settings and to update the firmware. We implemented a system based on the proposed approach, and evaluated the average time of the Network Membership Key (NMK) configuration in a network where three HPAV and three HPNA modems were connected respectively. The total time of the NMK configuration was 284.7 seconds, and we showed that the configuration time with the proposed approach was faster than the manual configuration. The evaluation of the configuration of QoS and update of the modem firmware will be our future directions, before the practical use.

Acknowledgments

We are indebted to Dr. Shigeyuki Akiba, President and Chief Executive Officer of KDDI R&D Laboratories Inc., for his continuous encouragement to this research.

References

1. HomePlug, AV: HomePlug AV White Paper (URL available in July 2007), http://www.homeplug.org/products/whitepapers/
2. Phoneline networking transceivers - Enhanced physical, media access, and link layer specifications: ITU-T G.9954 (2005)
3. Droms, R.: Dynamic Host Configuration Protocol. IETF RFC2131 (1997)
4. Droms, R.: DHCP: Options and BOOTP Vendor Extensions. IETF RFC2132 (1997)
5. Universal Plug and Play Forum: Universal Plug and Play Device Architecture (URL available in July 2007), http://www.upnp.org./specs/arch/UPnP-DeviceArchitecture-v1.0-20060720.pdf
6. IETF Zero Configuration Networking (zeroconf) (URL available in July 2007), http://www.zeroconf.org/
7. Williams, A.: Requirements for Automatic Configuration of IP Hosts. IETF draft-ietf-zeroconf-reqts-12.txt (2002)
8. Windows Connect Now (WCN): Windows Connect Now-UFD Specifications (URL available in July 2007), http://www.microsoft.com/whdc/rally/rallywcn.mspx
9. Wi-Fi Protected Setup (WPS): Wi-Fi Protected Setup White Paper (URL available in July 2007), http://www.wifialliance.com/wifi-protected-setup
10. AirStation One-Touch Secure System (AOSS): AirStation One-Touch Secure System White Paper (URL available in July 2007), http://www.buffalotech.com/technology/our-technology/aoss/
11. eBusiness Best Practices for All Industries, 2001 (URL available in July 2007), http://www.cisco.com/en/US/products/sw/custcosw/ps1844/products_implementation_design_guide_chapter09186a008057a628.html

Architecture of Context-Aware Integrated Security Management Systems for Smart Home Environment*

Seon-Ho Park, Joon-Sic Cho, Sung-Min Jung, Young-Ju Han, and Tai-Myoung Chung

Internet Management Technology Laboratory,
School of Information and Communication Engineering,
Sungkyunkwan University,
300 Cheoncheon-dong, Jangan-gu,
Suwon-si, Gyeonggi-do 440-746, Republic of Korea
{shpark,jscho,smjung,yjhan}@imtl.skku.ac.kr,
tmchung@ece.skku.ac.kr

Abstract. The smart home technology is a noticeable area of study in various studies on the ubiquitous computing technology. This paper presents the security management system to provide context-aware security service. The system supports the management and enforcement of context-aware security and the management of the context of users and resources. This system provides the security service functions such as the partial credential based user authentication and context-aware access control, and the security policy management functions. In addition, the system provides the policy enforcement simulation function to test and validate the security policy. This simulator can cover all security service in the smart home environment.

1 Introduction

Due to the rapid development of information and communication technology, the IT paradigm is rid of the past desk-based computing environment and entering the new IT paradigm such as a ubiquitous computing. The smart home technology is a noticeable area of study in various studies on the ubiquitous computing technology. For more than a decade, the term "smart home" has been used to introduce the concept of networking devices and equipments in the house[1]. The smart home environment(SHE) is based on various technologies such as the wireless network, the mobile communication technique, context-aware technique, and embedded software. The context-aware technique is specially important and interesting area of study for the SHE, since it is core technique for automatic intelligent computing service.

* This research was supported by the MIC(Ministry of Information and Communication), Korea, under the ITRC(Information Technology Research Center) support program supervised by the IITA(Institute of Information Technology Advancement) (IITA-2006-C1090-0603-0028).

S. Ata and C.S. Hong (Eds.): APNOMS 2007, LNCS 4773, pp. 102–111, 2007.

Ubiquitous computing technology, which is an important infra-technology of the SHE, have new security challenges, because it has different features from traditional computing environments. The core techniques of ubiquitous computing, such as context-awareness and wireless sensor network, require new security protocols that are different from the traditional security protocols. For example, the public key infrastructure cannot be applied to the wireless sensor network that has limited resources and the context-aware services require the enhanced security policies which can consider the context of users and resources[2,3].

Since the SHE includes various appliances that are networked through wired or wireless medium, in addition they interoperate with each other, security policies for the SHE is more complicated than typical security policies. In addition, the security policy for the SHE must be able to deal with the context that is related with users and resources and also, the security architecture that is able to enforce such policies is required.

This paper presents the security architecture to provide context-aware security service. The architecture supports the management and enforcement of context-aware security and the management of the context of users and resources. We implemented this architecture and call it the CAISMS(Context Aware Integrated Security Management System). The function of the CAISMS consists of a user authentication, an access control, a policy management, and a context management. The CAISMS includes the web-based policy UI for ease of policy management and also provides an additional UI for the simulation of the smart home environment. The policy administrator can test and validate the policy that s(he) configured using the simulation UI.

The remainder of this paper is composed as follows. In section 2 we explore related works. In section 3, we discuss an overview of the CAISMS. Section 4 illustrates detailed architecture, functions and operations of core components of CAISMS. Section 8 concludes the paper.

2 Related Works

The CASA[4] provides context-based access control using GRBAC [5] for security of Aware Home. GRBAC is an extension of traditional Role-Based Access Control. It enhances traditional RBAC by incorporating the notion of object roles and environment roles, with the traditional notion of subject roles. But GRBAC has some problems. First, GRBAC is not suitable for large and complex organizations, because of defining too many roles in the system. Second, RBAC loses its advantage of data abstractions by object role. RBAC abstract user-level information from system level information by notion of permission is relationship between objects and operations. In GRBAC, because object role violates this relationship, the data abstraction could not be achieved. In addition, this problem violates user/role and role/permission associations. Finally, GRBAC has an unnecessary overlapping that environment roles and object roles, because certain physical environmental things can be also objects.

Cerberus[6] enhances the security of ubiquitous applications that are built using Gaia. The security service of Cerberus is based on its Inference Engine. The Inference Engine performs two kinds of tasks which are managing the level of confidence for user authentication and evaluating queries from applications about access control. The access control provided by Cerberus considers context in making access control decisions and configuring policy. However, access control policy of Cerberus uses access control lists defined centrally by a system administrator. These access control mechanism is easy to maintain, but lack flexibility. The authentication of Cerberus uses a multi-level authentication scheme that has associated confidence values describing the trustworthiness of the authentication achieved. However, it is insufficient that the method for applying confidence value to authentication mechanism. The final defect of the Cerberus is that context-aware authentication such as location-aware authentication.

We presented the preliminary version of our CAISMS[7]. In the preliminary version, the CAISMS presents context-aware access control method using CRBAC, but context-role activation management is not supported so dynamic access control service is impossible. And We present the notion of AWL (authentication warrant level), but the calculation method of AWL is incomplete.

3 CAISMS Overview

The CAISMS provides a user authentication service based on the partial credential and context-aware access control service for users in the smart home. For ease of policy management, the CAISMS also provides web-based graphic user interface including the policy configuration function and the smart home simulation function.

The partial-credential based user authentication can enhance the level of security, since it gives differential trust degree to the principals based on confidence level of authentication methods. We used the fuzzy if-then rules to calculate the credentials of various authentication methods.

The access control part of the CAISMS is based on the context-role based access control(CRBAC) model for the context-aware access control. The model has all advantages of the RBAC and also enables the management of the context-based access control policy. The access control of the CAISMS provides secure user-role activation using the authentication credential value. The access control part includes the context-role activation manager(CRAM). The CRAM manages the activation and deactivation of context-roles.

The CAISMS includes the context server(CS) to manage the context information that is required for the context-role activation. The CS gathers various context information such as a location, a temperature, a humidity, and an illumination using the sensors. The communication between the CS and sensors is encrypted by skipjack algorithm.

The CAISMS provides convenient policy configuration function. The security administrator need not access the policy database directly, but can easily configure security policies through the web-based graphical user interface. In addition, the interface provides the policy enforcement simulation to test and validate the security policy. This simulator can cover all security service in the SHE.

4 CAISMS Architecture

This section describes detailed architecture, functions and operations of core components of CAISMS. Figure 1 is an overview of the CAISMS architecture.

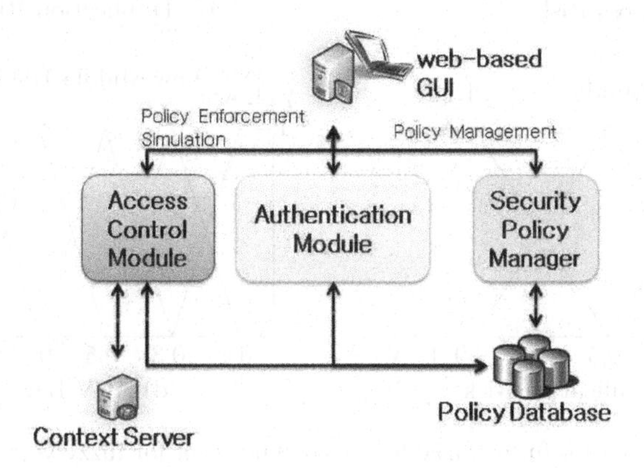

Fig. 1. The CAISMS Architecture Overview

4.1 User Authentication

The CAISMS applies different confidence levels to each user authentication mechanism using Authentication Credential Value (ACV). The ACV is a trust degree that differentiates authentication methods based on reliability.

User authentication mechanisms are various and have different features each other. User authentication in computing systems traditionally depends on three factors: something you have (e.g., a hardware token), something you are (e.g., a fingerprint), and something you know (e.g., a password). There are also some authentication failure factors which are loss risk, duplication risk and modification risk. A hardware token can be stolen or lost, so it has loss risk. A password can be forgotten by owner, or guessed or modified by attacker, so it has loss risk, duplication risk, and modification risk.

It is difficult to decide trust level of authentication methods, because the authentication failure factor is ambiguous value. It is impossible to get an exact numerical value which can decide quality and quantity of authentication failure factors. We used the fuzzy logic which is an extension of Boolean logic dealing with the concept of partial truth, as the solution of above ambiguities.

The model of auzzy logic control consists of a fuzzifier, fuzzy rules, fuzzy inference engine, and a defuzzifier. We have used the most commonly used auzzy inference technique called Mamdani Method[8] due to its simplicity. The process is performed in four steps.

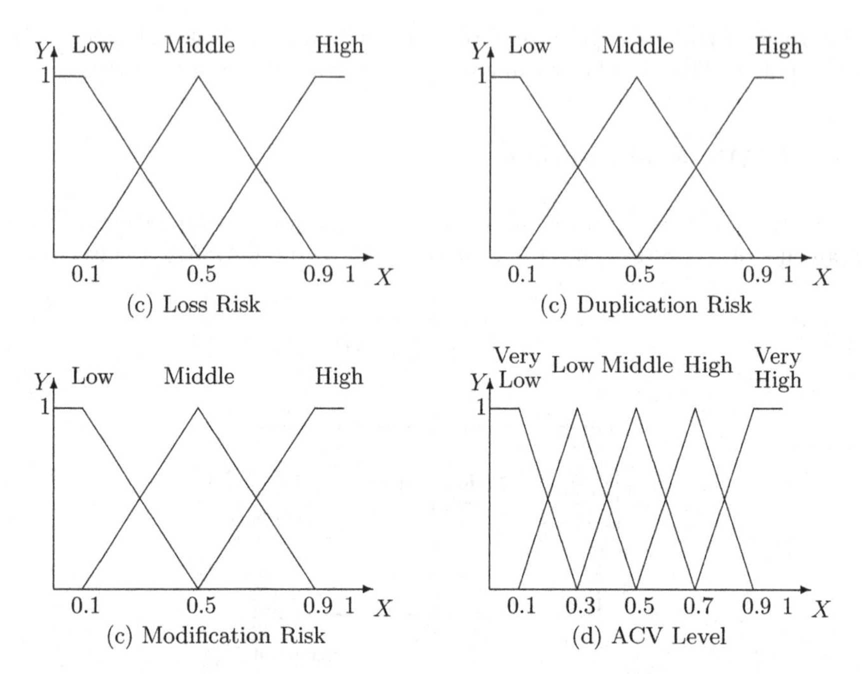

Fig. 2. (a) Fuzzy set for fuzzy variable loss (b) Fuzzy set for fuzzy variable duplication (c) Fuzzy set for fuzzy variable modification (d) Fuzzy set for fuzzy variable ACV Level

- Fuzzification of the input variables "loss risk", "duplication risk", and "modification risk" : taking crisp inputs from each of these and determining the degree to which these inputs belong to each of the appropriate fuzzy sets.
- Rule evaluation : tading the fuzzified inputs, and applying them to the antecedents of the fuzzy rules. It is then applied to the consequent membership function (Table 1).
- Aggregation of the rule output : the process of unification of the outputs of all rules.
- Defuzzification : the input for the defuzzification process is the aggregate output fuzzy set "ACV level" and the output is a single crisp number.

During defuzzification, it finds the point where a vertical line would slice the aggregate set "ACV level" into two equal masses. In practice, the COG (Center of Gravity) is calculated and estimated over a sample of points on the aggregate output membership function, using the following formula:

$$COG = (\sum \mu_A(x) * x) / \sum \mu_A$$

where, $\mu_A(x)$ is the membership function of set A.

The linguistic variables used to represent the three risks, are divided into three levels: low, medium and high, respectively. The outcome to represent the authentication credential value is divided into five levels: very low, low, middle, high, and very high. The fuzzy rule base currently includes rules like the following: if the “loss risk” is high and ”duplication risk” is low and “modification risk” is low then the “ACV level” is middle. Thus we used $3^3 = 27$ rules for the fuzzy rule base. We used triangle membership functions to represent the fuzzy sets medium and trapezoid membership functions to represent low and high in the three risk membership sets. The membership functions developed and their corresponding linguistic states are represented in Table 1 and Figure 2.

Table 1. Fuzzy if-then rule definitions

	Loss	**Duplication**	**Modification**	**Confidential Level**
Rule 1	Low	Low	Low	Very High
Rule 2	Low	Low	Middle	High
Rule 3	Low	Low	High	Middle
Rule 4	Low	Middle	Low	High
Rule 5	Low	Middle	Middle	High
Rule 6	Low	Middle	High	Middle
Rule 7	Low	High	Low	Middle
Rule 8	Low	High	Middle	Middle
Rule 9	Low	High	High	Low
Rule 10	Middle	Low	Low	High
Rule 11	Middle	Low	Middle	High
Rule 12	Middle	Low	High	Middle
Rule 13	Middle	Middle	Low	High
Rule 14	Middle	Middle	Middle	Middle
Rule 16	Middle	High	Low	Middle
Rule 17	Middle	High	Middle	Low
Rule 18	Middle	High	High	Very Low
Rule 19	High	Low	Low	Middle
Rule 20	High	Low	Middle	Middle
Rule 21	High	Low	High	Low
Rule 22	High	Middle	Low	Middle
Rule 23	High	Middle	Middle	Low
Rule 24	High	Middle	High	Very Low
Rule 25	High	High	Low	Low
Rule 26	High	High	Middle	Very Low
Rule 27	High	High	High	Very Low

4.2 Access Control

The access control component consists of the policy enforcement module(PEM), the user-role activation manager(URAM), and the context-role activation manager(CRAM). The PEM enforces the access control policies based on CRBAC models. Input and output of the PEM are well defined where the former is the user access requests and the latter is either allowing or denying those requests. The request contains the activated user-role, the object that needs to be accessed, operation that needs to be performed. When the PEM is received the reqeust, the PEM retrieves policy rules which is related with the request from the policy database. The policy rule is defined as a quadruple

$$PR = < ur, perm, cr, perm\_info >$$

where

- ur is a element of a set of user roles UR ($ur \in UR$),
- $perm$ is a element of a set of permissions P ($perm \in P$),
- cr is a element of a set of context roles CR ($cr \in CR$),
- $perm\_info$ is a decision information which is either allowing or denying the access control request.

The PEM checks that the cr in the retrieved policy rules is in the enabled context-role list, and if it is in the list the PEM makes the access control result based on the $perm\_info$ in the policy rule. The enabled context-role list contains context roles which can be activated currently, and is managed by the CRAM. If the request is allowed, the CRAM adds the cr which is used to the permission decision to the activated context-role list which is managed by the CRAM.

The CRAM provides the management function of enabled context-role list and activated context-role list, as well as the function of the context-role deactivation. The deactivation of context-roles occurs when the security session is closed and the current context is not satisfied the context condition which is related the context-role.

The URAM manages the activation/deactivation of user role and provides the role discovery function. The role discovery function supports the validation checking of mapping between the user-role and permission, and gives a suitable user-role set to the user if the user-role which is selected by the user is not valid to access the object which needs to be accessed.

4.3 User Interface

The user interface is a web-based graphic user interface, and consists of the security policy management part and the simulation part for a simulation of a policy enforcement in the SHE. The policy management part is an interface to the security policy manager component, and supports almost the functions to be needed for managing the policy database. The policy administrator inputs simply the policy information to the user interface, then the security policy manager component makes the policy rules using the information which is inputed through the user interface.

Figure 3 is a screenshot of the security policy management user interface. The figure shows the policy rule configuration to add the policy rule $< Guest, GateDoor\_OPEN, VACATION, Deny >$ to the policy database. All the other policy configuration functions are also implemented in the policy management user interface.

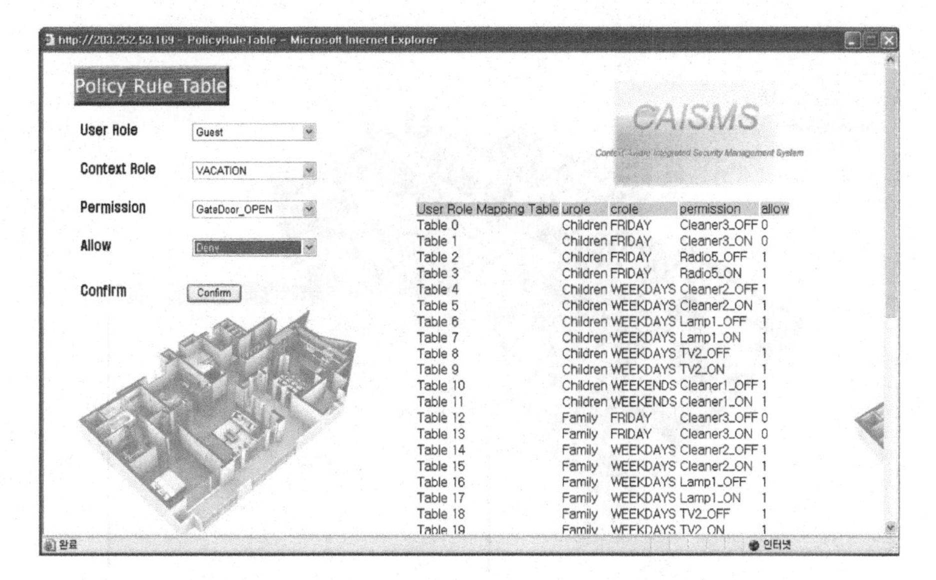

Fig. 3. A screenshot of security policy management UI

Figure 4 is the screentshot of simulating the policy enforcement in the SHE using our simulation UI. The administrator simulates the security services which are the partial credential based user authentication and access control based on the CRBAC model. The pictures indicates users in the home, and can be moved by the drag-and-drop. The movement of the picture simulates that the user which is related the picture moves in the home. The pictures are appeared when the user authentication is completed and successful. We use the RFID authentication as an user authentication, and also can select other authentication methods. The pop-up page is the interface for the request of the access control service. Input values are a user role, an object, an operation, and position information. The position information is the value which indicates the distance between the user and the object. The distance information is used for spatial context information. After the success of access control, the security session is made and maintained in the access control component. This security session can be closed by the context-role deactivation. If the security session is closed by the context-role deactivation, the access control component notify the fact that the security session is closed. The simulation UI can also express such security session changes.

4.4 Context Server

The context server(CS) gathers the context information and manages the gathered context. The context is gathered using the various sensors. Our system gathers the temporal context, the spatial context, and environmental context such as a temperature, a humidity, and an illumination.

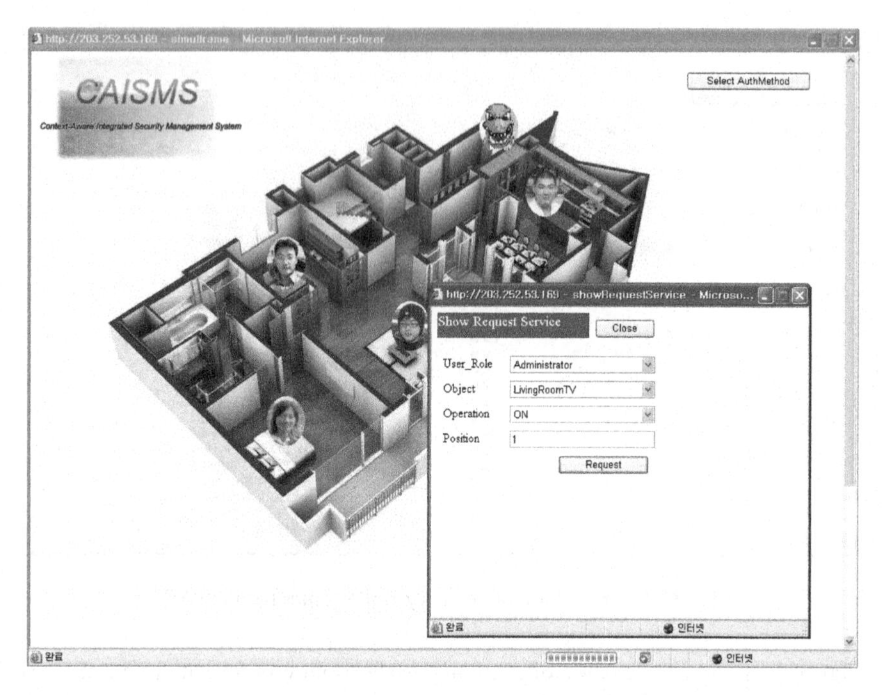

Fig. 4. A screenshot of the policy enforcement simulation UI

In the sensor network of our system, the sensors do not perform the in-network processing, because the environment is not spacious and the sensors need not to be densely deployed. The communication between the CS and sensors is encrypted using skipjack algorithm for secure communication.

5 Conclusion

The context-awareness feature of the smart home environment requires new security challenges. A paradigm shift in policy models is needed to move focus from the subject-centricity to the context. We presented the context-aware integrated security management system. The system provides the partial credential user authentication using authentication credential value and context-based access control using the context-role based access control model. We used the fuzzy if-then rule for calculating the authentication credential value. The CAISMS also provides the web-based graphic user interface which provides the security policy configuration function and the policy enforcement simulation function. The policy enforcement simulation function enables the policy administrator to easily test and validate the security policy.

Our future work is to study the theoretical validation methodology which can prove the level of security and privacy in our system.

References

1. Bierhoff, I., van Berlo, A., et al.: Towards an inclusive future. In: COST, pp. 110–156 (2007)
2. Park, S.-H., Han, Y.-J., et al.: Context-Role Based Access Control for Context-Aware Application. In: Gerndt, M., Kranzlmüller, D. (eds.) HPCC 2006. LNCS, vol. 4208, pp. 572–580. Springer, Heidelberg (2006)
3. Convington, M.J., Long, W., et al.: Securing Context-Aware Applications Using Environment Roles. In: SACMAT 2001, Chantilly, Virginia, USA, May 3-4, 2001 (2001)
4. Convington, M.J., Fogla, P., et al.: A Context-Aware Security Architecture for Emerging Applications. In: Proceedings of the 18th Annual Computer Security Applications Conference, p. 249 (2002)
5. Moyer, M.J., Ahamad, M.: Generalized role based access control. In: ICDCS 2001, Mesa, AZ (April 2001)
6. Al-Muhtadi, J., Ranganathan, A., et al.: Cerberus: A Context-Aware SEcurity Scheme for Smart Spaces. In: PerCom 2003, Dallas-Fort Worth, Texas (March 23-26, 2003)
7. Cho, J.-S., Park, S.-H., Han, Y.-J., Chung, T.-M.: CAISMS: A Context-Aware Integrated Security Management System for Smart Home. In: The 9th International Conference on Advanced Communication Technology (February 2007)
8. Negnevitsky, M.: Artificial intelligence: A auide to intelligent systems. Addison-Wesley, Reading, MA (2001)
9. Sandhu, R.S., Coyne, E.J., et al.: Role-Based Access Control Models. IEEE Computer 29(2), 38–47 (1996)
10. Ferraiolo, D.F., Sandhu, R., et al.: Proposed NIST Standard for Role-Based Access Control. ACM Transactions on Information and System Security 4(3) (2001)

Self-adaptability and Vulnerability Assessment of Secure Autonomic Communication Networks

Frank Chiang and Robin Braun

Faculty of Engineering, University of Technology Sydney,
NSW 2007, Australia
frankj@eng.uts.edu.au

Abstract. Risk and Vulnerability Analysis (RVA) aims at identifying the weaknesses of the networks that may be exploited to compromise the normal functions, such as service deployment, file system access permissions, applications activations and so on. Autonomic Communication Networks (ACNs) are recently proposed as business-objective driven high-level self-managed telecommunication networks with the adaptation capability to cope with increasing dynamics. Adaptation capability termed as adaptability becomes the premise of realizing thorough autonomy. As a theoretic foundation, we firstly propose an innovative Object-oriented Management Information Base (O:MIB). Secondly, a new information-theoretic security awareness strategy inspired from human immune system is proposed to reconfigure file access right, which has a direct relation with adaptability. The experimental results validate this methodology and find out a statistical bound for operators to set a vulnerability level of warning in practice.

1 Introduction

Self-protection and self-healing are two important attributes of Autonomic Communication Networks (ACNs), requiring the networks to not only reactively detect attackers but also proactively defend against potential security threats and recover from attacks autonomically. Currently associated research work in the literature includes design and implementations of Autonomic Defence Systems (ADSs) [1]. To deliver an efficient security defence, Vulnerability Analysis (VA) is strongly recommended to help with these requirements due to the fact that it aims at identifying the weaknesses of the networks that may be exploited to compromise the normal functions, such as service deployment, file system access operations and application instantiations.

Intensive research has been seen in recent years on Autonomic Communication Networks (ACNs) [2] since 2001. We have also proposed and proved learning and adaptation capabilities are two indispensible factors to the success of ACNs [3]. However, until now, there is a lack of literature to explore the comprehensive links between the vulnerability of autonomic systems and the adaptability in the society for autonomic communication networks, and to what extent the ACNs system is securely self-adaptable enough without breaching the maximum limits

S. Ata and C.S. Hong (Eds.): APNOMS 2007, LNCS 4773, pp. 112–122, 2007.

of safe content exposures to all usages (including illegal users and malicious attacks).

To the best of the authors' knowledge, this research could be the first attempt to tackle this important issue in depth for ACNs. In this paper, we carry out an information-theoretic analysis on the relation of *adaptability, autonomy,* and *vulnerability*, and propose a solution to breaking down these coherent links by finding a balance point for the tradeoff of maximizing systems' adaptability and reducing systems' vulnerability simultaneously. The research aims at yielding a tolerant bound for a desirable adaptation capability based on our proposed bio-inspired scheme, which enables the reduction of the vulnerability into a minimum set for ACN networks simultaneously. It is the belief of the authors that the analysis results in this paper will give operators a global view of what damage will occur under certain situations of information exposure due to security flaws. Consequently, operators can select a best timing to activate security mechanisms (e.g., Anti-virus software, Firewalls and Intrusion Detection System) to avoid further damages without performance compromises.

The remainder of the paper is organized as follows: Section 2 presents a detailed description of research questions. Section 3 presents our innovative O:MIB as well as discussing its links with conventional MIB in the sense of information modelling. Security awareness system (SAS) is discussed in section 4. Section 5 implements a validation simulation with a Java agent system interacting with O:XML. Experimental results show the efficiency of our proposed biological behavior-inspired vulnerability awareness system. Finally, we conclude the contribution of this paper.

2 Problem Statement

The research question arises from such statistical observations that reveal adaptability vs autonomy and adaptability vs vulnerability have strong similarity with respect to quantitative results, whereas, autonomy vs vulnerability are a pair of contradictory factors in practice. Adaptation capability termed as *adaptability* becomes the premise of realizing thorough *autonomy*. The statistical results are as follows: the system **vulnerability** is roughly proportional to the **rewritability**[1]. The dotted lines represents ideal **ADaptability (AD)**[2] without vulnerability bound concern.

It is statistically proven that the higher portion the system variables are rewritable, the more adaptable the system is to be, but at the same time, the more vulnerable the system is. Therefore, it is important to seek a way to maintain system robustness and adaptability at the same time so as to avoid this dilemma. To resolve these issues, we propose a biologically inspired awareness scheme that

[1] **ReWritability (RW)** of a system is defined as the ratio between the number of writable variables and total number of readable variables. Equation 1 and Eq. 2 define how to calculate writable and readable variables.

[2] The system **ADaptability (AD)** is defined as the mathematical integration of the **ReWritability (RW)**.

senses the security flaw and subsequently produce self-adaptive vulnerability results in the aim of activating the security system protection by reconfiguring *file/ variable access permissions.* This scheme presents an information-theoretic vulnerability analysis incorporating a number of concerns — (1) histogram of vulnerability and current vulnerability, (2) information contents exposure – rewritability, (3) traffic patterns, (4) disturbances and internal errors.

3 Information Modelling

3.1 CIM Schemas, SID Data Model vs. the Proposed O:MIB

The standard of Common Information Model (CIM) [4] developed by DMTF also produce an object-oriented scheme to organize the hierarchical data of Managed Elements (MEs) from different manufactures or sources. MEs includes devices and applications. With regards to its infrastructure specification and schema, the MEs are represented as *classes,* and the links between classes are desribed by associations. Most importantly, the *inheritance* is used to efficiently describe the common base elements and inherited sub-elements. UML notation is applied and CIM can be described in XML format in serveral ways. Share Information Data (SID) from Telecommunication Management Force (TMF) also shares some similarities. CIM standards present a conceptual schema to encapsulate all the MEs such as: services, devices, storages, computer systems, network system and applications. CIM essentially only provides an efficient *storage mangement* of various devices and applications so that system administrators and management programes can access it in an universal way. However, the methods for each class are not the main focus, the predefined methods are limited to static class methods. Moreover, the execution of corresponding actions are still not in need of a seperate high-level control (e.g., centralised control) to be delivered, which is independent from CIM structures.

In contrast with SNMP MIBS, CIM schema provides a better representation of information than *static* MIBs in terms of an OO structure for data modelling. Whereas SNMP MIBS have been used in industries for many years since ISO produced this hierarchically layered model. It represents informations of managed objects from a substantial view other than CIM models. A new O:MIB scheme is proposed in this paper as an alternative way to conventional MIB, also as a new attempt for the information modelling area by taking MIB as an example, it is expected the scheme proposed here could be extended to other sources of information modelling. It is known that conventional MIB is actually a table defined by OIDs to record the hierarchical object information without object-oriented principles. However, O:MIB is not only a data store, rather, it takes the object-oriented principle to manage the MIB objects and essentially divert CPU load into large amount of local CPU. The execution of corresponding actions can be invoked by the methods and algorithms through the local agent residing on each MIB variable dynamically, most of the decision-making taks can be done locally, the workload of system administrators or management programs in CIM standard could be furtherly reduced. Therefore, the nature of O:MIB structure designed specifically for distributed components with local execution capabilities determines its efficiency.

3.2 O:MIB vs. MIB

To fit the distributed scenario, we propose the object-oriented MIB (O:MIB) technique for distributed networks as an alternative to the conventional MIB for ACNs. It is implemented on the basis of *object-oriented XML* technology (O:XML[3]), and its integration with **Spring** application framework[4]. Figure 1 makes a chracteristic comparison between MIBs and O:MIBs. The proposed distributed approach based on O:XML can also be applied to other networks such as wireless ad-hoc networks, wireless sensor networks where peer-to-peer connectivity and network activities are supported.

Conventional MIB	Object-oriented MIB
Hierarchical Structure	Hierarchical Structure
Information Elements Stored	Information Elements + functions, Algorithms + Embedded Agent Semantics
Data Oriented	Object-oriented/On-demands
SMI/ASN.1- Standardized	O:XML-Enabled
Static/Fixed	Dynamic/Extensible/Reconfigurable

Fig. 1. Comparison Between Conventional SNMP MIB and O:MIB

4 Vulnerability Awareness System (VAS)

The main functions for VAS are described as follows: 1) Seeking the *right* timing to activate compact security mechanisms while maintaining performance, minimizing the use of system resources and operational costs. 2) Seeking the safety bound as the minimum vulnerability bottom line under which the system cannot maintain efficient autonomy anymore.

4.1 Information-Theoretic Mathematical Model

In this section, we propose a way of modelling MIB variables and a new way to apply the entropy metrics to measure the uncertainties of these random variables by applying Shannon's information theory. The readable and writable information contents are given in Definitions 1 and 2. We denote $\mathbf{E}$ as the elements in MIBs and $\mathbf{X}$ as the set of all the variables in MIBs, and we model the input of the *Awareness Strategy (AS)* in Figure 2 as set of random variables

[3] http://www.o-xml.org/

[4] http://www.springframework.org/

$\mathbf{X} = \{X_1, X_2 \cdots X_k, \cdots, X_n\}$ for elements, one element $\mathbf{E}_j$ could contain multiple variables; Each variable could have multiple symbols to describe itself, i.e., $X_k = \{x_1, x_2 \cdots x_i, \cdots, x_n\}$. Where $X_k = 1$ represents this variable is writable only; $X_k = 0$ represents this variable is readable only. And the access permissions of variables are *time-variant*, which means the MIB file variables accessibility could be altered due to the results from the output of awareness strategy functional block. The preliminary output of AS is also modeled as random variables $\mathbf{Y} = \{Y_1, Y_2 \cdots Y_i, \cdots, Y_n\}$ after acquiring all the variables statuses. The final output of AS would be a scalar parameter $\lambda \in [0, 1]$;

Proposition 1. *Given that the element $\mathbf{E}_j$ is of rewritable status, the larger the value of $\mathbf{IV}_{\mathbf{E}_j}$, the more significant the variable X_k for an element $\mathbf{E}_j$, and the more uncertain information content this variable takes, and hence, the more vulnerable this variable to the system will be.*

Some rules of the awareness strategy are related to this definition and remarks in this paper. We design a matrix form to represent all the associated information for MIB variables in matrices as follows. For example, we take 4 variables with up to 6 symbols for element j, the symbol set $\mathbf{S}_j$ at time t_k can be denoted as:

$$\mathbf{S}_j(t_k) = \begin{bmatrix} \mathbf{5} & \mathbf{1} & | & 1 & 1 & 1 & 1 & 1 \\ - & - & | & - & - & - & - & - \\ \mathbf{4} & \mathbf{1} & | & 1 & 1 & 1 & 0 & 1 \\ \mathbf{2} & \mathbf{0} & | & 0 & 0 & 0 & 1 & 1 \end{bmatrix}_{\mathbf{M}\times\mathbf{N}} \quad P_j(t_k) = \begin{bmatrix} \mathbf{5} & \mathbf{1} & | & 0.1 & \cdots & 0.15 & 0.15 \\ - & - & | & - & \cdots & - & - \\ \mathbf{4} & \mathbf{1} & | & 0.25 & \cdots & 0 & 0.25 \\ \mathbf{2} & \mathbf{0} & | & 0 & \cdots & 0.5 & 0.5 \end{bmatrix}_{\mathbf{M}\times\mathbf{N}}$$

where each row in the matrix represents each variable. Each value in the first column identifies the number of symbols for this variable in the current row. Each value in the second column shows the "writable" or "readable" for this current variable represented by this row. Value "0" of each row indicates the status of that variable is readable but not writable. Value "1" of each row indicates the status of that variable is writable and readable. The probability set $\mathbf{P}_j$ at time t_k of each symbol for the variable can be denoted as $P_j(t_k)$:

Hence, the average readable information content and rewritable information content for element j in a stream of bits can be calculated according to the above probability matrix.

Definition 1. *We define the writable information as follows:*

$$\mathbf{W}_{\mathbf{I}_j}(t) = -\sum_{m=0}^{\mathbf{M}-1} \sum_{n=2}^{[\mathbf{P}_j(t)]_{m,0}+1} \left\{ [\mathbf{P}_j(t)]_{m,1} \cdot [\mathbf{P}_j(t)]_{m,n} \cdot \log (\mathbf{P}_j)_{m,n} \right\} \tag{1}$$

Definition 2. *We define the readable information as follows:*

$$\mathbf{R}_{\mathbf{I}_j}(t) = \sum_{m=0}^{\mathbf{M}-1} \sum_{n=2}^{[\mathbf{P}_j(t)]_{m,0}+1} \left\{ \left[[\mathbf{P}_j(t)]_{m,1} - 1 \right] \cdot [\mathbf{P}_j(t)]_{m,n} \cdot \log \left(\mathbf{P}_j \right)_{m,n} \right\} \tag{2}$$

where $\mathbf{W}_{\mathbf{I}_j}(t)$ stands for the sum of *writable* information entropy; $\mathbf{R}_{\mathbf{I}_j}(t)$ stands for the sum of *readable* information entropy.

$$\mathbf{EI}_j(\mathbf{t}) = \boldsymbol{\lambda}_c \times \ [I_j(t) + f(\triangle(t)) + \gamma(t)] \tag{3}$$

where $\mathbf{EI}_j(\mathbf{t})$ is termed as the *Effective Information,* which is the ultimate probability of writable information contents in MIBs after considering disturbance, and mis-configuration errors. As shown in figure 2, λ_c is the normalized parameter, it is the output result of security awareness function, $\lambda_c \in [0, 1]$, where "0" shows *read only,* "1" represents *fully writable.* Any value between 0 and 1 represents the writable percentage of the MIB variables or files (e.g., due to the coded encryption for modifications of MIB varibables). λ_c will re-setup the writable levels of information contents based on threats level.

4.2 Integrated Security Awareness Framework (SAF)

Multiple factors are taken into consideration in the proposed integrated SAF. All these factors work together to affect the network security awareness status. The correlation of these factors is analyzed in this section too. The factors include (1) system vulnerability warning from readability and writability of the MIB variables, (2) information flow changes $\triangle f(x)$ during a time interval, which includes network traffic conditions, and real multiple services conditions (3) past memories of vulnerability towards attacks/virus conditions (4) disturbances effects and (5) internal mis-configurations errors influences (See figure 2). The awareness mechanism working to an awareness strategy is implemented with Self-organized Maps (SOMs) which are a kind of unsupervised ANN. The reason for selecting SOMs is its good classification and clustering performance as reported [5]. Furthermore SOMs have the ability to identify *new* vulnerability patterns when new threats or attacks increase. For details of SOMs used in this paper refer to the section on experiments. The proposed integrated SAF is assessed through our vulnerability analysis based on two main mathematical notations as follows:

Remark 1. **ReWritability** ($RW \triangleq \frac{ReWritableVariables}{Total\_ReadableVariables}$) of the file system is one of the main *necessary* conditions for evaluating the adaptability of the overall information system. $Total\_readableVariables$ consists of read only variables and rewritable variables. RW is used as a key parameter in determining the vulnerability because of the fact that we found that rewritability of *overall* MIB

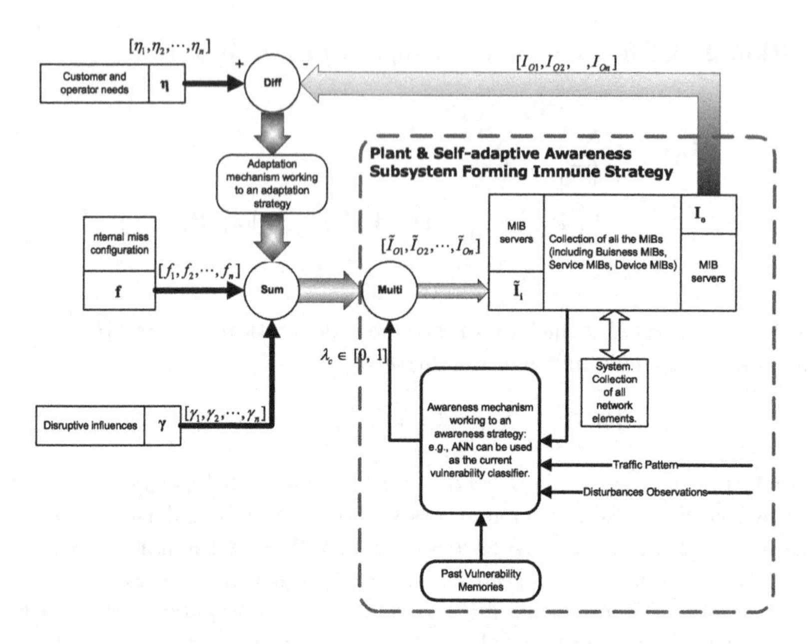

Fig. 2. Overview of Proposed Framework

variables for all system-wide MIB files is roughly *proportional* to the specifically predefined system vulnerable threshold values in the simulation[5]

Remark 2. The overall ReWritability of MIB variables in all MIB files can be used as a measurable parameter in terms of determining system vulnerability.

4.3 Our Algorithm: Adaptive Reconfiguration of O:MIB File/Variable Access Permission

Reconfigure file/variable access rights via Methods defined in O:XML which will modify its own symbols of O:MIB. Inspired by the cell mediated response model from human immune system, we propose a decentralised self-adaptive algorithm based on bio-inspired agent technology. Figure 3 lists the comparisons between the biological behaviors of human immune system and corresponding parts in our algorithm, where the biological inspirations can be clearly seen. The algorithm steps are as follows:

Step 1: When one agent finds some threats within radius r (r is the Euclidean distance between this agent and other agents) during one time interval t, it will warn the neighbors of those abnormal behaviors, hence, the system alarm level will be increased by $\triangle$. This process will be repeated every interval t.

[5] A Matlab function below is created as a proof: `function Ratio = ratio(U, WW, Min, Max, n)`.

Human Immune Network (Focusing on Cell Mediated Response)	Proposed Bio-inspired Algorithm
Cells (e.g., B-Cells and C-Cells)	Peer Agents
Pathogens	Network Threats/Attacks/Malicious codes
Paratope and Idiotope	Peer agents to find the best timing of stimulating awareness function based on (1) Level of threats/ damage (2) Other affinity (e.g., locations, service types, application categories,)
Information stored in Paratope and Idiotope	CSV(.) and H(.) used to calculate SV(.) according to Equation 11 and Eq. 12
Clustered Cells (e.g., Affinity)	Functional groups of agents (e.g., Functionality)

Fig. 3. Comparisons between Human Immune System and the Proposed Bio-inspired Algorithm

Step 2: Information theory is applied to reduce the dimensionality of related MiB variables.

Step 3: When the alarm level reaches the dynamic threshold, λ_c value is to be adjusted accordingly and consequently the protection mechanism will be activated, i.e., the rewritability of associated MiB variables will be reconfigured by increasing/decreasing $\triangle$. $\triangle$ is a dynamically varied data value.

Equation 1, Eq. 2 together with network traffic patterns are three important observations indicating abnormal network activities with respect to MIB information flows; and furthercoming actions become equivalent to hiding information against malicious attackers to limit their access. By manipulating the MIB variable access permission according to the threat information, we can achieve a security system for autonomic service management system.

5 Experiment

In order to evaluate the performance of the proposed vulnerability awareness strategy, an experiment is conducted based on the process described in Figure 2. Self-Organizing Maps (SOMs) are applied into the vulnerability awareness function as a vulnerability classifier in our experiments. To simply our experiments, the input of SOMs in this simulation are limited to the following three categories: (1) the changes of MIB information flows during a time interval, i.e., ReWritability variations; (2) network traffic conditions; (3) observed disturbances and internal errors if applicable. Compared to the higher dimensionality of input data, the output of SOMs is a lower dimension space. When the network is fully trained, it is ready to get the data clustered on the SOM map. After some tests by use of testing data, the 4 classified zones defined is found. In our simulation, a 2-dimensional space split into 4 classified zones, such as zone A, B, C and unacceptable zone, is the output of SOMs output neurons. These zones match with what we describe as safe zone A, risk zone B and high risk zone C.

Table 1. Parameters of Self-Organizing Map for Vulnerability Level

Number of Input Layer	3
Row of Output Layer	20
Column of Output Layer	20
Topology	Rectangular
Learning efficiency	0.9
Iteration Number	2000
error_limit	5E-12

Table 2. Part of Testing Results

Test Set	Desired	Winning Neurons
$\triangle f(x) + Disturbance\_30$	$\triangle f(x)$	153
$\triangle f(x) + Disturbance\_20$	$\triangle f(x)$	345
$\triangle f(x) + Disturbance\_10$	$\triangle f(x)$	127

Classified SV information is furthermore obtained by java agent simulator which invoke the call of the methods and algorithms predefined in the O:XML files. Afterwards, the **RW** of associated MIB variables is ready to be modified. It is shown the process of training SOMs and testing SOMs with the testing data, which is part of the training dataset generated by our Distributed Traffic Generator - D-ITG and ReWritability dataset which is generated by Monte Carlo Simulation where random number generator produce all the dataset.

The SOM structure we used in the simulation is a *rectangular lattice* network formed by a total of 400 neurons (20×20) is used. This size is chosen due to two reasons: firstly this size could give a clear visualization effects and secondly some literatures [5] suggest that the number of map cells should be lower than the number of input sets. Therefore, in order to give out a better visualization effects, the number of inputs samples are $400 > 300 > 17 \times 17$. It is acceptable to have 20 in that it is slightly larger than 17. In our simulation, we repeatedly generate groups of data every 300s, and the disturbance and internal errors are also generated arbitrarily, hence, in the future, we wish to adopt real network traffic data with real disturbance to train our SOMs.

5.1 Simulation Results

We use Distributed Traffic Generator (D-ITG) to generate real traffic patterns. The TCP traffic with Poisson distribution is adopted where packet size = 512 bytes, and average 1000 packets/sec. The traffic pattern we applied by use of bit rate figure for the generated traffic takes continuous simulation time = 700s, the same pattern is repeated in the input data sample to the SOMs. The disturbances are arbitrarily introduced during a period of time and are shown as the peak lines.

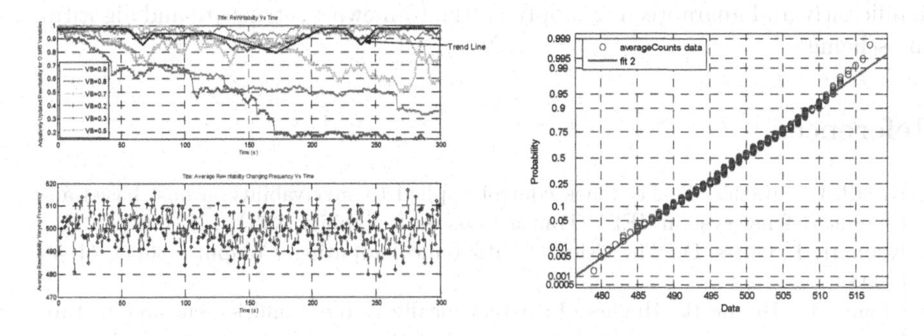

Fig. 4. Simulation Results

Fig. 5. Fitness of Average Counts

According to the vulnerability information from SOM, the methods and algorithms defined in O:XML will be invoked by java agents. The simulation results of our algorithm are shown in Figure 4, where we take the configuration parameters as VB=[0.1:0.1:0.9] and Step=0.1. This results aim at finding a bound used for guiding acceptable autonomy and vulnerability study.

We tested different system adaptability status under differently pre-setup VulneraBility Level (VBL), and found that with our algorithm with adaptive reconfiguring file/variable access permissions, in the long run, the system adaptability will come back to normal in the end, but considering the time efficiency, such as recovery time and awareness speed, we can see that when $VBL \leq 0.5$, (e.g., 0.2 and 0.3) it takes system an intolerable time ($\geq 300s$) to recover its adaptability. On the contrary, when the $VBL \geq 0.5$ (e.g.,0.5, 0.7, 0.8, 0.9), recovery time and response speed are both in tolerable range. And the trend line shown in this figure demonstrate the trend when $VBL = 0.8$ as an example. Therefore, we conclude that the tolerable region of VBL selection for operators are from 0.5 to 0.9. The best region for VBL selection in terms of awareness speed ($\leq \frac{\max imum \mathrm{Re}\, coveryTime}{totalPeriodCoverageTime} = \frac{20s}{60s} = 33\%$) and adaptability recovery (≤ 0.18) measured from the simulation result data is the value $VBL \geq 0.8$. A random number generator is used in this experiment, therefore, around 50% of the MIB variables would be writable or readable only (see Figure 5). This is a special case for real network scenario. The simulation results based on these data are still significant with valuable usage.

6 Conclusion

This paper proposes a general analysis on vulnerability and autonomy issues on the basis of our innovative O:MIB structure, this brand new structure is proved to be intuitively efficient and more attachable in future distributed peer-to-peers communication in ACNs. We argue that O:MIB can be as an alternative to replacing the current MIB used by industries. The simulation results based on the benchmark prototype - SOMs are promising and indicate a better performance. The self-protecting and self-healing features of desired ACNs can be improved significantly and guaranteed by applying the framework, structure and algorithmic scheme.

References

1. Kreidl, O., Frazier, T.: Feedback control applied to survivability: a host-based autonomic defense system. IEEE Transactions on Reliability 53(1), 148–166 (2004)
2. Kephart, J., Chess, D.: The vision of autonomic computing. Computer 36(1), 41–50 (2003)
3. Chiang, F., Braun, R., Hughes, J.: A biologically inspired multi-agent architecture for autonomic service management. Journal of Pervasive Computing and Communications 2(3), 261–275 (2006)
4. Sweitzer, J.W., Thompson, P., Westerinen, A.R., Williams, R.C., Bumpus, W.: Common Information Model: Implementing the Object Model for Enterprise Management. John Wiley and Sons, Chichester (1999)
5. Kohonen, T.: Self-organizing maps. Springer, Berlin, New York (2001)

Integrated OTP-Based User Authentication and Access Control Scheme in Home Networks

Jongpil Jeong, Min Young Chung, and Hyunseung Choo*

Intelligent HCI Convergence Research Center
Sungkyunkwan University
440-746, Suwon, Korea +82-31-290-7145
{jpjeong,mychung,choo}@ece.skku.ac.kr

Abstract. In this paper, we propose a new user authentication (UA) scheme based on one-time password (OTP) protocol for home networks. The proposed scheme is to authenticate home users identification who uses home devices. Several techniques such as biometrics, password, certificate, and smart card can be used for user authentication in the same environments. However, such user authentication techniques must be examined before being employed in the environment where home devices have low efficiency and performance. Here, we present the important security functions of home networks. The proposed authentication protocol is designed to accept the existing home networks based on one-time password protocol. Also, it is well suited solution and is quite satisfactory in terms of home networks security requirements, because of requiring low computation that performs simple operations using one-way hash functions. Our proposed scheme can protect illegal access for home services and devices and it does not allow unnecessary service access although they are legitimate users. Therefore, it allows the user to provide real-time privilege control and good implementation for secure home networks.

1 Introduction

Home networks provide remote access control over the connection between information home appliances and information devices on Internet [1,2,3,4]. And, it is possible to operate the bi-directional communication services that use the contents such as music, video and data. Therefore, it provides convenient, secure, healthy, pleasant and efficient life for home users through the future-directional home environments that possible to use several services regardless of devices, time and places. And, this can be realized based-on integrated IT technologies.

Home networks consist of several wired/wireless medium and its protocols, so it also has the existing security vulnerability. And it has problem that can be adapted current network-based cyber attacks. Home networks information appliances has low computing capabilities relatively, and they difficult to be built with security functions, so they can be used by cyber attacks and have many

* Corresponding author.

S. Ata and C.S. Hong (Eds.): APNOMS 2007, LNCS 4773, pp. 123–133, 2007.

possibility to be target of several attacks. Home networks services use many information related with personal privacy, they will provide direct-life services such as health-care service. Therefore, hacking attack for home networks can violate one person's privacy and threat life of home users ultimately, so security's countermeasure should be prepare urgently. OTP-based authentication scheme proposed for secure home networks that legitimate users only can use home services over the user authentication and access control verification in accessing home services.

There are two types of password-based authentication and public key based authentication for home users to provide security in wired/wireless networks. These authentication schemes are vulnerable for several attacks and have critical problem of processing overhead for home networks appliances, respectively. In this paper, the proposed scheme based on strong-password approach uses one-way hash functions to perform simple authentication operations, so it requires the low computational load. Also, it enhances the security level with the OTP technology that is variable authentication mechanism.

The proposed scheme protects replay attack of ID/PW very well, and uses OTP technique and one-way hash functions to reduce the processing overhead and then to operate overall procedures very fast. The OTP technique is based on mathematical cryptography for generating regular patterns, and it is the best technology for authentication paradigms because of ensuring its safety theoretically. The OTP transmits the input password that is different at each time. So it is impossible to reuse revealed value although its messages listened by attackers. In this paper, we propose a secure user authentication protocol based on the following OTP schemes such as S/Key [8], Lamport [7], Revised SAS [6] and SAS-2 [5], but more secure than them. It employs a three-way challenge-response handshake technique to provide mutual authentication. Also, computation in the user device is reduced, resulting in less power consumption in the mobile devices.

The rest part of the paper is organized as follows. In Section 2, related works is presented. An authentication protocol suitable for home network environments is proposed in Section 3. Finally, this paper is concluded, and future directions are noted in Section 5.

2 Related Works

Password-based authentication schemes are the most widely used methods for remote UA. Existing schemes could be categorized into two types [9]. One uses weak-password approach, while the other uses strong-password one. The weak-password authentication approach is based on El Gamal cryptosystem. The advantage of this scheme is that the remote system does not need to keep the userID-password table to verify the user. However, such weak-password authentication approach leads to heavy computational load on the whole system. Thus, it cannot be applied to home networks environment, as home appliances cannot afford the heavy computation. Unlike the weak-password approach, strong-password authentication is mostly based on one-way hash function [13] and exclusive-OR operations (XOR). It requires much less computation and needs only simple operations. With this in mind, this scheme may have advantages when it is applied to home networks environment.

Most of the existing UA schemes require high computation cost caused by exponentiation operations; and not suitable for mobile devices (e.g. PDAs, mobile phones, sensor nodes etc.). Lee et al. [11] also proposed an improved UA scheme with low computation cost by using smart cards and one-way hash functions. Only three phases are used here, namely, Registration Phase, Login Phase, and Authentication Phase. This scheme can resolve the attacks of forgery, replay, and modified login message. Our proposed solution in Section 3 makes use of the framework having three phases above; but adapts it to the home networks environment. Jeong et al. [19] proposed a mutual authentication scheme between 3-parties (user, authentication server, and home gateway server) for home networks, which uses pre-shared symmetric key based on two nonces between two servers and verifies the session key calculated by two nonces. However, authors do not consider the critical problem in home services in case of leaking user password to attackers. In this paper, we make up the weak points of [19], fulfilling the low computation load and security requirements for home networks. Here we employ the one-time password protocol to use the password between home appliances (mobile devices) and authentication server.

One form of attack on network computing systems is eavesdropping on network connections to obtain authentication information such as the login IDs and password of legitimate users. Once this information is captured, it can be used in a later time to gain access to the system. OTP system [14] designed to counter this type of attack, called a replay attack. A sequence of OTP produced by applying the multiple times of secure hash function gives to the output of the initial step (called S). That is, the first OTP to be used is produced by passing S through the secure hash function a number of times (N) specified by the user. The next OTP to be used is generated by passing S through the secure hash function $N-1$ times. An eavesdropper who has monitored the transmission of

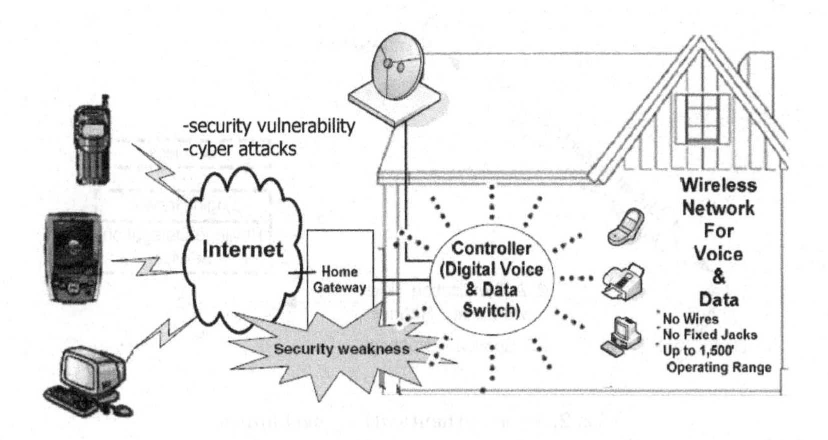

Fig. 1. Home Network architecture and Various attacks

OTP would not be able to generate the next required password because doing so would mean inverting the hash function. The OTP improves security by limiting the danger of eavesdropping and replay attacks that have been used against simple password system. The use of the OTP system only provides protections against passive eavesdropping and replay attacks. It does not provide for the privacy of transmitted data, and it does not provide protection against active attacks. The success of the OTP system to protect host system is dependent on the non-invert (one-way) of the secure hash functions used.

3 Integrated OTP-Based UA Scheme

Currently, home network is exposed to various cyber attacks through Internet, and has security vulnerability such as hacking, malignancy code, worm and virus, DoS (Denial of Service) attack, and communication network tapping as shown in Fig. 1. As a result, technical development of home network with respect to security mostly focuses on putting the security functions on the home gateway to cope with cyber attack. Home gateway needs countermeasures against the attacks on main resources through illegal device connection or possibility of leakage of main data. Especially, in the premise of home network, vulnerability of component and data security exists in the wireless part needing authentication for accessing the component and data.

Security function is preferred to be loaded into home gateway that provides a primary defense against the external illegal attacks as a entrance guard that connects public network out of the house to the home network. The representative security functions loaded in home gateway are firewall, VPN (Virtual Private Network), etc. However, they are not suitable to the HN because firewall allows data to enter the premise network if the destination is correct, and VPN is more suitable to a large network of high traffic.

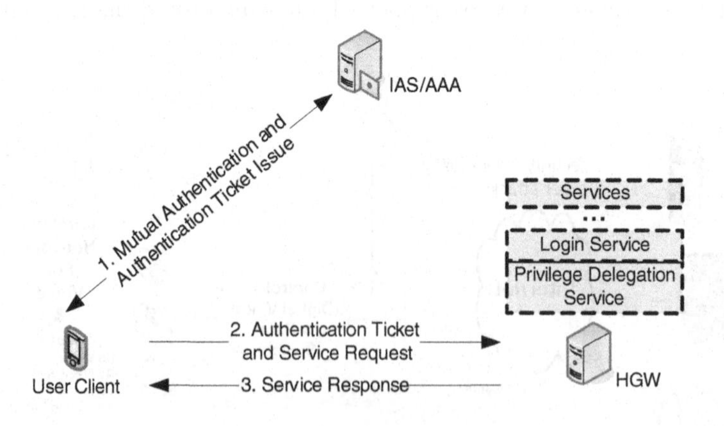

Fig. 2. User authentication mechanism

3.1 Preliminary and Notation

User (Mobile device) transmits information for OTP operation in login and verification phases to authentication server (IAS: Integrated Authentication Server) through the secure channel. User can select the own password in registration phase by separating registration phase and login/verification phases. [19] is based on public key infrastructure (PKI), so it causes the processing overhead for authentication messages between authentication servers and mobile devices. But, the proposed scheme has light-weighted overhead for home networks. Also it doesn't use password table per users but one-way collision-resistent hash functions for OTP mechanism.

Service subscribers require mutual authentication between IAS and home gateway server (HGW), in order to access home network services. In addition, they must be able to operate service access control when privilege services are granted. Users are authenticated through single-sign-on (SSO) and then, they can access other home services without additional authentication procedures. Fig. 2 illustrates the user authentication mechanism. And, it is assumed that IAS is located on the outside of the home network environment, manages the home gateway, and performs AAA functions of authentication, authorization, and accounting. A suitable user authentication protocol is proposed for home network environments, focusing on authentication for users receiving the home service and controlling the service privilege.

Table 1. Notation

Notation	Meaning
R_{Ui}	Number calculated by IAS using U_{ID} and Password
R_{Si}	Random Number generated by IAS
$F(), h()$	collision-resistant hash function
N	permitted number of login times
S_{key}	Shared Session Key between Client and HGW
U_{ID}	User's Identifier
IAS_{ID}	IAS's Identifier
$E_{IAS-HGW}(-)$	Encryption using Symmetric key between IAS and HGW
$E_K(-)$	Encryption using K
T	Timestamp to decide Session key's validation

The proposed scheme consists of three steps of registration, login, and authentication/service request phases. The proposed authentication scenario is described in Fig. 3, and 4, respectively.

3.2 Registration Phase

1. Registration phase is to store the message between authentication server and home devices for home devices's authentication request. Home users input the their identity and password information to home devices and then home devices generate the hashed-value using by one-time hash functions.

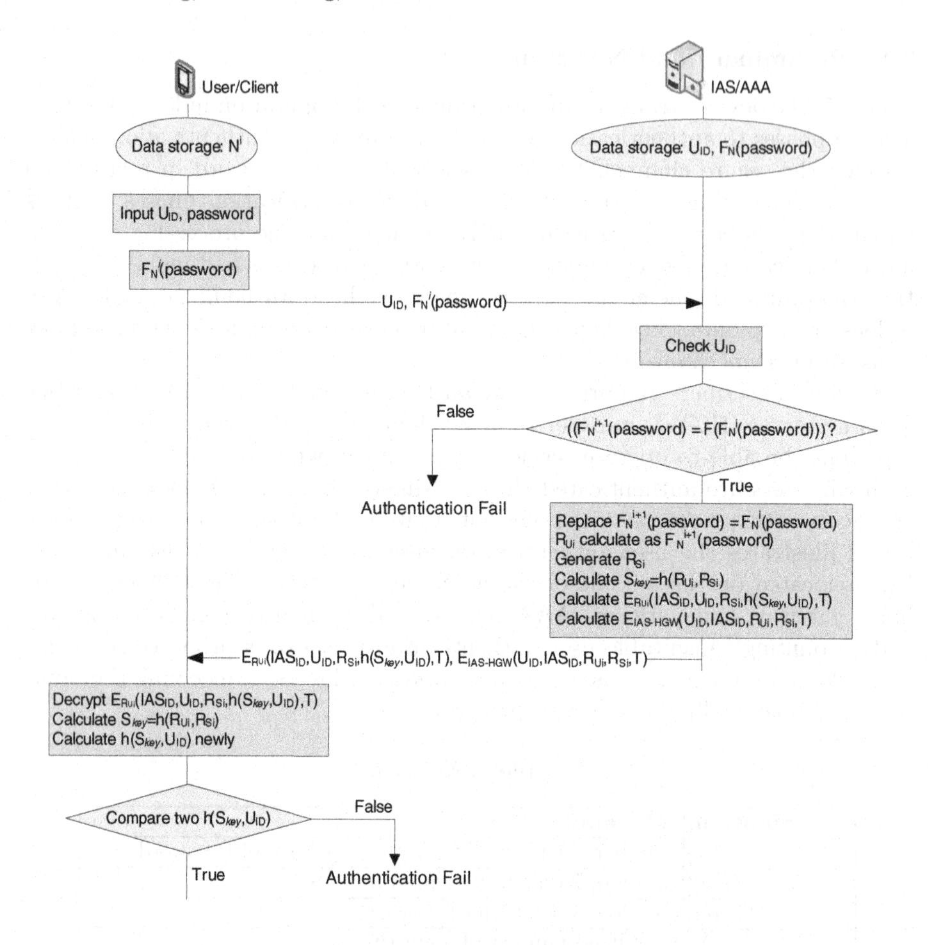

Fig. 3. *i*th Login and Authentication Phases

2. Home users transmit U_{ID} and $F_N(Password)$ to AAA-built authentication server over secure channel before the home-gateway server. Where, $F_N(.)$ represents the total number of hashing that accepted for home users.
3. Authentication server stores U_{ID} and password value calculated by F hash function.

3.3 Login and Authentication Phases

1. Client calculates result value of i-times hashing, $F_N^i(password)$ for i-th authentications and transmits U_{ID} and $F_N^i(password)$ to IAS/AAA.
2. IAS verifies U_{ID} and compare one more hashed value of $F_N^i(password)$ from client and one more hashed value, $F_N^{i+1}(password)$ of $F_N(password)$ on the server. If not equal, client reject.
3. IAS generates nonce value, R_{Si} and takes R_{Ui} as $F_N^i(password)$, and then calculates $S_{key} = h(R_{Si}, R_{Ui})$.

4. We assumed that IAS established the security association with home gateway server using symmetric key. IAS transmits the authentication ticket, $E_{IAS-HGW}(U_{ID}, IAS_{ID}, R_{Ui}, R_{Si}, T)$ encrypted with symmetric key between the two authentication servers, and this also includes messages for mutual authentication between user and IAS, encrypted with R_{Ui}. Here, this message is $E_{R_{Ui}}(IAS_{ID}, U_{ID}, R_{Si}, h(S_{key}, U_{ID}), T)$.
5. Legitimate user only can verify the stored password from the previous registration phase and one more hashed value for verification, and then calculate S_{key} on the IAS server. Therefore, legitimate user only can decrypt the $E_{R_{Ui}}$ $(IAS_{ID}, U_{ID}, R_{Si}, h(S_{key}, U_{ID}), T)$ message using $R_{Ui} = F_N^i(password)$. User decrypts the this message and acquires the R_{Si} and calculates the S_{key} and then verifies the requested authentication by checking $h(S_{key}, U_{ID})$.

3.4 Service Request Phase

1. Authenticated users request home services to home gateway server in home networks, according to the current available services for home users. User transmits two messages, $E_{IAS-HGW}(U_{ID}, IAS_{ID}, R_{Ui}, R_{Si}, T)$ authentication ticket and (U_{ID}, *Services*) encrypted with S_{key}, to home gateway server, and requests home services for the home user.
2. Home gateway server verifies the two U_{ID}, one is that decrypt the authentication ticket using symmetric key between IAS and home gateway server and the other is that decrypt the service request message, $E_{S_{key}}(U_{ID}, Services)$ using S_{key}, hashed value of R_{Ui} and R_{Si} messages.
3. HGW transmits R_{Ui} encrypted using S_{key} to the client, and then authenticates HGW implicitly.

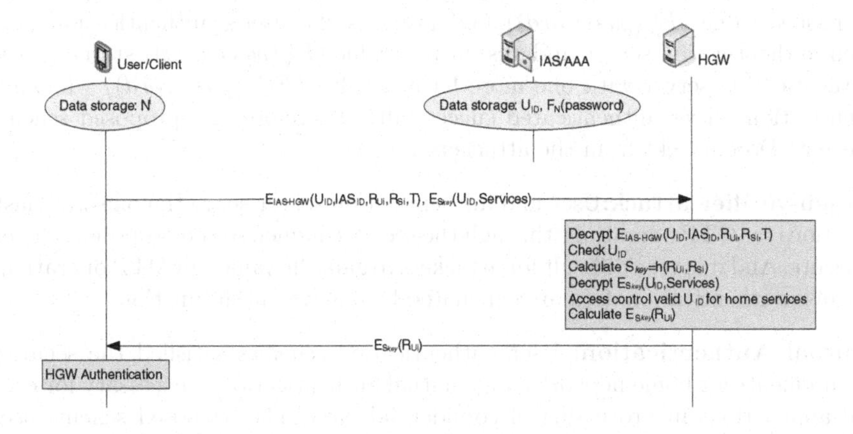

Fig. 4. Service Request Phase

4 Analysis

The proposed protocol is designed under the assumption that symmetric key is shared between IAS and HGW. In addition, it is assumed that IAS exists outside the home network, it manages the home gateway, authenticates users, grants privileges, and controls accounting as the home gateway operator. Another assumption is that service users trust IAS. Actually, the OSGi (Open Services Gateway Initiative) operator exists in OSGi framework, it is outside the home network as the home gateway manager for managing the home gateway and authenticating users.

Authentication between HGW and users employ the authentication ticket granted from the authentication server, and users can request and receive services with a valid authentication ticket after single authentication, there is no requirement to login each time when the users request services. Authentication ticket's validation can verify with its time-stamp, satisfied with authentication requirements. In addition, as U_{ID} is checked in authentication ticket after login, it can control whether having service privileges. ACL (Access Control List) is stored as table format for U_{ID} privileges list in HGW's policy file, and the purpose is to supply suitable services in response to user identification information.

Replay attack: One-time password that sends to authentication server is calculated by the one-way hash function($F()$), so attackers cannot replay the password to authentication server after intercepting U_{ID}'s password.

Man-in-the-middle attack: The main benefit in the proposed scheme based on one-time password protocol is that attackers cannot reuse the U_{ID}'s password because of changing U_{ID}'s password each time for login and authentication request to the authentication server. Furthermore, authentication data are transferred between clients and two servers, so attacks should be detected if it changed.

Denial of Service attack: The proposed scheme changes original password to hashed-value ($F_N^i(password)$) and protects the user's authentication, and then authentication server updates hashed-value ($F_N^i(password)$) stored in the authentication server with one more hashed value ($F_N^{i+1}(password)$) when authentication server authenticated successfully. Therefore, the proposed scheme prevents DoS attacks from the attackers.

Stolen-verifier attack: User and authentication server share the one-way hash functions for OTP operations through the secure channel, so the proposed scheme is secure. And it is very difficult for attackers to gain the values for OTP operations because authentication data are calculated by one-way hash function.

Mutual Authentication: User authentication schemes satisfied the security requirements for home networks, but mutual authentication is necessary for critical applications in processing of confidential data. The proposed scheme uses 3-way challenge-response handshake protocol to provide the mutual authentication. Authentication server transmits the authentication data (Authentication Ticket) to user, user checks the timestamp T and authentication server authenticated successfully by user if T value is allowed.

5 Efficiency

In this section, we summerize the performances and criteria for authentication schemes. For a protection mechanism for user authentication, the following criteria are crucial.
C1: *No verification table:* The remote system does not need the dictionary of verification tables to authenticate users.
C2: *Freely chosen password:* User can choose their password freely.
C3: *Mutual authentication:* Whether the users and the remote system can authenticate each other.
C4: *Lower communication and computation cost:* Due to hardware constraints of a home devices, it usually does not support power communication cost and higher bandwidth.
C5: *Session key agreement:* A session key agreed by the user and the remote system generated in every session.
C6: *No time synchronization:* Discard the timestamp to solve the serious time synchronization problem.

We made comparisons among the previous schemes and our proposed scheme. Table 2 shows that our scheme satisfies all criteria.

Table 2. Effectiveness comparisons among the previous schemes.

	C1	C2	C3	C4	C5	C6
Our scheme	Yes	Yes	Yes	Extremely low	Yes	Yes
Jeong et al. [19]	No	No	Yes	High	Yes	Yes
Wang and Chang [21]	Yes	Yes	No	Medium	No support	No
Yang and Shieh [20]	Yes	Yes	No	Medium	No support	Yes
Hwang and Li [23]	Yes	No	No	Medium	No support	No
Sun [22]	Yes	No	Yes	Extremely low	No support	No
Chien et al. [17]	Yes	Yes	Yes	Extremely low	No support	No
Hwang et al. [18]	Yes	Yes	No	Extremely low	No support	No

Without complicated cryptography algorithms, the proposed scheme has been successfully developed based on hash operations. In addition, the proposed scheme allows multiple-access requests with privacy protection. Furthermore, only a simple verification is required for a multiple-access request. We compare our proposed scheme with the previous schemes for six items in Table 2. It represents good performance for no verification table and freely chosen password items. Also, it shows significant benefits for communication and computation cost by using one-way hash functions and OTP mechanism, due to requiring low computation load by adopting the strong-password approach.

6 Conclusion

Home network is defined as environments where users can receive home network services for anytime and anywhere access through any device, connected with a wired/wireless network to home information appliances including the PC. In this environment, there are many security threats that violate users privacy and interfere with home services. In addition, the home network consists of several networks with each network being inter-connected, so network security for each network is required. This means that there are a number of security threats to other networks when a security threat occurs in any network. Also, users in home network need security mechanism, for receiving home services from attackers and sharing information between home information appliances.

In this paper, a user authentication mechanism between a home gateway and user is designed to accept existing home networks based on OTP mechanism. So, the proposed scheme provides low computation load and high security for secure home networks. In addition, it cuts off the illegal access for inside and outside home services and home devices. Therefore, our proposed scheme is possible to adopt for home networks at once, because it has a capability of real-time privilege control for legitimate users.

There is still progress in the standardization of home network architecture, which is required to be researched in the future. In addition, research regarding the integration of authentication servers for 3G-WLAN and authentication servers in home networks needs to be conducted.

Acknowledgment

This research was supported by Ministry of Information and Communication, Korea under ITRC IITA-2006-(C1090-0603-0046) and grant No. R01-2006-000-10402-0 from the Basic Research Program Korea Science and Engineering Foundation of Ministry of Science & Technology.

References

1. Sun, H.: Home Networking, Mitsubishi Electric Research Laboratories (2004), http://www.merl.com/projects/hmnt/
2. Choi, K., et al.: Trends of Home Networking Standardization in Korea. KETI Journal (2003)
3. Park, Y., et al.: Home Station Architecture based on Digital Convergence toward U-Home age. ETRI Journal (2003)
4. Lim, S., et al.: Home Network Protocol Architecture for Ubiquitous Communication. Journal of KIPS 10 (2003)
5. Shimizu, A.: A One-Time Password Authentication Method, Kochi University of Technology Master's thesis (January 2003)
6. Tsuji, T., et al.: Simple and secure password authentication protocol, ver.2 (SAS-2), IEICE Technical Report, OIS2002-30, vol. 102(314) (September 2002)
7. Lamport, L.: Password authentication with insecure communication. Communications of the ACM 24(11) (1981)
8. Bellcore, N.H.: The S/KEY One-Time Password System, Network Working Group (February 1995)
9. Das, M.L., et al.: A Dynamic ID-based Remote User Authentication Scheme. IEEE Transactions on Consumer Electronics 50(2) (2004)

10. Awasthi, A.: Comment on A dynamic ID-based Remote User Authentication Scheme. Transaction on Cryptology 01(02), 15–17 (2004)
11. Lee, C.Y., et al.: An Improved Low Communication Cost User Authentication Scheme for Mobile Communication. In: AINA 2005, Taiwan (March 2005)
12. El-Fishway, N., et al.: An Effective Approach for Authentication of Mobile Users. In: VTC (May 2002)
13. Schneier, B.: Applied cryptography, 2nd edn. John Wiley and Sons Inc, New York (1996)
14. Haller, N. et al.: A One-Time Password System, IETF RFC 2289 (February 1998)
15. http://www.faqs.org/docs/linux_network/x-082-2-firewall.attacks.html
16. Faisal, R.M., et al.: A Password Authentication Scheme With Secure Password Updating, Telematics System. Services and Applications 2004 (May 2004)
17. Chien, H., et al.: An efficient and practical solution to remote authentication: smart cards. Computers and Security 21(4), 372–375 (2002)
18. HWang, M., et al.: A simple remote user authentication scheme. Mathematical and Computer Modelling 36, 103–107 (2002)
19. Jeong, J., et al.: Secure User Authentication Mechanism for Digital Home Networks. In: Sha, E., Han, S.-K., Xu, C.-Z., Kim, M.H., Yang, L.T., Xiao, B. (eds.) EUC 2006. LNCS, vol. 4096, pp. 345–354. Springer, Heidelberg (2006)
20. Yang, W., et al.: Password authentication schemes with smart cards. Computers and Security 18(8), 727–733 (1999)
21. Wang, S., et al.: Smart card based secure password authentication scheme. Computers and Security 15(3), 231–237 (1996)
22. Sun, H., et al.: Cryptanalysis of password authentication schemes with smart cards. In: Information Security Conference 2001, pp. 221–223 (May 2001)
23. Hwang, M., et al.: A new remote user authentication scheme using smart cards. IEEE Transactions on Consumer Electronics 46(1), 28–30 (2000)

New Access Control on DACS Scheme

Kazuya Odagiri[1,3], Nao Tanoue[4], Rihito Yaegashi[2], Masaharu Tadauchi[3], and Naohiro Ishii[1]

[1] Aichi Institute of Technology, 1-38-1 Higashiyamadouri Chikusa-ku
Nagoya-city Aichi,Japan
odagiri@toyota-ti.ac.jp
ishii@in.aitech.ac.jp

[2] Shibaura Institute of Technology, 3-7-5 Toyosu Koutou-ku
Tokyo 135-8548
rihito@sic.shibaura-it.ac.jp

[3] Toyota Technological Institute, 2-12-1 Hisakata Tenpaku-ku
Nagoya-city Aichi Japan
tadauchi @toyota-ti.ac.jp

[4] Pasona tech, 3-6-1 Sakae Naka-ku,
Nagoya-city Aichi Japan

Abstract. Recently, social interest about network security becomes very high including the issue of information leak through a network. As one of the important technologies about network security, there is an access control for network service. There are some methods of access control: access control by packet filtering mechanism on the server side, access control by the communication control mechanism on the network course such as SSL-VPN, and access control by packet filtering mechanism on the server side such as personal firewall of quarantine network. In this paper, new access control is realized. This new access control is realized as the result of combining the above two methods and improving each other's problems.

Keywords: Access control DACS Scheme Packet filtering.

1 Introduction

Recently, social interest about network security becomes very high including the issue of information leak through a network. As one of the important technologies about network security, there is an access control for network service. In university networks, it is often necessary to improve security level for the purpose of making permitted users use network services. It is necessary to do so when data for individual users is handled. To be concreted, the case to access POP server and file server is considered. Moreover, because network can be managed in each laboratory, a computer management section may not manage a whole network. In that case, because a network administrator can not often change the system configuration of a network physically, the method that does not need to change the system configuration of the network is necessary. In addition, because there is the unspecified number of network

S. Ata and C.S. Hong (Eds.): APNOMS 2007, LNCS 4773, pp. 134–143, 2007.

services on the network, the access control for not only specific network services but also the unspecified number of network services needs to be performed.

As methods to perform access control of the communication that was sent from a client computer (client) to network services by a user unit, there are some methods as follows. First, there is a method to perform access control on the network server side. For example, access control is performed by making the communication between a client and a network service support VPN. Next, there is another method to perform access control by the mechanism for communication control (Communication Control Service) which is located on the network course. Moreover, there is the other method to perform access control by the packet filtering mechanism located on the client such as a personal firewall of quarantine network. In the method of access control on the network server side, it is unnecessary to change the system configuration of network physically. However, because access control for all communications is performed on the server side, there is the problem that processing load to occur on the server by access control becomes heavy. The method of access control by Communication Control Service on the network course is unsuitable to the university network because the system configuration of the network needs to be changed physically. In the method of access control on the client, it is unnecessary to change the system configuration of the network physically. However, when the client which does not possess the mechanism of packet filtering exists on the network, it is not impossible to perform access control for all communications form all clients. To do access control without changing the system configuration of the network, either the method to perform access control on the network server or the method to perform access control on the client needs to be used. But, both methods have problems.

Therefore, in this paper, new access control is realized. This new access control is realized as the result of combining the above two methods and improving and solving each other's problems. To be concreted, it is realized by functions of DACS (Destination Addressing Control System) Scheme. The communication from the client is supported by VPN and access control is performed on the server side by selecting whether the communication which is not supported by VPN is permitted or non-permitted.

The basic principle of DACS Scheme is as follows. Communication control by a user unit is realized by locating the mechanism of Destination Nat and packet filtering on the client, and a whole network system is managed through that communication control [1]. Then, the function was extended so that communication control by a client unit can coexist with communication control by a user unit for using in the practical network [2][3], and another function was extended by use of SSH so that communication is supported by VPN to solve security problem of DASC Scheme [4]. In addition, two kinds of functions of Web Service, which are realized on the network introducing DACS Scheme, are described as follows. The first function of Web Service is that, data which is stored in database dispersed on the network can be used efficiently [5]. The second function of Web Service is that, data which is stored in document medium such as PDF file and simple text file can be used efficiently [6]. Moreover, the information usage system for realizing the Portal which users can customize easily and freely is described [7].

In chapter 2, the existing method of access control is explained and compared with new access control proposed in this paper. The functions of DACS Scheme are explained in chapter 3. In chapter 4, the possibility of new access control is confirmed by verifying the movement of the DACS Scheme's prototype system.

2 Existing Research and New Access Control

The methods of access control for the communication from a client to a network server are described as follows from (1) to (3).

(1) The method of access control on the network server side.
(2) The method of access control on the Communication Control Service located on the network course
(3) The method of access control on the client

In the method of (1), there are some methods as follows from (a) to (c).

(a) The method of access control by the packet filtering mechanism which is located on the network server side when the communication from a client reaches a network server.
(b) The method of access control by use of authentication.
(c) The method of access control by supporting or not supporting VPN for the communication from a client to a network server.

In the method of (a), it is possible to perform access control only by an IP address and communication port unit. That is, it is impossible to perform access control by a user unit. The method of (b) can not applied to the unspecified number of network services, because authentication is performed by a network service unit individually. The method of (c) performs access permission control by supporting VPN for the communication from a client to a network server and permitting that communication. Then, access non-permission control is performed by denying the communication which is not supported by VPN. However, in this method, there is the problem that processing load at the server side becomes heavy because all access control is performed in the server side. In the method of (2), there are some methods such as follows from (d) to (e).

(d) The method of access control for the communication between LAN (Local Area Network) and external network by SSL-VPN [8] and Opengate [9][10].
(e) The method of access control for the communication from a client to a network server in the different network via Communication Control Service such as quarantine network with gateway [11] or authentication switch [12].

By use of the method of (2), it is possible to perform access control for the communication from a client to a network server by a user unit. However, because Communication Control Service needs to be located on the network course, the system configuration of existing network must be changed physically. Then, because communications from many clients concentrates, processing load on the Communication Control Service becomes heavy.

In the method of (3), there is a method to use the personal firewall of quarantine network [13][14]. The system configuration of the network does not need to be changed. But, because access control is performed by packet filtering mechanism on the client, it is impossible to perform access control for the communication from the client without packet filtering mechanism. Among the methods explained to here, there are two methods of access control by a user unit which does not need to change the system configuration of the network physically. Though it is the method of (1)-(c) or the method of (3), both methods have a problem. Therefore, each problem is improved and solved by combing these two methods.

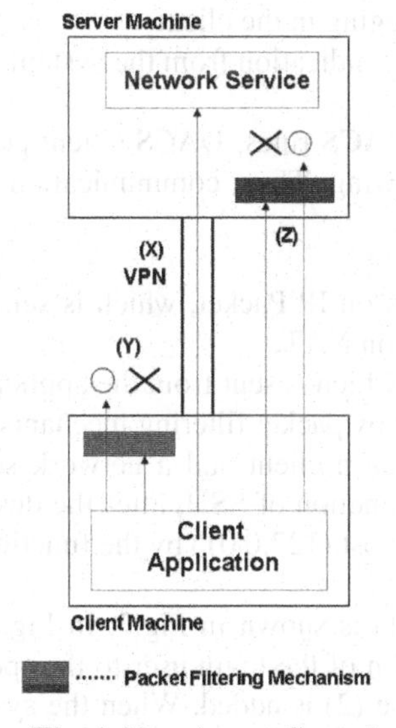

Fig. 1. New Access Control

In Fig. 1, new access control is shown. In this new access control, the packet filtering mechanism is located on the client, and access control for the communication from a client application is performed as shown in (Y) of Fig.1. When access is permitted on the client, access control on the server side in (Z) of Fig.1 is performed by supporting VPN for the communication between a client and a network server in (X) of Fig.2. To be concreted, access non-permission control is performed for the communication form the client without the packet filtering mechanism by denying the communication which does not support VPN on the server side. Access permission control is performed for the communication form the client without the packet filtering mechanism by permitting the communication which does not support VPN on the server side. This new access control is compared with the method of (1)-(c). Because access control is performed on the client, quantity of communications to perform access control in the server side decreases by just that much. As the result, processing load on the server side becomes low. The problem of the method of (1)-(c) is improved. Then, this new access control is compared with the method of (3). It is possible to perform access control for the communication from the client without the packet filtering mechanism on the server side. The problem of the method of (3) is improved.

3 Functions of DACS Scheme

First, summary of DACS Scheme is explained. Fig.2 and Fig.3 shows the functions of the network services by DACS Scheme. At the timing of the (a) or (b) as shown in the following, DACS rules (rules defined by user unit) are distributed from DACS Server to DACS Client.

(a) At the time of user's logging in the client
(b) At the time of a delivery indication from the system administrator

According to distributed DACS rules, DACS Client performs (1) or (2) or (3) operation as shown in the following. Then, communication control of the client is performed for every login user.

(1) Destination information on IP Packet, which is sent from application program, is changed by Destination NAT.
(2) Packet from the client, which is sent from the application program to the outside of the client, is blocked by packet filtering mechanism.
(3) Communication between a client and a network server is supported by VPN with the port forward function of SSH, after the destination of the communication is changed to localhost (127.0.01) by the function of (1).

An example of the case (1) is shown in Fig.2. In Fig.2, the system administrator can distribute a communication of the login user to the specified server among server A, B or C. Moreover, the case (2) is added. When the system administrator wants to forbid user to use MUA (Mail User Agent), it will be performed by blocking IP Packet with the specific destination information.

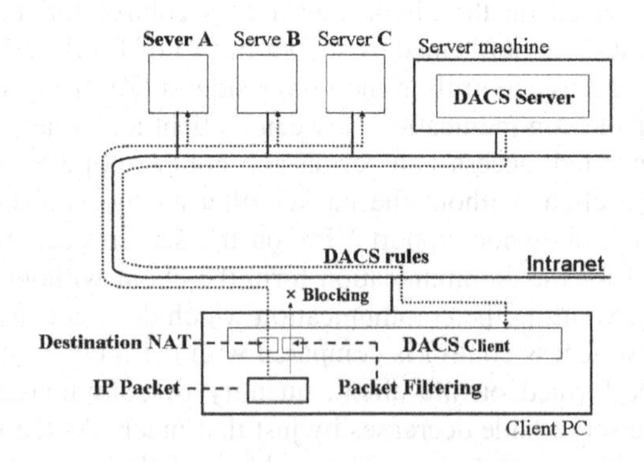

Fig. 2. Function of DACS Scheme (1)

An example of the case (3) is shown in Fig.3. The communication is supported by VPN, and the system administrator can distribute that VPN communication of the login user to the specific server A,B or C.

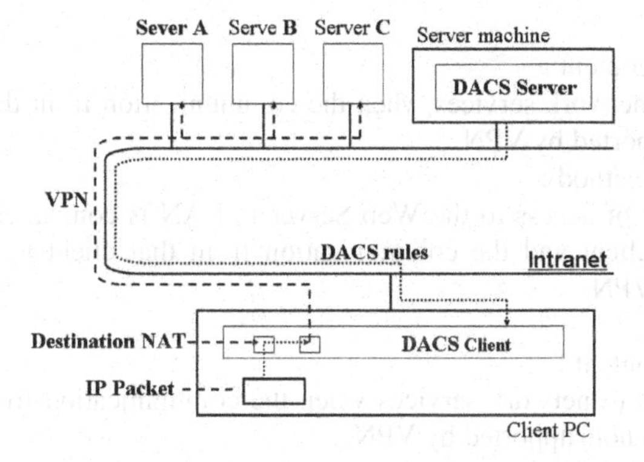

Fig. 3. Function of DACS Scheme (2)

In order to realize DACS Scheme, the operation is done by DACS Protocol as shown in Fig.4. DACS rules are distributed from DACS Server to DACS Client in (a) of Fig.4, and applied to DACS Control and DACS S Control in (b) and (c) of Fig.4. The steady communication control, such as a modification of the destination information or the communication blocking is performed at the network layer in (d) of Fig.4. Then, when communication is supported by VPN, communication is performed from (f) to (g) via (e). The VPN communication of (g) is sent by DACS S Control.

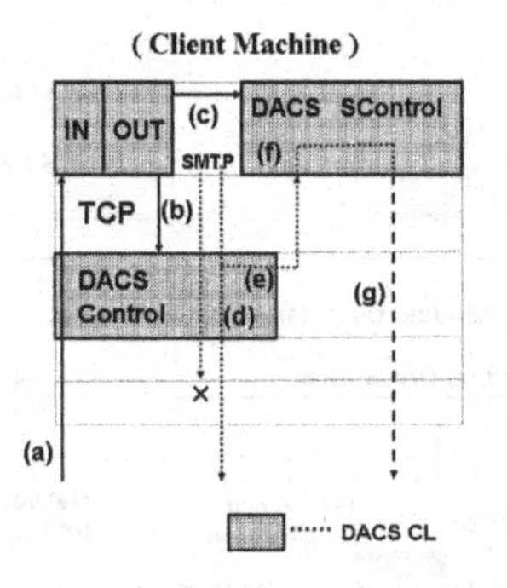

Fig. 4. Layer setting of DACS Scheme

4 Experimental Results

4.1 Range and Content of Movement Verification in Prototype System

To confirm the possibility of new access control, it is necessary to confirm the following Items. By confirming Item 1 and Item 2, it is confirmed that non-VPN supported communication is denied on the server side. By confirming Item 3, it is confirmed that non-VPN supported communication is permitted on the server side. By confirming Item 4, it is confirmed that access control on the client is realized.

(Item 1)

Confirmation content :

Access to network services when the communication from the client application is supported by VPN.

Confirmation method :

Permission of access to the Web Server in LAN is confirmed when one user logs in a client and the communication from that client application is supported by VPN.

(Item 2)

Confirmation content :

Non-access to network services when the communication from the client application is not supported by VPN.

Confirmation method :

Non-permission of access to the Web Server in LAN is confirmed when another user logs in a client and the communication from that client application is not supported by VPN.

(Item 3)

Confirmation content :

Access to network services when the communication from the client application is not supported by VPN.

(Item 4)

Confirmation content :

Access control for the communication from the client application on that client.

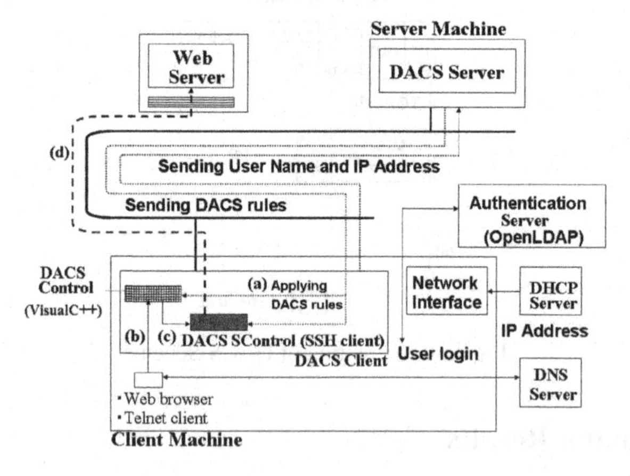

Fig. 5. Prototype System

Among these four Items, Item 3 does not need to be confirmed. This is because the communication method in Item 3 is the normal communication method when communication is not supported by VPN. Item 4 is a basic function of DACS Scheme, and has already been confirmed in the study of the conventional DACS Scheme. As the result, the possibility of this new access control is confirmed by confirming Item 1 and Item 2 in this experiment. The prototype system for Movement Verification is described in Fig.5. The details of system configuration is described in the following (1)-(3). This prototype system is located on LAN which is separated form external network, and one Web Server and one client are connected to the LAN. Therefore, it is confirmed that communication for the server is the communication from the user who sits before the client and logs in that client. The details of this prototype system are described as follows.

(1)Server Machine
CPU: Celeron M Processor340(1. 5GHz)
OS: FedoraCore3
Language: JAVA(DACS Server)
Database: PostgresSQL
Web Server: Apache
(2)Client Machine
CPU: Celeron M Processor340(1. 5GHz)
OS: Windows XP professional
Language: JAVA(DACS Client except DACS Control and DACS SControl)
Others: Visual C++ (DACS Control)
Putty (DACS SControl)
(3)Others
Authentication Server: OpenLDAP-2.1.22-8(FedoraCore1)
DHCP Server: Microsoft DHCS Server(WindowsNT4.0)
DNS Server: bind-9.2.2.P3-9(FedoraCore1)

4.2 Results of Movement Verification

In this chapter, experimental results of movement verification in prototype system are described. First, Item 1 is confirmed. The content is explained along the movement of prototype system. When the client with DACS Client is initialized, DACS rules are sent from DACS Server to DACS Client, and applied to DACS Control and DACS SControl as shown in (a) of Fig.5. When destination is changed by Destination NAT of DACS Control, that destination is port 80 of localhost (127.0.0.1) in the rectangular frame of Fig.6.

	destination before changing	destination after changing
DNAT	133.21.151.209 : 80	127.0.0.1 : 80
DNAT	133.21.151.209 : 8080	127.0.0.1 : 8080
DNAT	133.21.151.209 : 21	133.21.151.210 : 21
DNAT	133.21.151.209 : 110	133.21.151.210 : 110

Fig. 6. Results after the application of DACS rules to DACS Control

When destination is changed by DACS SControl, that destination is the IP address of Web Server in Fig.5. After DACS rules are applied, access to Web Server was performed through Web Browser. At that time, communicating server in URL is that Web Server. The destination is changed to port 80 of localhost by communication control of DACS Control. That communication is performed through the course from (b) to (c) in Fig.5. Then, by the port forwarding function of SSH Client as DACS SControl, the destination is changed to the Web Server and that communication is encrypted in (d) of Fig.5. As the result, response from Web Server to Web Browser was returned and Web page was displayed on that Web Browser. Because communication except the port of SSH (22) from the client is denied on Web Server and response from Web Server was returned, it is confirmed that communication is supported by VPN. At that point, communication records of the personal firewall on Web Server were confirmed. As the rectangular frame of Fig.7, communication quantity by SSH was increased from 0 byte to 1316 byte. It was confirmed that communication by SSH was performed.

```
Chain RH-Firewall-1-INPUT (2 references)
 pkts bytes target     prot opt in     out     source               destination
   14 31117 ACCEPT     all  --  lo     any     anywhere             anywhere
   13  1316 ACCEPT     tcp  --  any    any     anywhere             anywhere            tcp dpt:ssh
   24  4400 REJECT     all  --  any    any     anywhere             anywhere            reject-with
```

Fig. 7. Communication results on Network Server (1)

Then, Item 2 was confirmed. Different user logs in same client and performs access from Web Browser to Web Server in the state of stopping DACS Client (including DACS Control and DACS SControl). As the results, reply from Web Server was not performed and Web page was not displayed on Web Browser. As shown in Fig.8, it was confirmed that communication by HTTP was denied. Communication form the outside of Web Server is denied, and quantity of that communication was increased from 0 to 192byte. As the result, it was confirmed clearly that it was impossible to access Web Server when the communication is supported by VPN with SSH. That is, it was confirmed that it is impossible to access Web Server from the client without the packet filtering mechanism. From the results of the movement verification in prototype system, it was possible to confirm the possibility of new access control.

```
Chain RH-Firewall-1-INPUT (2 references)
 pkts bytes target     prot opt in     out     source               destination
    0     0 ACCEPT     all  --  lo     any     anywhere             anywhere
    4   192 REJECT     tcp  --  any    any     anywhere             anywhere            tcp dpt:http
```

Fig. 8. Communication results on Network Server (2)

5 Conclusion

In this paper, new access control was realized by combining the access control on the server side with the access control on the client side. When access control is performed on the sever side independently, access control for all communication form clients is performed on the server side. Processing load on the server side becomes heavy. In new access control, because access control is performed on the client side, quantity of communications to perform access control in the server side decreases by just that much. Processing load on the server side becomes low. When access control is performed on the client side independently, it is not always possible to perform access control for the communication from all clients. In new access control, because access control for the communication from these clients is possible on the server side, that problem is solved.

References

1. Odagiri, K., Yaegashi, R., Tadauchi, M., Ishii, N.: Efficient Network Management System with DACS Scheme: Management with communication control. International Journal of Computer Science and Network Security 6(1), 30–36 (2006)
2. Odagiri, K., Yaegashi, R., Tadauchi, M., Ishii, N.: New Network Management Scheme with Client's Communication Control. In: Gabrys, B., Howlett, R.J., Jain, L.C. (eds.) KES 2006. LNCS (LNAI), vol. 4252, pp. 379–386. Springer, Heidelberg (2006)
3. Odagiri, K., Yaegashi, R., Tadauchi, M., Ishii, N.: Efficient Network Management System with DACS Scheme. In: Proc. of International Conference on Net-working and Services, Silicon Valley, USA, July 16-19, IEEE Computer Society, Los Alamitos (2006)
4. Odagiri, K., Yaegashi, R., Tadauchi, M., Ishii, N.: Secure DACS Scheme. Journal of Network and Computer Applications (to appear)
5. Odagiri, K., Yaegashi, R., Tadauchi, M., Ishii, N.: New Web Service Based on Extended DACS Scheme. International Journal of Computer Science and Net-work Security 6(3), 8–13 (2006)
6. Odagiri, K., Yaegashi, R., Tadauchi, M., Ishii, N.: New Function for Displaying Static Document Dynamically with DACS Scheme. Int. Journal of Computer Science and Network Security 6(5), 81–87 (2006)
7. Odagiri, K., Yaegashi, R., Tadauchi, M., Ishii, N.: Free Information Usage System on the Network introducing DACS Scheme. In: Proc. of Int. Conf. on Internet and Web Applications and Services, Mauritius, IEEE Computer Society (to appear, 2007)
8. Shiraishi, Y., Fukuta, Y., Morii, M.: Port randomized VPN by mobile codes. In: CCNC, pp. 671–673 (2004)
9. Tadaki, S., Hirofumi, E., Watanabe, K., Watanabe, Y.: Implementation and Operation of Large Scale Network for User' Mobile Computer by Opengate. IPSJ Journal 46(4), 922–929 (2005)
10. Watanabe, Y., Watanabe, K., Hirofumi, E., Tadaki, S.: A User Authentication Gateway System with Simple User Interface, Low Administration Cost and Wide Applicability. IPSJ Journal 42(12), 2802–2809 (2001)
11. http://noside.intellilink.co.jp/product_se.asp#Inv
12. http://www.nec.co.jp/univerge/solution/pack/quarantine
13. http://www.ntt-east.co.jp/business/solution/security/quarantine/index.html
14. http://www.macnica.net/symantec_sygate/ssep.html

Design and Analysis of Hybrid On-Demand Multipath Routing Protocol with Multimedia Application on MANETs

Chuan-Ching Sue, Chi-Yu Hsu, and Yi-Cheng Lin

Department of Computer Science and Information Engineering
National Cheng Kung University, 701 Tainan, Taiwan
{suecc,p7895113,p7695175}@mail.ncku.edu.tw

Abstract. AODV and DSR are the two most widely studied on-demand ad hoc routing protocols because of their low routing overheads. However, previous studies have identified various limitations of these protocols. For example, whenever a link break occurs on the active route, they must both invoke a route discovery process. In general, on-demand protocols use query floods to discover routes. Such floods consume a substantial portion of the network bandwidth. To alleviate these problems, this study proposed an On-demand Hybrid Multipath Routing (OHMR) which extends AODV by equipping it with the capability to discover multiple routing paths. OHMR features two novel characteristics; it establishes multiple node-disjointed and braided routing paths between a source-destination pair and it reduces the frequency of route discoveries. The present results demonstrate that OHMR for multimedia communication can improve the performance of the fraction of decodable frames and achieve better performance in terms of video quality than the traditional node-disjoint multipath.

Keywords: MANET, On-demand routing, Hybrid, Node-disjointed, Braided multipath.

1 Introduction

A Mobile Ad hoc NETwork (MANET) is a collection of wireless mobile hosts forming a temporary network, with neither a fixed base station infrastructure nor centralized management function. Each node in the MANET acts both as a host and a router. If two hosts are out of radio range, all message communications between them must pass through one or more intermediate hosts. The hosts are free to move around randomly, and hence the network topology may change dynamically over time. Therefore, the routing protocols for a MANET must be adaptive and capable of maintaining routes as the characteristics of the network connectivity change. Designing an efficient and reliable routing protocol for such networks is a challenging issue. Ad hoc On-Demand Distance Vector (AODV) [9] and Dynamic Source Routing (DSR) [3] are two of the most widely studied on-demand ad hoc routing protocols, which are widely extended to implement on-demand multipath routing protocols. Multipath

S. Ata and C.S. Hong (Eds.): APNOMS 2007, LNCS 4773, pp. 144–154, 2007.

routing protocols can efficiently solve the limitations of the AODV and DSV by establishing multiple paths between a source and a destination in a single route discovery attempt. In these protocols, a new route discovery operation is invoked only when all of the routing paths in the network fail. This strategy reduces both the latency of route discovery and the routing overheads, e.g. both protocols build and rely on unipath routes for each data session. Whenever a link break occurs on the active route, both routing protocols invoke a route discovery process, which introduces significant latency and incurs considerable overheads. Clearly, the high route discovery latency induced by frequent route discovery attempts has a detrimental effect on the performance of dynamic networks. The Temporally Ordered Routing Algorithm (TORA) presented in [8] establishes multiple alternate paths by maintaining a "destination-oriented" directed acyclic graph (DAG) from the source. However, TORA does not provide a simple mechanism with which to evaluate the "quality" of these multiple routes in terms of which route is the shortest. Furthermore, TORA compares unfavorably with other on-demand protocols in that the overheads incurred in maintaining the multiple routes exceed the performance benefits obtained from maintaining route redundancy.

This paper proposes a new and practical routing protocol designated On-demand Hybrid Multipath Routing (OHMR). This protocol modifies and extends AODV to identify hybrid multipaths comprising multiple node-disjointed and braided routing paths with low routing overheads. The fundamental design principle of OHMR is to reduce the flood frequency. Controlling the frequency of network-wide flooding is important in maintaining the efficiency of on-demand protocols. To achieve this goal, OHMR searches for both multiple node-disjointed and braided routes using a single flooded query in order to provide sufficient redundancy at low cost. In this paper, OHMR with a multimedia traffic allocation strategy can efficiently support multipath multimedia communications in MANETs. The simulation results show that OHMR can reduce the frequency of route discoveries and achieve a higher packet delivery ratio. Our protocol uses a multimedia traffic allocation strategy to classify multimedia sub-streams among multiple paths according to different priority levels. The compressed multimedia stream can be segmented into several sub-streams. Each of these sub-streams takes a particular service class. The strategy is to allow braided paths to protect more important sub-streams, and node-disjoint paths to provide load balancing. Our experiments show that OHMR protocol for multimedia communication can improve the performance of the fraction of decodable frames and achieve better performance in terms of video quality than the traditional node-disjoint multipath.

The rest of this paper is organized as follows. Section 2 describes related work. Section 3 introduce proposed On-demand Hybrid Multipath Routing protocols. Section 4 shows a proposed transmission allocation strategy with multimedia traffic and simulation results. Conclusions are given in Section 5.

2 Related Works

Existing routing protocols in MANETs can be divided into two categories: on-demand and table-driven. On-demand protocols that are the most widely used protocols, in which nodes only compute when they are needed. Accordingly, each node maintains a routing table containing routes to all nodes in the network in table-driven protocols where the routing information must be exchanged periodically to keep routing information up-to-date. On-demand routing protocols include both ad hoc on-demand distance vector (AODV) [9] and dynamic source routing (DSR) [3] protocols, and DSDV [10] and OLSR [7] belong to table-driven routing protocols. Several investigations into the performance of ad hoc networks [4, 7] have shown that on-demand protocols incur lower routing overheads than their table-driven counterparts. However, these protocols also have certain performance limitations. In on-demand protocols, whenever a route is required, the route discovery process triggers a flooding process, in which the source node (or any node seeking the route) floods the entire network with query packets to search for a route to the destination. This flooding operation consumes a substantial amount of the available network bandwidth, which clearly is at a premium in wireless networks. A previous study [11] has identified several other limitations of the AODV and DSR protocols. For example, both protocols build and rely on unipath routes for each data session. Whenever a link break occurs on the active route, both routing protocols invoke a route discovery process, which introduces significant latency and incurs considerable overheads. Clearly, the high route discovery latency induced by frequent route discovery attempts has a detrimental effect on the performance of dynamic networks.

Multipath on-demand protocols alleviate these problems to a certain extent by establishing multiple paths between a source and a destination in a single route discovery attempt. In these protocols, a new route discovery operation is invoked only when all of the routing paths in the network fail. This strategy reduces both the latency of route discovery and the routing overheads. It has been shown that multipath routing yields significant benefits in wired networks. A lot of multipath routing protocols have also been proposed for ad hoc networks [2, 4, 6, 5, 11, 12], and these protocols are composed of different path selections, such as link-disjoint multipath, node-disjoint multipath, braided multipath, and non-disjoint multipath as shown in table 1. An example for node-disjoint multipath is shown in Fig. 1(a). The source S sends data to the destination node D, using one primary path which is S->c->d->D and two alternate paths which are S->a->b->D and S->e->f->D. Both of the alternate paths are node-disjoint with the primary path. Multiple paths in braided multipath are only partially disjoint from each other and are not completely node-disjoint. An example for braided multipath is shown in Fig. 1(b). The node S sends data to the destination D, using one primary path which is S->c->d->D and two alternate paths which are S->a->d->D and S->c->f->D. These two alternate paths are non-disjoint with the primary path. In addition, Temporally Ordered Routing Algorithm (TORA) presented in [8] establishes multiple alternate paths by maintaining a "destination-oriented" directed acyclic graph (DAG) from the source. However, TORA does not provide a simple mechanism with which to evaluate the "quality" of these multiple routes in terms of which route is the shortest. Furthermore, simulation studies [4, 7] have shown that TORA compares unfavorably with other on-demand protocols in that the overheads incurred in maintaining the multiple routes exceed the performance benefits obtained from maintaining route redundancy.

Table 1. The taxonomy of path selection in multipath routing

	Node-disjoint	Link-disjoint	Braided	Non-disjoint
Routing protcols	[6], [7], [8]	[6]	[10]	[5], [9]

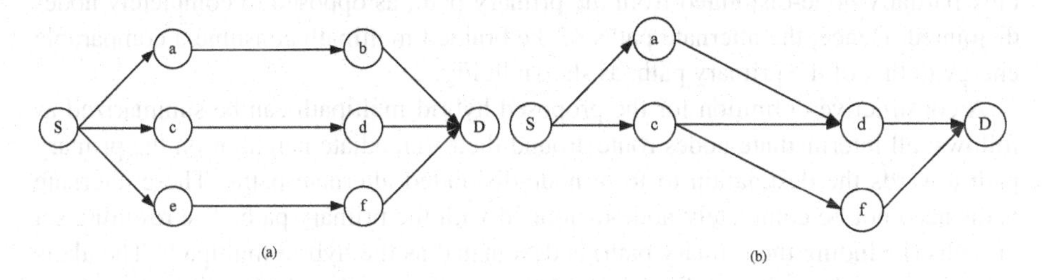

Fig. 1. The illustration of (a) link- and node-disjoint routes, (b) braided routes

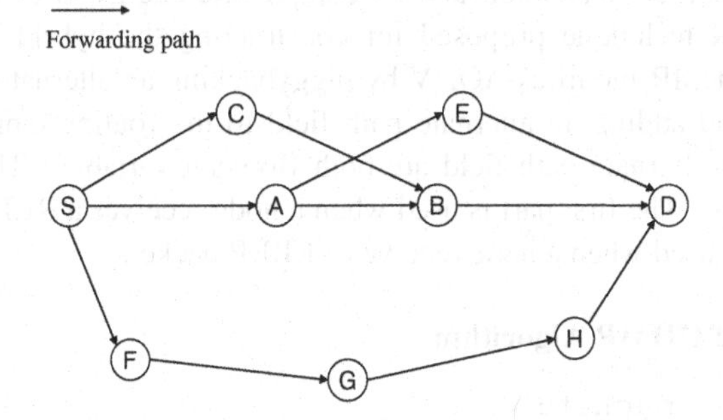

Fig. 2. The illustration of the proposed hybrid routes

3 The Overview of On-Demand Hybrid Multipath Routing Protocol

The basic principle of OHMR is to identify multiple paths during route discovery. OHMR is designed primarily for highly dynamic ad hoc networks in which link failures or node failures occur on a frequent basis. When a single path on-demand routing protocol such as AODV is used in such networks, a new route discovery must be launched each time a route break occurs. Each route discovery induces high overheads and latency. This inefficiency can be avoided by maintaining multiple redundant paths such that a new route discovery process is initiated only when all of the paths to the destination are broken. OHMR searches for hybrid multipaths comprising both node-disjointed and braided multipaths. The hybrid multipath approach combines the respective advantages of node-disjointed and braided multipaths. Alternate paths in the node-disjointed multipath are unaffected by failures along the primary path. Even though disjointed paths have attractive resilience properties, they can be energy inefficient in their transmission of data packets. Alternate node-disjointed paths tend to be longer, and therefore consume significantly more energy than the primary path. Because energy inefficiency can adversely impact the lifetime of ad hoc networks such as MANET, this study can construct both node-disjointed multipaths and braided multipaths simultaneously. The alternate paths in a braided multipath are only partially node-disjointed from the primary path, as opposed to completely node-disjointed. Hence, the alternate paths of the braided multipath consume a comparable energy to that of the primary path, as shown in Fig. 2.

A constructive definition for the proposed hybrid multipath can be summarized as follows: all intermediate nodes route around their immediate neighbor on the primary path towards the destination to form node-disjointed alternate paths. These alternate paths need not be completely node-disjointed with the primary path. The resulting set of paths (including the primary path) is designated as the hybrid multipath. The alternate paths can be either node-disjointed or non-disjointed with the primary path, or can be expected to be geographically close to the primary path. Therefore, data transmission on the alternate paths consumes a comparable energy to that on the primary path. The OHMR technique proposed for constructing the hybrid multipath is described below. OHMR modifies AODV by piggybacking an alternate path flag to the RREP packets and adding an alternate path field to the routing table. The alternate path flag and the alternate path field are both Boolean variables. The OHMR algorithm has two parts. The first part is used when a node receives a RREQ packet, while the second part is used when a node receives a RREP packet.

3.1 First Part of OHMR Algorithm

```
Received RREQ packet:
  STEP 1: if (the packet is received for the first time)
  STEP 1.a: Set up the first reverse path to the source
  using the previous hop of the packet as the next hop in
  the reverse route table
  STEP 1.b: if (the node is the destination)
  Generate a RREP and initialize its flag to "FALSE" and
  send the RREP to the next hop in the first reverse path
  STEP 1.c: else
  Broadcast the packet to the neighboring nodes
  STEP 2: else if (the packet is received for the second
  time)
  STEP 2.a: Set up the second reverse path to the source
  in the reverse route table
  STEP 2.b: if (the node is the destination)
  Generate a RREP and initialize its flag to "TRUE" and
  send the RREP to the next hop in the second reverse
  path
  STEP 3: else
  Discard the packet
```

Consider the case where a source initiates route discovery by flooding the network with a RREQ to search for a route to the destination. Any node receiving the RREQ packet for the first time sets up a reverse path to the source by using the previous hop of the RREQ as the next hop in the reverse path (Step 1.a). If the node receiving the RREQ is the destination node, it sends the RREP to the next hop in the first reverse path (Step 1.b.ii). Otherwise, it broadcasts the RREQ packet to its neighboring nodes (Step 1.c). When a node receives a duplicate RREQ packet (Step 2), it sets up the second reverse path to the source in its reverse route table (Step 2.a). If the node is the destination, it sets the alternate path flag of the RREP to "TRUE" and sends the RREP to the next hop in the second reverse path (Step 2.b). Duplicate copies of the RREQ packet received by the node are discarded (Step 3).

3.2 Second Part of OHMR Algorithm

```
Received RREP packet:
  STEP 1: if (the node is the source)
  Set up the forwarding path to the destination using the
  previous hop of the packet as the next hop in its rout-
  ing table
  STEP 2: else if (the packet is received for the first
  time)
  Save the packet's flag to the alternate path field in
  the routing table and send it to the next hop in the
  first reverse path
  STEP 3: else if(the packet's flag is "TRUE" and the al-
  ternate  path field is "FALSE")
  Reset the packet's flag to "FALSE" and return it to the
  previous hop
  STEP 4: else if(the packet's flag is "FALSE" and the
  alternate  path field is "TRUE")
  Reset the packet's flag to "TRUE" and send it to the
  next hop in the second reverse path
  STEP 5: else
  Discard the packet
```

If the node receiving the RREP packet is the source, it sets up a forwarding path to the destination by specifying the previous hop of the packet as the next hop in its routing table (Step 1). If a node other than the source receives the RREP packet for the first time, it saves the RREP's flag to the alternate path field in its routing table and sends the RREP to the next hop in the first reverse path (Step 2). If the conditions of Step 1 and Step 2 are not satisfied, the node checks the RREP's flag and the alternate path field in the routing table. If the RREP's flag is "TRUE" and the alternate path field is "FALSE", it resets the RREP's flag to "FALSE" and returns the RREP to the previous hop (Step 3). If the RREP's flag is "FALSE" and the alternate path field is "TRUE", it resets the RREP's flag to "TRUE" and sends the RREP to the next hop in the second reverse path (Step 4). If the conditions of Step 3 and Step 4 are not satisfied, the node does not propagate the RREP packet, but simply discards it (Step 5).

4 Transmission Allocation Strategy and Simulation Results

4.1 Transmission Allocation Strategy with Multimedia Application

Multimedia traffic allocation strategy in this paper is based on the popular standard MPEG encoding technique, where a video frame is encoded into three distinct types of frames, namely I-frame, P-frame, and B-frame. Reception of the I-frame or P-frame can provide low but acceptable quality, while reception of the B-frame can further improve the quality over the base layer alone, but the B-frame cannot be decoded without the I-frame and P-frame. When the I-frame, P-frame and B-frame are transmitted over multiple paths (e.g., two paths), we can design the traffic allocation scheme to send the packets of I-frame and P-frame on the primary path and the packets of B-frame on the node-disjoint alternate path.

We illustrate some examples when forwarding paths breaks. As shown in Fig. 3(a), when node A moves and the primary path breaks, node S uses the alternate path which the rt_mpath field of routing table is two to forward packets of I-frame and P-frame. And then, when node C moves and the braided path breaks, node S uses the alternate path which the rt_mpath field of routing table is three to forward I-frame and P-frame packets in Fig. 3(b). When node G moves and the node-disjoint path breaks, node S uses the alternate path which the rt_mpath field of routing table is one to forward packets of B-frame in Fig. 4(a). And then, when node B moves and the primary path breaks, node A uses the alternate path which the rt_mpath field of routing table is two to forward B-frame in Fig. 4(b).

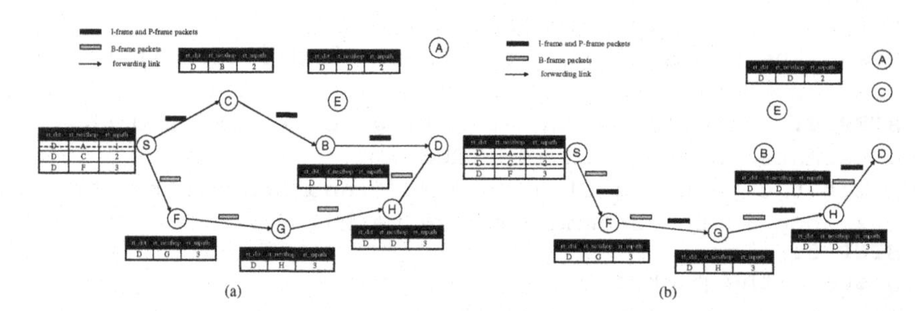

Fig. 3. Nodes move on I-frame and P-frame forwarding path

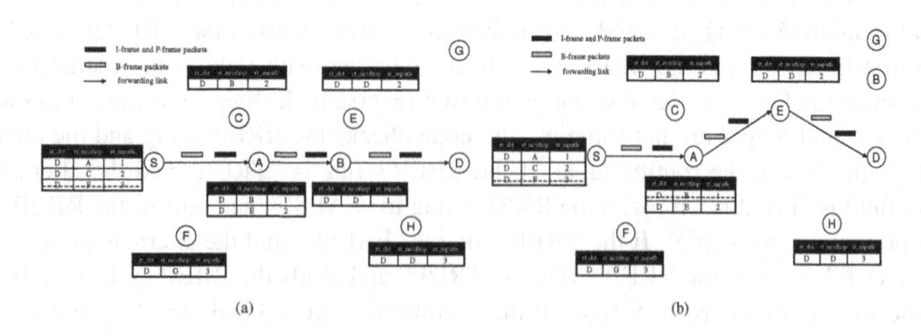

Fig. 4. Nodes move on B-frame forwarding path

Generally, a packet loss of I-frame or P-frame is likely to be experiencing a burst packet loss. OHMR finds an alternate braided path for each node on the primary path and a node-disjoint path for the primary path simultaneously to reduce the burst packet loss. Since the primary path and the alternate node-disjoint path are not correlated, source node uses the alternate node-disjoint path to provide load balancing in the beginning. That is, I-frame or P-frame is transmitted along the primary path while B-frame is transmitted along the node-disjoint path. To further improve the reliability when facing link failures, I-frame and P-frame packet transmission can use the alternative node-disjoint path to achieve higher successful delivery ratio instead of just using the alternate braided paths to forward packets when failures occur. Similarly, for B-frame packet transmission, the non-failed primary path can be used to improve the successful transmission probability before searching the braided path when failures occur.

4.2 Simulation with Multimedia Traffic

We use a simulation model based on NS-2 with CMU wireless extension [13]. In the simulation, the MANET consisting of sixteen mobile nodes are located inside a 600m × 600 m region. We only consider the continuous mobility case. Each mobile node has a continuous and random waypoint mobility model [9] (0s pause time) with a maximum speed of 5 meter/second. The radio propagation model is the two-ray ground reflection model for longer distance with omni-directional antenna. The shared radio media has a nominal bit rate of 2 Mbps. UDP is used as transport protocol. Ten UDP traffic flows are introduced as background traffics. Each of these flows has the traffic rate of four packets per second. The size of data payload was 512 bytes. The duration of these background flows are set randomly. Each node has a queue with size of 20 packets. These settings can be easily modified according to the requirements of applications. There are two format of video files, i.e. CIF (352 x 288) and QCIF (176 x 144). The difference between them is video frame size. Here we use the video file in CIF format to simulate. We decode the CIF video using an MPEG codec, at 30 frames per second provided by the tool set in [6]. Decoded video quality is measured in terms of the fraction of decodable frames and Peak-Signal-to-Noise-Ratio (PSNR).

The fraction of decodable frames. The fraction of decodable frames reports the number of decodable frames over the total number of transmitted frames. A frame is considered to be decodable if at least a fraction dt (decodable threshold) of the data in each frame is received. Furthermore, a frame is only considered decodable if and only if all of the frames upon which it depends are also decodable. Therefore, when dt=0.75, 25% of the data from a frame can be lost without causing that frame to be considered as un-decodable. In the simulations, we set the decodable threshold to one (dt=1).

PSNR (Peak Signal Noise Ratio). PSNR is one of the most widespread objective metrics to assess the application-level QoS of video transmissions. The following equation shows the definition of the PSNR between the luminance component Y of source image S and destination image D:

$$\text{PSNR(n)dB} = 20\log_{10}\frac{V_{peak}}{\sqrt{\frac{1}{N_{col}N_{row}}\sum\sum[Y_s(n,i,j)-Y_D(n,i,j)]^2}}$$

where n = frame number, Vpeak = 2k-1 and k = number of bits per pixel (luminance component). If the value of PSNR is bigger, it represents that the compressed video is much less undistorted than original one.

In the following we present a comparison study of the OHMR with a node-disjoint multipath for video streaming in MANETs. The video stream is segmented into two sub-streams based on the quality resolutions. The MPEG codec was used to generate two sub-streams. One of sub-stream is labeled as high priority (I-frame and P-frame), and the other is labeled as low priority (B-frame). In the experimental results presented, the performance of OHMR is compared with the performance of the node-disjoint multipath under the same topology and background traffic environment.

Table 2 shows that the packet delivery ratio for OHMR has better performance than that of the node-disjoint multipath. The node-disjoint multipath drops a larger fraction of the packets than that of the OHMR. It can be seen that the OHMR has higher reliability than the node-disjoint multipath. We observe that the peak signal-to-noise ratio (PSNR) drops when there is loss in packets. The deepest drop occurs if a large burst of losses occurs in I-frame and P-frame, accompanied with burst undecodable B-frame packets. The braided alternate paths in the OHMR can avoid communication failures. The PSNR curve in Fig. 5(a) has more frequent and a larger burst of frame loss than that in Fig. 5(b). Compared to the fraction of decodable frames and the average PSNR in Table 3, OHMR improves the performance by up to 16.89%, and OHMR achieves a significant 1.05 dB gain over the node-disjoint multipath in this experiment.

Table 2. Comparison of the packet delivery ratio between the OHMR and node-disjoint multipath

	OHMR	Node-disjoint multipath
Packets sent	4607	
Packets received	4558	3940
Packets lost	49	667
Packet delivery ratio	98.9%	85.5%

Table 3. Comparison of the fraction of decidable frames and average PSNR between OHMR and node-disjoint multipath

	OHMR	Node-disjoint multipath
Frames sent	2001	
Frames received	1976	1638
Not decoded frames	20	259
Frames miss	4	3
The fraction of decodable frames	98.75%	81.86%
Average PSNR	36.81	35.76

5 Conclusions

This study proposed a multipath extension to the AODV on-demand routing protocol. The proposed OHMR protocol searches for node-disjointed and braided routes using a single flooded query in order to provide sufficient redundancy with minimum overheads. The energy consumed by the alternate paths of the braided multipath is comparable to that consumed by the primary path. The alternate paths of the node-disjointed multipath are not affected by the node failures on the primary path. The key advantage of the proposed approach is a significant reduction in the frequency of route discovery flooding, which is recognized as a major overhead in on-demand protocols. The results show that OHMR with a multimedia traffic allocation strategy to classify multimedia sub-streams among multiple paths according to different priority levels can to allow more important sub-streams to travel over the primary path, and less important sub-streams to travel over the alternate node-disjoint and braided paths. In addition, the proposed protocol for multimedia communication can improve the performance of the fraction of decodable frames and achieve better performance in terms of video quality over the traditional node-disjoint multipath scheme.

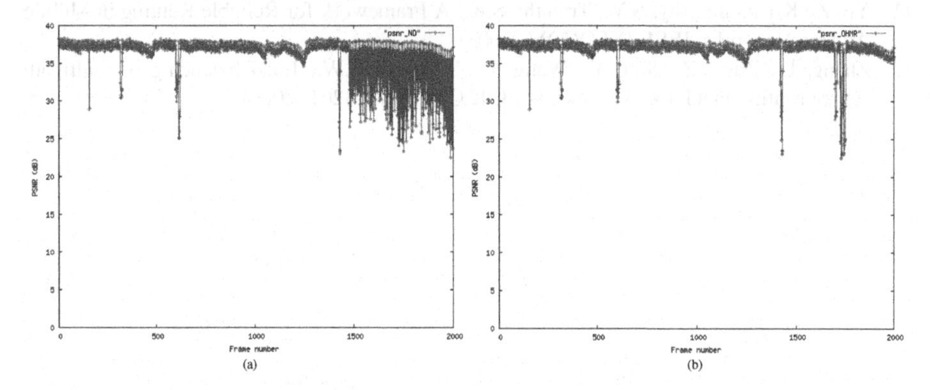

Fig. 5. The PSNRs of (a) the received video frames in the node-disjoint multipath, (b) the received video frames in the OHMR

Acknowledgments. The author is with the Department of Computer Science and Information Engineering, National Cheng Kung University, Taiwan, R.O.C. This research was supported by the National Science Council, Taiwan, R.O.C., under Grant NSC96-2221-E-006-093-MY2.

References

1. Broch, J., Maltz, D., Johnson, D., Hu, Y.C., Jetcheva, J.: A Performance Comparison of Multi-Hop Wireless Ad Hoc Network Routing Protocols. IEEE ACM MOBICOM, 85–97 (1998)
2. Ganesan, B., Govindan, R., Shenker, S., Estrin, D.: Highly-Resilient, Energy-Efficient Multipath Routing in Wireless Sensor Networks. Mobile Computing and Communications Review (MC2R). 4(5) (2001)

3. Johnson, D.B., Maltz, D.A.: Dynamic Source Routing in Ad Hoc Wireless Networks. Mobile Computing, 153–181 (1996)
4. Lee, S.J., Gerla, M.: Split Multipath Routing with Maximally Disjoint Paths in Ad Hoc Networks. In: IEEE International Conference on Communications. 10 (2001)
5. Li, X., Cuthbert, L.: Stable Node-Disjoint Multipath Routing with Low Overhead in Mobile Ad Hoc Networks. In: Proc. of the IEEE MASCOTS, IEEE Computer Society Press, Los Alamitos (2004)
6. Marina, M.K., Das, S.R.: On-demand Multipath Distance Vector Routing in Ad Hoc Networks. In: Proceedings of the International Conference for Network Procotols (2001)
7. Murthy, S., Garcia-Luna-Aceves, J.J.: An Efficient Routing Protocol for Wireless Networks. Mobile Networks and Applications. 1 (1996)
8. Park, V.D., Corson, M.S.: A highly adaptive distributed routing algorithm for mobile wireless networks. IEEE INFOCOM 1405–1413 (1997)
9. Perkins, C.E., Royer, E.M.: Ad-hoc On-Demand Distance Vector Routing. In: Proceedings of the 2nd IEEE Workshop on Mobile Computing Systems and Applications, IEEE Computer Society Press, Los Alamitos (1999)
10. Perkins, C.E., Bhagwat, P.: Highly Dynamic Destination-Sequenced Distance-Vector Routing (DSDV) for Mobile Computers. ACM SIGCOMM, 234–244 (1994)
11. Ye, Z., Krishnamurthy, S.V., Tripathi, S.K.: A Framework for Reliable Routing in Mobile Ad Hoc Networks. IEEE INFOCOM (2003)
12. Zhang, L., Zhao, Z., Shu, Y., Wang, L., Yang, O.W.W.: Load balancing of multipath source routing in ad hoc networks. IEEE ICC. 5, 3197–3201 (2002)

A Routing Scheme for Supporting Network Mobility of Sensor Network Based on 6LoWPAN*

Jin Ho Kim[1], Choong Seon Hong[1], and Koji Okamura[2]

[1] Department of Computer Engineering, Kyung Hee University
{jinhowin,cshong}@khu.ac.kr
[2] Computing and Communications Center, Kyushu University
oka@ec.kyushu-u.ac.jp

Abstract. Network Mobility (NEMO) and IPv6 over Low power Wireless PAN (6LoWPAN) protocols are the two significant important technologies in the current networking research areas and seem to be vital for the future ubiquitous environment. It can maximize the ripple effect of ubiquitous revolution due to the close correlation between NEMO and 6LoWPAN. In this paper, we propose an interoperable architecture between NEMO and 6LoWPAN. To accomplish our goal we have enhanced the routing protocol: An extended routing scheme for mobile routers to support mobility in 6LoWPAN sensor nodes. Enhanced routing scheme performs default gateway discovery and mobile network prefix discovery operations for packet forwarding, path optimization and backup route maintenance. Simulation shows that our mechanism is capable of minimizing the routing overheads, and end-to-end packet delay.

1 Introduction

WSN (Wireless Sensor Network) is one of the fastest growing segments in the ubiquitous networking today. Currently some sensor network protocols have non-IP network layer protocol such as ZigBee[1], where TCP/IP protocol is not used. However, future WSNs consisting of thousands of nodes and these networks may be connected to others via the Internet. Hence, efficient addressing mechanism will be needed to communicate with the individual nodes in the network. IPv6 can be the best solution for that. Also promising and suitable application is needed which can make effective use of IPv6 address. Accordingly, IETF (Internet Engineering Task Force) 6LoWPAN (IPv6 over Low power Wireless PAN) Working Group[2] is organized to define the transport of IPv6 over IEEE 802.15.4[3] low-power wireless personal area networks. So, external hosts in IPv6 Internet can directly communicate with sensor nodes in 6LoWPAN, and vice versa, as each sensor node will be assigned a global IPv6 address. With the help of IPv6 addressing, it is relatively easy to add mobility support in WSN. When node moves to other link from home link, there are many messages overhead e.g. BU (Binding Update) [4], BA (Binding Acknowledgement)

* "This work was supported by the Korea Research Foundation Grant funded by the Korean Government (MOEHRD)" (KRF-2006-521-D00394).

S. Ata and C.S. Hong (Eds.): APNOMS 2007, LNCS 4773, pp. 155–164, 2007.

messages[4]. As Mobile IPv6 protocol[4] is designed for host mobility support, nodes changing home link must exchange those messages for successful operation. As a consequence, Mobile IPv6 protocol is very inefficient for energy and computing constrained sensor nodes. We assume sensor network is homogeneous and mobility is as a unit of network (same like NEMO[5]). If NEMO is applied in sensor network, even though each node is not equipped with mobility protocol, it can maintain connectivity with the Internet through MR (Mobile Router) as a network unit. Also, sensor nodes should have IPv6 stack as NEMO protocol[5] is based on IPv6. Hence, the network mobility of the sensor nodes is supported by interoperable architecture technology between 6LoWPAN and NEMO.

In this paper, we propose an interoperable architecture between NEMO and 6LoWPAN. To the best of our knowledge this is the first work on interoperable architecture between NEMO and 6LoWPAN. To accomplish the inter operability, we have enhanced the routing protocol: An extended routing scheme for mobile routers to support mobility in 6LoWPAN sensor nodes. Enhanced routing performs default gateway discovery and mobile network prefix discovery operations for packet forwarding, path optimization and backup route maintenance.

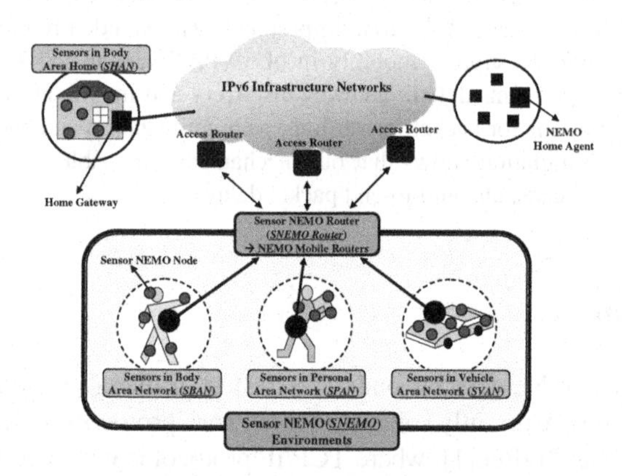

Fig. 1. Sensor NEMO (SNEMO) Environment

Figure 1 shows a Sensor Network Mobility (SNEMO) environment, which is integrated with NEMO and sensor network (6LoWPAN) protocols. This is the reference model used in this paper. SNEMO environment has three components: MR (Mobile Router), sensors, and AR (Access Router). MR use NEMO protocol that provides network mobility to each sensor node. An optimized route Internet connection is supported by mesh routing between the mobile routers. Sensors are equipped with IPv6 to be operable with NEMO environments. AR acts as the IPv6 default Internet gateway to the mobile routers.

Our paper is organized as follows. Section 2 introduces some related works such as NEMO and 6LoWPAN protocol. In section 3, we briefly described the operations of a suitable routing scheme for 6LoWPAN. Then we show the results and their performance evaluations in section 4. Finally, we conclude in section 5.

2 Related Works

Network Mobility. When a packet is sent to a fixed node in mobile network by using the Mobile IPv6 protocol, first it is transmitted to home network corresponding to the mobile network. But, home network does not know the information about the nodes moving with-in mobile network. The transmission is failed by routing loop occurring between the default router and the HA (Home Agent). To avoid the problem of the routing loop, we adopt explicit mode of NEMO protocol[5]. Prefix information of mobile network nodes are supplied to the mobile router, which is in charge of the mobile network. The mobile router sends a Binding Update message with mobile network prefix. Then, home agent for mobile router will be able to know the location of nodes moving with mobile router. Mobile network nodes that belong to the mobile router receiving the packets are directed to the current address for mobile router using tunneling. Thus, packets will reach to the final destination via mobile router. However, NEMO basic protocol does not solve the route optimization problem. The reason is that packets between CN and MNN would be forwarded through a bi-directional tunnel which is established between MR and its own HA. In case of a nested NEMO, packets would travel through the all of the HAs. Therefore, packet routing problem is more serious in nested NEMO. There are some solutions to solve route optimization problem such as ORC[6], RBU[7], ONEMO[8]. And MANEMO (Mobile Ad Hoc NEMO: integration of MANET and NEMO protocol)[9] solves the problem of route optimization in the nested case.

6LoWPAN. There are many problems in using existing IP-based infrastructure by adapting IPv6 mechanism to LoWPAN[10]. Specially, MTU for IPv6 is 1280 bytes, but PDU for 6LoWPAN is only 81 bytes. Accordingly, there is a need for Adaptation layer between IP layer and LoWPAN MAC layer[11]. Adaptation layer provides the fragmentation and reassembly process. 6LoWPAN should also support IPv6 automatic address configuration. In order to support multi-hop mesh network for 6LoWPAN, a routing protocol is required. AODV (Ad-hoc On-demand Distance Vector)[12] for MANET (Mobile Ad-hoc Networks) can be alternative plan for multi-hop routing protocol for 6LoWPAN. However, in order to adapt AODV for the 6LoWPAN, it should be adaptable for 6LoWPAN without fragmentation. LOAD (6LoWPAN Ad Hoc On-Demand Distance Vector Routing)[13] is the routing protocol which transmits the packet from 6LoWPAN to multi-hop node. LOAD, which is a simplified version of AODV, uses lightweight control and routing tables to adapt to 6LoWPAN using limited traffic and memory. Its basic operation is similar to AODV. First, it searches the path by broadcasting RREQ message and then decides the path after receiving RREP message. In order to adapt LoWPAN to IPv6 and IP-based networks, a gateway is needed. It is an important process for the gateway to convert the packet for LoWPAN into the IP-based packet.

3 Routing Scheme for Supporting Network Mobility of 6LoWPAN

In this section, we explain detailed operation of a suitable routing scheme for supporting network mobility of the sensor network based on 6LoWPAN.

3.1 Internet Gateway Discovery

If mobile network including 6LoWPAN sensor nodes cooperates with IPv6 Internet, 6LoWPAN default Internet gateway should exist between the 6LoWPAN network and the IPv6 Internet network. A mobile network can be consisting of hosts and mobile router. Mobile network communicates with the Internet gateway directly or using intermediate mobile routers. Each Internet gateway consists of one or more different subnets. Internet gateway has IEEE 802.15.4 interface to communicate with mobile network nodes and wired interface to communicate with IPv6 network. It performs packet routing function between mobile network and IPv6 Internet. Different mobile routers organize mesh network by IEEE 802.15.4 egress interface. When new mobile router moves to other link from home network, it sends a Binding Update message to home agent. However, unfortunately it is not sure of the route to the home agent, as mobile router doesn't know where the Internet gateway is located in the network. Thus, mobile router executes Internet gateway discovery in the current network. Figure 2 shows an Internet Gateway Request (IGREQ) and an Internet Gateway Reply (IGREP) messages format to discover the Internet gateway in the network.

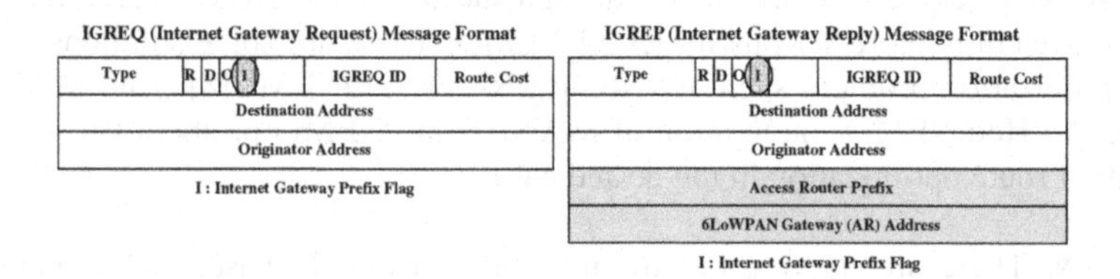

Fig. 2. Internet Gateway Request (IGREQ) and Internet Gateway Reply (IGREP) messages format

The mobile router sets the I-flag that means an Internet gateway prefix flag in the IGREQ message, and broadcasts it in the network (shown in figure 3(a)). At the same time neighbor mobile routers receiving the IGREQ message (as shown in figure 3(b)), will forward the message to the destination access router. Routing overhead of this method is reduced as compared to the other ad-hoc routing protocols such as AODV or LOAD, because IGREQ message is unicast rather than broadcast. As shown in figure 3(c), if the Internet gateway receives the IGREQ message, and sends an IGREP message to new mobile router immediately. The IGREP message includes a prefix and IPv6 global address (6LoWPAN Gateway address) of the access router. Mobile router constructs CoA (Care-of Address) in terms of the prefix information in the IGREP message. Each mobile router may receive multiple IGREP messages, which indicate the mobile router can access more than one Internet gateway. In the case of multiple Internet gateways, the mobile router establishes default Internet gateway according to the 'route cost' field of the received IGREP messages. Figure 3(c) displays the IGREP message route. New mobile router can receive prefixes and addresses information of AR1 as well as AR2, and the mobile router selects the best routing paths to the default Internet gateway. In this case the AR2 has been selected the Internet gateway.

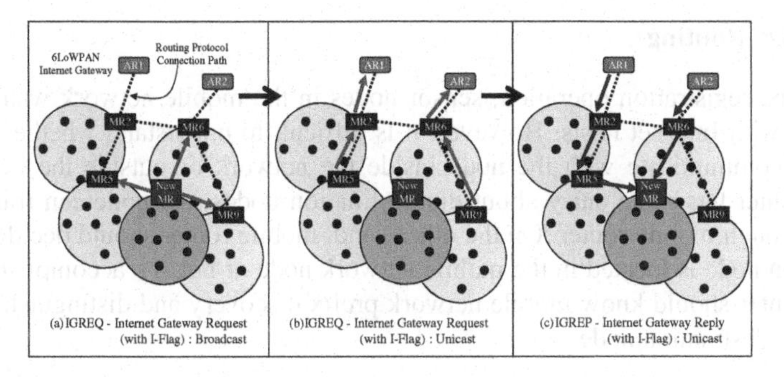

Fig. 3. Internet Gateway Discovery operation

3.2 Home Registration

If the mobile router detects its movement in the foreign network, the mobile router selects the path of the default Internet gateway, and has to perform home registration process in order to register the movement information. The mobile router sends Binding Update message to home agent as shown in figure 4. The Binding Update message includes mobile network prefix option, and home address option. In the figure 4, IPv6 header of source address is a CoA which is constructed by the access router prefix in the IGREQ message, and destination address is default Internet gateway. If the mobile router sends the Binding Update message, AR2 updates the mapping table entry (such as home address of mobile router, interface ID of CoA, and mobile network prefix). And then access router forwards it to home agent. If the Binding Update message is successfully transmitted to home agent, it will reply with Binding Acknowledgement message. The format is shown in the figure 4. The source address is home agent address and destination address is access router address. Then home agent sends the Binding Acknowledgement message to the mobile router. If this operation is successful, home registration process will be succeeded. Now sensor nodes can communicate with IPv6 Internet hosts successfully.

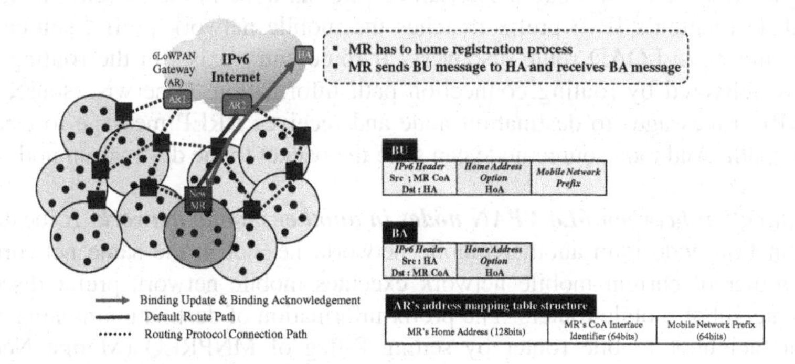

Fig. 4. Home Registration operation

3.3 Packet Routing

After home registration operation, sensor nodes in the mobile network want to communicate with Internet hosts. However, it is difficult to understand whether the node wants to communicate with the node inside the network or outside the network. If mobile router has route entry about the destination nodes, the node can transmit the packet through mobile router. On the other hand, mobile router should decide whether destination node is located in the mobile network node or not. To accomplish this, the mobile router should know mobile network prefix discovery and distinguish the location of the destination node.

Communication between 6LoWPAN nodes in another mobile network. Figure 5 shows the communication between 6LoWPAN nodes in the same mobile network.

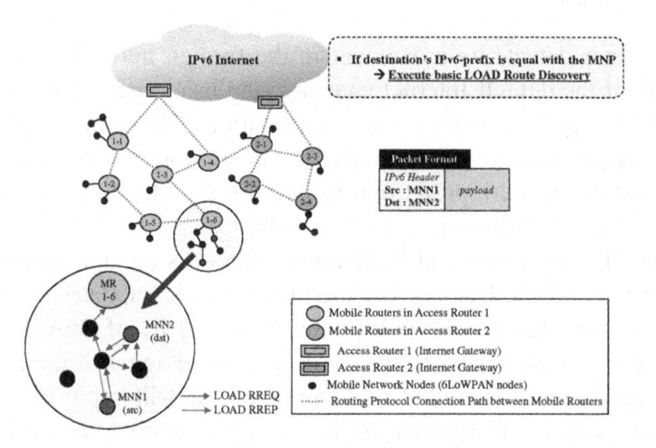

Fig. 5. The scenario of communication between 6LoWPAN nodes in same mobile network and its packet format

After considering the prefix of the destination node, source node will be able to know whether destination node is internal or external node in the current mobile network. If destination's IPv6 prefix matches the mobile network prefix, source node executes the basic LOAD route discovery. If route entry exists in the routing table, packet is delivered by routing connection path information. Otherwise source node sends RREQ messages to destination node and receives RREP message to establish the route path. And then source node can send the packet to the destination node.

Communication between 6LoWPAN nodes in another mobile network. If location of the destination node is in another mobile network, i.e. not in the same network, the mobile router of current mobile network executes mobile network prefix discovery between neighbor mobile routers. The prefix information of destination node is broadcasted to neighbor mobile router by setting P-flag of MNPREQ (Mobile Network Prefix Request) message. Upon receiving the message, mobile router sent MNPREP (Mobile Network Prefix Reply) message by setting P-flag. This indicates there is a mobile router corresponding to mobile network prefix of the destination node. If the mobile router successfully complete the mobile network prefix discovery, it is possible to communicate via an optimized route to the neighbor mobile router. If P-flag has been set 0 in the MNPREP message, the mobile router decides that destination node is an Internet host, so transmits the packet to the Internet gateway. Figure 6 shows the format of MNPREQ and MNPREP messages for mobile network prefix discovery.

MNPREQ (Mobile Network Prefix Request) Message Format

Type	R	D	O	P		MNPREQ ID	Route Cost
Destination Prefix							
Originator Address							

P : Mobile Network Prefix Flag

MNPREP (Mobile Network Prefix Reply) Message Format

Type	R	D	O	P		MNPREQ ID	Route Cost
Destination Address							
Originator Address (MR's CoA)							
Mobile Network Prefix							
Mobile Router Home Address							

P : Mobile Network Prefix Flag

Fig. 6. Mobile Network Prefix Request (MNPREQ) and Mobile Network Prefix Reply (MNPREP) messages format

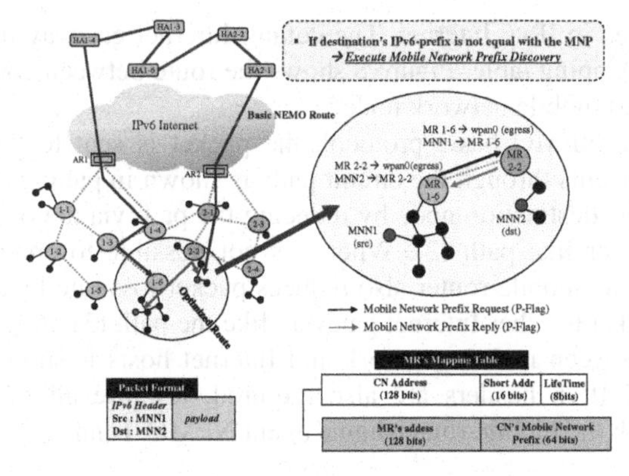

Fig. 7. The scenario of communication between 6LoWPAN nodes in another mobile network, its packet format, and mobile router's address mapping table

Figure 7 shows the communication between 6LoWPAN nodes in another mobile network. If destination's IPv6 prefix is not same as the mobile network prefix, the mobile router executes mobile network prefix discovery. As shown the figure 7, MR1-6 can transmit the packet to the destination node such as MNN2 after route path configuration. If the mobile network prefix discovery mechanism doesn't exist such as basic NEMO protocol or basic LOAD routing protocol, packet route leads to an inefficient result like detour route. In the case of the communication with another mobile network node, packet format and external mapping table of the mobile router appear in figure 7. And the mapping table stores mobile router's address and correspondent mobile network prefix's address.

Communication between mobile network node and Internet host. As mentioned above if destination's IPv6 prefix does not match the prefix of current mobile network, mobile router performs the mobile network prefix discovery. In particular, the mobile router receives MNPREP message without P-flag from default access router, the packets should be forwarded to the default Internet gateway. As the destination

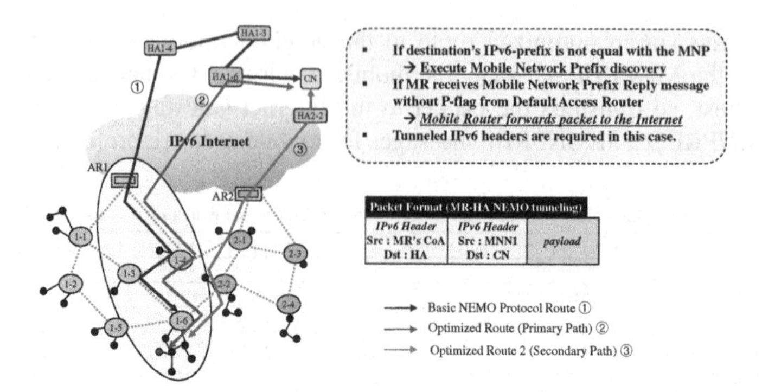

Fig. 8. The scenario of communication between mobile network node and Internet hosts, its packet format

node will be located in IPv6 Internet. The default Internet gateway makes a record in external address mapping table. Figure 8 shows the route between correspondent node in IPv6 Internet and mobile network node.

When operating NEMO basic protocol, the packet is sent to the destination by passing all home agents through the circuit path as shown in path (1) figure 8. But the packet is sent to the destination node by reducing the path via the only home agent of related mobile router like path (2). When it's not possible to connect with current Internet gateway, the mobile router also reduces packet loss rate by assuring the path to forward the packet to other Internet gateway like the path (3). Packet format for the communication between mobile network and Internet hosts is shown in figure 8. In addition, tunneled IPv6 headers are also required, because all of the packets are transmitted through the mobile router-home agent NEMO Tunnel.

4 Performance Evaluation

In order to measure performance of proposed routing mechanism, we configured the network topology as shown in figure 9(a). We measured routing packet overhead and end-to-end packet delay followed by the number of mobile router by using NS-2 (Network Simulator 2)[14]. We configured 100Mbps wired link among routers. AR network takes shape of mesh network by adapting AODV or LOAD routing protocol. Each mobile router includes three mobile network nodes. The parameters for simulation are stated in figure 9(b).

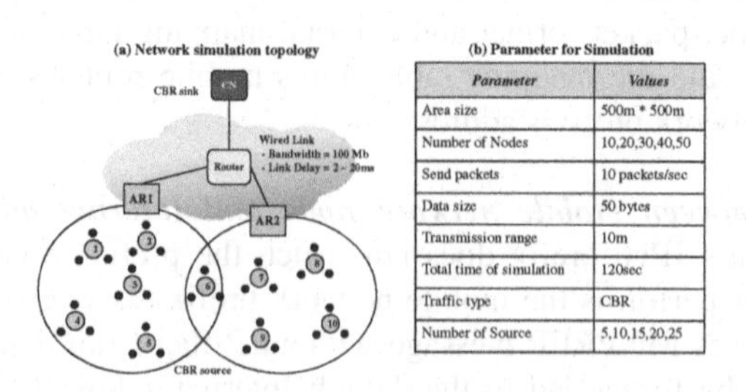

(b) Parameter for Simulation

Parameter	Values
Area size	500m * 500m
Number of Nodes	10,20,30,40,50
Send packets	10 packets/sec
Data size	50 bytes
Transmission range	10m
Total time of simulation	120sec
Traffic type	CBR
Number of Source	5,10,15,20,25

Fig. 9. Simulation environment

Figure 10(a) shows the measured routing overhead for varying numbers of network nodes. We compare our protocol with AODV, LOAD routing protocol without mobility scenario, NEMO with AODV and NEMO with LOAD routing protocol. It is found that our proposed packet routing method has least routing overhead. Comparing to our proposed protocol, adaptation of original AODV protocol results in the higher routing overhead. Overhead incurred by LOAD routing protocol is lower than AODV, because HELLO message is not used. And AODV with NEMO and LOAD with NEMO have lower overhead than original AODV and original LOAD routing protocol, because in NEMO protocol routing path is searched only by mobile router. In our protocol, packet is delivered with lowest routing overhead, because proposed method is improved suitably for NEMO environment with operating process searching the path by selecting lower overhead than LOAD routing protocol.

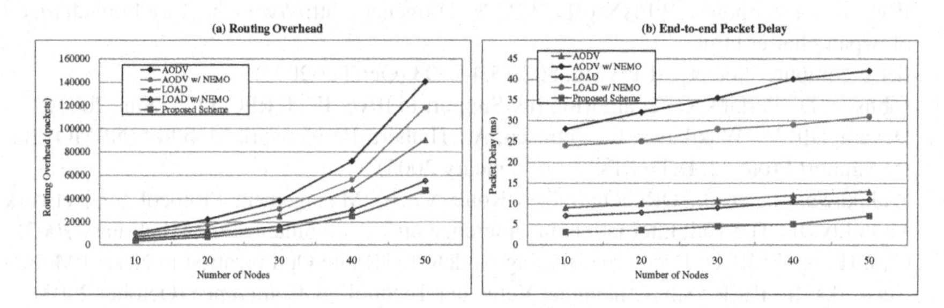

Fig. 10. Results of the routing overhead and end-to-end packet delay

The figure 10(b) shows the end-to-end packet delay experienced by CBR (Constant Bit Rate) traffic rate 10 packets per second, where each packet size of 50 bytes. Packet delay is much more in AODV with NEMO and LOAD with NEMO routing protocols than their original versions, the reason is packets are transmitted to the destination node via the home agent of each mobile router by NEMO basic protocol. In other words, delay time is increased because the routing path on Internet is extended. Packet delay is much lower in our proposed method as it transmits the packets only via the home agent of mobile router of source node. The above experimental results prove that proposed routing mechanism works efficiently than other proposed schemes in sensor network environment.

5 Conclusion

In this paper, we proposed the architecture for 6LoWPAN network mobility that also supporting communication between IPv6 node and sensor node. 6LoWPAN network is included in mobile router, although there's no mobility function for each sensor node. A sensor node works as a mobile network unit through mobile router with modified and defined LOAD routing protocol. Also, we have designed the mobile router architecture including the module of each layer. Mobile router detects its own default Internet gateway through the IGREQ (Internet Gateway Request) and the IGREP (Internet Gateway Reply) message, and after finishing this operation, it can communicate with Internet through IPv6 network and transmit Binding Update message to its own home agent. Therefore, mobile router can let home agent know its own current location. For optimized packet routing, through the MNPREQ (Mobile Network Prefix Request) and the MNPREP (Mobile Network Prefix Reply) message, it can exchange the prefix information between each other among neighbor mobile routers and can route the packet through optimized path between nodes. From simulation results we can conclude that the routing packet overhead and end-to-end packet delay are lower than other mechanisms.

References

1. ZigBee Alliance, http://www.zigbee.org/en
2. IPv6 over Low power WPAN (6LoWPAN) Homepage http://www.ietf.org/html.charters/6lowpan-charter.html
3. IEEE computer Society, IEEE Std. 802.15.4-2003 (October 2003)
4. Johnson, D., Perkins, C., et al.: Mobility Support in IPv6. IETF RFC 3775 (June 2004)
5. Devarapalli, V., Wakikawa, R., Petrescu, A., Thubert, P.: Network Mobility (NEMO) Basic Support Protocol, IETF RFC 3963 (January 2005)
6. Wakikawa, R., et al.: ORC:Optimized Route Cache Management Protocol for Network MObility. In: The 10th International Conference on Telecommunications (February 2003)
7. Cho, H., et al.: RBU: Recursive Binding Update for Route Optimization in Nested Mobile Networks. In: IEEE 58th Semiannual Vehicular Technology Conference (October 2003)
8. Watari, M., et al.: Optimal Path Establishment for Nested Mobile Networks. In: IEEE 62nd Semiannual Vehicular Technology Conference (September 2005)
9. Wakikawa, R., et al.: MANEMO Topology and Addressing Architecture. draft-wakikawa-manemoarch-00.txt, IETF (June 2007)
10. Kushalnagar, N., Montenegro, G.: 6LoWPAN: Overview, Assumptions, Problem Statement and Goals. draft-ietf-6lowpan-problem-06, IETF (November 2006)
11. Montenegro, G., Kushalnagar, N.: Transmission of IPv6 Packets over IEEE 802.15.4 Networks. draft-ietf-6lowpan-format-08, IETF (November 2006)
12. Perkins, C.E., Belding-Royer, E.M., Das, S.D.R.: Ad Hoc On-Demand Distance Vector (AODV) Routing. IETF RFC 3561 (July 2003)
13. Kim, K., Daniel Park, S., Montenegro, G., Yoo, S., Kushalnagar, N.: 6LoWPAN Ad Hoc On-Demand Distance Vector Routing (LOAD). draft-daniel-6LoWPAN-load-adhoc-routing-02, IETF (March 2006)
14. Network Simulator (ns) version 2, http://www.isi.edu/nsmam/ns

Cross Layer Based PBNM for Mobile Ad Hoc Networks with Vector Information in XML*

Shafqat-ur-Rehman[1], Wang-Cheol Song[1,**], Gyung-Leen Park[2], and Junghoon Lee[2]

[1] Department of Computer Engineering, Cheju National University, Jeju 690-756, South Korea
{shafqat,philo}@cheju.ac.kr

[2] Department of Computer Science and Statistics, Cheju National University, Jeju 690-756, South Korea
{glpark,jhlee}@cheju.ac.kr

Abstract. Management of mobile ad hoc networks poses new challenges due to the dynamic nature of the networks and the environment in which they operate. Management of these networks requires distributed systems which can adapt to the changing network conditions. An interesting approach to network management is Policy based Network Management (PBNM) which configures and controls the network as a whole. Lately, PBNM has become very popular but it has been mainly focused on large fixed networks and it is still evolving in the arena of mobile ad hoc networks. We present a PBNM framework utilizing the xml based cross-layer feed back. We propose solutions at the network layer, a cross layer feedback architecture and policy based management mechanism at the application layer. We also present initial analysis via ns2 simulations.

1 Introduction

With the miniaturization of personal computing devices, proliferation in their number and advances in the wireless communication technologies, mobile ad hoc networks (MANETs) have gained popularity worldwide. MANETs are self organizing and self configuring multi-hop networks wherein nodes act co-operatively to establish the network "on-the-fly". Communication between any two nodes might require the packets to traverse multiple hops and the constituent nodes act as both host and router. MANETs bear great application potential where wired infrastructure is not viable and temporary wireless network for instant communication is desirable such as disaster and emergency situations, battlefield communications, mobile conferencing, law enforcement operations and so on [1].

Mobile ad hoc networks have engendered many research areas in the recent years. In the management field of MANETs, data collection and monitoring have been the

* This research was supported by the MIC (Ministry of Information and Communication), Korea, under the ITRC (Information Technology Research Center) support program supervised by the IITA(Institute of Information Technology Advancement) (IITA-2006-C1090-0603-0040)

** Corresponding author.

S. Ata and C.S. Hong (Eds.): APNOMS 2007, LNCS 4773, pp. 165–174, 2007.

main topics for research, and research activities in the field of policy-based network management are gaining momentum in order to provide QoS mechanisms for MANETs [2, 3]. Unlike legacy network management, which generally involves configuring and managing each network entity individually, PBNM configures and controls the network as a whole, providing the network manager with a simplified, logically centralized and automated control over the entire network. However, so far, the work on policy-based network management has largely focused on large fixed networks [2, 3], e.g., enterprise networks, content provider networks, and Internet service provider (ISP) networks.

While there has been some previous research on deploying PBNM solutions for MANETs, our work introduces a novel organizational model specifically targeted to the characteristics of MANETs by incorporating cross-layer feedback and is dynamically adaptive to the continuously changing network conditions. The Framework consists of the following main components:-

- **Cluster Management/Topology Management:** Clustering offers scalability and manageability for large scale mobile ad hoc networks. Research has shown that clustering is not only important routing; it also paves way for scaleable management solutions at the management layer. The existing PBNM solutions manage the cluster formation and cluster maintenance at the application layer. But it is more logical in terms of control and architecture to manage the clustering at Network layer and make this information available via cross layer feed back. We therefore, propose an associativity-based clustering mechanism at the network layer. Information within the cluster is managed proactively.
- **Cross Layer Feedback:** Due to the dynamic nature of the MANET nodes and underlying wireless physical media, solutions for fixed wired networks don't perform optimally when deployed in ad hoc network environment. Recently there has been a growing interest in the design and implementation of cross layer solutions. We propose an XML based solution which to our knowledge has not been proposed before.
- **PBNM Application:** It is built on top of cross-layer feedback and network layer support for clustering, topology maintenance, service discovery etc. PBNM Application performs actual management of the network e.g., translation of high level policies into device level policies, adaptation of policies to the network conditions and policy distribution etc. Details are provided in section III.

The rest of this paper is organized as follows. After this brief introduction, section 2 reviews related work in the area. In Section 3 we present our proposed PBNM framework, focusing on issues such as the management model, cluster formation and cross layer feedback. Section 4 evaluates clustering part of our work using ns2 simulation, while Section 5 concludes the paper.

2 Related Work

Mobile ad hoc networks are significantly different from wireline networks and a lot of work has been generated on alternate network management paradigms. Network monitoring, data collection and QoS provisioning have been the focus of most of the research work on ad-hoc network management. The Ad-hoc Network Management Protocol (ANMP) [4] uses hierarchical approach to reduce the number of messages exchanged between the manager and the agents. It uses a SNMP compatible MIB (Management Information Base) to perform data collection, fault management, and QoS provisioning. MIBs are maintained at the clusterheads. However, constantly maintaining the MIB in the face of nodal mobility seems inefficient and unreliable. Furthermore, it is still not mature enough to fully qualify for network provisioning [5, 6].

A Guerilla Management Architecture [7] that employs a two-tier hierarchical infrastructure is proposed as an alternative to the conventional client-server architecture. It is based on the concept of mobile agents and focuses on adaptive network management. While this approach seems promising, considerable work is still required. Interoperability of this approach with the existing management systems (e.g., SNMP) needs to be better understood.

In recent years, policy-based management of mobile ad hoc networks has seen growing interest by the research community. In [8], authors discuss the adoption of policy-based management for mobile users within an enterprise network, while [9] proposes and demonstrates an agent-based architecture to provide policy-based personalized multimedia services to nomadic users. Both works only consider infrastructure-based mobile networks.

In [10], authors proposed a policy-based management framework which presents a policy provisioning architecture and a simulation-based evaluation of the architecture in different mobile ad-hoc networking conditions. All these protocols are implemented in a layered manner and may function inefficiently in mobile ad hoc environments. To our knowledge, a cross layer based PBNM solution has not been studied extensively for mobile ad hoc networks.

In this paper, we propose a PBNM architecture which employs cross layer feedback to get topology information, traffic load and track mobility of the nodes. The main contributions of this paper are:

- Clustering at the network layer
- XML based Cross layer feedback
- PBNM Framework which employs the above two approaches.

3 Cross Layer Based PBNM Framework with Vector Information in XML

3.1 Key Architectural Components of a PBNM Framework for Ad Hoc Networks

Before delving deep into the details of our architecture, we would like to provide a recap of the key concepts of a policy based network management framework. Typically, it consists of a Policy Management Tool (PMT), Policy repository, Policy Decision Point and PEPs. Policy Management tool provides the network manager with a UI to interact with the network. A network manager uses the PMT to define and edit various policies. The policies specified at the PMT are then stored in a policy repository commonly referred to as Policy Information Base (PIB). PIB is a data store that holds the policy rules. Policy Decision Point (PDP) or policy server translates the policies into a format that can be used to configure the Policy Enforcement Points (PEPs) or policy clients. PDP also monitors the policy changes that may take place at the PMT. PDP is responsible for the distribution of policies using COPS-PR. PEPs are network devices (end-hosts, routers etc.) where the policies are installed. PEPs are also responsible for reporting the local policy changes or status information to policy server(s). This facilitates the efficient network management [10].

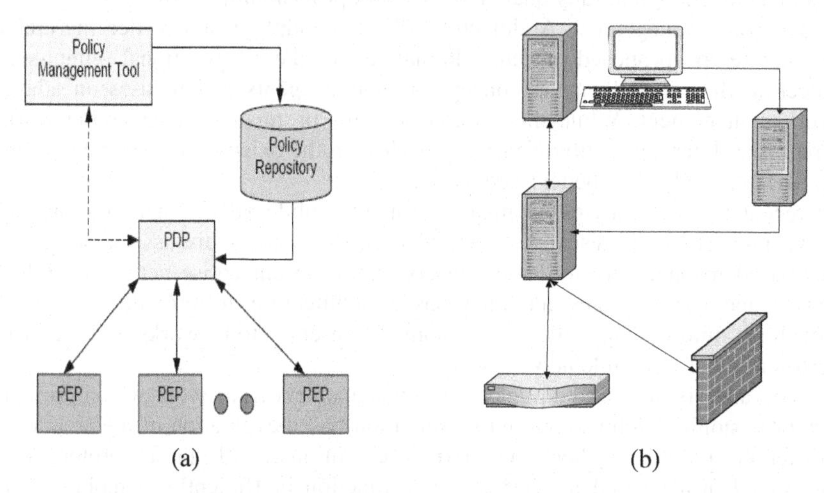

Fig. 1. (a) A Logical PBNM Architecture (b) An Example Deployment

3.2 Categories of PBNM Architectures

Based on the distribution model of policies we can divide PBNM architecture into three categories. First is outsourcing model wherein PEPs outsource the decision making to PDPs. Second is Provisioning model where PEPs are configured to take the decisions locally. Third and more flexible model is Hybrid model. In this model PEPs can make some of the decisions locally and for others they can outsource the decision making to PDPs.

There exist many protocols for policy distribution in the literature, e.g., CORBA, SNMP, Web Servers, COPS (Common Open Policy Service) etc. In our architecture we employ COPS-PR (Common Open Policy Service for PRovisioning) [11]. It is an extension of the COPS protocol. Some of the salient features of COPS-PR are as follows:

- COPS-PR supports hybrid policy distribution model which is the combination of both outsourcing and provisioning models.
- It is event driven.
- It supports transaction processing.
- Uses persistent TCP to transfer policy messages. Persistent TCP significantly improves the performance of data transfer.

3.3 Clustering Management

This is the one of the most distinctive features of our framework in that clusters are entirely managed at the network layer. We have implemented a clusterhead based clustering scheme called Associativity-based Adaptive weighted Clustering (AWC) [12]. Each node is assigned a nodal weight and selection of the clusterhead is based on this weight. Nodal weight is calculated as

Nodal Weight = Residual Battery Life + Nodal Associativity

$$W_v = |(P_b \times p) + I_v \times (1-p)|, \quad p \in \mathbb{R} : 0 \le p \le 1$$

where Nodal Associativity I_v is calculated based on the time averaged mobility rate and the nodal connectivity. It is given as

$$I_v = N_c - N_{topo}$$

Let X_i be the neighbor set of a node during i^{th} time interval then average nodal connectivity during h time intervals is

$$N_c = \frac{\sum_{i=0}^{h-1} |X_i|}{h}$$

The mobility rate averaged over h time intervals is calculated as

$$N_{topo} = \frac{\sum_{i=1}^{h-1} |(X_i \cup X_{i-1}) - (X_i \cap X_{i-1})|}{h-1}$$

The solution dynamically adapts the frequency of control messages to mobility pattern of the network. The size of the clusters is kept around an optimal value. This ensures energy and load balancing across clusters. The hop size of the cluster can vary between upper and lower boundaries. This helps eliminate the ripple effect of re-clustering. The stability of the clusters is further improved by improving Least Change Clustering (LCC) algorithm [15]. A timer Cluster Contention Interval (CCI) is used avoid unnecessary re-clustering when two clusters happen to pass by each other incidentally. If they remain within the radio range of each other longer than CCI, the one with the lowest W_v will give up its clusterhead role.

Each cluster maintains its topological information proactively. This information can be utilized by applications that assume self-adaptive hierarchical organization of the network e.g. hybrid routing protocols, network management tools. Solutions implemented at the upper layers can access this information through cross-layer interactions. Our proposed cross-layer mechanism is explained in the next section.

3.4 Cross Layer Feedback

The cluster wide topology information maintained at the network layer is exported to a shared repository called Shared Information Base (SIB). SIB is an Xml Database.

Structure of SIB Xml Database is specified through an Xml schema. Xml Database can be just an xml file or it can be part of a traditional Database Management System (DBMS). Xml Database concept has been incorporated into most of major commercial and open source database systems like Oracle [16], SQL Server [17], MySql [18] etc.

The layers employing cross-layer interactions communicate with SIB using Data Access service. The service implements xml data retrieval, modification and storage. With the popularization of Xml, there are several innovative technologies available for Xml processing. One such technology is Xml Query or XQuery which is built on top of XPath. XQuery is to Xml what SQL is to databases tables. As Xml is platform independent so it can be supported virtually on all the portable devices. XQuery is supported by all major database engines and it is a W3C recommendation. Architectural components of Data Access Service are demonstrated in Fig. 2

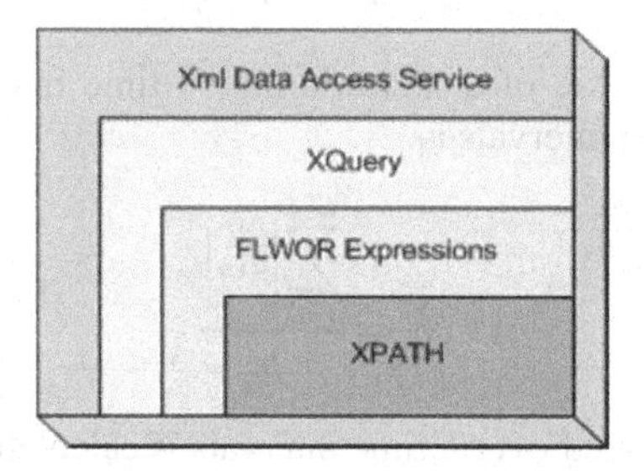

Fig. 2. Architecture of the Xml Data Access Service

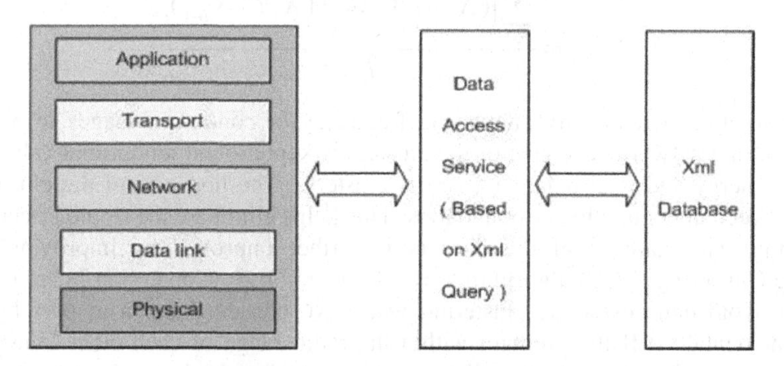

Fig. 3. Architecture for Cross Layer feedback

FLWOR (for, let, where, order by, return) expressions, pronounced "flower expressions" are the workhorse of XQuery language. They are similar to the SELECT-FROM-WHERE statements in SQL. Extended Backus-Naur Form (EBNF) of the FLWOR expressions can be found at [19]. For our research, implementation of only a subset of the FLWOR specification is sufficient. In addition to the implementation of XQuery, Xml Data Access Service also implements schema validation of the information retrieval and modification requests. The service converts the simple data requests from the layers into xml queries according to the schema and then runs these queries over the xml database.

The relevant information from all the layers can be stored in the xml database but currently in our scheme cross-layer interactions take place only at network layer and application layer.

3.5 Components of Our Architecture

The overall architecture is described as shown in figure 4. Topology Discovery is handled by the network layer and is based on the clustering mechanism explained in section 3.3. The information is retrieved by the application layer using the Xml Data access component. Clusterheads selected at the network layer are treated as PDP Servers by the PBNM framework at the application layer. QoS (Quality of Service) monitoring of the devices is performed using the COPS-PR protocol. The device capabilities are discovered using the device specific mechanisms. All this information

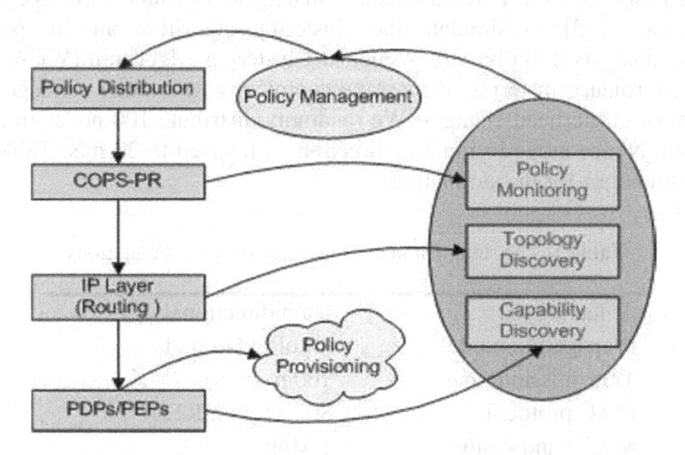

Fig. 4. Policy Based management for Ad hoc networks

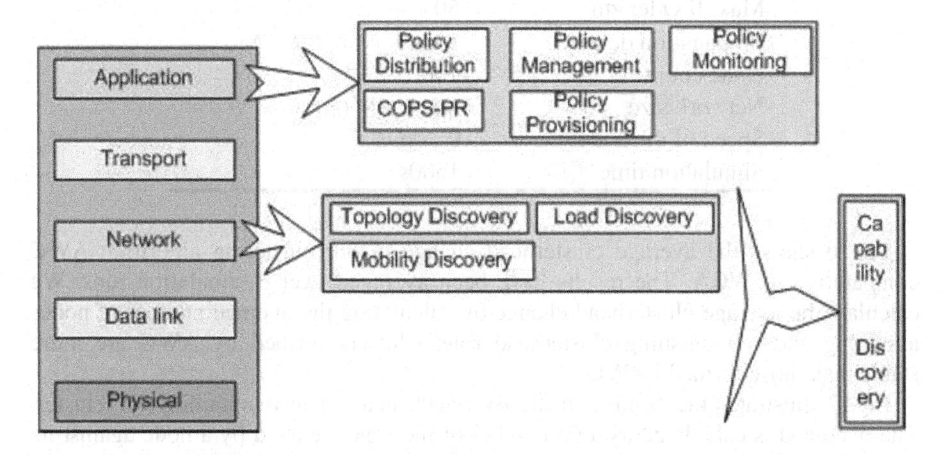

Fig. 5. Distribution of the constituent modules over layers of TCP/IP stack

is processed by the management tool to adapt the policies according to the network conditions. Then the policies are distributed to the PEPs using COPS-PR. The distinctive feature of the framework is network level support for clustering which can be further used as a backbone for service discovery. The other distinctive feature is that vector information is stored and exchanged using xml technologies which offer multitude of advantages over the legacy approaches. Fig 5 demonstrates the placement of various constituent components of our proposed framework at the respective layers of TCP/IP protocol stack.

4 Simulations

Currently, we have performed simulation analysis our clustering algorithm which is part of the proposed policy based network management framework. We have used NS-2 simulator [13] to simulate the clustering algorithm and to perform its comparative analysis initially with Weighted Clustering Algorithm (WCA) [14]. We measure performance in terms of message complexity or control message overhead and number of clusterhead changes. We randomly distribute 100 nodes in a space of 800m×800m. Nodes move in random direction with speed 0–20 m/s. Table 1 shows the setup parameters for the simulation.

Table 1. The setup parameters used for the ns-2 simulations

Antenna type	Omnidirectional
Propagation model	TwoRayGround
Transmission range	100m
MAC protocol	802.11 with RTS/CTS
MAC bandwidth	1 Mbit
Interface queue type	Drop-tail priority queue
Max. IFQ length	50
Propagation delay	1 ms
Node count	100
Network size	800m × 800m
Speed of the nodes	0-20m/s
Simulation time	1500s

Fig. 6 shows the average clusterhead change in our clustering algorithm AWC comparative to WCA. The results have been averaged over 3 simulation runs. We calculate the average clusterhead change by calculating the average number of nodes assuming and relinquishing clusterhead role. Clusters formed by AWC are more stable than those formed by WCA.

Fig. 7 illustrates the control traffic overhead incurred in maintaining the cluster. The overhead is calculated as total number of messages relayed by a node against its average speed. Due to the adaptation of AWC to the network mobility it is shown to perform better in a relatively static environment.

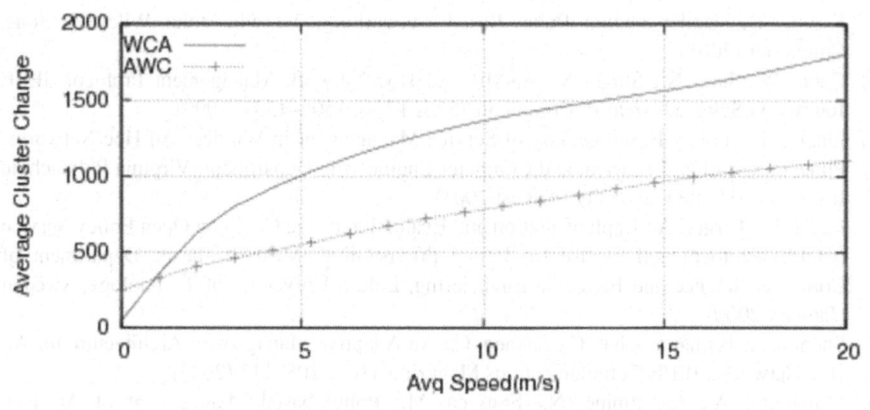

Fig. 6. Average Clusterhead Change

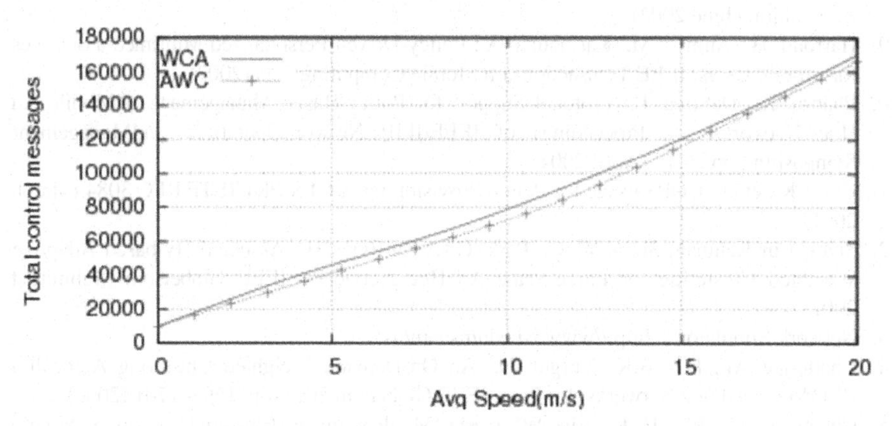

Fig. 7. Comparative Control message overhead

5 Conclusion

We have presented a PBNM framework which employs xml based cross-layer feed back for exchange of topology, mobility and traffic information. Network Layer offers more control and flexibility for clustering and management of topology information and hence improves the performance and scalability of the management solution. We have also proposed a novel technique for cross-layer feedback based on xml technologies which promises platform independence. The solution therefore can be deployed on a wide range of portable devices. At the first stage we have implemented our own clustering technique which facilitates the formation and management of more stable and load balanced clusters.

References

1. Perkins, C.E.: Ad Hoc Networking. Addison-Wesley, Reading (2001)
2. Verma, D.: Policy-Based Networking: Architectures and Algorithms, 1st edn. New Riders Publishing (2000)

3. Kosiur, D.: Understanding Policy-Based Networking, 1st edn. John Wiley & Sons, Chichester (2001)
4. Chen, W., Jain, N., Singh, S.: ANMP: Ad Hoc Network Management Protocol. IEEE Journal on Selected Areas of Communications 17(8), 1506–1531 (1999)
5. Phanse, K.: Policy-Based Quality of Service Management in Wireless Ad Hoc Networks, Ph.D. dissertation, Electrical and Computer Engineering Department, Virginia Polytechnic Institute and State University (August 2003)
6. Liden, E., Torger, A.: Implementation and Evaluation of the Common Open Policy Service (COPS) Protocol and its use for Policy Provisioning, Master's thesis, Department of Computer Science and Electrical Engineering, Lule°a University of Technology, Sweden (January 2000)
7. Shen, C., Srisathapornphat, C., Jaikaeo, C.: An Adaptive Management Architecture for Ad Hoc Networks. IEEE Communications Magazine 41(2), 108–115 (2003)
8. Munaretto, A., Agoulmine, N., Fonseca, M.: Policy-based Management of Ad Hoc Enterprise Networks. In: Proceedings of theWorkshop of the HP OpenView University Association (June 2002)
9. Harroud, H., Ahmed, M., Karmouch, A.: Policy-Driven Personalized Multimedia Services for Mobile Users. IEEE Transactions on Mobile Computing 2(1) (2003)
10. Phanse, K., DaSilva, L.: Protocol Support for Policy-Based Management of Mobile Ad Hoc Networks. In: Proceedings of IEEE/IFIP NetworkOperations andManagement Symposium, pp. 3–16 (April 2004)
11. Chan, K., et al.: COPS usage for Policy Provisioning (COPS-PR). IETF RFC 3084 (March 2001)
12. Shafqat-ur-Rehman, Song, W.-C., Park, G.L., Lutfiyya, H.: Associativity-based Adaptive Weighted Clustering for Large-Scale Ad Hoc Networks. IEEE GlobeCom (Submitted 2007)
13. Network Simulator 2. http://www.isi.edu/nsnam/ns/
14. Chatterjee, M., Das, S.K., Turgut, D.: An On-Demand Weighted Clustering Algorithm (WCA) for Ad hoc Networks. In: Proc. IEEE Globecom 2000, pp. 1697–1701 (2000)
15. Chiang, C.-C., Wu, H.-K., Liu, W., Gerla, M.: Routing in Clustered Multihop, Mobile Wireless Networks with Fading Channel. In: SICON 1997, Singapore (April 1997)
16. http://www.oracle.com/technology/tech/xml/xmldb/index.html
17. http://www.microsoft.com/learning/syllabi/en-us/2779bfinal.mspx
18. http://www.mysql.com/
19. http://www.w3.org/TR/xquery/#id-flwor-expressions

FECP Protocol for Energy Balanced Data Propagation in Smart Home Sensor Networks

Bao Nguyen Nguyen and Deokjai Choi

School of Electronic & Computer Engineering, Chonnam National University
Gwangju 500-757, Republic of Korea
nguyen_nguyen_bao@yahoo.com,
dchoi@chonnam.ac.kr
http://iat.chonnam.ac.kr/

Abstract. Nowadays Smart Home seems to be getting popular. An essential element of Smart Home is its sensor network to convey environment information to the control station. There have been many communication protocols for this sensor network. A problem from previous protocols is unbalanced energy consumption among sensor nodes. Even though they tried to overcome unbalanced energy consumption, they could not solve this problem completely yet. In this paper, we found that the basic reason for this unbalanced energy consumption is that there is normally one base node. Because of this architectural problem, the nodes around this sink node lose their energy more easily in relevance to other nodes. So, we propose the sensor network model with 2 and only 2 base stations with simulations, and its corresponding communication protocol.

1 Introduction

One of the essential things which make Wireless Sensor Networks (WSNs) different from the other types of networks is power consumption strategy. Unlike nodes in normal networks, each node in WSN is a sensor which has some restrictions in energy capacity. Recent extraordinary progress in sensor technologies has made it possible to build more powerful micro sensors in communication, data processing and storage capacities efficiently.

Various communication protocols such as [1], [2], [3], [4], [5], [6], [10], [11] and [12] have been proposed for WSNs in order to prolong the network lifetime and enhance data quality. The SPIN family of protocols in [1] and [2] utilize meta-data negotiations before any data is transmitted to assure that there is no redundant data sent throughout the network. Thus, it can save the network energy and bandwidth. In Directed Diffusion [3], C. Intanagonwiwat, et. al. proposed an approach using in-network data aggregation techniques to combine the data coming from different sources in order to minimize the number of transmissions. Following another approach, Heinzelman, et. al. [4] introduced LEACH, a hierarchical clustering protocol. In LEACH, cluster head (CH) nodes compress the data which arrives from nodes that belong to their respective cluster, and send an aggregated packet to the sink in order to reduce the amount of information that must be transmitted to the sink. Similar to

S. Ata and C.S. Hong (Eds.): APNOMS 2007, LNCS 4773, pp. 175–184, 2007.

LEACH, TEEN [5] and APTEEN [6] also utilize hierarchical clustering algorithms but with some enhancements. Cluster head sensors in TEEN and APTEEN send their members a hard threshold and a soft threshold to reduce the number of transmissions. This is done by allowing the nodes to transmit only when the sensed attribute satisfies such thresholds. However, these approaches could not maintain the balance of energy dissipation among the sensor nodes. Whether we want it or not, there are still some sensors that work harder and consume much more power than the others. For example, some nodes near the sink need to transmit data for nodes in further areas; also, cluster head nodes or leader nodes in hierarchical clustering protocols need to aggregate data coming from other nodes, and then send the final compressed packet to the sink. Such sensors may die very soon while others are still alive with high residual energy, and thus network performance and lifetime may still be low. Furthermore, a wireless sensor network protocol will perform better if it is application specific. Nowadays a prevalent protocol for a smart home environment has not been developed yet. It means that this area has strong potential as a research field. The protocols for home environment have some inherent characteristics, with some particular requirements.

In this paper, we shall therefore propose a novel wireless sensor network protocol, called Fair Energy Consumption Protocol (FECP), which is a highly balanced energy dissipation protocol appropriate to the home environment. Jose D.P. Rolim and Pierre Leone in [10], [11] and [12] also proposed an algorithm for energy balanced data propagation in WSNs but they used a statistics approach which is different from ours. Based on two base stations (BSs) topology, the key idea in FECP is to make energy dissipation in both horizontal and vertical axes more uniform, thus maximizing the network lifetime. Similar to the above approaches, our approach also adopts the data aggregation and threshold techniques, but they are improved in order to work better with the home environment.

The rest of this paper is organized as follows. Section 2 will briefly describes major reasons for the energy consumption in WSNs, and prove that unbalanced energy consumption is caused by one sink node. In section 3, we will present our new innovative protocol FECP. We will provide some metrics for simulations and compare our model with other models in section 4. Finally, section 5 will conclude the paper with a summary of our contributions and point out open research problems.

2 Energy Consumption in Wireless Sensor Networks

In this section, we intend to address some problems which frequently affect the sensor network lifetime. First of all, the energy of recent sensor nodes is still restricted whereas their processing and storage capacities have been increased many times. This problem certainly influences the overall network lifetime. Secondly, we encounter the problem of transmitting data from each sensor node. Such transmitting consumes much more energy than sensing, processing and storing data. If a sensor network has to transmit data too frequently, its energy might be drained rapidly. Similarly, when flooding techniques or broadcasting messages are overused, they would cause network congestion. That problem could also shorten the network lifetime. Thirdly, data collected by many sensors in WSNs is typically based on common phenomena, especially in overlapped sensing areas; hence there is a high probability that this data has some redundancy. Furthermore, there are usually some data values which could be unnecessary for some applications. Transmitting such redundant or unnecessary data would consume the network energy irrationally. For example, if the thermal application prefers to get the temperature greater than 50°C only in order to raise a fire alarm, then sending the value 23°C from every thermal sensor to the sink would waste the network energy. Finally, because the number of transmissions is different from each sensor, some sensors would die sooner than the others. Thus, the network lifetime would be affected.

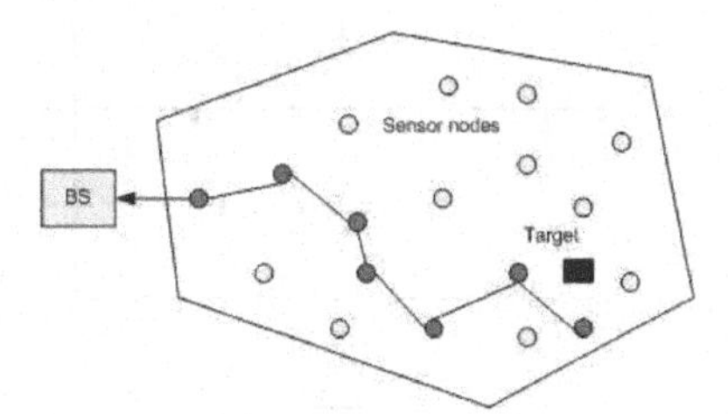

Fig. 1. One-base-station topology

Besides all of the above, there is still one more design problem which could significantly influence the network lifetime. Most previous protocols use one-base-station topology. In this topology, sensor nodes are usually scattered in a target field. They collect and route the data back to a particular external BS through other nodes, as in Fig.1. The sensor nodes are near the BS, therefore, would specifically have more pressure than the other nodes which are located further from the BS because they might transmit data to the BS not only for their own, but also for the further ones. Such sensor nodes would consume their energy much faster than the others. Some protocols such as LEACH, TEEN and APTEEN utilize a clustering technique to solve this problem. Data is transmitted inside each local cluster first. Then the cluster head nodes compress such incoming data, and send it to the BS. Thus, it might reduce the amount of information that must be transmitted to the BS. In addition, the cluster head role is rotated randomly on few sensor nodes to keep some certain nodes from draining too much energy for this role. However, the idea of dynamic clustering brings extra overhead, e.g. head maintenances, head changes, new clusters' advertisements, cluster collisions etc.

3 FECP in Details

3.1 Two-Base-Station Model

Our novel design idea is sharing the roles and responsibilities on every sensor equivalently. The sensor node which works hard will periodically change its phase to work lightly in its next turn, and conversely. Moreover, we try to avoid transmitting data to the same sensor node for a long time to extend the sensor network lifetime. Before sending data to the BS, each sensor node will choose its next hop based on the number of times that this hop is used. The hop with the least uses would be chosen as the next hop.

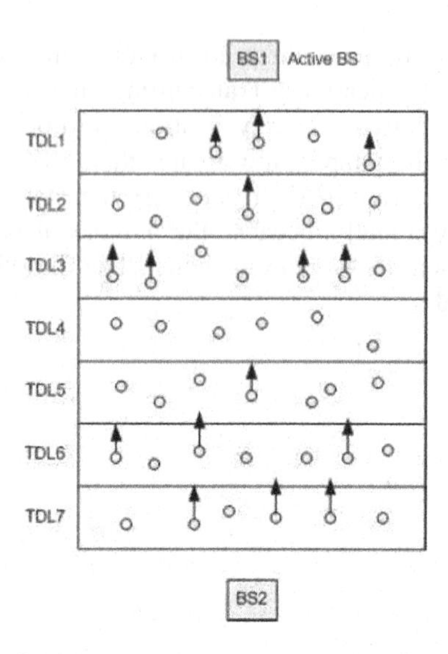

Fig. 2. Two-base-stations model with active BS1

We propose a two-base-stations model as Fig.2 and Fig.3. Sensor nodes are freely scattered in a smart house. Then we put two base stations at each end of the sensor field so that the sensor nodes should be considered a placement in the range of two BSs: BS1, BS2. We consider BS1 to be at top, and BS2 at the bottom of the network. The sensor nodes are classified into levels. There are two types of levels: Top-Down Level (TDL), which is numbered increasingly from BS1 toward BS2 (starting from 1), and Bottom-Up Level (BUL), which is numbered increasingly from BS2 toward BS1 (starting from 1) as shown Fig.2 and Fig.3. A TDL defines a number of hops from the sensor nodes to BS1. Similarly, a BUL defines a number of hops from the sensor nodes to BS2. Owing to the communication with 2 BSs, each sensor node should have both a TDL and a BUL. The sum of such two levels, TDL and BUL, should be equal to the total number of levels N in one direction (top-down or bottom-up) plus 1 as given by

$$TDL + BUL = N + 1 . \quad (1)$$

At first, BS1 works as an active BS (ABS). Every sensor node should transmit the collected data to this active BS in the bottom-up direction as shown in Fig.2. After a certain period of time, BS1 is deactivated and BS2 is activated vice versa. Then every sensor node should change the transmitting direction to the new ABS, top-down direction as displayed in Fig.3. It continues working like this periodically. In the first turn, the sensor nodes lying closer to BS1 tend to be utilized exhaustively, since all data passes through them. However, they will be the lightest ones in the next turn, since they just pass all data to their neighbors. It works similarly with the other sensors.

Besides indicating the number of hops to the BS, a level might also infer a transmission rate of the sensor nodes in this level. A low level sensor node should have a higher transmission rate than a high level sensor node. The sums of TDLl and BULl

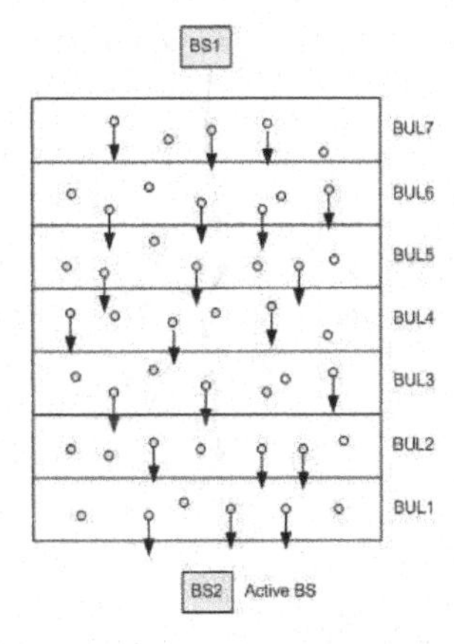

Fig. 3. Two-base-stations model with active BS2

of every sensor nodes are equal to each other with value of N + 1. Hence, any 2 sensors in different levels but in the same routing chain in Fig.4 should be intended to have similar transmission rate, and consequently consume mostly the same amount of energy. In other words, energy would be dissipated almost equally among all nodes in vertical axis. The picture of this transmission method is similar to an hourglass' working method.

Because the unbalanced energy consumption problem tends to occur mostly in vertical axis, we only need to add one extra base station to balance the energy consumption of each sensor in this axis. The use of more extra base stations might bring up complexity without any efficient help.

3.2 Transmission Counter to Choose the Lightest Neighbor

The above technique can only ensure the balance of energy dissipation among all nodes in the same routing chain. It can not balance the energy consumptions among the nodes in the same level, which means the horizontal axis. We shall therefore propose another technique, called Lightest Neighbor technique, to solve this problem. The nature of transmission of sensors shown in Fig.5 is broadcasting radial signals in a circular area.

Sensor A is transmitting data to sensor B, and sensor C is within the transmitting range of sensor A. Whether it wants to or not, the data from sensor A will reach sensor C also. In other protocols, sensor C will discard this packet since it is not a target of this packet. In our approach, however, before discarding the packet, sensor C will count up the number of transmissions of sensor A, called a transmission counter (TC). In our protocol, each sensor node should maintain the list of its closest neighbors, which are the reachable sensor nodes in lower and higher levels next to it, and also the TC corresponding to each neighbor. When a sensor wants to send a data to an ABS, it will look for a next hop in the lower neighbors list in the direction to this ABS. The neighbor which is chosen as a next hop should have the smallest TC. If there are more than two neighbors with the same smallest TC, the one which has highest energy

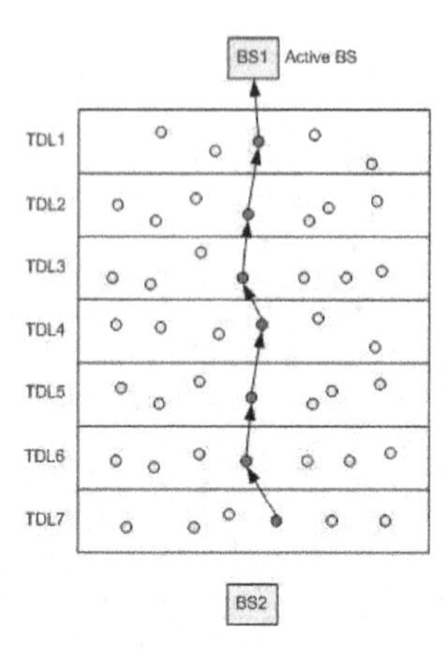

Fig. 4. Routing chain

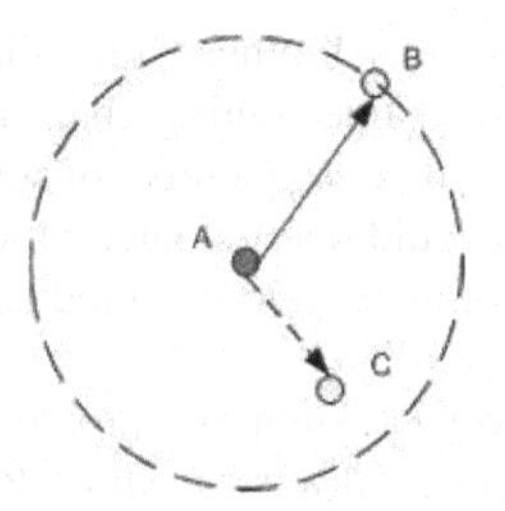

Fig. 5. Broadcasting in circle area

status will be chosen. The chosen neighbor is so called the Lightest Neighbor. By dynamically changing next hop and only choosing the lightest one, the energy dissipation tends to be equal among all nodes in the same level.

3.3 Threshold and Data Aggregation Techniques

Moreover, our approach also adopts the threshold and data aggregation techniques in order to reduce the number of transmissions. Our threshold technique is borrowed partly from TEEN protocol. Depending on applications, each sensor will have a hard threshold, which is the domain of the sensed attribute and a soft threshold, which is a small change in the value of the sensed attribute. The newest values of all attributes are cached in an internal array. The nodes will transmit data only when the following conditions are true: the current value of the sensed attribute is within the interested domain defined by the hard threshold; the current value of the sensed attribute differs from the newest value in the internal array by an amount equal to or greater than the soft threshold. After that, they update the newest value of this attribute by the current value. Unlike TEEN, we might not try to send periodically the thresholds to every node. We do the threshold function not only at sending nodes but also at receiving nodes. Moreover, broadcasting the thresholds is done by the BSs not by any nodes. Therefore, we can avoid broadcasting threshold announcement packets and move the overhead maintenance problems from sensor nodes to BSs. We also do not use any meta-data for negotiations, thus it might reduce the negotiating traffic.

3.4 Two-Base-Stations Setup Mechanism

At first, base station BS1 broadcasts a discovery message with a discovery level number (DL number) set to one while every node initially sets its Top-Down Level (TDL) to infinity (∞). Each node, upon receiving the discovery message originated at BS1, checks to see if it matches one of the following cases:

- The DL number is less than its TDL then the received node sets its TDL equal to DL number and resends the discovery message after increasing the DL number by one.
- The DL number is equal to the received node's TDL then it updates its routing table.
- The DL number is greater than the received node's TDL then it discards the message.

All discovery messages will be sent hop by hop until they reach the base station BS2. The Bottom-Up Level of each node is calculated by equation (1).

3.5 Regulation of Direction of Transmission

Two-base-stations model makes every nodes dissipate energy in a more balanced way. However, the regulation of direction of transmission between two base stations needs to be considered. In our mechanism, there are two triggers which cause a change in direction of transmission. The first one occurs at the active base station. After a predefined period T, such base station will send a message to the first level nodes to notify them a new direction. Then such nodes will update their current direction of transmission. The second trigger occurs at any member node if its rate of energy consumption larger than a predefined threshold. Then this node will update its current transmitting direction. The rate of energy consumption of one node is given by:

$$R = \frac{N_t \text{ x } E_t + N_s \text{ x } E_s}{\Delta t}. \tag{2}$$

N_t is the number of transmission and N_s is the number of sensing in a period Δt. E_t and E_s are the energy consumption unit for transmitting and sensing.

This mechanism regulates the transmitting directions dynamically depending on energy status of each node. Therefore, the network will spend its energy more wisely. Moreover, instead of broadcasting the notifying message, each node will update its transmitting direction when it receives a packet with a different direction. Hence, the control messages are reduced and the management cost is less.

4 Simulations and Comparisons with Other Protocols

We use Glonemo [9] as a simulation tool for our research because it has a model for energy consumption. In addition, we suggest some energy metrics to evaluate both the energy efficiency of overall network and energy consumption of each sensor.

1) Total Energy (e_{total}): total amount of energy consumed by the network for each transmitted packet.

$$e_{total} = e_{tx} + e_{rx} + (n-1)e_{re}$$

$e_{tx} + e_{rx}$: the amount of energy required by a node to transmit and receive a packet.

e_{re}: the amount of energy used to read only the header of the packet.

n: number of neighbors.

2) Energy Efficiency (e_{eff}): number of packets delivered to the sink for each unit of energy spent by the network.

3) Private Energy (e_{pr}): the probable amount of energy unit consumed by one sensor after the sources send p_{src} packets.

We perform simulations to observe the energy consumption rate at each local area and the residual energy of the overall network by the time. We simulate random static networks of 100 nodes having the same radio characteristics. At first approach, we just use the normal data aggregation technique with one base station and without Lightest Neighbor technique. In another way, we apply Two-Base-Station algorithm and Lightest Neighbor technique in a different approach. Our results show that such residual energy of the network in both approaches is almost the same during the observed time. However, in normal data aggregation technique, the energy consumption rate among each local area is not equivalent. The rate of the area which is near the sink is much higher than others. In FECP approach, the average rate of energy consumption of each area is similar. Therefore, the network which is applied Two-Base-Station algorithm consumes energy more balanced than that related to the former case.

We compare our proposed protocol FECP with some popular protocols such as the SPIN protocol, Directed Diffusion Protocol, LEACH protocol, TEEN and APTEEN protocols. This comparison is mainly based on the one performed in the surveys [7] and [8]. Table 1 lists the comparisons results.

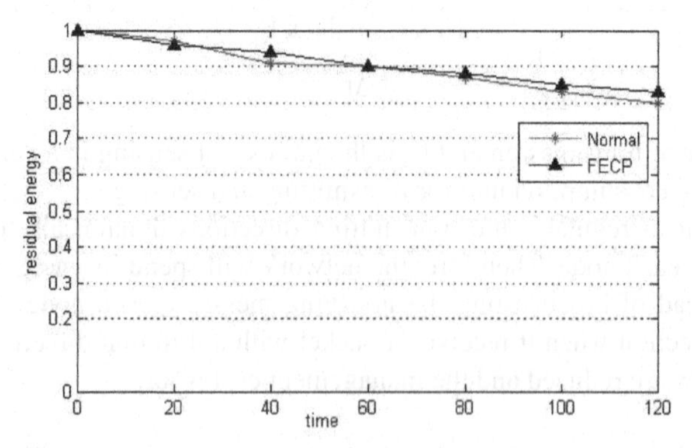

Fig. 6. Residual energy of overall network in both approaches

Table 1. Comparisons Results

	SPIN	Directed Diffusion	LEACH	TEEN & APTEEN	FECP
Complexity	Low	Low	CHs	CHs	Low
Balance of Energy Dissipation	Very low	Very low	Low	Low	Good
Number of BS(s)	All nodes	1	1	1	2
Radio Signal	No	No	No	No	Yes
Multi-path	Yes	Yes	No	No	Yes
Scalability	Limited	Limited	Good	Good	Average
Negotiation	Yes	Yes	No	No	No
Data Aggregation	Yes	Yes	Yes	Yes	Yes
Query Based	Yes	Yes	No	No	No
Threshold	No	No	No	Yes	Yes

The home environment is significantly appropriate for deploying our protocol. In such environment, sensor nodes are placed densely in a small area, and the number of deployed nodes is moderate. Each sensor node might have many candidates to choose its next hop. The distances from this node to its candidates are close, thus the energy consumption to transmit data to each of them is almost the same. Therefore, FECP might use these advantages to uniform the energy dissipation among all sensor nodes. We can obviously see that in our approach the higher density of sensor nodes, the better the balance of energy consumption. This feature, however, is a disadvantage for hierarchical clustering approaches. They will need to pay a higher cost for some heavy overheads such as dynamic clustering, cluster colliding, and cluster heads (CHs) maintaining. Besides, high density of sensor nodes is also appropriate for multi-path based routing. It can make a higher balance in power consumption by changing the route dynamically. Unlike LEACH, TEEN and APTEEN using single-path based routing, we propose a multi-path approach based on the Lightest Neighbor technique in order to uniform the power consumption in horizontal direction. Generally, by using our methodology the energy of an overall network would be consumed more efficiently than by using any other approaches, thus increasing the network lifetime.

In terms of comparing with data-centric protocols, we also have some adaptations that are more suitable for the home environment. The applications of such environment require data delivery to BSs continuously. Such data requirements will not change normally. Therefore, using metadata negotiation and query-based techniques like SPIN and Directed Diffusion should not be appreciated. It might be unnecessary to use extra negotiation packets and to query data from specific sensor nodes. Instead of that, FECP protocol would be intended to utilize threshold and data aggregation techniques in order to reduce the transmissions of redundant or uninterested data.

5 Conclusion

In this paper, we have described FECP, a two-base-stations protocol, which is near optimal for an energy dissipation balancing problem in WSNs, especially smart home environment networks. FECP outperforms the other prominent protocols such as SPIN, Directed Diffusion, LEACH, TEEN and APTEEN in terms of equivalence of energy consumption among all nodes. It can eliminate the overhead of dynamic cluster formation in hierarchical clustering protocols like LEACH, TEEN and APTEEN. It can also limit the number of transmissions among all nodes and it is much more suitable for home environment than data-centric protocols like SPIN and Directed Diffusion.

We also developed a regulation mechanism to synchronize the direction of transmission between two base stations and to enhance the energy awareness ability of the sensor networks. Our simulations show that FECP can keep the energy load being evenly distributed among all sensor nodes. Thus, the energy at a single sensor node or a small set of sensor nodes will not be drained out very soon. We hope that our novel protocol will bring out a new wireless sensor network model for the home environment.

References

[1] Heinzelman, W., Kulik, J., Balakrishnan, H.: Adaptive Protocols for Information Dissemination in Wireless SensorNetworks. In: Proc. 5th ACM/IEEE Mobicom Conference (MobiCom 1999), Seattle, WA, pp. 174–185 (August 1999)

[2] Kulik, J., Heinzelman, W.R., Balakrishnan, H.: Negotiation-based protocols for disseminating information in wireless sensor networks. Wireless Networks 8, 169–185 (2002)

[3] Intanagonwiwat, C., Govindan, R., Estrin, D.: Directed diffusion: a scalable and robust communication paradigm for sensor networks. In: Proceedings of ACM MobiCom 2000, Boston, MA, pp. 56–67 (2000)

[4] Heinzelman, W., Chandrakasan, A., Balakrishnan, H.: Energy-Efficient Communication Protocol for Wireless Microsensor Networks. In: HICSS 2000 (January 2000)

[5] Manjeshwar, A., Agarwal, D.P.: TEEN: a routing protocol for enhanced efficiency in wireless sensor networks. In: 1st International Workshop on Parallel and Distributed Computing Issues in Wireless Networks and Mobile Computing (April 2001)

[6] Manjeshwar, A., Agarwal, D.P.: APTEEN: A hybrid protocol for efficient routing and comprehensive information retrieval in wireless sensor networks. In: IPDPS 2002, pp. 195–202 (2002)

[7] Al-Karaki, J.N., Kamal, A.E.: Routing techniques in wireless sensor networks: a survey. IEEE Wireless Comm. 11, 6–28 (2004)

[8] Akkaya, K., Younis, M.: A survey on routing protocols for wireless sensor networks. Ad Hoc Networks, I (November 2003)

[9] Samper, L., Maraninchi, F., Mounier, L., Mandel, L.: GLONEMO, Global and Accurate Formal Models for the Analysis of Ad- Hoc Sensor Networks. In: InterSense: First International Conference on Integrated Internet Ad hoc and Sensor Networks, Nice (May 2006)

[10] Powell, O., Leone, P., Rolim, J.D.P.: Energy optimal data propagation in wireless sensor networks. J. Parallel Distrib. Comput. 67(3), 302–317 (2007)

Real-Time Multicast Network Monitoring

Joohee Kim, Bongki Kim, and Jaehyoung Yoo

Department of KT Network Technology Laboratory
463-1 Jeonmin-dong Yuseong-gu, Daejeon, 305-811, Korea
{jooheekim,bkkim,styoo}@kt.co.kr

Abstract. Through IP multicast has been deployed for more than a decade, several tools and systems for managing multicast network and service are developed. Most of them have their special propose to perform their tasks. That's not so comprehensive and easy-to-use customer-centric system. Especially today with the growth of multimedia service needs based on IP multicast such as IP-TV, it's more important to monitor and manage the network at real-time. In this paper, we present our multicast network management system 'Multicast NMS' we have built, aimed to monitor and manage our network for IP-TV service application delivery. we describe the design and implementing and experiences.

Keywords: IP Multicast, IP-TV, Network Management.

1 Introduction

IP multicast provides scalable and efficient mechanisms to support for multi-receiver network applications in the Internet. It is not only to save the access network resource, but also to reduce the load of the service servers.

IP-TV service, many ISPs are interested in today, consists of various services, and the live broadcasts and the applications are delivered over IP multicast. IP multicast is an appropriate method to transfer the one-to-many services though the network. So, management of multicast network has become more significant and the need of comprehensive management system is growing. The multicast management systems are required to detect and diagnose many multicast session problems such as reachability and quality at real-time. For deployment of the system on the enterprise network, the architecture is required to be scalable and flexible.

In this paper we present our system Multicast NMS, developed for monitoring IP-TV service application delivery over the KT network. we describe the architecture, implementing and experiences after deploying it. The reminder of this paper is organized as follows. Section 2 presents a brief overview of existing multicast management tools and the differences among them. Section 3 introduces Multicast NMS about key functions, the architecture and the implementation. Section 4 presents conclusions.

S. Ata and C.S. Hong (Eds.): APNOMS 2007, LNCS 4773, pp. 185–194, 2007.

2 Related Work

With the deployment of multicast services in global networks, many research proposals, experiments and tools were developed for monitoring multicast networks and operations. Here we briefly introduce existing multicast network management tools.

Mtrace [1] is one of earlier useful tools that shows the routers and loss statistics in the reverse multicast path from a given receiver to the source of a multicast group. but it does not show the entire forwarding tree and may cause congestion in the network. It is hard to provide real-time monitoring functionality.

MRM [3] is a protocol used to create active and passive multicast monitoring and measurement scenarios. It can measure the multicast loss.

SMRM [2] is a follow-up effort that incorporates MRM functionality into an SNMP-based framework monitoring packet loss,delay and jitter of multicast paths in a network.

RTPmon [4] was designed to monitor quality of service characteristics as observed at the receiver sites in a multicast application. but due to its dependence on RTCP, rtpmon can only be used for monitoring applications that use RTP as their transport mechanism.

Mhealth [5] uses mtrace and RTCP to display a real-time, graphical representation of a particular group's multicast tree including packet loss charactersistics of each link in the tree. but it also depends on RTP to perform its task and is not particularly scalable.

Sdr-monitor [6] observes the global reachability of sdr messages to group members by collecting feedback from multicast receivers. but it is solely based on sdr reports to collect the multicast information. Sdr reports provide a limited statistics and it suffers a potential scalability problem.

RMPMon [7] is SNMP-based system designed and developed to monitor end-to-end performance of multicast. It uses RTP MIB and RTP Sender MIB to perform the required monitoring.

Several systems developed to provide more comprehensive monitoring and management support for operational network environment are mentioned as follow.

MRMon [8] defines and uses several MIB modules to passively collect and monitor multicast service. The MIBs are Multicast Statistics Group MIB, Multicast Histoy Group MIB etc.

Mmon [9], developed by HP lab, was one of multicast monitoring and management systems from an operational network management point-of-view, and it utilizes standardized multicast routing MIBs to gather information. Mmon is very similar to Multicast NMS. The primary difference between them is the functionality of topology visualization. Mmon can provide each group forwarding path over the topology and Multicast NMS can provide all groups forwarding over the topology and also provide the forwarding path for a group too. The detail is described in the section 3.

3 Multicast NMS

In this section, we present Multicast NMS, developed for monitoring IP-TV service application delivery. It utilizes standardized multicast routing MIBs to gather information, and provide various functions for operators to be easy to control and manage the network. In the following, we describe the key functions of the system, present the architecture, and discuss our challenges with implementing, experiences.

3.1 Key Functions of Multicast NMS

Multicast NMS is designed to provide various functions of efficiently managing multicast network for the operators. we present the key functions of Multicast NMS as follow.

Discovery and visualization of the multicast topology: It is able to discover the multicast routers(and switches) and the multicast peering relationships between them. The display of multicast topology is important for operators to recognize how the multicast traffic flows though the path and which one is participant in the multicast delivery and provide whether there's any problem on an equipment or link of them.

Multicast protocol status monitoring: It monitors the status of multicast protocols and the availability of various elements in the multicast infrastructure.

Multicast traffic monitoring: It monitors the traffic and users can configure the thresholds so that when the traffic isn't normal it can warn the user.

Fault detection and alarm generation: Like any management system, an operational multicast system should generate alerts once it detect faults such as multicast traffic exceeding specific thresholds, failure of multicast peering relationship and router going down etc.

Visualization of forwarding tree: This helps the operator to monitor all of multicast channel flow, and also trace specific multicast data flows in the network. It is useful in both detecting and diagnosing non-reception of data by some group members.

Multicast groups forwarding history management: It saved the changes of groups(channels) flow on the interface of a router. It provides the history of the channels transiting on each interface so that users can get more information to deduce the reason about the changes.

Multicast traffic and fault history management: This is useful for post-analysis and capacity planning.

3.2 Software Architecture of Multicast NMS

Multicast NMS adopted 3-tier architecture in order to get performance and extensibility essential for management of huge networks. It consists of network

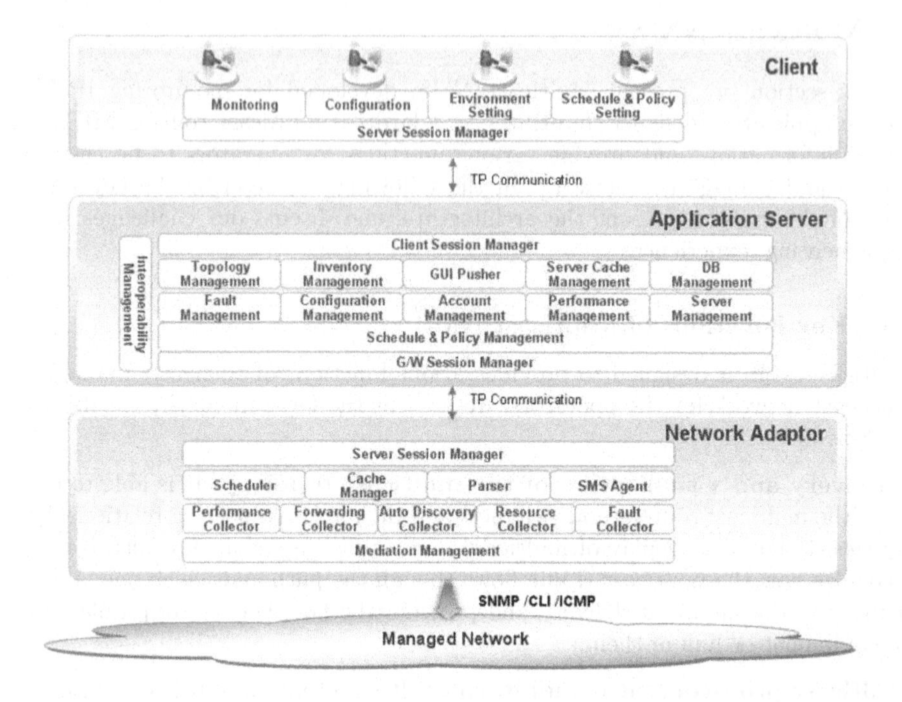

Fig. 1. Architecture of Multicast NMS

adaptor, application server and GUI(Graphic User Interface) client shown as fig 1.

Network adaptor use SNMP API to communicate with SNMP agents and also use CLI(Customer Line Interface) or ICMP(Internet Control Message Protocol) to get more information from the elements. Data collector modules such as forwarding collector, fault collector are controlled by the scheduler and they are polling data from the elements according to given schedules from the scheduler. Cache manager performs data caching, the parser is parsing the information when the data gathering via CLI.

Application server includes several data handing modules, such as topology management, fault management, account management,schedule management, policy management and so on. They deal with data gathered by network adaptors and put it into DB server stored. The schedule and policy management are managing the polling schedules and policies for each element. And operators can control the polling policy, configuration and other environment setting on GUI. The client communicates with application server via the session manager. Also for the communication of application server and network adaptor they use their own session managers. Among the 3-tier, they communicate with each other by TP(Transaction Processing) communication.

The architecture enables Multicast NMS to accept more additional devices easily.

3.3 Implementation

As architecture of our system mentioned, we have kept the 3 parts - the user presentation, data handling and data collection sperate. Users can control the data collecting policy over GUI. Scheduler module in network adaptor can get the policy from schedule management module in application server to know how to access the equipments and collect the data from them.

Multicast NMS frequently collects the data about multicast topology, multicast forwarding tree, multicast traffic, multicast protocol status from Multicast related MIBs.

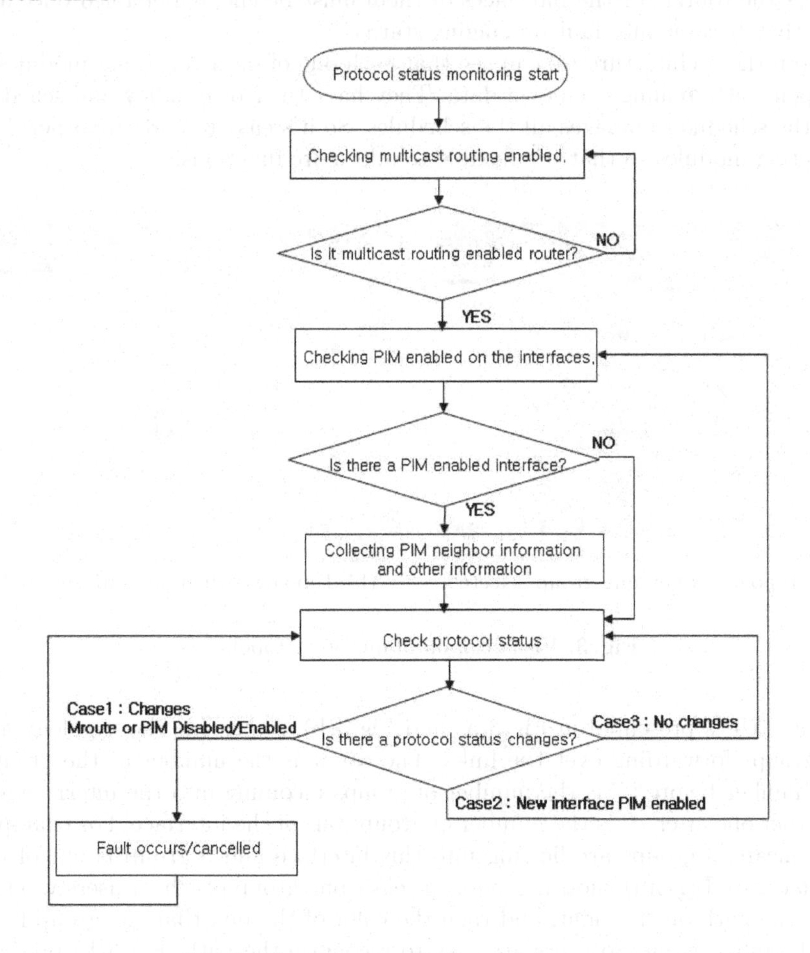

Fig. 2. Flow of protocol status management

Here we describe the process of managing equipment(Fig 2). At first, check the router(switches) whether it is multicast routing enabled. If yes, check the interfaces which are PIM(Protocol Independent Multicast) enabled. but if no, Multicast NMS has a frequency to check it again. When it is found to be an enabled one, it becomes managing status. And the PIM enabled interfaces on the equipment is found, then find the PIM neighbor interface to complete the connection between the equipments. When equipment is multicast routing enabled and there are some interfaces PIM enabled, then Multicast NMS frequently collects data not only about protocol status, but also about forwarding tree, multicast traffic and others from the equipment. During the managing status, if the protocol status is changes, it means a fault occurs. If an interface of a multicast enabled router is not checked it was a PIM enabled one, although the interface is checked disable while frequently collecting, it won't be a fault. That also indicates the routers or the interfaces of them must be checked enabled once and then they become into fault managing status.

From the architecture we can see that each one of data collecting modules is independently polling each own data. They have their own policy and schedule and the scheduler manages all the schedules. So it's easy to add any other data collecting modules so that it's flexible to add more functions.

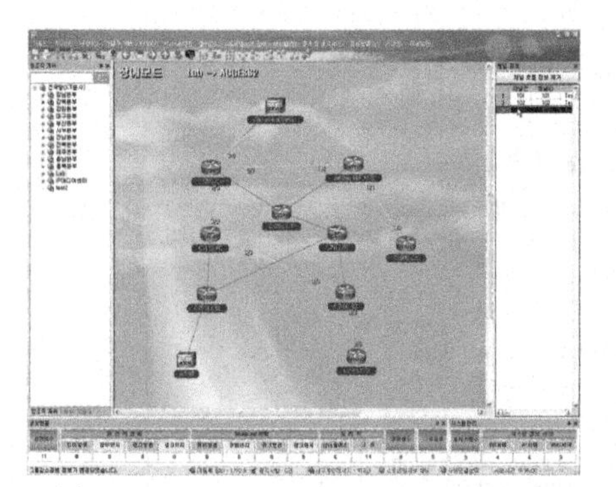

(a) Topology when one group selected

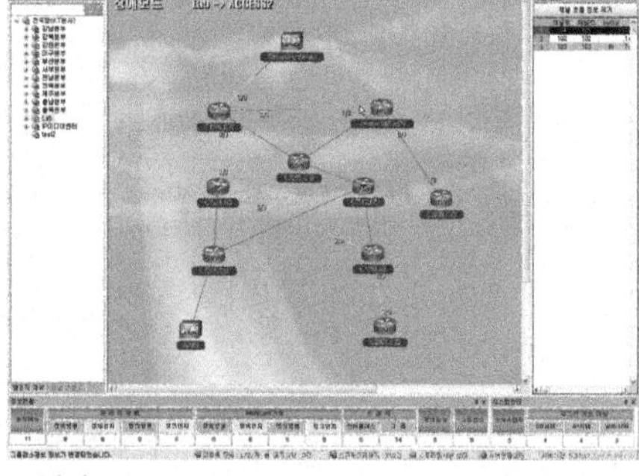

(b) Topology when one link selected

Fig. 3. Visualization of multicast topology

The GUI is presented in Fig 3(a) and Fig 3(b). It provide the topology and the groups forwarding over the links. The count is the number of the groups. The number before '/' is the number of groups incoming into the interface and the other one after '/' is the number of groups out of the interface. For example, '3/0' means 3 groups are flowing into this interface and 0 group is out of the interface. As Fig 3(a) shown, users can select one group of which users want to know the path on the right, and then the color of the link that the group flows over becomes green. So users are easy to recognize the path. Fig 3(b) presents the groups flow over the link user selected. The group list becomes orange color as you see.

Multicast NMS also provide forwarding tree trace to get the end-to-end path for each group. Users choose one equipment and one group first, and select the trace direction - 'top-down' or 'bottom-up', then Multicast NMS will present the forwarding path for the group. 'Top-down' means the forwarding path from

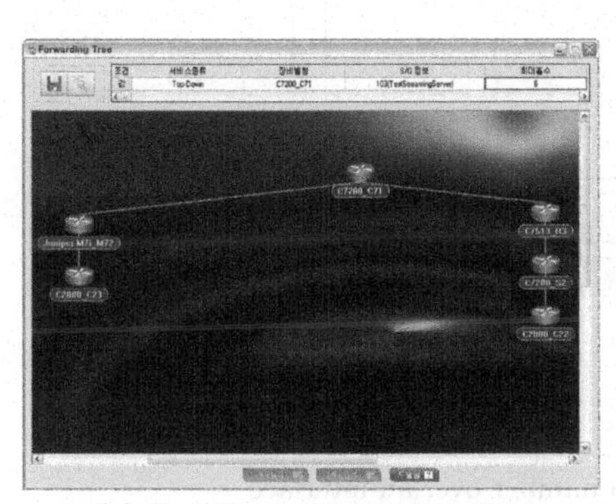

(a) Top-Down Forwarding Tree

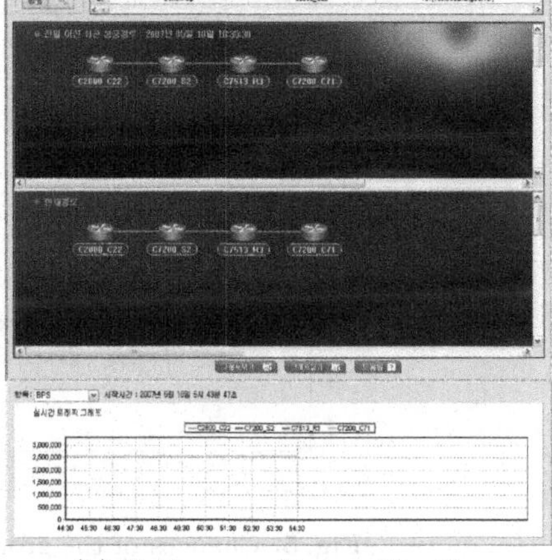

(b) Bottom-up Forwarding Tree

Fig. 4. Display of Forwarding Tree

the equipment to next hop equipments over that the traffic flows(Fig 4(a)) and 'bottom-up'means that from the selected equipment to the source(Fig 4(b)). In bottom-up forwarding tree, there're some more information provided. The first one is that users can compare the route now and the successful route one day before, so that users get more information about the path for the group to analyze the route changes. The second one is users can get the traffic information from each equipment on the path at real-time.

Multicast NMS also provide information about group changes on each interface and then users are easy to know which group is added or deleted. For example, the graph on the Fig 5 shows the count of groups flowing via the interface decrease(from 3 to 2). When click the graph, users can get more detail information about the groups. Here, it shows the result while clicking the time that's before the groups decrease, so the information about 3 groups are presented below the graph.

Multicast traffic monitoring is one of the important points. The multicast traffic statistics are also collected from the IP multicast routing MIB in the routers. This MIB has counters for measuring the multicast traffic on each interface as well as for each source and group pair transiting the router. The frequency of statistics collection and the threshold are user-configurable.This Fig 6 presents the group traffic including octets,pkts,bps and pps. The grid below the 4 graphs(octets,pkts,bps,pps) is present the real data collected and the red mark indicates "critical warning".

3.4 Experiences

Multicast NMS is deployed at KT network to monitor IP-TV service application delivery. More or less 450 equipments are managed now. With the extension of

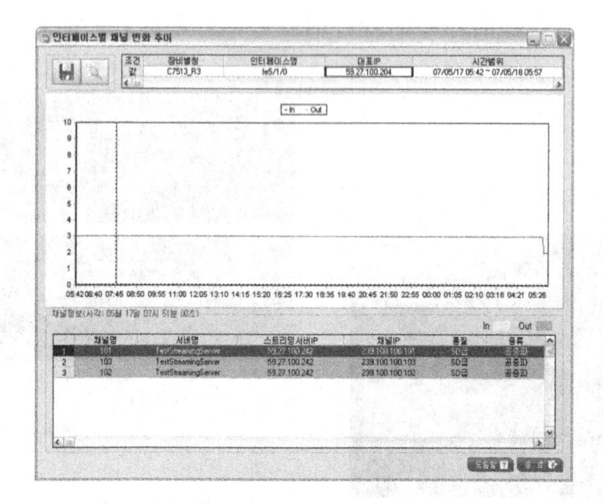

Fig. 5. Changes of the group number over an interface

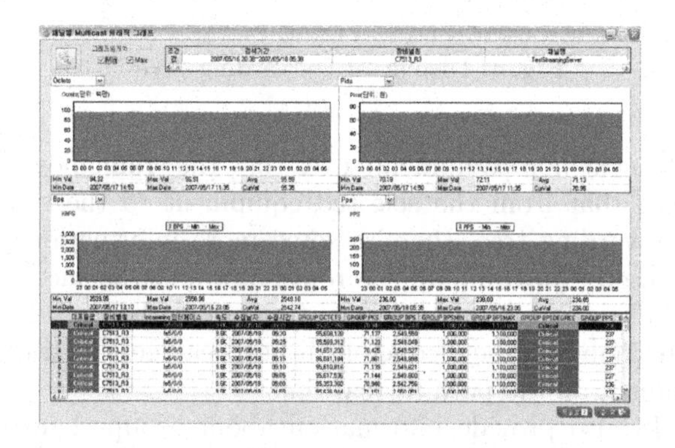

Fig. 6. Display of Multicast Traffic

the service coverage, the number of equipments to be managed will increase, especially access network. That is the remain of work for Multicast NMS.

In this section we discuss some issues we faced while implementing and testing Multicast NMS.

The first issue is about Multicast related MIBs we need. Although the IETF standard MIBs for multicast management are proposed, not all the vendors support the IETF standard MIBs.

KT network consists of QoS enabled backbone network, best-effort backbone network and access network and IP-TV service is provided over the QoS enabled network and access network. There're many routers and switches and they are produced by various vendors, especially those in access network.

Most in QoS enabled backbone network are the products of Cisco and Juniper and they supports the standard MIBs to be easy to manage. In order to monitor end-to-end network, managing the equipments in access network are important too. But in fact not all of the access network equipments can support the MIBs. So, vendors need to develop the standard MIBs for multicast management on switches and IGMP snooping MIB for L2 switches. Although IGMP snooping MIB is a private MIB, it gives useful data to analyze for customer service status.

Like this, if they don't support the MIBs, we must require to develop, or if the OS has a bug we must require to upgrade. For example, we found a bug from C6509 switches(OS version 12.2(18)SXF2) that can't provide exact multicast packet counts, and we found a bug from M320(OS version JUNOS7.2R2.10) that provides inaccurate active multicast groups. There's another case we found - the octets value we collected frequently is not the same to real data value. As we knows, switches produced by Cisco (sup720) support ipMroute-MIB instead of ipMroute-STD-MIB (GSR support ipMoute-STD-MIB) and ipMroute-MIB is an experimental MIB, just providing counter 32. So the it couldn't deal with so much data so that it gave the incorrect value. We changed to use Cisco-ipMroute-MIB that provides counter 64 to collect octets value.

The second one is to determine collecting period. We must consider the overhead of the equipments and also consider to realize the "real-time" monitoring. One thing we want to remind is the update period time of packets and octets in Cisco switches. It generally updates the value in every 90 seconds. That means we must collect the value every more than 90 seconds. If not that, the data may not be true.

Though implementing and testing we consider many things, so we have developed and deployed the system sufficiently managing KT network.

4 Conclusions

The implementation of Multicast NMS for KT has two objectives. The first one is to assist the managing and operating people for monitoring the service delivery over the whole network. The other is to provide multicast traffic information for designing and planning networks. Consequently, it makes the IP-TV service application delivery smooth and thus enhances customers' satisfaction.

It makes benefits in five aspect of ISP.

- Enables fast time-to-market by allowing auto-discovery of multicast information
- Provides useful statistical information for network planning or engineering
- Effective network management even during a network expansion
- Minimizes down time by monitoring before service alarm

We also plan to manage more equipments this year, so the following work is improving the performance and providing more information about IP-TV service management and customer management.

References

1. Fenner, W.: A traceroute facility for IP multicast. Internet Engineering Task Force (IETF), draft-fenner-traceroute-ipm-*.txt, Work in progress (December 2003)
2. Al-Shaer, E., Tang, Y.: SMRM: SNMP-based multicast reachability monitoring. In: IEEE/IFIP Network Operations and Management Symposium (NOMS), Florence, ITALY, pp. 467–482 (April 2002)

3. Almeroth, K., Wei, L., Farinacci, D.: "Multicast Reachability Monitor (MRM)", UInternet Engineering Task Force (IETF) Internet Draft (July 2000)
4. Swan, A., Bacher, D., Rowe, L.: rtpmon 1.0a7 (January 1997) University of California at Berkeley Available from `ftp://mmftp.cs.berkeley.edu/pub/rtpmon/`
5. Makofske, D., Almeroth, K.: Real-Time Multicast Tree Visualization and Monitoring. Software-Practice & Experience 30(9), 1047–1065 (2000)
6. Sarac, K., Almeroth, K.: Monitoring Reachability in the Global Multicast Infrastructure. International Conference on Network Protocol (ICNP)Osaka, JAPAN (November 2000)
7. Chesterfield, J., Fenner, B., Breslau, L.: Remote Multicast Monitoring Using the RTP MIB. In: IFIP/IEEE International Conference on Management of Multimedia Networks and Services. Santa Barbara, California (October 2002)
8. Al-Shaer, E., Tang, Y.: MRMON: Multicast remote monitoring. In: IEEE/IFIP Network Opeations and Management Symposium(NOMS), pp. 585–598 (April 2004)
9. Sharma, P., Perry, E., Malpani, R.: IP multicast operational network management: design, challenges, and experiences. IEEE Network 17(2), 49–55 (2003)

Monitoring SIP Service Availability in IPv4/IPv6 Hybrid Networks

Yung-Chang Wong[1] and Rhoda Chen[2]

[1] Department of Computer Science and Information Engineering,
[2] Computer and Communication Center,
Providence University, Taichung, 43301 Taiwan
{ycwong,rhoda}@ pu.edu.tw

Abstract. With the increasing deployment of IPv6, IPv6-based SIP applications draw more attention from service providers. To date, most work on SIP operations and management (OAM) is undertaken on IPv4 networks. In this paper we present a framework called MoSA for monitoring SIP service availability in heterogeneous IPv4/IPv6 networks. We also investigate the fault tolerance issue of our approach analytically. A prototype has been implemented in Providence University for validation.

Keywords: Voice over IP (VoIP), Session Initiation Protocol (SIP), IPv6, operations and management (OAM).

1 Introduction

Session Initiation Protocol (SIP) [1] is an application-layer signaling protocol for creating, modifying, and terminating multimedia sessions that works independently of underlying transport protocols. SIP defines a number of logical entities, namely user agents (UAs), proxy servers, and registrars. An UA is a logical entity that creates and sends a SIP request, and generates a response to a SIP request. A proxy server is used to route requests to the user's current location, authenticate and authorize users for services, and provide features to users. A registrar provides function that allows users to upload their current locations for use by proxy servers.

Most operational experience with SIP to date has been over the IPv4 network; however, deploying SIP over the IPv6 network [2] emerges. SIP benefits from IPv6 by having large address space and new features not found in IPv4, including security, plug-and-play, and Quality of Service (QoS). The transition to IPv6 will proceed gradually [3],[4]. During the transition period, SIP-based services coexist on both IPv4 and IPv6 networks.

Fig. 1 illustrates the flow of SIP call setup. In the Fig., Alice and Bob are assumed to be IPv6-based SIP UAs (SIPv6 UAs). Initially, Bob registers its location information with the SIPv6 proxy server by using a REGISTER message, so that new calls can be correctly redirected (step 1). The SIPv6 proxy server replies a 200 OK message (step 2). Afterward, Alice sends an INVITE message to Bob via the SIPv6 proxy server to setup a real-time transport protocol (RTP) [5] voice session (steps 3 and 4). The INVITE message contains the Session Description Protocol (SDP) [6]

S. Ata and C.S. Hong (Eds.): APNOMS 2007, LNCS 4773, pp. 195–204, 2007.

fields that include IPv6 address and port number information for RTP packets. When Bob receives the INVITE message, he sends a 200 OK message back to Alice via the SIP proxy server (steps 5 and 6). Upon receipt of the 200 OK, Alice responds with an ACK message (step 7). Then a RTP session is established between Alice and Bob.

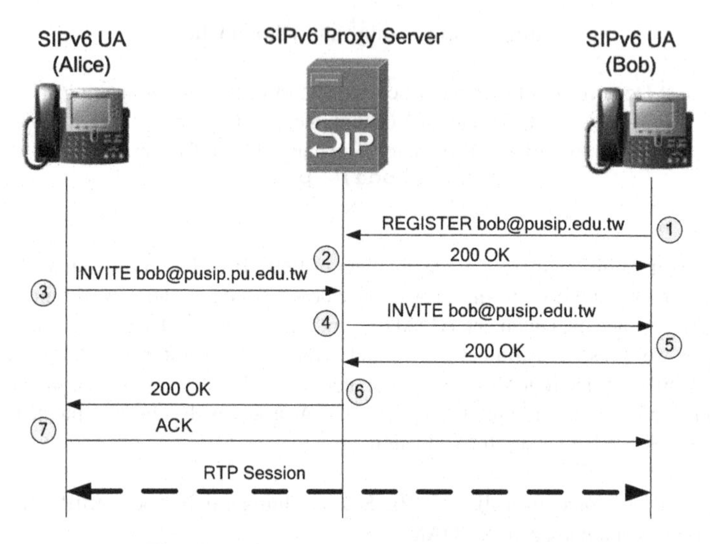

Fig. 1. SIP call setup between Alice and Bob

Most network management tools nowadays perform checks on SIP service using *port scan*. However, the scanned port is *open* or *accepted* does not necessarily imply that the SIP service listening on the port is still alive. Furthermore, these tools have limited support for the IPv6 protocol.

This paper addresses the regular surveillance over SIP service in a hybrid IPv4/IPv6 network. The contribution of this paper has been twofold. First, we have developed a tool instead of using port scan to test whether a particular SIP service is accessible across an IP network. Second, we have proposed a framework, known as Monitoring SIP-service Availability (MoSA), for periodically monitoring the specified services and warning about an abnormal situation. In MoSA, checks on SIP service availability can be achieved through two configurations: Configuration 1 (C1) and Configuration 2 (C2). C1 consists of a service monitoring server (SMS) and a service monitoring agent (SMA), and is suitable for the *IPv4 dominant* scenario where most services under SMS's monitoring are IPv4-based. On the other hand, C2 provides service monitoring for the *IPv6 dominant* scenario where most services under SMS's monitoring are IPv6-based.

This paper is structured as follows: Sections 2 and 3 describe, respectively, the architecture and the message flows of service monitoring for MoSA-C1 and MoSA-C2 configurations. Section 4 analyzes the reliability of SMS through an analytical model. In Section 5, we present the implementation of MoSA. Finally, we conclude this paper in Section 6.

2 MoSA-C1 for IPv4 Dominant Scenario

This section describes the MoSA-C1. We first present the designed network architecture and software architecture. Then, we elaborate on the message flows for monitoring SIP services.

2.1 Network Architecture

Fig. 2 displays the network architecture of MoSA-C1. The architecture consists of an IPv4 network and an IPv6 network. The IPv4 network connects to the IPv6 one through a dual-stack (DS) border router. The SIPv4 (SIPv6) proxy server is responsible for handling SIP sessions on IPv4 (IPv6) network area. Furthermore, the SIPv4 proxy server, the Web server, the Mail server, and the intermediate routers that reside in IPv4 network are *IPv4-only*. On the other hands, the SIPv6 proxy server and the intermediate routers that reside in the IPv6 network are *IPv6-enabled*; that is, they may not support IPv4 protocol.

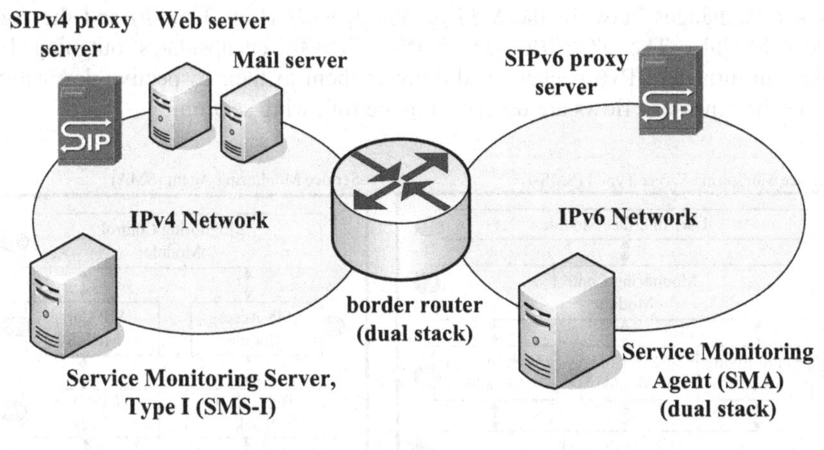

Fig. 2. MoSA-C1 network architecture

The core element of the network is the type 1 SMS (SMS-I). The SMS-I, an IPv4-only host, periodically checks on the specified services (in the example, the SIPv4 service, the Web service, and the email service) over IPv4 transport. Meanwhile, the SMS-I runs intermittent checks on the specified SIPv6 service through the corresponding SMA. The SMA is a dual stack agent that monitors the SIPv6 service over IPv6 transport while returns status information to the SMS-I over IPv4 transport. When problems are encountered, the SMS-I can send notifications out to administrators in a variety of different ways (such as email and instant message). In this scenario, we locate SMS-I in IPv4 network since most services under its monitor reside in IPv4 network.

2.2 Software Modules

Fig. 3 illustrates the respective software architectures for SMS-I and SMA. The SMS-I consists of six modules as shown in Fig. 3(a). The *User Interface Module* (Fig. 3(a)①) is responsible to display the status information, historical logs, and reports. The *Monitoring Control Module* (Fig. 3(a)②) is responsible for monitoring the specified services and collecting status information. For SIPv4 services, the Monitoring Control Module invokes the *SIP Core Module* (Fig. 3(a)④) to check on their availability. For SIPv6 services, the Monitoring Control Module asks the SMA, through message exchanges between the *Agent Access Module* (Fig. 3(a)③) and the *SMS Access Module* (Fig. 3(b) ③), to monitor them. When problems occur and get resolved, the Monitoring Control Module exercises the *Alarm-giving Module* (Fig. 3(a)⑤) to deliver notifications to administrative contacts. Finally, the *TCP/IPv4 Stack* (Fig. 3(a)⑥) uses the IPv4 socket to provide all packet-processing functions.

Fig. 3(b) shows the relationship among modules of SMA. The SMA relies on the *Monitoring Control Module* (Fig. 3(b) ①) to monitor the specified SIPv6 service. The Monitoring Control Module instructs the *SIP Core Module* (Fig. 3(b) ②) to check on the SIPv6 service, and sends the status information back to the SMS-I through message exchanges between the *SMS Access Module* (Fig. 3(b) ③) and the Agent Access Module. The *TCP/IPv6 Stack* (Fig. 3(b)④) encapsulates outgoing IPv4 packets in turn into IPv6 packets, and delivers them to their respective destinations. The detailed message flows are described in the following section.

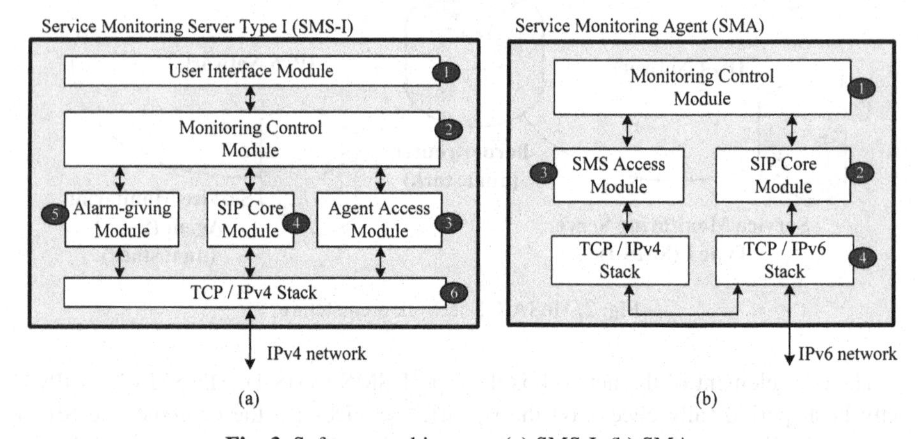

Fig. 3. Software architecture. (a) SMS-I. (b) SMA.

2.3 Message Flows

Fig. 4 shows the example message flow used by the SMS-I to monitor a SIPv4 proxy server, which includes the following steps.

Step 1. Base on the SIPv4 proxy server identity specified by the administrator, the Monitoring Control Module requests the SIP Core Module to initiate a check on the specified SIPv4 proxy server.

Step 2. The SIP Core Module performs Steps 2.1-2.5 (called the *SIP-availability-check* procedure) to obtain the status of the specified server.

Steps 2.1-2.2. At first, the *User Agent client* (UAC) initiates an INVITE message that is sent to the SIPv4 proxy server. The SIPv4 proxy server forwards the INVITE message to the *User Agent server* (UAS). The UAS responds to the INVITE message with a 200 OK.

Step 2.3. Upon receipt of the 200 OK, the UAC sends an ACK message to the UAS, which indicates that the tested proxy server operates properly.

Steps 2.4-2.5. Through the BYE and 200 OK message-pair exchange between the UAC and the UAS, the UAC tears the established RTP session.

Step 3. The SIP Core Module replies to the Monitoring Control Module with the status information.

Step 4. In the case of abnormality presence, the Monitoring Control Module invokes the Alarm-giving Module to send notifications out to administrators.

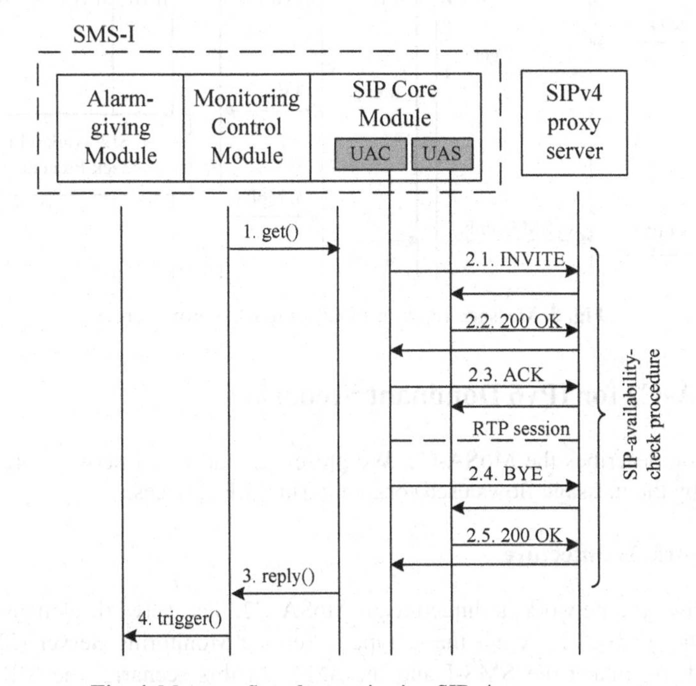

Fig. 4. Message flow for monitoring SIPv4 proxy server

Similarly, example message flow used by the SMS-I to monitor a SIPv6 proxy server is illustrated in Fig. 5 with the following steps.

Steps 1-3. Based on the SIPv6 proxy server identity specified by the administrator, the Monitoring Control Module on SMS-I sends a SMS-get request to the SIP Core Module on SMA to initiate the *SIP-availability-check* procedure.

The SMS-get request is firstly delivered to the border router. Then the border router forwards the request to the SIP Core Module through an IPv4-in-IPv6 tunnel.

Step 4. The SIP Core Module executes the *SIP-availability-check* procedure over IPv6 transport to obtain the status of the specified SIPv6 proxy server.

Steps 5-7. The SIP Core Module sends the status information back to the Monitoring Control Module along the path traveled by the SMS-get request.

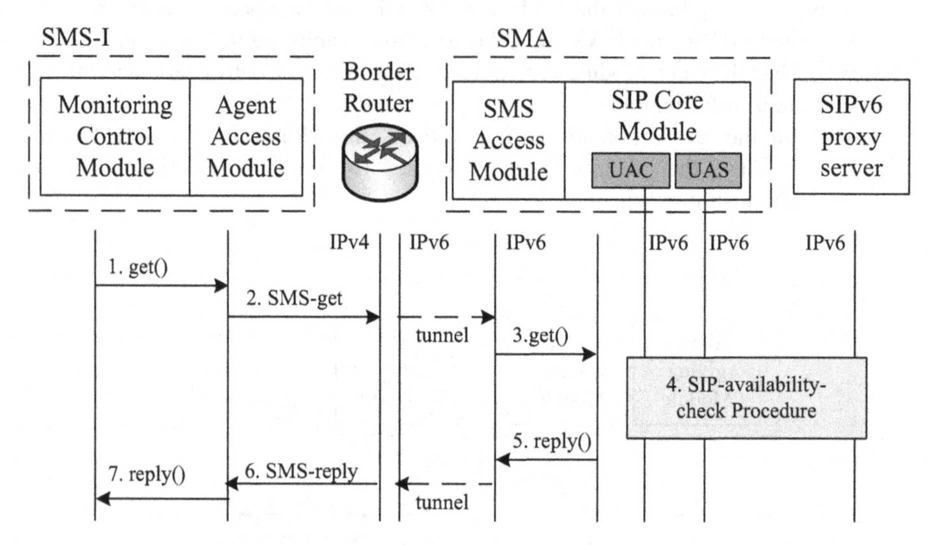

Fig. 5. Message flow for monitoring IPv6 proxy server

3 MoSA-C2 for IPv6 Dominant Scenario

This section describes the MoSA-C2. We present the designed network architecture, followed by the message flows used for monitoring SIP services.

3.1 Network Architecture

Fig. 6 shows the network architecture of MoSA-C2. The network elements are the same as that in Fig. 2, except that a type 2 Service Monitoring Server (SMS-II) is introduced to replace the SMS-I and the SMA. In this scenario, the SIPv4 proxy server and the intermediate routers that reside in IPv4 network are IPv4-only; the SIPv6 proxy server, the Web server, the Mail server, and the intermediate routers that reside in IPv6 network are IPv6-enabled. We locate SMS-II in IPv6 network since most services under its monitor reside in IPv6 network.

The SMS-II intermittently checks on the specified SIPv6 service, Web service, and email service over IPv6 transport while checks on the specified SIPv4 service over

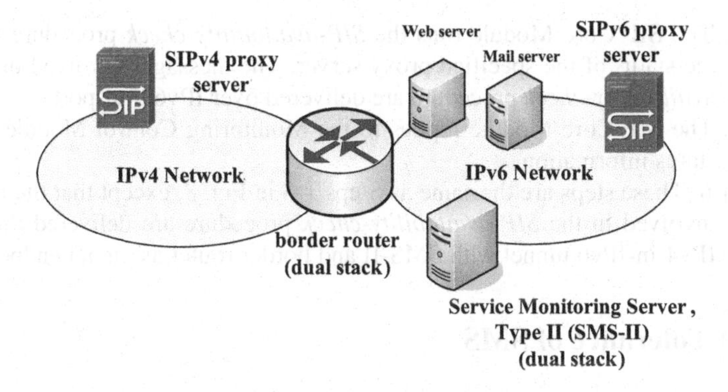

Fig. 6. MoSA-C2 network architecture

IPv4 transport. When problems are encountered, the SMS-II sends notifications out to administrators.

3.2 Message Flow

Fig. 7 shows the example message flow used by the SMS-II for monitoring the specified SIP proxy servers.

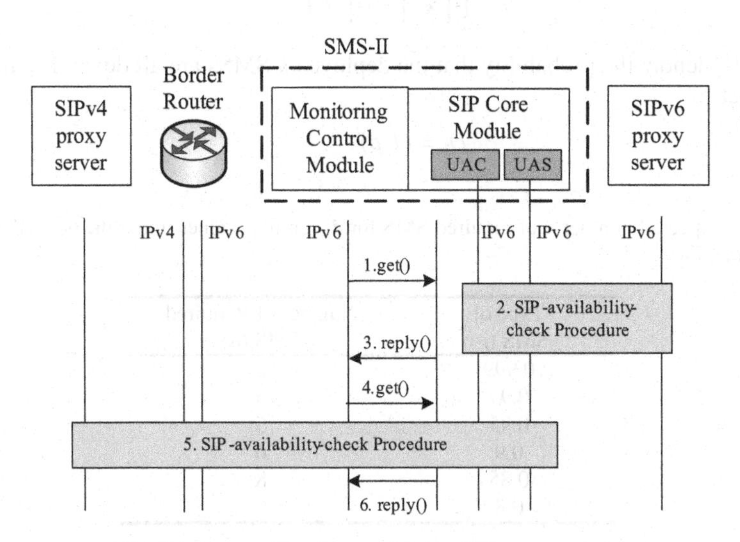

Fig. 7. Message flow for monitoring SIP proxy servers

In this example, the steps are as follows:

Step 1. Base on the SIPv6 proxy server identity specified by the administrator, the Monitoring Control Module requests the SIP Core Module to initiate the *SIP-availability-check* procedure.

Step 2. The SIP Core Module runs the *SIP-availability-check* procedure to obtain the status of the specified proxy server. The messages involved in the *SIP-availability-check* procedure are delivered over IPv6 transport.

Step 3. The SIP Core Module replies to the Monitoring Control Module with the status information.

Steps 4-6. These steps are the same as Steps 1-3 in Fig. 8, except that the messages involved in the *SIP-availability-check* procedure are delivered through an IPv4-in-IPv6 tunnel with SMS-II and border router as tunnel endpoints.

4 Fault Tolerance of SMS

Basically, fault tolerance of SMS can be achieved through deploying multiple servers of the same type. Since a SMS is either operational or down, the *operational readiness* of a SMS can be modeled by alternating renewal processes. Let X_i and Y_i denote the operational time and down time in the ith renewal cycle, $i \geq 1$, respectively. Assume that the random vectors (X_i, Y_i) are independent and identically distributed. If $E[X_i+Y_i<\infty]$ and X_i+Y_i is *nonlattice*[1], then the probability that a SMS is operational at time t, $t \rightarrow \infty$, is [7, Th. 3.4.4]

$$p = \frac{E[X_i]}{E[X_i]+E[Y_i]}. \tag{1}$$

Let D_k denote the probability that the deployed k SMSs are all down. From (1), we have that

$$D_k = (1-p)^k. \tag{2}$$

Table 1. Expected number n of required SMS for different p values under the desired 99.999% SMS reliability

Type of SMS (p)	Number of Required SMS (n)
0.999	2
0.99	3
0.95	5
0.9	6
0.85	8
0.8	9

To meet the carrier grade requirement, the SMS should achieve 99.999% reliability. Table 1 illustrates the effect of p on the number n of the required SMS to achieve 99.999% reliability. When $p=0.8$, nine SMSs are required to achieve the desired reliability. On the other hand, when $p=0.999$, only two SMSs are required.

[1] A nonnegative random variable is said to be *lattice* if it only takes on integral multiples of some nonnegative number.

5 Implementation

A MoSA prototype has been implemented in Providence University to validate the proposed framework. As SIP proxy server, the open-source implementation SER (SIP Express Router) [8], which supports for both IPv4 and IPv6, was chosen. For the automation of SIP testing, SMS was coded as a Nagios [9] plug-in. Nagios is a host and service monitoring tool licensed under the terms of the GNU General Public License. Finally, for SIP core module, SIPp [10], an open-source call flow generator for SIP, acts as UAC and UAS, respectively.

Fig. 8 and Fig.9 illustrate the status for the monitored SIP proxy server. At initial, the SIP proxy server functions properly (see Fig. 8). For demonstration purpose, we disconnect the server from the Internet. The server status transits from *OK* to *CRITICAL* shortly (see Fig. 9).

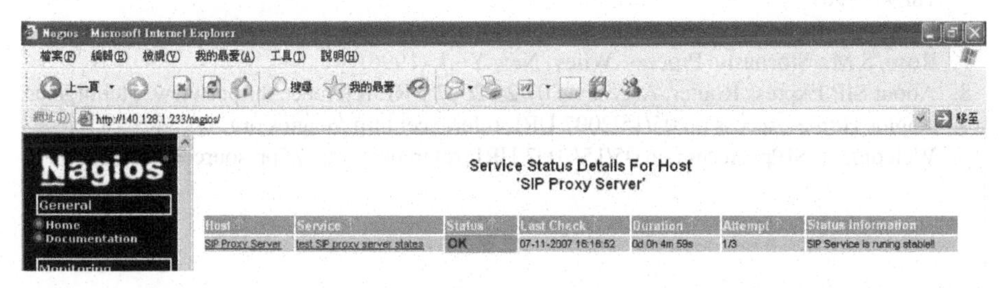

Fig. 8. Status for the monitored SIP proxy server: before disconnection

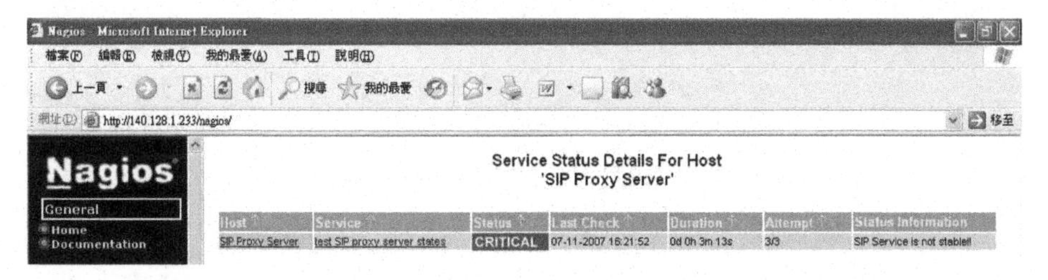

Fig. 9. Status for the monitored SIP proxy server: after disconnection

6 Conclusions and Future Work

In this paper, we described the design of MoSA, a framework for monitoring SIP service availability in hybrid IPv4/IPv6 environment. The detailed message flows for service monitoring and alarm giving are presented. We also suggested how to select the number of the required SMS to achieve telecom-grade performance. Furthermore, a MoSA prototype has been implemented using Nagios and SIPp for validation. MoSA can be easily extended to support checks on services besides SIP by adding their respective core control modules.

Future work includes a promotion for the alarm-giving module to deliver notifications to administrative contacts in cellular phone calls.

Acknowledgments. This work was supported by the National Science Council under Grand NSC 95-2745-E-126-003-URD.

References

1. Rosenberg, J., et al.: SIP: Section Initiation Protocol. IETF, RFC 3261 (2002)
2. Deering, S., Hinden, R.: Internet Protocol Version 6 (IPv6) Specification. IETF, RFC 2460 (1998)
3. Gilligan, R., Nordmark, E.: Transition Mechanisms for IPv6 Hosts and Routers. IETF, RFC 2893 (2000)
4. Tsirtsis, G., Srisuresh, P.: Network Address Translation - Protocol Translation (NAT-PT). IETF, RFC 2766 (2000)
5. Schulzrinne, H., et al.: RTP: A Transport Protocol for Real-Time Applications. IETF, RFC 1889 (1996)
6. Handley, M., Jacobson, V.: SDP: Session Description Protocol. IETF, RFC 2327 (1998)
7. Rose, S.M.: Stochastic Process. Wiley, New York (1996)
8. About SIP Express Router, Access on 07/25/2007 URL reference: http://www.iptel.org/ser
9. Nagios Home. Access on 05/15/2007 URL reference: http://nagios.org
10. Welcome to SIPp. Access on 05/15/2007 URL reference: http://sipp.sourceforge.net

Point of Reference in Perception of Network Performance by Active Probing

Myrvin Yap[1], Marat Zhanikeev[2], and Yoshiaki Tanaka[1,3]

[1] Global Information and Telecommunication Institute, Waseda University
1-3-10 Nishi-Waseda, Shinjuku-ku, Tokyo, 169-0051 Japan
[2] School of International Liberal Studies, Waseda University
1-21-1 Nishi-Waseda, Shinjuku-ku, Tokyo, 169-0051 Japan
[3] Research Institute for Science and Engineering, Waseda University
17 Kikuicho, Shinjuku-ku, Tokyo, 162-0044 Japan
myrvin@fuji.waseda.jp, maratishe@aoni.waseda.jp,
ytanaka@waseda.jp

Abstract. Reliability of active measurement results is an omnipresent issue in all existing measurement tools. Till this day, no reference point exists that would enable to compare measurement methods in efficiency and reliability. Partially, this is due to inability to attribute any particular probability model to both probing results as well as to the traffic probes interfere with. This paper proposed to analyze spectrum of measurement results in search for a reliable reference point. It is discovered that spectrum distribution of probing results are linearly dependent to linear changes in traffic.

1 Introduction

Many active measurement tools exist today [1]. They can be grouped by a particular network performance metric targeted by a tool, probe structure, and methods used to process measurement data. However, all active measurement methods are based on probes, which is a sequence of packets transmitted into the network in order to discover its performance by analyzing interference of probes with cross-traffic. Because of this, active measurement methods are often referred to as "*probing*" methods. In this paper we will use terms active measurement and probing interchangeably. Traffic participating in interference with probes is referred to as cross-traffic.

There is another common attribute of all probing methods besides the use of probes. They all share the uncertainty of probing results. The absence of any probability model attributed to traffic entails the same problems in processing probing results, which has to be done without a model attributed to them.

To give a simple illustration of this problem, let us suppose that a certain condition occurred in the network and is caused by some artifact in cross-traffic. At current levels of active measurement technology, there is no guarantee that probing will register this condition even if a probe is sent into the network during the condition in subject. The reason for this uncertainty has roots in traffic theory which is mostly probabilistic and assumes long periods of time, while probes operate only during a

S. Ata and C.S. Hong (Eds.): APNOMS 2007, LNCS 4773, pp. 205–214, 2007.

very short period of time. In simple words, it is therefore improbable that a probe will discover any short-term artifacts in traffic.

The above uncertainty problem has already been tackled by a number of research works, the most prominent of which are the ones in [2] and recently in [3]. Especially in [3], the authors attempt to apply traditional probability theory to probing, but performed only theoretical studies, which often does not hold in practice.

In this paper we attempt to find a reference point in the form of a metric that could have linear dependency on linear changes in traffic. Frequency or probability distribution cannot offer such a reference point as their changes are not consistent with changes in traffic. Instead, this paper proposes to use singular spectrum analysis (SSA) [4] to obtain distribution of spectrum in probing results. The analysis proves that spectrum distribution does have linear dependency on linear changes in the traffic, and thus can serve as a good reference point for deterministic network performance discovery.

2 Verification of Probing Results

In part, low reliability of probing results can be blamed on the lack of a mechanism of verifying them. In fact, any research proposing a new probing method has to make a choice between simulation and real network tests. Both environments have flaws described in the following chapters.

2.1 Basics of Probing

Basics of probing are graphically displayed in Fig.1. Probing always requires action at both ends of probing path even when a one-way metric such as one-way delay or one-way available bandwidth is concerned. Quite often probing is performed in round-trip fashion, there the receiving end of the path has to generate ACK packets for each probing packet it receives. In this way, information contained in the probe through interference with cross-traffic can be preserved in the group of space-preserving ACK packets. Interference with cross-traffic is normally due to cross-traffic packets queued

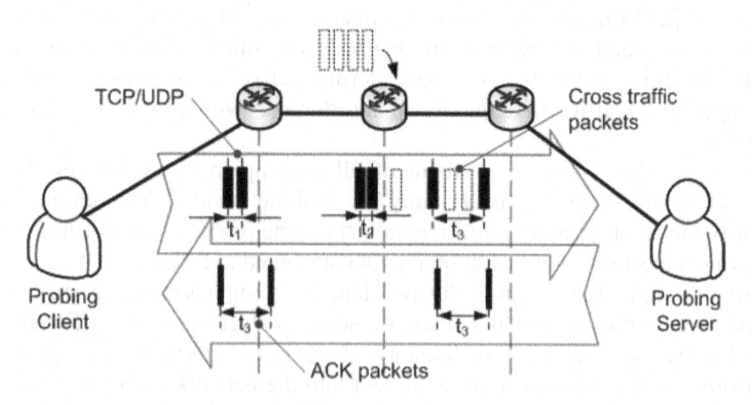

Fig. 1. Network diagram for end to end probing

before, after, or in between probe packets. This interference is, therefore, indicative of traffic intensity and subsequently of network performance.

In this paper we measure one-way delay to minimize unwanted inference. In one-way probing the interference is limited to forward direction and is easier to explore in quantitative terms.

2.2 Verification by Simulation

In a simulated environment it is very easy to verify performance of a given probing method. However, this results will be based on traffic generated in simulation and will directly depend on the traffic generation logic within the simulation. There are two well established general-purpose network simulators today, OPNET and NS. We do not reference them here since both tools have the same flaw described below. The information about them is freely available in the Internet.

In order to generate traffic, a simulator has to use a probabilistic model that will define packet size, time interval between packets, duration of ON/OFF periods, etc. There are many parameters that can be used to define a model of random traffic generation. This, however, does not solve the ultimate problem. Real traffic in real networks fail to adhere to any particular probability model and therefore cannot be simulated by a model-based automatic generation. This is the fundamental flaw of all simulations.

2.3 Active Measurement in Real Networks

Those probing methods that want to circumvent the fundamental flaw of simulation described above need to reserve verification through practical tests in real networks. Interference with authentic traffic is provided in this case, which seems to be the optimal solution.

However, this is not the case. While traffic inference is authentic, in real networks it is impossible to verify whether the probing results are true, thus undermining the goal of verification itself. Even if, for example, passive monitoring would be used to gather information on traffic directly from a network device at the time of probing, this data would still contain some error due to inability to synchronize probing with collection of physical data from a device. Besides, probing is done via a path rather than a single device, and collection of physical data from multiple devices will introduce even more error.

For the above reasons, comparison of existing measurement tools, such as that in [1], is done in a very primitive way – by visual comparison to determine if the results are realistic. In our opinion, such a comparison cannot constitute a sound scientific proof.

2.4 Real Traffic Traces in Simulation

In view of the two fundamental flaws above, the research in [5] tried to create a verification platform that would comprise useful features of simulation and real network tests in a single unified platform. As written in [5], if a simulation-based platform could import real traffic traces and generate cross-traffic in simulation while strictly following the time and size of packets found in the trace, such simulation could serve as a platform for verification, comparison, and practical analysis of probing methods. This paper uses the simulator in [5] to verify the results of power spectrum analysis. Traffic traces from several points in WIDE network [6] in Japan were used to generate cross-traffic.

3 Properties of Active Measurement Time Series

Before the proposal is made, some basic features attributed to active probing have to be considered. Before probing results are considered as conventional time series and submitted to scrutiny of data mining methods, one has to make a clear definition of contents of probing time series. The term "*probing time series*" in this paper stands for probing results represented as time series.

3.1 Trend Versus Residuals in Measurement Time Series

Normally, data mining attempts to discover a trend in time series and suppress noise (residuals). When probing time series are concerned, the target should be inversed. According to the fundamental assumption of all probing methods, information about network performance is inferred form interference of probing packets with cross traffic. Therefore, a trend in probing time series is a static value which is the best network performance, or, in other words, performance of the network without any traffic in it. The residual part of the signal, normally referred to by data mining as noise, is the main target of probing as it contains the valuable information of network performance.

In plain words, in probing time series, the residuals and not the trend is the target of mining.

3.2 Singular Value Decomposition for Model-Less Analysis

As mentioned above, analysis of time series results has to be done without a particular model attributed to data. One of the most efficient methods in model-less decomposition is Singular Spectrum Analysis (SSA) proposed and described in detail in [4]. Its closest rival is Principle Component Analysis (PCA), but it is well defined only for square matrices, while the matrices created from probing time series are rectangular MxN matrices, where M is normally smaller than N.

The core of SSA is Singular Value Decomposition (SVD), which is written as:

$$A = U \sum V^*, \quad (1)$$

where A is the original MxN rectangular matrix, U is an MxM unitary matrix containing orthonormal "input" basis vector directions for A, $\sum$ is an NxN diagonal matrix containing the singular values (spectrum components), and V^* is a NxN unitary matrix containing orthonormal "output" basis vector directions for A. In plain words, U is the input in the spectrum space of the signal, and $\sum V^*$ are eigenvalues and eigenvectors (in PCA terminology) that create the output.

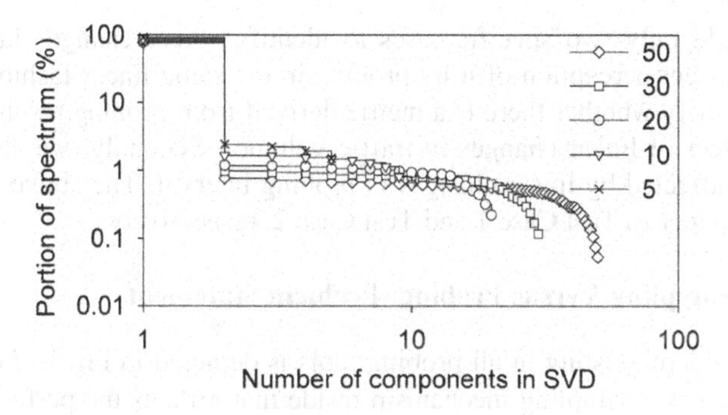

Fig. 2. Distribution of spectrum components in probing results

In practice, contents of diagonal matrix $\sum$ are of the main interest, as they can provide quantitative estimation on spectrum distribution in ordered components. Fig.2 displays distribution of spectrum components for 5, 10, 20, 30, and 50 components. An interesting feature in Fig.2 is the wide gap between the first component and all residual components. In fact, the first component alone is enough to reconstruct almost 100% of the signal. This proves the above assumption that trend is not of any value, as it represents zero interference encountered by the probe in cross-traffic. The distribution of all residual components starting from 2 is smooth and remains smooth over a wide range of the number of components.

Such a smooth distribution is indicative of possibly infinite number of components existing in signal with infinitely small share in spectrum, which, indirectly, proves the absence of any model that can be attributed to it. In fact, signal generated by deterministic models normally has abrupt changes in spectrum.

Before SVD can be performed on probing time series, single array of samples has to be converted into a matrix. This is done by using a lag matrix. Lag matrix uses a window of a given size to convert single-dimension array into a MxN matrix where M is the window size and $N = S - M + 1$, where S is the total number of samples. Each window is a column in the lag matrix where top element of each next column is the next sample in the original array. Naturally, if only first M rows from the top left corner of the matrix were used, this would be a symmetric matrix. However, since probing time series are longer than the size of the window, this is rarely the case.

Since spectrum decomposition is performed, naturally, the size of the window M directly affects the spectrum. Longer windows allow finer decomposition of signals. For probing time series, however, as was shown in Fig.2, a wide range of window sizes results in a similar smooth distribution.

All calculations related to SSA and SVD were performed by using procedures from the widely used LAPACK [7] library of scientific procedures that has an extensive set of tools for calculations using matrices and linear algebra in general.

4 Dependency Models for Probing Results

Coming back to the target the paper, to analyze the perception of network performance by probing, it is necessary to give a solid definition of the target. In this paper we tackle only two specific cases to identify where changes in the network performance affect perception of it by probing in the same linear fashion. First, it is important to verify whether there is a metric derived from probing results that change linearly in affect of linear changes in traffic volume. Secondly, we also study how this metric is affected by linear changes in probing interval. The above two cases are referred to in paper as Test Case 1 and Test Case 2, respectively.

4.1 Traffic Sampling Versus Probing: Problem Statement

Perception problem existing in all probing tools is depicted in Fig.3. Let us suppose that a network has a sampling mechanism inside that reflects the performance at any given time interval. The perception problem is then phased as follows. Given both sampling and probing are performed at identical regular time intervals, can it be stated that perception of network performance by probing is comparable to that by sampling?

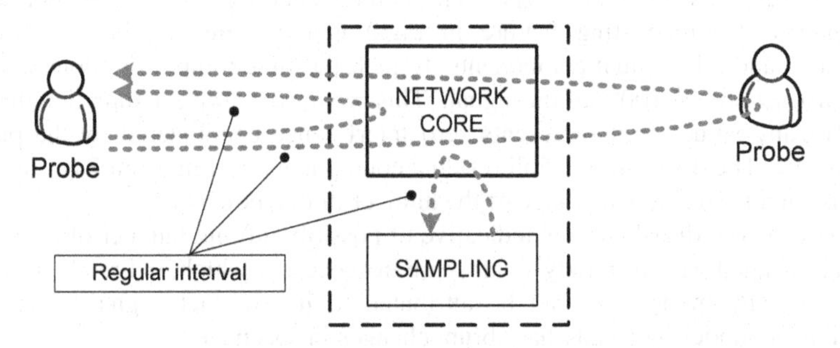

Fig. 3. Fundamental problem in perception of network performance

An answer to this question is rather simple. While sampling exploits cooperation by the hardware in the core of the network, it has direct acces to raw performance data. There is, therefore, no intrinsic error in it. Probes, on the other hand, have to traverse some path to reach the target network, continue to the other side of the path after the target network is exited, and often traverse the same path on the return if round-trip probing is used. So, the issue that needs to be verified by the proposed paper is whether or not probing results reflect changes in traffic or, interchangeably, network conditions, correctly as would be done by sampling. The role of sampling in this paper is played by rigid knowledge of the contents in traffic at any given point of time during simulation.

4.2 Fixed Measurement Model with Varying Traffic

The Test Case 1 is facilitated by applying fixed probing parameters to a set of varying traffic conditions. To provide linear changes in traffic conditions we apply a linear scaling parameter, which the multiplier for every space between packets in imported trace.

It can be argued that the change in the traffic is exponential, since a multiplier is used, but when we mention linear change we refer to scaling parameter. This paper focuses on quantitative analysis of change and does not care about qualitative essence of difference between addition and multiplication.

The use of scaling parameter also causes traffic to stretch in time. Since a fixed interval of 20 minutes is used to collect probing samples, scaling of 120% would mean that the last 20% of traffic will be lost. However, this is a tolerable effect since we analyze spectrum that will not be affected by this loss of traffic.

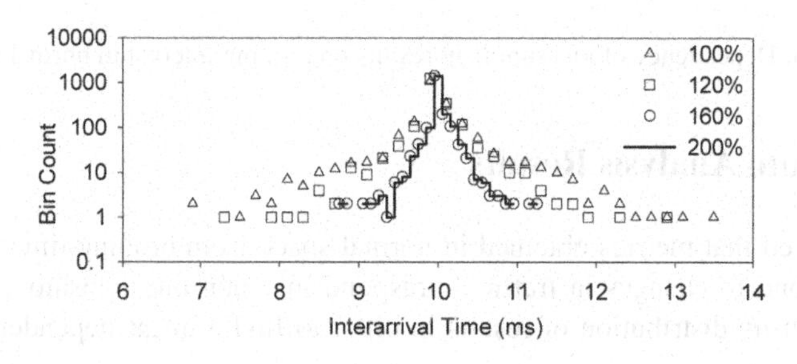

Fig. 4. Dependency of measurement results on scaling in normal space

Fig.4 contains histograms that visualize distribution of samples depending on scaling parameter ranging from 100% (original signal) to 200% (double length). Distribution is non-linear, as majority of samples are gathering around the main asymptote at 10ms, i.e. sampling interval. Also, minor interference (closer to the asymptote) appears to be less affected than major interference (farther from the asymptote), which stands for non-linear affect of change on distribution of samples. The behaviour of spectrum distribution depending on linear change in traffic will be studied later.

4.3 Varying Probing Models with Fixed Traffic

Fig.5 contains the histograms obtained from Test Case 2, i.e. varying probing interval with fixed traffic. As found in Test Case 1, frequency distribution in time series does not follow linear change in probing interval. Although due to fixed simulation time there are fewer probing samples to analyse, the distribution of frequency is similar with all three time intervals in Fig.5. That is, there are more samples closer to the asymptote and fewer samples further from it. Physically, this originates from common sense joined with probability. Probabilistically, it is stated that in a sufficiently long time period, there is a heavy-tailed distribution of traffic parameters, such as bulk size, interarrival time of packets, etc. Common sense helps by stating that this probability distribution is independent from probing interval, i.e. at any probing intervals, there is the same probability for the probe to encounter a major surge in volume of traffic. Distributions in Fig.2 prove it.

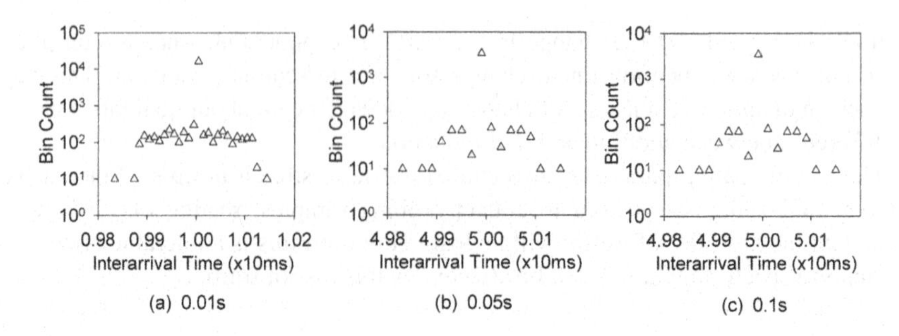

Fig. 5. Dependency of measurement results on probing interval in normal space

5 Spectrum Analysis Results

Having proved that metrics obtained in normal space from probing time series either do not respond to changes in traffic or respond in a non-linear fashion, this section studies spectrum distribution of results in order to find a linear dependency to linear change.

5.1 Static Probing and Varying Traffic

Fig.6 contains residual spectrum distribution from Test Case 1. The main trend fully represented by the first component is cut off. From Fig.6 we can see that distribution of the first 50 spectrum components shows linear dependency to traffic scaling parameter. More than that, the quantitative change is also reflected correctly, since average distance from 120% to 160% is more than from 100% to 120%, which is additional prove of linear dependency. Naturally, deeper in the ordered list of components there are some non-linear excursions, but the general trend in distribution is clearly linearly dependent on the scaling parameter.

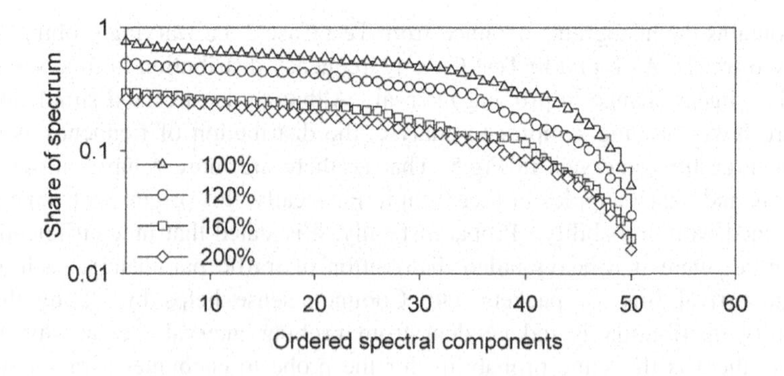

Fig. 6. Dependence of spectrum distribution on traffic scaling parameter

5.2 Static Traffic and Varying Probing

Fig.7 contains results for Test Case 2. In this case, the response to linear change in probing interval (~50ms increment) is not linear in spectrum distribution. However, the factor of change is the same for the total spectrum distribution, thus, maintaining the distance between trends. Naturally, the curve obtained from probing by 10ms interval is the smoothest due to the high number of samples. The last curve was calculated from 10 times less samples and therefore is less stable in spectrum distribution.

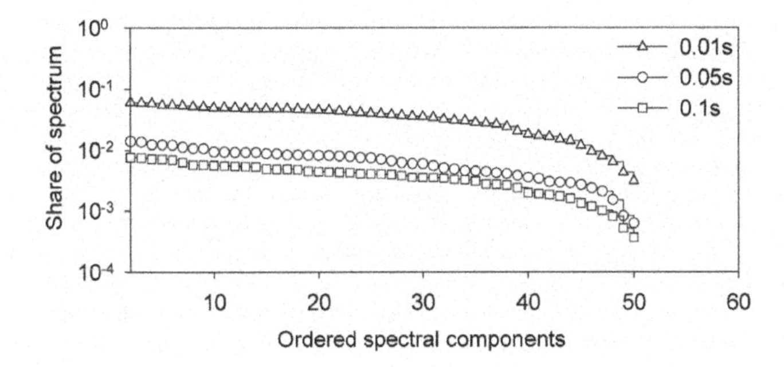

Fig. 7. Spectrum distribution as a function of probing interval

6 Conclusion

This paper analyzed spectrum distribution of probing results under a number of conditions. Test Case 1 tested spectrum distribution dependency by manually scaling traffic by a fixed factor. Test Case 2 studied dependency on probing interval. Both test cases proved that spectrum distribution obtained by using SSA shows linear dependency on the above parameters.

One of practical applications of the above findings would be obtaining a reference point for reliable perception of network performance. For example, with a fixed probing interval, such a reference point would be a global of a local best-case distribution of spectrum, which could be later used to detect changes in network performance based directly on changes in spectrum distribution. Since measurement results tend to have infinite number of components with very slow convergence rate, changes in spectrum distribution can easily take quantitative form as a distance between near-linear distribution curves. Quantitative analysis of spectrum changes is left for future study.

References

1. Montesino-Pouzols, F.: Comparative Analysis of Active Bandwidth Estimation Tools. In: PAM 2004, pp. 175–184 (2004)
2. Pasztor, A., Veitch, D.: A Precision Infrastructure for Active Probing. In: PAM (2001)

3. Liu, X., Ravindran, K., Loguinov, D.: What Signals Do Packet-Pair Dispersions Carry? IEEE INFOCOM 1, 281–292 (2005)
4. Golyandina, N., Nekrutkin, V., Zhigljavsky, A.: Analysis of Time Series Structure: SSA and Related Techniques. Chapman & Hall/CRC (2001)
5. Zhanikeev, M., Tanaka, Y.: Issues with Using Real Packet Traces in Simulated Environments. IEICE Technical Report on Telecommunication Management No.TM,2006-67 pp. 35–40 (2006)
6. MAWI Working Group Traffic Archive. Available at: http://tracer.csl.sony.co.jp/mawi/
7. LAPACK: Linear Library PACKage. Available at: http://www.netlib.org/lapack/

Real-Time Identification of Different TCP Versions

Junpei Oshio, Shingo Ata, and Ikuo Oka

Graduate School of Engineering, Osaka City University,
3–3–138 Sugimoto, Sumiyoshi–ku, Osaka 558–8585, Japan
{oshio@n.,ata@,oka@}info.eng.osaka-cu.ac.jp

Abstract. When multiple flows using different versions of TCP exist together on the network, they are influenced seriously between each other. One of methods to overcome this problem is to control such flows individually by identifying the TCP versions of each flow. Thus, this process needs a method to identify online which TCP version is used for each flow. In this paper, we first focus on the increase speed of the congestion window in two different versions of TCP, which are TCP Reno and HSTCP (High Speed TCP). We also show that increasing rate of congestion window has remarkable characteristics in each TCP version. We then propose a method to identify flow by using the characteristics. We show that TCP Reno and HSTCP can be completely identified by using our proposed method.

1 Introduction

TCP (Transmission Control Protocol) has been mainly used as a transport protocol, and TCP Reno, which is one of TCP versions, is widely used in the Internet. However, in recent years, the line speed of the Internet is increasing dramatically. Several researchers have shown that TCP Reno is not appropriate to communicate for high-speed network, because flows of TCP Reno cannot utilize the bandwidth of link efficiently. To solve this problem, new versions of TCP such as HSTCP (High Speed TCP) [1], FastTCP [2], STCP (Scalable TCP) [3] and BIC-TCP (Binary Increase Control-TCP) [4] have been proposed. These versions are made improvements to utilize the bandwidth of link efficiently. Therefore, in high-speed network, adoptions of new versions of TCP such as HSTCP are desired in order to achieve better performance.

However, even if the new version of TCP is quite reasonable in the future, it is difficult to migrate from TCP versions of current users in the Internet to new TCP versions at a time. During the migration, there is a phase that two or more versions coexist together on the same network. However, when multiple flows using different versions of TCP exist together on the same network, they are influenced seriously between each other. To avoid this situation, we need to identify TCP versions in real time, and control flows of new version with considering the influence on the current version [5].

S. Ata and C.S. Hong (Eds.): APNOMS 2007, LNCS 4773, pp. 215–224, 2007.

The researches to identify the TCP version exists so far. In [6], the authors have assumed Tahoe, Reno, and NewReno as TCP version used by flows. Next, they estimate the numbers of packets which will be sent in each versions. Then, they compare them with the number of packets which have been sent actually. As a result, each flow evaluated "violation" which indicates the degree how the measured number of packets exceeds the estimated number. The version which has the smallest value of violation is inferred as a version of a flow. In this method, they use rare events to identify the TCP versions, thus they cannot obtain high identification accuracy. Therefore, we need to identify the TCP versions in real time correctly. In [7], the authors cause the event such as the packet losses in communication with the Web server, and they identify the version used by Web server with TBIT (TCP Behavior Inference Tool). In this method, they can analyze in detail about TCP version used by Web server. However, we cannot use in actual environment, because this method have a influence to communication.

Therefore, in this paper, we propose a method to identify TCP version of flows at relay router accurately in real time. In TCP, the congestion window (cwnd), which is the number of packets which a sender can transmit at one time, is used to avoid network congestion. We note that increasing amount of cwnd differs depending on the TCP version. First, we calculate the cwnd in each of flows, and the rate of cwnd. We identify the version by using characteristics of the rate of cwnd. We simulate network environment that is constructed by nodes using Reno and HSTCP. We apply the proposed method to the trace data of this simulation, and show that the method enables identification of TCP version. Additionally, we apply the method to trace data of simulations in which we change the number of nodes using each version, and we investigate the amount of time needed for identifying TCP versions.

This paper is organized as follows. We describe about our identification method of TCP version in Section 2. We show our simulation environment, and describe the results that we apply proposal method to the trace data in Section 3. Finally we summarize this paper with future topics in Section 4.

2 Identification Method

In this section, we first describe outline of our proposal method. Next, we describe approaches which are used to estimate RTTs and rate of cwnd in our method. Moreover, we describe detailed algorithm of our method in the following subsection.

2.1 Outline of Identification Method

In each TCP versions, the cwnd is used to avoid network congestion. there are the characteristics that the increasing and decreasing amount of cwnd differs depending on the TCP version. Thus, in this paper, we use the characteristics of cwnd to identify TCP versions used by flows. However, the value of cwnd cannot be comprehended only by communicating end nodes, we cannot obtain the cwnd directly at relay router. To solve this problem, we use the nature that

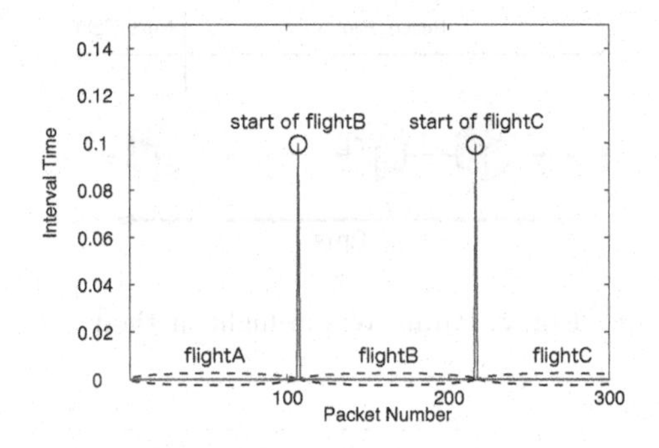

Fig. 1. Relations between flights and inter-arrival time of packets

the cwnd represent the number of packets sent in 1 RTT. We first estimate RTTs and regard the number of packets which reached in the RTT as the cwnd. Next, we estimate the rate of cwnd to obtain the characteristics which show the more obvious difference between each TCP versions. We identify TCP versions by using this difference. We describe about the method of estimating RTTs in Subsection 2.2, and about the rate of cwnd in Subsection 2.3. Moreover, we describe about process of proposed method circumstantially in Subsection 2.4.

2.2 The Method of Estimating RTTs

To estimate RTTs at relay router in real time, we use an estimation method called flight method [8]. Figure 1 depicts the relation between inter-arrival time of packets and flights. In this figure, the horizontal axis shows the number of packets, and the vertical axis shows the inter-arrival time of packets. This figure shows that there are packets making larger inter-arrival time and they make a leading edge of each flight. The flight is defined as a sequence of consecutive packets with nearly identical inter-arrival time followed by larger inter-arrival time. The flight method regards the time between the leading edge of a flight and the leading edge of the next flight as the RTT.

The procedure to compose flight is provided as follows. First, we consider a sequence of packets p_1, p_2, p_3, and inter-arrival times δ_1, δ_2 between the first and second pairs of packets respectively as in Figure 2. We define the ratio g which is given by

$$g(\delta_1, \delta_2) = \left| \frac{\delta_2 - \delta_1}{\delta_1} \right|. \quad (1)$$

Next, we compare g with a threshold value. If g is smaller than threshold, p_3 belongs to the current flight. If not, p_2 is the end of the current flight, and p_3 is the start of the new flight. According to [8], when we use ($\frac{1}{16}$, $\frac{1}{4}$, $\frac{1}{2}$, 1, 2, 4 and 8) as the threshold values, we have nearly identical results in the composing flight. Thus, we use 8 as the threshold.

In this way, the flight method regards larger inter-arrival time as separation of flights. Therefore, delayed arrival time of packets caused by occurrence of

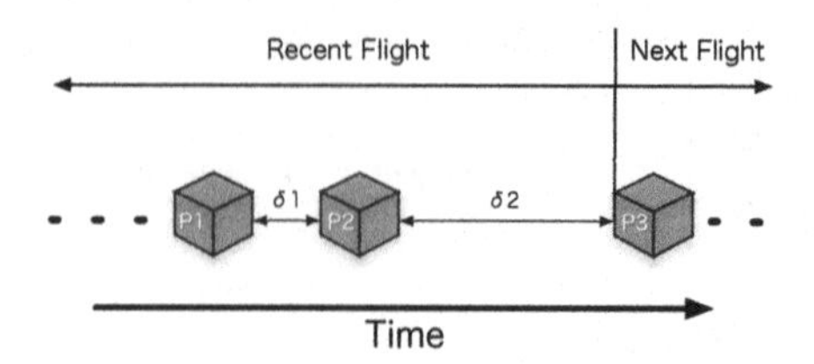

Fig. 2. Parameters in flight method

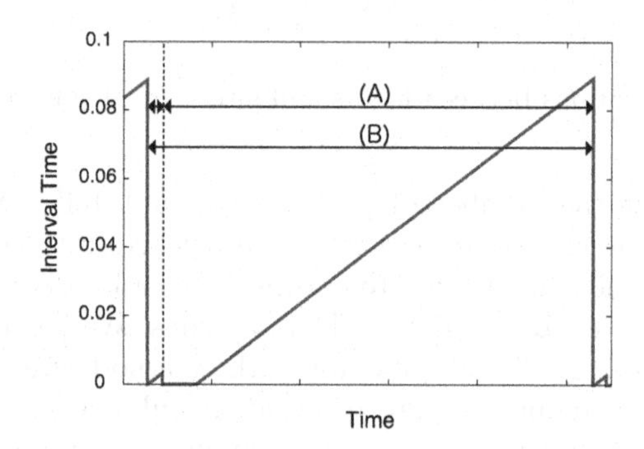

Fig. 3. Avoiding miss of classification in flight method

congestion is also regarded as separation. As a result, a flight which should be considered as one is divided into two shown as interval (A) in Figure 3. Thus, we cannot obtain a correct result when a delay occurred, because flight method is quite sensitive against the small variation of RTT. As a solution to this problem, first we calculate the average of RTTs until first drop of packet. Then, we compare estimated RTT with half of the average of the RTTs until first drop of packet, and if former is smaller than latter, we deal with current and next flight as one like interval (B) in Figure 3.

2.3 Rate of Congestion Window

We count the number of packets arrived in 1 RTT, in a word, the value of the cwnd is calculated by using estimated RTT. Next, we calculate R which is the differential ratio of cwnd between two flights of each flow. R is given by $R = \frac{w_i}{w_{i-1}}$, where w_i is the cwnd of i-th flight. Figure 4 shows the time dependent variation of R in Reno and HSTCP. In this figure, the horizontal axis shows the time, and the vertical axis shows the R. In Reno, the value is declining smoothly. This is because the increment of cwnd is fixed integer value in Reno. On the other hand, in HSTCP, the R is declining while repeating the increase and the decrease by little and little. This reason is as follows. The increment of cwnd is floating value, and the value is increased monotonically and gradually in HSTCP. Therefore, the R also varies smoothly. However, the number of packets to be sent is of course an integer value, and so the value after the decimal point

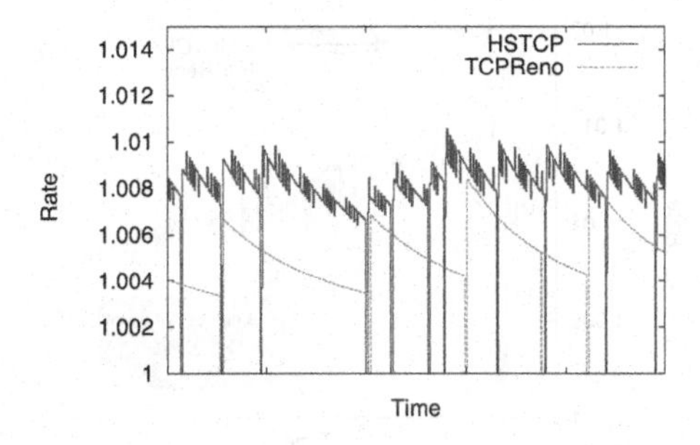

Fig. 4. Rate of cwnd in TCP Reno and HSTCP

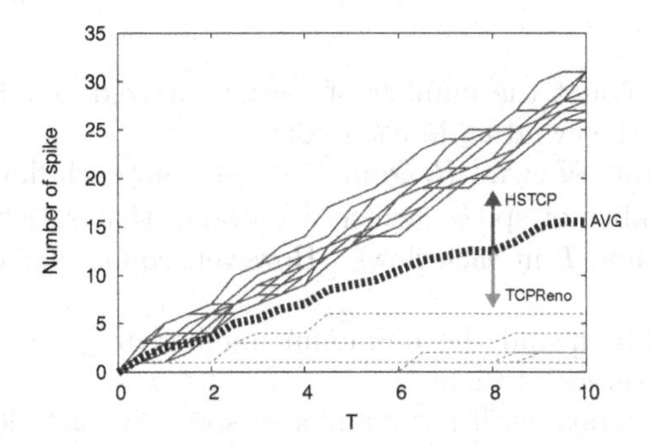

Fig. 5. Transition of the number of spike and the average

is rounded. Thus, the increment of cwnd lacks smoothness, and R is repeating the increase and the decrease as Figure 4. We identify TCP version by counting spikes of rate in observation time T. The spikes of rate means the events that $\frac{w_i}{w_{i-1}}$ is larger than $\frac{w_{i-1}}{w_{i-2}}$. The transition of the number of spike and the average of maximum and minimum number of them are shown in Figure 5. In this figure, the horizontal axis shows the identification time T, and the vertical axis shows the number of spikes. As T is increasing, the rate of the number of spikes in HSTCP is considerably larger than in Reno, and the average of maximum and minimum values can be used as the threshold between Reno and HSTCP. Figure 6 shows about the characteristics of rate in Reno and HSTCP. In this figure, the horizontal axis shows the time, and the vertical axis shows the rate of cwnd. In this figure, the distribution of the rate of Reno has one spike, while the distribution of the rate of HSTCP has eight spikes in T.

As above, the TCP versions have the different characteristics in the rate of cwnd respectively. We use the characteristics to identify TCP version.

2.4 Proposal Method

The procedure of our proposal method is shown as follows.

1. Estimate RTTs in each flows by using flight method.

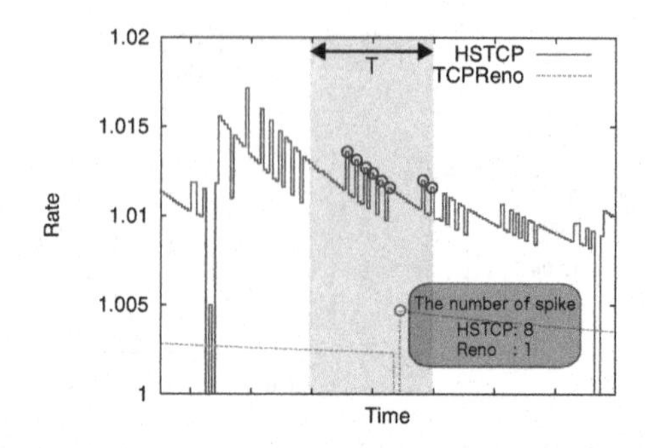

Fig. 6. Identification method by using the number of spike in T

2. For i-th flight, count the number of packets arrived in 1 RTT, and regard the number as the cwnd w_i in each flows.
3. Calculate the rate of cwnd R_i from $R_i = \frac{w_i}{w_{i-1}}$ for each flows.
4. Count the number of spikes, which are events that satisfy $R_{i-1} < R_i$, in identification time T in each flows. (However, count the consecutive spikes as one spike.)
5. Find the maximum and the minimum number of spikes in all flows, and calculate the average of them.
6. Compare the average with the number of spikes in each flows, and identify TCP versions as follows.
   (a) As Reno, if the number of spikes is smaller than the average.
   (b) As HSTCP, if the number of spikes is larger than the average.
   (c) As unknown, if they are equal.

Finally, to estimate the performance of our proposal method, we calculate identification rate I in each of TCP version. I means the value represented as $I = \frac{F_i}{F}$, where F is the actual number of flows using the version and F_i is the number of the flows identified accurately. For example, when the number of flows of Reno is 10, and 8 of those flows are identified as Reno accurately, I of Reno is 0.8.

3 Identification Results

In this section, we first describe about simulation environment we used in this paper. We then describe the results of applying the proposal method to the trace data. Additionally, we consider about influence of changes of identification time T, and we estimate the minimum time needed to identify TCP version by applying the method to various trace data.

3.1 Simulation Environment

We use ns-2 (ns-2.29.3) for simulation. The network topology we used is shown in Figure 7. We set the total number of flows of Reno and HSTCP to be 20,

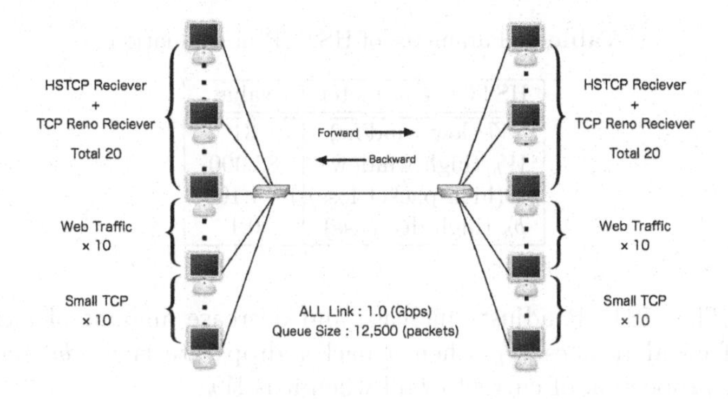

Fig. 7. Simulation environment

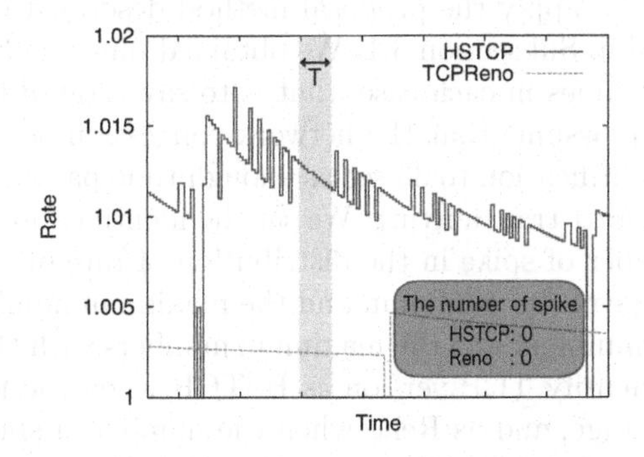

Fig. 8. Error of identification in a short T

and change the number of nodes of each version from 1 to 19 respectively in simulation. Thus, we obtain nineteen kinds of trace data. All traffic pass through a single (Bottleneck) link, whose capacity is 1 Gbps and transmission delay is set to be 50 msec. The capacity of the links connecting the nodes with router is 1 Gbps and the transmission delay of them is 5 msec. The size of router buffer determined by the bandwidth delay product are typically required to process TCP congestion control effectively [9]. Therefore we also set the size of router buffer to 8,333 packets. The simulation time is 50 sec, and the start of transmission of Reno and HSTCP is set from 0 to 5 sec in random order. The packet size is set to 1,500 bytes based on the standard Ethernet MTU (Maximum Transmission Unit), and all the Reno and HSTCP flows used forward direction. We enable a SACK (Selective ACK) option in both TCP Reno and HSTCP. We use FTP (File Transfer Protocol) as application. We setup small TCP flows and Web flows as back traffic.

The HSTCP parameters are set as shown in Table 1. The each of parameters means as follows. W_l is a threshold which specifies an algorithm of changing cwnd. The HSTCP uses the same algorithm as Reno when the current cwnd is smaller than W_l, and uses High Speed algorithm when current cwnd is larger

Table 1. Parameter of HSTCP in simulation

HSTCP parameter	value
W_l (low window)	31
W_h (high window)	83,000
p_h (high packet loss)	$1.0 * 10^{-7}$
b_h (high decrease)	0.1

than W_l. The HSTCP adjusts increase and decrease amount of cwnd as the average of cwnd reaches W_h when a packet drop rate is p_h. b_h specifies the decreasing proportion of current cwnd when it is W_h.

3.2 Result of Identification

In this subsection, we apply the proposal method described in Section 2 to the trace data obtained in Subsection 3.1. We obtained nineteen kinds of trace data with repeating five times in each case, that is to say total of trace data is 95.

In this paper, we assume that the network congestion occurs. Therefore, we set start time of identification to 30 sec as enough time passed since the nodes of Reno and HSTCP start transmitting. We set the identification time T to 10 sec, and count the number of spike in the distribution of rate of each flow in T. We calculate the average of the minimum and the maximum numbers, and compare the average of the minimum and the maximum numbers with the number of spike in each flow. We identify TCP version as HSTCP when the number of spike is larger than the average, and as Reno when the number of spike is smaller than the average. Moreover, we identify as unknown when the two numbers are equal. When we set T to 10 sec, we can obtain identification rate I as 1 in each TCP versions in all trace data.

3.3 Problem of Dependence on Parameter

We describe the results of identification by using proposal method in Subsection 3.2. However, the identification time T makes a difference to the results in this method. This is because, a situation often occurs in short T, which we are scarcely able to observe spikes of rate although in HSTCP. The example of this error is shown in Figure 8. As we have described so far, we identify TCP version by using the difference of number of spikes between Reno and HSTCP. Thus, there is not so much of a difference between them, we cannot identify them as shown in this figure. This error does not happen in long T, however a shorter T is more desirable. Therefore, we need to estimate the shorter and optimal T in which we can identify TCP version correctly. We describe this process in details as shown in Subsection 3.4.

3.4 Optimal Value of Identification Time T

To estimate the shorter and optimal T in which we can identify TCP version correctly, we apply the proposal method to them in T which is changed from

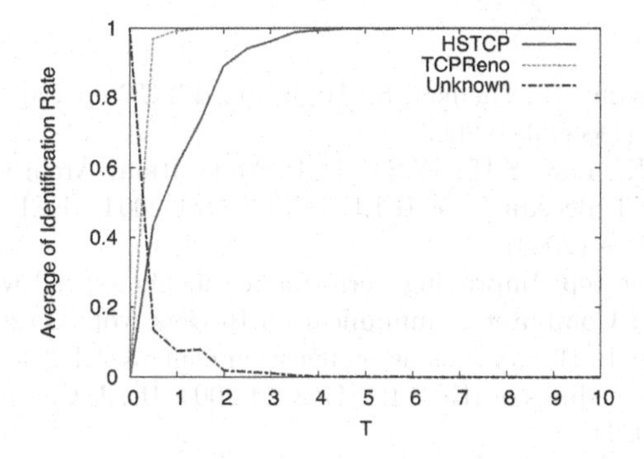

Fig. 9. Relations between identification rate and T

0.5 sec to 10.0 sec by 0.5 sec. As a result, we estimate T when we obtain I of Reno and HSTCP as 1 in the all of trace data, and we determine the T as optimal value.

Figure 9 shows relations between T and the average of I of Reno and HSTCP in all trace data. As shown in these figure, we can obtain I as 1 in all of trace data, when T is 6. The identification of TCP Reno requires 2 sec. It is thought that there is no difference between Reno and HSTCP because congestion is frequently generated in TCP Reno and it has taken time for identification. The identification of HSTCP requires 6 sec. It is thought that situation which we cannot observe very many spikes in the rate of HSTCP as shown in Figure 8.

We need to consider these factors and devise the method to be able to accurate identification at shorter T in the future.

4 Conclusion

In this paper, we have proposed identification method of TCP version at relay router in real time. This method uses the rate of the number of packets which reached in 1 RTT. We have applied this method to trace data that are obtained by simulation and use TCP Reno and HSTCP flows. We have shown that this method can identify TCP version accurately. Moreover, we have estimated the shorter and optimal value of T in which we can identify accurately. For future research topics, we need to propose the method not depending on the environment. In addition, we need to devise the method that can identify at shorter time.

Acknowledgment

This work was partially supported by the Grant-in-Aid for Young Scientists (A) (No. 19680004) from the Ministry of Education, Culture, Sports, Science and Technology (MEXT) of Japan.

References

1. Floyd, S., Ratnasamy, S., Shenker, S.: High Speed TCP for Large Congestion Windows. RFC 3649 (December 2003)
2. Jin, C., Wei, D.X., Low, S.H.: FAST TCP: Motivation, Architecture, Algorithms, Performance. In: Proceedings of IEEE INFOCOM 2004, IEEE Computer Society Press, Los Alamitos (2004)
3. Kelly, T.: Scalable tcp: Improving performance in high-speed wide area networks. ACM SIGCOMM Computer Communication Review, vol. 33 (2003)
4. Xu, K.H.L., Rhee, I.: Binary increase congestion control (BIC) for fast long-distance networks. In: Proceedings of IEEE INFOCOM 2004, IEEE Computer Society Press, Los Alamitos (2004)
5. Ata, S., Nagai, K., Oka, I.: Network support for tcp version migration. In: Proceedings of APNOMS, pp. 322–331 (September 2006)
6. Jaiswal, S., Iannaccone, G., Diot, C., Kurose, J., Towsley, D.: Inferring TCP Connection Characteristics Through Passive Measurements. In: Proceedings of IEEE INFOCOM 2004, Hong Kong, China, vol. 3, pp. 1582–1592 (2004)
7. Padhye, J., Floyd, S.: On Inferring TCP Behavior. ACM SIGCOMM Computer Communication Review 31, 287–298 (2001)
8. Shakkottai, S., Srikant, R., Brownlee, N., Broido, A., claffy, K.: The RTT Distribution of TCP Flows in the Internet and its Impact on TCP-based Flow Control, CAIDA Technical Report number tr-2004-02 (February 2004)
9. Bush, R., Meyer, D.: Some Internet Architectural Guidelines and Philosophy. RFC 3439 (December 2002)

End-to-End Flow Monitoring with IPFIX

Byungjoon Lee[1], Hyeongu Son[2], Seunghyun Yoon[1], and Youngseok Lee[2]

[1] ETRI, NCP Team, Yuseong-Gu, Daejeon, Republic of Korea
{bjlee,shpyoon}@etri.re.kr,
[2] Chungnam National University, Yuseong-Gu, Daejeon, Republic of Korea
{hguson,lee}@cnu.ac.kr

Abstract. End-to-End (E2E) flow monitoring is useful for observing performance of networks such as throughput, jitter and delay. Typically, E2E flow monitoring is carried out at end hosts with known tools such as iperf. However, the end-host approach may not be easily deployed in a large-scale network because of high cost and administrative overhead. Therefore, in this paper, we propose a new E2E flow monitoring method based on IP Flow Information eXport (IPFIX) that could provide QoS metrics such as throughput, retransmission rate, delay, and jitter for TCP flows and SIP-signalled RTP flows. We have extended the IPFIX templates for carrying QoS-related fields, and developed the E2E flow monitoring function with the open source that could be embedded into routers. From experiments, it was shown that the performance of TCP and RTP flows could be easily examined with the IPFIX-based approach.

1 Introduction

An E2E flow in the Internet could be regarded as a sequence of packet arrivals belongs to a web session, a file transfer, a video stream, or a VoIP call. Recently, a lot of performance questions on an end-to-end flow are being raised by users as well as Internet Service Providers (ISPs). Traditionally, a steady-state throughput of a long-lived TCP flow is considered to be a good indicator of the E2E flow performance, because most Internet traffic is associated with TCP protocol. Therefore, it is necessary to monitor the performance of TCP flows to correctly assess the end-to-end performance metrics of typical Internet applications such as Web downloading, email, and file exchanging.

On the other hand, although only the best-effort service is available within the current Internet, ISPs are willing to provide the QoS-guaranteed service for the specialized applications such as MMoIP(MultiMedia on IP) for maximizing the revenue. Most of the MMoIP services including VoIP and VoD have their own QoS requirements that are determined by the high-level Service Level Agreements (SLA) between the subscribers and the providers of the service. Therefore, it is essential for ISPs to monitor QoS parameters of MMoIP services for assessing whether the services are correctly delivered to the subscribers. Generally, it could be assumed that each QoS-critical Internet application is bound with a specific signaling mechanism like industry-proven and widely-accepted SIP or H.323 to collect charging-related information more easily.

S. Ata and C.S. Hong (Eds.): APNOMS 2007, LNCS 4773, pp. 225–234, 2007.

Hence, in this paper, we focus on monitoring QoS metrics of TCP flows and RTP sessions established by SIP signalling, because SIP is the most popular signaling protocol in current IP networks. For TCP, we figure out TCP throughput and the number of retransmitted packets of each TCP flow. For RTP sessions, we measure throughput, delay, and jitter of the RTP media streams created by SIP signaling messages.

For the observation of the E2E flow performance metrics, we employ the IPFIX architecture [1], where routers with IPFIX-compliant traffic monitoring functions will monitor and export E2E flow information to the collector. In general, the IPFIX architecture is more useful than the end-host approach in a large-scale network, because routers on the E2E path could monitor the E2E flow and export the flow information to the collector. In the end-host approach, each end-host with specialized software like iperf should collect the QoS information and send it to the collector, which may not be scalable. In addition, in the router-based IPFIX approach, the problematic segments of the path that an E2E flow is traversing could be easily identified.

In the IPFIX architecture, the performance information of each flow will be captured and computed by routers. The collected information is assembled into flow records, and exported to an external collector by IPFIX protocol [2]. The key concept of IPFIX protocol is the flexible and extensible template architecture that can be useful for various traffic monitoring applications. The template can be easily extended to export QoS measurement results.

Hence, in this paper, we propose new IPFIX templates which provide throughput and the retransmission rate for TCP flows, and throughput, delay, and jitter for SIP/RTP multimedia streams. We also propose an RTP stream detection method using SIP signaling information. Through the experiments with the prototype system for monitoring E2E flow performance, it is seen that our approach could easily provide the performance metrics of E2E flows.

The remainder of this paper is organized as follows. The related work is introduced in section 2. In section 3, the architecture of our system is explained. The results of the experiments with the prototype are introduced in section 4. Section 5 concludes this paper.

2 Related Works

In Internet2, there is a special working group called Internet2 End to End Performance Initiative (E2Epi) [3] to establish the architecture and to propose E2E monitoring applications. Similarly, there has been an Active Measurement Project (AMP) [4] that also addresses the issue of monitoring end-to-end performance using active measurement methods. However, with regard to the problem of detecting problematic segments in the network, the measurement granularity for these projects is too coarse to give detailed flow-based monitoring results.

An architecture to monitor end-to-end performance of selected flows has been proposed in [5]. Subscribers of an ISP service can check whether their flows are being serviced correctly by querying a database which stores per-flow information exported by each meter (i.e., routers) using Netflow v5 protocol. However, the suggested scheme cannot deliver the detailed measurement results of the QoS parameters because the Netflow v5 protocol can export only a limited set of performance parameters.

Cisco nBar [6] suggests a mechanism to detect RTP flows using payload classification. This system can mark the detected flows to be shaped or policed for satisfying certain QoS requirements. However, this system does not provide a solution for end-to-end performance monitoring.

A method to monitor end-to-end QoS metrics of media streams associated with SIP signaling has been presented in [7]. This method combines active and passive measurement methods. For active measurement, monitoring probes are injected to the network by agent-side packet generators. The measurement results are collected with SNMP and saved as flow files at the SIP proxy. Therefore, user agents and SIP proxy server software should be modified to use this scheme. Thus, this may be rather unrealistic because it is difficult in the real network to modify the software of the participating network entities for the specific measurement methods.

A QoS measurement method [8] for RTP flows has been proposed by using nProbe [9]. A new IPFIX template was introduced to export QoS monitoring results to an external collector. However, this work does not take SIP signaling into consideration. In this paper, we extend the work in [8] to extract RTP flow information from SIP SDP payloads.

3 The Proposed E2E Flow Monitoring Method

3.1 The IPFIX-Based Architecture

The system architecture of the proposed method is described in Fig. 1. As specified in [1], the IPFIX architecture consists of IPFIX devices and collector processes. IPFIX devices observe flows, meter flows, and export metered results for each flow to collector processes via the IPFIX protocol. Thus, we employ the IPFIX framework that consists of flow probes and a collector.

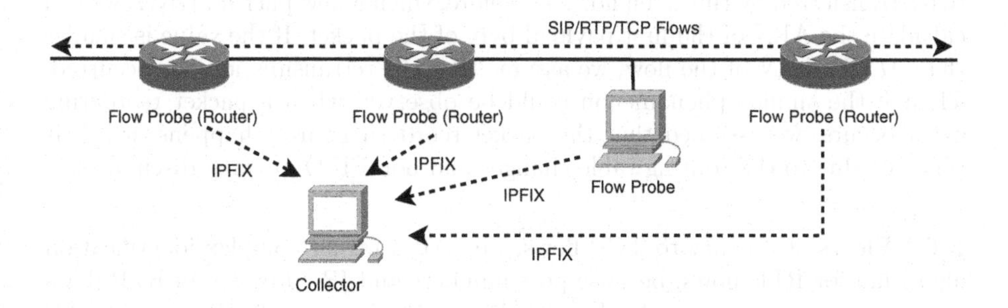

Fig. 1. The proposed IPFIX-based system architecture

As shown in Fig. 1, the flow probe can be deployed as either a function of a router or a specific system that captures packets and processes flows. A flow probe observes end-to-end flows using 5-tuples of packet headers, meters each flow, and exports the metered results to collector via the IPFIX protocol. The collector saves the exported information into DBMS. The saved information can be used later for further analysis or visualization.

As previously mentioned, the flow probe observes only TCP and SIP-signalled RTP flows. To identify RTP flows, the flow probe extracts 5-tuples of RTP flows associated with each SIP packet by inspecting SDP payloads of SIP INVITE/OK messages. The extracted information is used for RTP flow identification. For the TCP/RTP flows, the QoS metrics are also metered and exported.

3.2 The E2E Flow Monitoring Scenario

TCP Flows. In metering E2E performance parameters, it is assumed that every packet of a flow is captured without losses. Therefore, our flow monitoring scenario cannot be applied for the sampling-based measurement environment.

Given the assumption, it is straightforward to meter the throughput and the number of retransmissions of a TCP flow. We maintain a range of total bytes $[0, N]$ (N is integer, and $N > 0$) transferred for each flow. This range is tracked from TCP sequence numbers. Whenever a packet for the given flow is captured, we calculate the range of the payload bytes $[m, m + k]$ (m and k are integers, and $m, k > 0$) of the packet from its sequence number and payload size. If the range overlaps ($m < N$) with the current total range $[0, N]$, we conclude a retransmission has occurred. Therefore, we maintain following information for each TCP flow.

$$< ISN, WRAP\_CNT, MAX\_APN >$$

The initial sequence number (ISN) is captured from the TCP SYN message. $WRAP\_CNT$ is for counting how many times the sequence number of a flow wrapped around the maximum sequence number ($2^{32}-1$) to zero. MAX_APN is the maximum value of Absolute Position Number (APN). By shifting the ISN of a flow to zero, we know the absolute location of each byte (i.e., APN) on the total N bytes transferred by the given flow. Therefore, when a new packet arrives, we can calculate the APN of the first payload byte of the packet. If the value is smaller than MAX_APN of the flow, we assume that the retransmission has occurred. Though the similar phenomenon could be observed when a packet reordering event occurs, we assumed that the packet reordering rarely happens, which is typically due to the routing table changes and non-FIFO queuing disciplines.

RTP Flows. Contrary to TCP flows, we need a rather complex identification algorithm for RTP flows, because port numbers and IP addresses of RTP flows are not known in advance. As shown in Fig. 2, the 5-tuples of SIP-signaled RTP flows can be determined by inspecting the SDP payloads of SIP INVITE/OK packets exchanged between hosts [10]. The IP address of the party who sent the SDP can be found at the line starts with o=, and the port numbers can be found

SDP 1:
o=alice 2890844730 2890844731 IN IP4 *host.example.com*
m=audio 49170 RTP/AVP 0 8 97
m=video 51372 RTP/AVP 31 32

SDP 2:
o=bob 2890844526 2890844527 IN IP4 *host.biloxi.com*
m=audio 49174 RTP/AVP 0
m=video 52372 RTP/AVP 32

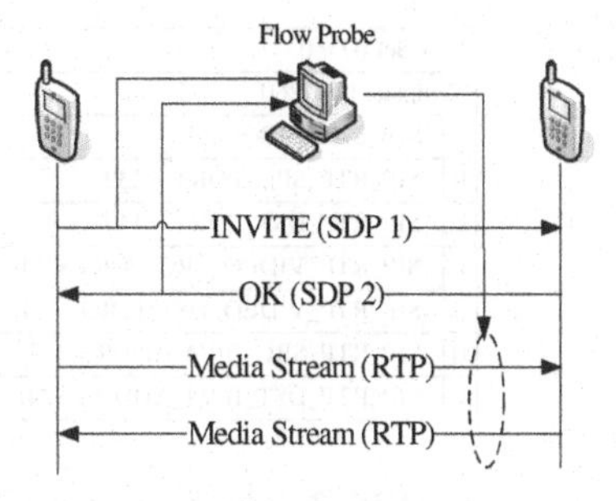

Fig. 2. The 5-tuples of the RTP streams are calculated from the SDP payloads

Algorithm 1. Calculating packet loss ratio of a RTP flow

1: $num\_of\_lost\_pkt \leftarrow num\_of\_lost\_pkt + current\_seq - prev\_seq$
2: $prev\_seq \leftarrow current\_seq$
3: $num\_of\_tot\_pkt \leftarrow prev\_seq + 1 - initial\_seq$
4: $pkt\_loss\_ratio \leftarrow (num\_of\_lost\_pkt / num\_of\_tot\_pkt) * 100$

at the lines start with m=. As a call possibly includes several media streams, multiple port numbers can be found in a SDP payload. The found 5-tuples are saved in an internal table. The QoS metrics are only metered for the RTP flows which have the matched entry in the table.

The packet loss ratio of a RTP flow is metered as shown in Algorithm 1. In the algorithm, *current_seq* and *prev_seq* are the RTP sequence numbers of the current and previous packets. *initial_seq* is the RTP sequence number of the first packet of the flow. The values of *num_of_lost_pkt* and *pkt_loss_ratio* are easily computed from the *current_seq* and *prev_seq*,

The jitter value of a RTP flow is metered by the Algorithm 2 defined in [11], where the *current_time* and *prev_time* are the timestamp values of the current and previous RTP packets. By tracking the value of variable *transit*, the delay is also monitored.

3.3 IPFIX Templates for Exporting E2E Flows

As the exporting information set differs by the protocol, we use different IPFIX flow templates for each of the protocol: SIP, RTP and TCP. The fields defined in TCP and RTP templates constitute a set of QoS parameters. The values of

Algorithm 2. Calculating jitter of a RTP flow

1: $transit \leftarrow current\_time - prev\_time$
2: $j \leftarrow |transit - last\_transit|$
3: $jitter \leftarrow (j - jitter)/16$
4: $last\_transit \leftarrow transit$

0 15	16 31	
FlowSet ID = 0		Length = 36
Template ID = 260		Field Count = 7
0	SIP_CALL_ID = 130	50
0	SIP_RTP_SRC_PORT = 139	2
0	SIP_RTP_DST_PORT = 140	2
0	SIP_RTP_VIDEO_SRC_PORT = 141	2
0	SIP_RTP_VIDEO_DST_PORT = 142	2
0	SIP_RTP_SRC_IPV4_ADDR = 143	15
0	SIP_RTP_DST_IPV4_ADDR = 144	15

Fig. 3. The proposed IPFIX template for SIP messages

0 15	16 31	
FlowSet ID = 0		Length = 60
Template ID = 261		Field Count = 13
0	LAST_SWITCHED = 21	4
0	FIRST_SWITCHED = 22	4
0	IN_BYTES = 1	4
0	IN_PKTS = 2	4
0	L4_SRC_PORT = 7	2
0	IPV4_SRC_ADDR = 27	4
0	L4_DST_PORT = 11	2
0	IPV4_DST_ADDR = 28	4
0	RTP_IN_JITTER = 154	4
0	RTP_IN_PKT_LOST = 156	4
0	RTP_IN_MAX_DELTA = 159	4
0	RTP_IN_MIN_DELTA = 161	4
0	RTP_IN_AVE_DELTA = 163	4

Fig. 4. The proposed IPFIX template for RTP messages

0 15	16 31	
FlowSet ID = 0		Length = 48
Template ID = 261		Field Count = 10
0	TCP_RETRAN_COUNT = 251	4
0	LAST_SWITCHED = 21	4
0	FIRST_SWITCHED = 22	4
0	IN_BYTES = 1	4
0	IN_PKTS = 2	4
0	L4_SRC_PORT = 7	2
0	IPV4_SRC_ADDR = 27	4
0	L4_DST_PORT = 11	2
0	IPV4_DST_ADDR = 28	4

Fig. 5. The proposed IPFIX template for TCP messages

the fields are calculated by the flow probe as presented in the previous section. The template used for SIP flows is shown in Fig. 3.

In Fig. 3, SIP_CALL_ID is an ID assigned by the flow probe to each SIP call. SIP_RTP_SRC_PORT and SIP_RTP_DST_PORT are source and destination port numbers assigned to the voice RTP flow. SIP_RTP_VIDEO_SRC_PORT and

SIP_RTP_VIDEO_DST_PORT are source and destination port numbers assigned to the video RTP flow. The rest of the fields are for source and destination IP addresses of the SIP session. The packets exported by this template are used only for correlating each call with its constituent RTP flows.

Fig. 4 is a IPFIX template for RTP flows. In this figure, LAST_SWITCHED and FIRST_SWITCHED are timestamp values demarking the start and end of a flow. IN_BYTE and IN_PKTS are counter values for the total volume and the number of packets observed. RTP_IN_JITTER is the jitter value of a flow. RTP_IN_PKT_LOST represents packet loss ratio of a flow. RTP_IN_MAX_DELTA, RTP_IN_MIN_DELTA, and RTP_IN_AVE_DELTA represent the maximum, minimum, and average delay.

Fig. 5 shows an IPFIX template for TCP flows. In this figure, TCP_RETRAN_COUNT is a counter for the number of retransmission. It is used to assess the transmission quality of a specific TCP flow.

4 Experiments

4.1 Environments

The experimental environment is shown in Fig. 6. We set up two end nodes for VoIP SIP soft-phones [12]. TCP flows are also tested between them. One of them is connected to the Internet through a commercial Wi-Fi access service, and the other node is connected to the CNU campus network. We have monitored the end-to-end flows at two observation points: one is the access router and the other one is the border router of the CNU campus. The mirrored traffic is captured by the flow probes. They IPFIX messages to the external flow collector.

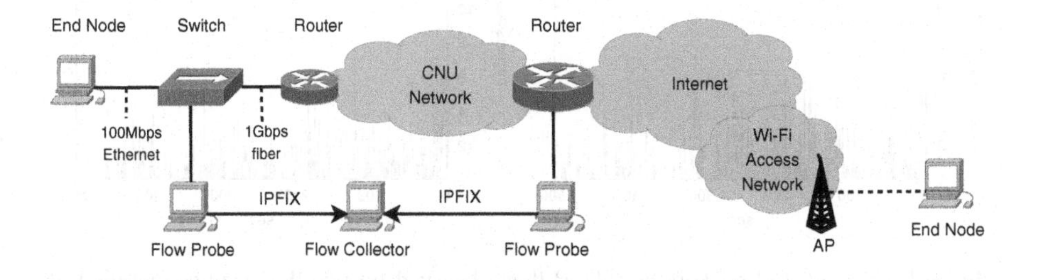

Fig. 6. Experimental Environment

The flow probes are plain Linux boxes equipped with our modified version of nProbe [9]. As the original implementation allows only one IPFIX template to be used for metering, we extended it to enable multiple IPFIX templates to be used at once, which is essential for SIP-based RTP monitoring. Moreover, we added some new fields to the SIP template included in the original implementation to support the monitoring of video RTP flows.

4.2 Results

Fig. 7 and 8 show the throughput and the number of retransmitted packets for a TCP flow observed at two flow probes. The congestion control behavior of the TCP sender is similarly seen because there are no queueing delays at the routers. The number of retransmitted packets is generally similar but not exactly same at two flow probes, because the retransmitted packets are not included by the same flow. On average, it is shown that the end-to-end QoS information of a TCP flow could be derived by our scheme.

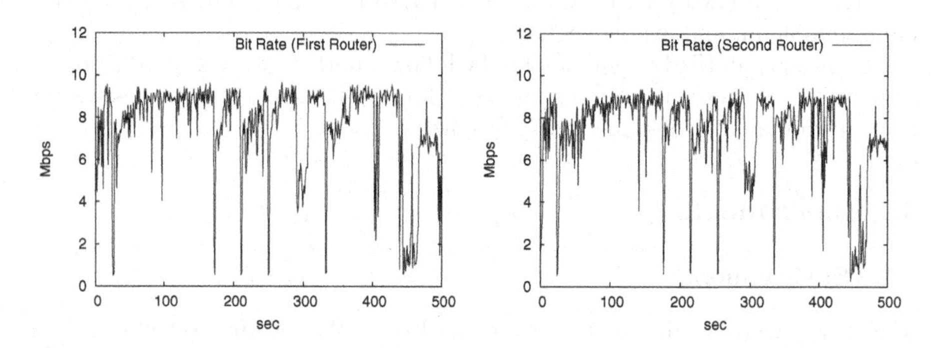

Fig. 7. Throughput (bit rate) of a TCP flow observed by two flow probes installed at routers

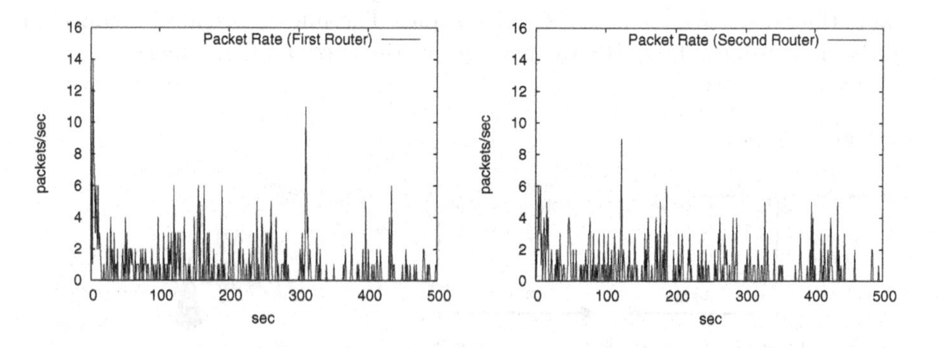

Fig. 8. Retransmitted packets of a TCP flow observed by two flow probes installed at routers

The representative QoS metrics of bit rate, delay, and jitter for a SIP/RTP flow are shown in Fig. 9, 10, and 11. The RTP stream generated by the PCM codec shows the constant bit rate of 81 Kbps. In the experiment, the observed QoS metrics imply the good quality of voice traffic. Specifically, delays and jitters shown in Fig. 10 and 11 are slightly different at each observation point, which caused by queueing delays at each router.

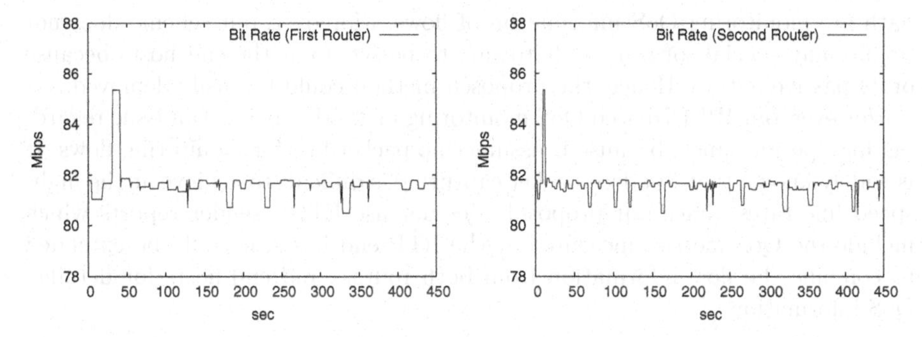

Fig. 9. Bit rates of a SIP/RTP flow observed by two flow probes installed at routers

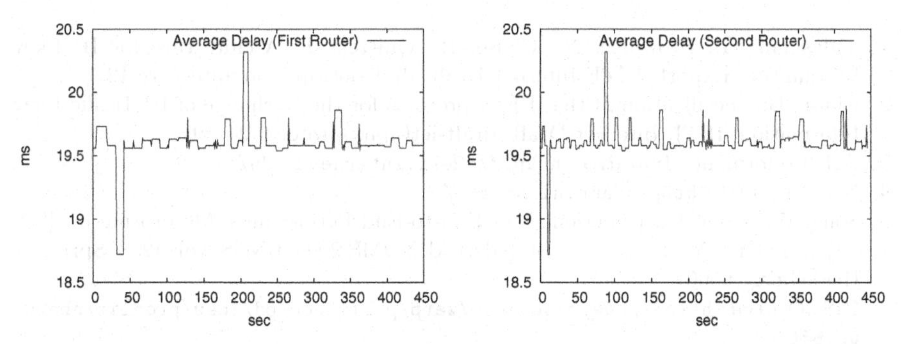

Fig. 10. Average delays of a SIP/RTP flow observed by two flow probes installed at routers

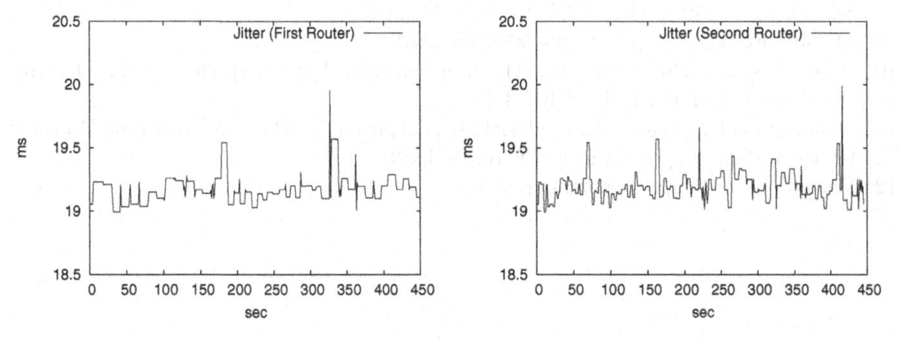

Fig. 11. Jitter values of a SIP/RTP flow observed by two flow probes installed at routers

5 Conclusion and Future Work

In this paper, we presented an E2E performance measurement scheme using IPFIX and its applications on TCP and SIP/RTP flows. Our method is useful for detecting problematic network segments because multiple flow probes on the path are monitoring QoS information of flows. Moreover, our scheme does not require any special software or hardware to be set up at the end hosts because of its passive nature. Hence, the proposed method could be easily deployed.

However, our IPFIX-based QoS monitoring method may have an issue regarding high performance, because it assumes no packet loss for monitoring flows. It is widely known that loss-free packet capturing is difficult to achieve at the high-speed line rates. Since our proposal does not use RTCP sender reports which include the QoS metrics measured at the RTP end hosts, it could be extended to combine the flow information from both routers and end-hosts for detailed QoS information.

References

1. Sadasvian, G., Brownlee, N., Claise, B., Quittek, J.: Architecture for IP Flow Information Export. IETF Internet-Draft, draft-ietf-ipfix-architecture-12
2. Claise, B.: specification of the IPFIX protocol for the Exchange of IP Traffic Flow Information. IETF Internet-Draft, draft-ietf-ipfix-protocol-24.txt
3. E2E Performance Initiative: `http://e2epi.internet2.edu/`
4. NLANR AMP: `http://amp.nlanr.net/`
5. Song, W., Choi, D.: Experiences in End-to-End Performance Monitoring on KOREN. In: Kim, Y.-T., Takano, M. (eds.) APNOMS 2006. LNCS, vol. 4238, Springer, Heidelberg (2006)
6. CISCO nBar: `http://www.cisco.com/warp/public/cc/pd/iosw/prodlit/nbarw_wp.pdf`
7. Lindh, T., Roos, E.: Monitoring of SIP-based Communication using Signalling Information for Performance Measurements. In: ICISP (2006)
8. Son, H., Lee, B., Shin, S., Lee, Y.: Flow-level QoS Monitoring of Multimedia Streaming Traffic with IPFIX. In: KNOM (2007)
9. nProbe: `http://www.ntop.org/nProbe.html`
10. Rosenberg, J., Schulzrinne, H.: An Offer/Answer Model with the Session Description Protocol(SDP). IETF RFC 3264
11. Schulzrinne, H., Casner, S., Frederick, R., Jacobsen, V.: RTP: A Transport Protocol for Real-Time Application. IETF RFC (1889)
12. WengoPhone: `http://openwengo.com/`

Advanced Scheme to Reduce IPTV Channel Zapping Time

Jieun Lee, Geonbok Lee, Seunghak Seok, and Byungdeok Chung

Network Technology Laboratory, KT, Daejeon, South Korea
{jieunlee,gblee,suksh,bdchung}@kt.co.kr

Abstract. IPTV just has been spread throughout the world. For the success of IPTV service, it is a key feature to satisfy QoS (Quality of Services) of total IPTV solution, such as tolerable channel zapping time. A related work reduces IPTV channel zapping time using a home gateway to take a role of IGMP-proxy. However, the related work considers the only case that a subscriber selects an adjacent channel of the current one. Additionally the related work requires a home gateway for each subscriber. In this paper, we propose a solution for the two problems using rating server without home gateway. Settop boxes instead of home gateways join or leave the adjacent channel groups for fast changes to the adjacent channels. Additionally settop boxes obtain an expected channel list by sending a query message to rating server and join the expected multicast groups in advance.

1 Introduction

As the border between broadcasting and IP communication is indistinct, the convergence between broadcasting and IP communication is getting faster and faster. IP communication companies recently attempt to advance to broadcasting market in corporation with broadcasting or movie companies. Broadcasting and cable companies are entering into high-speed internet service business. IPTV is the new convergence business to provide broadcasting through set-top box connected to internet. IPTV can support not only passive broadcasting service but also active, intelligent and bidirectional services such as VoD(Video on Demand), T-Commerce. Therefore IP communication companies expect that IPTV will be the representative service which can find a means of another huge earning. [1]

However to preoccupy and activate IPTV market, we should acquire multimedia QoS enough to satisfy subscribers. There are quality of video and audio media, fast channel zapping time, security and stability as QoS factors for IPTV service. Among them, intolerable service delay during channel change will be an obstacle to IPTV service expansion. According to a research, channel zapping time is increased rapidly as more subscribers are serviced and channel zapping time more 1 or 2 seconds lowers subscriber's satisfaction. [2]

There are many causes to occur service delay during channel change, which are settop box's processing time, encoding/decoding time, network delay etc. Among

S. Ata and C.S. Hong (Eds.): APNOMS 2007, LNCS 4773, pp. 235–243, 2007.

them, we focus on network delay time due to IGMP (Internet Group Management Protocol) [6] processing.

A previous work reduces channel zapping time using a home gateway to take a role of IGMP-proxy. A home gateway joins an adjacent channel of the current one in advance. Afterwards, if the subscriber changes IPTV channel to one of the adjacent channels, the home gateway can forward the multicast stream immediately and the subscriber changes new channel without network delay. However, the related work considers the only case that a subscriber selects an adjacent channel of the current one. If a subscriber uses hot-key on the remote controller or uses number button directly, the proposal doesn't work any more. As IPTV service is more activated, more random selection cases happen. Additionally the related work requires a home gateway for each subscriber and it is too inefficient. Therefore we need a solution to work well in both case of continuous channel change and random channel change without home gateway.

This paper is organized as follows. We present network topology and related protocols for IPTV services in section 2. And then we analyze a related work in section 3. In section 4, we propose advanced scheme to reduce IPTV channel zapping time using rating server without home gateway. In section 5, we compare the scheme with the previous work. Finally we come to a conclusion and offer future works in section 6.

2 Background

IPTV service platform is as like **Fig. 1**. It consists of IPTV head-end centre, backbone network, local network, access network and home network.

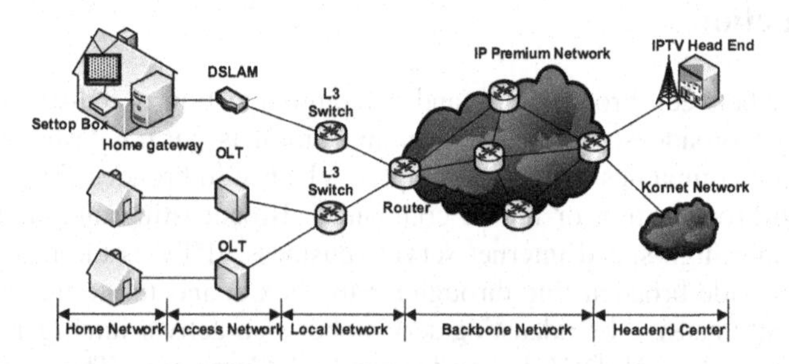

Fig. 1. IPTV Service Platform

IPTV head-end centre makes MPEG2-TS (MPEG2-Transport Stream) packets with video and audio and data from contents provider and sends them through IP premium network. It uses IP multicast protocol to forward packets because broadcasting service shares same data to many IPTV subscribers. A channel is assigned to one multicast group IP address and the address is used for source IP address. Routers which constructs IP premium network updates their routing table using multicast routing protocol such as PIM-SM (Protocol Independent Multicast - Sparse Mode) [3], PIM-DM (Protocol Independent Multicast - Dense Mode) [4], DVMRP (Distance Vector Multicast Routing Protocol) [5]. The routers forwards broadcasting data using multicast tree in their routing table. Broadcasting data is forwarded through backbone network, local network, access network and home network. Finally, settop box which is connected to home gateway can receive broadcasting data. Settop box decodes received IP packets, changes them into video and audio and sends it to IPTV.

L3 switch which is the closest equipment to home gateway uses IGMP [6] to manage subscribers who exist in the sub-network. When settop box receives channel change signal, it stops decoding current channel data and sends IGMP join message for a new selected channel to the upper network. When home gateway receives IGMP join message from settop box, it forwards the packet to the upper network. L3 switch which receives the packet from home gateway sends multicasting packet such as PIM-SM join to the upper network to join the channel group. When it finishes, L3 switch can receive multicast data packets for the new channel and sends them to sub-network.

Settop box which sends IGMP Join simultaneously sends IGMP leave message for current channel to home gateway. Home gateway received IGMP leave message forwards it to the upper network. When L3 switch receives the IGMP leave message, it sends IGMP Query message to sub-network to check whether any subscriber still watching the channel exists or not. If L3 switch doesn't receive response message for the query within configured maximum response time, it sends multicast packets such as PIM-SM leave message to the upper network to leave from the multicast group. Otherwise, it means that there is someone still watching the channel and therefore L3 switch don't need to do any action. IGMP operation for IPTV service is as like Fig. 2.

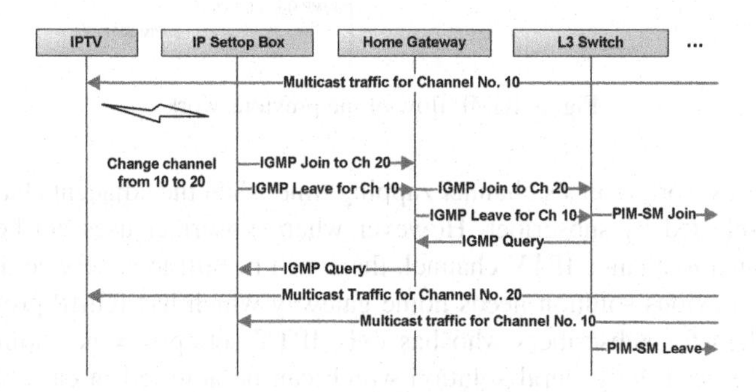

Fig. 2. IGMP operations for changing IPTV channel from 10 to 20

3 Previous Work

To reduce IPTV channel zapping time, a previous work [7] uses home gateway which has IGMP proxy function.

When home gateway receives IGMP Join or IGMP Leave message for a specific multicast group from IPTV settop box, it handles the IGMP Join or IGMP Leave operations for both the specific multicast group and the adjacent ones to the specific multicast group. Therefore, home gateway can receive both current channel traffic and the adjacent channels' traffic when IGMP Join operation is completed. In this situation, if subscriber change IPTV channel to the adjacent channel of current one, home gateway does not need to handle another IGMP Join operation and only forwards sub-network the adjacent channel traffic which has been being received since completion of IGMP operations for the adjacent channels. Afterwards there will be no IPTV service delay due to network delay in case of a change to one of the adjacent channels. When home gateway receives IGMP Leave message, it handles IGMP operations for both the related channel and the adjacent one of the related channel to reduce unnecessary channel traffics. The related message flow is as like **Fig. 3**.

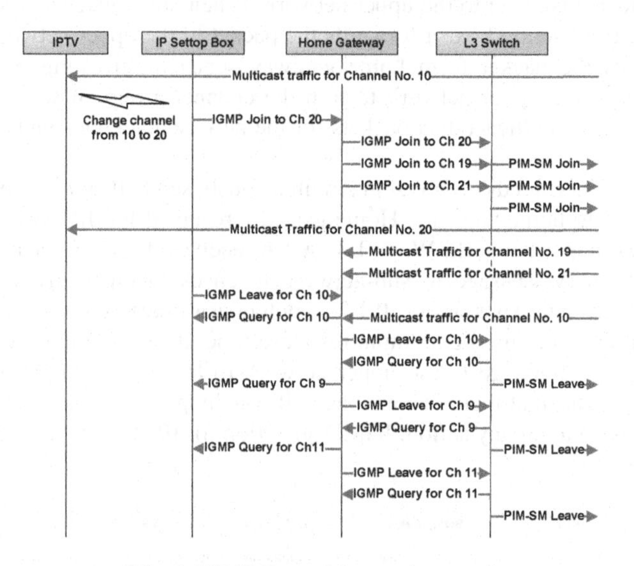

Fig. 3. IGMP flow of the previous work

The previous work reduces channel zapping time when the adjacent channel of current one is selected by subscriber. However when subscriber uses hot-key or select random channel to change IPTV channel, there will be still long service delay. Additionally, the previous solution needs home gateway which has IGMP proxy function and it's useless for subscribers who has only IPTV settop box not home gateway. Therefore, we need the general solution which can be adjusted in case of both continuous selection and random selection.

4 Proposal

This paper proposes a scheme to reduce channel zapping time in case of not only continuous channel selection but also random channel selection using rating server without home gateway. To reduce channel zapping time to adjacent channels, settop box instead of home gateway manages adjacent channel multicast group list and handles IGMP Join and Leave operation for adjacent channels. To reduce channel zapping time to random channels, settop box queries expected channel list to rating server and handles IGMP Join and Leave operation for the expected channels. Rating server gathers channel change event from settop boxes and manages statistics for each settop box. And if it receives a query message from a settop box, it obtains expected channel list from managed statistics and responses the list to the settop box. Settop box can obtain its expected channel information of specific time zone by sending a query message to rating server periodically.

1) Operations for rating server

In our proposal, rating server stores channel information sent from settop boxes and generates statistics from it. And if it receives a query message for expected channel list from a settop box, it obtains expected channel list from statistics and sends a response for the query to the settop box. Rating server locates at IP media centre in IP premium network.

First, rating server generates expected channel list from channel information received from settop boxes. When a subscriber changes his/her channel, the settop box sends only meaningful channel data to rating server. Meaningful data means that a subscriber picks up the channel and watches it more than 10 seconds. The channel data includes that settop box ID, child pin (subscriber's information), start time of audition, event generation time, remote controller action information (settop box ON/OFF, Change channel etc.), the previous channel, current new channel and so on. Using these channel data, rating server generates and manages channel statistics which supports daily audience share, monthly audience share, regional audience share and so on.

If rating server receives a query message for expected channel list from a settop box, it generates the list from channel statistics and sends a response message including the list to the settop box which requests it. The algorithm which generates expected channel list using channel statistics is out of this paper.

2) Operations for Settop box

In our proposal, a settop box plays a role of obtaining expected channel information and joining the related multicast groups in advance. There is two ways to obtain expected channel list.

First, a settop box manages adjacent channel list instead of home gateway in the previous work to prepare for the case of continuous channel change. The number of adjacent channels can be adjusted according to subscriber's available network bandwidth. The related work flow is as like **Fig. 4**.

Second, a settop box obtains expected channel list from rating server to prepare for the case of random channel change using hot-key or direct channel selection. A settop box queries expected channel list for the next time duration to rating server periodically. Rating server which received a query message for its expected channel list from a settop box extract expected channel list from channel statistics and sends the response to the settop box. The settop box obtaining expected channel list joins multicast groups related to the expected channel in advance using IGMP Join message. The related work flow is as like **Fig. 5**.

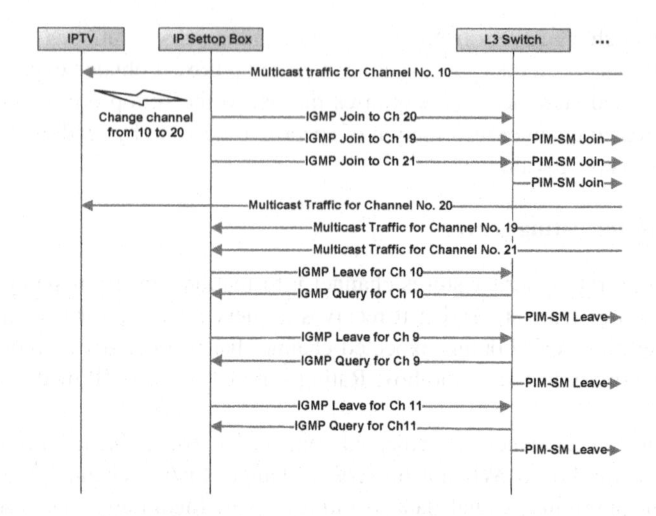

Fig. 4. Workflow of changing to adjacent channel

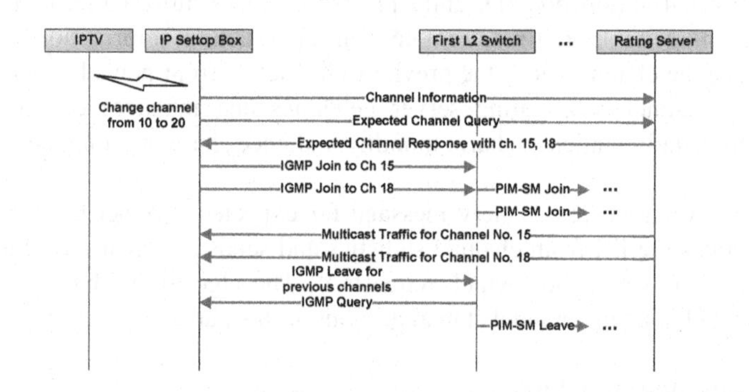

Fig. 5. Workflow using rating server

After a settop box sends IGMP join message for new channel and IGMP leave message for the current channel, it updates its local adjacent channel list and expected channel list. And it repeats IGMP operations for new added channels and unnecessary deleted channels in the lists.

3) Message format

Message formats between rating server and a settop box are as like **Fig. 6**, **Fig. 7** and.**Fig. 8** All messages are TCP packets and they contains application header like **Fig. 6**. Device type contains a code meaning home gateway, settop box, modem etc. Message ID stores identifier to map query and response message. Message Type is reserved field for future usage. Command type classifies Response and Query. Body length stores body field total length except for application header length.

Fig. 7 is a query message format from a settop box to rating server. SA-ID is identifier for a settop box and alphanumeric string. Child pin is identifier for an IPTV user

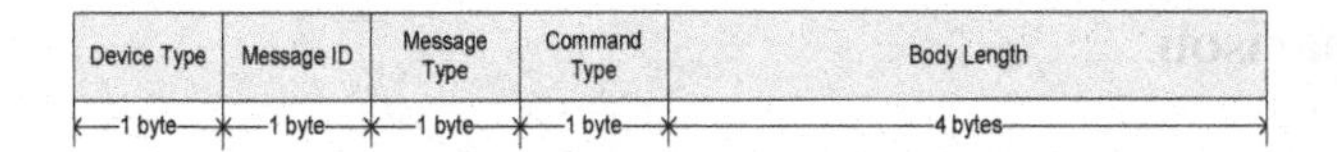

Fig. 6. Header format

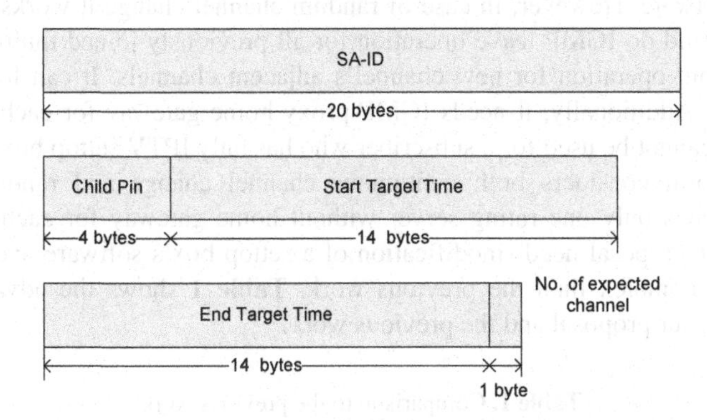

Fig. 7. Query message format

which consists of four numbers. Start Target Time and End Target Time is target time zone of expected channel. No. of expected channel means the number of requested channel. The more network bandwidth a subscriber has, the more expected channels one can obtain. It means if a subscriber has more available bandwidth, one can enjoy IPTV service with less channel zapping time.

Fig. 8 is a response message format from rating server to a settop box. SA-ID, Child pin, Start target time, end target time and No. of expected channel in a response message is the same as those in a query message. Ch No. is expected channel number and added it as many as No. of expected channel.

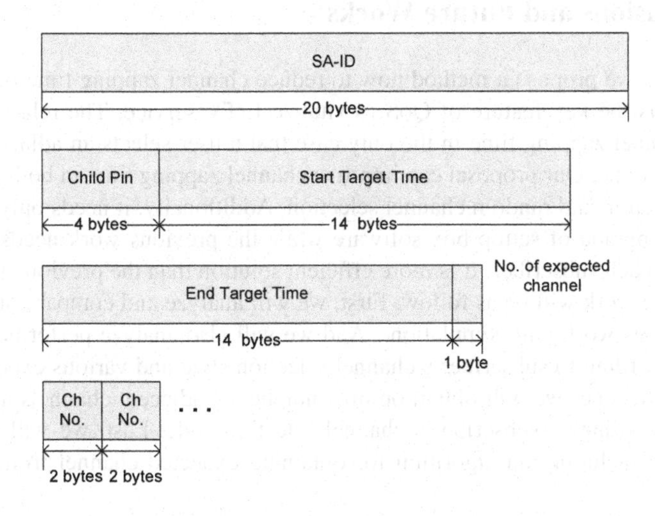

Fig. 8. Response message format

5 Comparison

The previous work reduced channel zapping time only in case of continuous channel switch. It operates simpler than our proposal and it doesn't need modification of a set-top box's software. However, in case of random channel change it works worse than before. It should do IGMP leave operation for all previously joined multicast groups and IGMP join operation for new channel's adjacent channels. It can lower overall performance. Additionally, it needs IGMP-proxy home gateway for each subscriber. Therefore it cannot be used for a subscriber who has only IPTV settop box.

Our proposal considers both continuous channel change and random channel change. It needs only one rating server without home gateway for each subscriber. However, our proposal needs modification of a settop box's software and the operation is more complex than the previous work. **Table 1** shows the advantages and drawbacks of our proposal and the previous work.

Table 1. Comparison to the previous work

	Previous work	Proposal
advantage	- simple settop box operation - no changes of IPTV settop box	- can be adjusted in case of contin uous & random selection - need only one rating server not one home gateway in one house
drawback	- cannot be adjusted in case of r andom selection - need one IGMP-Proxy home g ateway to a subscriber	- need changes of IPTV settop bo x - complex settop box operation

6 Conclusions and Future Works

In this paper, we propose a method how to reduce channel zapping time in IPTV service which is the key feature of QoS to vitalize IPTV service. The related work [7] reduces channel zapping time in the only case that a user selects an adjacent channel of the current one. Our proposal can improve channel zapping time in both continuous channel selection and random channel selection. Additionally, it needs only one rating server and upgrade of settop box software while the previous work needs one home gateway for each subscriber. It is more efficient solution than the previous work.

The future work will be as follow. First, we will analyze and compare our proposal to the previous work with simulation. And we will also analyze performance of our proposal according to subscriber's channel selection style and various expected channel hit rate. Second, we will obtain optimal number of adjacent channels and random channels according to subscriber's channel selection style. Last, we will implement our proposal including an algorithm for obtaining expected channel from rating information.

References

1. Son, J.: Triple Play Service: Broadcast TV and VoD over IP. Netmanias Network Newsletter 1(1) (2003)
2. Agilent Technologies. Inc. White Paper: Ensure IPTV Quality Of Experience (2005)
3. Estrin, D.: RFC2362: Protocol Independent Multicast – Sparse Mode (PIM-SM), IETF (June 1998)
4. Adams, A.: RFC3973: Protocol Independent Multicast – Dense Mode (PIM-DM), IETF (January 2005)
5. Waitzman, D.: RFC1075: Distance Vector Multicast Routing Protocol, IETF (November 1988)
6. Fenner, W.: RFC2236: Internet Group Management Protocol, Version 2, IETF (November 1997)
7. Cho, S.: Improvement of Channel Zapping Time in IPTV Services Using the Adjacent Groups Join-Leave Method. In: Advanced Communication Technology, 2004. The 6th international conference on, vol. 2 (2004)

XML-Based Policy Engineering Framework for Heterogeneous Network Management

Arjmand Samuel[1], Shahab Baqai[2], and Arif Ghafoor[1]

[1] School of Electrical and Computer Engineering, Purdue University
West Lafayette, Indiana 47906, USA
{amsamuel,ghafoor}@purdue.edu

[2] Lahore University of Management Sciences, Lahore, Pakistan
baqai@lums.edu.pk

Abstract. Network services, resources, protocols, and communication technologies give rise to multiple dimensions of heterogeneity in an enterprise, resulting in a complex administration and management operation. The administrative challenge is further exacerbated by the fact that multiple enterprises share resources and users, giving rise to semantic conflicts. While Policy-Based Network Management (PBNM) provides an effective means of automating and simplifying administrative tasks, it can also cause conflicts between policies meant for separate network entities, consequently giving rise to policy heterogeneity. In order to address issues of network and policy heterogeneity, we propose a policy engineering framework using the tried and tested system development methodologies from Software Development Life Cycle (SDLC) and apply it to PBNM language engineering. We present an XML based policy specification language, X-Enterprise, its corresponding UML meta-model along with a context sensitive and adaptive implementation framework. Use of XML and UML affords an open standard for cross-architecture implementation and use of existing UML tools for consistency and conflict analysis.

1 Introduction

Policy based networking has emerged as one of the most popular approaches to automating the management of large enterprise-wide networks [1]. The management tasks such as configuration, performance, security, fault restoration and service provisioning are complex, and have far reaching consequences [2]. This complexity is further exacerbated with the inclusion of context and consequent adaptation of network resources. Scalability and interoperability add yet another dimension to the administration and management problem. Policy Base Network Management (PBNM) has effectively addressed the above concerns and provides a seamless and cost effective means for managing vast, disparate and dispersed network resources.

The issue of heterogeneity across multiple networked enterprises can be further divided into semantic heterogeneity of services, resource heterogeneity (band width, time slots, frequency bands, buffers, OS support), protocol heterogeneity

S. Ata and C.S. Hong (Eds.): APNOMS 2007, LNCS 4773, pp. 244–253, 2007.

(TCP, IPv4, IPv6, ATM, CDMA, OFDMA) and communication technology heterogeneity (wired Vs wireless). The resolution of heterogeneity is important for the interconnection of a large number of disparate networks and to provide secure access to a wide variety of applications and services any where and any time. Scalability is yet another pertinent issue for the design of a large interconnected networked enterprise with vast number of end-nodes and applications.

In order to manage both interoperability and scalability of network resources at a multi-enterprise scale, we propose a Policy-Based Network Engineering approach. We employ a standardized methodology for the design of network policies using principles from resource management and software engineering. Specifically, we adopt the requirement specification methodology from Software Development Life Cycle (SDLC) and apply it to the engineering of network policies. We provide a UML-based meta-model of the network policy language thus exploiting the virtues of UML, ranging from expressive modeling to direct conversion from UML to XML schema [3]. We have successfully applied this approach to policy engineering for secure federated access management systems [4]. Combining these two efforts results in a uniform development and deployment framework of interoperable policies for Quality of Service (QoS) management and access control in heterogeneous networked enterprises.

The remainder of this paper is organized as follows. Section 2, presents the issues of network and policy heterogeneity. Section 3 presents the X-Enterprise policy language grammar. Section 4, discusses the UML-based X-Enterprise Metamodel and Section 5, presents the policy enforcement framework. Section 6 briefly outlines the experimentation conducted using the policy enforcement framework. Section 7, presents the related work and finally, Section 8, concludes the paper.

2 Policy Based Network Management

Network enterprise is a collection of heterogeneous services, resources, protocols and communication technologies. Policy based network management aims to create a collection of policies to effectively control this diverse set of entities. While each constituent policy controls its own stated domain of resources or services, the interaction between policies gives rise to another dimension of heterogeneity, namely *policy heterogeneity*. Conflicts of co-existence between policies can effectively cause the network enterprise to exhibit unexpected behavior. The range heterogeneity dimensions can best be addressed, on the one hand, by following the established system development paradigms such as Software Development Life Cycle (SDLC) and using the known verification and validation techniques, and on the other hand, by using architecture and system independent representation languages and formulations like XML and UML.

2.1 Network Heterogeneity

As mentioned above, sources of heterogeneity in a network enterprise are semantic heterogeneity of services, resource heterogeneity, protocol heterogeneity and communication technology heterogeneity. Network services are designed with specific end-requirements in mind. This gives rise to abundance of semantically diverse services which add to the problem of semantic heterogeneity. Physical resources range from gateways, routers, bridges, switches, hubs, repeaters, proxies, firewalls etc. Further, these resources are dispersed though out the enterprise. Route selection and allocation is also a task associated with optimal operation of the enterprise network and diverse routes offer dissimilar QoS behavior [5]. Issues of availability or non-availability of frequency bands, time slots, OS support etc also add to resource heterogeneity in a networked environment. Similarly, a network enterprise is home to a wide variety of protocols. Most of these protocols are designed with a specific task or QoS in mind and at times compete with each other on grounds of better QoS. The choice of UDP vs TCP is a typical example for protocol selection. TCP is capable of detecting congestion in the network and will back off transmission speed when congestion occurs. These features protect the network from congestion collapse. On the other hand UDP allows the fastest and most simple way of transmitting data to the receiver. There is no interference in the stream of data that can be possibly avoided. This provides the way for an application to get as close to meeting real-time constraints as possible. Likewise, algorithms also compete to provide services, and often, selection of an algorithm is based on QoS requirements thus contributing to protocol heterogeneity.

2.2 Policy Heterogeneity

As mentioned earlier, in order to control a heterogeneous set of network entities, a number of policies exist in an enterprise. Notable among them are the Resource Policy, Routing Policy, Monitoring Policy, User Policy and Re-configuration Policy. The resource policy is responsible to exercise control over all physical network resources such as routers, firewalls, switches, base-stations etc. The Routing policy states the conditions and parameters which affect the selection of a particular route. Monitoring policy lays down the framework for collection, processing and storage of environmental context. User Policy stores all SLA relevant information along with expected user credentials and any additional assignment constraints. The re-configuration policy provides a means for defining parameters and commands useful for the reconfiguration of network devices. While each of the above mentioned policies aim to exercise control over its own subservient entities, they indirectly interact with each other causing co-existence conflicts. Example of such a conflict is the route policy allowing a certain route to a user request, but resource policy disallowing the user use of the route. Clearly, any meaningful and coherent implementation of policies in an enterprise has to be verified for correctness and feasibility against the original user requirements and co-existence conflicts.

2.3 Proposed Approach

We propose a policy engineering mechanism which effectively addresses network and policy heterogeneity. Our approach offers a generic methodology for

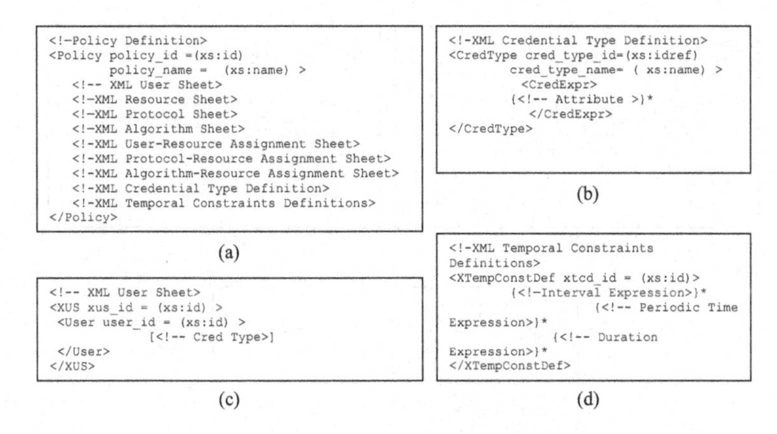

<!-Policy Definition>
<Policy policy_id =(xs:id)
policy_name = (xs:name) >
<!-- XML User Sheet>
<!-XML Resource Sheet>
<!-XML Protocol Sheet>
<!-XML Algorithm Sheet>
<!-XML User-Resource Assignment Sheet>
<!-XML Protocol-Resource Assignment Sheet>
<!-XML Algorithm-Resource Assignment Sheet>
<!-XML Credential Type Definition>
<!-XML Temporal Constraints Definitions>
</Policy>
(a)
<!-XML Credential Type Definition>
<CredType cred_type_id=(xs:idref)
cred_type_name= (xs:name) >
<CredExpr>
{<!-- Attribute >}*
</CredExpr>
</CredType>
(b)
<!-- XML User Sheet>
<XUS xus_id = (xs:id) >
<User user_id = (xs:id) >
[<!-- Cred Type>]
</User>
</XUS>
(c)
<!-XML Temporal Constraints Definitions>
<XTempConstDef xtcd_id = (xs:id)>
{<!-Interval Expression>}*
{<!-- Periodic Time Expression>}*
{<!-- Duration Expression>}*
</XTempConstDef>
(d)

Fig. 1. X-Enterprise Grammar (a) Top level policy definition. (b) Credential Type Definition. (c) User sheet. (d) Temporal constraint definition.

specifying all enterprise wide policies such as resource management policies, security management policies and network management policies. We employ an XML based grammar for the specification of system policies, and incorporate context parameters from the operating environment to adapt the system policies for optimal and desired operation. We address the issues of policy heterogeneity by defining a corresponding UML meta-policy, thus allowing standard UML based tools for detection and resolution of schema conflicts. This approach results in a modular language which can be extended and applied to a variety of systems and architectures. The policy thus created is configurable and maintainable using the same methodologies as other software. Since the policy is a software artifact and is in XML format, it can be deployed in a uniform manner across architectures and systems. Further, the policy can also be verified and validated using the standard software engineering approaches.

3 X-Enterprise Policy Language

We propose the X-Enterprise policy language for the specification of network behavior in an enterprise. The syntax of our proposed language is based on our previously proposed X-Grammar [4], used for policy engineering of secure federated access management systems. The policy language itself is based on a Backus-Naur Form (BNF) grammar with terminals and non-terminals. This allows X-Enterprise language to be accepted by an automaton. It also supports tagging of XML, which allows expressing attributes within the tags. We use the XML schema syntax as the type definition model. The non-terminals are expressed as $<!-'' non - terminal - name'' >$, whereas the terminals are the standard XML tags. Optional tags are placed within square brackets $''[]''$. Group portions of a production are included in curly brackets "{}", with the repeat count indicated by a subscript. The default count is one. A $''*''$ and a $''+''$ indicates a count of zero or more and one or more respectively, whereas a $''-''$ is

```
<!-XML Resource Sheet>
<XResS xress_id = (xs:id) >
  <Resource
   res_id = (xs:id)
   res_name = (xs:idref)
  [prop =(no_prop|first_level|cascade)] >
   <Object res_type_id=(xs:idref) />
   <Operation> (COPS commands)
   </Operation>
  </Resource>
</XResS>
```

(a)

```
<!-XML Protocol Sheet>
<XProtS xprots_id = (xs:id) >
  <Protocol
   prot_id = (xs:id)
   prot_name = (xs:idref)
    <Object prot_type_id=(xs:idref) />
   <Operation>
   </Operation>
  </Protocol>
</XProtS>
```

(b)

```
<!-XML Algorithm Sheet>
<XAlgoS xalgo_id = (xs:id) >
  <Algorithm
   algo_id = (xs:id)
   algo_name = (xs:idref)
    <Object algo_type_id=(xs:idref) />
   <Operation>
   </Operation>
  </Algorithm>
</XAlgoS>
```

(c)

Fig. 2. X-Enterprise Grammar (a) Resource Sheet. (b) Protocol Sheet. (c) Algorithm Sheet.

used to provide a range. A "|" indicates alternates within a production set, and exactly one can be chosen. The data types of the values of elements or attributes are indicated inside parenthesis "()" symbol.

The top level components of the X-Enterprise language are depicted in Figure 1(a). Credentials of the user request form the basis for allocating network resources to the user. The credential definition in X-Enterprise policy language is attribute-based and is modular, allowing fine grained control. Typical example of credentials of a user request is class of service of user, such as Gold User, Silver User etc. Inherent grouping provided by user request credentials allows scalability as well as fine grained control. The credentials are defined using grammar in the XML Credential Type Definition sheet (CredType) and are depicted in Figure 1(b). Each credential type bears an ID, a name (Gold, silver etc) and a set of attributes which corresponds to the environmental context parameters (e.g. originating from subnet xx.xx.xx.xx etc).

Users in the enterprise network are defined in the XML User sheet (XUS). The grammar is depicted in Figure 1(c). Each user ID has a set of credentials associated with it, which can be used to either define grouping of user or can be effectively used to add a fine grained user level control over user activity in the enterprise.

In order to provide fine grained control of enterprise resources in the temporal dimension, we also define temporal constraints using the XML Temporal Constraint Definitions (XTempConstDef) (Figure 1(d)). Modular definition of temporal constraints in a network enterprise aid in adaptation of network provisioning with the change in temporal context. Temporal constraints are particularly relevant as a user with certain credentials is only allowed to access resources for a given duration. Duration expressions are used to express the duration of usage of a certain resource irrespective of the time of activation.

```
<!-- XML User-Resource Assignment Sheet>
<XUResAS xuresas_id =(xs:id)>
  <UResA uresa_id=(xs:id) res_id=(xs:idref)>
   <AssignUsers>
     <AssignUser user_id=(xs:idref)>
            <AssignConstraint
[op =(AND|OR|NOT|XOR)]>
                  <AssignCondition
cred_type_id=(xs:idref)
                    [pt_expr_id=(xs:idref) |
d_expr_id=(xs:idref)] >
                       <LogicalExpr [op =
(AND|OR|NOT)]>
                     {<!-- Predicate>}+
   </LogicalExpr>
          </AssignCondition>
     </AssignConstraint>
  </AssignUser>
 </AssignUsers>
 </UResA>
</XUResAS>
```

(a)

```
<!-- XML Protocol-Resource Assignment Sheet>
<XPResAS xpresas_id =(xs:id)>
        <PResA presa_id=(xs:id)
res_id=(xs:idref)>
         <AssignProtocol>
      <AssignProtocol prot_id=(xs:idref)>
       <AssignConstraint
[op =(AND|OR|NOT|XOR)]>
                  <AssignCondition
cred_type_id=(xs:idref)
          [pt_expr_id=(xs:idref) |
d_expr_id=(xs:idref)] >
          <LogicalExpr
[op = (AND|OR|NOT)]>
                     {<!-- Predicate>}+
   </LogicalExpr>
          </AssignCondition>
     </AssignConstraint>
  </AssignProtocol>
 </AssignProtocol>
 </PResA>
```

(b)

```
<!-- XML Algorithm-Resource Assignment Sheet>
<XAResAS xaresas_id =(xs:id)>
        <AResA aresa_id=(xs:id)
algo_id=(xs:idref)>
         <AssignAlgorithm>
      <AssignAlgorithm algo_id=(xs:idref)>
       <AssignConstraint
[op =(AND|OR|NOT|XOR)]>
                  <AssignCondition
cred_type_id=(xs:idref)
          [pt_expr_id=(xs:idref) |
d_expr_id=(xs:idref)] >
          <LogicalExpr
[op = (AND|OR|NOT)]>
                     {<!-- Predicate>}+
   </LogicalExpr>
          </AssignCondition>
     </AssignConstraint>
  </AssignAlgorithm>
 </AssignAlgorithm>
 </AResA>
```

(c)

Fig. 3. X-Enterprise Grammar (a) User-Resource Assignment Sheet. (b) Protocol-Resource Assignment Sheet. (c) Algorithm-Resource Assignment Sheet.

Each network enterprise is composed of a set of resources such as base stations, routers, network storage devices, gateways, bridges, switches etc. In most cases, these resources are Active Network Components (ANC) and can be managed by running management software or scripts [6]. We propose to control the behavior of ANC devices by using fragments of X-Enterprise policy which is parameterized by context parameters. Each ANC device in the enterprise is defined in the XML Resource Sheet (XResS) as depicted in Figure 2(a). The definitions include the id and name of the resource together with the type (e.g. router, base station etc). XResS also provides definition of operations allowed on the resource in the form of COPS [7] commands.

In order to provide maximum flexibility to the user, X-Enterprise allows the selection of appropriate protocols for each component resource, depending on the context in which the resource is used. In X-Enterprise, these protocols are defined in the XML Protocol Sheet (XProtS) together with operations in which the protocol is to be used (Figure 2(b)). Example of two competing protocols in layer 3 are IPv4 and IPv6. Similarly, an example in layer 4 is UDP and TCP.

X-Enterprise provides an additional control over the network resources by enabling the enterprise to select the algorithm to be used by a resource in a particular situation. An example of competing choices is the packet scheduling algorithms in an IP router. A typical router in an enterprise has the choice to apply priority queuing, weighted fair queuing or custom queuing [8,9,10] besides

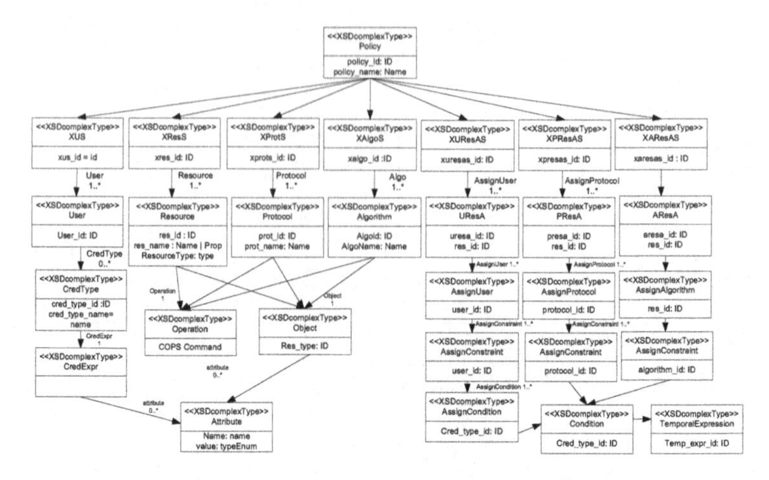

Fig. 4. UML Meta-Model of X-Enterprise policy language

other queuing algorithms. Custom queuing guarantees bandwidth per queue but does not adapt to changing network conditions. Priority queuing works well for a small number of traffic types and is not used for voice. Weighted fair queuing works well for data traffic and is not recommended for voice. However it can be used when no alternative is available, although it dynamically adapts to changing network conditions. Given the wide range of advantages and disadvantages of picking one packet scheduling algorithms over the other, the enterprise administrator only defines the conditions for each of the algorithms in X-Enterprise. Each algorithm is defined in the XML Algorithm Sheet (XAlgoS) and its grammar is depicted in Fig 2(c).

The flexibility to assign user requests to any resource in the enterprise is provided by the XML User-Resource Assignment Sheet (XUResAS). As depicted in Figure 3(a) XUResAS provides definition of user request with a resource ID (res_id) along with the assignment conditions. In order to afford maximum flexibility for the evaluation of the constraints logical expression such as AND, OR, NOT can also be defined here. Similarly, Protocols and Algorithms (as defined in XProtS and XAlgoS, respectively) are also assigned to resources by defining the XML Protocol-Resource Assignment Sheet (XPResAS) and XML Algorithm-Resource Assignment Sheet (XAResAS) respectively. These two sheets are also depicted in Figure 3(b) and (c).

4 UML Based X-Enterprise Meta-Model

The UML model of the proposed policy language depicted as a class diagram in Figure 4. UML is a standard approach for modeling of complex systems. By providing a UML model of our policy language, we also exploit the translation mechanisms available for translating UML models to readily usable program code. Our aim in this regard is to instantiate the UML meta-model using an

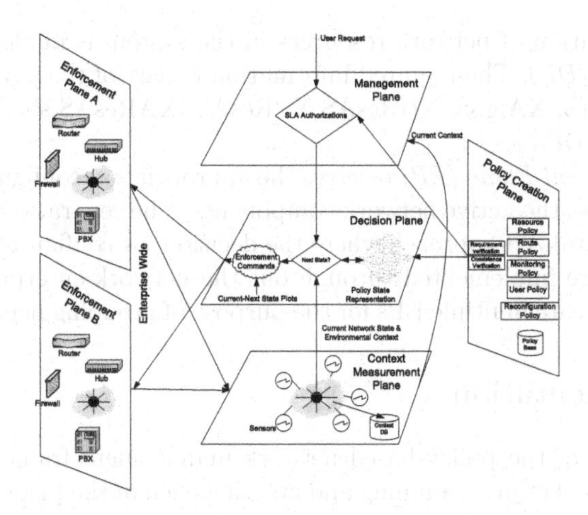

Fig. 5. X-Enterprise Enforcement Framework

XML schema [3] and use XML based policy for enforcement in an enterprise. In order to define our UML policy in XML we adopt a UML profile for XML Schema (UPX) [3].

5 Policy Enforcement Framework

In this section we present an enforcement framework for the X-Enterprise policy language in a network enterprise. The framework is depicted in Figure 5 and is guided by the IETF standards [11, 12]. The framework is broken down into planes, each of which is responsible for a set of tasks.

The *Policy Creation Plane (PCP)* provides an administrative interface for creation, transformation, XML policy instantiation, verification and storage of the policy. The transformation activity in PCP transforms the higher level user requirements to system level parameters [13]. Next, the XML policy as presented in Section 3, is instantiated with the resulting policy parameters. The resulting policy is then stored in the Policy Base.

The *Management Plane (MP)* interprets the user component of the policy (XUS) and implements the appropriate subscription levels for each user. The credentials provided by the user request are evaluated against the XUS definition of user. The current context parameters (e.g. time etc) are fetched from the context plane and form the basis for the assignment decision of the user.

The system environment is monitored by the *Context Plane (CP)*. Each network resource is connected to the CP and context parameters such as packet drop, end to end delay, jitter delay etc are monitored and converted to usable values by the Logical Context sub-module for onward storage in the context DB. The collection activity is periodic and is set by the administrators depending on the volatility of the network.

The configuration of network resources in the system is implemented by the *Decision Plane (DP)*. The required information is part of the policy in the form of XResS, XProtS, XAlgoS, XUResAS,XPResAS, XAResAS sheets and is made available to the DP.

The *Enforcement Plane (EP)* receives the appropriate reconfiguration instructions and controls the active network components. The separation of DP and EP ensures a decentralized approach where the decision for configuration is made at one point but are implemented through out the network enterprise. Also, each enterprise can create multiple EPs for the purpose of grouping network resources.

6 Experimentation

Implementation of the policy-based network management framework has been achieved using socket programming and creating each of the planes mentioned in Section 5. Policy files used in the experimentation as well as the complete documentation of the experiments can be accessed at http://web.ics.purdue.edu/ amsamuel/673/project1.htm.

7 Related Work

Notable among network policy languages is Clark's Policy Term [14], which provides a template based approach to define network policies. While this approach does provide a complete network policy, it is not attribute based and has issues of scalability. Strassner and Schliemier define a Policy Framework Definition Language [15] which provides transformation mechanism from business requirements to network device independent format. This approach does address the issues of translation between higher level requirements and lower level configuration, it does not provide consistency and verification analysis. Recently, [16] has proposed Ponder, a policy specification language for policy based network management. Ponder does provide an expressive language for definition of an effective network policy, but falls short of providing a mechanism for verification and conflict resolution. In an effort to standardize definition of network policy language design IETF has proposed framework details and information model of network policies in [11, 12]. Another helpful insight into policy based network management is presented by Lazar, Bhonsle and Lim [17] in the form of an Integrated Reference Model (IRM).

8 Conclusion

In this paper we have presented X-Enterprise, a policy language for policy based network management, its UML meta-model for conflict resolution and seamless conversion to XML, and its implementation framework for context sensitive network management. Our proposed software framework for the implementation of X-Enterprise is modular and scalable allowing easy deployment in a variety of network scenarios.

References

1. Lymberopoulos, L., Lupu, E., Sloman, M.: An adaptive policy-based framework for network services management. J. Network Syst. Manage. 11(3) (2003)
2. Chadha, R., Lapiotis, G., Wright, S.: Policy-based networking. IEEE Network. 16(2), 8–9 (2002)
3. Carlson, D.: Modeling xml applications with uml: Practical e-business applications. Addison-Wesley, Reading (2001)
4. Bhatti, R., Ghafoor, A., Bertino, E.: X-federate: A policy engineering framework for federated access management. IEEE Transactions on Software Engineering 32(5), 330–347 (2006)
5. Vogel, R., Herrtwich, R.G., Kalfa, W., Wittig, H., Wolf, L.C.: Qos-based routing of multimedia streams in computer networks. IEEE Journal on Selected Areas in Communications 14(7), 1235–1244 (1996)
6. Tennenhouse, D.L., Wetherall, D.: Towards an active network architecture. In: DANCE, pp. 2–15 (2002)
7. Chan, K., Seligson, J., Durham, D., Gai, S., McCloghrie, K., Herzog, S., Reichmeyer, F., Yavatkar, R., Smith, A.: Cops usage for policy provisioning (cops-pr). United States (2001)
8. Moon, S.-W., Rexford, J., Shin, K.G.: Scalable hardware priority queue architectures for high-speed packet switches. IEEE Trans. Computers 49(11), 1215–1227 (2000)
9. Bennett, J.C.R., Stephens, D.C., Zhang, H.: High speed, scalable, and accurate implementation of packet fair queueing algorithms in atm networks. In: ICNP, pp. 7–14 (1997)
10. Shaikh, F.A., McClellan, S., Singh, M., Chakravarthy, S.K.: "End-to-end testing of ip qos mechanisms. IEEE Computer 35(5), 80–87 (2002)
11. Strassner, J., Ellesson, E.: Policy framework core information model. Internet Draft draft-ietf-policy-core-schema-02.txt (February 1999)
12. Strassner, J., Ellesson, E.: Terminology for describing network policy and services. Internet Draft draft-ietf-policy-terms-00.txt (June 1999)
13. Beigi, M., Calo, S.B., Verma, D.C.: Policy transformation techniques in policy-based systems management. In: POLICY, pp. 13–22 (2004)
14. Clark, D.: Policy routing in internet protocols. IETF Network Working Group RFC 1102 (May 1989)
15. Strassner, J., Schleimer, S.: Policy framework definition language. Internet draft draft-ietf-policy-framework-pfdl-00.txt (November 17, 1998)
16. Damianou, N., Dulay, N., Lupu, E., Sloman, M.: The ponder policy specification language. In: POLICY, pp. 18–38 (2001)
17. Lazar, A.A., Bhonsle, S.K., Lim, K.-S.: A binding architecture for multimedia networks. In: COST 237 Workshop, pp. 103–123 (1994)

Autonomic Network Resource Management Using Virtual Network Concept

Myung-Sup Kim[1] and Alberto Leon-Garcia[2]

[1]Dept. of Computer and Information Science, Korea University, Korea
tmskim@korea.ac.kr
[2]Dept. of Electrical and Computer Engineering, University of Toronto, Canada
alberto.leongarcia@utoronto.ca

Abstract. Traditional telecommunications service providers are undergoing a transition to a shared infrastructure in which multiple services will be delivered by peer and server computers interconnected by IP networks. IP transport networks that can transfer packets according to differentiated levels of QoS, availability and price are a key element to generating revenue through a rich offering of services. Automated service and network management are essential to creating and maintaining a flexible and agile service delivery infrastructure that also has much lower operations expense than existing systems. In this paper we focus on the SLA-based IP packet transport service on a core network infrastructure and we argue that the above requirements can be met by a self-management system based on autonomic computing and virtual network concepts. We present a control and management system architecture based on this approach.

Keywords: Autonomic Computing, Virtual Network, SLA, Resource Management.

1 Introduction

Traditional telecommunications service providers are undergoing a transition to a shared infrastructure in which multiple services and applications will be delivered by peer and server computers interconnected by IP networks. IP/MPLS transport networks that can transfer packets according to differentiated levels of QoS, availability and price are a key element to generating revenue through a rich offering of services and applications. In this environment, Network Service Providers (NSPs) must address the challenge of how to deliver SLA-based network service to large customers. For example, a "customer" could be the Voice-over-IP (VoIP) organization of a service provider which requires that its aggregate flows be handled by the IP transport network according to specified delay, jitter, loss, and availability metrics. Another "customer" may be a third-party VoIP provider that purchases the packet transfer service. Clearly other service and applications can generate large customers, e.g., IPTV, VPNs.

Virtual network based resource management is necessary and efficient to guarantee the customer's diverse traffic demand [1, 2]. The customers' traffic must share the

S. Ata and C.S. Hong (Eds.): APNOMS 2007, LNCS 4773, pp. 254–264, 2007.

same physical core network infrastructure. The service provider should handle these traffics with different policies to meet each of their SLA requirements. Each customer's traffic consists of multipoint-to-multipoint flows that require a certain amount of network resources. Virtual network partitioning of network resources provides an approach to ensuring that the SLA requirements of large customers are met. A virtual network (VN) consists of sufficient network resources that are allocated to a customer to accommodate its flows and meet its SLA [3, 4].

Automated service and network management is essential to creating and maintaining a flexible and agile service delivery infrastructure that achieves much lower operations expense than existing systems. Autonomic computing as defined in [5, 6] is a natural approach to IP network resource management. Autonomic computing (AC) can address the delivery of multiple services with different SLAs using a shared set of computing resources. Effectiveness in service delivery is achieved by assigning to each service an autonomic manager that ensures that the service is delivered according to the prescribed SLA. The manager monitors and analyzes resource state and service metrics, and plans and executes appropriate actions to ensure that service metrics are met in the presence of faults and a changing environment. In the context of an SLA-based IP transport service, an autonomic manager is created for each virtual network.

The VN approach to network resource management is not new [3, 4]. The combining the VN and AC concept, however, brings greater clarity to the methodology of designing future network management systems that must attain greater degrees of scalability and manage higher levels of application diversity and adaptivity than required by networks in the past. In addition, with the focus of application delivery shifting to the peer/server computing infrastructure, it is our belief that future network management systems, to the extent possible, should incorporate resource management methodologies developed for information technology. In this paper, we propose a network service management methodology using the concept of virtual network and autonomic computing.

The organization of this paper is as follows. Section 2 describes the features of the SLA-based core network service and analyzes the requirements of this service in the perspective of customers and service provider. Based on the requirement analysis we developed a VN-based autonomic network resource control and management system named VNARMS. In section 3, we propose an overall architecture of VNARMS and describe the relationship among autonomic managers in VNARMS. Section 4 addresses the detailed design issues of our proposed system. In section 5, we conclude this paper with current status of our research and scheduled future work.

2 Requirement for SLA-Based Network Service

In this section, we consider requirements on the SLA-based network service from the perspective of the customers and the service provider. Based on this requirement analysis we design an autonomic network resource management system. The customers could be an enterprise (E), a content service provider, a 3rd party network reseller like a VPN provider, and application service providers, e.g. VoIP or IPTV. The network service provider (NSP) owns and operates the physical core network resources. The service instance delivered from the NSP to a customer is a virtual network (VN). The virtual network (VN) is an abstraction of a physical network that consists of a partition of network resources [1].

2.1 Customer Requirements

First, the customer requires the creation of a VN that can deliver a certain level of quality. The metric of VN quality can be end-to-end bandwidth and delay for multiple routes. When a customer negotiates a SLA for a VN, the VN topology may or may not be specified. Second, the customer wishes to monitor the current and historical status of his VN at various times. This VN usage data allows the customer to anticipate future VN usage and to plan for improved utilization. Third, the customer may wish to reconfigure the contracted VN capacity or to remove the VN. Fourth, the customer may choose to create one or more new VNs within its own VN capacity boundary and to resell these to other customers. This reselling activity should be performed independently from the NSP. Fifth, a VN control and management system should be provided to the customer when a new VN is created. Starting from the physical network infrastructure, a multi-level recursive creation of VNs and associated control and management system should be supported.

2.2 Service Provider Requirements

First, every service provider attempts to optimize the use of its resources (whether physical or virtual) to maximize its service revenue. In this general sense, the NSP that owns the physical network resources has the "root-VN" from which other VNs can be spawned. The "root-VN" is an abstraction of the whole network capacity of NSP core network. The VN customer in turn can be a service provider to other customers. Second, the operation expense of a VN should be minimized by automatically and efficiently carrying out the customer or service provider requests. The automatic configuration of VN is very important to reduce network operator's errors which are the cause for a significant amount of misuses and faults.

Third, the system should provide an autonomic way to handle fault and overload conditions on each VN and on its virtual network resources (VNR). For this autonomic operation, the system should have pre-defined autonomic routines to determine, isolate, and repair faults according to policy. Fourth, the service provider may wish to provide VN services with different levels of availability (such as 99.999% of availability) to different customers. For this goal, the system should provide appropriate levels of resource redundancy and appropriately fast fault recovery mechanisms. Fifth, when a new VN is created, the system should select an optimal topology, link bandwidths and call admission control mechanism. Finally, the ultimate goal of the NSP is to maximize its revenue. To do so, it must decide on the optimal mix of services and prices it should offer to its customers based on the available infrastructure and the forecasted demand.

3 Overall System Architecture and Operations

We now describe the overall architecture of VN-based autonomic network resource control and management system (VNARMS). The virtual network and autonomic computing concepts are applied to this architecture to make the service provisioning and operation efficient, scalable and cost-effective.

3.1 VNARMS Operation

Fig. 1 illustrates a high-level operation of VNARMS: construction of VN and VN spawning. We assume that the physical network is the IP-MPLS network with DiffServ capability, although the operation of VNARMS is not strictly dedicated to the IP-MPLS networks. The VNARMS first performs the abstraction of the physical network to construct a Root-VN which is controlled and managed by VNARMS. The VNARMS create a child VNs and provide them to customers according to the service provider's policy and customer's quality demand. A customer provided a VN by a service provider, in turn, could be a new service provider that sells some portion of its VN to another customer by VN spawning process. This recursive VN spawning process is one of the key features of our VNARMS system along with the autonomic features, which gives the network re-sellers or service organizers flexibility and controllability in operating their VNs.

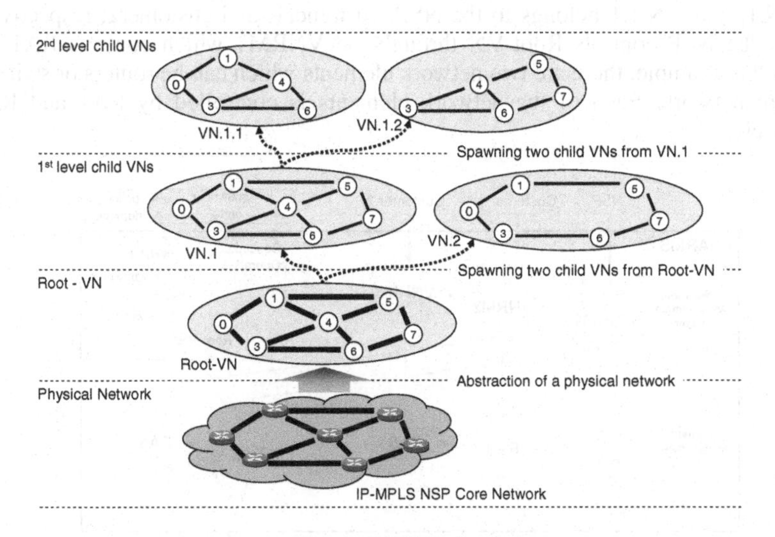

Fig. 1. VNRMS Operations

A VN is managed by two types of autonomic components in out VNARMS. Those are one virtual network resource manager (VNRM) and multiple resource agents (RA). The VNRM is responsible for the VN-level control and management, while the RA is responsible for the element-level resource control and management. In a single VN, each switching element is controlled by one RA. The VNRM manages its VN by controlling the operation of its RAs, while there is no interaction among RAs in a single VN. Multiple VNRMs can exist in our VNARMS in a hierarchical structure like that of the VNs. For example, if there are ten VNs created on a physical core network, there should be the same number of VNRMs, each of which controls one VN assigned to it. The RAs in different VNs run independently with little interaction, because, when a new VN is spawned from a parent VN, the parent RA creates a child RA when the parent VNRM creates the child VNRM.

The VNRM and RA are the autonomic managers that are arranged according to the autonomic control loop structure: monitor, analyze, plan, and execute [5, 6]. By monitoring their target objects, the autonomic managers diagnose the object's status. The managed objects of the VNRM and RA are the RAs and physical network elements, respectively. When a problem is determined, first, the manager tries to localize the problem and repair it by itself. Depending on the nature of the problem this may result in self-healing (in case of an error) or self-organizing (re-arrange the resources to attain optimum utilization) or self-configuring (adapt to the other changes). If the manager cannot handle the problem, the high-level autonomic manager is involved. Through this hierarchical autonomic problem handling, a hierarchical autonomic manager structure can manage and control large core networks.

3.2 Relationship Among VNRMS and RAs

Fig. 2 illustrates an example of VN spawning operation and the relationship among VNRMs and RAs. The conceptual VN-level operation in Fig. 2 shows that the Root-VN spawns the VN.1 and the VN.1 in turn spawns VN.1.1. The ownership of Root-VN, VN.1, and VN.1.1 belongs to the NSP, customer1, and customer2, respectively. Initially, the NSP controls Root-VN through the VNRM1 which controls RA1 and RA2. In this example, there are two network elements which can be routers or switches in a core network. Each of the network elements is controlled by RA1 and RA2, respectively.

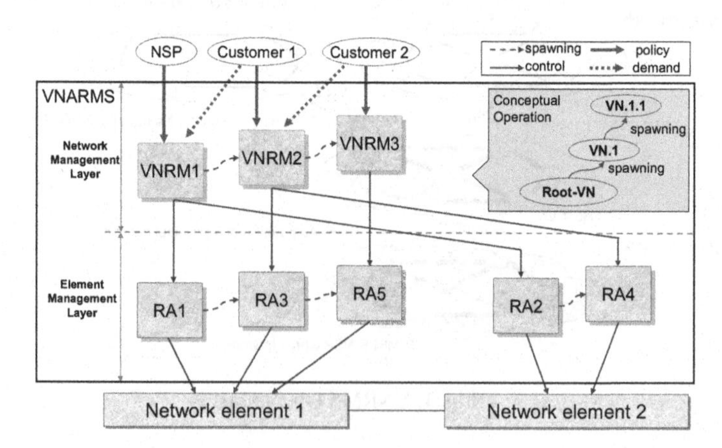

Fig. 2. VN spawning operation

When customer1 wants a VN (VN.1 in this example) he contacts VNRM1 and requests a new VN creation with a given SLA. After receiving a VN creation request, the VNRM1 calculates an optimal VN topology and effective bandwidth for each end-to-end route on this topology to create a new VN for the customer1. Note that the child

VN is a subset of the parent VN in terms of capability. After creating VN.1 for the customer1, the VNRM1 and its RAs (RA1 and RA2) perform additional actions to provide the customer1 with the controllability of VN.1. VNRM1 creates VNRM2, and RA1 and RA2 spawn RA3 and RA4, respectively. With the newly created VNRM2, RA3, and RA4, the customer1 can control its own VN.1. Similarly, customer2 contacts VNRM2 to create a new VN. VNRM2 and RA3 spawn VNRM3 and RA5, respectively, after they create VN3. The recursive spawning of VNRMs and RAs allows the customer of a VN to have controllability to its own VN, independently from other VNs and customers. Furthermore, this hierarchical processing of VNs provides scalability and extendibility.

All VNRMs and RAs are autonomic managers. The autonomic manager handles requests or notifications from external components in an autonomic way. A VNRM communicates with its parent VNRM, its child VNRMs, its owner service provider, customers, and its RAs. For example, the VNRM2 in Fig. 2 receives requests or notifications from VNRM1, VNRM3, customer1, customer2, RA3 and RA4. In case of RAs, the external components that generate requests or notifications are the VNRM, the parent RA and the physical network element. For example, RA2 in Fig. 2 receives the requests and notifications from VNRM2, RA1, and the physical network element 1. The contents of external requests are listed in Table 1, which is not exhaustive but important.

Table 1. Request from external component to VNRM and RA

From	To	Requests
Owner SP	VNRM	VN policy setup, VN monitoring, VN removal
Customer	VNRM	Customer subscription, Customer SLA negotiation, Child VN creation
Parent VNRM	VNRM	VN monitoring, VNRM termination
Child VNRM	VNRM	Child VN removal
VNRM	RA	RA policy setup, RA spawning, child RA removal, RA monitoring
Parent RA	RA	RA creation, VNR monitoring
RA	Network Element	monitoring, configuration

The high-level policy from a service provider is deployed automatically through the hierarchical autonomic manager structure and each autonomic manager configures itself based on the given policy (self-configuring) and deals with requests from external components according to the policy. This minimizes human intervention and leads to cost effective operation. Each autonomic manager tries to optimize its corresponding managed objects by forecasting future demand based on the policy. As indicated above, this hierarchical structure makes the system very scalable. The system can also adapt to the changes in network topology by adding new management components in the resource and network management layer.

4 Detailed Architecture of VNRMS

We now present the detailed architecture of VNRM and RA, the two types of autonomic managers in the VNARMS. We describe the functional building blocks of the VNRM and RA and show an example of autonomic operation of VNARMS.

4.1 Autonomic Manager Functional Blocks

First, we define generic functional building blocks for an autonomic manager as an extension of the IBM autonomic manager structure [6]. To achieve complete autonomic computing for a service, the four parts of the Fig. 3. should be executed automatically in each autonomic manager: customer control, policy control, service activation, and service maintenance. These four parts are tightly coupled and communicate to each other to achieve the ultimate goal: self-management of a service. This concept can be applied to all kinds of services, including application-layer services and network-layer services.

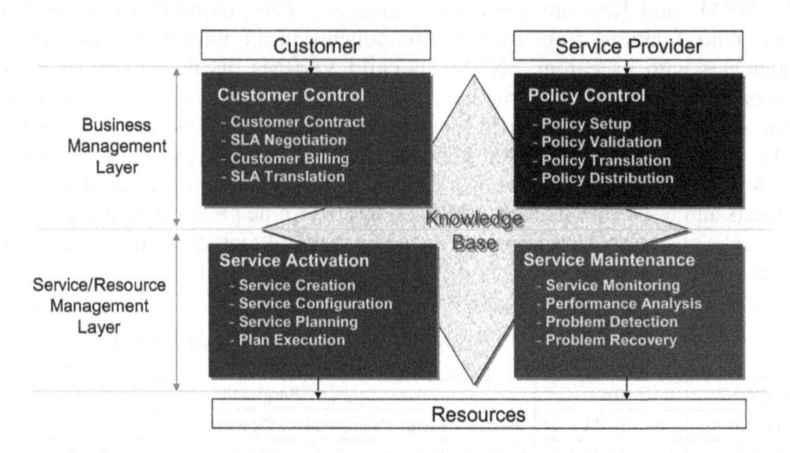

Fig. 3. The functional building blocks of an autonomic manager

The service provider uses the policy control building block to set up policy for the other three functional building blocks. The customer control building block is in charge of the interaction with customers. The customer contacts the system through the customer control block and negotiates the service quality. Based on the SLA information, the service activation building block creates a new service instance for the customer. The service maintenance building block controls and manages the resources during the lifetime of each service instance (problem detection, problem recovery, etc). The maintenance outcomes are sent to the customer control for customer care functionality, such as billing. There is some functional overlap among these parts. For example, the service planning to find optimal resource allocation could be processed in the service activation and service maintenance building blocks.

4.2 Detailed Architecture of VNRM

Fig. 4. illustrates the detailed ten functional building blocks of the VNRM. The VNRM has three different knowledge bases: PIB, CIB, and VNIB, which store static and dynamic information about policy, customer, and VN, respectively. The Policy Manager is responsible for policy translation, policy distribution, and policy validation. It receives the SP's policy and translates it into VN and VNR policy. The VN policy is stored in the PIB and VNR policy is distributed to all RAs through the Resource Manager. The policy for autonomic manager is one of our and others on-going research investigations [7]. In addition to the policy operation, the Policy Manager handles all other requests from the service provider, e.g. VN reconfiguration and VN removal requests. The Request Manager, on the other hand, handles all the requests from customers.

The Request Manager is responsible for customer contact and SLA negotiation. To create a new VN, a customer should subscribe itself to the VNRM and specify VN quality requirements through the Request Manager. SLA metric for VN creation and reconfiguration is also a part of our on-going investigation. The Request Manager also validates the customer VN requirements based on the pre-defined policy and its current VN status, and if appropriate it sends a VN create request to the VN Manager.

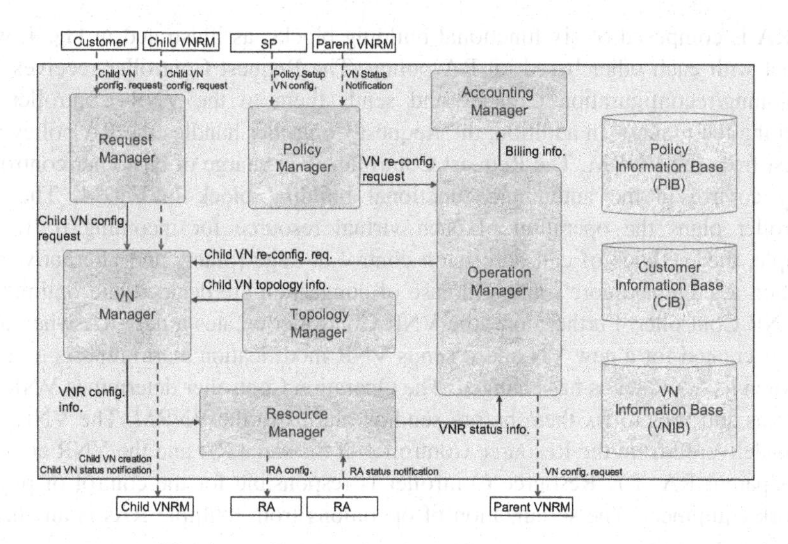

Fig. 4. Detailed Functional Architecture of VNRM

The VN Manager is responsible for the creation, modification, and removal of child VNs. On reception of a VN creation request, the VN Manager makes a topology request to the Topology Manager to find an appropriate VN topology. The VN Manager then makes a resource reservation request to the Resource Manager. In addition, the VN Manager handles the VN re-configuration and removal request from the Request Manager and the child VNRM. The VN Topology Manager is responsible for creating a set of optimal routes and calculating the corresponding effective bandwidth when a new VN is created, or when a VN is reconfigured. Much research [8, 9, 10, 11] has been devoted to route-level QoS, but extensions are required for VN-level QoS guarantees. The goal of VN manager is to find optimal VN allocation from the given network resources that maximize the resource utilization and service revenue, another area of on-going research.

The Resource Manager is responsible for the communication between the VNRM and the RAs. It translates the VN spawning message into a VNR partitioning message and distributes it to the corresponding RAs to create a new VN. It also sends monitoring requests to the RAs and receives status information from them. The Operation Manager is responsible for the analysis of each child VN and the overall VN status. It determines VN level problems and sends a VN reconfiguration message to the

VN Manager. In addition, it handles the SLA monitoring. The SLA analysis result is stored in the CIB, which is used for customer billing by the Accounting Manager.

The VNRM implements the autonomic control loop that involves the four main components: Resource Manager, Operation Manager, VN Manager, and Topology Manager. The Resource Manager does monitoring; the Operation Manager does monitoring and analysis; the VN Manager and Topology Managers do planning; and the Resource Manager performs execution. Furthermore, the seven functional components follow the four autonomic functional components described in Fig. 3.

4.3 Detailed Architecture of RA

The RA is composed of six functional building blocks, as illustrated in **Fig.** 4, which interact with each other based on RA policy. The Request Controller receives VNR partitioning/reconfiguration requests and sends them to the VNR Controller after validating the request. In addition, the Request Controller handles the RA policy setup request from the VNRM. The Request Controller is in charge of customer control and policy control in the autonomic functional building block in Fig. 4. The VNR Controller plans the operation of each virtual resource for incoming traffic. For example, the methods of call admission control at edge routers and alternative route selection at edge and core routers in case of congestion are decided and optimized in the VNR Controller. Furthermore, the VNR Controller creates a new RA when a new VNR is created for a new VN and it sends VNR modification alarm reports to a child RA when its VNR status has changed. The Operation Controller determines VNR level problems and tries to fix them before sending alarms to the VNRM. The VNR status data is delivered from the Resource Controller of the same RA and the VNR controller of the parent RA. The Resource Controller is responsible for the control of physical network equipment. The serialization of operations from multiple RAs is another on-going investigation.

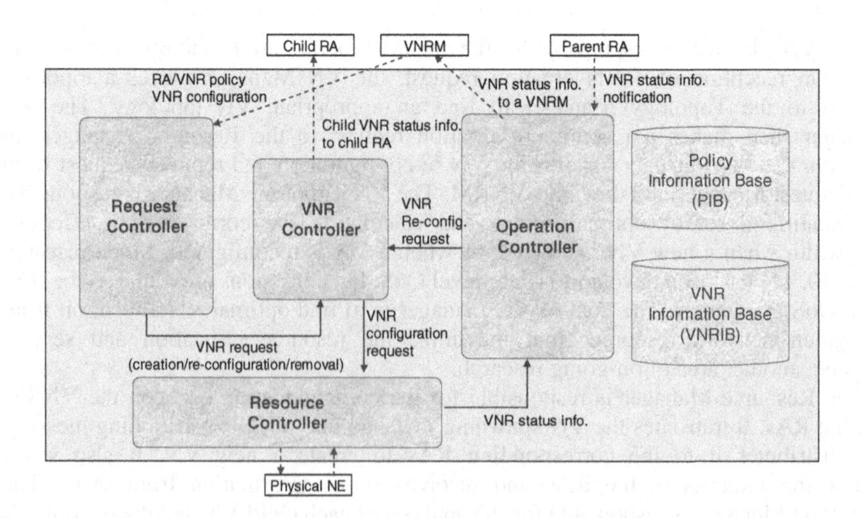

Fig. 4. Detailed Functional Architecture of RA

The Resource Controller, Operation Controller, and VNR Controller form the main autonomic control loop in the RA. The RA periodically retrieves status data of the VNR through the Resource Controller. When any fault is detected the Operation Controller tries to resolve the problem through the VNR controller. The Operation Controller coordinates the sequence of VNR recovery when more than one VNR are in an unstable state. The VNR Controller finds the optimal solution for the given problem based on the VNR policy and configures the physical network device through the Resource Controller. If the problem cannot be resolved by RA itself, the Operation Controller sends an alarm notification to the corresponding VNRM.

5 Conclusion

In this paper we presented requirements for SLA-based IP packet network service on a core network infrastructure, and we proposed a virtual network based autonomic network resource control and management system architecture. We argue that the VN-level QoS guarantee as well as link- or route-level QoS guarantees should be considered to meet customer various traffic demands on core network environments. In addition, providing a customer with the controllability for his virtual network is an important element to increase the service flexibility. This paper focused on the description of the architecture and operation of our proposed autonomic system. This is a work in progress and many issues remain to be addressed including adaptive algorithm development, system dynamics characterization, performance evaluation, and implementation of a proof-of-concept system.

References

1. Jun, A.D.-S., Leon-Garcia, A.: Virtual network resources management: a divide-and-conquer approach for the control of future networks. In: Proc. of Globecom 1998, Sydney, Australia, vol. 2, pp. 1065–1070 (1998)
2. Boutaba, R., Ng, W., Leon-Garcia, A.: Web-based customer management of virtual private networks. Journal of Network and Systems Management 9(1), 67–87 (2001)
3. Leon-Garcia, Mason, L.: Virtual network resource management for next-generation networks. IEEE Communications Magazine 41(7), 102–109 (2003)
4. Woodruff, G., Perinpanathan, N., Chang, F., Appanna, P., Leon-Garcia, A.: ATM network resources management using layer and virtual network concepts. In: IM 1997 (1997)
5. Kephart, J.O., Chess, D.M.: The vision of autonomic computing. IEEE Computer Magazine 36(1), 41–50 (2003)
6. IBM White Paper, An architectural blueprint for autonomic computing, (April 2003)
7. Appleby, K., Calo, S.B., Giles, J.R., Lee, K.-W.: Policy-based automated provisioning. IBM Systems Journal 43(1), 121–135 (2004)
8. Trimintzios, P., Griffin, D., Georgatsos, P., Goderis, D., T'Joens, Y., Georgiadis, L., Jacquenet, C., Egan, R.: A management and control architecture for providing IP differentiated services in MPLS-based networks. IEEE Communications Magazine 39(5), 80–88 (2001)
9. Bouillet, E., Mitra, D., Ramakrishnan, K.G.: The structure and management of service level agreements in networks. IEEE JSAC 20(4), 691–699 (2002)

10. Rosen, E.C., Rekhter, Y.: BGP/MPLS IP VPNs. IETF Internet Draft, draft-ietf-l3vpn-rfc2547bis-03.txt (October 2004)
11. Scoglio, C., Anjali, T., de Oliveira, J.C., Akyildiz, I.F., Uhl, G.: TEAM: A traffic engineering automated manager for diffServ-based MPLS networks. IEEE Communications Magazine 42(10), 134–145 (2004)

A New Heuristics/GA-Based Algorithm for the Management of the S-DRWA in IP/WDM Networks

Eduardo T. L. Pastor[1,2], H.A.F. Crispim[1], H. Abdalla Jr.[1], Da Rocha A. F.[1], A.J.M. Soares[1], and J. Prat[2]

[1] Universidade de Brasília, Dept. of Electrical Engineering, Campus Darcy Ribeiro CEP 70910-970, Brasília-DF-Brazil
tommy@unb.br, hcrispim@gmail.com, {abdalla, adson, martins}@ene.unb.br

[2] Universitat Politècnica de Catalunya, Dept. of Signal Theory and Communications, c/Jordi Girona 1-3 D-4, 08034 Barcelona-Spain
{tommy.lopez, jprat}@tsc.upc.edu

Abstract. This work presents the design and validation of a new hybrid Heuristic-Genetic Algorithm (HGA) for the management of the Survivability Dynamic Routed and Wavelength Assignment (S-DRWA), based on lightpath protection sharing, and applied on a IP/WDM transport network.

Keywords: RWA, Survivability (S-DRWA), IP/WDM networks, Genetic Algorithms (GA).

1 Introduction

Service provisioning issues, especially routing, wavelength assignment (RWA) and survivability problems, frequently require NP type algorithms, which are difficult to undertake with exact methods of mathematic programming because of the large time process [1]. On the other hand, evolutionary algorithms, as Genetic Algorithms (GA) [2], have drawn hard attention to its application in the solution of complex problems and the optimization in different fields, including telecommunications, even though it is not practical in some applications due to the high execution time of the computational process.

This work proposes an alternative based on the design and validation of a novel hybrid Heuristic/GA algorithm, named HGA, for the management of the dynamic routing and wavelength assignment with survivability (S-DRWA) in IP/WDM transport networks. This hybrid algorithm is based on sharing lightpath protection.

The requirements associated with the heuristic mechanism searching for the bests working path with its respective protection-paths are given. This information determines the search space for the Genetic Algorithm, which selects the "best" working path and the "best" sharing protection path. The provision of the wavelength is implicitly determined in the selection process.

For the validation of this proposal, the HGA was implemented directly on the control plane of the SIMOMEGA testbed [3], and tested over the OMEGA [4] and NSFNet [5] network topologies.

S. Ata and C.S. Hong (Eds.): APNOMS 2007, LNCS 4773, pp. 265–275, 2007.

2 Hybrid Heuristic/GA Algorithm (HGA) Design

The proposed mechanism is based on heuristics and on a genetic algorithm (GA), establishing an alternate-adaptive mixed routing scheme [6].

2.1 Heuristics

1) A set of primary routes for every *s-d* pair is pre-established, using a fixed alternated routing algorithm based on the Dijkstra algorithm [7].
2) For every primary route, disjoint paths are selected to stablish a set of protection paths. An algorithm based on search trees is used [8].

The heuristics are executed only once, or when a change in the topology or network configuration exists. The main goal of the heuristics is to originate a searching space for the GA, to remove the random recursive creation of the first generation in the GA, so optimizing the computation time.

2.2 Genetic Algorithm

The initial population given by the heuristics is filtered in terms of the *s-d* pairs, obtaining in this way a specific population P. It evolves by the action of the genetic operators (crossover and mutation). These chromosomes are once again evaluated depending on its fitness. The process is repeated until the best chromosome is selected or the stopping condition has been reached. The GA pseudo-code is presented in Table 1.

Table 1. GA pseudo-code

```
{Population given for the heuristics}
t=0
Select population P to the lightpath request
Fitness evaluation
C= best cost between the s-d nodes
while (t<G AND did not had reach the o criterion of optimization)
 do crossover and fitness evaluation of the children
 do mutation and fitness evaluation of the children
 Select P fittest to next generation
 C=C+1;
 t=t+1;
end while
Lightpath assignment
Table of Codes and Lightpaths actualized
```

In the algorithm proposed, a chromosome is represented by a code symbolized by an integer number, where each number identifies a lightpath node. Each code is formed by a primary path and by a backup path, and these paths are disjoint. Fig.1 shows, for example, the chromosome 2-1-5-2-4-3-5 and, implicitly the Primary-Backup pair.

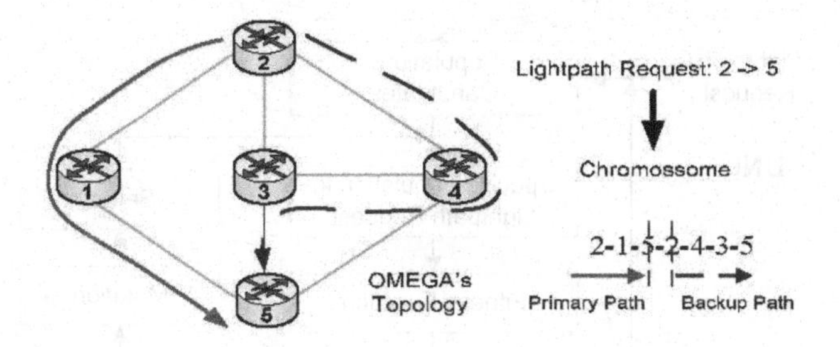

Fig.1. Chromosome: implicit representation of the pair Primary-Backup path

2.3 Fitness Evaluation

The chromosome is evaluated as a function of its cost. The chromosome with the smallest number of hops and smallest number of wavelengths will have the lowest cost. The following definitions are used in this work:

a) **Cost of the Primary Lightpath (Cw):** Defined by the number of hops, assuming that they have at least one available wavelength, otherwise Cw = ∞.

b) **Cost of the Protection Lightpath (Cp):** Considering the link L and the cost of each link with wavelength *w*.
Cpw: Cost of the protection lightpath candidate with λ=w, (Eq. 1)
CL,w: Cost the protection lightpath with λ=w of the link L; then:

$$C_{pw} = \sum_{L \in Path} C_{L,w} \tag{1}$$

$C_{L,w} = 1$ → if the λ in the link has never been used before in the same link.
$C_{L,w} = 0$ → if the λ in the link was already been used and can be shared, with the primary *lightpath* in different SRLG.
$C_{L,w} = \infty$ → otherwise;
If the C_{pw} is the best candidate → $C_{pw} = C_p$.

c) **Total Cost:** then, the total cost (C_T) and fitness function is:

$$C_T = C_w + C_p + h_w(I) \tag{2}$$

h_w = number of hops of the primary lightpath;
I = number of nodes of the candidate chromosome;

The term $h_w(I)$ intends to privilege the chromosomes with less number of hops. Eq. 2 shows a linear relationship, with low complexity, resulting in lower execution time.

In the GA, the non-linear restrictions introduced by sharing of the protection links between SRLG will be considered. A block diagram of the proposed algorithm is shown in Fig. 2.

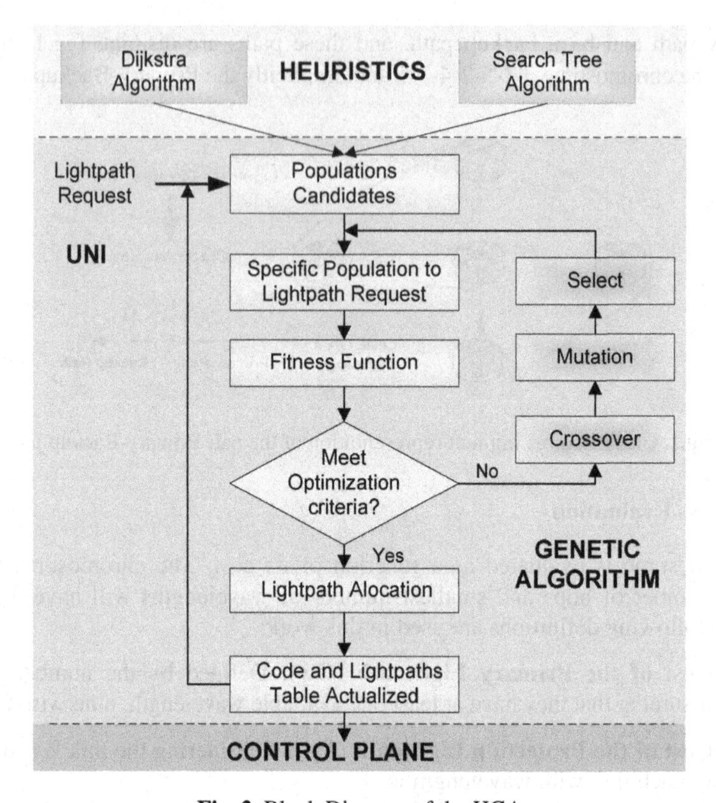

Fig. 2. Block Diagram of the HGA

If the optimization criterion is not reached, the following genetic operators will be used: crossover, mutation and selection. These operators are described as follows.

2.4 Crossover

The crossover operator is applied to pairs of chromosomes, leading to changes in the genetic characters. The crossover point is selected based on the first meeting with the source node. Therefore, the information to be crossed corresponds to the part of the lightpath protection code (Fig. 3).

2.5 Mutation

The mutation operator causes random changes in the genetic characteristics of a chromosome, leading to obtain individuals with new characteristics. Here, a node of the candidate lightpath, named "n", is randomly selected (Fig. 3). The part of the lightpath between the source node and the "n" node remains unchanged. The part of node "n" that goes up to a hop before the destination (part of the protection way) is then recreated. The mutation is applied to some chromosomes with fitness values below the average aptitude (in this work, only for 2 individuals) from all individuals at the crossover stage. The probability associated with this operator is low.

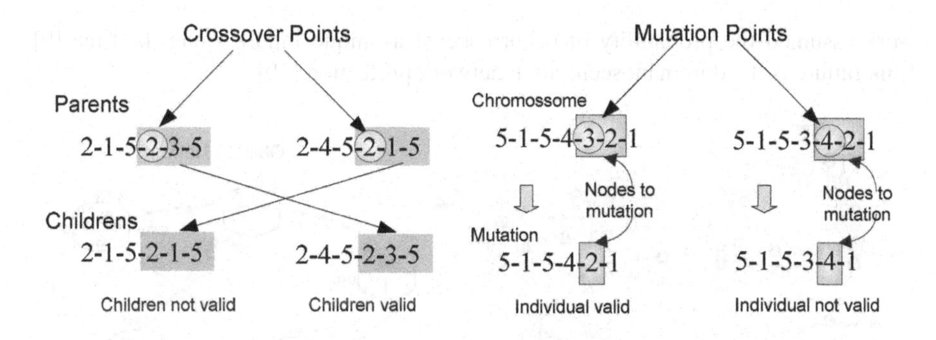

Fig. 3. Examples of Crossover and Mutation operations

2.6 Selection

After the crossover and mutation, the selection stage chooses the P individuals with greater aptitude between parents and children, which become part of the population of the next generation. This process will be repeated until the stopping condition is reached and the best individual is selected.

Finally, the lightpath with required protection will be provided, updated to the table of code and lightpaths, and the information to the control plane for the provisioning process will be provided.

3 Development Scenario

The environment IDE used for programming the mechanism HGA was the KDevelop KDE/C++ 3.1, Linux, FEDORA 5, Fig. 4. A client can access the services of the system via telnet sessions.

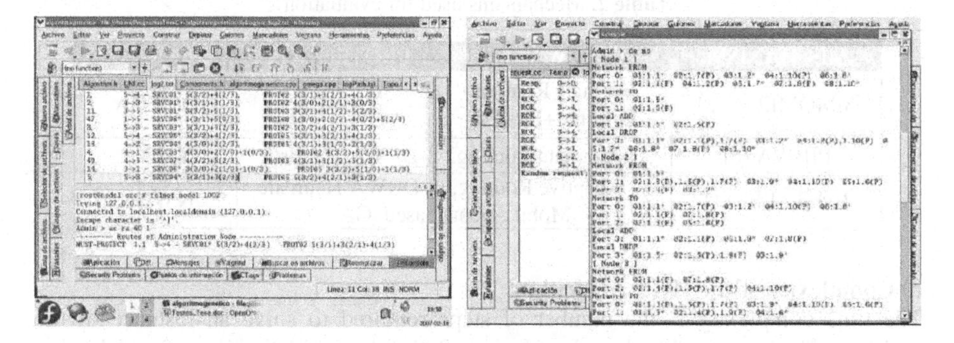

Fig. 4. The environment IDE used for programming the HGA mechanisms

The HGA was implemented directly over the UNI and control plane of the SIMOMEGA testbed, and tested over the OMEGA and NSFNet network topologies, (Fig. 5), using a PC with processor Pentium III 447 MHz and 384 MB of RAM. This work assumed the probability of occurrence of a unique failure along the time [9]. Link failure is the dominant scenario in network protection [10].

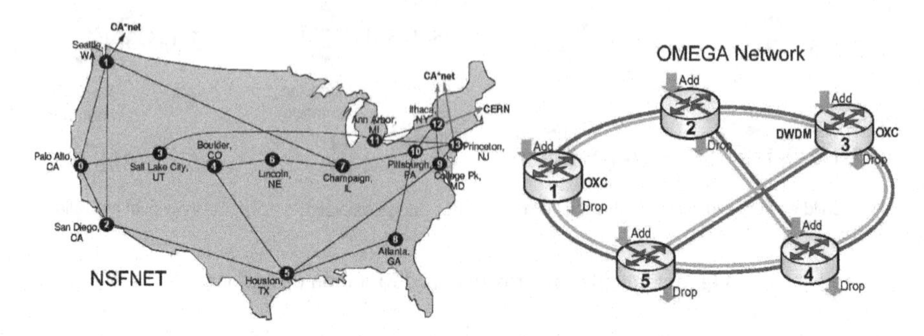

Fig. 5. Topologies used for the proposed evaluation

4 Network Experiments and Results

A generator of connection demands was used to simulate requests according to a probabilistic model. The interaction generator-control plane is a Client/Server model. The following parameters were evaluated:
a) Complexity of the algorithm;
b) Mean blocking probability;
c) Network redundancy ratio and sharing capacity;
d) Mean execution time.

The HGA proposal was compared to other Heuristics mechanisms which are using alternate or adaptive routing [3][6][11]. The Heuristics mechanisms from references [3][6][11] are resumed in the Table 2.

Table 2. Mechanisms used for evaluation

MECHANISM	RWA	PROTECTION	TOPOLOGY
SIMOMEGA [3]	Alternate Routing, Dijkstra +FirstFit	Type 1:N	OMEGA
PIBWA [6]	Alternate Routing,	S-DRWA Heuristic	NSFNet
Hybrid-LE [11]	Adaptive Routing, Mobil Agents based–GA	S-DRWA Heuristic	NSFNet

a) Complexity of the HGA Algorithm

The time complexity is the number of steps required to solve an instance of the problem. The process for obtaining the population by heuristics is performed using O(N+N). This happens only once for all the requests from the client; therefore, its computational costs will not have a large impact. Thus, the complexity of the HGA mechanism will be the time complexity of the genetic algorithm. In comparison with the other proposals, a relatively low temporal complexity, $O(Gx(PxWxN^2 + P^2xN))$, was obtained (Table 3), where K is the number of possible primary lightpaths, with H hops, N is the number of nodes of the networks, W is the number of wavelengths per fiber, G is the maximum number of generations, and P is the population size.

Table 3. Complexity HGA vs. PIBWA vs. Hybrid-LE

MECHANISM	COMPLEXITY
PIBWA	$O(K_2xH_2xW_2)$
Hybrid-LE	$O(Gx(PxWxN_2+P_2xN)$
HGA	$O(Gx(PxWxN_2+P_2xN))$

b) Mean blocking probability (P_b)

The calculus of the mean blocking probability is performed in terms of the number of blocked connections (RNOK) over the total number of requested connections (Eq.3).

$$P_b = \frac{\sum RNOK}{\sum Total\,\mathrm{Re}\,q} \tag{3}$$

Blocking Probability as a function of G and P

The block probability (P_b) as function of G and P parameters for a 30 Erlangs load, for both P=4 and P=P_{max} (maximum population gave by heuristics) is presented at Fig. 6. You can observe that P_b improves when G and P grow. This analysis is important to define the criterion for stopping the algorithm. The best performance was found for P=P_{max} and G=3, being this values used as a stopping criterion.

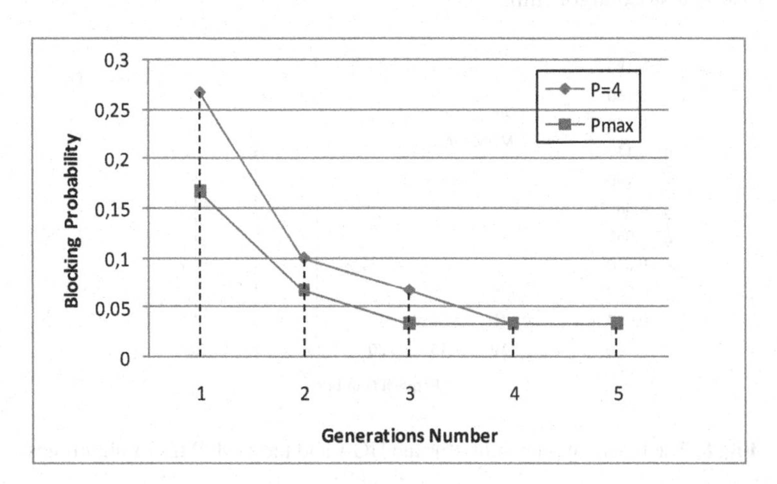

Fig. 6. Blocking Probability as a function of G and P

Individuals assignment probability for generation

The individuals assignment probability by generation for a P_{max} is shown in Fig. 7. We can observe that more than 80% of the lightpaths can be allocated in the first searching, which shows that in many cases a "good" lightpath can be found in the first interaction for the minimum cost, without using the genetic operators and, thus, with a lower execution time.

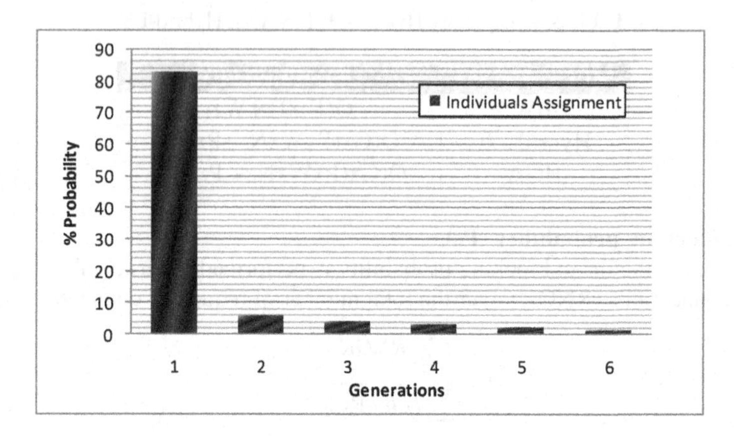

Fig. 7. Individuals Assignment Probability x Generations on the HGA

Blocking Probability: HGA vs SIMOMEGA, PIBWA and Hybrid-LE
Fig. 8 shows a comparison between the mean blocking probability obtained with HGA and with the SIMOMEGA testbed algorithm. We observe a good performance of the HGA, with better values (about a 50% lower) than the ones obtained by the SIMOMEGA testbed algorithm.

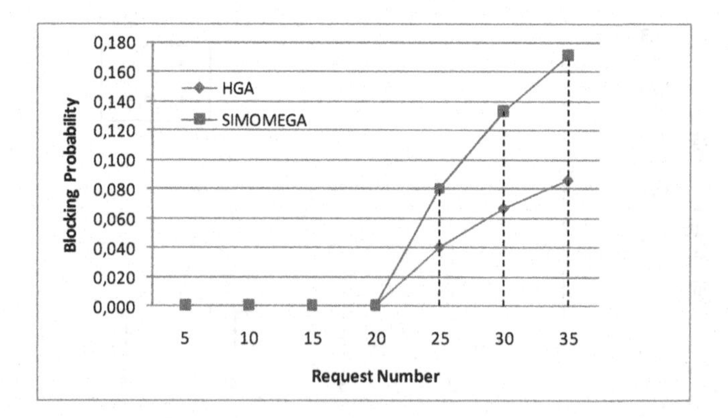

Fig. 8. The blocking probability for the HGA and the SIMOMEGA algortihms

Fig. 9 shows the values of $\mathbf{P_b}$ of the PIBWA and Hybrid-LE algorithms, obtained in [6] and [11] respectively, and these values are compared with the HGA, in an NSFNet topology. Our algorithm had a performance about 20% better.

c) Network Redundancy Ratio ($\mathbf{R_r}$) and sharing capacity
The R_r is defined as the ratio of the total spare capacity over the total working capacity [12] (Eq. 4).

$$R_r = \frac{\sum_{(ij)\in L} B_{ij}}{\sum_{(ij)\in L} A_{ij}} \tag{4}$$

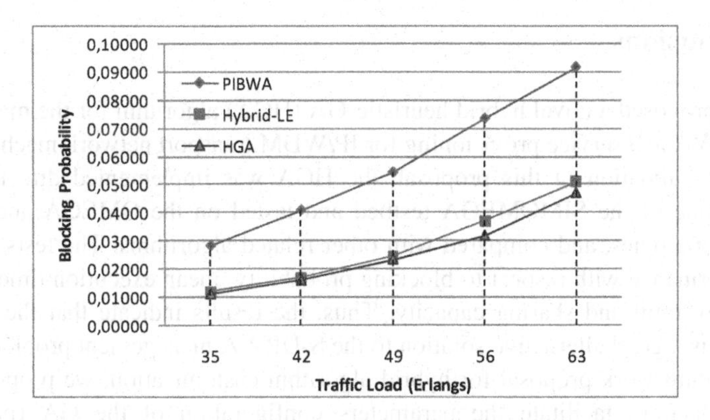

Fig. 9. Blocking Probability evaluation: PIBWA, Hybrid-LE and HGA

Redundancy depends on the network topology as well as on the used algorithm. Without explicitly considering SRLG constraints, the shared path protection scheme proposed in [13] obtained 74% to 87% for 32 nodes. With HGA, we obtained a redundancy rate of 58%, an optimal value compared to the results in [13]. In [12], similar values have been obtained, but with a static S-RWA algorithm. The sharing capacity reached by HGA (the capacity of re-using protection paths) is about 39%.

d) Mean execution time

The mean execution time is calculated as the ratio between the execution time of a series of requests and the number of calls of such series (Eq. 5).

$$T_m = \frac{\sum Request\ Execution\ Time}{\sum Request\ Number} \tag{5}$$

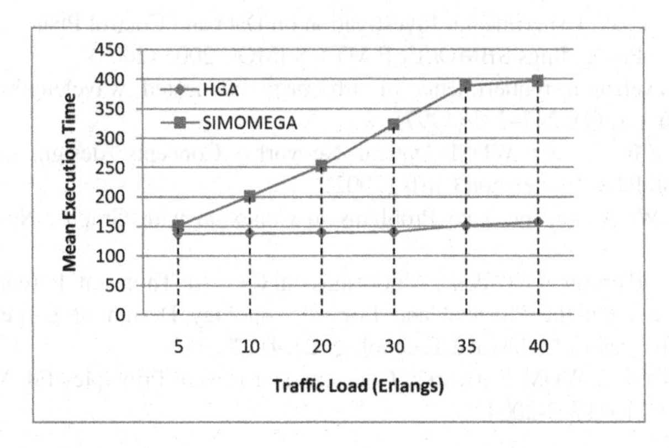

Fig. 10. Mean execution time: HGA vs. SIMOMEGA

Fig. 10 presents its comparison of the HGA and SIMOMEGA testbed algorithm, for different traffic load values. In every case, a very low t_m is obtained with our proposed model (up to quicker 60%); as the measurement of traffic increases, the characteristic of backup sharing of our algorithm is more effective.

5 Conclusions

We have proposed a novel hybrid heuristic-GA (HGA) algorithm for the managing of the S-DRWA and service provisioning for IP/WDM transport network mechanisms.

For the validation of this proposal, the HGA was implemented directly on the control plane of the SIMOMEGA testbed and tested on the OMEGA and NSFNet network topologies, and compared with other related algorithms. The tests showed a high performance with respect to blocking probability, mean execution time, network redundancy ratio and sharing capacity. Thus, the results indicate that the proposed algorithm is a good alternative solution to the S-DRWA management problem.

As a future work proposal for hybrid algorithm customization, we propose a new user interface to facilitate the parameters configuration of the GA (population, generations number, stopping criteria, etc) in order to achieve best functionality.

Acknowledgments. This work was partially supported by the CNPq-Brasil and supported by the Alban program, scholarship No. E05D056349BR.

References

1. Proestaki, A., et al.: Impact of topology on wavelength and switch_port requirements in alloptical hierarchical networks. In: Proc. IEEE GLOBECOM-1999 - RJ – Brazil (1999)
2. Goldberg, D.E.: Genetic Algorithms in Search, Optimization and Machine Learning. Addison-Wesley, Reading (1989)
3. Crispim, et al.: IP/WDM Optical Network Testbed: Design and Implementation. In: Kim, Y.-T., Takano, M. (eds.) APNOMS 2006. LNCS, vol. 4238, pp. 27–29. Springer, Heidelberg (2006)
4. Sachs, A.C., et al.: Experimental Investigation on Data and Control Planes of the OMEGA Test Bed. In: Proceedings SBMO/IEEE MTT-S IMOC 2003 (2003)
5. Baroni: Wavelength requeriments in arbitrarily connected wavelength-routed optical networks. JLT 15(2), 242–251 (1997)
6. Murthy, C.S.R., et al.: WDM Optical Networks: Concepts, design, and algorithms. Prentice Hall PTR, Englewood Cliffs (2002)
7. Dijkstra, E.W: A note on Two Problems in Conexion with Graphs. Numerische Math (1959)
8. Drozdek, A.: Estrutura de dados e Algoritmos em C++. In: Thomson, P. (ed.) SP (2002)
9. Xin, J., et al.: On the Physical and Logical Topology Design of Large-Scale Optical Networks. Journal of Lightwave Technology 21(4) (2003)
10. Zheng, J.: Optical WDM Networks: Concepts and Design Principles Ed. Wiley IEEE, p. 312 ISBN 0471671703-2004

11. Le, V.T., et al.: A Hybrid Algorithm for Dynamic Lightpath Protection in Survivable WDM Optical Networks. In: ISPAN 2005 (2005)
12. Zhou, B., et al.: Spare capacity planning using survivable alternate routing for long haul WDM networks. In: Proceedings ISCC – 2002 (2002)
13. Alanyali, et al.: Provisioning algorithms for WDM optical networks. IEEE/ACM Trans. Net. 7(5), 767–778 (1999)

Providing Consistent Service Levels in IP Networks

Solange Rito Lima, Pedro Sousa, and Paulo Carvalho

University of Minho, Department of Informatics, 4710-057 Braga, Portugal
{solange,pns,pmc}@di.uminho.pt

Abstract. The use of Internet as an ubiquitous communication platform puts a strong demand on service providers regarding the assurance of multiple service levels consistently. Designing flexible and simple service-oriented management strategies is crucial to support multiconstained applications conveniently and to obtain a deployable and sustainable service quality offer in multiservice IP networks. In this context, this paper proposes the use of a self-adaptive QoS and SLS management strategy sustained by a service-oriented traffic admission control scheme to ensure the negotiated quality levels. A proof-of-concept of the proposed strategy is provided, illustrating its ability to self-adapt and control efficiently distinct QoS requirements in multiservice IP networks.

Keyword: Multiservice Networks, Admission Control, Service Quality.

1 Introduction

Providing service integration in IP networks assuring, at same time, consistent levels of service quality tend to require the adoption of specific service models and traffic control mechanisms to handle traffic with multiple requirements. The challenge is increased when considering end-to-end Quality of Service (QoS) delivery, involving multiple heterogeneous domains with negotiated Service Level Specifications (SLSs) between them to be fulfilled. In fact, the end-to-end QoS panacea will not be based on a single network service model attending to the diversity of business goals and technologies available. This means that the network control tasks, when in place, should be flexible enough to accommodate heterogeneity and service integration efficiently. At same time, simplicity is a major design goal and a key aspect for their deployment in real networks.

This paper addresses the problematic of efficient and versatile QoS/SLS management, proposing a service-oriented and self-adaptive management architecture to improve QoS guarantees and enforce SLSs fulfillment in multiservice networks. In this architecture, we identify and structure the main issues and tasks subjacent to the definition, building and control of network services both intra and interdomain. Attending to the key role of Admission Control (AC) in preventing QoS instability and service degradation, we specify a service dependent AC criteria adjusted both to explicit and implicit AC scenarios, widening the diversity of services supported.

S. Ata and C.S. Hong (Eds.): APNOMS 2007, LNCS 4773, pp. 276–285, 2007.

The contents of this paper is organized as follows: current related work on SLS definition and management is debated in Section 2; an overview of the proposed QoS and SLS management architecture and operation is given in Section 3; the specification of the service-oriented AC criteria, including the proposed AC rules, is included in Section 4; the proof-of-concept and performance results are provided in Section 5; finally, the main conclusions are summarized in Section 6.

2 SLS Definition and Management

An SLS defines the expected service level, QoS related parameters and traffic control issues. The definition of a standard set of SLS parameters and semantics, apart from being a key aspect for QoS provisioning, is crucial for ensuring end-to-end QoS delivery and for simplifying interdomain negotiations [1]. Several working groups are committed to SLS definition and management [2,3]. Although a large combination of quality, performance and reliability parameters is possible, service providers are expected to offer a limited number of services. To define SLSs for quantitative and/or qualitative standard services adapted to different application types is, therefore, a major objective [4, 5].

For SLS management and control, solutions based on Bandwidth Brokers (BBs) (RFC 2638) centralize service information required to perform control tasks such as AC, removing them from the network core. However, this involves a large amount of information to manage and a functional dependence on a single entity. To improve reliability and scalability in large networks, several approaches consider distributed service control tasks with variable control complexity depending on the QoS guarantees and predictability required. To provide guaranteed services, existent proposals tend to require significant network state information and, in many cases, changes in all network nodes [6]. To provide predictive services, control tasks based on network measurements performed node-by-node [7,8] and end-to-end [9,10] have deserved special attention. These solutions lead to reduced control information and overhead, but eventually to QoS degradation. To control elastic traffic, for efficient network utilization, implicit strategies i.e., without requiring explicit signaling between applications and the network, have also been defined [11].

In these proposals, detailed in [12], aspects such as the trade-off between service assurance level and network control complexity for a scalable and flexible support of distinct service types and corresponding SLSs, intra and interdomain, are not covered or balanced as a whole. This grounds the motivation for the present proposal.

3 Multiservice QoS and SLS Management Architecture

The proposed self-adaptive architecture for managing multiple service levels involves different tasks related to service definition, monitoring and control, interrelated as illustrated in Fig. 1.

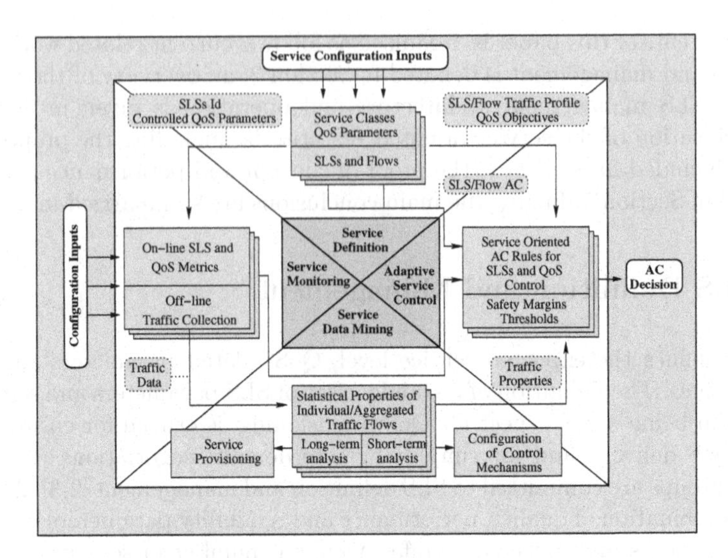

Fig. 1. Multiservice management architecture

Service definition involves the definition of basic service classes oriented to application with different requirements, the definition of relevant QoS parameters to control within each service type and the definition of SLSs' syntax and semantics. *Service monitoring*, performed on-line, keeps track of QoS and SLS status in the domain through a set of well-defined service metrics, providing feedback to drive *Self-adaptive service control* tasks such as AC. Traffic aggregates may also be collected for subsequent off-line analysis and characterization. *Service data mining* allows to determine the statistical properties of each class as a result of traffic aggregation so that more realistic service-oriented control mechanisms and service provisioning can be established. The knowledge resulting from interrelating these areas provides the basics for defining a multiservice management architecture and corresponding *AC decision criteria.*

In order to pursuit design goals such as flexibility, scalability and easy deployment, the service control strategy illustrated in Fig. 2 comprises: (i) distributed control between edge nodes; (ii) no control tasks within the network core; (iii) reduced state information and control overhead; (iv) measurement-based self-adaptive behavior regarding network dynamics. This model, oriented to accommodate multiple services, may perform AC irrespectively of the applications' ability to signal the network.

A primary idea of the AC strategy is to take advantage of the need for on-line QoS and SLS monitoring in today's networks and use the resulting monitoring information to perform distributed AC. This monitoring process, carried out on a per-class and edge-to-edge basis, allows a systematic view of each service class load, QoS levels and SLSs utilization in each domain, while simplifying SLSs' auditing tasks. An additional and crucial characteristic of the devised AC strategy is to consider a service-dependent degree of overprovisioning in order to achieve a simple and manageable multiservice AC solution. These levels

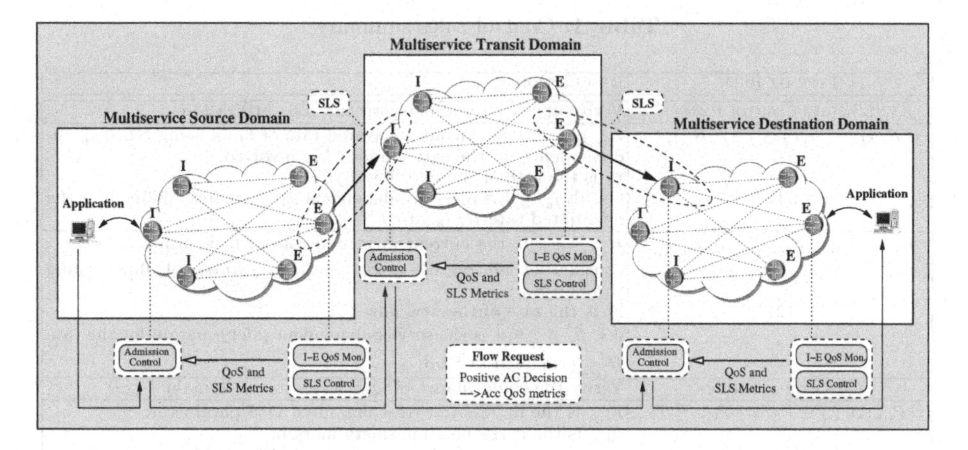

Fig. 2. Distributed monitoring-based AC approach

of overprovisioning, embedded in the AC rules, allow to relax the AC process widening the range of service types covered by a monitoring-based AC solution.

As shown in Fig. 2, in the model's operation, while ingress nodes perform implicit or explicit AC resorting to service-dependent rules for QoS and SLS control (see Section 4), egress nodes collect service metrics providing them as inputs for AC. When spanning multiple domains, collecting and accumulating the QoS measures available at each domain edge nodes will allow to compute the expected end-to-end QoS. This cumulative process is consistent with the cascade approach for the support of interoperator IP-based services, which has the merit of being more realistic, i.e., in conformance with the Internet structure and operation, and more scalable than the source-based approach [1].

4 Specifying the Multiservice AC Criteria

For controlling both the QoS levels in the domain and the utilization of existing SLSs, the following rules have been defined: (i) rate-based SLS control rules; (ii) QoS parameters control rules; (iii) end-to-end QoS control Rules. The specification of these rules, following the notation in [13], is presented in Table 1. The conformance of the defined rules determine the acceptance of a new flow F_j. Note that Eq. (3) is not flow dependent, i.e. it is checked once during Δt_i to determine $AC\_Status_{\Delta t_i}$. An $AC\_Status_{\Delta t_i}$ =`accept` indicates that the measured QoS levels for SC_i are in conformance with the QoS objectives and, therefore, new flows can be accepted. An $AC\_Status_{\Delta t_i}$ =`reject` indicates that no more flows should be accepted until the class recovers and restores the QoS target values, which will only be checked at Δt_{i+1}. For a service class SC_i under *implicit AC*, as flows are unable to describe r_j, traffic flows are accepted or rejected implicitly according to the value of $AC\_Status_{\Delta t_i}$.

In order to increase the scalability of the control strategy, it is the QoS parameter target value for the service class that bounds the corresponding SLS's

Table 1. Control rules summary

TYPE OF RULE	DESCRIPTION
SLS Rate Control Rules	Verify upstream and downstream SLSs utilization
$\tilde{R}_{i,(I_n,*)} + r_j \leq \beta_{i,I_n} R_{i,I_n}$ (1)	$\tilde{R}_{i,(I_n,*)}$ is the current measured rate of flows using SLS_{i,I_n} independently of the egress nodes E_m involved; r_j is the rate of the new flow F_j; $0 < \beta_{i,I_n} \leq 1$ is a service-dependent safety margin defined for the negotiated rate R_{i,I_n} of SLS_{i,I_n}.
$\tilde{R}^+_{i,(*,E_m)} + r_j \leq \beta^+_{i,E_m} R^+_{i,E_m}$ (2)	$\tilde{R}^+_{i,(*,E_m)}$ is the current measured rate of flows using SLS^+_{i,E_m}, considering all ingress-to-E_m estimated rates of flows going through E_m; r_j is the rate of the new flow F_j; $0 < \beta^+_{i,E_m} \leq 1$ is the service-dependent safety margin for the rate R^+_{i,E_m} defined in SLS^+_{i,E_m}.
QoS Control Rules	Verify the conformance of QoS levels in the domain
$\forall (P_{i,p}, \beta_{i,p}) \in P_{SC_i} : \tilde{P}_{i,p} \leq T_{i,p}$ (3)	$\tilde{P}_{i,p}$ is the ingress-to-egress measured QoS parameter; $\beta_{i,p}$ is the corresponding safety margin; $T_{i,p}$ is the parameter's upper bound or threshold, given by $T_{i,p} = \beta_{i,p} P_{i,p}$, used to set the acceptance status for Δt_i.
End-to-end Control Rules	Cumulative computation and verification of e2e QoS
$\forall P_{j,p} \in P_{F_j}$: $(\mathsf{op}_1\ (P^{acc^-}_{j,p}, P_{i,p}))\ \mathsf{op}_2\ (\gamma_{j,p} P_{j,p})$ (4)	$P_{j,p}$ is a flow's QoS parameter, allowing a tolerance factor $\gamma_{j,p}$; $P^{acc^-}_{j,p}$ is the parameter's cumulative value when crossing upstream domains; $P_{i,p}$ the corresponding target value in present domain.

expected QoS value and respective flows. Depending on the semantics of each parameter, $P_{i,p}$ can either be an upper or lower bound. Embedding the expected SLS parameters values in the respective class parameter target values is of paramount importance as QoS and SLS control in the domain is clearly simplified.

5 Proof-of-Concept

To evaluate the performance of the service control strategy regarding its ability to manage multiple service commitments in a multiclass environment, a simulation prototype was set using NS-2. This prototype implements three functional interrelated modules - Automatic Source Generation Module, AC Decision Module, and QoS and SLS Monitoring Module. Fig. 3 presents a simplified diagram of the simulation model architecture, including the relation between these modules and the main underlying functions and variables. The two recursive modules represented in gray are responsible for the dynamic behavior of traffic source generation and monitoring.

Taking into consideration current service configuration guidelines [14], three initial service classes have been defined. Table 2 summarizes the service classes implemented, highlighting AC and QoS monitoring parameters used to configure the AC rules specified in Table 1. Three downstream SLSs have been considered, one per service class, with a negotiated rate (R^+_{i,E_m}) defined according to the traffic load share intended for the corresponding class in the domain. As shown, the parameterization of the AC rules is service-dependent and larger safety margins β^+_{i,E_m} and tighter thresholds $T_{i,p}$ are defined for more demanding classes.

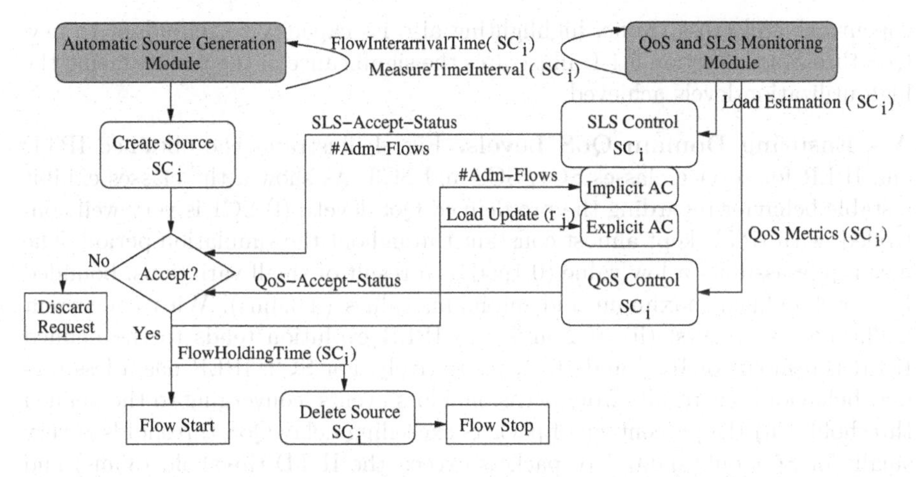

Fig. 3. Simulation model diagram

Table 2. Configuration of service classes SC_i

SC_i	Serv. Type	AC Type	R^+_{i,E_m}	β^+_{i,E_m}	$P_{i,p}$	$T_{i,p}$	Example	Traffic Src
SC1	guaranteed (hard-RT)	explicit and conservative	3.4Mbps (10% share)	0.85	IPTD ipdv IPLR	35ms 1ms 10^{-4}	VoIP Cir.Emulation Conv. UMTS	Exp.or Pareto on/off (64kbps,pkt=120B on/off=0.96/1.69ms)
SC2	predictive (soft-RT)	explicit and flexible	17Mbps (50% share)	0.90	IPTD IPLR	50ms 10^{-3}	audio/video streaming	(256kbps,pkt=512B on/off=500/500ms
SC3	best-effort	implicit	13.6Mbps	1.0	IPLR	10^{-1}	elastic apps.	FTP (pkt=512B)

For instance, a $\beta^+_{i,E_m} = 0.85$ corresponds to impose a safety margin or degree of overprovisioning of 15% to absorb load fluctuations and optimistic measures.

The network domain consists of ingress routers I_1, I_2, a multiclass network core and an edge router E_1. I_2 is used to inject *concurrent* or *cross* traffic (referred as CT-I2), allowing to evaluate concurrency effects on distributed AC and to assess the impact of cross traffic on the model performance. The test scenarios with cross traffic allow to evaluate the presence of unmeasured traffic within the network core. This type of traffic, likely to occur in real environments, impacts on domain's QoS and load without being explicitly measured by E_1 SLS rate control rule (Eq. (2)). The domain routers implement the three service classes according to a hybrid Priority Queuing - Weighted Round Robin (2,1) scheduling discipline, with RIO-C as AQM mechanism. The domain internodal links capacity is 34Mbps, with a 15ms propagation delay. Δt_i is set to 5s.

5.1 Simulation Results

This section intends to assess the self-adaptive behavior and effectiveness of the proposed service control strategy in keeping QoS levels and SLSs share consistently and efficiently controlled. For this purpose, Section 5.1-A presents results illustrating the solution's ability to ensure domain QoS levels in presence of concurrent and cross traffic, highlighting also its capacity to self-adapt to new QoS thresholds; Section 5.1-B illustrates the significance of the results facing the high utilization levels achieved.

A - Ensuring Domain QoS Levels. Fig. 4 illustrates the obtained IPTD and IPLR for service classes SC1, SC2 and SC3. As shown, the classes exhibit a stable behavior regarding the pre-defined QoS levels: (i) SC1 is very well controlled, with IPTD kept almost constant throughout the simulation period. The mean ipdv assumes a low value (0.1ms) as a result of small variations, bounded by a well-defined maximum and minimum values (±0.4ms). With concurrent traffic no loss occurs; (ii) SC2 and SC3 IPLR evolution tends to the defined IPLR thresholds of 10^{-3} and 10^{-1}, respectively. For SC2, IPLR has a less regular behavior as it results from occasional loss events, converging to the defined threshold; (iii) the percentage of packets exceeding delay QoS thresholds is very small: for SC1 only 0.007% of packets exceed the IPTD threshold (35ms) and 0.0005% the ipdv threshold (1ms). For SC2, 2.95% of packets exceed the delay threshold (50ms). Note that, exceeding a QoS threshold does not necessarily imply a service QoS violation, as the defined concept of threshold comprises a safety margin to the QoS parameter target value (see Eq. (3)). These results are particularly encouraging attending to the high network utilization obtained (see Section 5.1-B).

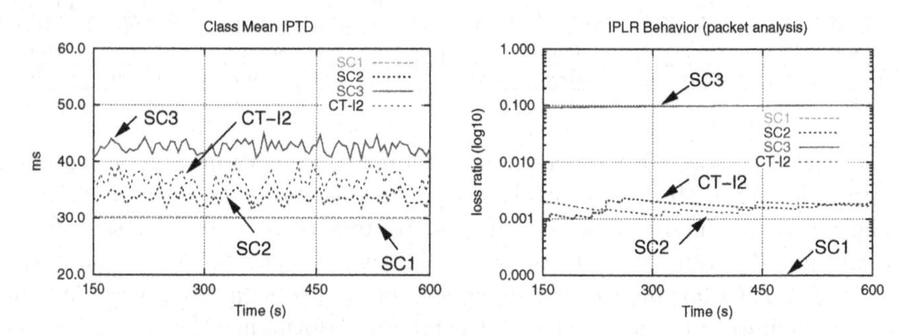

Fig. 4. IPTD and IPLR evolution

Impact of Cross Traffic and Adaptation to New Thresholds. This new test scenario intents to illustrate the model's ability to self-adapt to distinct QoS thresholds, in particular, to control new delay and loss bounds. In addition, the traffic conditions are now more demanding as the traffic submitted to ingress I2 is cross traffic. Fig. 5 presents a multimetric analysis of the IPTD and IPLR obtained for each service class in each Δt_i.

As shown, when a tighter IPTD threshold of 35ms is set for SC2, AC is effective in bringing and maintaining IPTD controlled around that value. Simultaneously, considering a new IPLR threshold of 0.05 for SC2 and SC3 (more relaxed and tighter than the previous one of 10^{-3} for SC2 and 10^{-1} for SC3), it is notorious that the control strategy has been able to bring IPLR to the new value. It is also evident that IPLR is more difficult to keep tightly controlled,

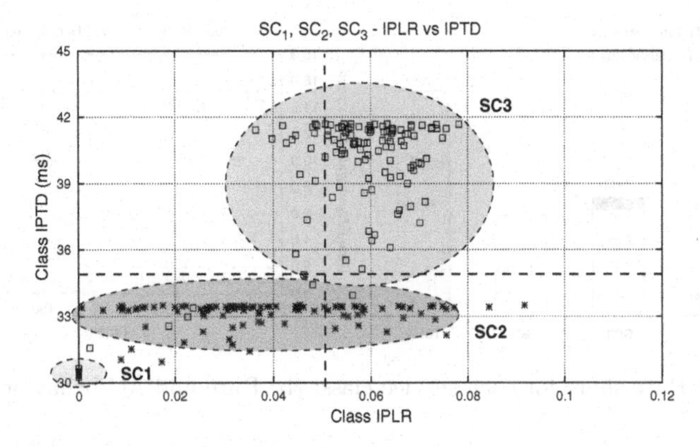

Fig. 5. Adaptation to new QoS thresholds in presence of cross traffic

however, a consistent behavior around 0.05 is achieved. In the presence of cross traffic, the main rule determining AC decisions is the QoS control rule Eq. (3), with $AC\_status_{\Delta t_i} = reject$ activated mostly by IPLR threshold violations. This rule by itself maintains the QoS levels controlled, as shown in Fig. 5.

From these set of experiments, the relevance of the defined AC rules becomes evident for assuring service commitments in the domain. While the rate control rule (Eq. (2)) assumes a preponderant role for service classes SC1 and SC2 to control the traffic load and indirectly QoS, particularly in situations involving concurrent traffic, the QoS control rule (Eq. (3)) is decisive to assure the domain QoS levels in presence of unmeasured cross traffic. In real environments, where the two type of situations are likely to occur simultaneously, the two AC rules will complement each other to increase the domain capabilities to guarantee service commitments.

B - Controlling SLSs Share. Fig. 6 (a) illustrates the obtained share of each class under the concurrent traffic test scenario. Note that, for the safety margins (β^+_{i,E_m}) and the SLSs rate share (R^+_{i,E_m}) defined in Table 2, the utilization target for SC1, SC2 and SC3 is 8.5%, 45% (SC2+CT-I2) and 40%, respectively. As shown, the share configured for each class and SLS is well accomplished and the global utilization is kept very high. SC2 and CT-I2 obtain a similar behavior and share and SC3 exceeds its share slightly. This occurs due to the adaptive nature of traffic within SC3, the more relaxed implicit AC criterion and the work conserving nature of the scheduling algorithm in use, which allows SC3 to take advantage of unused resources. The per-class and global utilization with cross traffic decreases slightly comparing to the concurrent case. This decrease is a consequence of the effect of cross traffic on queues occupancy increasing loss events and triggering the QoS control rule more frequently. However, the rate share of each class is also well accomplished and the global utilization very high.

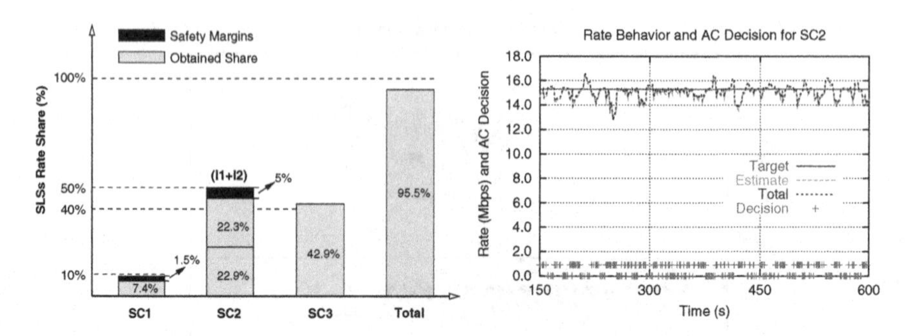

Fig. 6. (a) Rate share for each service class; (b) Rate and AC behavior for SC2

When observing the behavior of the AC rules for SC2 and the resulting AC decision along time (Fig. 6 (b)[1]), it can be seen that, although the rules are effective in blocking new flows when QoS degradation or an excessive rate is sensed, the effect of previously accepted flows may lead to some episodes of rate estimates above the target line. In fact, traffic fluctuations reflecting a low estimation in Δt_{i-1} may lead to over acceptance during Δt_i. Despite the fast time reaction of the control rules, these situations evince the advantage of using protective safety margins.

6 Conclusions

In this paper, a self-adaptive service management strategy has been proposed to improve QoS guarantees and enforce SLSs fulfillment in multiservice networks. The solution relies on service-dependent AC rules which allow a versatile and consistent control of QoS levels and SLS usage both intra and interdomain. The proof-of-concept has demonstrated that the self-adaptive behavior inherent to on-line measurements combined with the proposed AC rules is effective in controlling the quality levels of each service class. Under demanding traffic conditions, the relevance of the two defined AC rules became evident complementing each other to increase the domain capabilities to guarantee service commitments.

References

1. Georgatsos, P., et al.: Provider-level Service Agreements for Inter-domain QoS delivery. In: ICQT 2004 (September 2004)

[1] In Fig. 6 (b), *Target* line represents the value $\beta^+_{i,E_m} R^+_{i,E_m}$ above which AC rejection occurs, *Estimate* line represents the estimated rate or load of SLS^+_{i,E_m}, i.e., $\tilde{R}^+_{i,(*,E_m)}$, and *Total* line reports to the previous estimate by adding the new flow rate r_j. *Decision* dots represent a positive (dots above the x-axis) or negative (dots overlapping the x-axis) AC decision, considering also the QoS control rule evaluation.

2. Morand, P., et al.: Mescal D1.2 - Initial Specification of Protocols and Algorithms for Inter-domain SLS Management and Traffic Engineering for QoS-based IP Service Delivery and their Test Requirements. Mescal Project IST-2001-37961 (January 2004)
3. Diaconescu, A., Antonio, S., Esposito, M., Romano, S., Potts, M.: Cadenus D2.3 - Resource Management in SLA Networks. Cadenus Project IST-1999-11017 (May 2003)
4. Seitz, N.: ITU-T QoS Standards for IP-Based Networks. IEEE Communications Magazine 41(5) (2003)
5. Miras, D., et al.: A Survey of Network QoS Needs of Advanced Internet Applications. Internet2 Working Document (November 2002)
6. Stoica, I., Zhang, H.: Providing Guaranteed Services Without Per Flow Management. In: ACM SIGCOMM 1999 (October 1999)
7. Jamin, S., Danzig, P., Shenker, S., Zhang, L.: A Measurement-Based Call Admission Control Algorithm for Integrated Services Packet Networks (Extended Version). IEEE/ACM Transactions on Networking, 56–70 (1997)
8. Breslau, L., Jamin, S.: Comments on the Performance of Measurement-Based Admission Control Algorithms. In: IEEE INFOCOM 2000 (March 2000)
9. Cetinkaya, C., Kanodia, V., Knightly, E.: Scalable Services via Egress Admission Control. IEEE Transactions on Multimedia 3(1), 69–81 (2001)
10. Elek, V., Karlsson, G., Rnngren, R.: Admission Control Based on End-to-End Measurements. In: IEEE INFOCOM 2000 (2000)
11. Mortier, R., Pratt, I., Clark, C., Crosby, S.: Implicit Admission Control. IEEE Journal on Selected Areas in Communication 18(12), 2629–2639 (2000)
12. Lima, S.R., Carvalho, P., Freitas, V.: Admission Control in Multiservice IP Networks: Architectural Issues and Trends. IEEE Computer Communications Magazine 45(4), 114–121 (2007)
13. Lima, S.R., Carvalho, P., Freitas, V.: Self-adaptive Distributed Management of QoS and SLSs in Multiservice Networks. In: IEEE/IFIP International Conference on Integrated Management (IM 2005), Nice, France, IEEE Press, Los Alamitos (2005)
14. Babiarz, J., Chan, K., Baker, F.: Configuration Guidelines for Diffserv Service Classes. RFC 4594, Informational (August 2006)

A Visual Component Framework for Building Network Management Systems

Ichiro Satoh

National Institute of Informatics
2-1-2 Hitotsubashi, Chiyoda-ku, Tokyo 101-8430, Japan
ichiro@nii.ac.jp

Abstract. This paper presents a component framework for rapidly building and operating the visual interfaces for network management. It can easily and dynamically assemble visual components that can monitor and control their target network devices or sub-systems into an active document as a visual interface for the whole network. Since each visual component is an autonomous programmable entity, it can define network management protocols in addition to its visual interface inside it. End-users or administrators can manually customize smart spaces through user-friendly GUI-based manipulations for editing documents. This paper presents the design for this framework and describes its implementation and a practical application already being used for sensor networks in the real world.

1 Introduction

Network management is needed in various systems. For example, a ubiquitous computing environment consists of many heterogeneous computing devices connected through wired or wireless networks in a house or office. The requirements of applications in such environments tend to depend on their targets, e.g., users, houses, or offices. In fact, one of the most typical and popular applications of pervasive computing is context-aware services. To support such services, ubiquitous computing systems must be able to know the context and process this in the real world, e.g., in terms of people, locations, and time. However, most ubiquitous computing environments often lack professional administrators unlike other network systems. Therefore, end-users themselves are required to customize their own ubiquitous computing environments to their individual requirements and applications.

We propose a component framework to rapidly and easily build visual interfaces for network management systems. The framework is constructed based on a compound-document framework, called *MobiDoc*, developed by the author [9,10]. It enables one document to be composed of various visible parts, such as text, image, and video created by different applications, like other compound-document frameworks, e.g., COM/OLE [2], OpenDoc [1], CommonPoint [7], and Bonobo [3]. Compound-document technology is useful for constructing visual interfaces for network management, because it enables end-users to easily and dynamically assemble visual components into a seamless interface. However, existing compound document frameworks, including MobiDoc, have not been designed for network management. In fact, the framework themselves cannot monitor and control network systems. It enables visual components to define

S. Ata and C.S. Hong (Eds.): APNOMS 2007, LNCS 4773, pp. 286–295, 2007.

network protocols as well as the GUIs inside them, so that they can communicate their target network devices or sub-systems through the devices' or sub-systems' favorite controlling and monitoring protocols. It can assemble multiple components into one document or component by using GUI-based manipulation. Therefore, end-users can customize visual interfaces to monitor and manage their networks.

2 Background

This framework was inspired by our experiences with network management tasks in real systems. We have constructed and operated sensor networks for tracking the positions of visitors in several museums, e.g., at the National Science Museum (at Ueno) in Tokyo to support user/location-aware visitor assistance systems.[1]. As exhibitions are often changed in museums, even curators who have no knowledge about pervasive computing systems should be able to easily and naturally configure the systems, e.g., the topology of sensor networks and filtering the data measured by sensors, according to changes in the museums without the need for any professional administrators.

There have been a few attempts to enable end-users to easily manage their networks. Nevertheless, networks for ubiquitous computing environments in houses and offices are usually administered by end-users, who may have no knowledge or experience of networks. User-friendly interfaces are therefore needed so that end-users can easily manage networks in their ubiquitous computing environments. Several commercial or academic systems for network management provide visual interfaces for professional administrators rather than end-users. Furthermore, they explicitly or implicitly assumed that they were being used without any other network management systems. That is, their visual interfaces cannot coexist with those of other systems. We need to be able to seamlessly unify visual interfaces for different network management systems.

Although commercial or academic web-based tools for network management have recently been used, they cannot always monitor and control their target network systems in a real-time manner. This is because they periodically query the target systems and update their visual interfaces displayed within Web browsers executed on client-side computing devices through http-based protocols. Ajax technology may be able to reduce latency between the target systems and the visual interfaces, but most network devices or sub-systems do not support the technology, because they support only simple or particular protocols.

Administrators or users must support heterogeneous network devices, sub-systems, or services that are different. Our framework needs to be independent of any network management systems and open to various network management protocols. Networks in ubiquitous computing environments are evolving in the sense that network devices and services are dynamically being added to and removed from them. Therefore, a user-friendly management system supports the dynamic evolution of network systems or ubiquitous computing systems.

There have been several mechanisms for automatically generating GUIs to controlling devices [6,5]. Most existing approaches can provide GUIs for individual devices and can support them being dynamically generated for devices that may be added. They

[1] The system is presented later in detail.

assume the target devices, which they should provide visual interfaces for, can support their own control protocols for visual interfaces. However, most network devices do not support such protocols. Instead, they can be controlled and monitored through their own favorite protocols, e.g., SMNP, HTTP, and telnet, in addition to management-specific protocols. Therefore, the framework presented in this paper is required to support various protocols. Network devices and ubiquitous computing devices only have limited resources, such as restricted amounts of CPU power and memory. However, client-side devices, i.e., PCs and PDA, have sufficiently resources with their input/output devices, e.g., display, keyboard, and mouses. Therefore, visual interfaces should be managed and executed as much as possible in external systems, including client-side devices, rather than in network devices.

3 Component Framework for Network Management

This section presents a component framework for building and operating the visual interfaces for network management systems by using compound document technology.

3.1 Basic Approach

Like other compound document frameworks, the framework presented in this paper enables components to maintain their own contents within them and to be dynamically assembled into one document or component. It also supports GUI-based manipulations to edit individual components and to layout components on GUI windows or control panels so that end-users can create GUIs for their networks. Unlike other existing frameworks, e.g., COM/OLE, OpenDoc, CommonPoint, and Bonobo, it provide each component with its own program code to view and edit its content within the component. Therefore, such a component itself can implement network management protocols to communicate with its target network device, sub-system, and service. For example, when a user wants to manage a new network device in his/her networks, he/she drags and drops the visual component that can define the network protocol to monitor and control the device as well as the visual interfaces on his/her control panel.

Although the framework inherits many features of our previous compound document framework, `MobiDoc`, it has been extended to support network management. It consists of two parts: visual component and component runtime system. The former is defined as a collection of Java objects and the latter is executed on the Java virtual machine (VM). Since the Java VM and libraries abstract away differences between underlying systems, e.g., operating systems and hardware, components and runtime systems can be executed on different computers, whose underlying systems may be different.

3.2 Component Runtime System

Each runtime system governs all the components within it and provides them with APIs for the components in addition to Java's classes. It assigns one or more threads to each component and interrupts them before the component terminates, or is saved. Each component can request its current runtime system to terminate and save itself and its

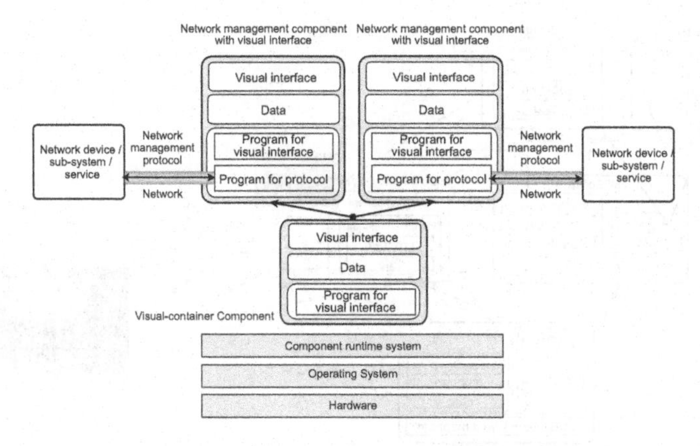

Fig. 1. Visual component for network management

inner components in secondary storage. This framework provides each component with a wrapper, called a *component tree node*. Each node contains its target component, its attributes, and its containment relationship and provides interfaces between its component and the runtime system (Fig. 2). When a component is created in a runtime system, it creates a component tree node for the newly created component. When a component migrates to another location or duplicates itself, the runtime system migrates its node with the component and makes a replica of the whole node.

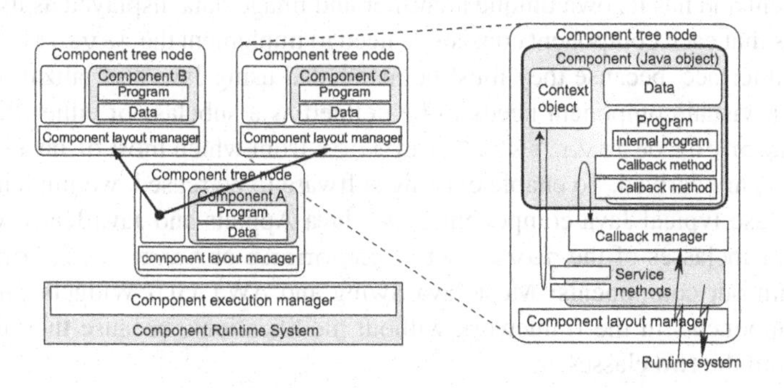

Fig. 2. Component hierarchy and structure of components

A hierarchy is maintained in the form of a tree structure of component tree nodes of the components (Fig. 3). Each node is defined as a subclass of `MDContainer` or `MDComponent`, where the first supports components, which can contain more than one component inside them and the second supports components, which cannot contain any components. For example, when a component has two other components inside it, the nodes that contains these two inner components are attached to the node that wraps the container component. Component migration in a tree only occurs as a transformation of the subtree structure of the hierarchy. When a component is moved over a network, on the other hand, the runtime system marshals the node of the component, including

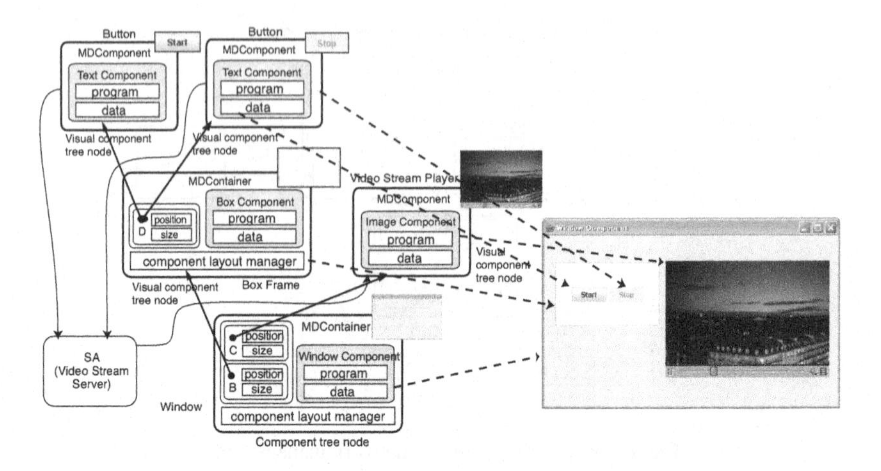

Fig. 3. Component Hierarchy

the nodes of its children, into a bit-stream and transmits the component and its children and the marshaled component to the destination.

3.3 Visual Component

As we can see from Fig. 2, each component is a collection of Java objects wrapped in a component and has its own unique identifier and image data displayed as its icon. All the objects that each component consists of need to implement the `java.io.Serializable` interface, because they must be marshaled using Java's serialization mechanism. Each visual component needs to be defined as a subclass of either the `java.awt.Component` or `java.awt.Container` from which most of Java's visual or GUI objects are derived. To enable existing software to be reused, we implemented an adapter to use typical Java components, e.g., Java Applets and JavaBeans, which are defined as subclasses of the `java.awt.Component` or `java.awt.Container` class within our components[2] Most Java Swing and AWT GUI Widgets can be used as our components in the framework without modifications, because they have been derived from the two classes.

The runtime system can invoke specified callback methods defined in components when they are created, relocated, and terminated and can assign more than one active thread to the components. We can define program codes to communicate with network systems or devices within these callback methods. In fact, we constructed several components for network management through basic protocols. For example, an HTTP server (or client) component plays the role of an HTTP server (or client) to monitor and control network devices as HTTP clients (or servers) and a Telnet component can connect to its target device through a telnet protocol. Since these components are defined as abstract classes, we can define visual interfaces for network management by

[2] This is not compatible with all kinds of Applets and JavaBeans, because some of these existing applications manage their threads and input and output devices depreciatively.

using Java Swing or AWT Widgets. We also developed various components, e.g., a text viewer/editor component and a JPEG, GIF, MPEG viewer component and audio-player component. Note that visual components allow their content to be in arbitrary as well as standard formats, because they have codes for viewing and modifying content. Components can support further application-specific protocols. For example, the Video Stream Player component as shown in Fig. 3 supported Real-Time Protocol (RTP) to received a video stream and displayed the stream on its visual rectangle with a GUI control panel to stop, play, forward, back and pause.

3.4 Component Manipulation

Each component can display its content within the rectangular estate maintained by its container component. The node of the component, which is defined as a subclass of the `MDContainer` or `MDComponent` class, specifies attributes, e.g., its minimum size and preferable size, and the maximum size of the visible estate of its component in the estate is controlled by the node of its container component. These classes can define their new layout manager as subclasses of the `java.awt.LayoutManager` class.

This framework provides an editing environment for manipulating the components for network processing, as well as for visual components. It also provides in-place editing services similar to those provided by OpenDoc and OLE. It offers several value-added mechanisms for effectively sharing the visual estate of a container among embedded components and for coordinating their use of shared resources, such as keyboards, mice, and windows. Each component tree node can dispatch certain events to its components to notify them when certain actions happen within their surroundings. `MDContainer` and `MDComponent` classes support built-in GUIs for manipulating components. For example, when we want to place a component on another component, including a document, we move the former component to the latter through GUI manipulations, e.g., drag-and-drop or cut-and-paste. When the boundary of the visible area of a component is clicked, the component is *selected* and eight rectangular control points for moving it around and resizing it are displayed (Fig. 4 (a)). The user can resize the selected component, move it to another, save it, and terminate it by dragging its handles (Fig. 4 (b)).

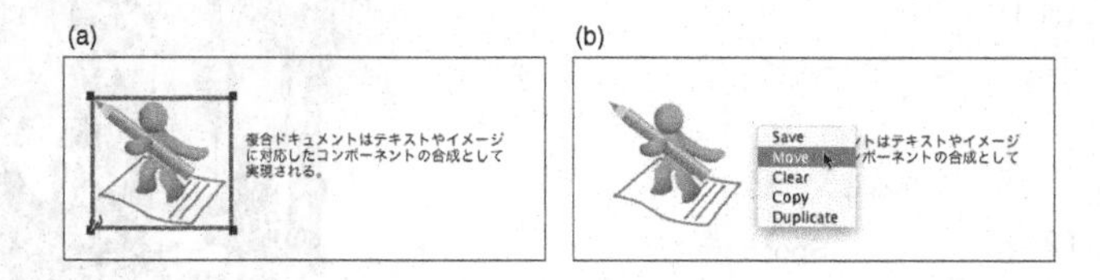

Fig. 4. Editing layout for components and popup menu for controlling components

3.5 Component Persistence

By using mobile agent technology, the runtime system can save or duplicate a component, its children, and information about their containment relationships and visual layouts into a bit-stream and can then later unmarshal the components and information from the bit-stream. When a component is saved or duplicated, its code and state, e.g., instance variables, can be marshalled by using the Java object serialization package. The package does not support the capturing of stack frames of threads. Consequently, our system cannot marshal the execution states of any thread objects. Instead, the runtime system (and the Java virtual machine) propagates certain events to components before and after marshaling and unmarshaling them. The current implementation of our system uses the standard JAR file format for passing components that can support digital signatures, allowing for authentication.

4 Application

The framework has already been used in a management system for context-aware user-assistant services in public museums, e.g., at the National Science Museum in Tokyo and the Museum of Nature and Human Activities in Hyogo, Japan. The system was constructed as a sensor network where RFID tag readers were connected through a wireless network. These readers were located at specified spots in several exhibition spaces at these museums (Fig. 5). Visitors were provided with active RFID-tags to track their locations. When they came sufficiently close to various objects, e.g., zoological specimens and fossils, located at the spots, they could listen to sound content that annotated the objects. The RFID-tag readers identified all the visitors within their range of coverage, i.e., a 2-meter diameter and sent the identifiers of their detecting RFID tags to a service-provider computer through TCP sessions. All service-provider computers had databases for storing audio contents and selected and play content according to their guest's knowledge and interests when they receive the identifiers of tags from readers. Fig. 6 shows a screenshot of the visual interface for the management system. The interface enables users to deploy services at areas by using drag-and-drop manipulations. For example, the exhibition had more than 200 visitors daily and the system

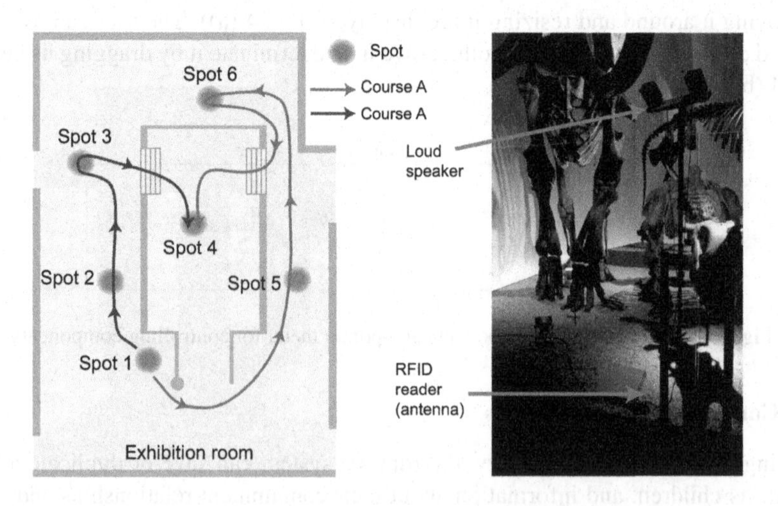

Fig. 5. Spots in museum and networked RFID tag reader at spot

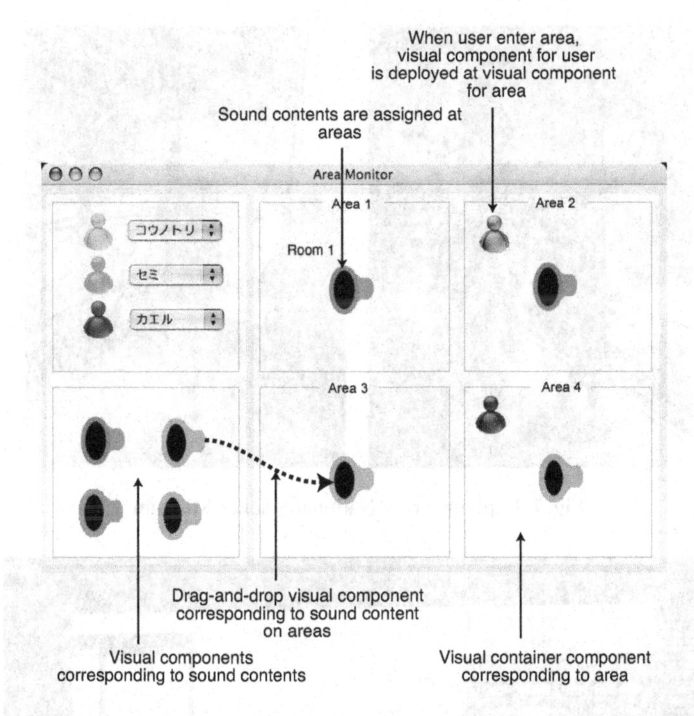

Fig. 6. Screenshot of monitor system window for user/location-aware visitor assistance system with sensor network

continued to monitor and manage RFID-tag readers and location-aware services for a week without experiencing any problems (Fig. 7).

The interface consisted of four visual components that monitored four RFID-tag readers located at spots throughout the exhibition consisting of four spots. As we can see from Fig. 6, the window was implemented as a window component that contains six components corresponding to the frame-boxes in it. The four of the box-frame components represented the spots and had programs that communicated with their readers through TCP sessions to monitor the presence of tags within their coverage areas, where RFID readers could work as TCP servers to send the identifiers of such tags. Fig. 8 shows the visual interface for user/location-aware systems used at the National Science Museum. An image view component drew a map of the exhibition room and it contained six management components for RFID readers connected through a network.

When a visitor with an RFID-tag entered a spot, the component corresponding to him or her was deployed at the component corresponding to the spot. We could dynamically add/remove location-aware services to/from spots. We deployed software to define the service at the component corresponding to a spot by dragging-and-dropping the visual component of the software on the visual components corresponding to the places in which the services should be provided. Curators, who may have no knowledge about ubiquitous computing systems, can therefore easily and intuitively change audio-based assistance services at exhibitions.

Fig. 7. Experiments at National Science Museum

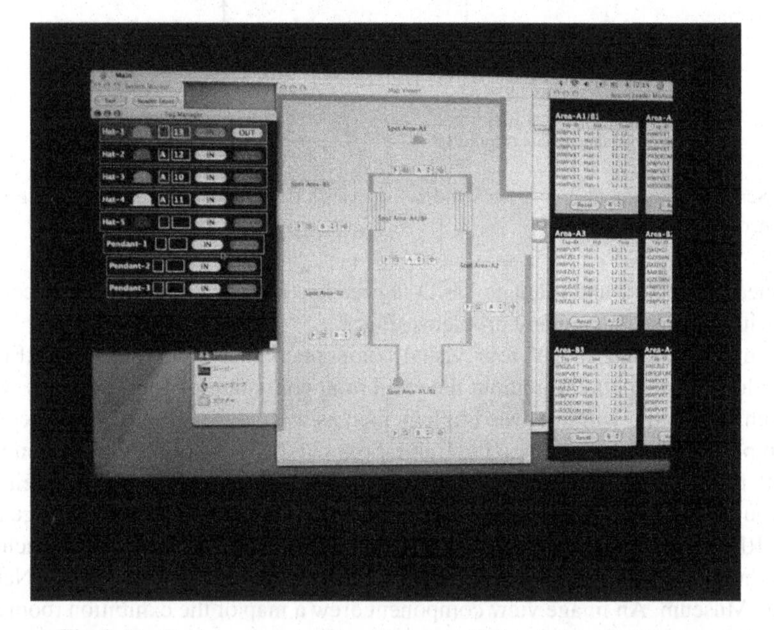

Fig. 8. Monitoring computer for six spots at National Science Museum

5 Conclusion

We presented a component framework for rapidly building and operating the visual interfaces for network management. The framework can dynamically assemble visual components into a visual interface. It enable components to communicate with their target network sub-systems through the sub-systems' favorite protocols, since these components contain their own program code in addition to the content inside them. Since it provides GUI-based manipulations for editing visual components, end-users can easily add and remove the visual interfaces for network management on their control panels. We designed and implemented a prototype-system based on the framework and demonstrated its effectiveness in a network management system for sensor networks in public museums.

In concluding, we would like to identify further issues that need to be resolved. As the framework is general, it has not only been designed for visual components for monitoring and controlling network management systems. Even though the current implementation relies on Java's security manager, we are interested in additional security mechanisms for these components.

References

1. Apple Computer Inc. OpenDoc: White Paper, Apple Computer Inc. (1994)
2. Brockschmidt, K.: Inside OLE 2. Microsoft Press (1995)
3. The GNOME Project, Bonobo (2002), `http://developer.gnome.org/arch/component/bonobo.html`
4. Hamilton, G.: The JavaBeans Specification, Sun Microsystems (1997), `http://java.sun.com/beans`
5. Hodes, T.D., Katz, R.H., Servan-Schreiber, E., Rowe, L.: Composable ad-hoc mobile services for universal interaction. In: MobiCom 1997, pp. 1–12 (1997)
6. Nichols, J., Myers, B.A., Higgins, M., Hughes, J., Harris, T.K., Rosenfeld, R., Pignol, M.: Generating remote control interfaces for complex appliances. In: UIST 2002, pp. 161–170. ACM Press, New York (2002)
7. Potel, M., Cotter, S.: Inside Taligent Technology. Addison-Wesley, Reading (1995)
8. Rogerson, D.: Inside COM. Microsoft Press (1997)
9. Satoh, I.: Network Processing of Documents, for Documents, by Documents. In: Alonso, G. (ed.) Middleware 2005. LNCS, vol. 3790, pp. 421–430. Springer, Heidelberg (2005)
10. Satoh, I.: A Component Framework for Document Centric Network Processing. In: SAINT 2007, IEEE Computer Society, Los Alamitos (2007)

The Primary Path Selection Algorithm for Ubiquitous Multi-homing Environments*

Dae Sun Kim and Choong Seon Hong

Department of Computer Engineering, Kyung Hee University,
Seocheon, Giheung, Yongin, Gyeonggi, 449-701 Korea
{dskim,cshong}@khu.ac.kr

Abstract. The multi-homing technology can provide an extended coverage area via distinct access technologies. Also, it is able to redirect a flow from one interface to another without reinitiating the flow. However, there is no suitable multi-homing technology for ubiquitous network environment at the moment. To provide multi-homing schemes into the ubiquitous environment, various related researches should be studied. Therefore, we proposed primary path selection algorithm which can provide a ubiquitous access and Flow redirection.

1 Introduction

Currently, multi-homing issues[1] are one of the most important factors in ubiquitous network environment. Multi-homing technologies can establish connection more efficiently and reliably between nodes those communicate each other. To adopt multi-homing technologies into the ubiquitous networks, various algorithms such as source address selection, primary path selection and failure discovery must be defined. An efficient algorithm for address selection and failure discovery has been articulated in [2]. However, existing works do not provide suitable path selection algorithms for multi-homing technology in ubiquitous networks which is very important considering the performance of the network. So in this paper our main focus is on primary path selection algorithm to improve the network performance. Our paper is organized as follows. Section 2 introduces related works about multi-homing issues. In section 3, we described the primary path selection algorithm for Ubiquitous network environment. After that we show the simulation results in section 4. Finally, we conclude in section 5.

2 Related Works

The monami6 working group in IETF has defined scenarios so that multiple interfaces and multiple CoA (care of addresses) registration [3] can be used. According to those scenarios, mobile node [5][6] can use multiple types of access technologies such as

* This work was supported by the Korea Research Foundation Grant funded by the Korean Government (MOEHRD)" (KRF-2006-521-D00394).

S. Ata and C.S. Hong (Eds.): APNOMS 2007, LNCS 4773, pp. 296–305, 2007.

802.11.b, WiBro and Cellular network in order to maintain ongoing communication, especially when it is moving out of coverage area of a specific technology. Moreover, these scenarios described how the mobile node can dynamically redirect flows [3] from one type of access technology to another and how it can also select the best access technology according to traffic characteristics or preferences. In multiple care of addresses registration [2], binding unique identifiers sub option for registration of multiple CoA [5][8] is defined. Although, this group defined well about multi-homing, still there are problems to solve such as source address selection and primary path selection algorithm. These algorithms should be defined as because mobile node must select one of the access technologies and it can change path according to user preference, applications and traffic quality. Therefore, this research focused mainly on primary path selection when mobile node make connection with its' home agent and correspondent node. Moreover it defined the path selection algorithm that can dynamically change primary path according to network state.

3 Path Selection Scheme for Ubiquitous Network Environment

In this section the primary path selection algorithm, binding update with multiple care of address, administration of binding cache and operation of proposed scheme are described.

3.1 Primary Path Selection Algorithm

The primary path selection algorithm is a method that can select primary path by Home Agent and Correspondent Node. When mobile node selects primary path in multi-homing environment, it can allocate primary path using the following algorithm. There are three different factors; which are signal strength, data rate and bandwidth utilization used in this algorithm. Our research considered these factors to compare the most appropriate paths. If signal strength and data rates have higher value, it means network can provide high quality of service to a mobile node. In addition, according to primary path selection algorithm, network states can be separated in eight different cases as shown figure 1. The figure 1 shows the results to measure each signal strength, data rate and bandwidth utilization by mobile node at the same time.
The graphs are analyzed as follows:

① Time 1: The mobile node should select A as a primary path because it has higher value for both signal strength and data rates than B

② Time 2: Although A's signal strength and data rate are higher than the B's, bandwidth at interface a ('a' is connected with access router A)

③ Time 3: If bandwidth utilization of interface 'a' and 'b' are over threshold value, the mobile node can select the A as a primary path which have higher value for data rate and signal strength.

④ Time 4: The mobile node should select the B as a primary path even the A has higher data rate than the B. This is because of the signal strength of A falls under

Table 1. Pseudo code for primary path selection algorithm

```
If signal strength  A>B
      If data rate A>B
           If  A's signal strength ≠ threshold
                If  interface a of MN's bandwidth utilization = enough
                          A is selected primary path by mobile node
                Else if  interface b of MN's bandwidth utilization = enough
                          B is selected primary path by mobile node
                Else // 'a' and 'b' = threshold
                          A is selected primary path by mobile node
           Else // A's signal strength = threshold
                B is selected primary path by mobile node
      Else // data rate A<B
           If  B's signal strength ≠ threshold
                If  interface b of MN's bandwidth utilization = enough
                          B is selected primary path by mobile node
                Else if  interface a of MN's bandwidth utilization = enough
                          A is selected primary path by mobile node
                Else // 'a' and 'b' = threshold
                          B is selected primary path by mobile node
           Else // B's signal strength = threshold
                A is selected primary path by mobile node
                A is selected primary path by mobile node
End
* A and B are Access Router
* 'a' is a interface of mobile node to connect with A
* 'b' is a interface of mobile node to connect with B
* threshold is a  minimum value which can connect with access router by a MN
```

the threshold. It means mobile node moves far from access router so that it can prepare a handover or its' connection can be disconnected with access router.

⑤ Time 5: Although B's signal strength is lower than the A's, the mobile node should select the B as a primary path because the B's data rate is higher than the A's data rate. It means the B can provide a better quality of service than the A.

⑥ Time 6: It has same condition with the Time 5 except bandwidth utilization. In this timing, the B's bandwidth utilization increases over the threshold. Which means it can not provide efficient throughput through the interface 'b' at that time. Therefore, the mobile node should select A as the primary path.

⑦ Time 7: It has similar condition with the time 3. Even though the A's signal strength is higher than the B, but B's data rate is higher than A. So the mobile node should select the B as a primary path.

⑧ Time 8: The mobile node should select A as a primary path because A's signal strength falls under the threshold.

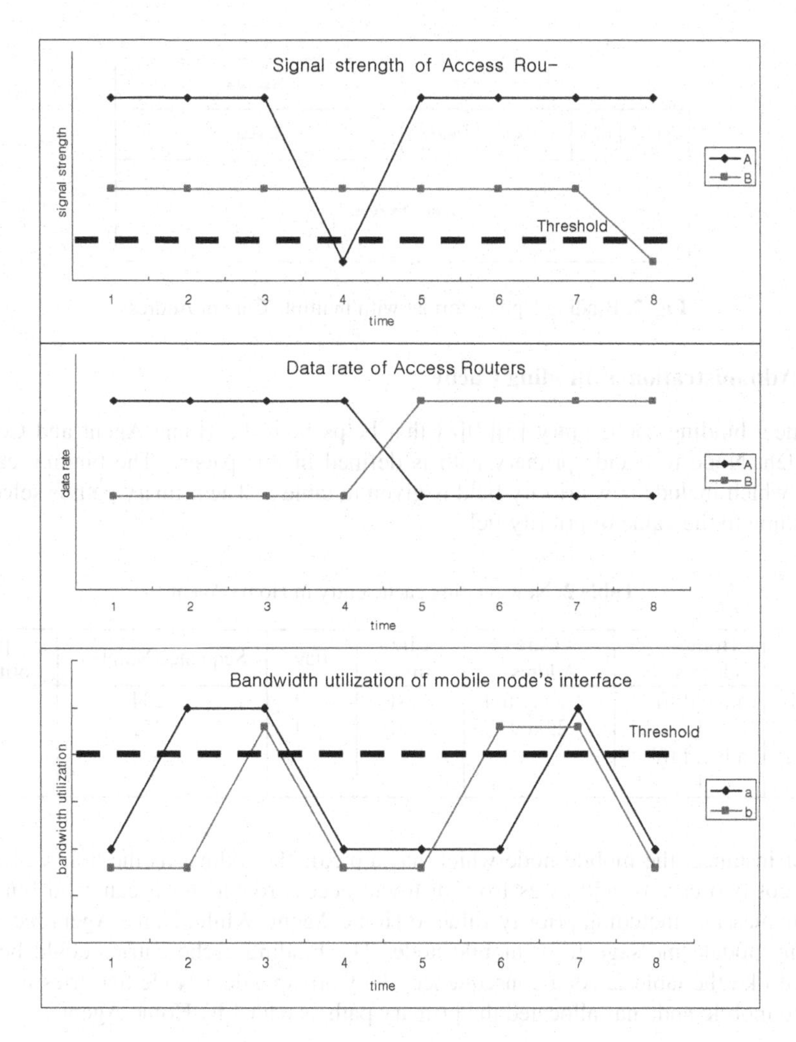

Fig. 1. Eight cases for primary path allocation

3.2 Binding Update with Multiple Care of Address

New binding update [5] fields are defined to accommodate multiple binding and priority value registration. The following additional fields are required in the binding update message. Figure 2 depicts the new binding update message format

- 'M' flag: Mobile node sends multiple care of address in binding update
- 'Priority' field: The priority value is calculated using path selection algorithm by mobile node. It is used by Home Agent and Correspondent Node to select a primary path.

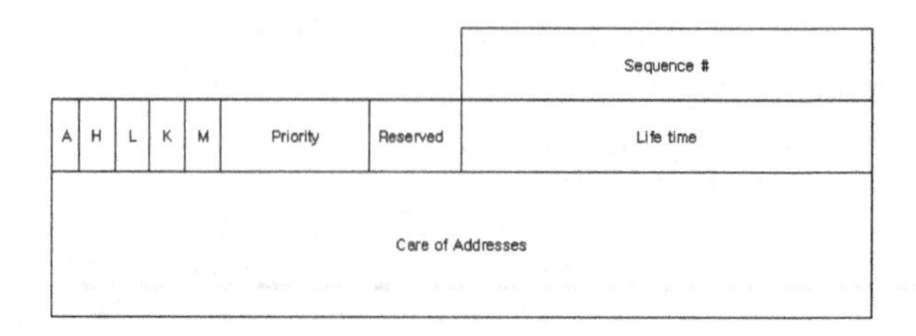

Fig. 2. Binding Update format with multiple Care of Address

3.3 Administration of Binding Cache

The new binding cache entry [5][7][8] that helps both the Home Agent and Correspondent Node to decide primary path is defined in this paper. The binding cache entry which include new priority field is given in table 2. The primary path is selected according to the value of priority field.

Table 2. New binding cache entry in Home Agent

	Home address	Care of Address	Life time	flag	Sequence Number	priority
Path1	a:b:c:d:fff1	1:b:c:fff1	10	1	234	1
Path2	a:b:c:d:fff1	2:b:c:fff1	10	1	252	2

For instance, the mobile node which has a:b:c:d:fff1 address as the home address have got two care of addresses from different access router. So it can send binding update message including priority value to Home Agent. While, Home Agent receives binding update message from mobile node. The binding cache entries could be recorded like the table 2. As a consequence, the Correspondent Node that tries to connect to mobile node has allocated the primary path as path 1 by Home Agent.

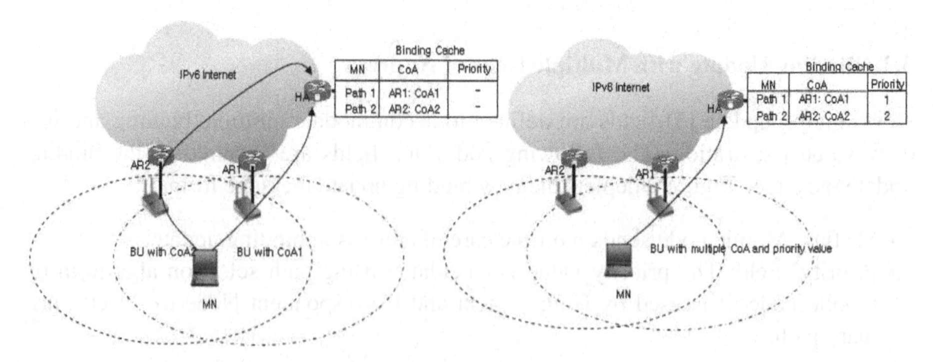

Fig. 3. Normal binding procedure and Binding Update with multiple CoA and Priority values

The mobile node can get binding acknowledgements for each binding update. Then it measures signal strength and data rate for each access router. As a result, it sends binding update which includes multiple CoA and priority value that is calculated by primary path selection algorithm as shown in figure 3. The following figure 4 shows the header format when the mobile node sends binding update to home agent through the AR1.

IPv6	MIPv6: M='1', Priority='1', Address: CoA1	MIPv6: M='1', Priority='2' Address: CoA2

Fig. 4. Packet which is sent to Home Agent through the Access Router 1

The binding cache database is updated by the priority value. It provides two options to Home Agent. First option is that Home Agent can separate traffic according to the traffic class value in the IPv6 header. While, Second option is Home Agent allocates most suitable paths for each correspondent node.

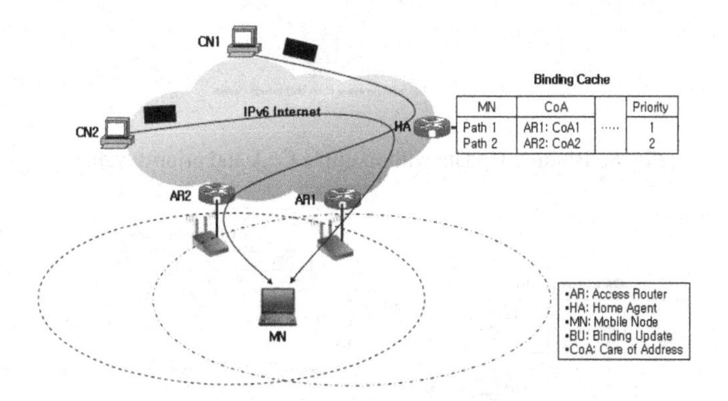

Fig. 5. Path selection according to the traffic specific

Figure 5 shows path selection procedure of Home Agent.

Following figure 6 and 7 shows the header information [4] of each correspondent node.

IPv6 Header traffic class='high value'	data

Fig. 6. Packet which is sent to Home Agent through the Access Router 1

IPv6 Header traffic class='low value'	data

Fig. 7. Correspondent node 2's header

When home agent receives packets from correspondent nodes, it checks the traffic class value in the IPv6 header. Then it allocates primary path for each correspondent node. For instance, when the correspondent node 1 sends packet having high value in

IPv6 traffic class field to the mobile node, the home agent allocates path 1 to correspondent node 1 according to a traffic class value. In contrast, when the correspondent node 2 sends packet having low value in IPv6 traffic class field to the mobile node, the home agent allocates path 2 to correspondent node 2 according to a traffic class value.

The mobile node can send the binding update with multiple CoA and priority value after it receives packets of correspondent nodes from the home agent as shown figure 8. After that the correspondent nodes update their binding cache database like figure 9.

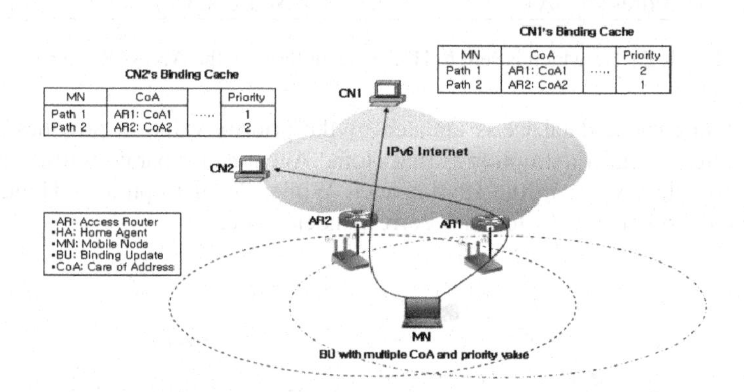

Fig. 8. Binding Update with multiple CoA and priority value

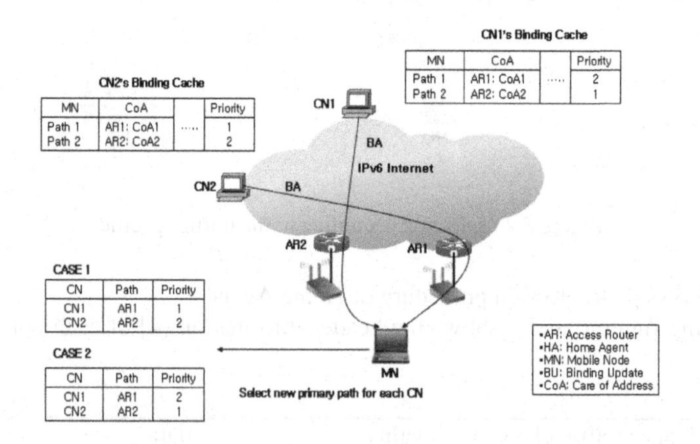

Fig. 9. Primary path selection for each correspondent node

When mobile node receives binding acknowledgements from each correspondent node, it executes primary path selection algorithm and the result could be like figure 8. There are two expected results as shown in case 1 and case 2 in figure 9. Figure 9 shows that, although, the primary path is selected as AR2 for correspondent node 1, the mobile node can change primary path as AR1 for correspondent node 1 according to the result of primary selection algorithm.

The following algorithm describes the process of registering a care of address to home agent:

① Mobile node creates care of addresses which are based on prefix information received from Access Router 1 and Access Router 2.
② MN sends binding update messages through AR 1 and AR 2.
③ When Home Agent receives binding update messages from the mobile node, care of addresses are registered on the Home Agent's binding cache database after address validation.
④ Home Agent then, sends binding acknowledgement messages to each care of address.
⑤ When the mobile node receives binding acknowledgement messages from Home Agent, it decides the primary path using primary path selection algorithm.
⑥ Then, the mobile node sends binding update message which includes multiple care of addresses through the primary path.
⑦ When the Home Agent receives binding update message, it marks the primary path value into the primary care of address's record.

3.4 Operation of Correspondent Node

The following procedure describes the alteration of primary path by the mobile node:
① The correspondent node 1 sends data having high flow level value in the IPv6 field to mobile node.
② In contrast, the correspondent node 2 sends data having low flow level value in the IPv6 field to mobile node.
③ When HA intercepts each packet which is sent by correspondent nodes, it decides correspondent node 1's path and correspondent node 2's path according to the primary path value. Therefore, correspondent node 1's packet sends to mobile node through the primary path and correspondent node 2's packet sends to mobile node through the secondary path.
④ After returning from routing and binding procedure, mobile node can make decision to allocate different primary paths for each correspondent nodes using primary path selection algorithm.

4 Simulation

In this section, we describe the simulation environment and show the simulation results. For the simulation, we used the Network Simulator version 2 (NS-2) [9].

4.1 Simulation Environment

For the simulation, we configured the network topology as shown in figure 10. There are four wired nodes which are connected with 5Mbps data rate and 200ms link delay and one mobile node which are connected with node 2 and node 3. Node 2 and node 3 are connected through a wireless link (path 1) of 0.5Mbps and 600ms delay. While

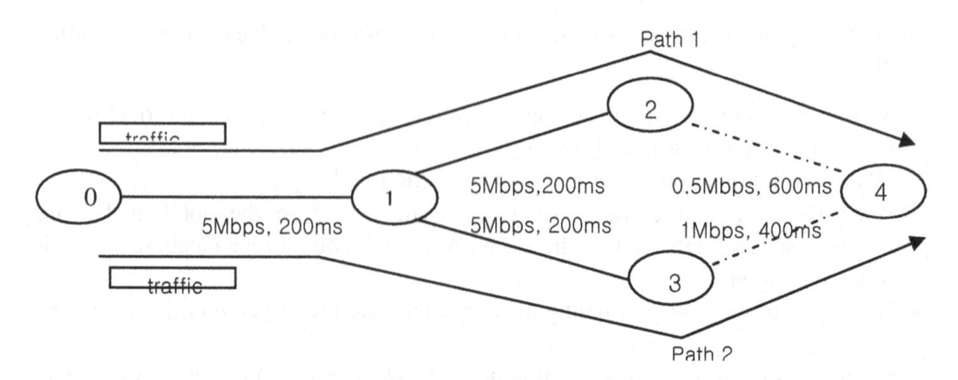

Fig. 10. Simulation environment

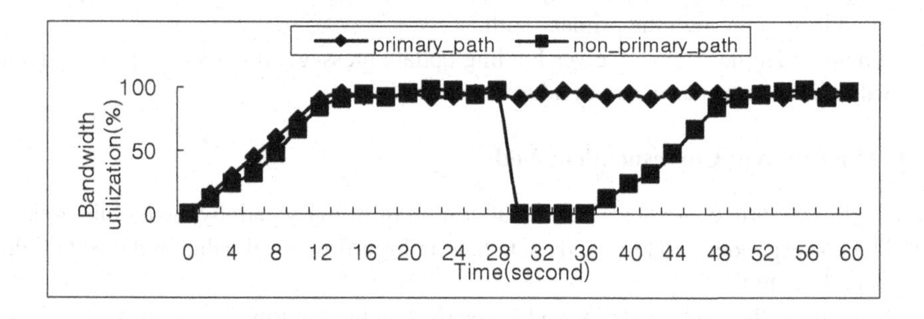

Fig. 11. Bandwidth utilization

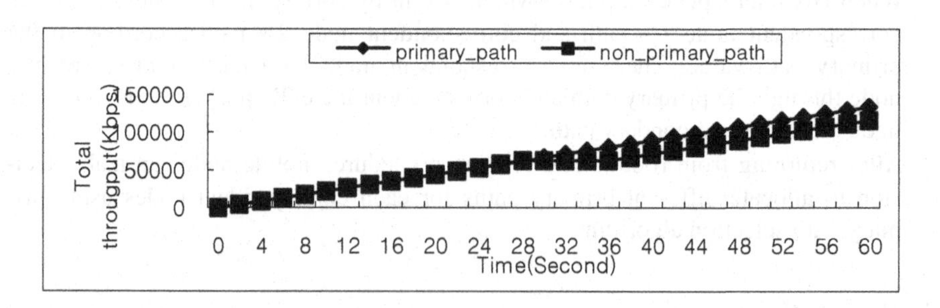

Fig. 12. Throughput

node 3 and node 4 has a wireless link (path 2) of 1Mbps and 400ms delay between them. In order to measure the performance of proposed scheme, we have generated two different size traffics at the node 0. The traffic which has 0.5Mbps rate is sent to the node 4 through the path 1 by the node 0. Other traffic which has 2Mbps rate is sent to the node 4 through the path 2 by the node 0. The simulation scenario is that node 0 sends two different size traffics to node 4 through the node 2 and node 4 for 60 seconds. During the simulation, we have configured that mobile node moves near to node 3 so that connection of the mobile node and the node 2 is disconnected for 6 seconds. And then, the mobile node returns to original position. Figure 11 shows bandwidth utilization of the mobile node measured during simulation. From the figure we conclude that, primary path scheme and non-primary path scheme both have same bandwidth utilization except disconnected time. Also, we have got total throughput as shown the figure 12. It shows that the proposed scheme's total throughput is better than normal scheme's because when the mobile node is disconnected from an access router, the mobile node can make new path through other access router immediately.

5 Conclusion

Multi-homing has become one of the most crucial issues in ubiquitous network environment. As a consequence, recent literature has enriched by a number of proposed multi-homing schemes. Most of the schemes have come up with abstract concepts and scenarios of ubiquitous network. To provide an extended coverage area via distinct access technologies anywhere, anytime and anyone, we propose the primary path selection algorithm for ubiquitous multi-homing environment. We also defined new binding update and binding cache entry. Therefore, a mobile node can select suitable path for data transmission. In addition, it can change primary path when wireless network environment is changed such as a mobile node moves from one place and another place and a mobile node find better access router than current access router. From the simulation results we can conclude that primary path selection algorithm improves the network performance.

References

1. Montavont, N., et al.: Analysis of Multihoming in Mobile IPv6. draft-ietf-monami6-mipv6-analysis-01 (June 26, 2006)
2. Ernst, T., et al.: Motivations and Scenarios for Using Multiple Interfaces and Global Addresses. draft-ietf-monami6-multihoming-motivation-scenario-00 (February 2006)
3. Wakikawa, R., et al.: Multiple Care-of Addresses Registration. draft-ietf-monami6-multiplecoa-00 (June 12, 2006)
4. Deering, S., Hinden, R.: Internet Protocol, Version 6 (IPv6)Specification, RFC 2460 (December 1998)
5. Perkins, C., et al.: Mobility Support in IPv6, RFC 3775 (June 2004)
6. Manner, J.,, Kojo, M.: Mobility Related Terminology, RFC 3753 (June 2004)
7. Ernst, T., Lach, H.: Network Mobility Support Terminology, draft-ietf-nemo-terminology-05 (February 2006)
8. Devarapalli, V., Wakikawa, R., Petrescu, A., Thubert, P.: Network Mobility (NEMO) Basic Support Protocol. IETF RFC 3963 (January 2005)
9. Network Simulator (ns) version 2, http://www.isi.edu/nsmam/ns

Design of Location Management for Heterogeneous Wireless Networks*

Li-Der Chou, Chang-Che Lu, and Chyn-Yen Lu

Department of Computer Science and Information Engineering,
National Central University
Taoyuan County 32001, Taiwan (R.O.C.)
cld@csie.ncu.edu.tw

Abstract. As the rapid development of mobile broadband access network technologies in recent years, 802.11 Wi-Fi, GPRS and 3G access networks have been widely used. Users have the characteristic of mobility in these wireless heterogeneous networks. As users are roaming, How to report rapidly the current location of users to network management center, so that the pagers are able to obtain the right user location for establishing communication, has become an important issue. we focus on the architecture and policies of location management for wireless heterogeneous networks. we propose the Geographical Location Registration (GLR) mechanism for location updating and paging, where the location information in wireless heterogeneous networks is maintained to rapidly page other users and to reduce the overall cost. Compared to traditional mechanisms, the proposed GLR mechanism is capable of improving the paging cost by 79%. Besides, as the moving speed of users increases, the paging cost will not increase substantially.

Keywords: heterogeneous access network, network management, location update, paging.

1 Introduction

The development of wireless network is close to our life. We can use different wireless access networks connecting to internet. In other words, we could connect to internet with multiple wireless access networks at one time. These multiple wireless access networks are also called the heterogeneous wireless networks. The heterogeneous wireless networks will not be integrated into one wireless access network in short time because of many reasons. For example, the cost and purpose are two of these reasons. The users hope to use the heterogeneous wireless networks for the wide coverage, mobility, and lower cost. Therefore, the heterogeneous wireless network will become the master wireless network in the future.

---

* This research was sponsored in part by National Science Council under grants NSC 95-2627-E-008-002 and NSC 95-2221-E-008-031, and by National Central for High-Performance Computing.

S. Ata and C.S. Hong (Eds.): APNOMS 2007, LNCS 4773, pp. 306–315, 2007.

Fig. 1 (a) is the traditional heterogeneous wireless network management architecture. The parameters i and j indicate the different wireless network. The message transmission and paging will make cost and overhead in the heterogeneous wireless network. The traditional heterogeneous wireless network is based on switching packets in IP network and the communication with telecommunication network is based on using telecommunicating equipments to switch packets in IP network. In this paper, we using Location Register Gateway (LRG) [1] to decrease the broadcasting packets and cost when switching packets and paging in the heterogeneous wireless network. [2] give the definition of network management architecture and how to reduce transmission cost in telecommunicating network. There are some problems that the user roams to another wireless access network but he actually does not have any movement. In this paper, we propose that the GLR records the moving and switching information between these two the heterogeneous wireless network i and j, like Fig. 1. (b). It could decrease the load of Home Location Register (HLR) and the paging cost. In order to increases the success of paging.

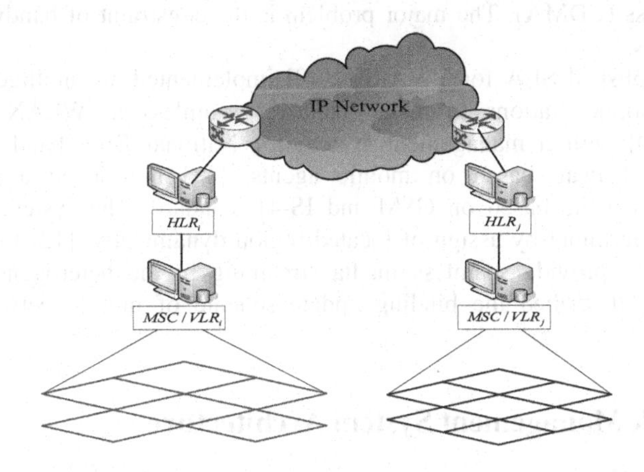

(a) The traditional heterogeneous wireless network management architecture

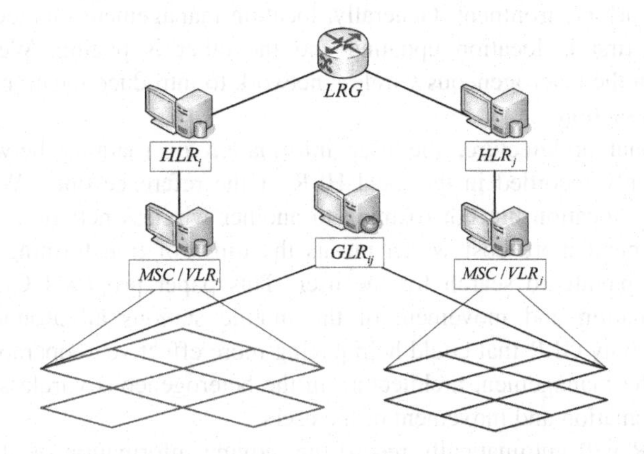

(b) The GLR heterogeneous wireless network management architecture

Fig. 1. Traditional and GLR heterogeneous wireless network management architecture

Section 2 is the related works about 802.11, 802.16, and location management. Section 3 is the proposed architecture that the location updating and paging happen in the heterogeneous wireless network. In section 4, we take advantage of NS2 to prove our proposed. Section 5 is conclusion.

2 Related Work

Wu proposed frame-based adaptive multirate transmission to expand the capacity in IEEE 802.11 [3]. [4] introduced IEEE 802.11 protocol and contrasted some 802.11 products. [5] discussed how to modify IEEE 802.11 to become high capacity wireless network and proposed IEEE 802.11 Extension up to 22Mb/s. [6] used DiffServ in wireless LAN distributively to apply classification of services, monitor, and control. [7] discussed the trend and challenge of wireless broadband access technology nowadays, included local multipoint distribution systems (LMDS) and code-division multiple access (CDMA). The major problem is the constraint of bandwidth and the size of packets.

[8] accomplished SLA for TWAREN. [9] implemented an intelligent agent that located in mobile stations to make handover seamless in WLAN and cellular networks. [10] built a management system for National Broadband Experimental Networks in Taiwan based on mobile agents. [11] introduced a new location management scheme based on GSM and IS-41 standard. The system could locate every mobile terminal by assign of located region dynamically. [12] took advantage of multicast to provide seamless media streaming in the heterogeneous wireless network. [13] described the binding update scheme of mobile networks in IPv6 environment.

3 Network Management System Architecture

Because of the mobile users, location management is more complicated than fixed in wireless network environment. Generally, location management includes two topical subjects, the first is location updating and the other is paging. We propose the architecture in the heterogeneous wireless network to introduce location updating and paging in this section.

In traditional architecture, the user information of roaming between wireless networks is only recorded in the local HLR or the reference ones. When a user is actually in the location but not roaming to another wireless network, the traditional paging will spend more cost which means the times of transforming signal in the procedure of paging to search for the user. This paper proposed GLR that could record information and movement of the mobile stations all around the system regions. We apply GLR that could help paging more effective cooperating with LRG to establish the management architecture in the heterogeneous wireless network and maintain information and movement of the users.

Every GLR will automatically record the moving information of all users in the heterogeneous wireless network. While a user move to the heterogeneous wireless network, GLR will also record the location of the user. When there is a demand of paging for the user, HLR will directly query to GLR that where the user is if HLR could not search for the user in the local MSC / VLR. By thy way, HLR does not need to search for the user around all the network.

(1) Location Updating

The network management center must know well and keep the most new location of the users in order to search for them in any time. The information of the users are stored in the location management database. Hence, the location of the database is very important with regard to the updating speed. We propose the architecture in this paper including the homogeneous or heterogeneous wireless network and identical or different LRG. So, we can divide four conditions in our proposed. The users are in homogeneous wireless network with identical LRG, in the homogeneous wireless network with different LRG, in the heterogeneous wireless network with identical LRG, and in the heterogeneous wireless network with different LRG. The following describes the procedure flow.

The former four conditions will be the same location updating procedure like Fig. 2 These steps are as follows:

1. The user moves from MSC/VLR_i to MSC/VLR_j and transfers the location updating information to MSC/VLR_i.
2. MSC/VLR_i forwards the updating demand to HLR_i and GLR_{ij} simultaneously to update database.
3. HLR_i transfers ACK back to the user after updating database.
4. The user will transfer Cancel Message to HLR_j to cancel the old record after receiving ACK.
5. HLR_j forwards Cancel Message to MSC/VLR_j to cancel the old record after receiving information.

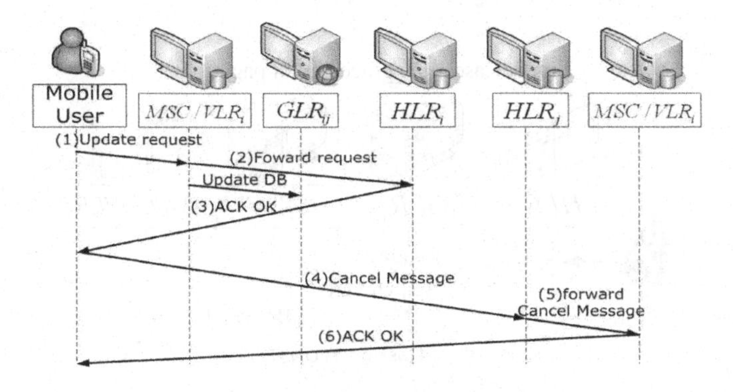

Fig. 2. Location updating flow

(2) Paging

Paging is that someone would like to search for a certain user in the heterogeneous wireless network. The network management center must search for the user in all of these base stations to apply the connection of service. In this paper, we proposed three possible conditions of paging according to the location of the user. The description is as follows:

Case (1): The caller is located in the same LRG and under the same subnet system.

Fig. 3 is the procedure of paging, steps as follows:

1. The caller who has registered on HLR_i just right transfers a paging to HLR_i, then HLR_i directly forwards the request to MSC/VLR_i.
2. MSC/VLR_i searchs for the user location after receiving the request, then assigns Temporary Location Directory Number （TLDN） back to the caller through HLR_i.

Case (2): The caller is located in the same LRG and under the different subnet system.

Fig. 4 is the procedure of paging, steps as follows：

1. The caller who has registered on HLR_i just right transfers a paging to HLR_i, then HLR_i directly forwards the request to MSC/VLR_i.

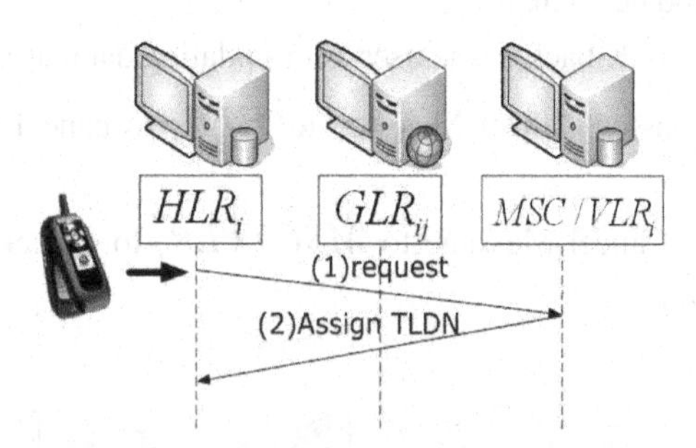

Fig. 3. Case 1 the procedure of paging

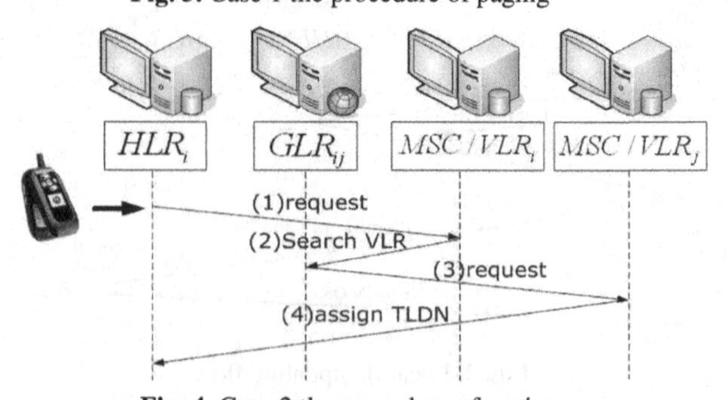

Fig. 4. Case 2 the procedure of paging

2. MSC/VLR_i could not search for the target in the local database, then forwards the request to GLR_{ij}.
3. GLR_{ij} searchs for the database that owned by MSC/VLR_j, and forwards the request.
4. MSC/VLR_j searchs for the user location after receiving the request, then assigns TLDN back to the caller through HLR_i.

Case (3): The caller is located in different LRG.
Fig. 5 is the procedure of paging, steps as follows:

1. The caller who has registered on HLR_i just right transfers a paging to HLR_i, then HLR_i directly forwards the request to MSC/VLR_i.
2. MSC/VLR_i could not search for the target in the local datebase, then forwards the request to GLR_{ij}.
3. GLR_{ij} searchs for these database that owned by MSC/VLR_j, and forwards the request.
4. MSC/VLR_j searchs for the user location but failed after receiving the request, then transfers the failed message back to HLR_i.
5. HLR_i could not search for the user in any MSC/VLR of the region, then, forwards the request to the neighbor GLR_{mn}.
6. The neighbor GLR_{mn} forwards the request to MSC/VLR_m of the region.
7. MSC/VLR_m searchs for the user location after receiving the request,then assigns TLDN to the caller through HLR_i.

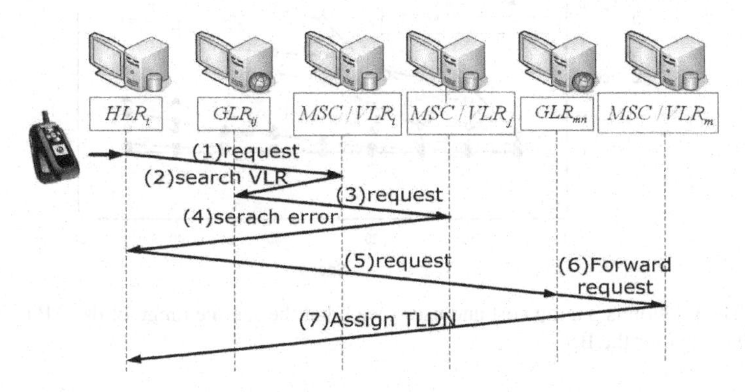

Fig. 5. Case 3 the procedure of paging

4 Simulations and Discussions

In this paper, we take advantage of Network Simulator (NS-2) for the evaluation. The environment is a PC (Pentium-4, 2 GHz, 512 MB main memory). And the operating system is Red Hat Linux 9.0. The proposed the GLR heterogeneous wireless network management architecture is compared with traditional one. The compared factors include paging cost which means the times of transforming signal in the procedure of paging, location updating rate, the range of base stations, and the user moving speed. Table 1. is the setting of the simulation environment that includes the paged mobile users. We assume that the numbers of AP with different service range are constant and compare with the three kinds of number of BS with different service range. The times of location updating and paging are as follows 10:1, 2:1, 1:2, and 1:10. So we let GLR_10:1 represent the times of location updating and paging to be 10:1 in GLR. Equally, noGLR_1:10 represents the simulation results that the times of location updating and paging to be 1:10 in traditional architecture.

Table 1. The setting of the simulation environment

Elements	Parameter
Field Range	3000(m)*3000(m)
Access Points number	400
Access Points Range	100(m)
Base Stations number	100,25,4
Base Station Range	200,500,2000(m)
Mobile User number	500

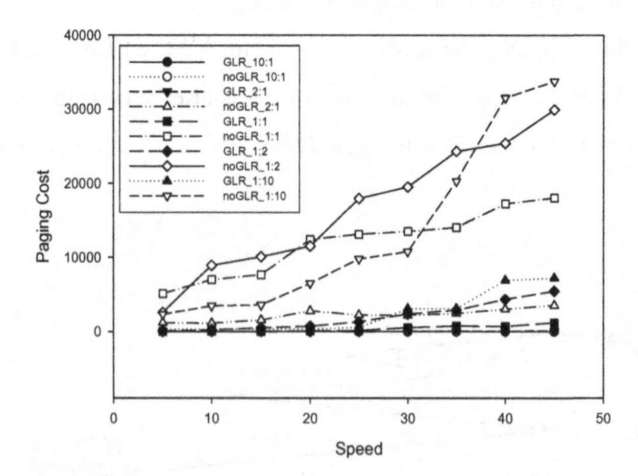

Fig. 6. The relation is paging cost and user speed that the service range of the AP is 100 meters and 500 meters for the BS

Fig. 6 is the relation of paging cost and user speed that the service range of the AP is 100 meters and 500 meters for the BS. The result shows that the moving speed of users increases, the paging cost will not increase substantially. The proposed architecture has 78% improvement than the traditional one when the fewer times of location updating. Besides, we can pay attention to the result of noGLR_1:10. The paging cost will increase substantially while the speed greater than 32m/s. The mobile nodes moved continued with the increase of moving speed to roam within AP or BS but there is no increase of location updating times. So, the cost will increase while the rise of moving speed.

Fig. 7 is the relation of paging cost and user speed that the service range of the AP is 100 meters and 2000 meters for the BS. The proposed architecture has 79% improvement while location updating times and paging is 1:10. Compare with Fig. 6, the result of noGLR_1:10 has the same case in the speed 32 m/s.

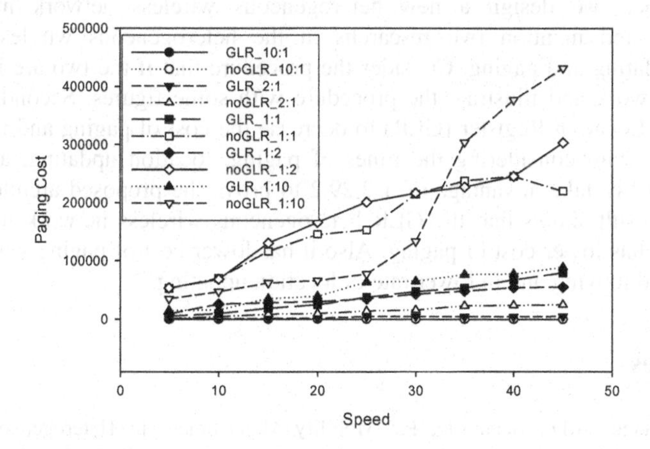

Fig. 7. The relation is paging cost and user speed that the service range of the AP is 100 meters

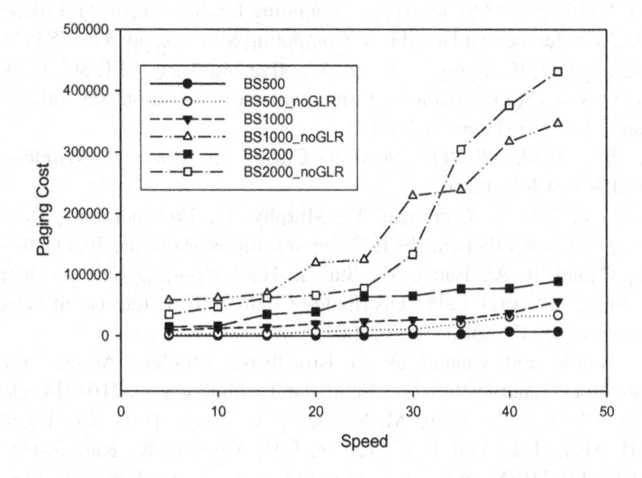

Fig. 8. The relation is paging cost and user speed that location updating and paging is 1:10 in BSs with different service range

Fig. 8 is the comparison that the times of location updating and paging are 1:10 in BSs with different service range. The result shows that the cost will increase while the increase of service range, but the cost of our proposed architecture is still lower. There is lower cost while low location updating rate. We can obtain that the cost will rise suddenly in speed 33 m/s in the traditional architecture and the service range of BS is 2000 meters. Compare with that the service range of BS is 1000 meters. The reason is that the mobile user roams to other BSs, then results in spending more cost to search for the mobile user.

5 Conclusion

In this paper, we design a new heterogeneous wireless network management architecture and mention two researchs in the heterogeneous wireless network: location updating and paging. Consider the procedure that if the two are in the same wireless network and illustrate the procedure with some figures. Secondly, propose Geographic Location Register (GLR) to decrease the cost of paging and improve the performance with considering the times of paging, location updating, and moving speed. Then, we take advantage of ns2.29.2 to prove the proposed architecture. The simulation result shows that the GLR heterogeneous wireless network management architecture has lower cost of paging. Also it has lower cost of paging when the user in high speed moving and in lower rate of location updating.

References

1. Assoumaand, A.D., Beaubrun, R.: Mobility Management in Heterogeneous Wireless Networks. IEEE Journal on Selected Areas in Communications 24(3), 638–648 (2006)
2. Choi, S.-J., Baik, M.-S., Hwang, C.-S.: Location Management & Message Delivery Protocol in Multi-region Mobile Agent Computing Environ- ment. In: Proceedings of 24th International Conference on Distributed Computing Systems, pp. 476–483 (March 2004)
3. Wu, J.-L.C., Liu, H.-H., Lung, Y.-J.: An Adaptive Multirate IEEE 802.11 Wireless LAN. In: Proceedings of 15th International Conference on Information Networking, Beppu City, Oita, Japan, pp. 17–23 (February 2001)
4. Stallings, W.: IEEE 802.11: Moving Closer to Practical Wireless LANs. IT Professional 3(3), 17–23 (2001)
5. Heegard, C., Coffey, J., Gummadi, S., Murphy, P., Provencio, R., Rossin, E.: High Performance Wireless Ethernet. IEEE Communications Magazine 39(11), 64–73 (2001)
6. Veres, A., Campbell, A., Barry, M., Sun, L.-H.: Supporting Service Differentiation in Wireless Packet Networks Using Distributed Control. IEEE Journal on Selected Areas in Communications 19(10), 2081–2093 (2001)
7. Sari, H.: Trends and Challenges in Broadband Wireless Access. Proceedings of Symposium on Communications and Vehicular Technology, pp. 210–214 (October 2000)
8. Chou, L.-D., Chen, Y.-W., Hong, M.-Y., Lin, Y.-C., Liang, H.-H., Chen, J.-M., Yen, J.-J., Chen, S.-H., Hsu, H.-J., Tsai, L.-C., Hsueh, J.-F., Yang, C.-N., Kuo, S.-H.: Service level agreement for TWAREN networks. Journal of Internet Technology 7(2), 145–152 (2006)

9. Chou, L.-D., Lai, W.-C., Lin, C.-H., Lin, Y.-C., Huang, C.-M.: Seamless handover in WLAN and cellular networks through intelligent agents. Journal of Information Science and Engineering 23(4), 1087–1101 (2007)
10. Chou, L.-D., Shen, K.-C., Tang, K.-C., Kao, C.-C.: Implementation of mobile-agent-based network management systems for National Broadband Experimental Networks in Taiwan. In: Mařík, V., McFarlane, D.C., Valckenaers, P. (eds.) HoloMAS 2003. LNCS (LNAI), vol. 2744, pp. 280–289. Springer, Heidelberg (2003)
11. Chen, R., Yuan, S., Zhu, J.: A Dynamic Location Management Method of Personal Communication System. E-Tech, pp. 1–9, (July 31, 2004)
12. Chou, L.-D., Chen, J.-M., Kao, H.-S., Wu, S.-F., Lai, W.: Seamless streaming media for heterogeneous mobile networks. ACM Springer Mobile Networks and Applications (MONET) 11(6), 873–887 (2006)
13. Chen, Y.-W., Shih, J.-M.: Binding Update Scheme of Mobile Networks in IPv6 Environment. Journal of Internet Technology (to be published in Aug 2007)

Network Architecture and Fast Handover Scheme Using Mobility Anchor for UMTS-WLAN Interworking*

Incheol Kim, Sungkuen Lee, Taehyung Lim, Eallae Kim, and Jinwoo Park

Korea University, Anam-dong Seongbuk-ku, Seoul 136-713, S. Korea
{ickim,food2131,autonome,earhead,jwpark}@korea.ac.kr

Abstract. UMTS-WLAN interworking has attracted many research efforts because each of UMTS and WLAN can maximize its capability in service provisioning. Through UMTS-WLAN internetworking, users can be offered the most suitable services according to service area and service provider can reduce network building and maintenance costs. In this paper, we propose a network architecture and fast handover scheme using Mobility Anchor (MA) for UMTS-WLAN interworking. MA is provided at the boundary between GGSN and PDG, under the 3GPP-proposed interworking reference model. Such MA can enable authentication and session establishment before L2 handover of the mobile node, so that seamless and fast vertical handover could be possible. Through computer simulation using OPNET for the performance measurement, the efficiency and validity of the proposed scheme has been examined.

1 Introduction

The recent introduction of commercial High Speed Downlink Packet Access (HSDPA) services based on Universal Mobile Transmission System (UMTS) has made mobile wideband wireless data services available to the public. By embedding the HSDPA technology, UMTS offers data services at rates up to 14 Mbps at an average of 2-3 Mbps with a wide service area at a high service fee. On the other hand, The Wireless Local Area Network (WLAN) provides a higher speed data rate and lower price, but covers only small areas and allows limited mobility.

Therefore, a UMTS-WLAN interworking approach can make the best use of the advantages of each type of network and can eliminate the stand-alone defects of the two services. This calls for efficient interworking mechanisms between UMTS and WLAN networks. Several approaches have been proposed for UMTS-WLAN interworking networks. The Third Generation Partnership

* This work was partly supported by the IT R & D program of MIC/IITA [2006-S058-01, Development of Network/Service Control Technology in All-IP based Converged network] and the MIC, Korea, under the ITRC support program supervised by the IITA (IITA-2006-(C1090-0603-0005)).

S. Ata and C.S. Hong (Eds.): APNOMS 2007, LNCS 4773, pp. 316–325, 2007.

Project (3GPP) has specified six interworking scenarios and UMTS-WLAN interworking reference model [1], [2]. The European Telecommunications Standards Institute (ETSI) specifies two generic approaches for UMTS-WLAN interworking: loose coupling and tight coupling [3]. The main difference between tight coupling and loose coupling is whether the user's traffic is delivered through the core network of UMTS or not. That is, in case that UMTS and WLAN are tightly coupled, traffic from WLAN flows into the core network of the UMTS and flows out to the external PDN via SGSN and GGSN. On the other hand, in the loosely coupled case, WLAN doesn't share any core network nodes of UMTS except AAA functionality. When WLAN and UMTS are loosely coupled, each network operates independently. Therefore, networks don't need to change their network architectures or protocol stacks. But loosely coupled network cannot support service continuity to other access network during handover, thus loosely coupled scheme has long handover latency and packet loss. When WLAN and UMTS are tightly coupled, the users from WLAN can access the UMTS services with guaranteed QoS and seamless mobility. At the same time, there are issues to be resolved including a standard of new interface and dedicated emulator must be developed, as well as heavy burden imposed to the core UMTS network. Generally speaking, loose coupling is more preferable compared with tight coupling due to the simplicity and less reconfiguration. However, loose coupling is worse in terms of seamless mobility, QoS provision, and network security [4].

Therefore, we propose a new interworking network architecture and fast vertical handover scheme using MA for UMTS-WLAN interworking. This scheme that guarantee independent UMTS and WLAN operations, allow easy network implementation, and provide seamless service. The proposed scheme is an effective network architecture that utilizes the advantages of the two UMTS-WLAN interworking approaches (loose coupling and tight coupling) and offers fast vertical handover.

The paper consists of the following. Chapter 2 describes the proposed interworking network architecture and fast handover scheme using MA for UMTS-WLAN interworking. In Chapter 3, performances were compared between the proposed scheme and conventional scheme based on OPNET analysis to verify the advantages of the proposed model. Finally the conclusion of the paper is presented in Chapter 4.

2 Proposed UMTS-WLAN Interworking Network Architecture and Vertical Handover Scheme

In order to minimize changes to the conventional UMTS and WLAN and provide a seamless service during vertical handover, this paper proposes a new interworking scheme that uses a telecom-based node called the mobility anchor (MA) on the loose coupling foundation using MIP.

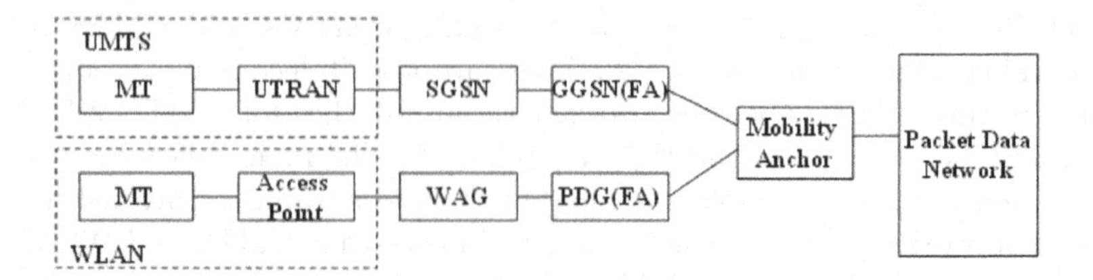

Fig. 1. Proposed UMTS-WLAN interworking network architecture

2.1 Proposed Interworking Network Architecture

The proposed interworking network architecture integrates the MIP and the Telecom emulation scheme as shown in Fig. 1

The proposed interworking network architecture offers the following features.

- 1) The proposed network architecture for UMTS-WLAN interworking is based on interworking network models of 3GPP standards documents [2].
- 2) The proposed scheme uses the MIP approach to manage the MN's mobility, which enables two independent service operations, simplifies interworking network implementation and provides flexibility into other services.
- 3) MA is added in place of the conventional HA to perform mobility management. And MA is located on the boundary between GGSN and PDG within the service provider network and has hierarchical layer architecture.
- 4) The interworking between MA and GGSN(FA)/PDG(FA) allows authentication for MN prior to L2 handover as well as early session establishment, minimizing the handover latency.
- 5) The network architecture involves integration of the tight coupling on top of loose coupling scheme.

The proposed interworking network architecture features the following new objects and functions.

- 1) GGSN(FA)/PDG(FA): Based on the 3GPP interworking reference network model, Foreign agent (FA) function is added to GGSN and PDG to register the MN's Home address (HoA) and Care of address (CoA) to the MA.
- 2) Similar to tight coupling network, MA should be added to interwork between UMTS and WLAN network. The functions of MA are as following:
- MA contains the GGSN/PDG mapping table that corresponds to the Access point name (APN), or the corresponding GGSN/PDG can be found from the DNS server.
- MA manages the HoA and CoA of MN. It also acts as the gateway to external network for transmitting traffic from users that wish 3GPP IP access in the WLAN.
- MA can communicate with the AAA server/HSS. The result of authentication is referenced to either authorize or reject a handover request from a MN. The authentication result is relayed to the FA to enable pre-authentication.

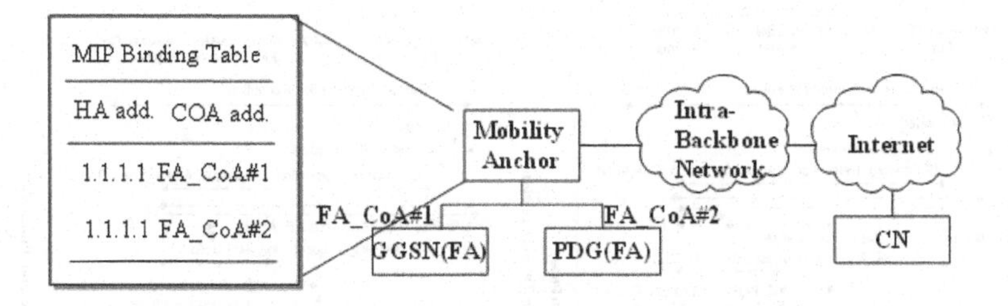

Fig. 2. Hieratical architecture of proposed network model

- MA interworks with FA(GGSN/PDG) to establish a session required for MN' handover prior to L2 handover.
- MA has a packet-buffering capability at the handover starting point as well as a packet-forwarding function upon the completion of a handover.

As shown in Fig. 2, MA is located at the boundary of the service provider network, and it can be considered as a node for mobility and bearer traffic management for UMTS-WLAN users who are subscriber UMTS-WLAN interworking service. When MA receives packets from CN, MA finds related FA address (FA-CoA) through table look-up and delivers packets to FA through tunneling between MA and FA. And the FA delivers packets to the final destination MN through tunneling between FA and MN.

A remarkable feature of the proposed scheme is that binding is updated by a MA which is located in a service provider network instead of a public network. Therefore we can reduce the load to the public network. Moreover, MA establishes a session required for a MN prior to L2 handover and relays authentication information between MN and AAA server/HSS as well as MN's profile information for pre-authentication, so that the proposed scheme can improve the handover latency without increasing overhead in public network. Moreover, the proposed scheme can minimize packet losses by packet-buffering and forwarding functions.

2.2 Proposed Vertical Handover Schemes

This section describes the proposed fast vertical handover schemes in the proposed network model. Fig. 3 shows the handover procedures when a MN moves between WLAN and UMTS, respectively. Unlike the conventional scheme, the proposed vertical handover scheme consists of two phases; Preparation phase and Execution phase. Once a MN detects a need for a handover with an L2 trigger and sends handover request message (MN-HO-Request) to oFA(GGSN/PDG), and the oFA(GGSN/PDG) relays it to MA in order to notify MN' handover. When MA received this message, MA performs pre-authentication between the MN and the AAA server/HAA and establishes a session in the new wireless network before L2 handover, and oFA(GGSN/PDG) finally sends approval message (MN-HO-Accept) to the MN. These procedures is defined as the Preparation phase (Fig. 3(a) 1-10, Fig. 3(b) 1-8). The Execution phase involves performing

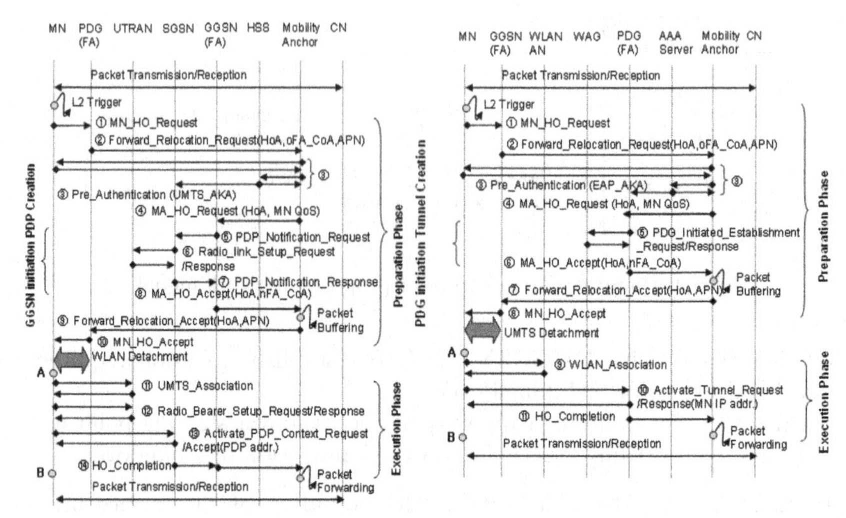

Fig. 3. Proposed handover procedures

association from the wireless networks, activating the pre-established session between nFA and the MN and exchanging handover completion message between MA and nFA. The handover procedures of the proposed scheme are as follows.

1) Preparation Phase

- Step 1: The MN detects a need to change the access network based on handover decision algorism and notifies handover request to oFA. And oFA relays the information to MA (Fig. 3(a,b) 1,2).
- Step 2: MA requests authentication to the MN and performs pre-authentication with the AAA server/HSS based on the information received from the MN to determine acceptance or rejection (Fig. 3(a,b) 3).
- Step 3: MA notifies to nFA the MN handover and establishes the necessary session. Acceptance or rejection is determined based on available resources in the new wireless network. If the request is accepted by nFA, MA starts packet-buffering the corresponding session data for a MN. (Fig. 3(a) 4-9, Fig. 3(b) 4-7)
- Step 4: The oFA notifies the acceptance for handover to the MN (Fig. 3(a) 10, Fig. 3(b) 8).

2) Execution Phase

- Step 5: The MN detaches the channel of the current access network and performs wireless channel association with the new wireless access network (Fig. 3(a) 11, Fig. 3(b) 9).
- Step 6: Activation is requested/authorized for an established session during preparation phase (Fig. 3(a) 12-13, Fig. 3(b) 10-11).

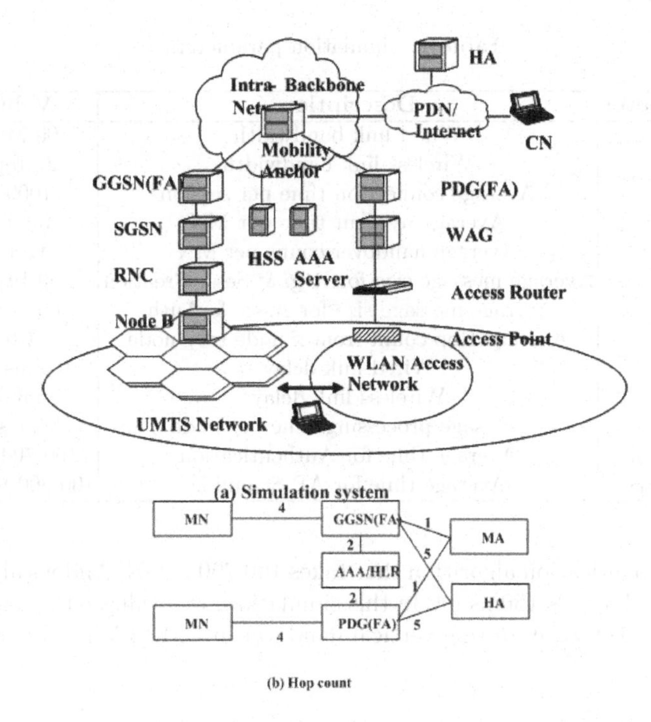

Fig. 4. Simulation environments

– Step 7: Handover completion is notified to the final MA, which stops packet-buffering and forwards the packets of the CN including the stored packets to nFA to complete the handover process (Fig. 3(a) 14, Fig. 3(b) 11)

3 Simulation Results and Analysis

This chapter analyzes the handover performance of the proposed scheme based on computer simulation and compares the result with the conventional scheme.

3.1 Environment for Simulation

In order to verify the performance of the proposed interworking network architecture, simulation was conducted using OPNET. The simulation environment is displayed in Fig 4, and Table 1 shows the simulation parameters [5]-[9]. The WLAN service used 802.11b, and it is assumed that WLAN access network located in UMTS network and the MN is moved freely around between UMTS and WLAN network. For the purpose practical simulation environment, the CN generates Constant bit rate (CBR) packets based on the User datagram protocol (UDP) at intervals according to the Poisson distribution. And simulation was conducted for this paper under the assumption that authentication method uses

Table 1. Simulation parameters

Parameter	Description	Value
B_w	Wired link bandwidth	100 Mbps
B_{wl}	Wireless link bandwidth	2 Mbps
t_s	Average connection time per session	1000 s
t_r	Average resident time per MN	5-60 s
N_H	Average handover count per MN	t_s/t_r
L_R	Average message size for Reg. & Sess. Creation	50 Byte
L_A	Average message size for Asso. & Auth.	60 Byte
N_{x-y}	Average Hop count from x node to y node	1-6
D_w	Wired link delay	0.5 ms [8]
D_{wl}	Wireless link delay	2 ms [8]
P_x	Message processing time per node	5 ms
T_{AUTH}	Average time for Authentication	100-700 ms
T_{ASSO}	Average time for AP Scanning	100-200 ms [9]

a fast re-authentication algorithm that takes 100 700 ms [7]. And we also assume that average T_{Asso} is 150ms [9]. In this simulation, we evaluate the handover latency and packet losses during vertical handover in UMTS-WLAN interworking network.

3.2 Performance Analysis Results

Handover latency in a conventional MIPv4 scheme can be considered as a sum of L2 handover duration for performing association time (T_{Asso}) and authentication time (T_{AUTH}), registration update time (T_{REG}), and the time required for establishing a session between the MN and FA (T_S) as expressed in Equation 1.

$$T_{HO-MIPv4} = T_{ASSO} + T_{AUTH} + T_{REG} + T_S, \tag{1}$$

in which T_{HO} is Overall handover latency for vertical handover of a MN.

In the case of the proposed handover schemes, the handover latency is determined by L2 handover time and activation for pre-established session delays. Because the proposed scheme performs authentication, registration and session establishment before L2 handover, the handover latency only includes the L2 handover time and activation of a pre-established session delay. Therefore, Handover latency of the proposed scheme can be expressed as Equation 2.

$$T_{HO-Proposed} = T_E, \tag{2}$$

in which T_E is average time required for a MN between L2 handover and activation of a pre-established session for the proposed scheme. Therefore, Handover latency can be expressed as Equations 3-6.

$$\begin{aligned} T_{HO-WtoU}(MIPv4) = {} & T_{ASSO} + T_{AUTH} + 7(L_R/B_{wl}) + 7D_{Dwl} + 13D_w \\ & + 11(N_{MN-FA}(L_R/B_w)) + 2(N_{FA-HA}(L_R/B_w)) \\ & + (N_{MN-FA} + N_{FA-HA} + 2)P_x \end{aligned} \tag{3}$$

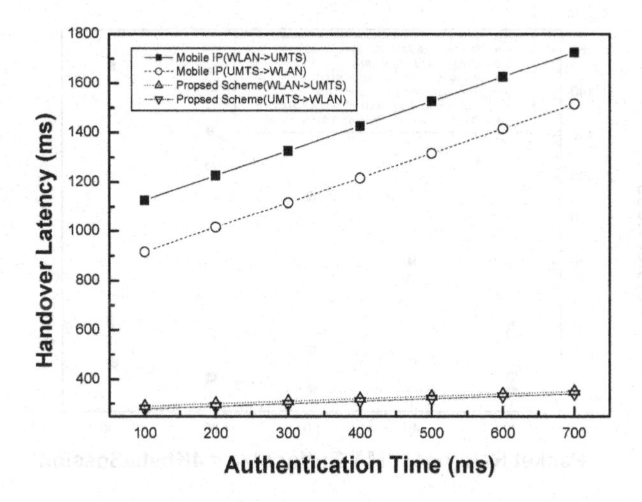

Fig. 5. Handover latency according to Authentication time

$$\begin{aligned} T_{HO-UtoW}(MIPv4) = & T_{ASSO} + T_{AUTH} + 5(L_R/B_{wl}) + 5D_{Dwl} + 9D_w \\ & + 7(N_{MN-FA}(L_R/B_w)) + 2(N_{FA-HA}(L_R/B_w)) \\ & + (N_{MN-FA} + N_{FA-HA} + 2)P_x \end{aligned} \quad (4)$$

$$\begin{aligned} T_{HO-WtoU}(Proposed) = & T_{ASSO} + 4(L_R/B_{wl}) + 4D_{Dwl} + 3D_w \\ & + 2N_{MN-FA}(L_R/B_w)) + (N_{FA-MA}(L_R/B_w)) \\ & + (N_{MN-FA} + N_{FA-MA} + 2)P_x \end{aligned} \quad (5)$$

$$\begin{aligned} T_{HO-UtoW}(Proposed) = & T_{ASSO} + 2(L_R/B_{wl}) + 2D_{Dwl} + 3D_w \\ & + 2N_{MN-FA}(L_R/B_w)) + (N_{FA-MA}(L_R/B_w)) \\ & + (N_{MN-FA} + N_{FA-MA} + 2)P_x \end{aligned} \quad (6)$$

Because the session creation algorithm of UMTS is a little more complicate, more latency is required for movement from WLAN to UMTS.

Fig. 5 presents handover latency versus the authentication time. With respect to handover latency, the authentication procedure is a significant portion in the conventional scheme. But the proposed scheme performs authentication, registration, session establishment before L2 handover, the interval for which it is disconnected does not contain the authentication processing time, registration time and session establishment time only includes the L2 handover delay in pure latency, providing much improved performance.

Fig. 6 presents packet losses versus packet transmission rates of CN. It can be showed that the packet loss rate increases with high data transmission rate from CN. As indicated in performance analysis, the proposed scheme offers better packet loss rate compared to the conventional scheme. Moreover, when the 4

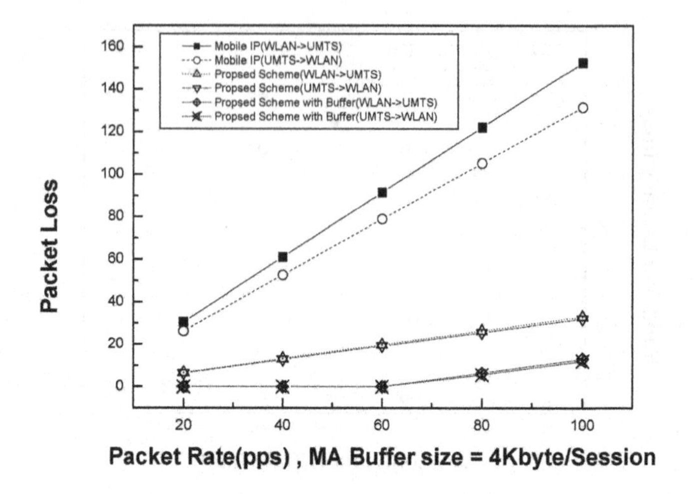

Fig. 6. Packet loss according to Packet rate of CN

Kbyte per session packet-buffering/forwarding capability is activated at MA with the proposed scheme, data loss does not occur until 60 pps. It is also expected that the packet loss can be eliminated beyond 60 pps if the packet buffer size is adjusted according to network characteristics.

In conclusion, the simulation results prove that the proposed scheme outperforms the conventional schemes in several aspects. Therefore, the proposed scheme is more suitable for UMTS-WLAN interworking network. Table 2 shows the comparison of the proposed and conventional scheme.

4 Conclusion

This paper proposed a new interworking network architecture and fast handover scheme that guarantee independent UMTS and WLAN operations, allow easy network implementation, and provide seamless service. The proposed scheme is an effective network architecture that utilizes the advantages of the two UMTS-WLAN interworking approaches (loose coupling and tight coupling) and offers fast vertical handover. The proposed scheme can be regarded as an intermediary phase of the evolution of the System architecture evolution (SAE) by 3GPP and IMS integrated network, and the proposed MA can be re-used into the SAE Anchor within 3GPP SAE in the future [10].

References

1. 3GPP TR 22.934, Feasibility Study on 3GPP System to WLAN Interworking (Release 6), (December 2002)
2. 3GPP TS 23.234 v1.3.0, 3GPP system to Wireless Local Area Network (WLAN) Interworking; System Description, Rel.6 (December 2004)

3. ETSI BRAN TR 101 957 V1.1.1, HIPERLAN Type2: Requirements and Architectures for Interworking Between HIPERLAN/2 and 3rd Generation CellularSystems (August 2001)
4. Floroiu, J.W., et al.: Seamless Handover in Terrestrial Radio Access Networks: A Case Study. Communications Magazine 41(11) (2003)
5. Choi, H.H, Song, O.S, Choi, D.H.: Seamless Handoff Scheme Based on Pre-registration and Pre-authentication for UMTS-WLAN Interworking. Wireless Personal Communications (2006) (September 2006)
6. Casals, O., Cerda, L., Willems, G., Blondia, C., Van den Wijngaert, N.: Performance evaluation of the post-registration scheme, a low latency handoff in MIPv4. In: IEEE ICC 2003, vol. 1, pp. 522–526 (May 2003)
7. Kwon, H.Y., Cheon, K.Y., Roh, K.Y, Park, A.S: USIM based Authentication Test-bed for UMTS-WLAN Handover
8. Seshan, S.: Low-Latency Handoff for Cellular Data Networks Ph.D dissertation (1995)
9. Shin, S.H, Forte, A.G., Schulzrinne, H.: Seamless Layer-2 Handoff using Two Radios in IEEE 802.11 Wireless Networks
10. 3GPP TR 23.882 v0.11.0, 3GPP System Architecture Evolution: Report on Technical Options and Conclusions (Release 7) (February 2006)

Implementation of 802.21 for Seamless Handover Across Heterogeneous Networks

Wonseok Lee, Mun-Seok Kang, and Mi-Sook Lim

17, Woomyeondong, Seocho-gu, Seoul, 137-792, Korea
{alnova2,mskang,misook}@kt.co.kr

Abstract. MIH (Media Independent Handover)[1] is the standard technology of IEEE 802.21. It gives seamless handover to mobile terminals that have multiple wireless interfaces. Recently, many mobile terminals having multiple wireless interfaces are emerging on the market. Thus, the handover technology, which makes it possible to hand-over between heterogeneous networks, is attracting the attention of network operators. In this article, we explain IEEE 802.21 framework and its laboratory implementation and compare between IEEE 802.21-based handover and non-802.21-based handover. We performed laboratory experiments on handover between IEEE 802.11 (WiFi) access networks and IEEE 802.3 (Ethernet) using MIH Functions. We also analyzed and compared 802.21-assisted handover and non-802.21-assisted handover.

Keywords: IEEE 802.21, MIH, Seamless handover.

1 Introduction

Recently, there have been an increasing number of products on dual-mode terminals that have CDMA and WLAN access interface. However, handover between heterogeneous access network is not automatically operated.

Seamless handover is an indispensable capability of the next-generation network. As Figure 1 shows, in NGN, there are many access networks and there is one core network. So, if we want to offer voice services in NGN, it is very important to support seamless handover between many heterogeneous access networks.

It is very hard to offer such seamless handover between heterogeneous access networks because many heterogeneous access networks have different mobility mechanisms and they have various QoS quality and security requirements. In addition to that, applications such as VoIP and streaming services have tight performance requirements like end-to-end delay and packet-loss.

Handover decision is very important for mobility management. We can reduce delay using pre-retrieved network information and a well-ordered handover policy. In this case, inter-operation between network and terminal and another network component is needed. The IEEE 802.21[1] standard defines the framework that is needed to exchange information between handover participants to provide mobility. It also defines functional components doing handover decision.

S. Ata and C.S. Hong (Eds.): APNOMS 2007, LNCS 4773, pp. 326–333, 2007.

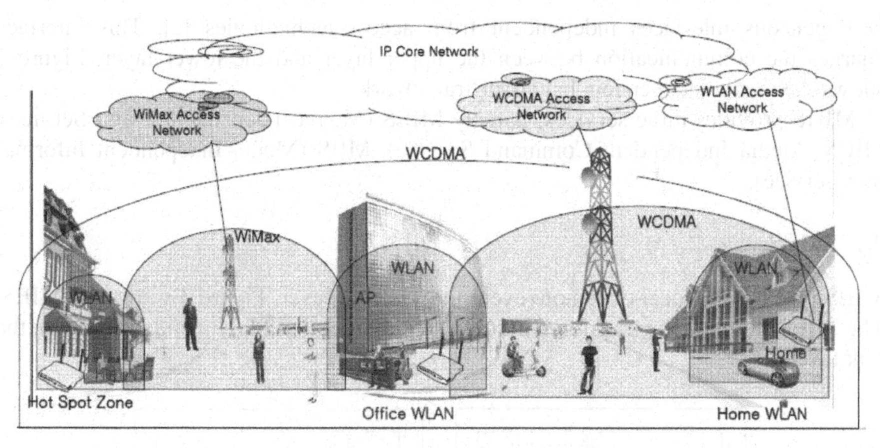

Fig. 1. Network architecture of NGN

In this article, we discuss the IEEE 802.21 framework and its laboratory implementation. We also discuss our experimental results on handover between 802.11 (WiFi) access networks and between 802.16 (WiMax) access networks and 802.11 access networks.

2 802.21 Framework

The IEEE 802.21 framework offers the method and procedure for handover between heterogeneous networks. This handover procedure uses the information from both mobile terminal and network infrastructures. The IEEE 802.21 framework informs the available network nearby of the mobile terminal and helps the mobile terminal to detect and select the network. This information includes link layer information.

The core of 802.21 lies in MIHF (Media Independent Handover Function). MIHF consists of intermediate functionalities residing between layer2 and layer3. It presents

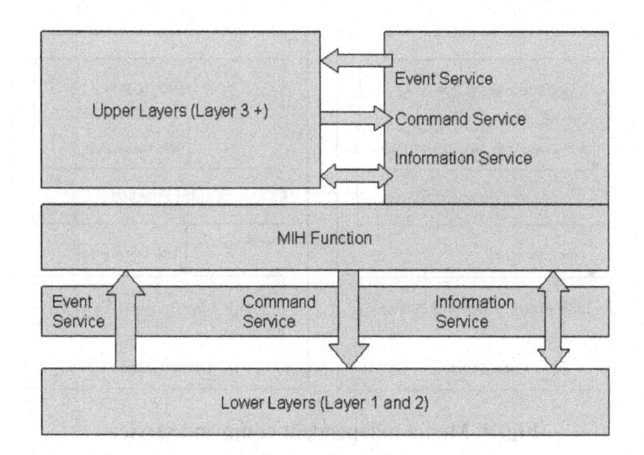

Fig. 2. Media independent handover framework

homogeneous interfaces independent from access technologies [2]. This interface charges the communication between the upper layer and the lower layer. Figure 2 shows the media independent handover framework.

MIHF provides three services; namely MIES (Media Independent Event Service), MICS (Media Independent Command Service), MIIS (Media Independent Information Service).

2.1 MIES

MIES proffers the local or remote event to the upper layer. Figure 3 shows the MIES. The MIES supports the transfer, filtering and classification of dynamic changes on the link layer.

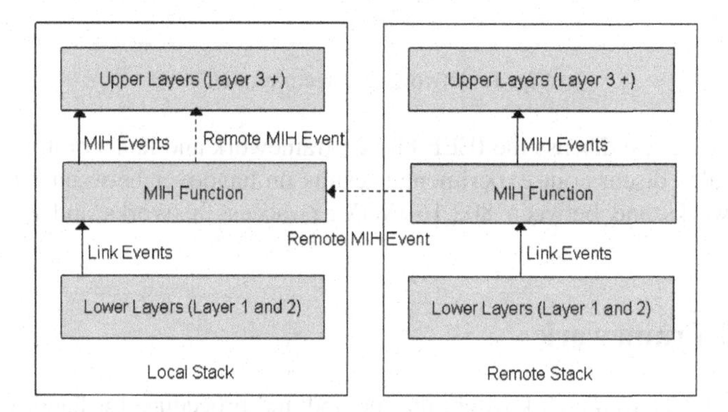

Fig. 3. Media independent event service

2.2 MICS

The MICS offers the functions for managing and controlling the link layer. If the MIH application wants handover and mobility, it can control the MAC layer by using MICS. Figure 4 shows the MICS block.

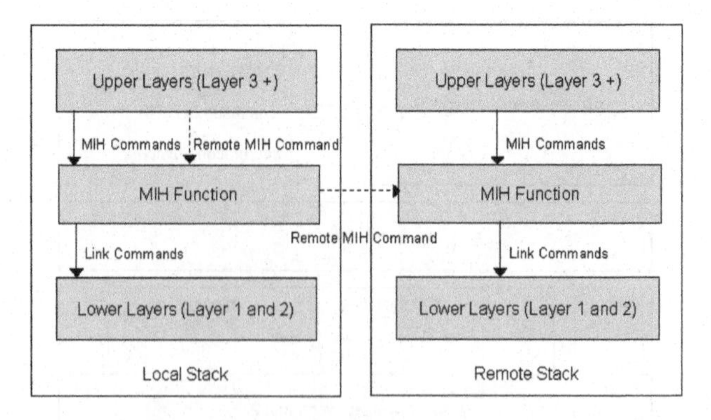

Fig. 4. Media independent command service

2.3 MIIS

The MIIS offers information that is needed to perform the handover. This information includes nearby available access networks. Using this information, the mobile terminal makes a decision on the handover. Figure 5 shows an examples of MIIS message.

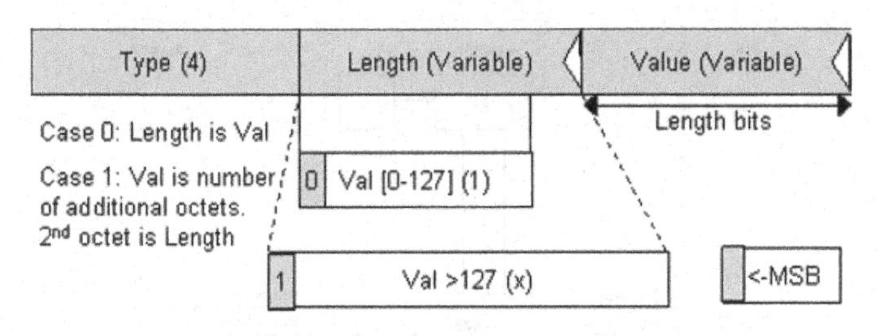

Fig. 5. TLV messages (octets)

2.4 SAP

SAP is an abbreviation for Service Access Point. It is an API that makes it possible to communication between the lower and upper layers. Figure 6 shows the 802.21 reference model on various heterogeneous networks.

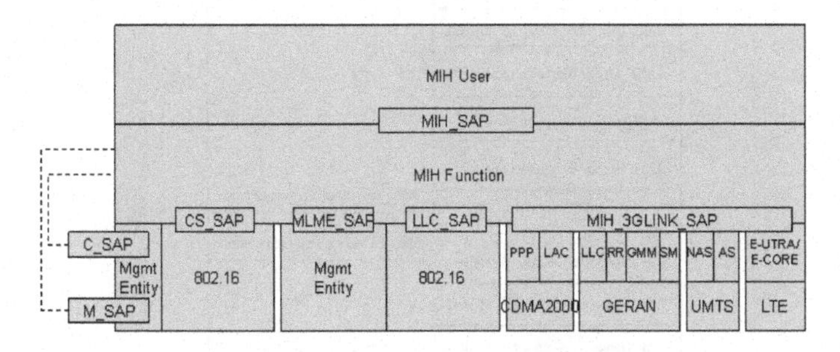

Fig. 6. IEEE 802.21 SAP reference model

3 Implementation of 802.21

In this paper, we implemented 802.21 based on the Linux platform. We made the 802.11 access point and information server using a Linux box. We used a notebook PC as the mobile terminal that can access the 802.11 and 802.16 networks. We experimented on handover between 802.11 (WiFi) networks and we also experimented on handover between 802.11 and 802.16 (WiMax) access networks.

3.1 Implemenation of PoA MIHF

PoA (Point of Attachment) is the function that makes it possible for a mobile terminal to access wireless media, and it is the component of the network structure. PoA MIHF performs the transportation of messages passing among mobile terminals and between the mobile terminal and other PoA MIHF or Information server and analyzes the messages passing them.

Figure 7 shows the 802.11 PoA MIHF block diagram and Figure 8 shows the messages between the Mobile node and the PoAs.

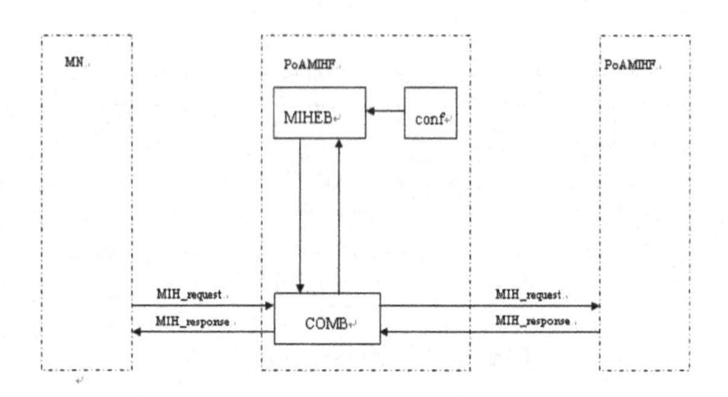

Fig. 7. Block diagram of PoA MIHF

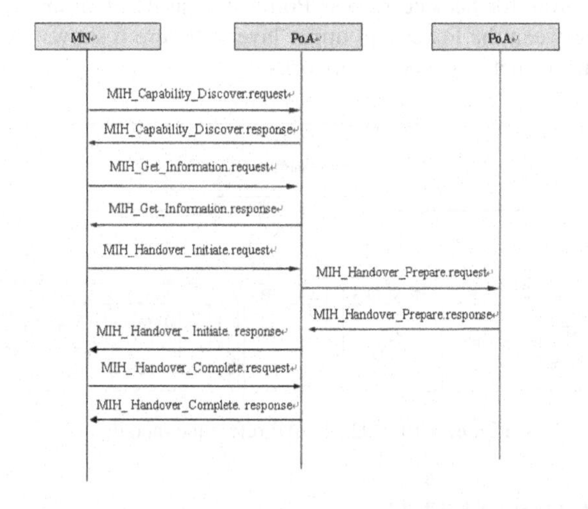

Fig. 8. Flow of MIH messages

3.2 Implemenation of MN MIHF

MN (mobile node) MIHF support handover linked to MjSIP[3] and Linux device driver interface. Figure 9 shows the interaction between MjSIP (application) and MN MIHF through the Linux device driver interface and other PoA MIHF.

3.3 VoIP Application

In this paper, we performed an experiment using MjSIP that enables video/audio communication based on SIP. MjSIP processes session creation and manages dialog.

It can support video/audio communication using RTP. We chose this application to experiment on handover when we communicated voice on an ip network. MjSIP calls or receives the MIH messages based on JNI. Using this interface, MjSip can access the MIHF driver as described in 3.2

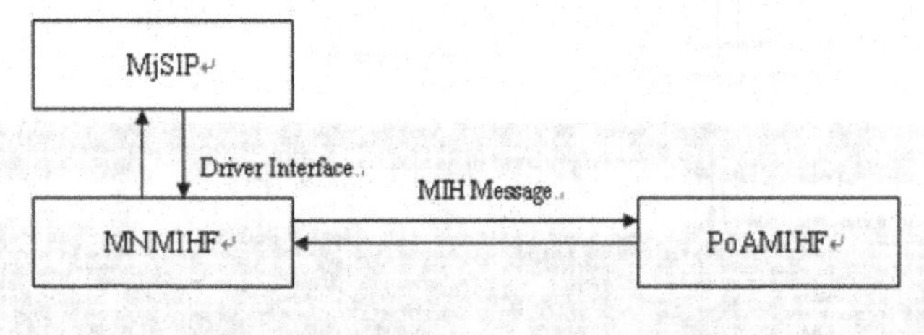

Fig. 9. Message flow between MIHF

3.4 Handover Experiments

Figure 10 shows the experimental environment that performs handover between 802.11 access networks.

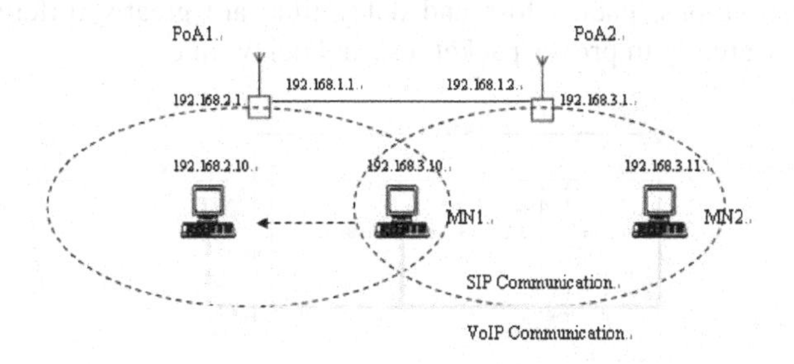

Fig. 10. Service architecture using IP network

Mobile node 1 communicates with Mobile node 2 using VoIP. MN1 has two WLAN interfaces (WLAN0, WLAN1). At first MN1 communicates with MN2 using WLAN0. As MN1 escapes from PoA2's access area, it queries if there is an available access network near PoA2. Its procedures are shown in figure 7. After receiving MIH_GET_INFORMATION_RESPONSE, MN1 activates its other WLAN interface (WLAN1) and sets the necessary parameters, which are received from the previous MIH response to access PoA1's access network. When the preparations for accessing PoA1's network is complete, MN1 sends "RE-INVITE" to the SIP proxy server using ip that is set for PoA1's access network. It updates MN1's ip binding. And then the media session is re-established. At last, it shuts down WLAN0.

Figure 11 shows the experimental environment for testing handover between an 802.11 access network and an 802.3 access network.

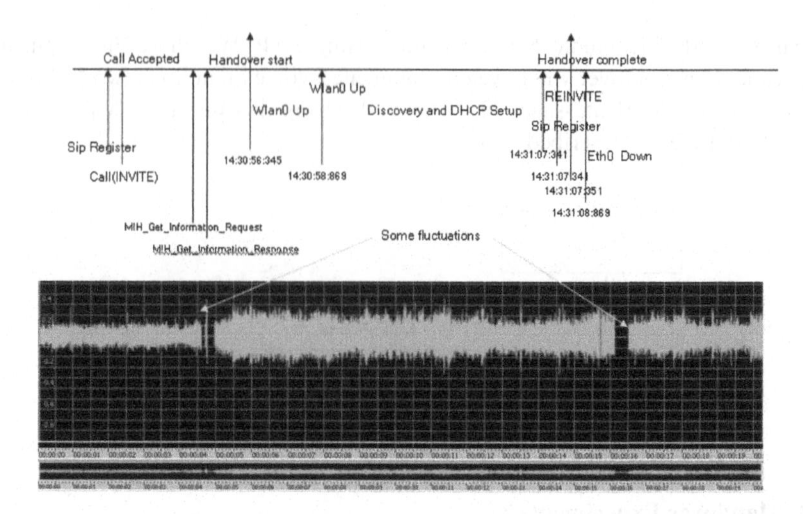

Fig. 11. Experimental result of 802.21-assisted handover

Figure 12 shows the experimental result for 802.21-assisted handover and Figure 12 shows the experimental result for non-802.21-assisted handover. As can be compared from these graphs, packet loss and delay time are greatly different. 802.21-assisted handover greatly improves packet loss and delay time.

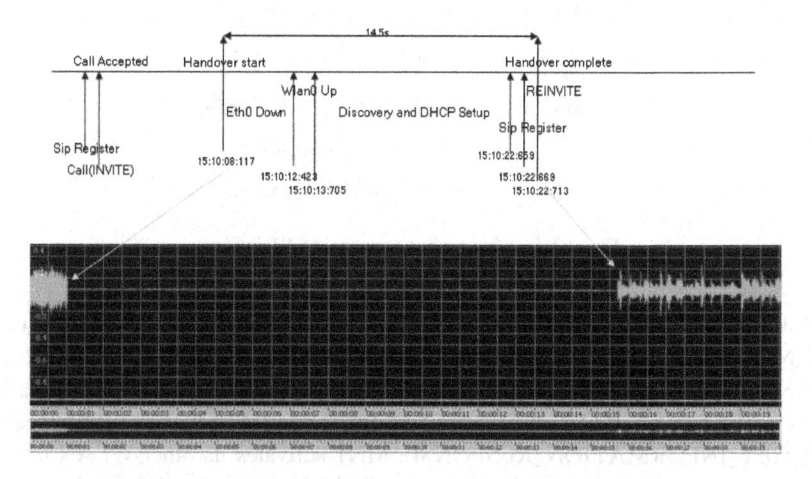

Fig. 12. Experimental result of non- 802.21-assisted handover

4 Summary and Conclusion

As communication networks evolve to NGN, communications based on ALL-IP networks is emerging. It is necessary to support seamless handover between heterogeneous networks for service quality. In addition to that, there will be more multi-mode terminals and the demand for mobile VoIP will be increasing. So, the functions of

802.21 that support seamless handover in heterogeneous access networks are essential. In this paper, we discussed several functional components of the framework and their roles in supporting handover. We explained the laboratory implementation of 802.21 and handover experiments using 802.21. We also compared 802.21-assisted handover and non-802.21-based handover. As the experimental results show, 802.21-assisted handover provides seamless handover by reducing delay and packet loss during handover to a level that is acceptable for mobile VoIP and streaming traffic.

In this paper, we analyzed 802.21-assited handover. But handover policies and other techniques that improve handover packet loss and delay such as fast-handover [4] will affect packet loss and delay time. Another important issue is how to make the information server and gather the geographical network information.

In the future, we will conduct experiments on handover between various wireless access networks, and we will measure the delays at each handover step to create an optimized handover policy. These experimental results will be an important data for seamless mobile ip communication.

References

1. Draft IEEE Standard for Local and Metropolitan Area Networks: Media Independent Handover Services. IEEE P802.21/D01.00 (March 2006)
2. IEEE 802.21 Wikipedia General overview of 802.21 (April 2006), http://en.wikipedia.org/wiki/IEEE802.21
3. http://www.mjsip.org
4. Dutta, A., et al.: Fast handover schemes for application layer mobility management. In: PIMRC (2004)

FPGA-Based Cuckoo Hashing for Pattern Matching in NIDS/NIPS

Thinh Ngoc Tran and Surin Kittitornkun

Dept.of Computer Engineering, Faculty of Engineering,
King Mongkut's Institute of Technology Ladkrabang, Bangkok, 10520 Thailand
tnthinh@dit.hcmut.edu.vn, kksurin@kmitl.ac.th

Abstract. Pattern matching for network intrusion/prevention detection demands exceptionally high throughput with recent updates to support new attack patterns. This paper describes a novel FPGA-based pattern matching architecture using a recent hashing algorithm called *Cuckoo Hashing*. The proposed architecture features on-the-fly pattern updates without reconfiguration, more efficient hardware utilization, and higher throughput. Through various algorithmic changes of Cuckoo Hashing, we can implement parallel pattern matching on SRAM-based FPGA. Our system can accommodate the newest Snort rule-set, an open source Network Intrusion Detection/Prevention System, and achieve the highest utilization in terms of SRAM per character and Logic Cells per character at 15.63 bits/character and 0.033 Logic Cells/character, respectively on major Xilinx Virtex FPGA architectures. Compared to others, ours is more efficient than any other Xilinx FPGA architectures.

Keywords: NIDS, NIPS, Cuckoo Hashing, FPGA, Pattern Matching.

1 Introduction

In recent years, Network Intrusion Detection/Prevention Systems (NIDSs/NIPSs) are more and more necessary for network security. Normally, traditional firewalls only examine packet headers to determine whether to block or pass the packets. Due to busy network traffic and smart attacking schemes, firewalls are not as effective as they used to be. NIDSs/NIPSs are designed to examine not only the headers but also the payload of packets to match and identify intrusions.

To define suspicious activities, most modern NIDSs/NIPSs rely on a set of rules which are applied to matching packets. At the heart of almost every NIDS is pattern matching algorithms. For example, Snort [1] is an open source network intrusion prevention and detection system utilizing a rule-driven language, which combines the benefits of signature, protocol and anomaly based inspection methods. Snort uses a set of rules to filter the incoming packets. As the number of known attacks grows, the patterns for these attacks are made into Snort signatures (patterns or strings). The simple rule structure allows flexibility and convenience in configuring Snort. However, unfortunately, checking every byte of every packet to see if it matches one

S. Ata and C.S. Hong (Eds.): APNOMS 2007, LNCS 4773, pp. 334–343, 2007.

of a set of thousand strings becomes a computationally intensive task as the highest network speed increases to several gigabits/second.

To improve the performance of Snort, various implementations of FPGA-based systems have been proposed. These systems can simultaneously process thousands of rules relying on native parallelism of hardware so their throughput can satisfy current gigabit networks. However, the drawback of hardware-based systems is the flexibility. With emergence of new worms and viruses, the rule set must be frequently updated. Although SRAM-based field programmable gate array (FPGA) can be reconfigured; the process of recompiling the updated FPGA design can be lengthy. For recently proposed FPGA-based NIDSs/NIPSs, adding or subtracting any number of rules requires recompilation of some parts or the entire design. The compilation process takes several minutes to several hours to complete. Today, such latency in compilation may be not accepted for most networks when new attacks are released at a high frequency. It is necessary to update pattern database faster to reduce down time.

Based on Cuckoo Hashing [2], we implement a novel architecture of variable-length pattern matching best suited for FPGA. New patterns can be added to or removed out of the Cuckoo hash tables. Unlike most previous FPGA-based systems, the proposed architecture can update the rule set on-the-fly without reconfiguration thanks to Cuckoo Hashing. Our contributions also include parallel Cuckoo hashing, better hardware utilization and high matching throughput reaching multiple gigabits per second.

The paper is organized as follows. In section 2, some previous FPGA implementations of pattern matching and Cuckoo Hashing are presented. Section 3 proposes the architecture of FPGA-based Cuckoo Hashing. Next, the FPGA implementation of Cuckoo Hashing for multiple pattern matching and its experimental results are discussed in section 4 and 5, respectively. Finally, future works are suggested in the conclusion.

2 Background and Related Works

2.1 FPGA Implementations of NIDS

For a line speed of gigabit network, many previous FPGA approaches of NIDS are proposed. Some of them as [3, 4] implement regular expression matching (NFAs/DFAs) on FPGA. Other approach [5] uses content addressable memory (CAM). While the processing speed is fast, they suffer the two scalability problems such as too many states consume too many hardware resources and FPGA device has to re-program every time patterns be changed. Furthermore, incoming characters are broadcasted to all character matchers. This requires the use of extensive pipelined trees to achieve a high clock rate. The clock frequency of these architectures tends to drop gradually as the number of patterns increases.

Another hardware approach implements hash functions [6-10] to find a candidate of pattern match. Dharmapurikar proposed to use Bloom Filters to do the deep packet inspection [6]. Unlike other hardware approaches mentioned above, this method does not require reprogramming of FPGA for patterns added. Nevertheless, the blooming filter method could generate a false positive match, which requires extra cost of hardware for the match rechecked.

Dionisios et al. proposed CRC hashing named HashMem [8] system using simple CRC polynomials hashing implemented with XOR gates that can use efficient area resource of FPGA than before. For the improvement of memory density and logic gate count, they implemented the V-HashMem [9]. However, these systems have some drawbacks: 1) To reduce the sparse of memory and avoiding collision, CRC hash functions have to be chosen carefully depending on specific pattern groups, 2) since pattern set is dependent, probability of redesigning the system and the reprogramming the FPGA is very high every time patterns are changed.

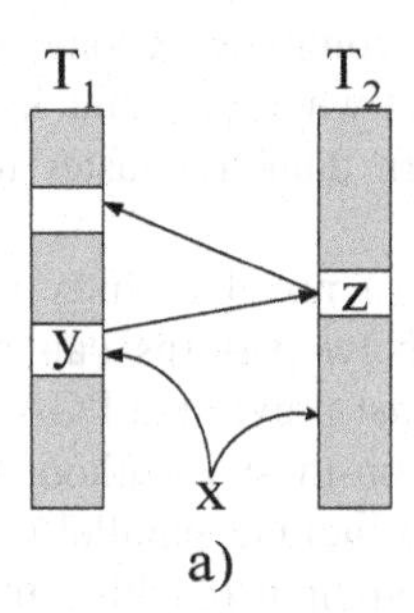

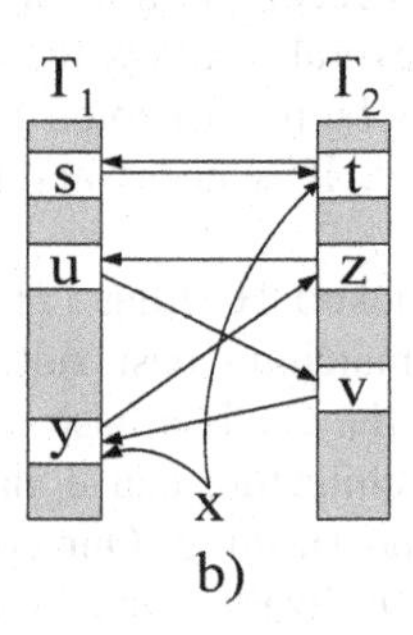

Fig. 1. Original Cuckoo Hashing [2], a) Key x is successfully inserted by moving y and z, b) Key x cannot be accommodated and a rehash is necessary

2.2 Cuckoo Hashing

Cuckoo hashing is proposed by Pagh and Rodler [2] as an algorithm for maintaining a dynamic dictionary with constant lookup time in the worst case scenario. The algorithm utilizes two tables T_1 and T_2 of size $m = (1+\varepsilon)n$ for some constant $\varepsilon > 0$, where n is the number of elements (strings). Cuckoo hashing guarantees $O(n)$ space and does not need perfect hash functions that is very complicated if the set of elements stored changes dynamically under the insertion and deletion. Given two hash functions h_1 and h_2 from universe U to $[m]$, one maintains the invariant that a key x presently stored in the data structure occupies either cell $T_1[h_1(x)]$ or $T_2[h_2(x)]$ but not both. Given this invariant and the property that h_1 and h_2 may be evaluated in constant time, lookup and deletion procedures run in worst case constant time.

Pagh and Rodler described a simple procedure for inserting a new key x in expected constant time. If cell $T_1[h_1(x)]$ is empty, then x is placed there and the insertion is complete; if this cell is occupied by a key y which necessarily satisfies $h_1(x) = h_1(y)$, then x is put in cell $T_1[h_1(x)]$ anyway, and y is kicked out. Then, y is put into the cell $T_2[h_2(y)]$ of second table in the same way, which may leave another key z with $h_2(y) = h_2(z)$ nestless. In this case, z is placed in cell $T_1[h_1(z)]$, and continues until the key that is currently nestless can be placed in an empty cell as in Figure 1(a). However, it can be seen that the cuckoo process may not terminate as Figure 1(b). As a result, the number of iterations is bounded by a bound *MaxLoop* chosen beforehand. In that case everything is rehashed by reorganizing the hash table with new hash functions h_1 and h_2 and newly inserting all keys currently stored in the data structure, recursively using the same insertion procedure for each key.

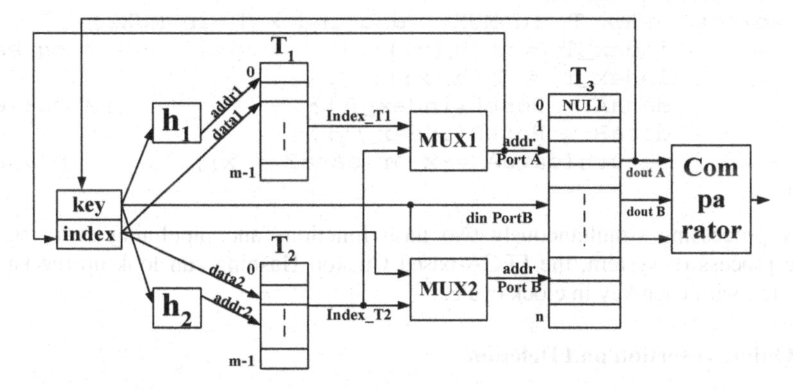

Fig. 2. FPGA-based Cuckoo Hashing. Tables T_1 and T_2 store the key indices; Table T_3 stores the keys.

3 FPGA-Based Cuckoo Hashing

To apply FPGA-based Cuckoo Hashing for variable pattern lengths, the memory efficient for storing patterns in hash tables is required because of the limited numbers of hardware resources. In the original Cuckoo Hashing, the width of table must be equal to the longest pattern in the rule set. Thus, the remaining short patterns will waste so much memory. In order to increase memory utilization, we build up the hashing module for each length of pattern and use indirect storage. Small and sparse hash tables contain indices of keys which are the addresses of a condensed pattern-stored table.

The architecture of a FPGA-based Cuckoo Hashing module as shown in Figure 2 includes three tables: two index tables (hash tables) T_1 and T_2 are single-port SRAMs and a pattern-stored table T_3 is the double-port SRAM for concurrent processing. Hash functions are any universal hashes that can change if they are required to rehash. Two multiplexers are used to select addresses for two ports of T_3. The output of first multiplexer (*MUX1*) is the address of *port A* that is the read-only port. *MUX1*'s inputs are the output value of T_1 (*index_*T_1) and T_2 (*index_*T_2). The output of second multiplexer (*MUX2*) is the address of *port B* that is both reading and writing port. *MUX2* selects *index_*T_2 as lookup function or *index* of key as insertion function.

3.1 Parallel Lookup

A lookup function of element (key) x can be divided into three phases. In each phase, instructions can be processed simultaneously. In the first phase, x is hashed by two hash functions in parallel. The values of two hash functions are used as the address for reading data of two index tables. In the second phase, output data of two index tables are used as the address for two ports of T_3 for reading. In the third phase, the data outputs of T_3 are compared with the incoming character to determine the match. Following is the pseudo-code of parallel Cuckoo lookup function.

```
function lookup(x)
  select index_T1 in MUX1  and index_T2 in MUX2;
        index_T1 = T1(h1(x));                        // phase1
        index_T2 = T2(h2(x));
        dataA = PortA(index_T1);                     // phase2
        dataB = PortB(index_T2);
        return(dataA = x or dataB = x);              // phase3
end
```

By processing simultaneously two hash functions and pipelining every step of whole process of system, the FPGA-based Cuckoo Hashing can look up the keys in streaming with each key in clock cycle.

3.2 Online Insertion and Deletion

When a key insertion occurs, we consider both key and its index. The key is used as an input for two hash functions and stores in table T_3. Its index is used for storing in table T_1 or T_2 and also as the address for lookup the space of key in T_3.

The insertion of element x as description in C-like pseudo-code below is only started after the lookup process failed. If one of outputs of two index tables is empty (NULL), *x's* index is inserted into T_1 or T_2. This is an improvement as compared with the original Cuckoo Hashing. The original one always inserts the index into T_1 without referring to the value of T_2. Otherwise, we consider both tables to reduce the insertion time. If both of the outputs of T_1 and T_2 are not NULL, we insert the key index into T_1. At the same time, the data from T_3 and its address, *index_*T_1, will be written into the key storage and the index storage for starting of cuckoo process. Then, the *MaxLoop* is decreased and the key value is hashed by hash function h_2. The output data will be checked for whether the value is NULL. If it is NULL, the process ends with successful insertion. Conversely, the process is continued by taking in turns hashing from h_2 to h_1.

The worst case happens when the *MaxLoop* decreases to zero. Hence, rehashing is required. Two new hash functions h_1 and h_2 are issued by a pseudo-random number generator. As rehashing cost can be expensive, the choice of good hash function has to be discussed in the next subsection.

For deletion, it is as simple as the lookup process. If the lookup succeeds, the deletion resets the key value to become NULL and the key index to become index of either table T_1 or T_2. We then write key value to table T_3. After that, we reset the key index to NULL and write to the appropriate index table T_1 or T_2. Rehashing for deletion can be required when table T_3 is too sparse. However, this rehashing does not require new hash functions.

```
procedure insert(x)
    if (lookup(x)) return;
    select index_T1 in MUX1  and index in MUX2;
    PortB(index) = x;
    if(index_T1 == NULL){
        T1(h1(x))= index;
```

```
        return;}
    else if(index_T2 == NULL){
        T2(h2(x))= index;
        return;}
    loop MaxLoop times
        if (select index_T1 in MUX1){                    //phase1
          key = PortA(index_T1);
          index = index_T1;
          select index_T2 in MUX1;}
        else{ // (select index_T2 in MUX1)
          key = PortA(index_T2);
          index = index_T2;
          select index_T1 in MUX1;}
        index_T1 = T1(h1(x));                            //phase2
        index_T2 = T2(h2(x));
        if (select index_T1 in MUX1){                    //phase3
          if(index_T1 == NULL) return;
          dataA = PortA(index_T1);}
        else{ // (select index_T2 in MUX1)
          if(index_T2 == NULL) return;
          dataA = PortA(index_T2);}
    end loop
    rehash(); insert(x);
end
```

3.3 Selection of Hash Functions

The choice of hash functions greatly affects the performance of the system. Moreover, the probability for rehashing in Cuckoo Hashing is also based on the randomized property of hash functions. In Cuckoo Hashing, the authors use the Siegel's universal hashing [11] that has a constant evaluation time. However, this constant time is not small and complex in practice. In this section, we discuss some other simple and fast hash functions for string and chose the best one that easily be implemented on hardware.

The universal class of hash functions [12] has the good performance which can be guaranteed independent of the input keys by randomly selecting hash functions from the family. An example of such construct is modular hash functions. However, it is not suitable for hardware because of the complexity of the prime modulo operation. A fast way of generating a class of universal hash function without the modular operation and hardware-friend, tabulation based hashing method [12], is defined as follows:

$$H_t(x) = a_t[0][x_0] \oplus a_t[1][x_1] \oplus \ldots \oplus a_t[n-1][x_{n-1}] \quad (1)$$

A randomized table contains a 2-D array of random numbers in the hashing space. A key is string n characters $x_0x_1..x_{n-1}$ and the hash process is calculated by bit-wise exclusive-or ($\oplus$) a sequence of values $a_t[i][x_i]$, which is indexed by each byte value of x_i and position of i in the string. The drawback of this method is that the size of random table is very large and depends on the length of key.

Another class of simple hash function for hashing character strings named *shift-add-xor* (SAX) [13] is proposed by Ramakrishna et al. The function utilizes only the simple and fast operations of shift, exclusive-or and add.

$$H_i = H_{i-1} \oplus (S_L(H_{i-1}) + S_R(H_{i-1}) + c_i) \quad (2)$$

Two operators S_L and S_R denote the shift left and right, respectively. The symbol c_i is the character i^{th} of string and H_i is an intermediate hash value after examination of i characters. The initial value H_0 can be generated randomly. The authors have shown that the class is likely to be universal. Good performance can be achieved in practice by randomly choosing functions from this class. Moreover, the main advantages of SAX over random-table are a very small space of hardware, and the simple architecture achieving high clock frequency hence the system is faster. To generate the new SAX hash function in case of rehashing, we only need to change the value of H_0 by simple pseudo-random circuit LFSR [14]. Therefore, the SAX is the best choice for our system. The practical performance will be shown in the next section.

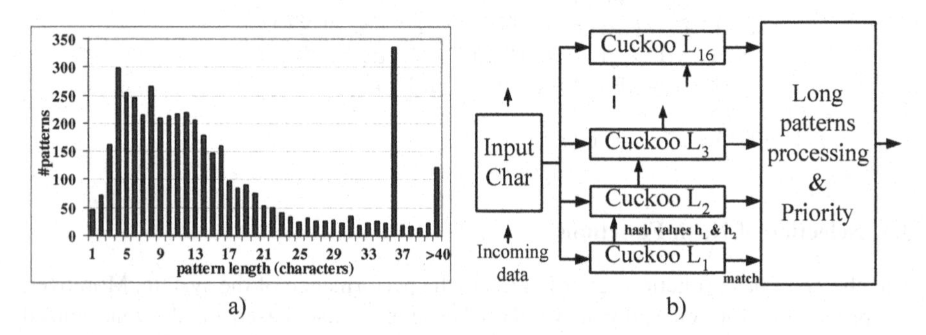

Fig. 3. a) Pattern length distribution of pattern set of SNORT in Dec 2006. b) Overview of FPGA-based Cuckoo Hashing in NIDS.

4 Implementation of FPGA-Based Cuckoo Hashing for Variable-Length Pattern Matching in NIDSs/NIPSs

In Dec 2006, there are 4,748 unique patterns which contain 64,873 characters in Snorts rule set. Figure 3(a) shows the distribution of the pattern lengths in the Snort database. For the pattern matching in the NIDS, the patterns are searched on incoming packets. The matched pattern can occur anywhere as the longest substring. In order to process at the network speed in Gigabits per second, we have to construct Cuckoo Hashing module for every length of pattern from 1 up to 109 characters. Fortunately, in Figure 3(a), we can see that 65% of total numbers of patterns are up to 16 characters long. Therefore, we build the Cuckoo Hashing modules for patterns which are less than or equal 16 characters according to this fact. For longer patterns, we can break them to shorter segments so that we can insert those segments to the Cuckoo modules of short patterns. We then use simple address link-lists to combine these segments later. Figure 3(b) shows the overview of our NIDS system.

As discussed in the above section, SAX hash function is the best choice of our system. Now, we implement SAX and random table hash with patterns of the length from 2 to 16 characters for practical comparison. Figure 4 shows the number of insertion times of Cuckoo Hashing with two methods: random tables and SAX function in which the size of each index table is 512. The lines *SAX_par* and *Tabulation_par* are the hash functions whose architectures are changed as in section 2 of this paper with keys inserted in parallel. The results show that the parallel systems have the insertion time less than the original systems by 20% and the performance of SAX hash function is close to that of random table.

Moreover, we can reduce significantly large amount of hardware area resource by accumulative characteristic of SAX hash function. From (2), to calculate hash value of pattern of length *i* characters in the hash module i^{th}, the requisite inputs are hash value of *i-1* characters calculated beforehand in the hash module $(i-1)^{th}$ and the ith character. Therefore, the value of previous hash module can be reused for the next hash module. This property of the hash functions results in a regular and less resource-consuming hash function. In comparison with previous implementations [6-10], the module i^{th} requires all *i* characters of the input string for every calculation and no reused of previous hash values at all. Their weak points lead to the consumption of a remarkable the number of logic gates for implementation of hash functions.

To reduce the number of memory blocks in FPGA, we can implement two index tables in the same block RAM; the T_1 is in a low addresses part and the T_2 is in a high addresses part. The block RAM of Xilinx FPGA can be configured as dual-port mode that can be accessed concurrently. Therefore, the performance of system still remains the same.

With improvements in the architecture, we can reduce a large amount of hardware resources. In next section, we will illustrate the experimental results on FPGA.

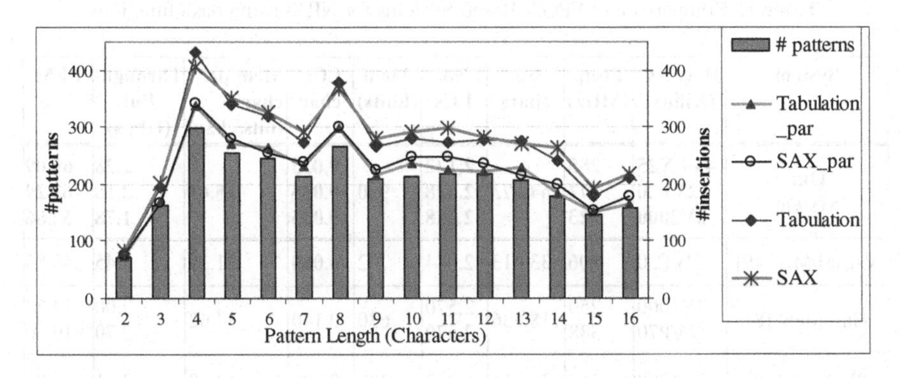

Fig. 4. The number of insertions of various hash functions vs. pattern lengths. Bar graphs are the numbers of patterns. The lines are the numbers of insertions. Index table size is 512. The number of trials is 1,000.

5 Experimental Results

Our design is developed in Verilog hardware description language and Xilinx's ISE 8.1i for hardware synthesis, mapping, and placing and routing. The target chip is a

Virtex4 XC4VLX25 which has 24,192 logic cells and 72 RAM memory blocks. Based on our parallel Cuckoo Hashing pattern matching system described earlier, the numbers of required memory blocks are 39 for pattern storage and 15 for index memories at 1k x 18bit for each block. Besides, we need 1 more memory block to match with the narrow patterns of one character. Totally, we use only 2,142 logic cells and 55 RAM blocks to fit 64,873 characters of entire rule set in the XC4VLX25 FPGA chip. The throughput of a design is calculated by multiplying the clock frequency with the data width (8-bit) of incoming characters. For a design running at 285 MHz clock frequency, the throughput is 2.28 Gigabits per second (Gbps).

Table 1 shows the comparison of our system with other hashing systems. For ease of comparison, we also implement the system on other FPGA chips as Virtex2 XC2V2000 and VirtexPro XC2VPro20. Two metrics, Logic Cells per character (LCs/char) and SRAM bits per character (bits/char), are used to compare hardware NIDS designs. LCs/char is determined by dividing the total number of logic cells used in a design by the total number of characters programmed into the design. SRAM Bits/char is the ratio of memory blocks in bits per total number of characters. With only about 0.033-0.035, our LCs/char is twice smaller than the best one, V-HashMem [9], of previous systems. With 15.63 bits/char, the memory usage of our architecture is of very high density and is acceptable in comparing to other systems.

Another metric used to compare hardware NIDS designs is the Performance Efficiency Metric (PEM) that is the ratio of throughput in Gbps to the Logic Cells per pattern character. PEM of our system is at 62.29 for Virtex2Pro and 69.09 for Virtex4 devices, the best one among all of FPGA hashing systems.

Table 1. Comparison of FPGA-Based Systems for NIDS using hash functions

System	Dev.-XC (Xilinx)	Freq. (MHz)	No. chars	No. LCs	Mem (kbits)	LCs/ char	Mem per char (bits/char)	Through-Put (Gbps)	PEM
Our System	**4VLX25** **2VP20** **2V2000**	**285** **272** **223**	**64,873**	**2,142** **2,328** **2,328**	**990**	**0.033** **0.035** **0.035**	**15.63**	**2.28** **2.18** **1.78**	**69.09** **62.29** **50.86**
V-HashMem [9]	2VP30	306	33,613	2,084	702	0.060	21.39	2.45	40.83
HashMem [8]	2V1000 2VP70	250 338	18,636	2,570 2,570	630	0.140	34.62	2.00 2.70	14.50 19.60
PH-Mem [10]	2V1000	263	20,911	6,272	288	0.300	14.10	2.11	7.03
ROM+Coproc[7]	4VLX15	260	32,384	8,480	276	0.260	8.73	2.08	8.00

6 Conclusion and Future Works

A novel FPGA-based pattern matching based on Cuckoo Hashing for NIDSs/NIPSs is proposed. In addition, selections of practical hash functions are also discussed. As a result, the most suitable one is shift-add-xor (SAX). According to the implementation results, the utilization of our system is the best when compared with other previous systems and the achievable throughput can be up to 2.28 Gbits/s. One of remarkable features of our system is dynamic pattern insertions and deletions with no FPGA reconfiguration. For future work, we will improve our system by reducing the size of hash tables. In addition, IP header matching will be combined to complete the system.

Acknowledgments. We would like to acknowledge AUN-SeedNet Program of JICA for the scholarship and Xilinx, Inc. for donating the software tools.

References

1. SNORT: The Open Source Network Intrusion Detection System. http://www.snort.org
2. Pagh, R., Rodler, F.F.: Cuckoo hashing. Journal of Algorithms. 51, 122–144 (2004)
3. Moscola, J., Lockwood, J., Loui, R.P., Pachos, M.: Implementation of a content-scanning module for an internet firewall. In: Proceedings of the 11th IEEE Symposium on Field-Programmable Custom Computing Machines (FCCM), pp. 31–38. IEEE Computer Society Press, Los Alamitos (2003)
4. Clark, C.R., Schimmel, D.E.: Scalable pattern matching for high speed networks. In: Proceedings of the 12th IEEE Symposium on FCCM, pp. 249–257. IEEE Computer Society Press, Los Alamitos (2004)
5. Sourdis, I., Pnevmatikatos, D.: Pre-decoded cams for efficient and high-speed NIDS pattern matching. In: Proceedings of the 12th IEEE Symposium on FCCM, pp. 258–267. IEEE Computer Society Press, Los Alamitos (2004)
6. Dharmapurikar, S., Krishnamurthy, P., Spoull, T., Lockwood, J.: Deep Packet Inspection using Bloom Filters. In: Hot Interconnects, pp. 44–51 (2003)
7. Cho, Y.H., M-Smith, W.H.: Fast reconfiguring deep packet filter for 1+ gigabit network. In: Proceedings of the 13th IEEE Symposium on FCCM, pp. 215–224. IEEE Computer Society Press, Los Alamitos (2005)
8. Papadopoulos, G., Pnevmatikatos, D.: Hashing + memory = low cost, exact pattern matching. In: Proceedings of the 15th International Conference on Field Programmable Logic and Applications, pp. 39–44 (2005)
9. Pnevmatikatos, D., Arelakis, A.: Variable-length hashing for exact pattern matching. In: Proceedings of the 16th International Conference on Field Programmable Logic and Applications, pp. 1–6 (2006)
10. Sourdis, I., Pnevmatikatos, D., Wong, S., Vassiliadis, S.: A reconfigurable perfect-hashing scheme for packet inspection. In: Proceedings of the 15th International Conference on Field Programmable Logic and Applications, pp. 644–647 (2005)
11. Siegel, A.: On universal classes of fast high performance hash functions, their time–space tradeoff, and their applications. In: Proceedings of the 30th Annual Symposium on Foundations of Computer Science, pp. 20–25. IEEE Computer Society Press, Los Alamitos (1989)
12. Carter, J.L., Wegman, M.N.: Universal classes of hash functions. Journal of Computer System Sci. 18, 143–154 (1979)
13. Ramakrishna, M.V., Zobel, J.: Performance in Practice of String Hashing Functions. In: Proceedings of the Fifth International Conference on Database Systems for Advanced Applications, vol. 6, pp. 215–224 (1997)
14. Xilinx Application Note. http://www.xilinx.com/bvdocs/appnotes/xapp211.pdf

ATPS – Adaptive Threat Prevention System for High-Performance Intrusion Detection and Response

Byoungkoo Kim, Seungyong Yoon, and Jintae Oh

Security Gateway System Team, Electronics and Telecommunications Research Institute, Gajeong-dong, Yuseong-gu, Daejeon, 305-700, Korea
{bkkim05,syyoon,showme}@etri.re.kr

Abstract. The fast extension of inexpensive computer networks has increased the problem of unauthorized access and tampering with data. Many NIDSs are developed till now to respond these network attacks. As network technology presses forward, Gigabit Ethernet has become the actual standard for large network installations. Therefore, software solutions in developing high-speed NIDSs are increasingly impractical. It thus appears well motivated to investigate the hardware-based solutions. Although several solutions have been proposed recently, finding an efficient solution is considered as a difficult problem due to the limitations in resources such as a small memory size, as well as the growing link speed. In this paper, we propose the FPAG-based intrusion detection technique to detect and respond variant attacks on high-speed links. It is possible through novel pattern matching mechanism and heuristic analysis mechanism that is processed on FPGA-based reconfiguring hardware. Most of all, It was designed to fully exploit hardware parallelism to achieve real-time packet inspection, to require a small memory for storing signature. The technique is a part of our proposed system, called ATPS(Adaptive Threat Prevention System) recently developed. That is, the proposed system has hardware architecture that can be capable of provide the high-performance detection mechanism.

Keywords: intrusion detection, header lookup, pattern matching.

1 Introduction

In the last decade, networks have grown in both size and importance. In particular, TCP/IP networks have become the main means to exchange data and carry out transactions. But, the fast extension of inexpensive computer networks also has increased the problem of unauthorized access and tampering with data[1]. As a response to increased threats, many Network-based Intrusion Detection Systems (NIDSs) have been developed to serve as a last line of defense in the overall protection scheme of a computer system. These NIDSs have two major approaches; misuse intrusion detection and anomaly intrusion detection[9][10], but most of existing NIDSs, such as Snort[6], NFR[7], and NetSTAT[8], only employs the misuse detection approach for reducing a lowering of performance to the minimum. Also, most of NIDSs based on misuse detection approach has concentrated on catching and

S. Ata and C.S. Hong (Eds.): APNOMS 2007, LNCS 4773, pp. 344–353, 2007.

analyzing only the audit source collected on Fast Ethernet links. However, with the advancement of network technology, Gigabit Ethernet has become the actual standard for large network installations. However, existing NIDSs have problems of a lowering of performance as ever, such as bottleneck, overhead in collecting and analyzing data in a specific component. Therefore, the effort of performing NIDS on high-speed links has been the focus of much debate in the intrusion detection community, and several NIDSs, such as Real-Secure[3], ManHunt[4], and CISCO IDS[5], that is run on high-speed links actually has been developed. But, these NIDSs is still not practical because of technical difficulties in keeping pace with the increasing network speed, and real-world performance also will likely be less. Therefore, there is an emerging need for security analysis techniques that can keep up with the increased network throughput. This paper presents our Gigabit IDS, called ATPS(Adaptive Threat Prevention System), to detect and respond attacks on the high-speed network. It is possible through FPGA(Field Programmable Gate Array)-based intrusion detection technique.

The remainder of the paper is structured as follows. The next section presents related works about early studies of NIDS. Then, section 3 presents the architecture of our proposed system, and describes the efficient detection techniques that are run in FPGA-based reconfiguring hardware. Section 4 introduces our prototype that we have developed, and briefly shows the experimental results. Finally, we conclude and suggest directions for further research in section 5.

2 Related Work

Basically, IDS is classified into host-based IDS and network-based IDS[9]. Audit sources discriminate the type of IDSs based on the input information they analyze. Host-based IDS analyzes host audit source, and detects intrusion on a single host[11]. With the widespread use of the Internet, IDSs have become focused on network attacks. Therefore, most IDSs employed network-based IDS. NIDS uses the network as the source of security-relevant information. Consequently, NIDS moves security concerns from the hosts and their operating systems to the network and its protocols.

Besides, IDS is classified into two major approaches based on the detection method they operate; misuse intrusion detection and anomaly intrusion detection[12]. The first, misuse intrusion detection is based on the detection of intrusions that follow well-defined patterns of attack exploiting known vulnerabilities. Therefore, this approach is based on the pattern of known misuse or abnormal behavior. This approach is very efficient, but this is hard to detect new intrusion patterns. Also, this approach is possible to draw false negative detection. The second is based on the detection of anomalous behavior or the abnormal use of the computer resource. This approach is based on the database of normal behavior. Therefore, it costs a great deal, but this approach is capable of detecting unknown intrusions. Also, It is possible to draw false positive detection, but hard to set a threshold value. Like this, these approaches all have each advantages and disadvantages. However, anomaly intrusion detection approach does not apply to operate real-time intrusion detection, since it tends to be computationally expensive because of several maintained metrics that are updated after every system activity. Therefore, most IDSs employed misuse detection approach only because of performance and availability consideration.

Primary approaches to misuse detection might be implemented by on the following techniques: Expert Systems, State Transition Analysis, Model based Approach and so forth[12]. But, these techniques do not present the definite mechanism, and sometimes contain complex and ambiguous concepts. Also, these approaches are not suited as a speedy detection mechanism in high-speed network environments. Therefore, most IDSs focus on more speedy and exact pattern matching algorithm and detection mechanism about Denial of Service(DoS) attacks and Port Scan attacks.

3 ATPS – Adaptive Threat Prevention System

In this section, we briefly introduce the architecture of our system and components of the architecture called ATPS. And then, we present efficient detection techniques for high-speed intrusion detection and response. Through these techniques, our proposed system can perform the real-time operations as inline-mode.

3.1 System Architecture and Components

Our system is aimed at real-time network-based intrusion detection based on misuse detection approach. As shown in the figure 1, the proposed system consists of two parts; Application Task for policy management, alert management and system management, and Security Engine Board for wire-speed packet forwarding, packet preprocessing, high-performance intrusion detection and response. Most of all, for detecting network intrusions more efficiently on high-speed links, Security Engine Board is composed of several sub FPGA Logics. Here, communication between Security Engine Board and Main CPU is run through PCI interface. Communication through PCI interface is divided into two channels. One is a channel for policy enforcement. The other is a channel for alert information transmission. Through the interoperability of these components, our system analyzes data packets as they travel across the network for signs of external or internal attack. That is, the major

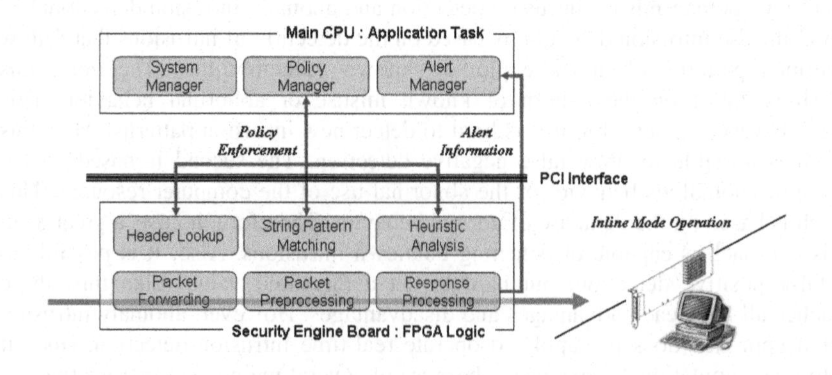

Fig. 1. System Architecture and Components

functionality of our system is to perform the real-time traffic analysis and intrusion detection on high-speed links. Therefore, we focus on effective detection strategies applied FPGA Logics.

3.2 Security Engine Board

As shown in the figure 2, Security Engine Board is composed of three FPGA chips and one FPGA chip for PCI interfacing.

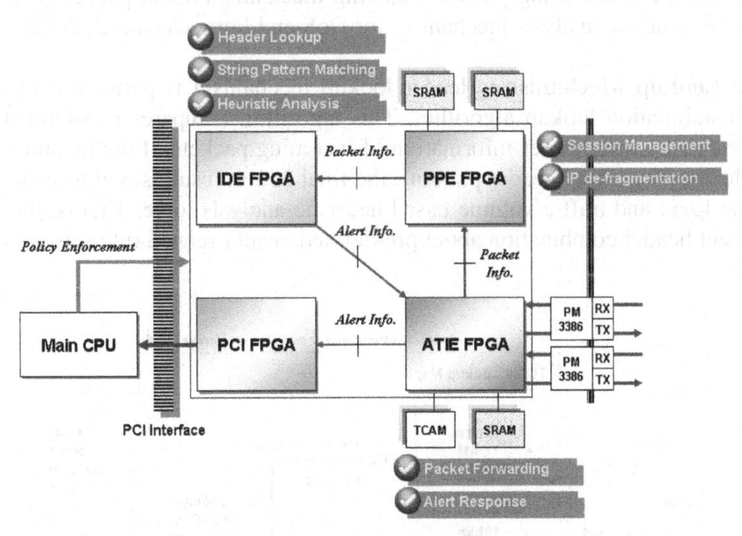

Fig. 2. The FPGA Chips Arrangement of Security Engine Board

First, ATIE(Anomaly Traffic Inspection Engine) FPGA chip uses the XILINX XC2VP70 FPGA chip. And, it is connected to the PM3386 for incoming packet forwarding. Also, it uses external TCAM and SRAM for incoming packet scheduling and management. Briefly, the main function of ATIE is the wire-speed packet forwarding and response coordinating such as alert message generation and packet filtering. Basically, incoming packets from PM3386 is sent to PPE FPGA chip, and if it is determined as attack according to the analysis result from other FPGA chips, alert information is sent to the main CPU through FPGA chip for PCI interfacing. Second, PPE(Pre-Processing Engine) FPGA chip uses the XILINX XC2VP50 FPGA chip. And, it uses two external SRAM for session management, IP de-fragmentation and TCP reassembly. Briefly, the main function of PPE FPGA chip is to process the before steps for intrusion detection. The preprocessing function supports the SPI (Stateful Packet Inspection) based intrusion detection and IDS evasion attack detection. Finally, IDE(Intrusion Detection Engine) FPGA chip uses the XILINX XC2VP70 FPGA chip. It uses three mechanisms for high-performance intrusion detection; Flexible Header Combination Lookup Algorithm for packet header pattern matching, Linked Word based Store-less Running Search Algorithm for string pattern matching about packet payload, and Traffic Volume based Heuristic Analysis

Algorithm for DoS and Port-scan attack detection. Through these mechanisms, it executes the high-performance intrusion detection without packet loss.

3.3 High-Performance Intrusion Detection Techniques

The detection mechanism of our system is mainly run on IDE FPGA chip. For effective high-performance intrusion detection, our system has three detection mechanisms. One is the header lookup mechanism for flexible header combination lookup, another is the string pattern matching mechanism about packet payload, and the other is heuristic analysis mechanism for DoS and Port-scan attack detection.

Header Lookup Mechanism: Header lookup mechanism is performed by flexible header combination lookup algorithm. This algorithm compares pre-defined header related rule-sets with header information of incoming packets. If the incoming packet is matched with existing header patterns, the final 256bit result is sent to string pattern matching logic and traffic volume based heuristic analysis logic. That is, the number of different header combination about pre-defined header related rule-sets is limited to 256.

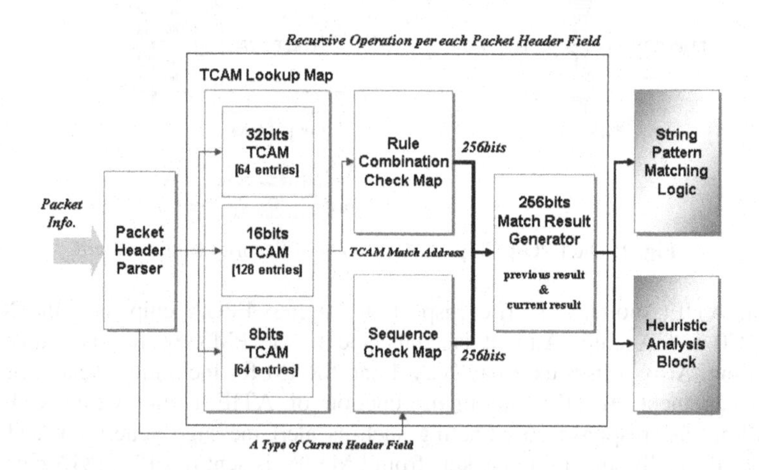

Fig. 3. Header Lookup Mechanism

As shown in the figure 3, this algorithm uses three memory maps; internal TCAM(Ternary Content Addressable Memory) lookup map, rule combination check map and sequence check map. First, internal TCAM lookup map is composed of three TCAM; 8bits lookup map for 8bits header fields such as ICMP type and TCP flags, 16bits lookup map for 16bits header fields such as service port value, and 32bits lookup map for 32bits header fields such as IP address field. The match address of these lookup maps is used by rule combination check map. Second, rule combination check map is composed of combination map of 256*256. That is, the 256bit result of rule combination check map, which is indicated by match address from TCAM lookup map, presents the rule-set information of current matched field. For example, if the match result of ICMP type field is "{255{2'b0}, 2'b1}", the first rule-set(header combination) is to be matched. Finally, sequence check map is composed of sequence map of 32*256, and includes don't care information of current matching field. Here, '32' means the number of header field kinds. For example, if don't care information of ICMP type field is "{255{2'b0}, 2'b1}", the first rule-set is always to be matched irrespective of TCAM matching. The 256bit result of this map is combined with result of rule combination check map. Basically, this mechanism is performed recursively about all packet header fields of incoming packet. The final 256bit result is referred by logics for string pattern matching and traffic volume based heuristic analysis.

String Pattern Matching Mechanism: String pattern matching mechanism about packet payload is performed by linked word based store-less running search algorithm. This algorithm compares pre-defined packet payload related rule-sets with packet payload information of incoming packets. If the incoming packet is matched with existing payload patterns, alert message is generated in combination with the header lookup result. For this operation, this algorithm uses the pattern reconstruction technique. As shown in the figure 4, reconstruction substring length of each pattern has boundary length of 5 or 7 because of the limit of block memory in FPGA chip. The first 5byte substring of "/bin/echo" pattern is equal to the first 5byte substring of "/bin/kill" pattern and "/bin/chmod" pattern. Therefore, "/bin/" substring of these patterns is stored in the same memory space. Like this, other patterns are also reconstructed. Through a pattern reconstruction like this, our system can have about 2,000~3,000 rule-sets in the limited memory storage on FPGA chip.

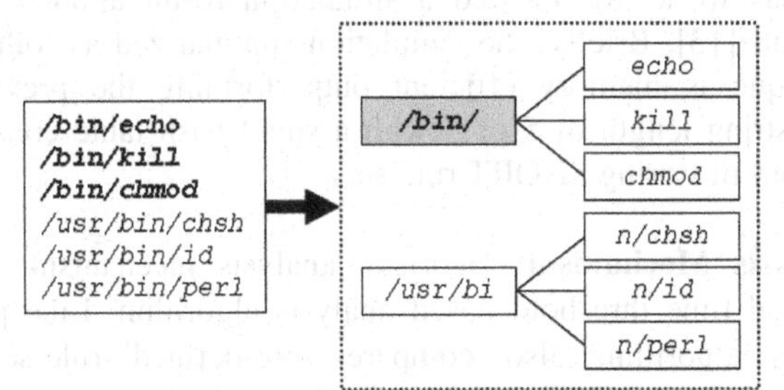

Fig. 4. Payload Pattern Reconstruction

After above pattern reconstruction, linked word based store-less running search algorithm is performed as string pattern matching mechanism. This algorithm uses the spectrum dispersion technique as shown in the figure 5. The spectrum dispersion technique is method to calculate unique hash value about reconstructed substrings. For example, "/etc/" substring of 5byte length has the 9bit hash value by a given hash formula. Hash values calculated about each substring are used as a storing memory address for each substring. The pattern reconstruction and hash memory allocation like this is performed by main CPU when the system booting is run in advance. After system booting, IDE FPGA logic performs the hash value calculation about the incoming packet to the unit of byte. If the payload string in incoming packet is matched with the substring in memory pointed by the calculated hash value, then it is checked out which the related reconstructed substrings are matched or not. If all connected substrings of one pattern are matched with incoming packet, alert message is generated in combination with the header lookup result. Through these operations, our system performs the string pattern matching operation without lowing of performance and packet loss.

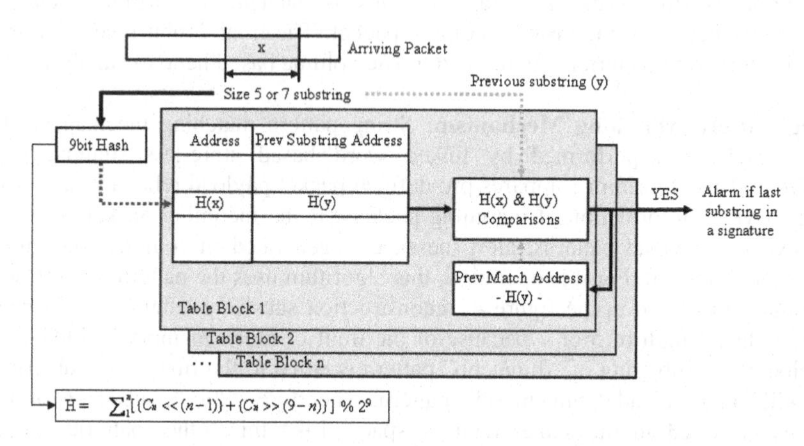

Fig. 5. Linked Word based Store-less Running Search Algorithm

In the previous work, we showed a simulation result about our string pattern matching technique[13]. Briefly, the simulation summarized as following; First, the proposed technique is memory efficient outperforming the previous techniques. Second, the substring length of 5 or 7 with a small hash table consumed minimum amount of memory in storing SNORT rule-sets.

Heuristic Analysis Mechanism: Heuristic analysis mechanism is performed by traffic volume and time threshold based analysis algorithm. Like pattern matching mechanism, this algorithm also compares pre-defined rule-sets with packet information of incoming packets. However, this mechanism generates alert message by traffic volume within time threshold. If the incoming packet is matched with existing rule-set, then count value of the rule-set is increased, and count threshold and time threshold is checked out. If the count threshold is exceeded within time threshold, alert message is generated. Through this mechanism, our system is capable of detecting the DoS and Port-scan attacks.

4 Implementation and Experimental Results

We have developed Gigabit IDS based on our architecture, called ATPS. Our system is implemented in programming languages that is best suited for the task it has to perform. Basically, application tasks of our system are implemented in C programming language. FPGA Logic of our system is implemented in verilog HDL(Hardware Description Language) that is best suited for high-speed packet processing. Most of all, our system focus on FPGA logic for real-time intrusion detection on high-speed links. Also, we employed inline mode capable of effective response by using four Gigabit Ethernet links. That is, our system has developed in the side of improvement in performance for packet processing.

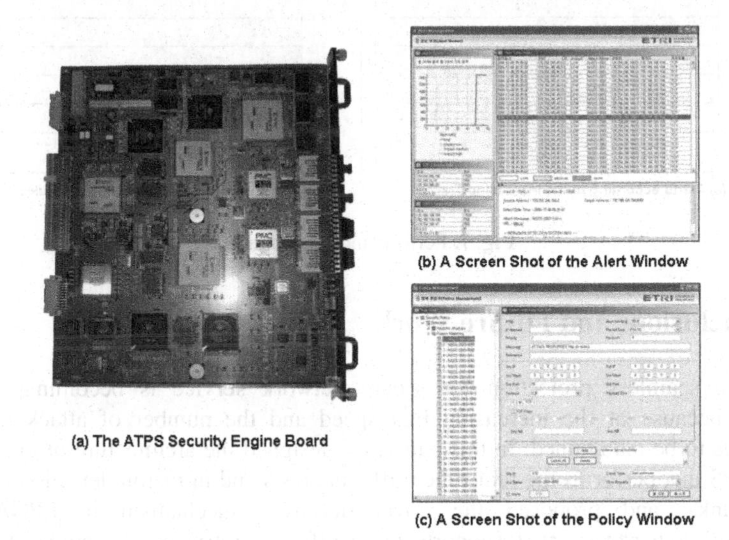

(a) The ATPS Security Engine Board
(b) A Screen Shot of the Alert Window
(c) A Screen Shot of the Policy Window

Fig. 6. Adaptive Treat Prevention System

As shown in the figure 6 (a), our system was implemented in a XILINX Virtex-II Pro flatform FPGAs. Most of all, The IDE FPGA device, XC2VP70, has 74,448 logic cells and 5.9Mbits of on-chip SRAM, which is a configurable block select memory. In the above figure, it is marked with the red square. Also, the screen shots were captured during experiments to validate the performance of the prototype. The screen shot (b) shows that web-related attacks were detected. The next screen shot (c) shows that rule-sets for intrusion detection and response were applied.

We applied the snort rule-sets for performance evaluation of our system. And we used IXIA Traffic Generator[14] for background traffic generation, IDS Informer Attack Tool[15] and Nessus Vulnerability Scanner[16] for attack traffic generation. The first, we observed the rate of alert generation when background traffic generated by IXIA increase gradually. That is, we measured the decrease in effectiveness of the detection when the traffic rate increases. As shown in the figure 7 (a), increasing traffic rate hasn't an effect on detection rate. The second, our experiment was run with a constant traffic rate of 100Mbps and an increasing number of signatures. The experiment starts with only the 200 rule-sets that are needed to achieve maximum detection for the given attacks. As shown in the figure 7 (b), increasing number of signatures also hasn't an effect on detection rate of our system. The previous two experiments using Snort sensors are performed by Kruegel et al.[2]. Compared with Snort sensor, our system showed a consistent performance in traffic level and had nothing to do with increasing number of signatures applied.

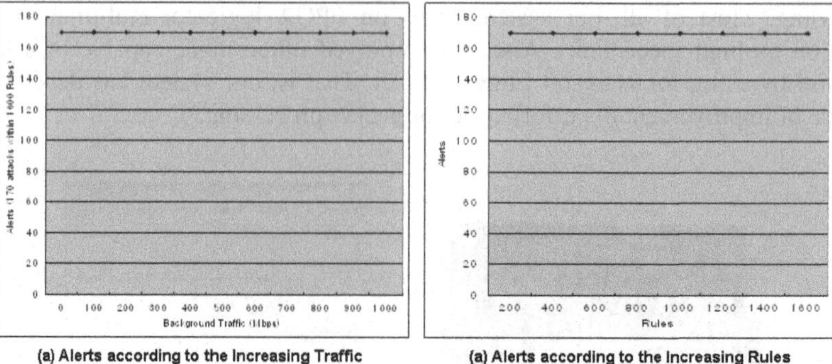

(a) Alerts according to the Increasing Traffic   (a) Alerts according to the Increasing Rules

Fig. 7. Performance Evaluation

5 Conclusions and Future Work

Providing seamless protection for secure network service is becoming difficult primary because of the increasing link speed and the number of attack patterns, signatures to be maintained. In this paper, we designed the architecture of our system call ATPS that performs the real-time traffic analysis and intrusion detection on high-speed links, and proposed the novel detection mechanism in FPGA-based reconfiguring hardware that supports more efficient intrusion detection. Also, we have developed the prototype of our system for the analysis of the traffic carried by a Gigabit link. Most of all, our system focus on reducing a lowing of performance caused by high-speed traffic analysis to the minimum. Therefore, it is run by the FPGA logic proposed for improvement in performance. Also, it has the advantage that is capable of supporting the effective response by using inline mode monitoring technique on four Gigabit links. However, the current prototype is very preliminary and a thorough evaluation will require experimentation in a real-world environment. In future, for resolving the problem derived from the verification of implemented system, we will go and consider on system performance, availability, faults-tolerance test with prototype. Also, we will keep up our efforts for improvement in performance of detection mechanism on high-speed links. Finally, we will implement and expand our designed system and give more effort to demonstrate effectiveness of our system.

References

1. Kim, B.-K., Kim, I.-K., Kim, K.-Y., Jang, J.-S.: Design and Implementation of High-Performance Intrusion Detection System. In: Laganà, A., Gavrilova, M., Kumar, V., Mun, Y., Tan, C.J.K., Gervasi, O. (eds.) ICCSA 2004. LNCS, vol. 3046, pp. 594–602. Springer, Heidelberg (2004)
2. Kruegel, C., Valeur, F., Vigna, G., Kemmerer, R.: Stateful intrusion detection for high-speed networks. In: Proceedings of the IEEE Symposium on Security and Privacy, pp. 266–274. IEEE Computer Society Press, Los Alamitos (2002)
3. ISS. RealSecure Gigabit Network Sensor (September 2002), http://www.iss.net/products_services/enterprise_protection/rsnetwork/gigabitsensor.php

4. Symantec. ManHunt (2002), http://enterprisesecurity.symantec.com/products/products.cfm? Product ID=156
5. CISCO. CISCO Intrusion Detection System. Technical Information (November 2001)
6. Roesch, M.: Snort-Lightweight Intrusion Detection for Networks. In: Proceedings of the USENIX LISA '99 Conference (November 1999)
7. Ranum, M.: Burglar Alarms for Detecting Intrusions, NFR Inc. (1999)
8. Ptacek, T., Newsham, T.: Insertion, Evasion, and Denial of Service: Eluding Network Intrusion Detection, Secure Networks Inc. (1998)
9. Debar, H., Dacier, M., Wespi, A.: Research Report Towards a Taxonomy of Intrusion Detection Systems, Technical Report RZ 3030, IBM Research Division, Zurich Research Laboratory (June 1998)
10. Kumar, S., Spafford, E.: A pattern matching model for misuse intrusion detection. In: Proceedings of the 17th National Computer Security Conference, pp. 11–21 (October 1994)
11. Anderson, D., Frivold, T., Valdes, A.: Next-generation intrusion detection expert system(NIDES), Technical Report SRI-CLS-95-07 (May 1995)
12. Kumar, S.: Classification and Detection of Computer Intrusions, Phd, Purdue University (1995)
13. Yi, S., Kim, B.-k., Oh, J., Jang, J., Kesidis, G., Das, C.R.: Memory-Efficient Content Filtering Hardware for High-Speed Intrusion Detection Systems. In: Proceedings of the 2007 ACM Symposium on Applied Computing, Seoul, Korea, March 11-15, pp. 264–269. ACM Press, New York (2007)
14. http://www.ixiacom.com
15. http://www.blasdesoftware.net
16. http://www.nessus.org

A Practical Approach for Detecting Executable Codes in Network Traffic

Ikkyun Kim[1], Koohong Kang[2], YangSeo Choi[1], Daewon Kim[1], Jintae Oh[1], and Kijun Han[3]

[1] Information Security Research Division, ETRI, Gajeong-dong, Yuseong-gu, Daejeon, 305-700, South Korea
{ikkim21,yschoi92,dwkim77,showme}@etri.re.kr

[2] Dept. of Information and Communications Engineering, Seowon University 231, Mochung-Dong, Chongju, 361-742, South Korea
khkang@seowon.ac.kr

[3] Dept. of Computer Engineering, Kyungpook National University 1370 Sankyuk-dong, Buk-gu Daegu 702-701, South Korea
kjhan@knu.ac.kr

Abstract. The research on the detection of zero-day network attack and the signature generation is highlighted as an issue according to the outbreak of the new network attack is faster than a prediction. In this paper, we propose a very practical method that detects the executable codes within the network packet payload. It could be used as the key function of the signature generation against the zero-day attack or the high speed anomaly detection. The proposed heuristic method in this paper could be expressed in terms of visually classifying the characteristic of the instruction pattern of executable codes. And then we generalize this by applying the discrete parameter Markov chain. Our experimental study showed that the presented scheme could find all types of executable codes in our experiments.

Keywords: zero-day attack, signature generation, malicious code.

1 Introduction

As it is mentioned in the vulnerability's law, new vulnerabilities are released with every year and the infection rate of new network attack is faster than our prediction. That is, whilst it activates the automatic patch function, the threats of the zero-day network attack are increased as times go by. As the gap between the release time of the vulnerability and the outbreak time of the attack using the vulnerability gradually decreases, the necessary time to response its vulnerability reached zero-day and zero-time. Actually in case of Slammer worm in 2003, after the server buffer overflow vulnerability was released, its exploit occurred in 185 days (called 185-day attack). And in case of Sasser worm in 2004, it took 19 days (called 19-day attack). At last, the attack (zero-day attack) using the vulnerability which is not released was occurred in the Mozilla Firefox application

S. Ata and C.S. Hong (Eds.): APNOMS 2007, LNCS 4773, pp. 354–363, 2007.

in 2005. Furthermore, recently the research on the detection of zero-day network attack and the signature generation is highlighted as an issue. In the classification point of view of the typical network intrusion detection methodology, we can consider the "zero-day" worm problem as the extension of the anomaly detection methodology. However, as the zero-day network attack became more sophisticated and faster in spreading across network, it differs from the existing anomaly detection methodology and researches. In the initial of this research of a field, it was initiated from the detection and signature generation method using the content prevalence model which considers the propagation of the super worm including Code-Red, Slammer, etc. But the false rate of the detection methods which are based on it is a little high and the accuracy of the signature generation is relatively low. Moreover, it is not suitable to detect exploits and generate the signature, if we look into the recent trend of new network attacks. For example, after Sasser worm occurred in 2004, the network attack of the similar type markedly decreases. And malicious software mainly spread by using E-mail, downloader, dropper, and etc. Therefore, as to researches [1,2] using the property of the similarity or the repeatability of the network traffic, the effectiveness decreases, while some static or dynamic analysis method of network packet have gotten the attention in detecting the malicious software. In this paper, we propose a very practical method that detects the executable part within the network packet payload. The proposed method can be used for a pre-processing module for detecting the malicious executable codes showing up in the network packet. In order to find the executable part within a payload, we propose the instruction spectrum analysis of which each instruction set is represented its color and then the whole sequence of instructions is analyzed by the translated color pattern. We assume two things; (i) there is a tendency which is not so with the tendency to be consecutively repeated with the instruction set, and (ii) as the context between an instruction group and the other instruction group, there is any correlation pattern. So if these relations are expressed with a color, a pattern toward the correlation of the executable instruction group can become visible and we can found the executable instruction spectrum. The instruction spectrum analysis method of this paper could be expressed by the form visually classifying the characteristic of the instruction pattern of executable codes. It generalized this by applying the discrete Markov model, and a correlation probability between the current instruction group and previous instruction group is found and we determine the executable instruction sequence. The method proposed in this paper can be used as the key function of the signature generation against the zero-day attack or the high speed anomaly detection. The rest of this paper is organized as follows. In Section 2, we summarize the work related to ours. The definition of the problem considering in this paper is presented in Section 3, and the basic idea using the instruction code distribution is introduced in Section 4. In Section 5, we describe about the discrete-parameter Markov chain for generalizing the basic idea. In Section 6, we show our experiment results, and finally, we conclude in Section 7.

2 Related Works

As the method for the detection of the unknown zero-day exploit, there are the signature generation, the static analysis method, and the dynamic analysis method of emulation technique and etc. Honeycomb [1] was the first attempt which automatically generates the intrusion detection signature using honeypot. Autograph [3]and Earlybird [4] can be said to be the beginning of the method for automatically generating a signature against the worm which is delivered through TCP by using the prevalence feature of a content. Furthermore, the Polygraph [2] and Hamsa [5] can be given as the research for the signature generation of the polymorphic worm. Many literatures about the malware detection and prevention on the host level have provided the insight about the execution environment and malware form. Recently, PolyUnpack [6] proposed a method which automatically detects and extracts the executable code hiding in the unpacked malware, and Christodorescu [7] proposed a method which detects the polymorphic malware by using the semantic behavior models called as the templates. But, these two proposals also have the disadvantage not to provide the method for the self-modifying decryption loop. As the widely used disassembly technology in this domain, the liner sweep method and recursive traversal method can be given. Because the liner sweep method successively decoding bytes is unable to distinguish embedded data and real instructions, there is the disadvantage not to prepare for the attack using the data injection attack and instruction intersection. And there is the advantage of the recursive traversal method decoding bytes according to the control flow of the program. But, this method requires the start address of the program in order to carry out the analysis, because in fact, the destination address of the branch instruction is not always statically determined, so linear sweep method can make more valid instructions. In order to distinguish program instruction more accurately, Kruegel [8] proposed the method using the control flow graph of the program and the statistical technologies. As the hybrid method of the network-level for the exploit code detection, there are static-analysis [9,10,11,12] and emulation method [13]. The validity of the static-analysis is determined by the method for distinguishing the program-like payload from non-code data and non-exploit code. The exploit code usually hides the start position which is not clear in the network traffic and can be scattered in invalid data. The Intel instruction set (IA-32) is so simple architecture that its codes cannot be distinguished from data bytes by using only the disassembly method. The static-analysis methods [9,10,11,12] which have been researched down to date can detect only some portion of the problems earlier illustrated. Due to unspecific start position of code, these methods cannot confront the exploits attack using the static analysis-resistant technology like a self-modifying and indirect control instructions. The methods proposed in [11] and [9] focused on the NO-OP sled can miss an execution codes in the progressed exploit code. The method which is proposed in [10] to extract the control flow of an exploit has the disadvantage not to prevent the data injection attack. The work of Wang et al [12] proposed the code

	Group Name		Instruction examples
0	Arithmetic		AAA, AAM, DAA, DAS, ADD, XADD, ADC, SUB, SBB, DIV, ...
1	Transfer		MOVS, MOVSX, MOVZX,
2	Logical		AND, OR, TEST
3	Load/Store		STOS, LODS
4	Stack		PUSH, POP
5	Control		JMP, CALL, LOOP, RET
6	NOP		NOP
7	Interrupt		INT
8	Float		FADD, FIADD, FSUB, FISUBR
9	MMX		MMX instructions
10	SSE		
11	Privileged		Privileged Instruction
12	Others		DB

Fig. 1. Instruction group classification

abstraction method which extracts the valid instruction from the instruction sequence in order to detect the exploits.

3 Problem Definition: Network-Level Executable Code Detection

In the process to detect the malicious code on the network, the key element is to determine the presence of the executable code within a payload. The method to distinguish the program-like payload from non-code data and non-exploit code, that is, in order to decide the presence of the executable code, we assume that all of payload are executable. Because Intel instruction set (IA-32) is simple structure, its codes can't be distinguished from data bytes by using only the disassembly method. If the static disassembling is performed about all byte data, the complete assembly code can be generated, and if the emulation of the execution level is not performed, it is very difficult to determine that it is executable code. Because the Intel IA32 OP-Code uses the CISC format, all OP-code encodings can be expressed as 1byte (256 Encodings) size. Therefore all kinds of data, which are the ASCII format or the binary, can be translated into the assembly language with the IA-32 static disassembler. It means that it is very difficult to determine whether it is executable or not through the syntax examination of the disassembled code. The challenge of this paper is to look for the method for determining the executable part within network payload without the emulation of the execution level.

4 Basic Idea: Spectrum Analysis

In order to find the executable part within a payload, we propose the instruction spectrum analysis of which each instruction set is represented its color and then the whole sequence of instructions is analyzed by the translated color pattern.

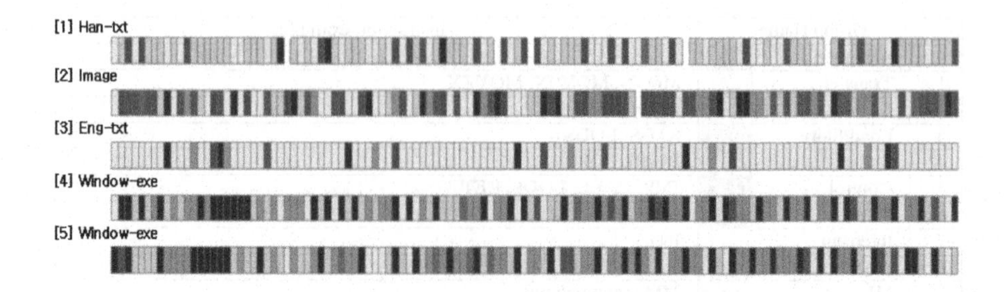

Fig. 2. Instruction spectrum for five different types of payload

We assume two things; (i) there is a tendency which is not so with the tendency to be consecutively repeated with the instruction set, and (ii) as the context between an instruction group and the other instruction group, there is any correlation pattern. If these relations are expressed with a color, a pattern toward the correlation of the executable instruction group can become visible and we can found the executable instruction spectrum. The instruction spectrum analysis method of this paper could be expressed by the form visually classifying the characteristic of the instruction pattern of executable codes. We classify the several hundreds of IA-32 OP-codes into 13 instruction groups as shown in figure 1, and every instruction group is represented by its corresponding color. In order to check the feasibility of basic idea, we disassembled the some kinds of files and we confirmed whether the correlation of the each instruction group is discernable visually.

Figure 2 shows an example of instruction spectrum for five different types of payload. In fig. 2, a slot indicates an instruction cycle and the color represents its corresponding instruction group. In the first case of Hangul text file (Han-txt), the slots of the orange-red color showing the transfer instruction group are mainly repeated. In the second case of image JPG file (image), there are many repetitions of the red and yellow slots showing arithmetic and logical calculations respectively. In the third case of text file (Eng-txt), there are many repetitions of a bunch of yellow slots. And in case of three non-executable files, there is the tendency that the instruction belonging to the float calculation or the others group periodically show up. However, in case of Windows execution PE files, the fourth and the fifth, we can see that the instruction sets of control, transfer, stack, logical, arithmetic, and etc are distributed in a rate, and we note that their color spectrums are visually different from ones of the previous three non-executable data files.

As shown in simple example, we can find some different points of instruction spectrum between executable code and non-executable code. So we can say that the basic idea of this paper could be feasible. However there are some challenging points how to generalize the visual differences in the instruction spectrum and how to automatically determine the executable region.

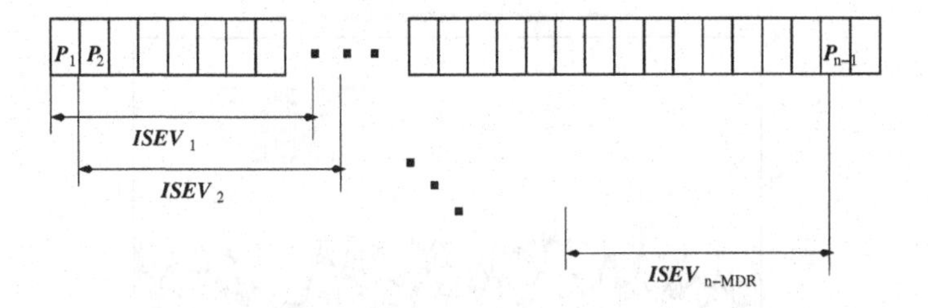

Fig. 3. The expectation value of instruction sequence

5 Generalization: Discrete-Parameter Markov Chain

In order to generalize the instruction spectrum analysis, we introduce the discrete parameter Markov chain. That is, we assume that every execution time of each instruction is same and we choose to observe the instruction group at a discrete set of times. Moreover, we further assume that the "future instructions" only depends on the "current instruction". Now we drive a matrix of the instruction transition probability (ITPX) as follows,

$$\mathbf{p^1} = \begin{pmatrix} p_{0,0} & \cdots & p_{0,12} \\ \vdots & \ddots & \vdots \\ p_{12,0} & \cdots & p_{12,12} \end{pmatrix}$$

where, $p_{i,j}, i, j = 0, \ldots, 12$ is the transition probability from instruction group i to instruction group j. For the determination about in which the instruction sequence of the executable code exists in the partial region of specific payload, we define the minimum decision range (MDR) and the instruction spectrum expectation value (ISEV) as follows,

- *Definition 1.* MDR : The number of minimum instructions required for the determination in which the instruction sequence of the executable code exists.
- *Definition 2.* ISEV: The summation value of the transition probability of each instruction within the MDR section.

$$ISEV_n = \sum_{k=n}^{n+MDR-1} p_k,$$

where n is from 1 to $(n - MDR)$ and $p_k$is the instruction transition probability. It determines that the executable code can exist in the corresponding area of the MDR size if ISEV exceeds the threshold (C). If $ISEV_n > C_{threshold}$, then n^{th} chunk is executable.

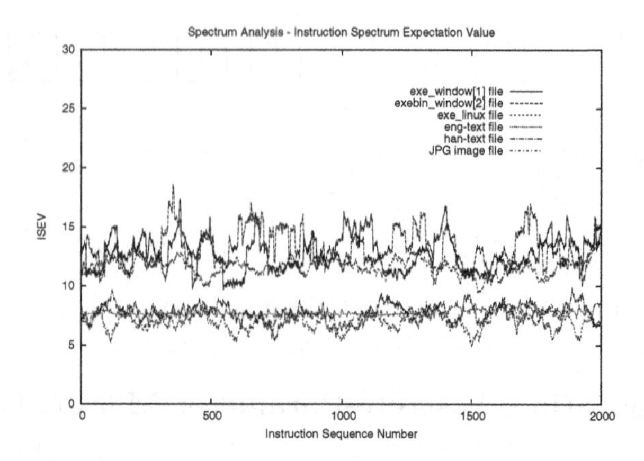

Fig. 4. Instruction spectrum expectation values for six sample files

6 Experimental Results

Our implementation architecture has two interfaces to handle the input data for the instruction spectrum analysis. The one is for learning the instruction transition probability of the executable code from various kinds of files, and the other is for deciding an executable part in the payload of the network packet from network interface. We use the libdasm [14] disassembler of the linear sweep type as a decoding tool for the input data of each interface. The learning matrix is made by the learner using the executable code which is collected by the execution file. The learning matrix maintains the value of the instruction transition probability going with each instruction sequence. The instruction spectrum analyzer calculates the ISEV according to a MDR with the sliding window mechanism on the network packet and determines whether its ISEV exceeds the threshold or not.

In order to get the reference ITPX of the instruction spectra of executable codes in network traffic, we used 400 Windows PE files which are executable format in Windows. Here we ran 2000 instructions in the execution code area (.txt) of each PE file and finally got the reference ITPX.

We measured the ISEVs for six sample files by using reference ITPX - (i) two execution files of the Windows, (ii) one Linux execution file, (iii) one general text file, (iv) one Hangul text file, and (v) one image file. As the MDR size of this basic test is 50, the sum of the ITP (Instruction Transition Probability) of consecutive 50 instructions was used as the ISEV of the corresponding instruction location. As shown in Figure 4, the values of ISEV of the Windows and Linux execution files fluctuate between 10 and 17 while the ones of non-executable files stay between 6 and 10. We believe that the obtained values of ISEV in Fig. 4 show the typical values for executable and non-executable codes. Moreover, from this figure we can estimate the threshold value that determines whether the current payload has an executable or not. From Fig. 4 the threshold value could be 10 when MDR equals to 50. We know that this threshold value gives an important factor on the false rate of the determination. We measured the ISEV values

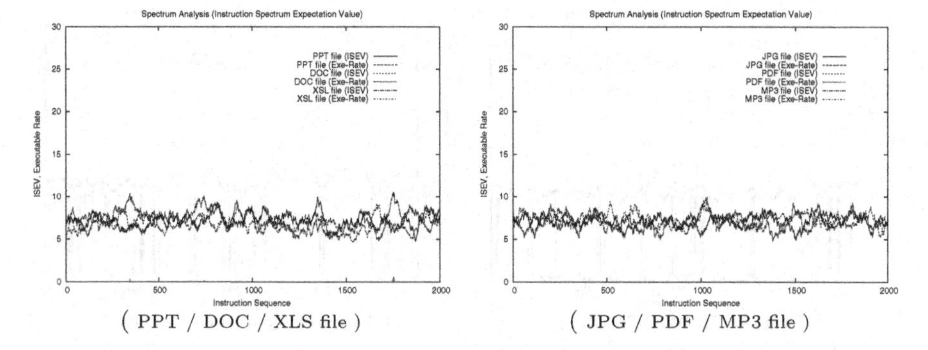

Fig. 5. ISEV and executable probability for non-executable files

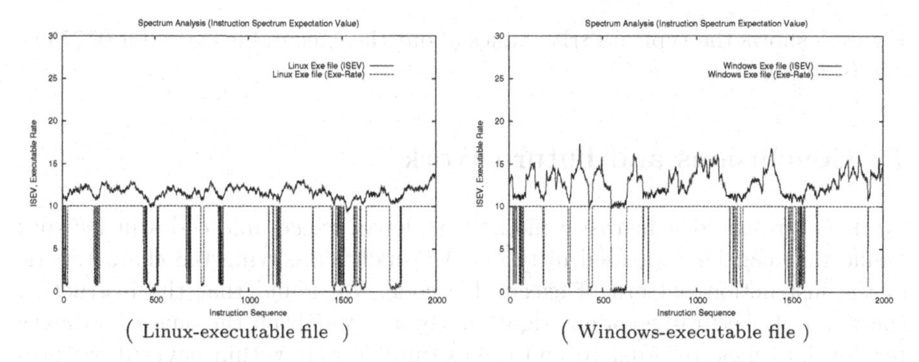

Fig. 6. ISEV and executable probability for executable files

for diverse non-executable files such as the Powerpoint, Word, Excel files of the Windows and JPG, PDF, and MP3 files when MDR=50 and the threshold value is 10. As shown in Figure 7, we note that the ISEV values below 10 as we can expect.

On the other hand, in the execution files of Linux and Windows, the region exceeding a threshold is considerable, and it shows that each executable probability of the Window execution file is higher than one of the Linux execution file. This is why the execution file of the Windows was used as learning data in the learning process of ITPX. This shows the significance of learning data used in IPTX learning process, so a proper learning data for each goal and environment need to be selected. The reason in which the executable probability intermittently decreases is that the data region like DB is used in the section of the specific instruction or there can be a reason including the division of a segment, and etc.

Recently, the prevalent types of malware are mainly the trojan-horse, worm, dropper types, and the polymorphism by the new execution compression tool. In this paper, we measured the ISEV of Dropper - two samples 0777.exe and 0420.exe - that installs the Trojan horse or worm. As we can expect,

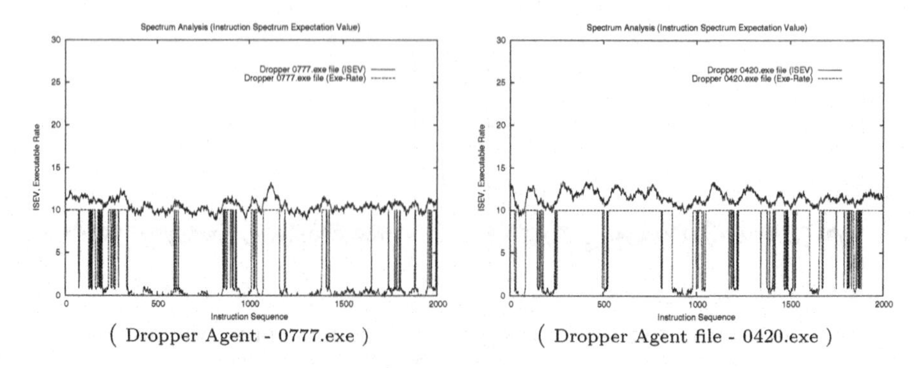

Fig. 7. ISEV and executable probability for Dropper malware

Figure 8 shows the typical ISEV values about the executable codes for 0777.exe and 0420.exe.

7 Conclusions and Future Work

We have presented a heuristic method to detect executable codes in network traffic that can be expressed in terms of visually classifying the characteristic of the instruction pattern of executable code. We found that the instruction spectrum of executable code is significantly different from the ones of ordinary text or data files. In order to find the executable part within payload, we proposed the instruction spectrum analysis that generalizes this by applying the discrete parameter Markov model. The method proposed in this paper can be used as the key function of the signature generation against the zero-day attack or the high speed anomaly detection. However, the basic experiments which are carried out earlier are just about the non-executable, executable files, and the sample of the malware. The various experiments are required about the adaptability of the network-level which is originally an object of this paper, and the deep consideration about the false rate of the executable code determination is necessary.

References

1. Kreibich, C., Crowcroft, J.: Honeycomb - Creating Intrusion Detection Signatures Using Honeypots. In: Proceedings of the Second Workshop on Hot Topics in Networks (Hotnets II), Boston (2003)
2. Newsome, J., Karp, B., Song, D.X.: Polygraph: Automatically generating signatures for polymorphic worms. In: IEEE Symposium on Security and Privacy, pp. 226–241. IEEE Computer Society, Los Alamitos (2005)
3. Singh, S., Estan, C., Varghese, G., Savage, S.: Automated worm fingerprinting. In: OSDI, pp. 45–60 (2004)

4. Singh, S., Estan, C., Varghese, G., Savage, S.: The EarlyBird system for realtime detection of unknown worms. Technical Report CS2003-0761, UC San Diego (2003)
5. Li, Z., Sanghi, M., Chen, Y., Kao, M.Y., Chavez, B.: Hamsa: Fast signature generation for zero-day polymorphicworms with provable attack resilience. In: S&P, pp. 32–47 (2006)
6. Royal, P., Halpin, M., Dagon, D., Edmonds, R., Lee, W.: Polyunpack: Automating the hidden-code extraction of unpack-executing malware. In: ACSAC 2006. Proceedings of the 22nd Annual Computer Security Applications Conference on Annual Computer Security Applications Conference, pp. 289–300. IEEE Computer Society, Washington, DC, USA (2006)
7. Christodorescu, M., Jha, S., Seshia, S.A., Song, D.X., Bryant, R.E.: Semantics-aware malware detection. In: IEEE Symposium on Security and Privacy, pp. 32–46. IEEE Computer Society Press, Los Alamitos (2005)
8. Kruegel, C., Robertson, W., Valeur, F., Vigna, G.: Static disassembly of obfuscated binaries. In: SSYM'04. Proceedings of the 13th conference on USENIX Security Symposium, Berkeley, CA, USA, p. 18. USENIX Association (2004)
9. Akritidis, P., Markatos, E.P., Polychronakis, M., Anagnostakis, K.G.: Stride: Polymorphic sled detection through instruction sequence analysis. In: SEC, pp. 375–392 (2005)
10. Chinchani, R., van den Berg, E.: A fast static analysis approach to detect exploit code inside network flows. In: Valdes, A., Zamboni, D. (eds.) RAID 2005. LNCS, vol. 3858, pp. 284–308. Springer, Heidelberg (2006)
11. Toth, T., Krügel, C.: Accurate buffer overflow detection via abstract payload execution. In: Wespi, A., Vigna, G., Deri, L. (eds.) RAID 2002. LNCS, vol. 2516, pp. 274–291. Springer, Heidelberg (2002)
12. Wang, X., Pan, C.-C., Liu, P., Zhu, S.: Sigfree: a signature-free buffer overflow attack blocker. In: USENIX-SS'06. Proceedings of the 15th conference on USENIX Security Symposium, Berkeley, CA, USA, p. -16. USENIX Association (2006)
13. Polychronakis, M., Anagnostakis, K.G., Markatos, E.P.: Network-level polymorphic shellcode detection using emulation. In: Büschkes, R., Laskov, P. (eds.) DIMVA 2006. LNCS, vol. 4064, pp. 54–73. Springer, Heidelberg (2006)
14. jt: Libdasm. http://www.klake.org/jt/misc/libdasm-1.4.tar.gz

A Visualized Internet Firewall Rule Validation System

Chi-Shih Chao

Department of Communications Engineering
Feng Chia University, Taiwan 40724, ROC
cschao@fcu.edu.tw

Abstract. For the security consistency, firewall rule editing, ordering, and distribution must be done very carefully on each of the cooperative firewalls, especially in a large-scale and multi-firewall-equipped network. Nevertheless, a network operator is prone to incorrectly configuring the firewalls because there are typically thousands or hundreds of filtering/admission rules (i.e., rules in the Access Control List file; or ACL for short) which should be setup in a firewall, not mention these rules among firewalls which affect mutually can make the matter worse. For this reason, our work is to build a visualized validation system for checking the security consistency between firewalls' rule configuration and the demands of network security policies. The system collects the filtering/admission rules (or ACL rules) from all of the firewalls (and routers if they are ACL-configured) in the managed network and then checks if these rules meet the demands of the global network security policies. The checked/analyzed results would later be visualized systematically by our system with different viewpoints for error debugging or anomaly removal. Currently, part of the firewalls' configuration of our campus network has being used to demonstrate our system's implementation.

Keywords: Firewall security consistency, Rule anomalies, Policy-based network security management, System visualization.

1 Introduction

A firewall is a combination of hardware and software that isolates an organization's internal network from the Internet at large, allowing some packets to pass and blocking others. It functions to avoid unauthorized or illegal sessions established to the hosts in the network areas it protects. Basically, several of firewalls can be deployed in the proper positions of the managed network for cooperative and integrated network security purpose [1]. However, in a large and complex network equipped with plenty of firewalls, it is very possible for a network manager to make mistakes while setting the firewall rules (i.e., ACL rules) since maintaining the security consistency between firewalls' rule configuration and the demands of network security policies is always time-consuming, laboring, and error-prone [2], [3], [4]. Sometimes, the matter can even go worse where there are several of managers assigned to do this job collectively.

The security inconsistency typically can be revealed by the occurrence of (1) anomalies between the firewall rules and (2) demand-mismatching of network

S. Ata and C.S. Hong (Eds.): APNOMS 2007, LNCS 4773, pp. 364–374, 2007.

security policies [5]. E. Al-Shaer and H. Hamed formally define an anomaly as a duplicate or multiple rule-matching for a packet in a rule set. Based on the concept, they further define 5 intra-firewall anomalies and 4 inter-firewall anomalies among the firewall rules [6-9]. However, a duplicate (or multiple) rule-matching for a packet may not cause misbehavior of the managed network. Yet, there is no mechanism offered in their work to check whether the setting of these rules can make the managed network behave as expected or not. Y. Bartal, et al. [10], [11] design a high-level global policy complier to reduce the burden and anomalies while setting the firewall rules. The compiler can transfer the global security policy specification written in its own high-level object-oriented MDL (Model Definition Language) into several individual but cooperative ACLs, and put them on the proper locations of the network. But, this compiling system lacks real deployment examples to demonstrate its performance, especially in large and complex networks with a number of firewalls.

In our work, an Internet firewall policy validation system is developed to help network managers ease the maintenance of the security consistency among firewalls. Our system can not only pin-point the anomalies among the firewall rules, but also figure out if the rule setting can satisfy or match the demands of the network security policies. In addition, a visualized tool to facilitate the job is also developed and integrated by systematically showing where and what inconsistency is. The rest of this paper is organized as following: Section 2 defines what security consistency for a managed network is and describes how the security inconsistency can be revealed. Section 3 details our three-tier visualization model built in a circle-based fashion which can effectively provide the information regarding security inconsistency and ease the labor-intensive work of the maintenance of security consistency. Our prototype system and its system implementation are displayed in Section 4 as a performance demonstration of our visualization approach, while Section 5 specifies our future works as an end of this paper.

2 Inconsistency Analysis

Internet firewalls typically use packet filtering to achieve the protection/prevention of the managed network from attacks and threats. Packet filtering operates by first parsing traffic datagram headers and then sequentially checking filtering rules against the datagram being inspected from an administrator-specified rule set (called an ACL) to determine whether to drop the datagram or let the datagram pass. The first rule matching the datagram determines the action taken [12]. In general, an ACL consists of a number of filtering rules in the form of (<order>, <protocol>, <source_IP>, <source_port>, <destination_IP>, <destination_port>, <action>) to do packet filtering on some specific interface of a firewall with direction indications (for outbound or inbound traffic filtering). Filtering decisions are typically based on (1) IP source or destination address, (2) TCP or UDP source and destination port, (3) ICMP message type, and (4) connection-initialization datagram using the TCP SYN or ACK bits.

For a computer network, the setting of firewall filtering rules can be said to be consistent if it can conform to the demands of global security policies defined by the manager(s), and vice versa (say security inconsistency). Our system in this work is developed to check if the consistent relationship remains; if not, the inconsistent rule setting in firewalls should be pointed out. According to our study in [5], the inconsistent rule setting of firewalls can be revealed via (1) anomalies between the firewall rules or (2) mismatching of the desired (and constraint) network behavior.

2.1 Rule Anomalies and Behavior Mismatching

For the anomalies between the firewall rules, they are defined completely by E. Al-Shaer et al and classified into two types: anomalies within one single ACL (or referred to as intra-ACL anomalies) and anomalies among different ACLs (or called inter-ACL anomalies) [6], [8]. Regarding anomalies within one single ACL, there are five different categories:

(1) S*hadowing anomaly*: Rule *x* is call being shadowed by Rule *y* because Rule *x* will become useless at all on packet filtering/matching while the position of Rule *y* is prior to Rule *y*.
(2) *Correlation anomaly*: Rule *x* and Rule *y* are defined as being abnormally correlated where 1) the values of their <Source_IP> and <Destination_IP> fields are mutually contained and 2) they have different actions to take.
(3) *Generalization anomaly*: Rule *x* with a lower filtering order/priority than Rule *y* in the ACL generally contains or covers Rule *y*, except the value of the action field.
(4) *Redundancy anomaly*: Rule *x* is fully redundant with the existence of Rule *y*.
(5) *Irrelevance anomaly*: Rule *z* doesn't function at all due to its irrelevant setting on source/destination addresses for the managed network.

As to anomalies among different ACLs, there are four different categories:

(1) *Shadowing anomaly*: It is defined as some specific traffic permitted to pass through the downstream filtering rule is prohibited by the upstream rule.
(2) *Spuriousness anomaly*: It is defined as some specific traffic prohibited to flow through the downstream filtering rule is permitted by the upstream rule.
(3) *Redundancy anomaly*: Rule *x* in ACL_i is completely redundant with the existence of Rule *y* in ACL_j, where $i \neq j$.
(4) *Correlation anomaly*: It is similar to the definition of the intra-ACL correlation anomaly, yet the two abnormally-correlated rules are positioned in different ACLs.

Anomalies between firewall rules (or ACL rules) might induce the security inconsistency. For example, if the managed network is supposed to disallow the WWW traffic flowing through (as one of the demands of global security policies). However, a network manager incautiously put the corresponding rule *x* below rule *y* which can completely shadow rule *x*. This mis-configuration will result in that Rule *x* can not do the designated job any more. For this reason, in our system, rule anomalies between firewall rules would be pointed out for security consistency.

Still, the security inconsistency can arise without the existence of any rule anomaly. For example, due to the erroneous port number configuration on an ACL rule, the P2P traffic originally be blocked can flow through the managed network, whereas the anti-virus code update traffic may be blocked. This mis-configuration of port number in ACL rules could hardly be found by the check/analysis of rule anomalies, but found by the behavior of the managed network which is beyond (or against) the desired demand(s) of global security policies. In our work, to explore the causality of behavior mismatching, two kinds of errors would be reviewed:

(1) *Incorrectly blocking error*: With the incorrect setting of rule contents, the traffic that should originally pass through is blocked. For example, to block the traffic of illegal P2P file-sharing applications, some of well-known ports could accordingly be set "blocked" by the firewall since the network managers have little knowledge of Internet services. Please notice that shadow anomalies could also result in the incorrectly blocking errors.
(2) *Incorrectly admitting error*: With the incorrect setting of rule contents, the traffic that should originally be blocked is allowed to pass through. For example, it is possible that some warned ports are forgotten to be shut down simply due to the oversight of managers. It can easily lead to hackers using some sophisticated tools, such as port-scanning or port-mapping, to see the vulnerabilities of the managed network.

As mentioned above, the security insistency of firewall rule setting could not only cause chaos, but also create security breaches of the managed network. Our work in this paper is to develop a system that can help managers point out each of problematic rules to keep the security consistency.

3 System Visualization

To facilitate the validation of the security consistency of firewall rule setting, visualization is employed as part of our system implementation. In our work, a visualization approach is developed to systematically present the validating results to the managers and direct them to retain the security consistency.

3.1 Three-Tiered Visualization Hierarchy

As shown in Fig. 1, a three-tiered visualization hierarchy is established for our systematic visual presentation of the validating results of security consistency. At the bottom tier of this hierarchy, the detailed system configuration of the managed network as well as the firewall rules (or ACLs) will be exposed for the sake of raw-data inspection, if needed. At the middle tier of the hierarchy is a logical firewall topology created for effectively visualizing the security inconsistency occurring within/between firewalls (ex., the inter-ACL anomalies between two different firewalls) as well as the topological relationships among them. For this, a circle-based logical firewall network is developed, rather than the typical network topology with some traditional shape, like tree, to achieve a scalable visualization of network topology.

At the middle tier in Fig. 1, our logical firewall topology evenly puts the nodes, standing for the firewalls (and ACL-configured routers) in the managed network, on the edge of a fit-sized circle. And then, by making use of this logical topology, we put the visualized validating results (i.e., rule anomalies and behavior-mismatching errors) on the circle to effectively present the allocation of security inconsistency.

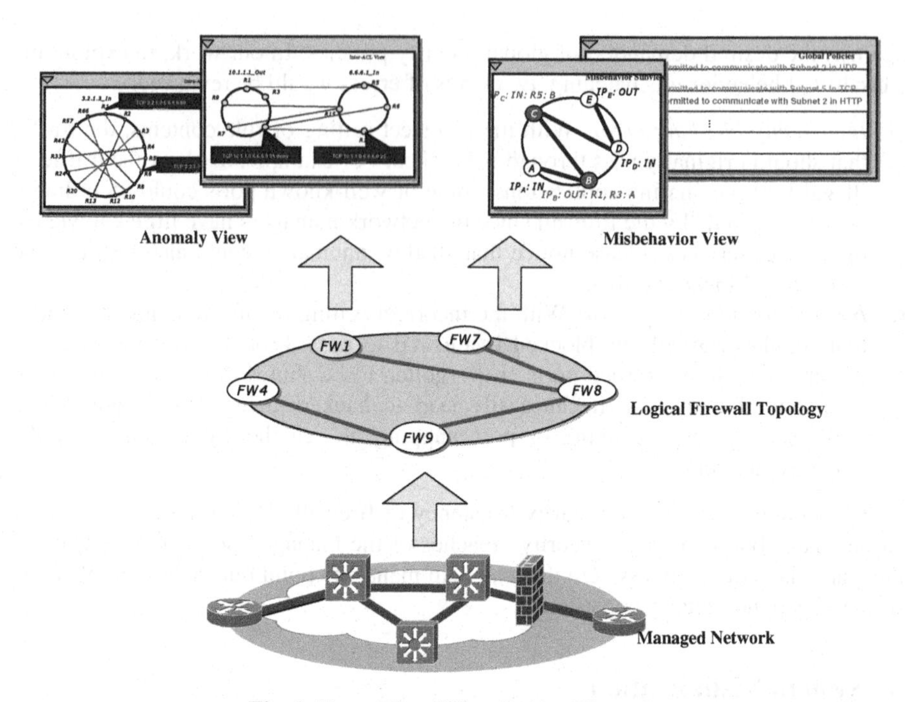

Fig. 1. Three-Tiered Visualization Hierarchy

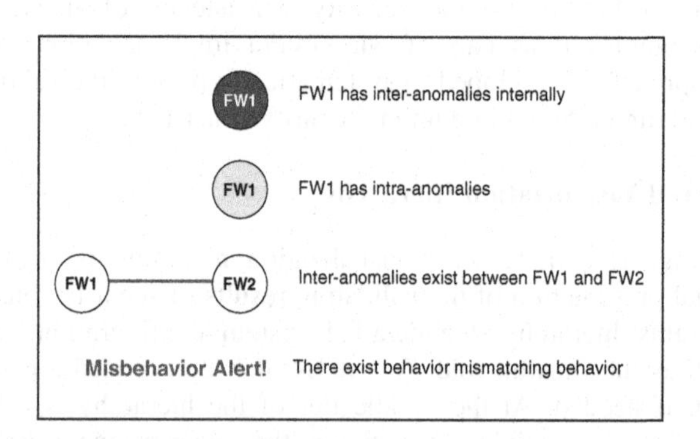

Fig. 2. Visual Presentation in the Logical Firewall Topology

To effectively present the validating information of security inconsistency on the logical firewall topology, a set of visualization rules is specified as shown in Fig. 2:

(1) If there are intra-ACL anomalies within a firewall, the color of the iconic node representing the problematic firewall will become yellow whereas a node with white color means the node is in the 'normal' status. Also, if there are inter-ACL anomalies within a firewall (namely, there are more than one ACL in the designated firewall), then the node color of the firewall will become red.

(2) If inter-ACL anomalies exist between two of the iconic firewall nodes on the circle edge, there would be a red link which interconnects them where the color of these two nodes are white.
(3) If there are behavior-mismatching errors, the alert with red bold letters "Misbehavior Alert" will flash at the corner of the window canvas.

According to the visualization information presented by the logical firewall topology, network managers can recognize what problems they have for the security inconsistency of firewall rule configuration. At that time, they can further utilize two offered subviews (Anomaly View and Misbehavior View: will be described at Sections 3.2 and 3.3, respectively) which are located at the top tier of the three-tiered visualization hierarchy, for more detailed look-at about the validating results; in the Anomaly View, intra- and inter-ACL anomalies are displayed, and in the Misbehavior View, two different behavior-mismatching errors along with the involved ACLs as well as the global policies are indicated.

3.2 Anomaly View

As described in Section 2.1, rule anomalies consist of two different types: intra- and inter-ACL anomalies. To systematically visualize each of them, two separate widows for the Anomaly View which employ the similar circle-based visualization approach are provided (Fig. 3). In the Intra-ACL Anomaly Subview window, all the problematic filtering rules within the ACL which is configured at the firewall (or router) selected from the logical firewall topology are put evenly on the edge of a

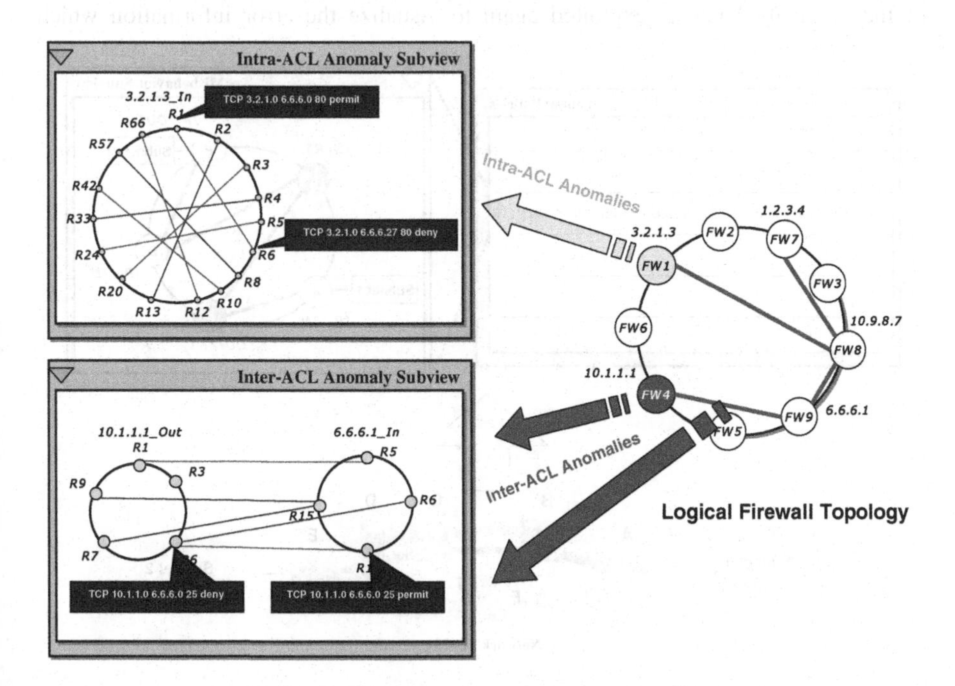

Fig. 3. Visual Presentation for Rule Anomalies

circle. If an anomaly exists between two rules of the ACL, there would be a line interconnecting the corresponding iconic nodes of these two rules where the color of the line represents what category of anomalies between them is. For example, as shown by the Intra-ACL Anomaly Subview window in Fig. 3, rules R1 and R6 in the ACL which is configured on the interface with IP address 3.2.1.3 of firewall FW1 for filtering inbound traffic have a red line interconnecting them. It means there is an Intra-ACL shadowing anomaly between these two rules.

For the inter-ACL anomalies, the Inter-ACL Anomaly Subview window employs two separate circles to represent two problematic ACLs, respectively. If an inter-ACL anomaly exists, there would be a line which runs across the two circles and interconnects the iconic nodes of the two rules causing the anomaly where the color of the line depicts the category of the anomaly. For example, as shown by the window of Inter-ACL Anomaly Subview in Fig. 3, rules R6 in the ACL configured on the interface with IP address 10.1.1.1 of firewall FW4 for filtering outbound traffic and R1 in another ACL configured on the interface with IP address 6.6.6.1 of firewall FW9 for filtering inbound traffic have a red line connecting them. This means there exists an inter-ACL shadowing anomaly between them. Please notice that if the IP addresses shown above the two circles are the same, it means there is an inter-ACL anomalies occurring between two rules which are in the different ACLs, but on the same firewall.

3.3 Misbehavior View

As to the Misbehavior View, a circle-based visualization approach analogous to that of the Anomaly View is exploited again to visualize the error information which

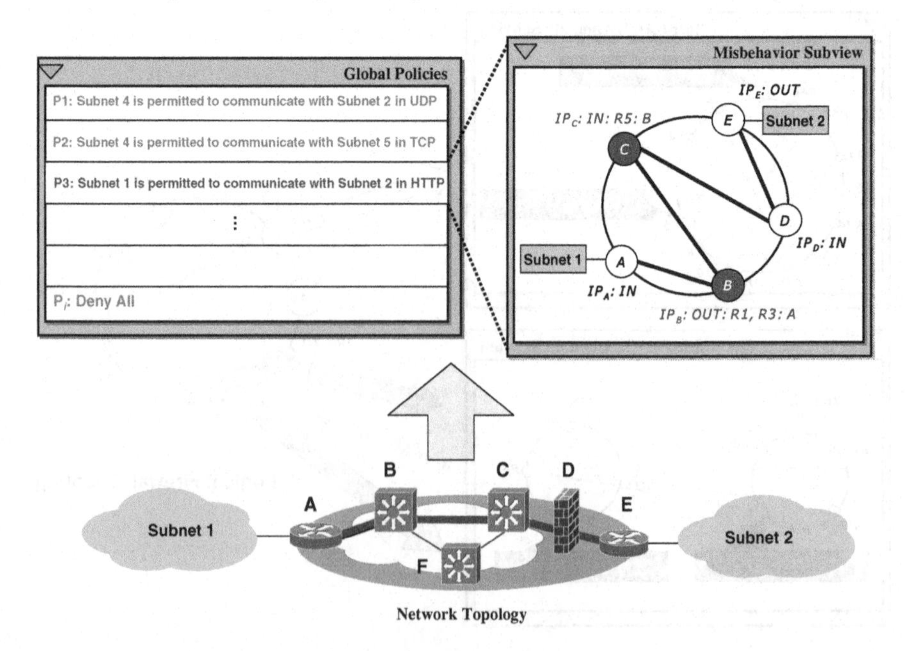

Fig. 4. Visual Presentation for Behavior-Mismatching Errors

comes from the inconsistency between ACL rules and global security policies. If there is a behavior-mismatching error in the managed network, in our work, the involved global policy would be pointed out first in red letters within the Global Policies window, as the global security policy P3 shown in Fig. 4. After clicking on the textual line of "P3: Subnet 1..." in the Global Policies window, the Misbehavior Subview window will pop up to show all the routers/firewalls on the route from subnet 1 to subnet 2 (that is, routers A, B, C, D, and E), along with all of the ACL rules' number involved in the security inconsistency with P3.

Following the same example of security inconsistency in Fig. 4, the Misbehavior Subview window evenly puts all the firewalls/routers along the path from subnet 1 to subnet 2 on the edge of a circle with their corresponding iconic nodes, where solid straight lines interconnecting these iconic nodes stand for the logical connectivities among them. One node in the window labeled IP_i: IN/OUT represents the interface with IP address IP_i of that node on which an ACL is configured for filtering the inbound/outbound traffic flowing from subnet 1 to subnet 2. As to the color of the iconic nodes, red/white represents the nodes with/without misbehavior rule configuration in the corresponding firewalls or routers. For example, in Fig. 4, IP_A: IN labeled beside node A means the ACL which is configured on the interface of router A with IP address IP_A for inbound traffic filtering is error-free with respect to the security inconsistency example of Fig. 4. However, rules R1 and R3 in the ACL which is configured on the interface with IP address IP_B of router B for filtering outbound traffic have the incorrectly admitting error which is inconsistent with the global policy P3. For this, IP_B: OUT: R1, R3: A would be labeled along with node B in red color where letter A at the end field of this label stands for incorrectly *a*dmitting error. Likewise, IP_c: IN: R5: B denotes rule R5 in the ACL which is configured on the interface with IP address IP_c of router C for the inbound traffic filtering has the incorrectly *b*locking error, which is also inconsistent with the global policy P3. With the assistance of our subviews' windows mentioned in this section, the security inconsistency while configuring the firewall rules can be revealed easily.

4 System Implementation and Future Work

The system prototype has been completed as a demonstration of our visualization approach. The ACL configurations of some of the routers in our campus network are offered and utilized as the input of our system implementation. Figures 5~10 show the system execution results for these rule configurations. A network manager can first click "Physical Network Topology" on the menu bar of the main window (Fig. 5) to obtain the physical connectivities of routers (and firewalls, if exist) in the managed network. And then, he can click "Start Validation" to display the Logical Firewall Topology with the entire validation results. As shown in Fig. 6, there are five Inter-ACL rule anomaly events (one in Router G with a red circle and the other four are indicated by four red lines) and one intra-ACL rule anomaly event in Router H (the yellow circle).

On the basis of the information provided by the Logical Firewall Topology in the main window, the manager can click on the yellow circle of Router H to have a detailed look-at about the Intra-ACL rule anomalies occurring within Router H (shown in Fig. 7). Also, he can click on these red lines as well as the red circle of Router G on the Logical Firewall Topology to further investigate the Inter-ACL anomalies (Fig. 8), where the line colors no matter in Intra- or Inter-ACL Anomalies Subview window stand for the categories of rule anomalies. In addition, in Fig. 6, a textual line "There is an Incorrectly Admitting Error!!!" flashing at the window's bottom. The manager can follow the warning and click on the textual line to launch the Global Policies window for checking which global policy is involved in the warned error. According to those shown in Fig. 9, the manager can recognize global policy P3 has been involved in the error. Thus he can click on the line of P3 in the Global Policies window to launch the Misbehavior Subview window (Fig. 10) and find which ACL rules, as well as their locations (indicated by red circles), on the routing path with respect to P3 (through Routers F, E, H, D, and B, sequentially) are involved in the error. In this manner, a network manager can continue the checking procedure just described in this subsection to make the proper corrections on the rule configurations in the managed network until the security inconsistency (i.e., behavior mismatching errors and rule anomalies) does not exist.

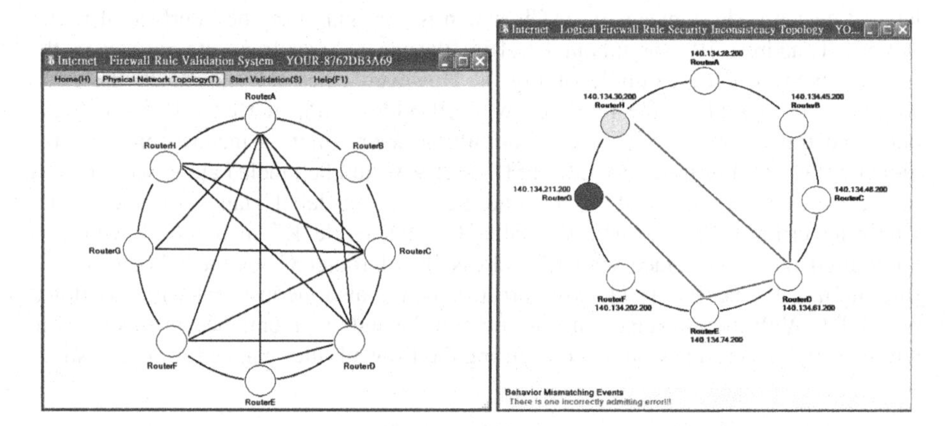

Fig. 5. Physical Network Topology

Fig. 6. Logical Firewall Topology

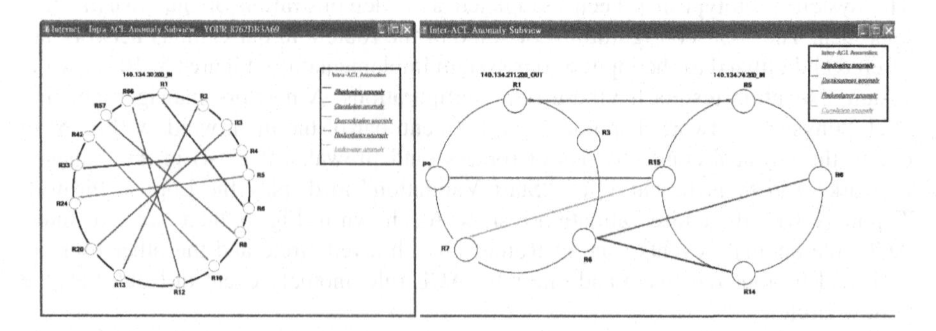

Fig. 7. Intra-ACL Rule Anomalies

Fig. 8. Inter-ACL Rule Anomalies

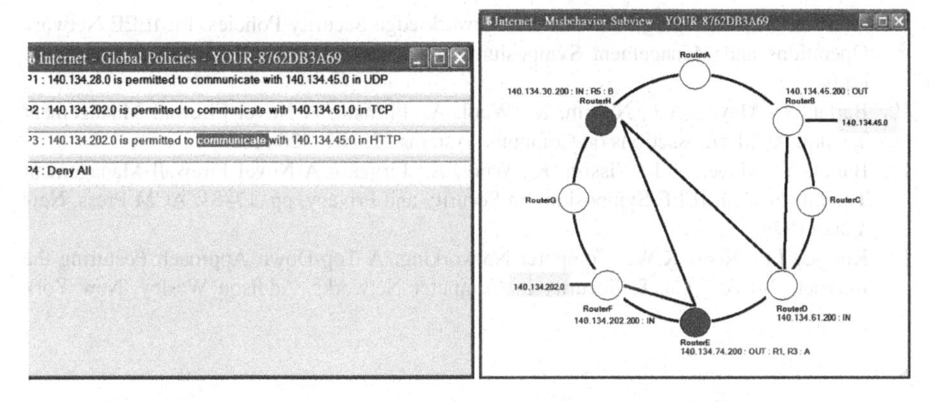

Fig. 9. Global Policies Window

Fig. 10. Behavior Mismatching Error

Right now the first version of our prototype system has been finished and being tried to be applied to our campus network. Still, a number of works are in our hands for this project in the coming future, especially for system scalability and expansibility. Below lists two of them as reference: 1) Develop a faster computing method for the check of the behavior mismatching errors, and 2) provide users effective and practical suggestions of how to correct the behavior mismatching errors.

References

1. Ioannidis, S., Keromytis, A., Bellovin, S., Smith, J.M.: Implementing a Distributed Firewall. In: ACM Computer and Communication Security, pp. 190–199. ACM Press, New York (2000)
2. Eppstein, D., Muthukrishnan, S.: Internet Packet Filter Management and Rectangle Geometry. In: 12th Annual ACM-SIAM Symposium on Discrete Algorithms, pp. 827–835. ACM Press, New York (2001)
3. Hari, B., Suri, S., Parulkar, G.: Detecting and Resolving Packet Filter Conflicts. In: IEEE INFOCOM, vol. 3, pp. 1203–1212. IEEE Press, New York (2000)
4. Guttman, J.D.: Filtering Postures – Local Enforcement for Global Policies. In: IEEE Symposium on Security and Privacy, pp. 120–129. IEEE Press, New York (1997)
5. Chen, I.D.: An Internet Firewall Policy Validation System. Feng Chia University, Taiwan (2006)
6. Al-Shaer, E., Hamed, H.: Discovery of Policy Anomalies in Distributed Firewalls. In: 23rd Annual Joint Conference of the IEEE Computer and Communications Societies, vol. 4, pp. 2605–2616. IEEE Press, New York (2004)
7. Al-Shaer, E., Hamed, H.: Firewall Policy Advisor for Anomaly Discovery and Rule Editing. In: 8th International Symposium on Integrated Network Management, pp. 17–30. IEEE Press, New York (2003)
8. Al-Shaer, E., Hamed, H., Boutaba, R., Hasan, M.: Conflict Classification and Analysis of Distributed Firewall Policies. IEEE J. on Selected Areas in Communications 23(10), 2069–2084 (2005)

9. Al-Shaer, E.: Managing Firewall and Network-edge Security Policies. In: IEEE Network Operations and Management Symposium, vol. 1, pp. 926–932. IEEE Press, New York (2004)
10. Bartal, Y., Mayer, A.J., Nissim, K., Wool, A.: Firmato: A Novel Firewall Management Toolkit. ACM Transactions on Computer Systems 22, 381–420 (2004)
11. Bartal, Y., Mayer, A.J., Nissim, K., Wool, A.: Firmato: A Novel Firewall Management Toolkit. In: 20th IEEE Symposium on Security and Privacy, pp. 17–31. ACM Press, New York (1999)
12. Kurose, J.F., Ross, K.W.: Computer Networking: A Top-Down Approach Featuring the Internet, 3rd edn. Ch. 8, Security in Computer Networks. Addison Wesley, New York (2005)

A Secure Web Services Providing Framework Based on Lock-Keeper

Feng Cheng, Michael Menzel, and Christoph Meinel

Hasso-Plattner-Institute, University of Potsdam,
P.O.Box 900460, 14440, Potsdam, Germany
{feng.cheng,michael.menzel,christoph.meinel}@hpi.uni-potsdam.de

Abstract. A general model for securing widely deployed Web Services has been recommended in which the security of Web Services is divided into three layers: network security, host security and the security of Web Service message, also called SOAP message security. According to principles of this model, we propose a new secure Web Services Providing Framework based on the Lock-Keeper technology, which is a high level security solution implementing the basic security concept, "Physical Separation". In the proposed framework, the internal Web Services provider and its network are protected well by being physically isolated with the external world. At the same time, trusted Web Service message based communications can be performed smoothly and securely with the guard of a "SOAP Verification Module", which is integrated in the Lock-Keeper system. The SOAP Verification Module realizes general functionalities of both "Trust Management" and "Threat Prevention" that have been specified by most common WS-Security standards. Experiments demonstrated in this paper show that our proposed framework, which can simultaneously guarantee all the three layers of Web Services security, is feasible, applicable and secure.

1 Introduction

Web Services (WS) technology that effectively facilitates the integration of web applications as well as offering interoperable machine-to-machine interactions, has been widely deployed in modern businesses [1]. Using standardized Web Service message (i.e. XML-based SOAP message) communication to build rich federated environments that enable complex business-to-business scenarios and allow organizations to expose powerful line of business applications is tremendously exciting [2]. However, to provide Web Services for customers or partners, enterprises have to open their internal networks and expose their private resources in loosely-coupled public network infrastructures, such as the untrusted Internet, which causes more and more possibilities for the internal network and its hosts, especially the sensitive WS provider, to be exploited [3]. To simultaneously guarantee the security of the internal network, the WS provider and the WS communications, has been a main and serious problem to extend the utilization of WS technology [1], [4].

S. Ata and C.S. Hong (Eds.): APNOMS 2007, LNCS 4773, pp. 375–384, 2007.

In recent years, many new security models, solutions and even standards, have been proposed to solve the addressed problem. Traditional perimeter security mechanisms, such as packets-filtering firewalls [5], are integrated into the common Web Services providing model to enhance the security of Web Services network. However, such network-level security measures make no sense for preventing attacks, which are targeted on Web Services message itself, since the application-level content of passing Web Services traffic can not be thoroughly examined. Malicious or illegitimated users are able to easily intrude the normal Web Services conversations and abuse the provided Web Services and other internal resources. On the other hand, just solely securing Web Service messages, e.g., applying Web Services related security standards or specifications such as WS-Security, WS-Policy and WS-Trust, etc [6], to achieve end-to-end security, is not sufficient too. The permitted network channel, i.e. opened ports and supported transport or message layer protocols such as HTTP and SMTP, required by regular Web Services interactions, also offers a convenient way for hackers to successfully penetrate the network and fully occupy the internal hosts, including the Web Services provider. Therefore, a general security module has been proposed for piloting the design of holistic Web Services security approaches in [7]. According to this model, the Security of Web Services is divided into three layers: network security, host security and the security of Web Service message, also called SOAP message security.

Motivated by implementing this Web Security model, we propose a Lock-Keeper based secure Web Service Providing Framework in this paper. Based on the simple security concept of "Physical Separation" that "the ultimate method to secure a network is to disconnect it", the Lock-Keeper technology has been convinced to be an efficient approach to guarantee high-level network security and prevent online attacks by physically separating the protected hosts or network from the external world [8]. The benefit of Lock-Keeper on providing "Physical Separation" is exactly identical with requirements of the first two layers of the above mentioned general Web Security model, i.e. securing network and securing host. In addition, the Lock-Keeper architecture that consists of three independent normal PCs, named INNER, OUTER and GATE respectively, is flexible to be seamlessly integrated with other security methods, e.g. Third-Party anti-virus programs, etc, to prevent offline attacks on the application level. Most existing Web Services security standards can be easily realized in the Lock-Keeper system so that the third layer of that WS security model, i.e. the SOAP message security, can be achieved as well. Based on this analysis, we believe that Lock-Keeper is a suitable alternative for implementing enhanced Web Services security solutions.

From another point of view, to reach the high-level network security, normal network usage has to be restricted. For this reason, almost all the existing "Physical Separation" implementations, including our Lock-Keeper technology, have prohibited most of network applications that are supported by real time application level protocols so that just a limited number of specific applications, e.g., email exchange, file exchange and database replication, etc, can be proffered. Fortunately, the XML based SOAP communication is perfectly propitious to the working principle of Lock-Keeper. It provides us a new field of vision for implementing Lock-Keeper applications. Therefor, we expect that the Lock-Keeper based Web Services Framework would be a promising Lock-Keeper application. Based on this framework, many other Lock-Keeper application scenarios can be revealed in practices which can significantly improve the usability of the "Physical Separation" technology.

The proposed framework mainly includes three components: *Web Services Consumer*, *Web Services Provider* and *Lock-Keeper Web Services Application Module* that broadly refers to "Web Services Proxy" and "Web Service Routing Module" on OUTER, "SOAP Verification Module" on GATE and "Web Service Invocation Module" on INNER. *Web Services Consumer* and the *Web Services Provider* play the same roles as the normal "Consumer" and "Provider" in the original Web Services model [1]. That means, the message processing operated on the Lock-Keeper is transparent to both the Web Services client and the Web Services server. "Web Services Proxy", "Web Service Routing Module" and "Web Service Invocation Module" are responsible for the Web Services parsing, forwarding, routing and invoking. On GATE, the SOAP message will be examined by the "SOAP Verification Module", which realizes functionalities of both "Trust Management" and "Threat Prevention" specified by common WS-Security standards. Experiments on an implemented prototype of this framework show that the Web Services interactions work well over Lock-Keeper and the security of Web Services can be practically enhanced by combining the software-based WS-Security technologies and the hardware-based Lock-Keeper technology.

The remainder of this paper is organized as follows: next section introduces the general security module for Web Services. Section 3 gives a short review of the Lock-Keeper technology. The new Lock-Keeper based secure Web Services Providing Framework is presented in section 4. Some experiments are illustrated in section 5. Section 6 concludes the paper with a summary of major contributions and possible future works.

2 Security Model for Web Services

A general security model of Web Services has been proposed in one of the Microsoft Web Appliction guides, named *Improving Web Applications Security: Threats and Countermeasures*[7]. It was originally called as *"a holistic approach to security of general applications"*. Here, we just specify the term "application" to "Web Services". The figure is reformulated in Figure 1. According to requirements of this model, all efforts to secure Web Services should be embedded in an environment considering three layers: *Network Security*, *Host Security* and *Web Service Message Security*. To prevent potential vulnerabilities of each layer, different security measures have to be applied to different layers accordingly.

1. **Securing Your Network:** As stated before, a secure Web Service relies upon a secure network infrastructure. The role of securing network is not

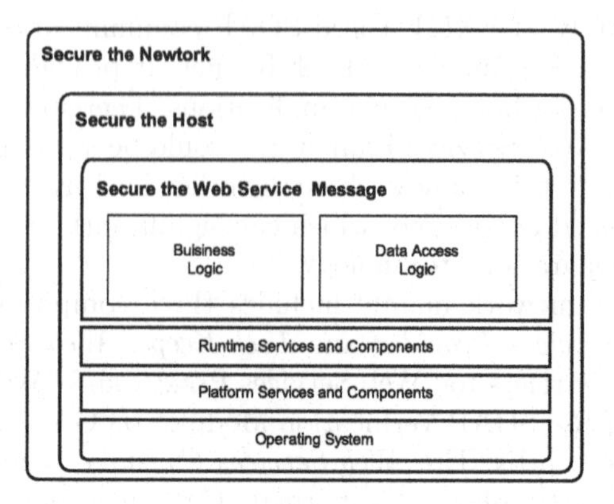

Fig. 1. Web Services Security Layers

only to protect the network itself from TCP/IP-based attacks, but also to implement countermeasures such as secure administrative interfaces and strong passwords. The secure network is also responsible for ensuring the integrity of the traffic that it is forwarded [5]. If it is known at the network layer about ports or protocols that may be harmful, those potential threats must be countered at this layer. The mostly used network security measures include *Router*, *Switch*, *Gateway*, *Firewall*, *IDS/IPS*, etc.

2. **Securing Your Host:** The "Host" indicated here is not just the internal Web Services server (also called provider), it also refers to other application server, database server, or normal workstation in the internal network. Any vulnerabilities or security holes on any single host can be utilized by hackers to exploit other hosts or even the whole network. The popular operations to enhance the host security comprise *regularly Performing Patches and Updates*, *disabling unused User Accounts, Services, Protocols and Ports*, *Security Auditing and Logging*, *carefully setup of File Shares*, *strictly Controlling of Files and Directories*, etc. To a certain extent, the security of a host depends mostly on the reliability of its underlying Operation System (OS) and the effectiveness of locally deployed security tools such as personal firewalls [5].
3. **Securing Web Service Message:** For securing Web Service message, traditional network centric security approaches are not sufficient since most vulnerabilities and threat for Web Service message take place at the application layer, such as *Web Service Interface Probing*, *Recursive, Oversized or Malicious XML Document*, *SOAP Flooding Attack*, *Interaction Replay Attack*, *"Man-in-the-middle" Attacks*, etc [3]. To prevent these attacks, a large number of security related Web Service standards and specifications have been proposed and implemented (see [2], [3], [4], [6], [9], and [10]). The functionalities of these approaches are focused on *Trust Management* and *Threat Protection*. *Trust Management* ensures that a consumer can trustedly call a Web Service by confirming both authentication and authorization of the

sender. WS-Security tokens, e.g., X.509 certificates [11], User name token, XML-encoded Kerberos tickets [12], etc, are usually used to realize the *Trust Management. Threat Protection* addresses threats associated with the provided Web Methods and their content, since attacks can be performed by abusing legitimated requests.

3 Review of the Lock-Keeper Technology

As mentioned before, Lock-Keeper is an implementation of the "Physical Separation" concept which aims at finding a way to transmit data between two different networks without having to establish a direct physical connection. Besides Lock-Keeper, there are also many other tools emerged to implement the idea of "Physical Separation", such as Microsoft e-GAP-based Intelligent Application Gateway (IAG) [13], Nump technology proposed by U.S. Naval Research Laboratory [14], etc. The commercial version of the Lock-Keeper technology has already been developed and is now under the marketing extension by our industrial partner, Siemens [15]. Here, we just use a SingleGate Lock-Keeper system as an example to briefly explain what the Lock-Keeper is and how it works. More detailed information on Lock-Keeper and other "Physical Separation" related works can be available from [8] and [15]. As shown in Figure 2,

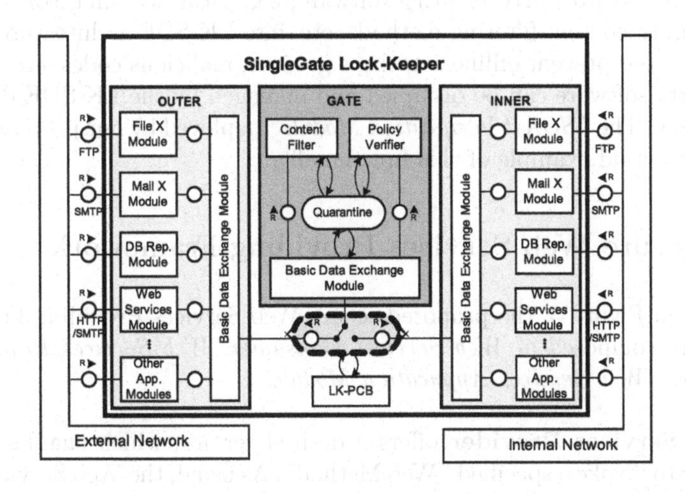

Fig. 2. Conceptual Architecture of the SingleGate Lock-Keeper

a SingleGate Lock-Keeper system consists of three independent Single Board Computers (SBCs): INNER, OUTER and GATE, which are connected using a patented switch unit. The switch unit restricts the connection so that GATE can just be connected with only one partner, either INNER or OUTER and there are no ways to directly establish the connection between INNER and OUTER at any time. The switch mechanism is realized by a hardware-based Lock-Keeper

Printed Circuit Board (LK-PCB) which is the core component of the Lock-Keeper system. There is not any software, even assembler programs, running on the LK-PCB. It works automatically when the system starts.

Besides these hardware components, Lock-Keeper Secure Data Exchange (LK-SDE) software is required to facilitate the data exchange between the separated networks. As important components of the LK-SDE software, several application modules are deployed on INNER and OUTER on both sides to work as interfaces for outside users and provide popular network services. Besides the Lock-Keeper Web Services Module proposed in this paper, there are other three implemented application modules, i.e. File Exchange (File-X) Module, Mail Exchange (Mail-X) Module, and Database Replication (DB-Rep.). Standard communication protocols, such as SMTP and FTP, are stopped and analyzed by respective application modules. Then, several unified file-based Lock-Keeper Message Containers (LKMC) can be created to carry the data content of services. These LKMCs will be transferred by "Basic Data Exchange Module" that is another important component of LK-SDE software. The secure Lock-Keeper "Basic Data Exhange Module" is responsible to manage the data exchange inside the Lock-Keeper and works based on the "Pull-and-Push" mechanism [8], which means all the data transfer operations are initially started by GATE. It is one of important criteria for implementing "Physical Separation".

In addition, because GATE is also a normal computer, it is possible to integrate some Third-Party security software, e.g. virus scanning software, mail analysis tools, content filtering methods, etc, into LK-SDE architecture to check data traffic and prevent offline attacks, e.g. virus, malicious codes, etc. All these Third-Party software can be occupied and managed by the LK-SDE data scanning module. The *"SOAP Verification Module"* deployed by our proposed framework is an actual example of this functionality.

4 A Secure Web Services Providing Framework

As shown in Figure 3, the proposed secure Web Services Providing Framework is basically composed of *Web Services Consumer*, *Web Services Provider* and *Lock-Keeper Web Services Application Module.*

1. **Web Services Provider** offers a desired service, which enables the consumer to invoke a specified "Web Method". As usual, the Web Services Server receives the request, triggers the Web Method and creates the response. In our framework, the Web Services provider, supposed to be a sensitive internal host, has been physically isolated by Lock-Keeper.
2. **Web Services Consumer** acts as a legally registered Web Services user who has necessary security credentials required by the security operations performed on GATE (i.e. by "SOAP Verification Module"). Normally, it is realized by a specially designed or standard client program, such as a web browser (for SOAP over HTTP/HTTPS), or an email client (for SOAP over SMTP), etc, by which the consumer can create, edit, and send out the Web

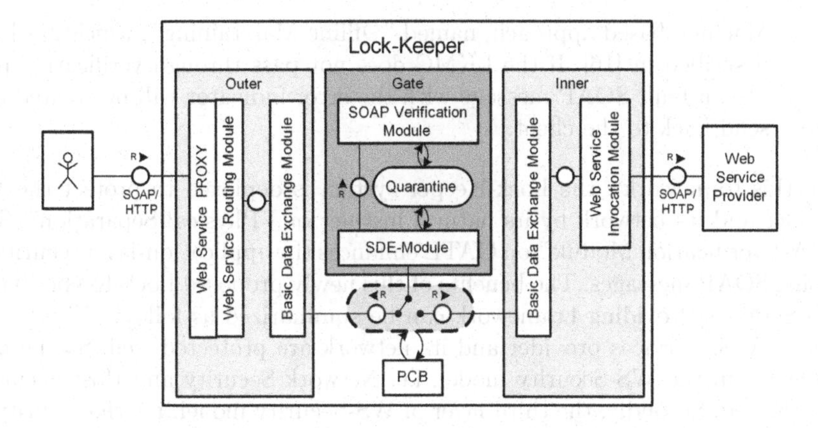

Fig. 3. The Secure Web Service Model using Lock-Keeper

Service requests and later receive responses. On the client side, consumer should also be able to create and manage different security policies by adding registered certificates, encrypting the Web Service request or executing any other security approaches.

3. **Lock-Keeper Web Services Module** is used to facilitate the Web Services communication on the special "disconnected" connection provided by Lock-Keeper as well as performing SOAP Verification for applying WS-Security standards on GATE as stated before. It consists of "Web Services Proxy" and "Web Service Routing Module" on OUTER, "SOAP Verification Module" on GATE and "Web Service Invocation Module" on INNER.
   (a) *Web Services Proxy* exposes the "Web Method", hosted by the internal provider, to the external network. Similar to other Lock-Keeper application modules, the incoming SOAP requests over HTTP or SMTP are accepted by the "Web Services Proxy" to then translated into the LKMC. When a new message is received, a unique identifier is generated and assigned into the message header. The identifier facilitates the mapping between requests and responses since the response will also be verified using the same mechanism on its way back to prevent insider attacks.
   (b) *Web Service Routing Module* on OUTER handles the communication between "Web Services Proxy" and "Basic Data Exchange Module".
   (c) *Web Service Invocation Module* on INNER invokes the desired method based on information saved in header of the received LKMC message. It is also responsible for forwarding the response back after resolving the unique WS-LKMC identifier.
   (d) *'SOAP verification module'* on GATE enforces the concrete "Trust Management" and the "Threat Prevention". To perform security actions and make decisions, a "Policy Store" with all accepted or legitimated policies or policy combinations and a "Certificate Store" to keep certificates or private keys are required to be set up on GATE in advance. The late updating or configuration changing can be realized by a new Virtual

Machine based approach, named "Offline Maintaining", which has been described in [16]. If the LKMC does not pass through verification process, a fault SOAP message with the error indicator will be created and send back to the client.

In this framework , the Lock-Keeper system is deployed to protect the WS provider and its network by its natural instinct on "Physical Separation". The "SOAP Verification Module" on GATE enhances the application-layer security of passing SOAP messages. The benefits of this newly proposed Lock-Keeper based Web Services Providing Framework can be summarized as follows. Firstly, the internal Web Services provider and its network are protected well. So the first two layers in the WS-Security model, i.e. Network Security and Host Security, are achieved. Secondly, the third layer of WS-Security model, i.e. the security of WS messages, can be realized by the "SOAP Verification Module" on GATE. Moreover, the proposed framework provides a completely new Lock-Keeper application Module, which significantly improves the usability of the Lock-Keeper.

5 Experiment Results

To verify the feasibility, applicability and security of the proposed Lock-Keeper based secure Web Services Providing Framework, we build a testing environment according to the requirements specified in Section 4. Through the Test Client, *Web Services Consumer* can create, edit, sign, encrypt the SOAP request and check the response. The required policy can be imported or composed as well. As shown in Figure 4, a policy named "secure" is composed by checking accessible security methods, such as the encryption named "gate" and a signature named "client", both realized by X.509 certificates. Besides, there is another policy named "unsecure" without checking any security options. We set up a *"Web*

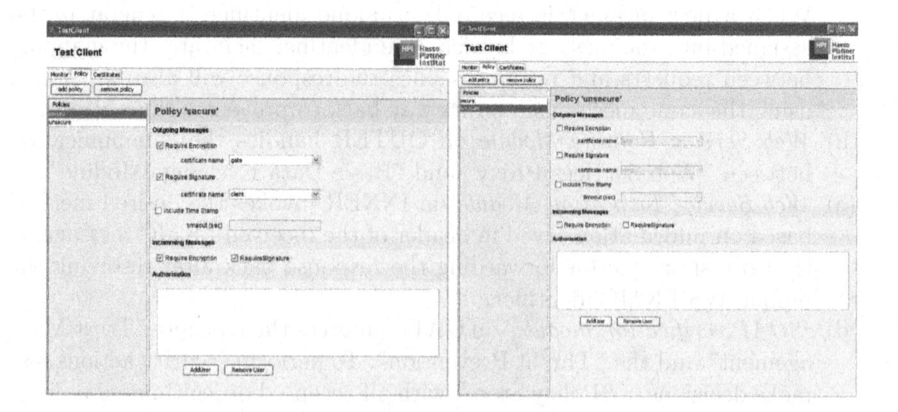

Fig. 4. Policy "secure" and Policy "unsecure" created by the WS Client

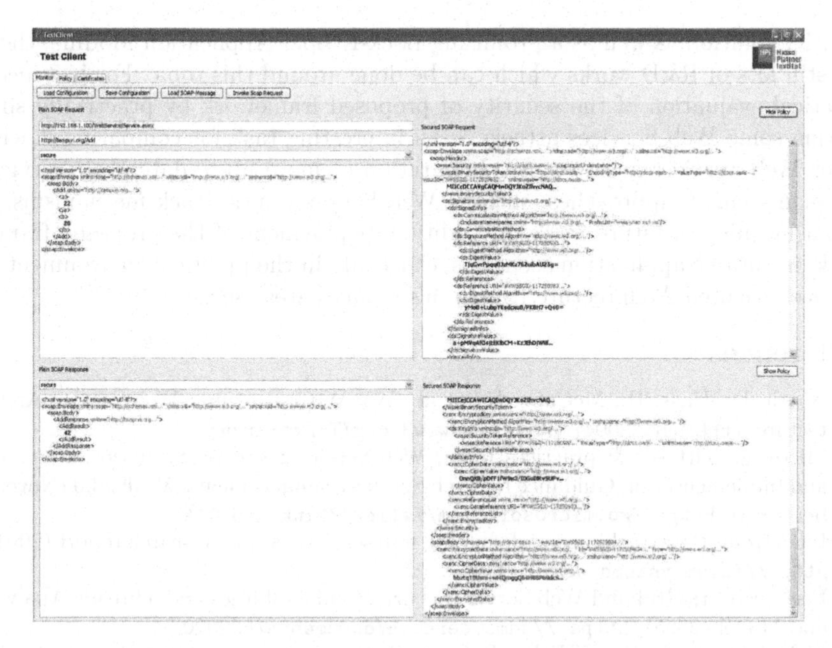

Fig. 5. A Successful SOAP Response from WS Provider

Services Provider to provide a simple Web Method "Add", which performs the addition of two numbers. The Consumer and the Provider are located in different networks which have been physically separated by Lock-Keeper.

As indicated in the Figure 5, The original SOAP message, shown in the top-left dialog box, embeds the "Add" document, which represents a SOAP request. After executing the necessary security operations (e.g. checking the "secure" policy), which would be required later by the SOAP Verification process on GATE, a secured SOAP request (top-right) is generated and sent to the Web Services Provider. After a couple of seconds (the required Round-Trip-Time mostly depends on the length of the Lock-Keeper switch interval), the secured response can be received in the bottom-right box. The plain result (i.e. 42, which is gotten from 22 + 20) is displayed after being decrypted by the same policy "secure". In the case of "unsecure" scenario, we can just get a fault SOAP response with the indication that "The Request is Illegal" because the user policy is not identical with any policies available in the "Policy Store" on GATE.

6 Conclusions

In this paper, we propose a secure Web Services Providing Framework based on the "Physical Separation" device, Lock-Keeper. With the new framework, the three layers of Web Services security: Network Security, Host Security, SOAP Message Security, can be achieved at the same time. The popular Web Services communication can also be facilitated over the Lock-Keeper which significantly improve the usability of the Lock-Keeper technology. As a new Web Services security solution as well as a promising Lock-Keeper Application Module, there are still lots of R&D works which can be done around this topic. For instances, practical evaluation of the security of proposed framework by practically simulating some Web Services attacks is an interesting task for comparing it with other Web Services architecture. Integration of more up-to-date Web Services message security approaches, such as Web Services anti-attack mechanisms, is also a meaningful future work. In addition, deployment of the proposed framework in suitable application scenarios, especially in the practical environment of Service Oriented Architecture (SOA), also makes great senses.

References

1. Booth, D., Haas, H., McCabe, H., et al.: WWW Consortium: Web Service Architecture. (February 2004), `http://www.w3.org/TR/ws-arch/`
2. Microsoft patterns & practices team: Web Service Security: Scenarios, Patterns, and Implementation Guidance for Web Services Enhancements (WSE) 3.0 (November 2005), `http://go.microsoft.com/fwlink/?LinkId=57044`
3. Lindstrom, P.: Attacking and defending web services, a spire research report (2004), `http://forumsystems.com/`
4. Eege project: Grid and Web Service Security: Vulnerabilties and Threads Analysis and Model (2005), `https://edms.cern.ch/documents/632020/`
5. Meinel, C., Sack, H.: WWW - Kommunikation, Internetworking, Web Technologien. Springer-Verlag, Berlin, Heidelberg, New York (2004)
6. Nadalin, A., Kaler, C., Hallam-Baker P., Monzillo, R.: Web Services Security: Soap Message Security 1.1 (WS-Security 2006): Oasis standard 200602 (March 2006), `http://docs.oasis-open.org/`
7. Curphey, M., Scambray, J., Olson, E., Howard, M.: Improving Web Application Security: Threats and Countermeasures. Microsoft Press, Redmond, Washington (2003)
8. Cheng, F., Meinel, C.: Research on the Lock-Keeper Technology: Architectures, Applications and Advancements. International Journal of Computer & Information Science 5(3), 236–245 (2004)
9. Della-Libera, G., Gudgin, M., et al.: Web services security policy language (ws-securitypolicy) (July 2005), `ftp://www6.software.ibm.com/`
10. McIntosh, M., Austel, P.: Xml Signature Element Wrapping Attacks and Countermeasures. In: Proceedings of the ACM (2005)
11. Housley, R. , Ford, W., Polk, W., Solo, D.: Internet X.509 Public Key Infrastructure Certificate and CRL Profile. IETF - Network Working Group, The Internet Society, RFC 2459 (January 1999)
12. Neuman, C., Yu, T., Hartman, S., Raeburn, K.: The Kerberos Network Authentication System (RFC4120). IETF- Network Working Group, The Internet Society (July 2005), `http://www.kerberos.info/`
13. IAG 2007 website in Microsoft: (2006-2007), `http://www.microsoft.com/iag/`
14. Kang, M.H., Moskowitz, I.S.: A Pump for Rapid, Reliable, Secure Communication. In: CCS1993. Proceedings of 1st ACM Conference on Computer & Communications Security, Fairfax, VA, ACM Press, New York (1993)
15. Lock-Keeper WebSite in Siemens Switzerland: (2005-2007), `http://www.siemens.ch`
16. Cheng, F., Meinel, C.: Deployment Virtual Machines in Lock-Keeper. In: Lee, J.-K., Yi, O., Yung, M. (eds.) WISA 2006. LNCS, vol. 4298, Springer, Berlin, Heidelberg (2006)

Measurement Analysis of IP-Based Process Control Networks*

Young J. Won[1], Mi-Jung Choi[1], Myung-Sup Kim[2], Hong-Sun Noh[3], Jun Hyub Lee[3], Hwa Won Hwang[4], and James Won-Ki Hong[1]

[1] Dept. of Computer Science and Engineering, POSTECH, Korea
{yjwon,mjchoi,jkbon,jwkhong}@postech.ac.kr
[2] Dept. of Computer and Information Science, Korea University, Korea
tmskim@korea.ac.kr
[3] Electric &Control Maintenance Dept., POSCO, Korea
[4] Technical Research Laboratories, POSCO, Korea
{vishnu4,mujigae,hwawon}@posco.co.kr

Abstract. This paper presents a measurement study of the traffic traces from the industrial process control IP networks. We present some interesting and unique traffic characteristics of the IP networks which support the control of manufacturing and precision-control machines. Understanding their traffic behaviors would help us to operate the fault-tolerant control IP networks, where the cost of network malfunctioning is far more severe than ordinary IP data networks. We observe rather steady and cyclic traffic patterns in the collected traces between the control IP network entities, mainly PLCs and process controllers.

Keywords: passive measurement, traffic analysis, process control IP networks.

1 Introduction

Process control networks must sustain the robust communications between controlling devices and controlled devices in a manufacturing environment. There are several categories of control networks, namely Building Automation (BA), Factory Automation (FA), and Process Automation (PA) [1]. These networks are deployed in mission critical operations which require a maximum level of network stability. However, we know little about the traffic behavior of such networks, simply because we are dealing with very secure and closed networks which entail company's confidentiality.

Existing process control network technologies (e.g., FOUNDATION Fieldbus [3], PROFIBUS [4], MODBUS [5], BACnet [6], Lon Works [7], etc.) are developed separately from the relatively recent emergence of Ethernet and IP technology. The

* This research was supported by the MIC (Ministry of Information and Communication), Korea, under the ITRC (Information Technology Research Center) support program supervised by the IITA (Institute of Information Technology Assessment)" (IITA-2006-C1090-0603-0045).

S. Ata and C.S. Hong (Eds.): APNOMS 2007, LNCS 4773, pp. 385–394, 2007.

newer versions of them have adapted Ethernet (e.g., Industrial Ethernet) and IP for low cost, high scalability, and easy maintenance purposes. A few studies [1, 2] have introduced some important issues of moving towards the all IP-based control networks, such as QoS requirements of the control network and existing IP network security concerns. Nonetheless, the decision of such change in a manufacturing plant environment is beyond the technical superiority of Ethernet and IP because it involves a huge investment.

Unlike IP data networks (e.g., Internet, enterprise networks), the traffic characteristics of control IP networks have not been studied in-depth previously. Due to significant differences of traffic nature, the general OA&M guideline running IP data networks may not coincide fully with process control IP networks. Understanding the traffic behavior helps us to operate fault-tolerant control IP networks where the cost of network malfunctioning is far more severe than the IP data networks. We present a measurement study of the traffic traces from the real industrial process control IP networks. To our knowledge, this is the first work to provide an empirical traffic analysis in such networks.

The remainder of this paper is organized as follows. Section 2 provides an overview of our measurement environment. In Section 3, we present the results of empirical analysis of the collected traffic traces. Finally, we summarize our findings and discuss possible future work in Section 4.

2 Measurement Environment

A typical process control network is illustrated in Figure 1. We have captured traffic traces from three different measurement points (illustrated as A, B and C in Figure 1). The process control network elements follow a hierarchical model where the controller at the top triggers actions in one or more controlled devices. The following are brief descriptions of the component elements and their role.

- *Process Controller (PC) – This is a part of the software and hardware package provided by the PLC vendors. It is process control software on a computer*

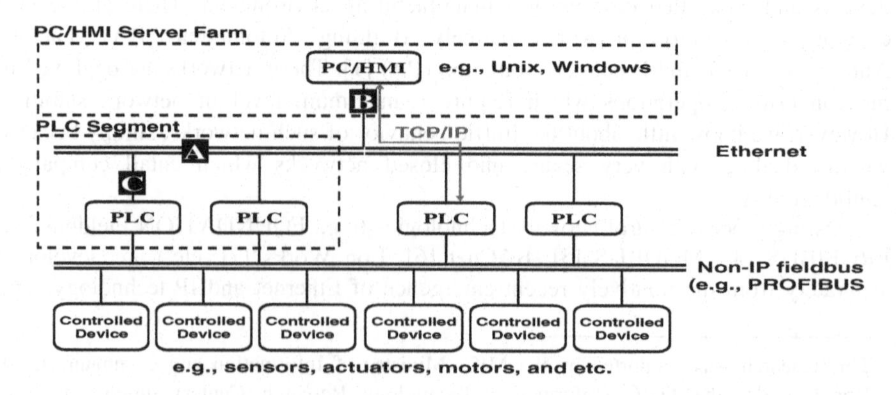

Fig. 1. Control IP network topology & Measurement points – A, B, and C

running UNIX or Windows that can remotely access PLCs. Custom built or vendor provided server applications are placed in a PC to communicate with PLCs. It also communicates with the machines running Human Machine Interface (HMI) solutions which provide graphical presentation of real-time process status monitoring. The communication with PLCs is established over reliable TCP/IP.

- *Programmable Logic Controller (PLC) - It is a microprocessor computer for process control attached to a process control network. A complex sequence control of machinery (or low end controlled devices on factory assembly line) is handled by the custom-built software programs running in PLC. The PCLs in our monitoring environment are equipped with two separate interfaces: Ethernet (e.g., RJ-45) and PROFIBUS (e.g., EIA-485).*
- *Controlled Devices – These machineries refer to sensors, actuators, motors, etc. They receive the command signal from PLCs via the embedded interface.*

Two measurement domains exist in process control networks: PC to PLC network and PLC to controlled device network. It is a combination of IP-based and non-IP based control technologies; in fact, PLC acts as a gateway accessing the lower end devices, which are on non-IP based networks. In this study, we focus on the first half of the networks (i.e., PC to PLC). A PLC segment refers to a group of PLC networks at the edge.

We have collected the traffic traces from a various points of the process control networks using the standard libpcap [8]. Points A, B, and B in Figure 1 refers to the top of PLC segment, the process control backbone network, and the nearby end-host (PLC device), respectively. These representative locations are carefully selected to provide a precise and unbiased snapshot of the network.

3 Empirical Data Analysis

The experimental data set we used was collected at the process control networks of POSCO, the world's fourth largest iron and steel manufacturer [11]. It operates a number of plants world-wide and a single operational site consists of about 40 manufacturing plants, equally 40 process control networks where they are organized in a synchronous and sequential order. Each process control network is a group of edge network segments. In simpler terms, a number of networks are working together to interconnect the machineries running continuously on a conveyor belt.

Our assumption is that no other data traffic is injected into the monitoring PLC networks. This was achieved by the complete isolation of PLC network from the Internet or any other enterprise networks. The private IPs and dedicated Ethernet links were assigned to PLC devices.

3.1 Traffic Summary

Table 1 illustrates the traffic summary of the collected traces. We have analyzed a week-long trace at one of the edge segments, a typical working-hour trace at the plant backbone, and short (e.g., 5 minutes) traffic traces at the end-host segment – Segment A, B, and C, respectively. The actual name of each process segment is undisclosed

Table 1. Summary of the datasets

Set	Date	Duration	Byte	Packets	Flows	TCP%	Util.%	Location
Segment-A	2006-09-29	170 hr (7 days)	63.5 GB	542 M	48 K	98	1	Edge
Segment-B	2007-02-27	10 hr (12:00~20:00)	74 GB	122M	25 K	99	19	Backbone
Segment-C	2006-05-11	5 min (13:15~13:19)	22 MB	84 K	48	99	0.57	Edge

due to security reasons. Most traffic is exchanged using TCP and the average utilizations yield a very low percentile in the 100 Mbps physical links environment. We observed the total traffic volume of Segment B regardless of the three times much as the packet counts of Segment B. It implies a low yielding traffic volume and a major occupancy of light-size packets. Interestingly, in Segment B, only 533 flows are TCP flows and responsible for 121M packets. In fact, a small number of identical sessions with fixed amount of hosts (PLC and PC) are continuously observed and generate the traffic.

Figure 2 (a) illustrates a long-term traffic pattern of particular PLC segment – Segment A. The graph shows that the bandwidth consumption is not bound to the time-of-day effect in the typical IP data networks where more traffic is generated during the day or working-hours. Its bandwidth usage is very steady and predictable throughout the course of monitoring period. A few sudden drops in the graph, the shade regions, are observed which indicates an instant shutdown of the process due to scheduled or unscheduled maintenance purpose. Its bandwidth consumption is strictly proportional to the number of devices in the network. Thus, the network planning for control IP network can rely on a very precise projection of bandwidth growth model which is almost impossible in other type of IP networks.

The Segment B traces are a collection of multiple instances of the Segment A level of traffic. Figure 2 (b) shows a short-term bandwidth measurement, but much larger traffic volume at the backbone. Its pattern closely coincides with the behavior as in Figure 2 (a). Indeed, the microscopic view of traffic behavior can reflect the

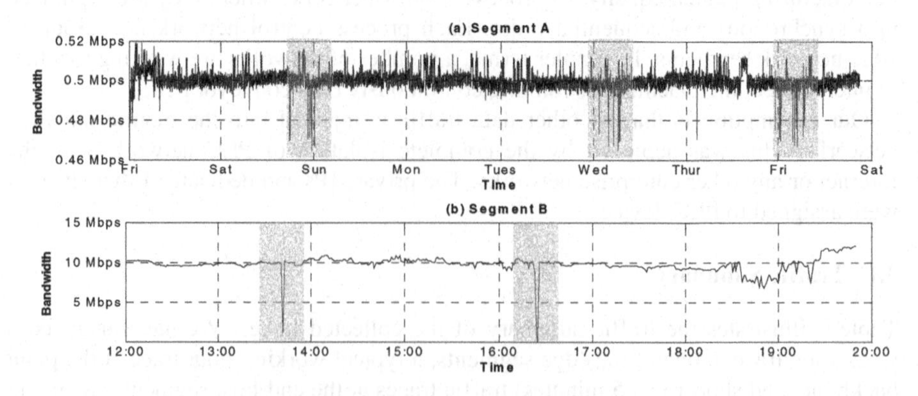

Fig. 2. Traffic Volume – Segment A & B

long-term behavior without loss of generality in control IP networks because there exists a fixed pattern in every session occupying the control IP networks. More details will be covered in the following sections.

3.2 Traffic Cycle

Figure 3 illustrates the four representative packet arrival patterns in PC-PLC sessions. The upper plane (above 0 on y-axis) of the graph indicates a unidirectional packet size arrival sequence according to the packet inter-arrival time. The lower plane is a packet size arrival sequence of the corresponding reverse transmission. Thus, a single graph represents two bidirectional flows. The packet inter-arrival time is measured in time granularity of millisecond. The dense region of graph implies that the packet inter-arrival gap is reduced. In all four graphs, we observe a unique and regular cycle of dense and sparse region occurrences over a short time period as well as the packet size distributions. For example, Figure 3 (a) has about 12-second cycle of two dense regions followed by a sparse region in both upper and lower planes. In the lower plane, it also shows a regular cycle of 1000 bytes packet transmission in every 1.5 second. In a similar fashion, the rest of the graphs can be expressed as a traffic pattern candidate for general PLC-PC sessions. All the sessions in our measurement belong to one of the pattern shapes shown in the graphs of Figure 3. Figure 3 (b), (c), and (d) illustrated the session traffic patterns for different PLC vendor solutions or sessions involving possibly in different processes.

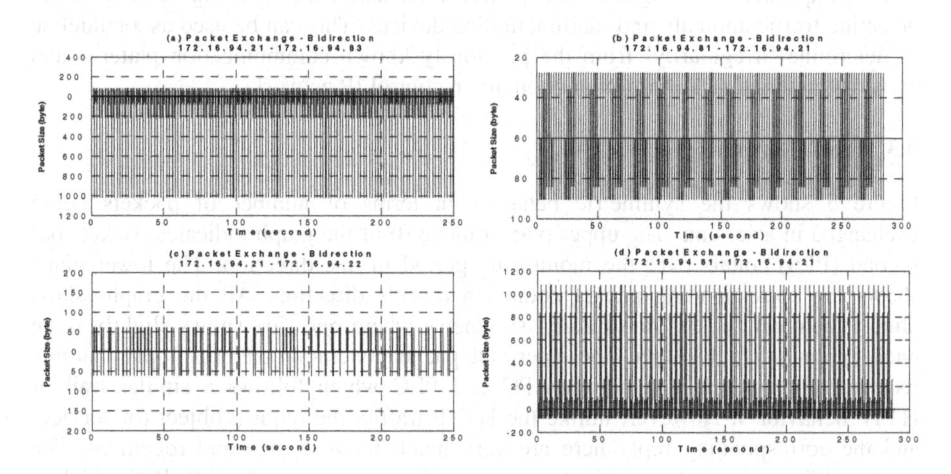

Fig. 3. Bidirectional packet-size arrival sequence patterns of PC-PLC sessions

The average inter-arrival time of these four sessions ranges from 120 ms to 1.5 s. Note that, the selected sessions were operational without any performance irregularities at the time of monitoring period. A matter of hundreds of millisecond or above may not be acceptable a set of values for a packet delay in the IP data networks. It is a unique characteristic where relatively longer packet delays are acceptable.

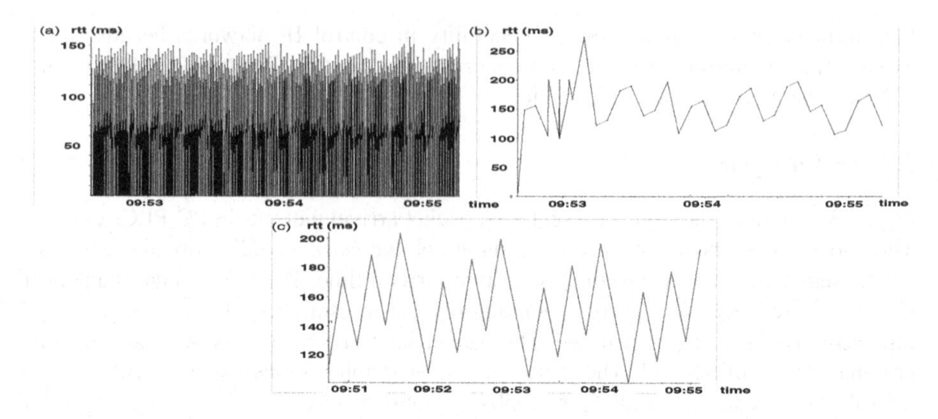

Fig. 4. RTT measurement of PC-PLC sessions

Figure 4 illustrates the three sample RTT measurement graphs over the monitoring periods. The RTT values of each session are measured using tcptrace [10]. These PLC-PC sessions show clear periodic patterns which distinguish them from normal IP sessions. Their maximum RTT values also ranges from 150 ~ 250 msec, coinciding with the packet inter-arrival values of the sessions in Figure 3. The fixed packet arrival sequence is again apparent through the recurring shape of the graph.

It is important to recognize such pattern information of on-going sessions when detecting traffic anomaly and malfunctioning devices. This can be used as a guideline to determine 'irregularity' from the previously known communication patterns and avoid the ambiguous definition of anomaly in control IP networks.

3.3 Traffic Symmetry

Figure 5 shows the symmetric behavior in terms of number of packets being exchanged in a session. The upper plane (above 0) of the graph indicates packets per second (PPS) counts over the monitoring period in one direction. The lower plane shows the corresponding packet counts in reverse direction. All the graphs show almost identical packet transmission symmetry where one side shows slightly more packet counts than the other. These periodic packet generation patterns imply a simple request-response behavior between PC and PLC which follows a similar trail of HTTP behavior 9. However, unlike the HTTP traffic, the request object (or service) and the corresponding reply here are very much fixed in size and repetitive. The average PPS count of the sessions in control IP networks is below 10 PPS which is quite low compared to the Internet traffic.

3.4 Packet Size Distribution

The purpose of transmitting packets in control IP networks can be classified into the following: Signaling purpose for PLC operations, HMI display purpose, and management purpose (e.g., SNMP). At the time of data collection, we were assured that no management traffic was injected into the network. Figure 6 (a) and (c)

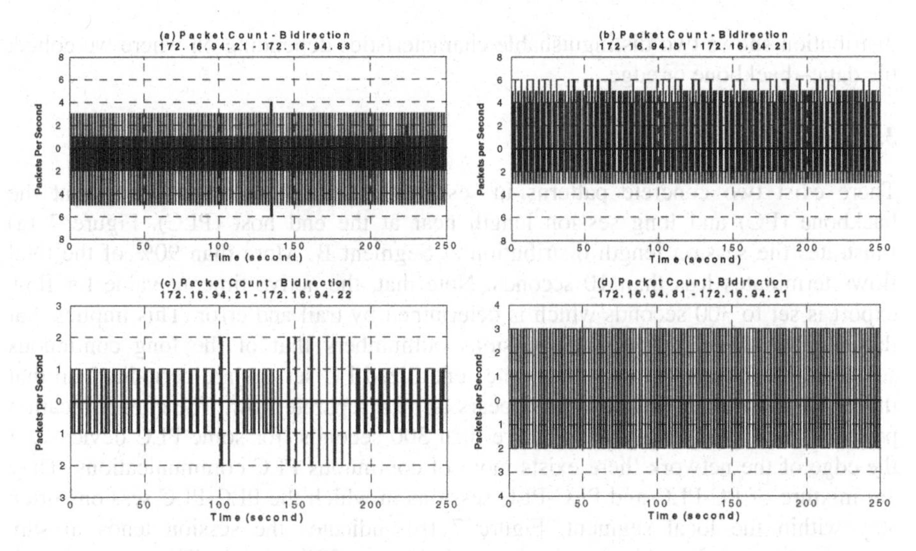

Fig. 5. Bidirectional packets per second counts in PC-PLC sessions

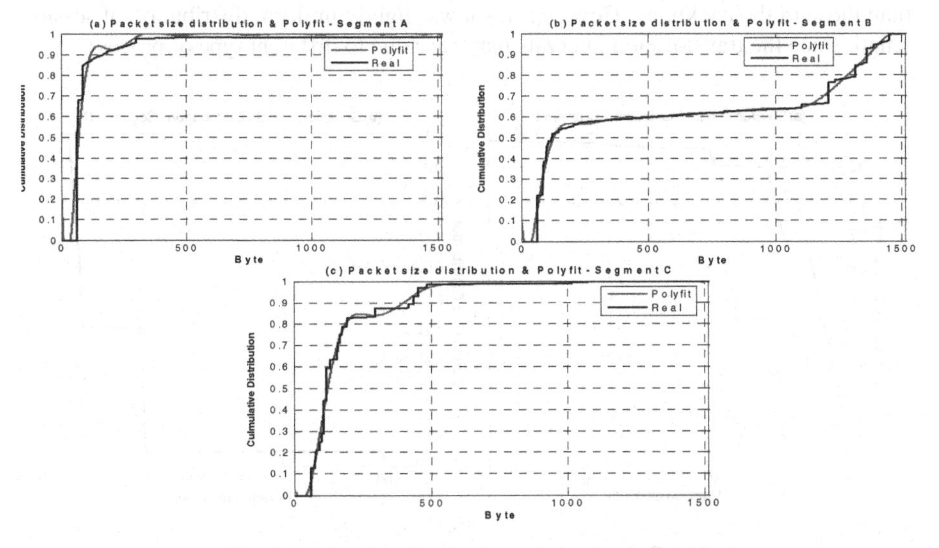

Fig. 6. Packet size distribution – Segment A, B & C

indicates the packet size distribution of packets mainly for signaling purpose. They show that over 90% packets are less than 100 bytes. In fact, their size ranges from 60 to 80 bytes which is just sufficient enough to contain TCP header information and a very few bytes for the actual payload. It explains for low bandwidth consumption in the control IP networks.

Figure 6 (b) shows the increase of average packet size compared to Figure 6 (a) and (c) where they are collected at near the edge PLC segment. At Segment B, we collect the HMI request/reply packets along with the PLC packets. The packet size distribution shows two distinguishable characteristics depending on where we collect the data – backbone or edge.

3.5 Session Length Distribution

There exist two concrete patterns in session length: Short session length at the backbone (PC) and long session length near at the end host (PLC). Figure 7 (a) illustrates the session length distribution at Segment B. More than 90% of the total flows terminates less than 10 seconds. Note that, the active timeout value for flow export is set to 300 seconds which is determined by trail and error. This implies that the number of short periodical sessions outnumbers that of the long continuous sessions. The communications with the centralized PC servers are periodical, so that the flows from the identical hosts appears to be a separate flow. The communication period is relatively long, at least more than 300 seconds, for some PLC devices. At the edge of the network there exists more of continuous PLC communications. They are mixture of PC-PLC and PLC-PLC sessions in which the PLC-PLC sessions often stay within the local segment. Figure 7 (b) indicates the session tends to stay connected during the entire monitoring period, about 250 seconds. The session length distribution from another edge, Segment C, shows relatively longer session length than those in the backbone. However, it shows slightly uniform distribution of session length since the transmission intervals may vary due to different type of process.

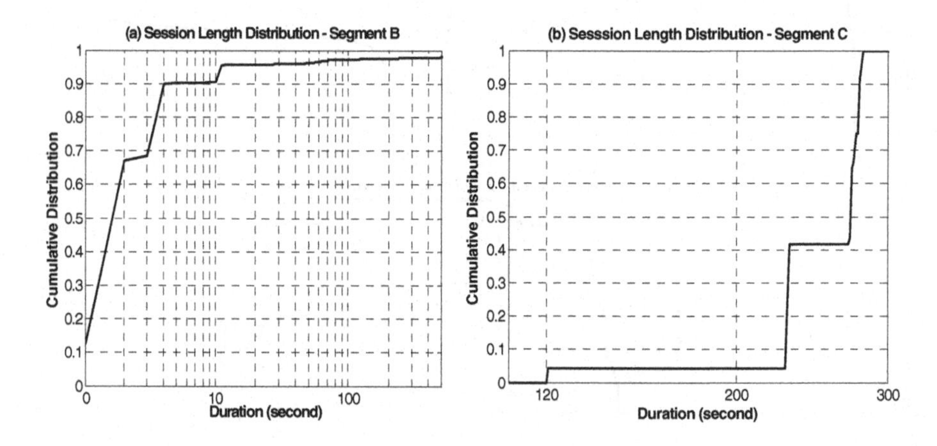

Fig. 7. Session length distribution - Segment B & C

It is generally believed that the longer and continuous sessions occupy the control IP network because its operations often involve continuous and repetitive tasks over long hours (e.g., production line built on conveyor belt.) Overall, it follows a typical action-trigger behavior; we observe lighter commands (e.g., small packets) from the top of the hierarchy and corresponding actions at the bottom which triggers a complex sequence of communications.

3.6 Packet Reordering

We measure the following two categories to detect any packet reordering in a session: Out-of-sequence packets and retransmission packets. It refers to any out-of-order packet delivery. Figure 8 illustrates the ratio of the flows experiencing the out-of-sequence and retransmission packets. In the Segment B traces, we observe the total 21,539 TCP flows, and 92% of them experience one or more retransmission packets while online. The backbone traces have shown the sign of improper packet delivery compared to the rest of the traces. Despite of the high retransmission ratio, the process networks operates without any problem at the time of data collection. For future work, it is worthwhile to investigate whether such characteristic is bound to this particular case.

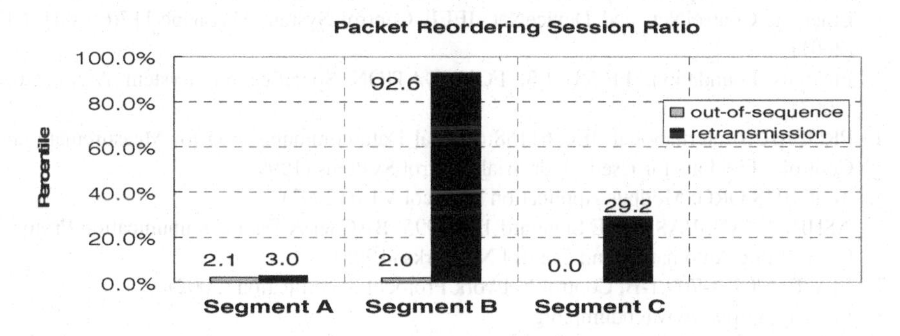

Fig. 8. TCP sessions experiencing out-of-sequence and retransmission

4 Concluding Remarks

Despite of its large deployment in the real world, the area of monitoring and analyzing process control IP networks has not yet been focused very much by the network measurement and management research community. We have known very little details about the traffic behaviors of process control IP networks. These networks are deployed in mission critical operations which entail a maximum level of network stability. Understanding the traffic behavior helps us to operate fault-tolerant control IP networks where the cost of network malfunctioning is far more severe than the IP data networks.

In this paper, we have presented a measurement study of the traffic traces from the real industrial process control IP networks. The traces for analysis are carefully selected to provide a precise snapshot of the network from different perspectives, the traces from the backbone, the edge network, and the end host. We have summarized the following unique characteristics.

- Low-yielding and steady bandwidth usage regardless of monitoring periods
- Periodic traffic cycle in terms of packet arrival sequence and inter-arrival time
- Traffic symmetry
- Occurrence of small signaling packets
- Session length distribtuion patterns
- High packet reordering ratios at the backbone

Based on the primilnary analysis of traffic characteristics, we plan to develop a control IP network traffic model which can be used for network planning and anomaly detection. The comparison study between the process control networks and the ordinary IP networks will be also helpful to identify and formulate the systematical differences of process control networks.

References

1. Okabe, N.: Issues of Control Networks When Introducing IP. In: Proc. of Symposium on Applications and the Internet Workshops, vol. 00, pp. 80–83 (2005)
2. Lian, F.-L., Moyne, J.R., Tilbury, D.M.: Performance Evaluation of control networks: Ethernet, ControlNet, and DeviceNet. IEEE Control System Magazine 117(6), 641–647 (2001)
3. Fieldbus Foundation. FF-581-1.3, FOUNDATION Specification: System Architecture (2003)
4. PROFIBUS International. IEC 61158, Digital Data communication for Measurement and Control – Fieldbus for Use in Industrial Control Systems (1999)
5. MODBUS.ORG: Modbus Application Protocol V1.0 (2002)
6. ASHRAE. ANSI/ASHRAE Standard 135-1995, BACnet A Data Communication Protocol for Building Automation and Control Networks (1995)
7. EIA. EIA/CEA-709.1-B, Control Network Protocol Specification (2002)
8. Libpcap, http://www.tcpdump.org/
9. Kunz, T., Barry, T., Zhou, X., Black, J.P., Mahoney, H.M.: WAP Traffic: Description and Comparison to WWW Traffic. In: ACM Workshop on Modeling, Analysis and Simulation of Wireless and Mobile Systems (August 2000)
10. Tcptrace, http://jarok.cs.ohiou.edu/software/tcptrace/
11. POSCO, http://www.posco.com

On the Use of Anonymized Trace Data for Performance Evaluation in IP Routers

Yusuke Toji, Shingo Ata, and Ikuo Oka

Graduate School of Engineering Osaka City University
3–3–138 Sugimoto, Sumiyoshi-ku, Osaka, 558–8585, Japan
{toji@n.,ata@,oka@}info.eng.osaka-cu.ac.jp

Abstract. For IP routers, it is important for realistic performance evaluation of address lookup algorithms to consider both routing table and trace data obtained from target network. However, most of trace data available are published after anonymizing personal information. Thus, the published trace data cannot be applied directly to estimate performance of routers. In this paper, we propose a new method for predicting more realistic router performance by using anonymized trace data. For our motivation, we analyze correlations of address space usage between real trace data and routing table. Based on analytic results, we also propose a method which transforms IP addresses in trace data by using statistics of address space usage in routing table. Through trace-driven simulation we show that our method can predict the routers' performance closer to the actual one.

1 Introduction

Demands for high-speed and high-performance relay routers have been increasing by growth and advancement of the Internet. One of the main functions that affect performance of the routers greatly is IP address lookup. IP address lookup searches an appropriate entry for the destination address in the IP packet from the routing table which manages forwarding ports for each address. Many techniques to improve the forwarding capability have previously been proposed and evaluated.

For the performance evaluation of address lookup technologies, it is important to evaluate in the actual environment, because address lookup techniques depend heavily on the network environment where routers are deployed. It is well known that the performance of routers often loses touch with theoretical figure in the actual environment. Predicting actual performance of routers in real networks has efficacy for enhancing the reliability of the router performance.

It is necessary to evaluate performance of routers using both routing table and real trace data obtained from the target networks, however, it is commonly difficult to obtain these data together. Most of ISP (Internet Service Providers) deny to publish the monitored traffic from the viewpoint of security and privacy. In particular, it is factually impossible that vendors of routers use trace data.

S. Ata and C.S. Hong (Eds.): APNOMS 2007, LNCS 4773, pp. 395–404, 2007.

In addition, there are several world-wide projects on the Internet data analysis and/on traffic measurement. CAIDA [1] is one of the organizations which publishes traffic data observed in the networks. However, the trace data available in such projects are often anonymized due to security and privacy reasons. Therefore, the published trace data cannot be applied directly to estimate performance of routers which depend on the real destination addresses.

From above background, in previous works, performance evaluation on routers' address lookup algorithms are performed following approaches. Literatures [2,3] use randomly generated addresses for the evaluation. On the other hand, in [4], a small amount of the traced data is used. In such ways, performance evaluation may be different from actual performance of routers. Alternatively, [5] uses single anonymized trace directly for its performance evaluation. However, since the anonymized trace data has no relative information with the routing table, the results obtained from the anonymized trace lack the reliability. Another approach is proposed in [6] which introduces the address generation method of arriving packets by taking the time-dependent behavior of packet arrivals into consideration. However, this method still requires an amount of real trace data to obtain a set of IP addresses appeared in the destination IP field of packets. In addition, performance comparisons among different address lookup algorithms are quite difficult because there is no unified benchmark environment today. To provide a compatitive field in the area of IP routing, both trace data and routing table are needed to be available in public.

Our motivation in this paper is to provide reliable and simple performance evaluation method for IP address lookup as well as packet classification algorithms. Unlike others, our approach does not use the anonymized trace data directly, but transform IP addresses in data to realer ones. We believe that our method realizes a common benchmark scheme in IP routing area, and hope it would help ISPs to publish anonymized trace data for various evaluations on IP related research.

In this paper, we analyze address space usage of both trace data (captured at campas network) and the routing table. We show that there is strongly relationship on the address space usage between the trace data and the routing table. By using this phenomenon, we propose a method to transform IP addresses in the anonymized trace data into most likely ones correlated to the prefixes in the routing table. Moreover, we show that our method can predict near actual performance with a trace-driven simulator. We also insist that our method has a generality by applying a routing table in a core network.

Hereinafter, in Section 2 we briefly describe prefix-preserving anonymization and problems of using anoynmized trace data. In Section 3, we present a method to analyze real trace data and routing tables in a campus network. In Section 4, we present a transform method between anonymized trace data and routing table from the analysis. In Section 5 we evaluate address-lookup performance of trace data applied our method, and evaluate the performance. Finally, in Section 6 we remark a summary and future topics.

2 Prefix-Preserving Anonymization

Our method exploits nature of prefix-preserving anonymization which is used to anonymize most of trace data in public domain. In this section, we first give a brief summary of prefix-preserving anonymization schemes, and describe problems of using anonymized trace data for predicting performance of routers.

2.1 Brief Description of Anonymization

Packets in trace data have IP header information, including destination and source IP addresses. Published trace data in public domain have been anonymized destination and source addresses. A straightforward approach is to translate each IP address to a random 32-bit address, however, such translation removes the relation of prefix information among packets. For example, two packets having 192.168.1.2 and 192.168.1.3 are considered that packets are from/to the same network. If we apply the random translation, two IP addresses are randomly changed. As the result, we cannot identify that the packets are from/to the same network by checking the translated address. To solve the problem, prefix-preserving anonymization is proposed which keeps the relationship of address prefixes. Prefix-preserving anonymization use a following anonymization function. If two IP addresses which share k-bits prefix ($0 \leq k \leq 32$), a and b are defined as

$$a = a_1 a_2 \cdots a_{32} \quad (1)$$

$$b = b_1 b_2 \cdots b_{32} \quad (2)$$

$$a_1 a_2 \cdots a_k = b_1 b_2 \cdots b_k, \text{and}, a_{k+1} \neq b_{k+1}. \quad (3)$$

Then, the anonymization function F is given by

$$F(a) = a_1' a_2' \cdots a_{32}' \quad (4)$$

$$F(b) = b_1' b_2' \cdots b_{32}' \quad (5)$$

$$a_1' a_2' \cdots a_k' = b_1' b_2' \cdots b_k', \text{and}, a_{k+1}' \neq b_{k+1}'. \quad (6)$$

That is, if two original IP addresses share a k-bit prefix, their anonymized addresses will also share a k-bit prefix. We use `tcpdprive` (`http://ita.ee.lbl.gov/html/contrib/tcpdpriv.html`) which is an implementation of the prefix-preserving anonymization.

2.2 Problems of Using Anonymized Trace Data

Though, if the prefix relationship is saved, we cannot use those anonymized trace directly for performance evaluation of IP lookups. Commonly, the performance of IP lookup strongly depends on the value of IP address and the routing table. The performance would be significantly different when we use the anonymized trace data compared to the real data. We suppose a binary trie for instance. A trie is a tree-based structure allowing the organization of prefixes and make it possible to search any address up to 32-comparisons. we explain this example

Table 1. Example of Rouitng table

No	Prefix
1	80.0.0.0/8
2	100.168.0.0/16
3	175.168.10.0/24

Table 2. Summary of routing table

Network	Campus	Core (RouteView)
Number of prefixes	29,765	8,936,822
Number of unique prefixes	28,500	209,972

Table 3. Summary of trace data

Network	Campus	Core (CAIDA)
Measurement period	44 min 24 sec	1 hours
Number of addresses	46,989,248	117,855,519
Number of unique addresses	112,585	785,795

with a routing table shown in Table 1. When actual address is 80.93.174.60, it matches the first entry in the routing table. However, if the real address is anonymized to 175.168.10.90, the anonymized address matches the third entry in the routing table. In such case, the number of comparisons should be 8, however, it would be 24 if we apply the anonymized trace directly.

3 Analysis of Relation Between Trace Data and Routing Table

We propose a method for predicting more realistic router performance based on anonymized trace data. We first exploit nature of prefix-preserving anonymization which is used for anonymization of public trace data. In this paper, we focus on the relation of the usage of addressing space between IP addresses appeared in trace data and entries of the routing table. For example, volume of traffic to the organization having the prefix 10.0.0.0/8 is expected to be larger than the smaller organization having 11.11.11.0/24, because the organization has 2^{16} times as many nodes as another one. Thus, observed packets that first 8 bits of IP addresses are 10 should be more in trace data. Prefix-preserving anonymization translates IP addresses to the non-relative value, however, their prefix information are preserved. Namely, though the first 8 bits are translated from 10 to 25, for example, the volume of traffic having IP addresses beginning 25 in the anonymized trace would be the same as the one having IP addresses beginning 10 in the real trace data. Inversely, if we found the first 8 bits in which many of packets in anonymized trace have (e.g., 25), we can expect that this first 8 bits should be re-translate to 10.

From the above-mentioned conception, we analyze distribution of address space in real trace data and a routing table. We use two types of routing tables shown in Table 2 and use the campus trace as shown in table 3.

Here we define *p-bit aggregate address* as the first p bits part of addresses. After converting to p-bit aggregate address, the address is represented as the integer value (from 0 to 2^p). We aggregate both the destination addresses in the trace data and entries in the routing table into p-bit aggregate addresses. We also define *p-bit aggregate address distribution* as distribution of the number of addresses/prefixes having the same p-bit aggregate addresses (we refer as *size* in this paper). For example, in case of 8-bit aggregation, two IP addresses 172.16.4.10 and 172.19.34.10 are aggregated to the same 8-bit aggregate address 172, and size of 172 is two. We show the derivation of p-bit aggregate address distribution according to the following procedure.

1. Transform destination addresses of the real trace data and entries in the routing table to p-bit aggregate addresses.
2. Count the size of each p-bit aggregate address.
3. Sort p-bit aggregate addresses according to its size in ascending order, and give a rank.
4. Normalize the ranking index to make the bottom ranking to be 1. Let r_i be the rank of p-bit aggregate address i, and L be the bottom ranking. The normalized rank N_i is given by

$$N_i \equiv \frac{r_i}{L}. \tag{7}$$

5. Obtain the p-bit aggregate address distribution from the relationship between the value of p-bit aggregate addresses and its normalized rank.

In the Step. 2, we should consider the width of the address space covered by the prefix in case of the routing table. For example, if the prefix is 192.0.0.0/8, it should be translated to 256 16-bit aggregate addresses from 192.0 to 192.255.

Figure 1 compares 8, 16 and 24-bit aggregate address distribution between campus trace data and campus routing table. From this figure, 8 and 16-bit aggregate addresses show that there is a correlation between the trace data and routing table. However, in 24-bit aggregate address distribution, the correlation is quite weak, because the most of the entries in the routing table have 24 bits prefix. In 24-bit aggregation, entries of the routing table are sparsely distributed in the whole addressing space. Therefore, the size of each 24-bit aggregate address is a few. Therefore, we consider that using 8 or 16-bit aggregate address is appropriate to clarify the correlation.

We next divide the normalized rank into *segments*, and investigate the ratio satisfying that the same p-bit aggregate addresses in the trace data and the routing table are in the same segment. We define f_i as the normalized rank of the p-bit aggregate address in the trace data, where i is the integer value of p-bit aggregate address. Let g_i be the normalized rank of the p-bit aggregate address in the routing table. We the define Z as the number of segments, and M as number of aggregate addresses which satisfy Eq. (8)

$$[f_i Z] = [g_i Z]. \tag{8}$$

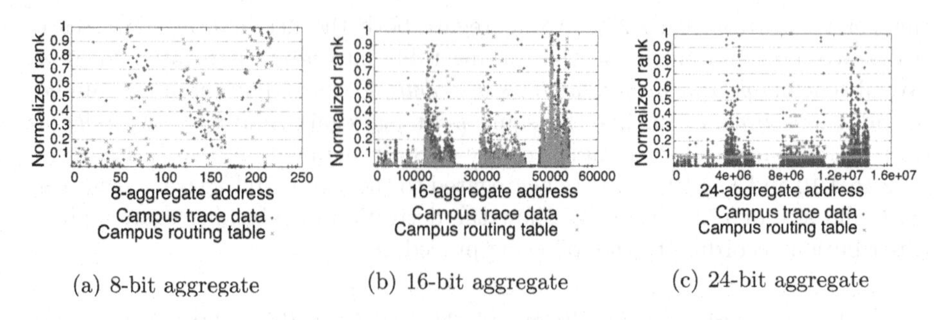

(a) 8-bit aggregate (b) 16-bit aggregate (c) 24-bit aggregate

Fig. 1. Aggregate address distribution

We also define N as the total number of aggregate addresses, Let R_z be the ratio of coexistence in the same segment, which is given by

$$R_z = \frac{M}{N}. \tag{9}$$

In addition, we show the correlation coefficient with the normalized ranking in the trace data and the routing table. The correlation coefficient R_c is defined as

$$R_c \equiv \frac{\frac{1}{N}\sum_{0}^{N} f_i g_i}{\sqrt{\frac{1}{N}\sum_{0}^{N} f_i^2}\sqrt{\frac{1}{N}\sum_{0}^{N} g_i^2}}. \tag{10}$$

Figure 2 and Table 4 show R_z and R_c respectively by changing the granularity of aggregation p.

As shown in Figure 2, R_z of the 8, 12 and 16-bit aggregate address is higher than other aggregate address. From Table 4, there is strongly correlation between aggregate address in the trace data and the routing table in 8-bit aggregate, but the correlation does not appear in 16-bit aggregate.

4 Transform the Anonymized IP Addresses

In this section, we propose the method which is considered the ratio of co-existence in the same segment, because the transform using 16-bit aggregate addresses is able to re-transform to the address closer to actual ones rather than using 8-bit aggregate addresses. The method consists the following procedures.

1. Aggregate addresses in both the anonymized trace and the routing table into p-bit aggregate addresses according by following steps of aggregation described in Section 3.
2. Obtain the normalized rank of each p-bit aggregate addresses by Eq. (7).

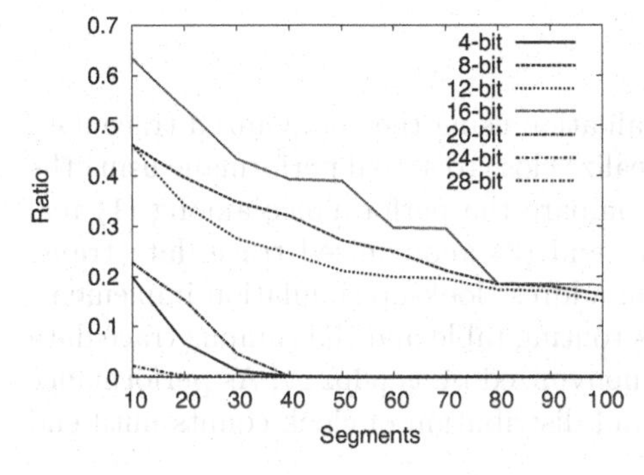

Fig. 2. Relation between number of segments and R_z

Table 4. Correlation between campus trace and campus routing table

p-bit	R_c
4-bit	0.95501
8-bit	0.87920
12-bit	0.55959
16-bit	0.33316
20-bit	0.14868
24-bit	0.12200
28-bit	0.07506

3. Classify the associate segment of p-bit aggregate addresses in both the anonymized trace and the routing table. Let Z be the number of segments. After classifying p-bit aggregate addresses are grouped by Z segments, the p-bit aggregate address whose normalize rank is N_i belongs to the segment $\lceil N_i * Z \rceil$.
4. Create the mapping table which stores a pair of (p-bit aggregate address in anonymized trace, p-bit aggregate address in the routing table) as follows. Let S_t be the segment of the p-bit aggregate address t in anonymized trace.
   – If p-bit aggregate addresses in the routing table exist in segment S_t:
     (a) Pick randomly the p-bit aggregate addresses in the routing table in segment S_t.
   – Else:
     (a) Increase j until the segment $S_t + j$ or $S_t - j$ has at least one p-bit aggregate address in the routing table.
     (b) Pick randomly the p-bit aggregate addresses in the routing table in segment $S_t + j$ or $S_t - j$.
5. Based on the mapping table, transform first p-bit of IP addresses in anonymized trace into the picked p-bit aggregate addresses.

We show an example as follows. We use 16-bit aggregate address for the anonymized trace and the routing table, and separate the routing table into the 10 segments. Suppose the address in the anonymized trace is 180.10.10.10, the 16-bit aggregate address is 180.10. We also suppose that the normalized rank of the aggregate address 180.10 is 0.34, we randomly pick an aggregate address in routing table from Segment 3 ($\lceil 0.34 * 10 \rceil$). If the picked aggregate address is 120.90, first p-bit of the anonymized address of 180.10.10.10 is transformed into 120.90. The IP address of the anonymized address is re-transformed into 120.90.10.10. Our method is able to transform p-bit prefix of addresses in the anonymized trace data into it of addresses in the real trace data.

5 Simulation Results

We show the results of performance evaluation using the anonymized trace data applied our transform method, which realize close to actual performane using the real trace data. For the purpose, we compare the performance among (1) real trace data, (2) anonymized trace data, and (3) anonymized trace data transformed by proposed method through an address look-up simulation implementing a trie structure. We use the campus routing table and the campus trace data shown in Table 3, and the trace data anonymized by `tcpdpriv`. As performance metrics, we use memory access times and distribution of clock counts until end of search for each addresses.

5.1 Memory Access Times

We divide the normalized rank in the trace data and the routing table into 100 segments, and simulate the memory access times. The result is shown in Figure 3(a). "8 mapped" and "16 mapped" are the results using the anonymized trace data transformed by 8-bit aggregate address and 16-bit aggregate address in the routing table. "Campus Trace Data" is the result using the campus trace data and the campus routing table. For the comparison, we also show the results by using anonymized trace data directly (labeled by "Anonymized Trace Data"). That is, the nearer the distribution of "Campus Trace Data", the closer estimate of performance in actual environment. Figure 3(a) shows that performance using the addresses transformed by 8-bit aggregate addresses is closer to performance in the actual environment. However, in case of using the addresses transformed by 16-bit aggregate address, the performance is greatly differ from the actual performance. The reason is because there are address areas losing the correlation of aggregate address distributions in the campus trace data and the routing table due to the aggregation of routing prefixes which is performed to reduce the number of prefixes in the routing table (see RFC 3765[7]).

Therefore, instead of campus routing table, we use the full-route routing table obtained at core network which dose not have much aggregated routes. We use the full-route routing table obtained from RouteView Project [8]. The full-route routing table is shown in Table 2. We show the result in Figure 3(b). The performance using the addresses transformed by the full-route routing table represents more precisely the performance in the real environment than using the addresses transformed by the campus routing table. From this result, aggregate address distribution in the campus trace data correlates stronger with it of the full-route routing table than it of the campus routing table. The fact indicates that distribution characteristic of aggregate address which we mentioned in this paper is not local, but more generic. In other words, if trace data of campus scale have the similar distribution characteristic regardless locations, our performance evaluation using only the trace data in a specific campus network can obtain the general (i.e., core networks) performance. In addition, we can use our performance evaluation in a core network because of the stronger correlation between the campus trace data and the full-route routing table.

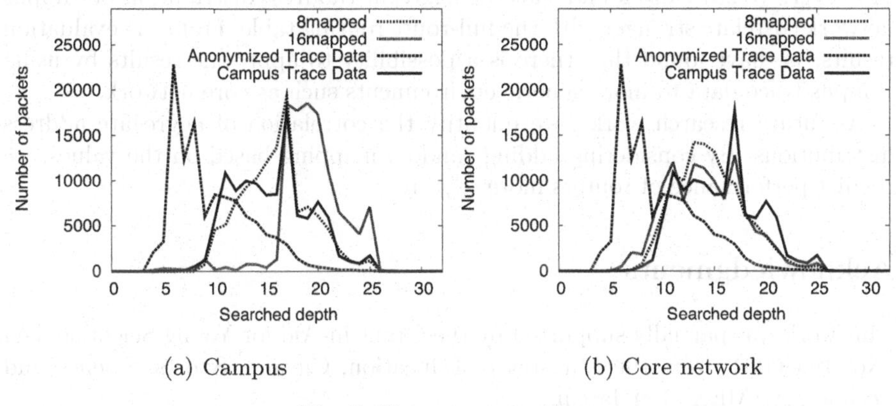

(a) Campus (b) Core network

Fig. 3. Memory access times

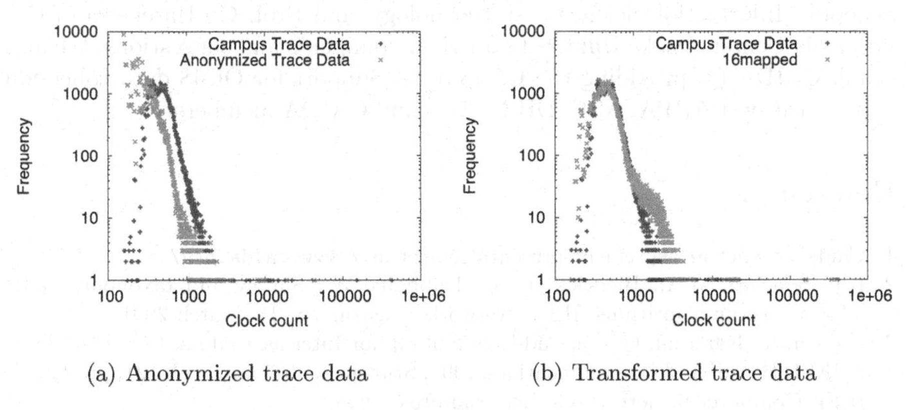

(a) Anonymized trace data (b) Transformed trace data

Fig. 4. Distribution of clock counts

5.2 Distribution of Clock Counts

Figure 4 shows the distribution of clock counts until end of search addresses in the same data shown in Section 5.1. The clock counts of the campus trace data and the anonymized trace data are shown in Figure 4(a), and also those of the campus trace data and trace data transformed by the full-route routing table are shown in Figure 4(b). As shown in Figure 4, in case of using the anonymized trace data, most of addresses are searched at lower clock counts than actual clock counts. Otherwise, in case of the trace data applied our method, most of addresses are searched at close to actual clock counts. In this result, our method can also evaluate search time for IP-lookup.

6 Conclusion and Future Work

We have proposed the method to re-transform anonymized addresses into close to actual addresses, and obtain the near performance in real environment.

Moreover, we have also found that the aggregate address distribution in campus network correlate stronger with the full-route routing table. From the evaluation results, we have found that there is a possibility to apply the results by using campus trace data to more generic environments such as core network.

As future research works, we quantify the correlation of aggregate address distributions. By considering adding random mapping based on the values, we predict performance of routers more exactly.

Acknowledgments

This work was partially supported by the Grant-in-Aid for Young Scientists (A) (No. 19680004) from the Ministry of Education, Culture, Sports, Science and Technology (MEXT) of Japan.

Additionally we would like to thank Prof. Masayuki Murata of Graduate School of Information Science and Technology, and Prof. Go Hasegawa of Cybermedia Center, Osaka University for their comments and suggestions. We also thank CAIDA for providing the OC48 data. Support for OC48 data collection is provided by DARPA, NSF, DHS, Cisco and CAIDA members.

References

1. Caida internet measurement presentation, http://www.caida.org/
2. Ruiz-Sanchez, M.A., Biersack, E.W., Dabbous, W.: Survey and taxonomy of IP address lookup algorithms. IEEE Network Magazine, 8–23 (March 2001)
3. Nilsson, S., Karlsson, G.: Fast address look-up for Internet routers. In: Proceedings of IEEE Broadband Communications 1998, Stuttgart, Germany, vol. 121, pp. 11–22. IEEE Computer Society Press, Los Alamitos (1998)
4. Gupta, P., Prabhakar, B., Boyd, S.: Near-optimal routing lookups with bounded worst case performance. In: Proceedings of IEEE INFOCOM 2000, Tel Aviv, Israel, pp. 1184–1192. IEEE Computer Society Press, Los Alamitos (2000)
5. Narlikar, G., Zane, F.: Performance modeling for fast IP lookups. In: Proceedings of ACM SIGMETRICS 2001, vol. 29, pp. 1–12. ACM Press, New York (2001)
6. Kawabe, R., Ata, S., Murata, M.: Address generation method based on traffic characteristics and its application to performance predictions of routers. IEICE technical report. Information networks, vol. 101, no. 716, 201–208 (2002)
7. Huston, G.: Nopeer community for border gateway protocol (bgp) route scope control (2004)
8. University of oregon route views project, http://www.routeviews.org/

10Gbps Scalable Flow Generation and Per-flow Control with Hierarchical Flow Aggregation & Decomposition Using IXP2800 Network Processors*

Djakhongir Siradjev[1], JeongKi Park[1], Taesang Choi[2], Joonkyung Lee[2], BongDae Choi[3], and Young-Tak Kim[1,]**

[1] Dept. of Information & Communication Engineering, Graduate School, Yeungnam University, 214-1, Dae-Dong, Gyeongsan-Si, Gyeongbook, 712-749, Korea
[2] BcN Division, ETRI, Kajong-Dong, Yusong-Gu, Daejeon, 305-350, Korea
[3] Dept. of Mathematics & Telecommunication Mathematics Research Center, Korea University, Korea
m0446086@chunma.yu.ac.kr, jk21p@ynu.ac.kr, {choit,leejk}@etri.re.kr, queue@korea.ac.kr, ytkim@yu.ac.kr

Abstract. With the growth and development of the Internet, understanding the composition and dynamics of network traffic is greatly important for network engineering, planning, design, and attack detection. This paper proposes 10Gbps dynamic flow monitoring and control using IXP2800 network processors with flow dynamic rate-based flow aggregation that ensures scalability of the system. We propose fast and scalable 6-tuple TCAM-based classification, which provides three levels of aggregation and flow control that includes flow metering, marking, queuing and scheduling. Using flow control allows network operators to adjust QoS-level dynamically and restrict malicious activity in the network. The proposed implementation design is based on Radisys ATCA-7010 processing module, containing two IXP2800 network processors and TCAM module for highspeed classifications. Flow information collector can be either implemented on another processing blade, or on external computer. The proposed architecture supports up to 125,000 flows in total.

Keywords: flow monitoring, control, classification, network processors.

1 Introduction

The content of traffic passing through Internet Service Providers (ISPs) is becoming increasingly diverse because various applications, such as, WEB, peer-to-peer, and VoIP applications, are widely employed. Therefore ISPs must now

* This research is supported by the MIC, under the ITRC support program supervised by the IITA (IITA-2006-(C1090-0603-0002)).

** Corresponding author.

S. Ata and C.S. Hong (Eds.): APNOMS 2007, LNCS 4773, pp. 405–414, 2007.

take up the challenge of real-time monitoring of traffic classified into flows to make their networks more reliable. Flow monitoring allows network operators to view the detailed, time- and application-based usage of a network. This information can be used in problem detection and troubleshooting, network planning, denial of service and other attacks detection, accounting and billing and traffic engineering.

Rapidly increasing network rates in access and core networks complicates the problem of dynamic flow monitoring, since the time given to process a packet is reduced, thus complicating performing additional processing on network nodes. Another severe challenge in flow monitoring is continuous increase of number of various network applications that results in high number of active flows. Recent measurements in [1] using a variety of traces shows the number of flows between end host pairs in an one hour period to be as high as 1.7 million. Recent estimates show about 1 million concurrent flows in network backbone, and further growth is expected. This fact creates problem for the algorithms having the storage complexity of $O(N)$, due to limitations of majority of SRAM and TCAM sizes. Currently available commercial TCAM module sizes are 9 Mbits and 18 Mbits.

Nowadays Cisco NetFlow [2] is the de-facto standard for flow monitoring in most of environments. In order to reach high processing rate packet with limited processing power, packet sampling technique is used. The simplest packet sampling technique has 2 main drawbacks: (1) the re-normalization of flow volume statistics and (2) there may be flows for which no packet is sampled. Although intelligent techniques proposed in [3,4] can improve the correctness of received results, still sampling can not be used for error-critical operations like billing. Fortunately, current speed of TCAMs can provide lookup rates enough to process packet on the rates below OC-192. The NetFlow performs quite well in majority of current environments, but the rapid increase of number of networking applications used by a single host, can degrade its performance. Authors of [5] propose adaptive aggregation scheme, but that scheme requires more complex processing due to its software-based nature. Also, the approach does not have dynamic expansion of aggregated flows, so once grouped the flow always remain as aggregated until it is expired.

This paper proposes an architecture for scalable and flexible flow monitoring and control system with rate-based dynamic aggregation and expansion of flows. The main idea of our approach is to start measurements on aggregated flows, expanding them when their rate goes beyond predefined threshold, in order to provide the network operator with real-time data and allow network operators to know the details about high rate flows. Another advantage of the proposed scheme is that it does not require complex processing and can be used on high-speed backbone links. Additionally our scheme includes flow control modules, that allow network operators to apply specific policy on the network traffic, based on traffic observations. The design of packet processing pipeline and overall application architecture is presented, including details of proposed TCAM-based classification engine. The architecture is in implementation stage on Radisys

ATCA-7010 [6] packet processing module and has the goal of 10Gbps processing rate and support of 125,000 flows with the use of Intel IXA SDK Framework [7].

The rest of this paper is organized as follows. Section II describes the related work on Cisco Netflow, flow aggregation techniques, Intel IXP 2800 [8] and Intel IXA SDK Framework. In Section III, we describe the design and implementation of scalable flow monitoring and control system with dynamic flow aggregation and expansion, and details of implemented functional modules. We cover the TCAM-based classifier implementation plan in details, as one of the focuses of this paper. Section IV analyzes current performance and theoretical final performance of the proposed architecture, and finally section V concludes the paper.

2 Related Work

2.1 Cisco NetFlow

Cisco Netflow[9] is the dominant industry standard for exporting per-flow information, and its evolution is becoming an IETF standard (IPFIX, see Ref. [10]). Cisco routers that have the Netflow feature enabled generate netflow records, which are exported from the router and collected using a netflow collector. The routers use flow cache that stores currently active flows information. A flow record is maintained within the NetFlow cache for each active flow. Each flow record is created by identifying packets with similar flow characteristics and counting or tracking the packets and bytes per flow.

NetFlow accounts for every packet (non-sampled mode) and provides a highly condensed and detailed view of all network traffic that entered the router or switch. Besides periodical export of flow entries specific events, like timeout, flow cache overflow may trigger export procedure. If the size of flow cache is less than number of concurrent flows, additional export procedure will be triggered every time on new flow detection. This may cause link overload between the flow monitor and collector. Although Cisco equipment can use additional aggregated cache, the aggregation property is set statically for flows and it results in poor performance in dynamic environment.

2.2 Studies on Flow Aggregation

Measurements of variety traces show continuous exponential increase of flows in network backbone, which is expected to be continued [1,11]. Considering the slower growth of TCAM and SRAM sizes, this fact causes scalability problem for flow monitoring systems. Authors in [4] proposed a method of adapting the sampling rate to traffic. They use dynamical decrease of the sampling rate until it is low enough for the flow records to fit into memory. This algorithm guarantees a stable flow cache and export bandwidth even under severe DDoS attacks. But under DDoS attacks the sampling rate will decrease to a very low level, which results in poor overall accuracy in per flow counting including legitimate flows.

Authors in [5] propose adaptive flow aggregation algorithm, which detects traffic clusters in real-time and aggregates large amount of short flows into a few flows. The algorithm has high accuracy in gathering flow information, but entails high complexity of cluster detection mechanism that requires sorted list of flow ID nodes. The clustering mechanism used for aggregation perform list sorting into a sorted list, which makes it hard to use the approach on high-speed links, when the number of IP addresses, having same hash key is high. Adaptive flow aggregation algorithm was designed to protect NetFlow system from memory overflows Comparing to the abovementioned approach, our proposed scheme is designed to provide realtime measurement data to the network operator, and uses reverse strategy for aggregation. It starts measurements of aggregated flows, and expands it when its rate goes beyond predefined threshold. Also it exploits TCAM module for fast prefix-based lookups.

3 10Gbps Scalable Flow Generation and Per-flow Control with Hierarchical Flow Aggregation and Decomposition Using IXP2800 Network Processors

3.1 Architecture of Flow Monitoring and Control System

The flow monitoring and control system provides flow information measurements. The architecture includes 3 major components: (i) data plane processing module; (ii) collector and (iii) the control application. The interactions and functional modules of components are shown on Fig. 1. The usual data plane processing module performs additional flow generation and flow control functions. The flow generation includes classification, flow monitoring and exporting procedures. Flow control includes traffic policing by metering/marking, active queue management and scheduling mechanisms. Periodical flow exports are done using export interface. Active timeout of flow export is set depending on the goals of monitoring system usage and channel capacity used between data plane module and collector. To get realtime flow information, another network processor module connected to switching fabric can be used. If the time period between two consecutive flow information exports is not critical, usual PC can be connected to data plane module using network interface. Control application is used to make the decisions based on the observations either automatically or semi-automatically. Once the decision is made, control application connects to data plane module using its control interface, and sets required rules for specific flows. Collector may also perform automated flow control if it has analyzing rules for flow information. The flexibility of scheme allows using it in various environments. In the simplest case collector and control application can be merged in a single module.

This paper covers only the implementation of data plane processing module, since collector and control application do not differ from other usual applications used in network monitoring. The implementation is done on Radisys ATCA-7010 [6] processing module, which includes 18 Mbit TCAM module and 2 IXP2800[8]

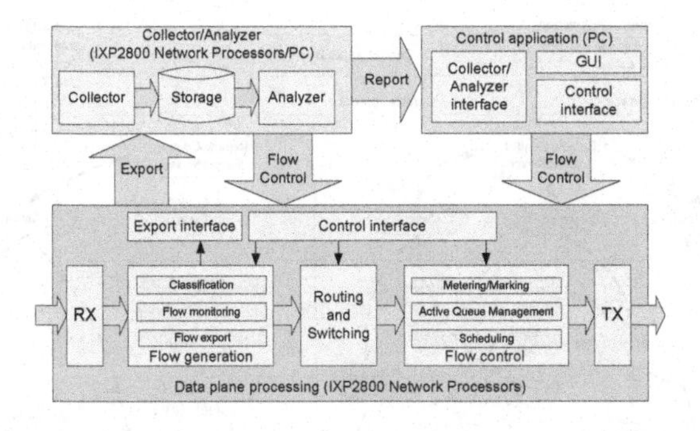

Fig. 1. Architecture of Flow Monitoring and Control System

Network Processors. Application implementation uses Intel IXA SDK Portability Framework [7]. The IXA Portability Framework [7] is a network application framework and infrastructure for writing modular and portable code, which can save time by providing robust infrastructure software and APIs, re-configurable building blocks, and an ideal structure for third-party plug-in application modules. Every functional module running on network processor is usually split into core component and microblock. Microblock handles usual packets processing and has limitation in time required for processing a packet. Core component is responsible for complex packets processing and performing control functions of each module.

Implementation scheme is shown on Fig. 2. The application has 4 layers and is also split into 2 sides: ingress and egress. Ingress and egress side separation allows application to remain scalable in case of using multiple processing blades. The lowermost layer shows microblocks implemented on IXP2800 microengines. Rx microblock receives data from network interface card, initializes packet related information and sends data to layer 2 decapsulation microblock, which checks the MAC type field of packet header. The classifier that receives IP packets from L2 decapsulation microblock, uses TCAM to check whether the flow cache entry exists for the packet. If the entry does not exist packet key fields and other packet related information is sent to the classifier Core Component (CC) running on XScale processor, else flow information is updated. After classification packet goes through usual IPv4 and flow control processing. Once a new message is received, classifier CC initiates flow addition procedure, which is explained in section 3.2.

3.2 TCAM-Based Classification with Dynamic Aggregation and Expansion

The proposed classifier is built with the assumption that network operator has low interest for low rate flows detailed information, and single aggregated entry can be used for them. Flows in proposed classifier can be divided in 3 categories, based on level of details: (i) network-to-network aggregated flows; (ii) host-to-host aggregated flow; (iii) process-to-process aggregated, or non-aggregated

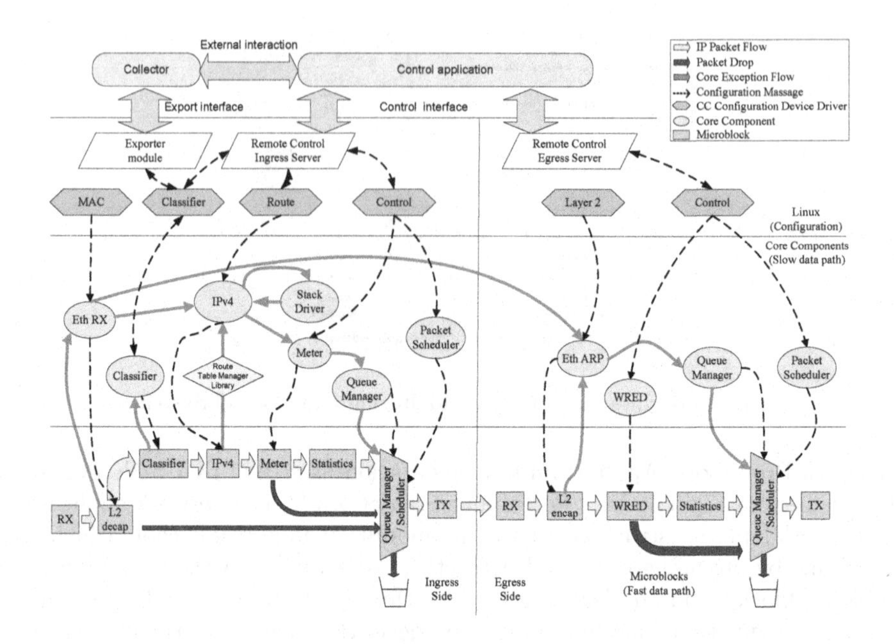

Fig. 2. Flow Monitoring and Control System Implementation Architecture

flows. Authors of [12] show that network-to-network aggregation results in 48%, and host-to-host aggregation results in 60% reduction of required flow entries in their experiment with real packet traces of the Internet Gateway. Since these 2 aggregation schemes give more compression ratio, comparing to others they are used in our proposal. The proposed classifier uses following 7 tuples as a flow key: source IP address, destination IP address, source port number, destination port number, DSCP, IP protocol type, incoming interface number.

The implementation of classifier module includes classifier CC and microblock. In order to avoid producing a bottleneck in data plane processing the classifier microblock performs only lookup function, and detects whether the flow that the packet belongs to the existing flows in cache or not. If the packet belongs to new flow classifier microblock generates a message, containing key fields of packet header, related packet information, like packet size and TCP flags, and sends the message to the CC. The flow cache management was intentionally moved to CC, because treating timeout and aggregation/expansion of flow cache entries requires sophisticated data structure handling, which is hard to implement on microengines, since that requires complex critical section implementation due to parallel execution of the application on several microengines.

In order to provide flow aggregation flow cache must be built with new data structure, simplifying flow aggregation and expansion procedures. The data structure used for flow aggregation is shown in Fig. 3. Each TCAM entry has two associated pointers. *Next entry* pointer is used to point the next entry on same aggregation level. This link is used for periodical traversal done for timeout detection. *Expanded list* pointer is used to point the list head of expanded flows,

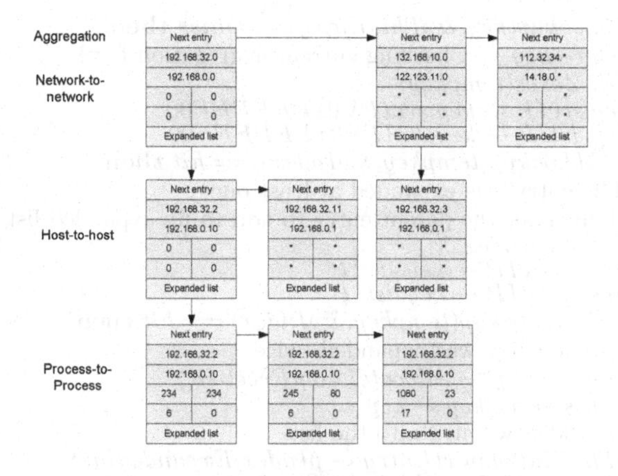

Fig. 3. Data structure of flow cache entries

belonging to the aggregate. Aggregate flows have specific values of some key parts. For aggregated flows using network-to-network aggregation, last octets of IP address numbers and, fields except IP addresses are set to *NULL*. For host-to-host flows, all fields except IP addresses are set to *NULL*.

Upon receiving the message from microblock, classifier CC initiates flow cache entry creation procedure. The procedure pseudocode is shown on the Fig. 4. CC checks whether the entry does not exist because there might be multiple messages sent from microengines to the CC during entry creation. Also CC periodically traverses the data structure and checks flows elements for timeouts and rate exceed of threshold. Both of rate thresholds can be set by control application.

3.3 Flow Control Operation

For the flow control system provides flow policing using Single Rate Three Color Marker (SRTCM) [13] and Two Rate Three Color Marker (TRTCM) [14]. Metering algorithm included into IXA SDK Framework [15] is used in our implementation. When the control application makes the decision to set some policy on the flow, it sets the parameters of token buckets and marks the flow as static. In case of using SRTCM, Committed Information Rate (CIR), Committed Burst Size (CBS), and Excess Burst Size (EBS) must be defined. Packets that conform CIR have the lowest drop precedence, while packets exceeding CIR have higher drop precedence. CBS defines allowed burstiness of conforming traffic and EBS defines the total amount of allowed data that exceeds CIR. In case of using TRTCM, CIR, CBS, Peak Information Rate (PIR) and Peak Burst Size (PBS) must be defined. TRTCM allows defining the maximum traffic rate and burstiness for the exceeding traffic, in contrast to SRTCM that defines only the amount of exceeding traffic.

Static flows can not be aggregated or expanded, based on their rate. They will present in the system until they are explicitly removed by control application. This is done to eliminate the situation, when a new flow under same

```
1: if TCAMLookup(key, &pThisEntry) == miss then
2:   tempkey ← key // Checking current aggregation level
3:   FillWithZeros(tempkey)
4:   tempkey.srcIP ← key.srcIP&0xFFFFFF00
5:   tempkey.dstIP ← key.dstIP&0xFFFFFF00
6:   if TCAMLookup(tempkey, &pIndex) == hit then
7:     // The entry was expanded at least once
8:     // Remember the parent entry to add to its expanded list
9:     parent ← pIndex
10:    tempkey.srcIP ← key.srcIP
11:    tempkey.dstIP ← key.dstIP
12:    if TCAMLookup(tempkey, &pIndex) == hit then
13:      // The entry was expanded twice
14:      pThisEntry ← AllocateFromFreelist()
15:      pThisEntry.key ← key
16:      // Add new entry into the list
17:      pThisEntry.nextEntry ← pIndex.ExpandedList
18:      pIndex.ExpandedList ← pThisEntry
19:    else
20:      // There was only one expansion
21:      FillWithWildCards(tempkey)
22:      tempkey.srcIP ← key.srcIP
23:      tempkey.dstIP ← key.dstIP
24:      pThisEntry ← AllocateFromFreelist()
25:      pThisEntry.key ← tempkey
26:      // Add new entry into the list
27:      pThisEntry.nextEntry ← parent.ExpandedList
28:      parent.ExpandedList ← pThisEntry
29:    end if
30:  else
31:    // There was no expansion
32:    FillWithWildCards(tempkey)
33:    tempkey.srcIP ← key.srcIP/24 // Put wildcards to last octet
34:    tempkey.dstIP ← key.dstIP/24
35:    pThisEntry ← AllocateFromFreelist()
36:    pThisEntry.key ← tempkey
37:    // Add new entry into the list
38:    pThisEntry.nextEntry ← TopList.Head
39:    TopList.Head ← pThisEntry
40:  end if
41:  UpdateFlowInfo(pThisEntry, packetinfo) // Update flow information
42: end if
```

Fig. 4. Pseudocode for new flow addition

ID gets the policy applied to removed flow. Also control application can assign specific traffic class to each flow. Each class has its own WRED parameters, and scheduling priority or weight. For the WRED and scheduler microblocks IXA SDK Framework modules are used.

4 Performance Evaluation and Analysis

Currently we have implemented our approach on Intel IXDP2400 Development Platform, that contains two IXP2400 Network Processors and has target processing rate equal to 4Gbps. Although the overall application is in the debugging

Table 1. Maximum flow inter-arrival rates for each aggregation level

Aggregation type	Avg. Time Taken, ms	Max. number of flow per second
Network-to-network	0.0156	64,103
Host-to-host	0.0175	57,143
Process-to-process	0.0194	51,546

stage, we present some performance analysis and discussions available now. We used Smartbits SMB-6000 traffic generator to test the performance of fast data path. Since we moved new flow generation procedure from usual data path to XScale core the classifier operation on data path will only include one TCAM lookup, thus processing time of classifier microblock running on fast data path has very low variation. Overall performance of data path in the worst case that includes classification, IP forwarding, and token bucket metering is 5.2 Mpps, which allows aggregated processing rate of 3.5 Gbps in the worst case.

It is obvious that bottleneck of the system is new flow arrival processing. We have measured the processing latencies for flow generation to find out the maximum flow inter-arrival rate that the system can handle, without any loss of information. We have to mention that in most of cases that flow statistics are not affected strongly, when flow inter-arrival rate exceeds the maximum supported rate, because information of several packets have low weight in overall flow statistics, especially when flow is long. Processing, required for each aggregation level, differs in complexity. Table 1. shows corresponding results for each flow aggregation type.

Our approach exploits TCAM operation and makes use network addresses and port numbers that will never occur in the packet, like A.B.C.0 IP address and 0 port number. The list traversal complexity for timeout detections does not increase even though it becomes 2-dimensional, and requires the same $O(N)$ memory accesses, like simple NetFlow. Rate threshold exceed checking is merged with timeout checks, so it also does not increase the proposal complexity. According to our estimation, implementation of the proposed architecture on Radisys ATCA-7010 packet processing module can provide flow monitoring and control on 10Gbps link rate.

5 Conclusion

NetFlow is the traffic measurement solution most widely used by ISPs to determine the composition of the traffic mix in their networks. However, increasing diversity of applications may cause the situation, when NetFlow will have to export entry on every new flow arrival. Most of solutions countermeasures have the goal of protection of NetFlow cache from DDoS attack, that reduces flow locality on several magnitudes. In this paper we propose an architecture for high-speed scalable and flexible flow monitoring and control system with rate-based dynamic aggregation and expansion of flows. Our proposal uses 3 aggregation levels that allow flexible and dynamic flow aggregation or expansion. Proposed approach does not suffer from majority of DDoS attacks, since most of current DoS and DDoS attacking flows are short and are mostly directed to attack certain protocol level or CPU resource of victim, rather than the bandwidth. Another advantage of the proposed scheme is that it does not require complex processing and can be used on high-speed backbone links. Implementation of the proposed architecture on Radisys ATCA-7010 allows combining flexibility and high processing speed, while maintaining scalability. Integrated control functions allow network operator to respond the changes in network behaviour of some flows. Our future work includes: finishing implementation of the proposed architecture, getting various performance results and comparing them to other approaches. Also, we plan to improve algorithm to enable fast DDoS attack detection by applying more complex and sophisticated aggregation schemes.

References

1. Fang, W., Peterson, L.: Inter-AS traffic patterns and their implications. In: Proceedings of IEEE GLOBECOM, Rio, Brazil (December 1999)
2. Introduction to Cisco IOS NetFlow - A Technical Overview, Cisco Systems (February 2006)
3. Patcha, A., Park, J.M.: An adaptive sampling algorithm with applications to denial-of-service attack detection. In: Proceedings of 13th ICCCN (October 2004)
4. Estan, C., Keys, K., Moore, D., Varghese, G.: Building a better NetFlow. In: Proceedings of ACM SIGCOMM, ACM Press, New York (2004)
5. Hu, Y., Chiu, D.-M., Lui, J.C.: Adaptive flow aggregation - a new solution for robust flow monitoring under security attacks. In: Proceedings of IEEE/IFIP NOMS 2006, Vancouver, Canada (April 2006)
6. ATCA-7010 Packet Processing Module: Hardware Reference, Radisys Corporation (November 2005)
7. Intel Internet Exchange Architecture Portability Framework Developer's Manual, Intel Corporation (November 2004)
8. Intel IXP2800 Network Processor Hardware Reference Manual, Intel Corporation(August 2004)
9. Cisco Netflow, `http://www.cisco.com/warp/public/732/Tech/netflow/`
10. Sadasivan, G., Brownlee, N., Claise, B., Quittek, J.: Architecture for IP flow information export. IETF, draft-ietf-ipfix-architecture-12 (September 2006)
11. Thompson, K., Miller, G., Wilder, R.: Wide-area internet traffic patterns and characteristics. IEEE Transactions on Networking, 10–23 (November 1997)
12. Dressler, F., Munz, G.: Flexible flow aggregation for adaptive network monitoring. In: Proceedings of IEEE LCN Workshop on Network Measurements 2006, Tampa, Florida, USA (November 2006)
13. Heinanen, J., Finland, T., Guerin, R.: A single rate three color marker. IETF, RFC 2697 (September 1999)
14. Heinanen, J., Guerin, R.: A two rate three color marker. IETF, RFC 2698 (September 1999)
15. Intel Internet Exchange Architecture Software Building Blocks Developer's Manual, Intel Corporation (November 2004)

Quantitative Analysis of Temporal Patterns in Loosely Coupled Active Measurement Results

Marat Zhanikeev[1] and Yoshiaki Tanaka[2]

[1] School of International Liberal Studies, Waseda University
1-21-1 Nishi-Waseda, Shinjuku-ku, Tokyo, 169-0051 Japan
[2] Global Information and Telecommunication Institute, Waseda University
1-3-10 Nishi-Waseda, Shinjuku-ku, Tokyo, 169-0051 Japan
Research Institute for Science and Engineering, Waseda University
17 Kikuicho, Shinjuku-ku, Tokyo, 162-0044 Japan
maratishe@aoni.waseda.jp, ytanaka@waseda.jp

Abstract. With constantly increasing complexity of active measurement methods, the issue of processing measurement results becomes important. Similarly to traditional pattern discovery, temporal patterns found in active measurement samples should be provided effective storage and means to compare to other samples. Traditional time series data mining is not applicable to temporal patterns in active measurement time series. This paper proposes a pattern discovery method based on unique features of active measurement results. The method is implemented in form of a database and is used in the paper to verify the proposed method.

1 Introduction

Active measurements are becoming more and more important in network operation. Often referred to as probing, active measurement methods are now used in network performance monitoring and management, traditionally an area restricted to a limited set of passive technologies operating directly on network hardware.

An active measurement method normally consists of two parts. One part defines the structure of the actual active probe, which is simply a sequence of packets transmitted into the network in order to infer characteristics of the network path based on interference of probing packets with traffic existing in the path at the time of probing. Another part is of equal importance and consists of procedures which process the results of active measurements in order to produce an intelligent output.

Presently, most of existing active measurement tools invest considerable research into the first part, i.e. the structure of the probe and probing technique, while underestimating the need to process results. A good overview on a number of currently existing active measurement tools can be found in [1].

This paper focuses specifically on processing of measurement results and does not propose any original probing method. Instead, single-packet probes are sent at regular intervals to obtain one-way delay samples which are then processed for

S. Ata and C.S. Hong (Eds.): APNOMS 2007, LNCS 4773, pp. 415–424, 2007.

temporal patterns using the proposed method. The term "measurement series" used in this paper will refer to active measurement results, specifically, a set of time/one-way delay pairs ordered by time. An index of measurement series is a set of patterns in numeric form that describe measurement series and can be stored and retrieved from a database. The word "signal" may also be occasionally used to refer to measurement series.

Section 2 of this paper contains an overview of traditional methods of mining time series and consideration of measurement series in traditional context. Section 3 contains the proposal for pattern discovery in measurement series based on unique properties of active measurement. Section 4 describes implementation of the method as a database with table-based and graphical search functionality. Section 5 compares the performance of the proposal with traditional data segmentation. A conclusion is finally drawn in Section 6.

2 Traditional Pattern Analysis in Time Series

According to [2], traditional data mining in time series has two fundamental approaches. One approach splits data into segments using a given criteria in such a way that original series can be reconstructed based on its segments. The other approach transforms data into phase space by analyzing patterns in changes over a number of consecutive samples. Both approaches create an approximated version of the original time series based on patterns discovered in it. However, they face a number of difficulties with measurement series, which are discussed in this section.

2.1 Segmentation and Phase Space Analysis

Given a time series $X = \{x_i : i = 1, 2, ..., N\}$, it can be split into k segments $\{x_{ij} : i \in N, j \in N, i < j\}$ based in a given criteria in such a way that $X \to K$, i.e. original time series can be represented by values of its segments, i.e. $X = \sum_{i=1}^{k} X_i$, where X_i is a metric for segment i specified by the approximation method concerned. When identifying trends in time series, the segment-based metric is often the plain average of all the samples in the segment.

In phase space analysis, instead of splitting samples into shorter segments, time series is analyzed for change patterns. Change patterns are defined as differences among k consecutive samples in time series. Accordingly, the notion of dimensionality of phase space is introduced and stands for the "depth" of phase space, i.e. the number of consecutive samples that occupy the phase space at any given point of time. Let us consider a time series $X = \{x_i : i = 1, 2, ..., N\}$. If the phase shift of j is denoted as s_j and stands for quantitative difference between x_i and x_{i-j}, where $i = j, j+1, j+2, ..., N$, then the time series can be represented as a set of vectors $\sum_{i=k}^{N} \sum_{j=i}^{i+k} s_j$ in phase space.

In practice, dimensionality of phase space rarely exceeds 2 or 3, which allows relatively easy visual presentation of patterns found in time series. A 2-dimensional phase space, for example, grasps instant changes in signal and allows to separate them from minor changes.

There are more complicated methods for extraction of patterns in time series based on the two approaches. Most current collection of research in this area can be found in [3].

2.2 Reality of Active Measurement Time Series

Measurement series is special because it is the opposite of traditional time series. If traditionally the main trend in time series is the main target, in active measurement time series it conveys no information and should be cut off. Similarly, the noise, i.e. short-term variations in measurement series contain information on network performance and should be analyzed, while in traditional processing noise would be removed.

These unique features in measurement series relate to the main theory of active measurements, which states that network performance can be inferred by analyzing how probing packets interfere with traffic existing in the network at the time of probing. Naturally, when a path in question has no traffic, measurement results would be static with zero variance. The more traffic in the path, the more probing packets interfere with it, thus, creating variance in measurement samples.

Another major limitation for measurement series is the negligibly small role of the absolute value of samples in it. For example, let us suppose that time series obtained from a path from Japan to Europe is compared to a path from Japan to USA. Given the aggregate physical lengths of both paths are comparably equal, there is a good chance that absolute one-way delay would be the same in both cases. Naturally, a direct comparison would say that both time series are similar because the main trends in them are near equal.

Similarly, even if the main trend in the two above time series is removed completely, it would not guarantee that signals would be different. Variance in measurement time series tends to have certain patterns, such as very short-term surges, white noise originating from network equipment, etc. Practical experience proves that both absolute and relative values of measurement results are not sufficiently reliable to facilitate extraction of patterns or effective indexing.

Active measurements also face the problem of loose coupling. There is no guarantee that two measurements performed from two different points in a network are perfectly synchronized. Therefore, a method that extracts and uses temporal patterns from measurement data has to offer a certain degree of flexibility in time.

Based on the above factors, basic assumptions about measurement series should be defined before one could apply traditional statistical methods. These assumptions can be declared as follows:

- a measurement series always has a static trend, which does not carry any valuable information;
- any excursion from the main trend has finite duration and will always return back to the main trend (referred to as "excursion");
- since measurements cannot be synchronized, excursions in two dependent time series can occur at slightly different times.

This paper proposes a method for discovery of temporal patterns in measurement series based on unique properties of active measurement results.

3 Temporal Patterns Based on Active Measurement Properties

This section proposes a method for discovery of patterns in measurement series. The method is based on the intrinsic properties of time series originating from probing and cannot be applied as a general data mining solution, just as traditional data mining is not applicable to measurement data. However, it can still be used in similar cases, i.e. where change in time series is more important than the trend.

3.1 Reduction of Dimensionality Based on Excursions from Minimum

In any measurement series there is a global minimum provided communication path remains unchanged during probing. This assumption does not hold when load balancing or other dynamic routing techniques are used, but can still be satisfied by calculating a global minimum at regular intervals.

This paper uses one-way delay, which has a global minimum for a static path. However, this can be generalized for other performance metrics. For example, interarrival time between packets has a globally static mean, and therefore can be split into two series containing mean with positive change and inversed mean with negative change. Similar linear transformation can be performed for other metrics as well.

If global minimum in measurement series T is found to be L, two major events can be defined in a reliable way. First, for any measurement series $X = \{x_i : i = 1, 2, ..., N\}$ there is a set of points in time $t \in T$, where the following condition is true:

$$S = \{t : x_{t-1} = L, x_t > x_{t-1}, \forall t > 1\}. \quad (1)$$

Similarly,

$$F = \{t : x_t = L, x_{t-1} > x_t, \forall t > 1\}. \quad (2)$$

Sets in (1) and (2) are called start and end sets, respectively. Provided a subset of measurement series from position i to position j is denoted as $x_{i,j}$, a set of excursions from the global minimum, created from k start and k finish sets can be defined as:

$$E = \{x_{i,j} : i \in S, j \in F, j - 1 < i < j < i + 1\}. \quad (3)$$

Naturally, there are as many excursions as there are elements in the end set F, however there may be one less end sets than start sets as the last excursion may not reach the end by the end of the measurement series. However, this is a minor and infrequent condition, and the number of elements in both sets is normally considered equal.

3.2 Temporal Patterns as Sequences of Excursions

Absolute or relative values are not reliable in measurement series, which is why from the set of excursions in (3) only duration of excursion $d_{i,j} = t_j - t_i$ is used. It is, however, desirable to introduce additional dimensions to improve quality of pattern discovery.

If $D = \{d_{i,i} : i \in S, F\}$ is the set of durations of excursions $E = \{x_{i,i} : i \in S, F\}$, where notation $x_{i,i}$ reads "samples from position i in S to position i in F", then, a time gap between excursions can be written as $G = \{d_{i,i-1} : i \in S, F\}$, i.e. the time between the end of the previous excursion and the beginning of the current one. Notation $d_{i,i-1}$ reads "absolute value of time interval between time at $i \in S$ and $i-1, i \in F$", and is insensitive to order.

Additional dimension can be quantitatively defined as $D - G$ or $\sum_{i=1}^{k}(d_{i,i} - d_{i,i-1})$, where i is the sequence number of an excursion. We refer to this space as **capacity**, and distinguish it from **time** dimension, which is used to keep correct sequence of excursions in time.

This paper uses two distinct cases to deal with capacity of patterns. Case 1 takes into account the gap between excursion by applying a "gap penalty", i.e. decreasing the capacity. Case 2 ignores the gaps and concentrates only on excursion durations, which is why the capacity in this case constantly increases. Case 1, on the other hand can have negative values for capacity. Both increments and decrements create lines at 45 degree angles to horizontal line, because time is incremented by the same value.

Obviously, Case 1 is better at retaining temporal sequences of excursions, while Case 2 is better at identifying the total capacity of excursions in measurement series. Both methods are used in this paper to compare the quality of pattern discovery.

4 Implementation of Database Storage and Search

This section provides implementation of the proposed method. Simulation was used for verification to provide high precision of timestamps in packets.

4.1 Network Topology and Traffic

A ring topology in Fig.1 is used to obtain measurement data. Five measurement paths from P1 to P5 are defined in such a way that some paths share topology, and, therefore, share patterns in measurement series. Some paths are completely independent and should have different sets of patterns. Specifically, P1 and P3, P2 and P4 and P4, P4, P5 have partially shared topology. The other combinations are completely independent.

To create interference, three flows of cross-traffic were created using real packet traces from different links within the WIDE network in Japan [4]. Details on how to use real packet traces in simulation to generate authentic traffic conditions can be found in our previous study in [5].

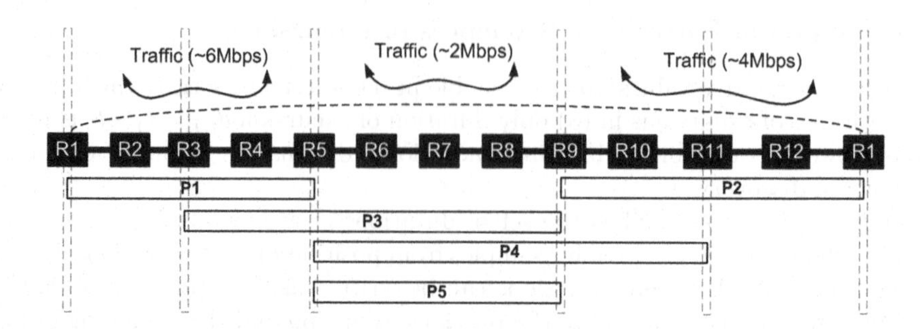

Fig. 1. Network model with paths sharing topology

4.2 Table-Based and Graphical Search Functionality

The proposed method was implemented in form of a database that was used to store measurement series and patterns discovered in it. The database allows to query its indices using another table, with graphical output in Fig.2.

Fig.2(a) is the result of comparison between a pair of existing database and query indices, which automatically generates the plot visualizing differences in patterns. Fig.2(b) is a plot of distances that are displayed as gray lines between patterns in Fig.2(a). Visually, Fig.2(b) is helpful for analyzing dynamics of compared indices as a function of time. Finally, Fig.2(c) is another automatically generated plot that displays the volume of distance between indices. Although it might seem that distance between indices and volume of this distance are mutually exclusive, they convey different messages. In Fig.2, for example, although distance between indices grows with time, volume of this distance in relatively small. In fact, the shorter the gap or excursion duration, the smaller is the affect of distance between indices on distance volume. Distance and distance volume can also be presented as error and volume of error.

Logically, while flexibility of patterns in time should be provided, it is natural that the longer the gaps or excursions, the higher is the potential error in comparison. Therefore, long excursions or gaps introduce higher penalties or rewards into the total capacity of the pattern. On the other hand, very small excursions created by white noise hardly affect distance volume. This makes the capacity dimension a natural weighting solution insensitive to random noise in measurement series.

Finally, Fig.2(d) is pair-based quantitative comparison between indices in base and query tables. Graphics and tables are provided for each of several metrics used in the search, but for simplicity we will only consider distance as the most practical indicator. When such a search is performed in practice, database should naturally return the closest time series based on the distance. As output is different for each pair and is calculated on the fly, it is difficult to come up with a global numbers for "relativity" or "hit ratio", as is the case of text-based search, so distance or distance volume are the only means of providing searches.

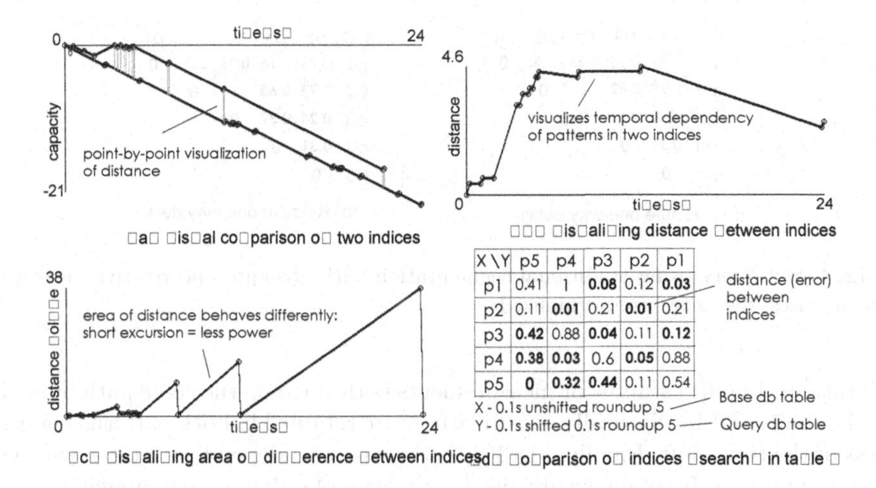

X\Y	p5	p4	p3	p2	p1
p1	0.41	1	**0.08**	0.12	**0.03**
p2	0.11	**0.01**	0.21	**0.01**	0.21
p3	**0.42**	0.88	**0.04**	0.11	**0.12**
p4	**0.38**	**0.03**	0.6	**0.05**	0.88
p5	**0**	**0.32**	**0.44**	0.11	0.54

Fig. 2. Graphical and table representation of search results by comparing pairs of base and query indices

4.3 Cases Proving Robustness of Proposed Indices

Fig.3 contains several cases learned by experience using the implementation of the proposed method. Fig.3(a) is the case of different length of indices. Visual analysis in this case is enough to tell that two time series are completely independent. Fig.3(b) stands for indices of the same length originating from shared topology, hence the similarity of time sequence of patterns. Capacity may be negative and means that aggregate gaps in patterns are longer than aggregate excursion

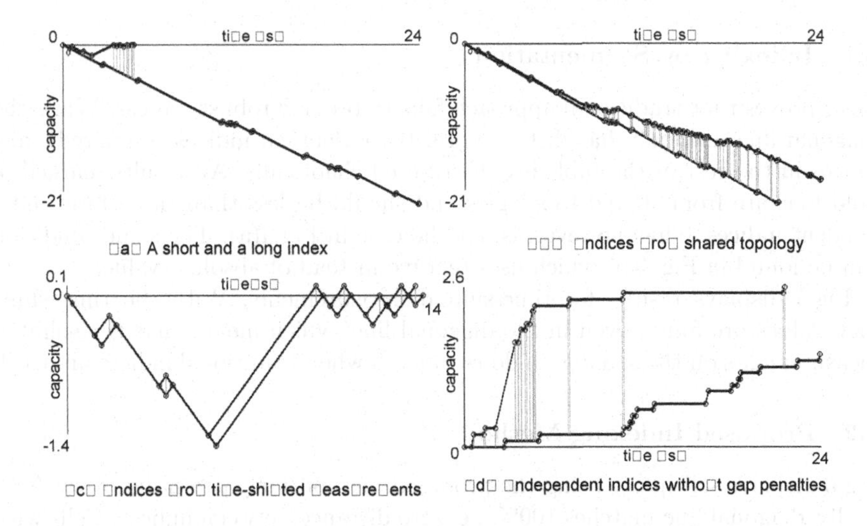

Fig. 3. Visualization of various cases experienced in searches

X \Y	p5	p4	p3	p2	p1
p1	0.31	0.28	**0.26**	0.78	**0**
p2	0.99	**0.82**	1	**0**	
p3	**0.4**	**0.51**	**0**		
p4	**0.31**	**0**			
p5	**0**				

(a) Absolute one-way delay

X \Y	p5	p4	p3	p2	p1
p1	0.26	0.16	**0.26**	0.78	**0**
p2	0.93	**0.63**	1	**0**	
p3	**0.24**	**0.37**	**0**		
p4	**0.31**	**0**			
p5	**0**				

(b) Relative one-way delay

Fig. 4. Self-query using traditional segmentation with absolute and relative measurement series, distances are normalized

durations. Fig.3(c) stands for measurements performed on the same path by with a time shift of 0.1s. Naturally, most features are retained in both patterns regardless of the time shift. Finally, Fig.3(d) is the case of independent indices with no penalty for gaps between excursions, i.e. the case of only positive capacity.

5 Performance Analysis of Pattern Discovery

This section proves the robustness of the proposed method in discovery of patterns versus traditional segmentation approach. For verification, two common cases of shared topology and loosely coupled probing are studied. In the former case self-queries are performed by studying similarities between paths in the same table, and the latter case studies two separate tables created by measurement results with 0.1s time shift. Time shift refers to measurements conducted on the same path but with a time shift between each probe in the series, and allows to explore resilience of patterns to loose synchronization between probes.

5.1 Indexing by Segmentation

Fig.4 proves that traditional approach fails to provide robust indices. While the diagonal line naturally has distance of 0 (two identical indices compared), distances in the rest of the table are distributed chaotically. All results marked in bold font are from shared topologies and should be less than those form independent indices. This, however, is not the case in Fig.4(a). The same condition can be found in Fig.4(b) which uses relative instead of absolute values.

Fig.5 displays results for time-shifted measurements. With the time shift, high values are found even in the diagonal line, which means that time-shifted measurements on the same path do not match when traditional indices are used.

5.2 Proposed Indexing Method

Fig.6 displays self-query using the proposed method for pattern discovery. Naturally, diagonal line matches 100%, i.e. zero distance between indices. Cells with partially shared topologies (bold) contain smaller distances compared to the rest of the table, which is above 70% on the normalized scale. This enables division

X \ Y	p5	p4	p3	p2	p1
p1	0.31	0.28	0.27	0.77	0.02
p2	0.99	0.82	1	0.19	0.78
p3	0.4	0.51	0.21	0.99	0.28
p4	0.31	0.02	0.52	0.8	0.28
p5	0	0.31	0.26	0.97	0.32

X - interval 0.1s unshifted
Y - interval 0.1s shifted by 0.1s

(a) Absolute one-way delay

X \ Y	p5	p4	p3	p2	p1
p1	0.26	0.15	0.27	0.77	0.02
p2	0.93	0.64	1	0.19	0.78
p3	0.24	0.37	0.21	0.99	0.28
p4	0.31	0.02	0.38	0.62	0.16
p5	0	0.31	0.07	0.92	0.26

X - interval 0.1s unshifted
Y - interval 0.1s shifted by 0.1s

(b) Relative one-way delay

Fig. 5. Query of segmentation-based indices using measurements shifted by 0.1s in time, distances are normalized

X \ Y	p5	p4	p3	p2	p1
p1	0.48	0.85	0.12	0.81	0
p2	0.57	0.05	0.69	0	
p3	0.18	0.13	0		
p4	0.21	0			
p5	0				

(a) With gap penalty

X \ Y	p5	p4	p3	p2	p1
p1	0.48	0.85	0.16	0.81	0
p2	0.59	0.05	0.45	0	
p3	0.16	0.18	0		
p4	0.11	0			
p5	0				

(b) Without gap penalty

Fig. 6. Self-queries using excursion space with and without gap penalty

X \ Y	p5	p4	p3	p2	p1
p1	0.48	0.86	0.17	0.82	0.03
p2	0.79	0.04	0.49	0.03	0.53
p3	0.18	0.14	0.03	0.76	0.1
p4	0.12	0.03	0.14	0.08	1
p5	0	0.13	0.12	0.76	0.48

X - interval 0.1s unshifted
Y - interval 0.1s shifted by 0.1s

(a) With gap penalty

X \ Y	p5	p4	p3	p2	p1
p1	0.48	0.86	0.17	0.82	0.03
p2	0.79	0.04	0.49	0.03	0.53
p3	0.12	0.09	0.03	0.76	0.1
p4	0.11	0.03	0.12	0.08	1
p5	0	0.11	0.09	0.76	0.48

X - interval 0.1s unshifted
Y - interval 0.1s shifted by 0.1s

(b) Without gap penalty

Fig. 7. Normalized distances in queries using measurement results shifted by 0.1s

of indices into groups based on distance, i.e. perfect matches, partially shared patterns, and independent patterns.

Fig.7 contains the time-shifted case with indices generated by the proposed method. Diagonal line has negligibly small values and is, therefore, resilient to time-shift. Remaining part of the table can also be easily split into two distinct groups using 0.5 as a threshold, which will allow to separate shared topology cases from independent measurement series.

6 Conclusion

This paper proposed a novel method for effective pattern discovery in measurement series. Due to specifics of measurement results, traditional data mining methods cannot be used to find patterns. Instead, the proposed method exploits properties specific to active measurement to discover temporal patterns.

The paper also introduced a database application developed by authors to store and query measurement series, as well as to provide online visualization of patterns and query results. The patterns discovered by the proposed method are compared with traditional segmentation approach and proved to be reliable and resilient to lack of synchronization while results by using traditional approach were mostly inconsistent with topology. Proposed patterns were sufficient to discriminate among identical, partially shared and independent topologies both with and without time shift. The use of capacity metric attributed to the proposed patterns proved that patterns are mostly dependent on duration of excursions than on time gaps between them.

Although the paper used only one-way delay, virtually any network performance metric can be used in a similar fashion since any performance metric either has a global minimum by default or can be linearly transformed to meet this condition. These particular cases will be considered in future research.

References

1. Shriram, A., Murrey, M., Hyum, Y., Brownlee, N., Broido, A., Fomenkov, M., Claffy, K.: Comparison of Public End-to-End Bandwidth Estimation Tools on High-Speed Links. In: Passive and Active Measurements (PAM), pp. 306-320 (2005)
2. Last, M., Kandel, A., Bunke, H.: Data Mining in Time Series Databases, vol. 57. World Scientific, Singapore (2004)
3. Zhu, Y., Shasha, D.: High Performance Discovery in Time Series: Techniques and Case Studies. Springer, Heidelberg (2003)
4. MAWI Working Group Traffic Archive. available at `http://tracer.csl.sony.co.jp/mawi/`
5. Zhanikeev, M., Tanaka, Y.: Issues with Using Real Packet Traces in Simulated Environments. IEICE Technical Report on Telecommunication Management No.TM2006-67, pp. 35–40 (2007)

Constella: A Complete IP Network Topology Discovery Solution

Fawad Nazir[1,2,4], Tallat Hussain Tarar[3,4], Faran Javed[4], Hiroki Suguri[5], Hafiz Farooq Ahmad[4,5], and Arshad Ali[4]

[1] National ICT Australia (NICTA), Australia
[2] University of New South Wales (UNSW), Australia
fawad.nazir@nicta.com.au
[3] European Oeganization For Nuclear Research (CERN), Switzerland
tallat.hussain@cern.ch
[4] NUST Institute of Information Technology (NIIT), Pakistan
{43faran,arshad.ali}@niit.edu.pk
[5] Communication Technologies, Japan
{farooq,suguri}@comtec.co.jp

Abstract. Network topology discovery for the large IP networks is a very well studied area of research. Most of the previous work focus on improving the efficiency in terms of time and completeness of network topology discovery algorithms and less attention has been given to the deployment scenarios and user centric view of network topology discovery. In this paper we propose a novel network topology discovery algorithm and a flexible architecture. The silent features of our work are *loosely coupled architecture, network boundary aware architecture, discovering the transparency of dumb/incorporative elements, flexible network Visualization, and intelligent algorithm for quick response to user discovery request.* To the best of our knowledge no existing solution has focused on the above mentioned requirements. After several years of research experience in developing a complete, flexible and scalable solution for network topology discovery we propose to divide it into three loosely coupled components: topology discovery algorithm, topology object generation and persistence, and topology visualization. In this paper we will present our proposed integrated complete network topology discovery solution, discuss the motivation of our proposed architecture, the efficiency and user-friendliness of our work. Our results show that the average accuracy of our algorithm is 92.4% and takes one second to discover 100 network elements.

Keywords: Network Monitoring and Management, Simple Network Management Protocol (SNMP).

1 Introduction

Network topology is a representation of the interconnection between directly connected peers in a network. A physical topology corresponds to many logical topologies each at a different level of abstraction. At the IP level, peers are hosts or routers one IP hop away from each other and at the workgroup level the peers are

S. Ata and C.S. Hong (Eds.): APNOMS 2007, LNCS 4773, pp. 425–436, 2007.

hosts connected by a logical link. Network topology constantly changes as nodes and links join a network, personnel move offices, and network capacity is increased to deal with added traffic. Keeping track of network topology manually is a frustrating and often impossible job. Network topology knowledge including the path between endpoints, can play an important role in analyzing, engineering, and visualizing networks. Most of the previous work [7][9][10][11][13][14] focus on improving the efficiency in terms of time and completeness of network topology discovery algorithms and less attention has been given to the deployment scenarios and user centric view of network topology discovery. The goal of our work is to automatically discover network topology and visualize it. We have been doing research on network topology discovery for the last few years now [1][2][3][5][6] and this paper is summary of our earlier work in order to propose an integrated complete IP network topology solution. In addition to that in the next section we explain the novel objectives of our work.

2 Contribution of This Paper

The motivation of this paper comes from the problem of discovering network topology at European Oeganization for Nuclear Research (CERN), Switzerland. At CERN there are thousands of machine, hundereds of routers/L3 switches/L2 switches, hundereds of network printers and huge clusters of computer for data analysis. Each department at CERN has its own network management team. These teams differ in the type of network management and monitoring polocies/requirements. Therefore there is a need for an autonomous network topology discovery solutions that is able to support multiple clients at the same time, flexible enough to discover the network topology given a network segments as a referecne start and management solutions and support different views for different administrator clients. Currently, to the best of our knowledge there is no solutions thats supports these type of features. With these challenges in mind, we introduce *Constella* that encompasses the following advantages:

Loosely coupled architecture: Loosely coupled architecture is an innovative, challenging and an important concept in order to discover the network topology and share it with large numbers of clients. We address this problem through decoupling the topology discovery and topology visualization algorithms by adding an additional layer "topology object generation and persistence" in between. More detail about decoupling is discussed in our previous work[5].

Boundary aware architecture: Boundary awareness is an important feature of network topology discovery for distributed management and probing restriction. To facilitate distributed network topology discovery, we have developed a boundary aware network topology discovery algorithm using which we can discover networks within a predefined management scope/domain.

Transparency of Dumb/incorporative elements: Networks contain Hubs that do not participate in switching protocols and, thus, are essentially transparent to switches and bridges in the network. Similarly, the network may contain switches from which no address-forwarding information can be obtained either because they do not speak

SNMP or because SNMP access to the switch is disabled. Clearly, inferring the physical interconnections of hubs and "uncooperative" switches based on the limited AFT information obtained from other elements poses a non-trivial algorithmic challenge.

Efficient User friendly Visualization of Large Networks: Visualization of network topology after the complete network discovery is what most of the software's support and in most of the current solutions network visualization is tightly coupled with the discovery algorithm. In our solutions the network discovery is what is done at the server side (could be one or multiple) and the network visualization is completely separate from network discovery and this is done on the client side (at CERN we need more than 100 clients at one time). The communication between discovery machine and visualization machine is done by the topology object.

Intelligent Algorithm for Quick Response to Discovery Request: Sending ICMP Echo request to all the possible IP addresses is not feasible to determine the availability of in large networks. Therefore there is a need to devise an intelligent algorithm that generated a list of IP addresses having a high probability of being assigned to devices in the network. We devised an algorithm that utilizes the entries in the ARP cache of the router to generate the list of IP addresses to be tested [3].

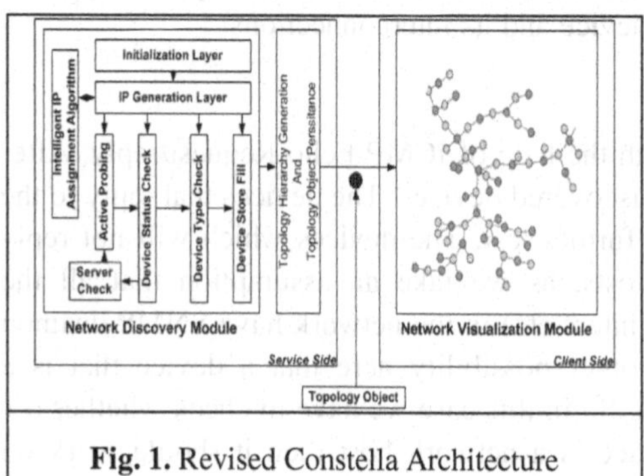

Fig. 1. Revised Constella Architecture

3 Constella: IP Network Topology Solution

This section will discuss two different proposed architectures of Constella. This section will discuss the functionality and architectural importance of three main modules of our topology discovery algorithm i.e. topology discovery algorithm, topology object generation and persistence, and topology visualization Furthermore, two revisions of this architecture are discussed.

3.1 Constella System Architecture

The basic Constella Architecture is discussed in details in[5]. Here we discuss in detail the revised architecture which has minor changes from the our previously proposed architecture [5]. The major revision in this architecture is that in the first architecture we need SNMP Agent to be installed on the machine which is running Constella but in the second architecture we don't need SNMP agent on the machine running Constella. This makes Constella more deployable and mobile. In (Fig-1) this architecture we get rid of the ARP-MAC mapping layer which was used to get the ARP information from the local host using SNMP requests. Furthermore we introduce another layer "Server Check" layer. This layer is responsible of identifying DNS,

Mail etc servers in the network and get useful topology or active host information from them and feed it in to the Active probing layer.

3.2 Constella Algorithms

In this section we will discuss all the individual algorithms and their relationship with each other in order to discover the network topology in a collaborative way.

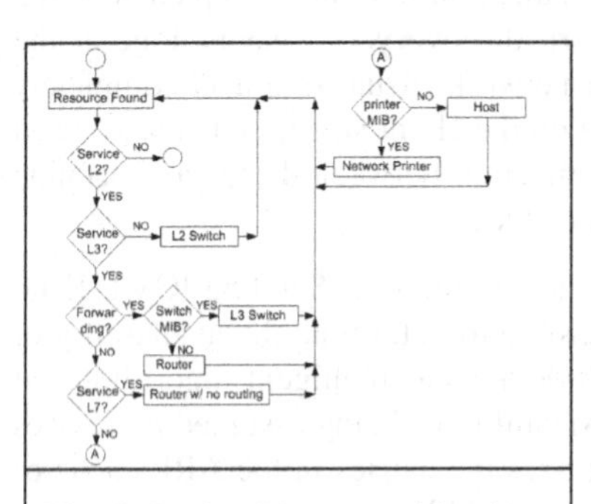

Fig. 2. Device Discovery Algorithm Flowchart

3.2.1 Device Discovery Algorithms

In this paper (Fig 2) we discuss the discovery of mainly five type of devices. These devices are network router, network printer, L2/L3 switches, computers/servers/host and network HUBs / unmanaged switches. The two main protocols used for device discovery are Internet Control Message Protocol (ICMP) Echo request/reply and Simple Network Management Protocol (SNMP). ICMP is used to discover the device availability/connectivity and SNMP is used to identify the type of device and its interconnections.

A. Host Discovery

First of all we discover the devices with the help of ICMP Echo Request/reply. After that we send SNMP request to all the discovered devices. The devices that reply to the SNMP request will be considered for further test. The devices which will not reply will be considered to be Network Hosts, as we take an assumption that all the managed devices (routers, switches, printers etc) in the network have SNMP deamon running. However, there could be another possibility here that a device that is a network host is also an SNMP daemon. So in this case we have to check whether it's a host or some other device. If a device is a network host then it should work on Layer 7.

Now we will discuss the tests which will be performed to actually see whether a device is a network host or not.

```
bash-2.05b$ snmpwalk -c public 10.10.21.8 .1.3.6.1.2.1.1.7.0
SNMPv2-MIB::sysServices.0 = INTEGER: 76
```

In the above command we have mentioned an OID .1.3.6.1.2.1.1.7.0 which is used to check the service provided by the specific host. To check a device whether its providing services on Layer 7 and 4 should have a value of sysServices as 72.

$$L4 \ \& \ L7 \ services: (2 \wedge (4-1) + 2 \wedge (7-1)) = 72$$

This means if a network devices has services value greater than $2 \wedge (7 - 1) = 64$ than it's a network host.

B. Switch Discovery

The switch discovery contains discovery of Layer 2, Layer3 and un managed switches. In the following section we will discuss about the discovery of all these switch types.

a. Managed L2 Switch Discovery

The layer 2 switches work on Layer 2 and Layer 1. The figure bellow shows an snmpwalk commands to retrieve sysServices. As shown in the output of this command the sysService value for a L2 switch (10.10.6.99) is 3. That means that the L2 switch works on Layer 1 and Layer 2.

```
bash-2.05b$ snmpwalk -c public 10.10.6.99 .1.3.6.1.2.1.1.7.0
SNMPv2-MIB::sysServices.0 = INTEGER: 3
```

This can also be proven mathematically as:

L1 & L2 services: (2 ^ (1 – 1) + 2 ^ (2 – 1) = 3

In this equation we have shown that *L1 mean 2 ^ (1 – 1) = 1* and *Layer 2 means 2 ^ (2 – 1).* Another way to identify the L2 switch is to convert the 3 into binary and upto 7 digits and then check which bits are on. The layer 2 switch will be working on those layers. The sysServices value for L2 switch is 3 and if we convert this to binary it makes 1100000. So the binary representation shows that the L2 switch works on Layer 1 and Layer 2. In this we can identify the L2 switches.

1	1	0	0	0	0	0
Layer 1	Layer 2	Layer 3	Layer 4	Layer 5	Layer 6	Layer 7

b. Managed L3 Switch Discovery

Layer 3 switches as the name states works on Layer 3 and as it's a switch so it also work on Layer 2 and Layer 1. The layer 3 switches have the capability of working on the IP Layer and switching. It can also be discovered by checking its sysServices parameters.

L1 & L2 & L3 services: (2 ^ (1 – 1) + 2 ^ (2 – 1) + 2 ^ (3 – 1) = 7

Now looking at the binary conversion method. Converting 7 to binary will be 1110000. This clearly shows that this device is a L3 switch and it works on Layer 1, Layer2 and Layer 3.

1	1	1	0	0	0	0
Layer 1	Layer 2	Layer 3	Layer 4	Layer 5	Layer 6	Layer 7

c. Un-Managed Switch Discovery

Besides SNMP-enabled bridges and switches that are able to provide access to their AFT, a switched network can also deploy "dumb" elements like hubs to interconnect switches with other switches or hosts. Hubs do not participate in switching protocols and, thus, are essentially transparent to switches and bridges in the network. Similarly, the network may contain switches from which no address-forwarding information can be obtained either because they do not speak SNMP or because SNMP access to the switch is disabled. This was also a challenge to discover the unmanaged switches or hubs. The algorithm which we have developed can not find out the hierarchy of the unmanaged switches or hubs but at least it can identify that the link which is attached to unmanaged device or devices. It can also identify that with a particular port of certain device an unmanaged switch/switches are connected.

Now we will talk about identifying an unmanaged switch which is connected to some managed switch. One Ethernet card has one MAC address and if one network card is attached to one port of the switch then in the Address Forwarding table of the switch only one MAC will be mapped with one Port. Similarly if we have a switch connected to certain port of another managed switch then you can see in the Address forwarding table (AFT) it will display multiple MAC for a single port. So if we see some multiple MAC for a single port this means that there is some switch connected to that port.

We have successfully discovered that switch is connected to certain port of the managed switch. Now we have to discover that this switch is a managed or an unmanaged switch. To discover this the MAC address of the managed discovered switch is to be compared with the all MAC address connected to a particular switch port. If the MAC address is there then this means the switch connected is a managed switch other wise if we don't find any MAC this mean it's a unmanaged switch.

C. Router Discovery

In this section we will discuss how to discover the routers in the network. In the case of the router we have to check two things. First of all we will check that whether forwarding is enabled or not. This can be checked by using SNMP IP-MIB. We check the ipForwarding parameter of IP-MIB. If the ipForwarding is one then this means it a router and if its 0 then this means it a switch.

```
bash-2.05b$ snmpwalk -c public 10.10.22.3 .1.3.6.1.2.1.4.1
SNMPv2-MIB::sysServices.0 = INTEGER: 3
```

The OID for ipForwarding is .1.3.6.1.2.1.4.1.0. Now we have discovered that the IP is a router or not. Now we also have to check the services it provides to check whether it's a soft router or a proper hardware router. For this we have to check the services it provides with the help of sysServices parameter.

$$L1 \ \& \ L2 \ \& \ L3 : 2 \wedge (1-1) + 2 \wedge (2-1) + 2 \wedge (3-1) = 7$$

The equation above shows that's it's a router and its works on Layer 1, Layer 2 and Layer 3. Now we have to check one more thing this machine can be a Layer 3 switch. As the layer 3 switches also works on Layer 3. Now to check whether is a Layer 3 switch to a router we have to do one more check. This check is for switching services. To check that a device is providing switching services or not we have to check for the Bridge-MIB. To check that a devices has implemented bridge MIB or not. We will send a request to a device for OID : .1.3.6.1.2.1.17.1 . If the machine repyles then that means it a L3 switch otherwise it's a hardware router.

3.2.2 Single Subnet Discovery Algorithm

Single subnet means devices that are under one interface of a router. The single subnet discovery is different from multiple subnet discovery, as in the case of multi subnet discovery router posses many limitations to the protocols. The MAC level

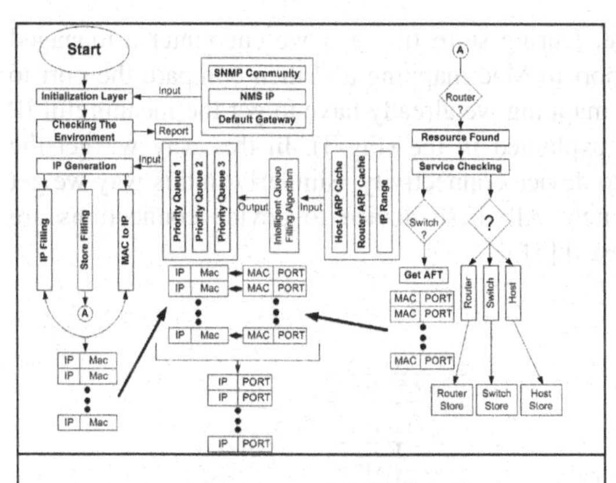

Fig 3. Single Subnet Topology Discovery Algorithm Flowchart

broadcast are blocked at the router so ARP information from multi-subnets cannot be retrived. On the other hand single subnet can utilize RARP and ARP information for topology discovery. In multi-subnet case we can not use the ICMP echo request/ reply broadcast message as the devices that are beyond the first router will not reply because router block IP level broadcasts. Due to these reasons we choose to separate algorithms for single subnet and multiple subnets. In this section we will limit our discussion to the discovery of a single subnet and then in the next sections we will extend our discussion to the discovery of multiple subnets. In the single subnet discovery we will discuss discovery of hosts, stub router, managed switches & unmanaged switches that are all inside a single subnet. Our single subnet discovery algorithm only requires one input i.e. the SNMP community string. This input is given at the initialization layer (Fig-3). Other inputs to this layer are extracted automatically by extracting information from the local operating system like NMS (Network Management Station) IP address and the gateway address of the NMS. The community string is used by SNMP to query the devices in the network. The NMS IP is used to get the range of the IP in that network. The default gateway information is used to get the ARP cache of the router in order to guess the active IP ranges in the network. Furthermore, we also get the next hop information form the router.

The next layer is the "checking the network environment" layer. In this layer we will check the network and the NMS environment that whether it's feasible for topology discovery or not. In this layer we check the following things, IP address is assigned to the NMS, ping and ARP commands are available in the NMS, gateway is assigned to the NMS, is the gateway device up, is SNMP enabled on the gateway router or not etc. This layer checks all these questions if some thing is missing then it gives report to the user otherwise it continues toward topology discovery interface. The next layer is the IP generation layer. This layer generates the IP's in different priority queues. The priority is given to IP's that are most likely to be assigned and discovered in the network. The algorithm used in this layer is discussed in detail in a our previous work[3]. This layer use information from multiple sources like ARP cache of the gateway router, ARP cache from the NMS and the IP range of the NMS. After this information is gathered by the IP generation layer this layer initiates three more layers IP Filling, Store Filling and MAC to IP Layer. These three layers are three different threads that will work in parallel. The IP Filling layer is responsible to check for alive computers and then store them into a data structure. The MAC to IP layer builds a MAC to IP mapping of each of the resource discovered. The third layer that is the store filling layer is responsible of identifying the type of resource and inserting it into the particular store. During store filling if we encounter a managed switch our algorithm will get the port to Mac mapping and then compare the port to Mac mapping with the IP to MAC mapping we already have to get the meaningful IP to Port mapping. This process in explained in the (Fig-3). In this way we get the information about the switch port to device connectivity. Similarly in this way we get the connection of router with switch. All these device to device connections are discussed in detail in our previous work[5].

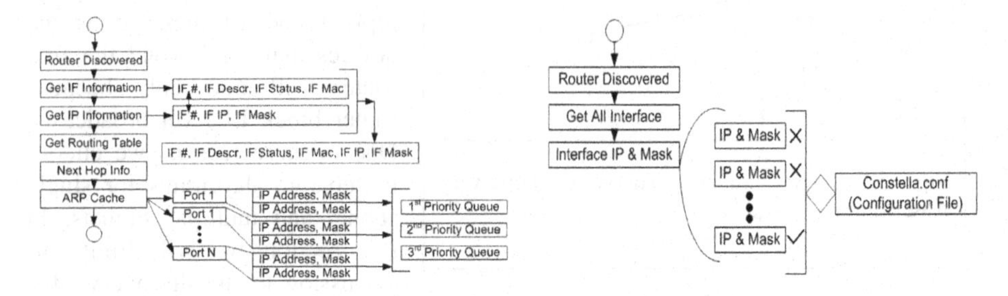

Fig. 4. Multi-Subnet Topology Discovery Algorithm Flowchart

Fig. 5. Boundary Aware Topology Discovery Algorithm Flowchart

3.2.3 Multi-subnet Discovery Algorithms

In multiple subnet topology discovery we have to discover subnets beyond a single router interface. In this case the most useful information is the routing information and *ARP* information from the routers. Now lets look (Fig 4) at how we can utilize this information from the routers to discover multiple subnets. First of all we will discuss how to discover all the subnets that are attached to a particular router. To start our algorithm we get the first router from the router store. Then get the *IF* (interface) information from the *IF-MIB*. The *IF-MIB* will contain the information about the Interface number which is a sequence number of all the interfaces in the router, the interface description that explains what type of interface it is, the interface status which tell about the interface status that whether its up or down and the interface MAC address. Now for mapping the interface to an IP-address and subnet mask we have to get the information from the *IP-MIB*. From the IP-MIB we will get information about the Interface number, the associated IP address to that interface and the subnet mask. The interface number will be used as an index between IF-MIB and IP-MIB. Therefore after mapping the IP-MIB to the IF-MIB using IF number we can get some useful information like Interface number, interface description, interface status, interface MAC address, interface IP address and the interface MASK of a particular interface. In this way you will get the IP address of all the interfaces and their subnet masks. In order to get the next Hop information we will utilize the default gateway information in the routing table of that router. Now we will discuss about the process of topology discovery of each subnet that is attached to a router. As discussed earlier that the input to our topology discovery for single subnet is decided by the IP generation layer. Originally for single subnet the IP generation layer get the input as ARP cache from the gateway router, ARP cache from the NMS and IP and Mask from the NMS. Now in the case of multiple subnets this information will be a bit different. In this case we will be getting information from the router about the specific ARP cache of each interface. In addition to that we will also have the IP and subnet mask of that interface. All this information about an interface will become an input to our single subnet topology discovery algorithm. Now similarly, iteratively we will be able to discover the topology of the whole network.

3.2.4 Boundary Aware Topology Discovery

In general, a network is owned and administrated by an enterprise, organization, or college. The network within an administrative domain is usually also the scope of network management softwares. Our Complete solution for boundary aware network topology discovery is discussed in[1]

3.2.5 Device to Device Connections

These are given in detail in out earlier work[5].

3.2.6 Alive Device Detection Algorithm

Sending ICMP Echo request to all the possible IP addresses in a network is not feasible to determine the availability of large networks. Therefore there is a need to devise an intelligent algorithm that generated a list of IP addresses having a high probability of being assigned to devices in the network. We devised an algorithm that utilizes the entries in the ARP cache of the router to generate this list of IP addresses. First of all we retrieve the IP addresses in the ARP cache of the router using SNMP e.g.

IP_{R} = {10.10.5.2, 10.10.0.1, 10.10.6.99, 10.10.5.1, 10.10.5.3, 10.10.100.1}

We then remove the last octet.s.

IP_{RP} = {10.10.5., 10.10.0., 10.10.6., 10.10.5., 10.10.5., 10.10.100.}

Then remove duplicates.

IP_{RP} = {10.10.5., 10.10.0., 10.10.6., 10.10.100.}

For all the above elements in the set now we generates 254 IP addresses by appending values from 1 to 254 in the fourth octet resulting in a list of High Priority IP addresses.

*The number of High Priority IP addresses = N * 254 (where N is the distinct first 3 octets in the routers ARP cache)*

IP_{HP} = {(10.10.5.1->10.10.5.254), (10.10.0.1 -> 10.10.0.254), (10.10.6.1->10.10.6.254), (10.10.11.1->10.10.11.254)}.

For every entry generated as a result of *Step 3* it will create more entries, One entry each by subtracting 1, subtracting 2, adding 1 and adding 2 to the third octet (if it results in a valid entry). An entry is not entered if it is already present in the list or if this address range is already in the High Priority IP address list.

IP_{ll} = {10.10.4, 10.10.3, 10.10.7, 10.10.1, 10.10.2, 10.10.8, 10.10.99, 10.10.98, 10.10.101, 10.10.102}.

For every entry, created as a result of *step 5,* it generates , 254 IP addresses by appending values from 1 to 254 in the fourth octet resulting in a list of Low Priority IP addresses

IP_{lp} = *{(10.10.4.1 ->10.10.4.254), (10.10.3.1->10.10.3.254), (10.10.7.1-> 10.10.7.254), (10.10.1.1 ->10.10.1.254), (10.10.2.1->10.10.2.254), (10.10.8.1->10.10.8.254), (10.10.99.1->10.10.99.254), (10.10.98.1->10.10.98.254), (10.10.101.1->10.10.101.254), (10.10.102.1->10.10.102.254)}*

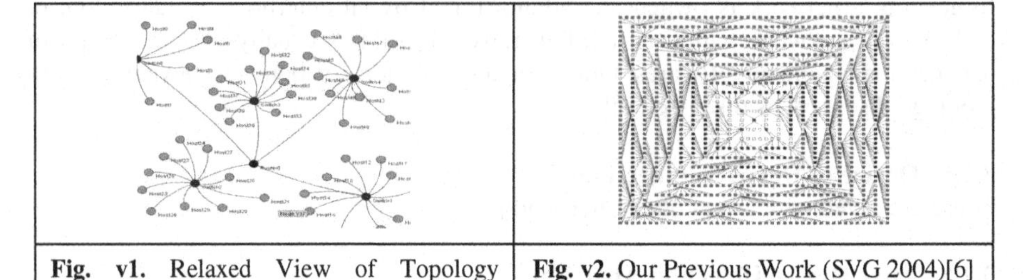

Fig. v1. Relaxed View of Topology Discovery

Fig. v2. Our Previous Work (SVG 2004)[6]

4 Network Visualization

In this section due to lack of space we will give a brief overview of our network visualization approach and present some snapshots (Fig v1-v2) from our implemented system. As our system is implemented in Java, therefore we use a java based visualization tool i.e. Java Universal Network and Graph (JUNG) Studio for network visualization. Moreover, we have also proposed a network visualization technique using Scalar Vector Graphics in our SVG 2004 paper[6]. The visualization using JUNG takes network topology information as input in a file and plots the nodes in the best possible manner. It also has a facility to locate a node or path between two nodes.

5 Performance Evaluation and Results

In this section we will discuss the performance evaluation and testing that we performed in order to validate the functionality and performance of our proposed topology discovery algorithm. Constella has been tested as a whole (Black Box Testing). Results have been drawn from the testing of each module. We have tested Constella in five different networks that have separate network configuration and devices. The algorithm was tested in Comtec Japan, CERN, Switzerland, KHU Korea, NUST research lab and NIIT, Pakistan. The accuracy of the algorithm is measured by the following formula:

$$Accuracy(\%) = \sum_{i=0}^{n} X(i)/N$$

Here n=the total number of subnets discovered and N is the total number of nodes in the network. X(i), is a single instance of a network subnet.

Table 1. First Five Graphs show the trends in nodes discovery and tables shows the algorithm accuracy

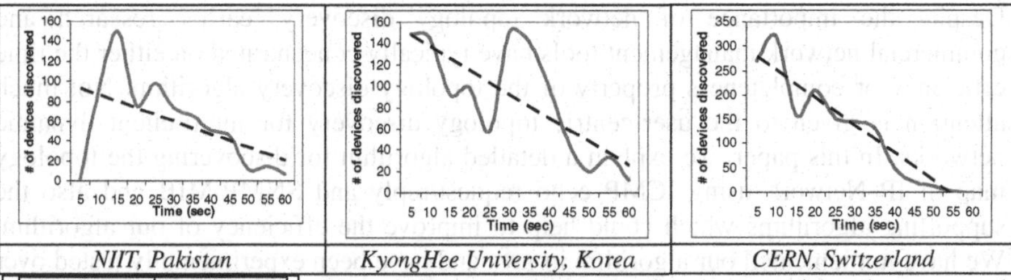

NIIT, Pakistan	KyongHee University, Korea	CERN, Switzerland

Site Name	Algorithm Accuracy
COMTEC	100%
CERN	~82%
Korea	~95%
NIIT Network	~85%
NIIT Research Lab	100%

In the above graphs dotted lines shows the node discovery trend-line and stable lines shows the actual nodes discovered.

Now a days, we have very large networks to be managed. To start discovering networks from first IP of the networks to the last IP would make it very slow and irritating for the network administrators. Currently all the software that discover the topology takes a lot of time to give initial response to the network administrator. The purpose of our approach was to discover the network in a way that large percentage of the devices in the network are displayed first and then rest of the network is discovered accordingly. The results in Table-1, show that our algorithms discovers the large percentage of the devices in the network initially. We tested the Constella algorithm for one hour in NUST Institute of Information Technology (NIIT) working network. The following are the results that we got from the log file generated. These results clearly show some very interesting trends (Fig-A). Every bar in the graph has two parts. The bottom part show the number of nodes discovered, the above part shows the nodes not discovered and the whole bar shows the total number of nodes checked/tested. This clearly shows that number of nodes discovered is directly proportional to number of nodes tested).

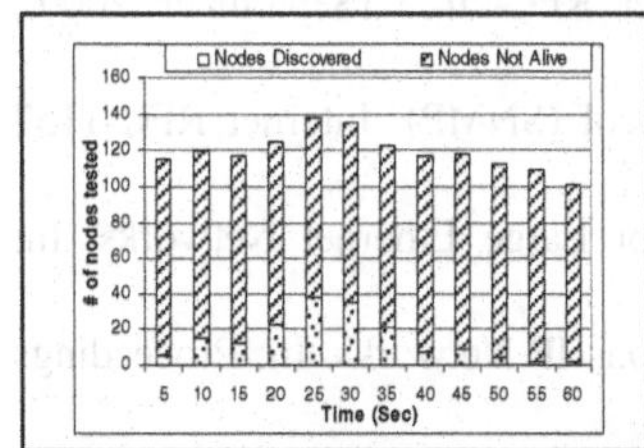

Fig-A. Relationship between nodes discovered, nodes not alive and nodes tested at NIIT.

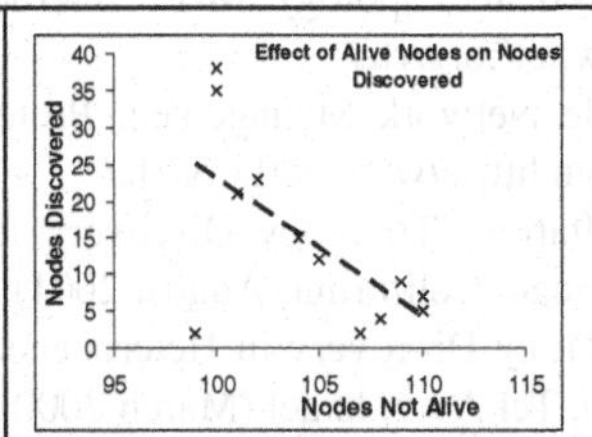

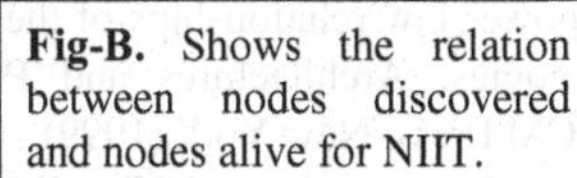

Fig-B. Shows the relation between nodes discovered and nodes alive for NIIT.

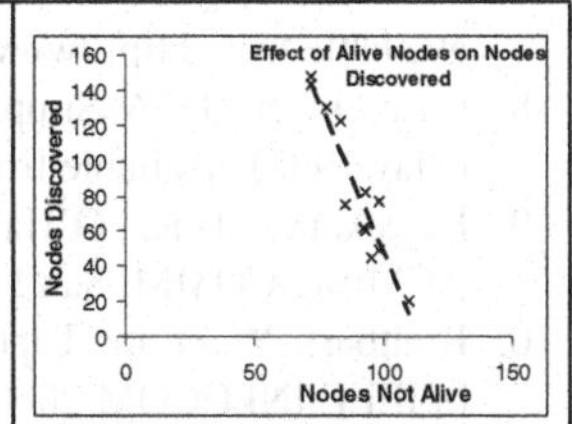

Fig-C. Shows the relation between nodes discovered and nodes alive for KHU.

The following are the conclusion from the results in Fig A,B and C.

$n \propto 1/{\sim}A$. *This equation shows, numbers of nodes discovered is inversely proportional to the number of nodes not alive(A). That mean if number of nodes not alive is less then number of nodes discovered will be more. (Fig A,B,C,D)*
$n \propto T$ *This equation shows, number of nodes discovered is directly proportional to number of nodes tested (T). (Fig A,B,C,D).* ${\sim}n \propto 1/T$. *This equation shows, number of not nodes discovered is inversely proportional to number of nodes tested. (Fig A,B,C,D) (T).*

6 Conclusion

Despite the importance of network topology discovery, earlier research and commercial network management tools have typically concentrated on either the time efficiency or completeness property of the topology discovery algorithm. Not much attention is given to the user-centric topology discovery for multi-client dynamic networks. In this paper, we explain a detailed algorithm for discovering the topology map of IP Network using ICMP echo request/reply and SNMP-MIB and also the supporting algorithms which could help to improve the efficiency of our algorithm. We have implemented our algorithm in java and have been experimentally tested over different production networks. The results clearly validate our methodology, demonstrating the accuracy and practicality of the proposed algorithms and architecture. The average accuracy of our algorithm is 92.4%.

References

1. Ali, A., et al.: Restriction of Network Topology Discovery within a Single Administrative Domain. In: CIC'05, USA, June27-30, pp. 222–225 (2005)
2. Nazir, F., et al.: Standardizing IP Network Topology Discovery through MIB Development. In: CIC'05, USA, June 27-30, pp. 226–230 (2005)
3. Najeeb, Z., Nazir, F., et al.: An Intelligent Self-Learning Algorithm for IP Network Topology Discovery. In: LANMAN 2005, Greece (September 2005)
4. Javed, F., et al.: Loosely Coupled Architecture for Integrating Network Topology Discovery Algorithms. In: HONET 2005, Pakistan (2005)
5. Nazir, F., Jameel, M., et al.: Efficient Approach Towards IP Network Topology Discovery for Large Multi-subnet Networks. In: ISCC 2006, Sardinia, Italy (2006)
6. Ali, A., et al.: Single Snapshot Exploratory Approach for Visualizing Very Large Network Topologies. In: SVG Tokyo, Japan, September 7-10, 2004 (2004)
7. Bierman, Jones, K.: Physical Topology MIB. Internet RFC-2922 (September 2000), available from http://www.ietf.org/rfc/
8. Case, J., et al.: A Simple Network Management Protocol (SNMP). Internet RFC-1157 (May 1990), available from http://www.ietf.org/rfc/
9. Lowekamp, D.R., O'Hallaron,: Topology Discovery for Large Ethernet Networks. In: ACM SIGCOMM, San Diego, California (August 2001)
10. Breitbart, Y., et al.: Topology Discovery in Heterogeneous IP Networks. In: Proceedings of IEEE INFOCOM 2000, Tel Aviv, Israel (March 2000)
11. Faloutsos, M., et al.: On power-law relationships of the Internet topology. In: Conference on Applications, Technologies, Architectures and Protocols for Computer Communications, pp. 251–262. ACM Press, New York (1999)
12. Claffy, K.C., McRobb, D.: Measurement and Visualization of Internet Connectivity and Performance. http://www.caida.org/Tools/Skitter/
13. Stott, T.: Snmp-based layer-3 path discovery. Tech. Rep. ALR-2002-005, Avaya Labs Research, Avaya Inc. Basking Ridge, NJ (2002)
14. Govindan, R., et al.: Heuristics for Internet map discovery. In: INFOCOM 2000, IEEE, Los Alamitos, CA (2000)

What Are Possible Security Threats in Ubiquitous Sensor Network Environment?

Marie Kim[1], YoungJun Lee[1], and Jaecheol Ryou[2]

[1] RFID/USN Middleware Research Team, Telematics&USN Research Division, ETRI, 161 Gajeong-dong, Yuseong-gu, Daejeon, 305-350, Korea
[2] Information Security Research Lab, CNU
220 Goong-dong, Yuseong-gu, Daejeon, 305-764, Korea

Abstract. USN infrastructure means the logical and physical space which consists of all kinds of networks, which provides useful sensing information surrounding human being. It includes wireless sensor network, RFID, Mobile RFID, IP-USN, and wired sensor network. These kinds of network are sensing information providers for human beings and use different technologies to collect information from the environment and provide those information to information consumers. USN middleware provides USN infrastructure abstraction, complicated and integrated sensing information processing, Open Application Interface to the information consumer. The importance of USN middleware security lies in this role of USN middleware. It manages sensor networks and collects sensor data on the behalf of the sensing information consumer. Therefore a malicious consumer can disrupt the sensor networks through USN middleware and collect sensitive sensing information through USN middleware. Or, malicious sensor networks may provide contaminated sensing values to consumers through USN middleware. In this case, all information provided to consumers may be contaminated and destroyed. In this paper, we analyze possible security threats in USN environments. Based on this analysis, possible counter-threat measures can be taken.

Keywords: USN, middleware, security threat.

1 Introduction

USN (Ubiquitous Sensor Network) is very hot research and development topic in Korea these days. From the various kinds of sensor devices, various kinds of sensor networks and USN applications are studied and implemented. Leading by MIC (Ministry of Information Communication Republic Korea), many USN test-beds are established to verify the related technologies and their feasibility. These are the Automatic Management of ammunitions in the Ministry of National Defense, the Monitoring of Rivers for Management of water quality, the Control of the u-City Fundamental Facilities, etc. The result briefing meeting was held in March, 2007 and they are all working well. The point we have to consider is that they work well by themselves without integrating with other projects. From a Ubiquitous point of view, all kinds of information providers have to be shared and managed in a certain unified

S. Ata and C.S. Hong (Eds.): APNOMS 2007, LNCS 4773, pp. 437–446, 2007.

way. By doing so, redundant investment cost can be reduced and interoperability among products can be brought in. For this purpose, the USN Middleware project has been going on in ETRI (Electronics and Telecommunication Research Institute) since 2006. We expect that USN middleware plays an important role in building u-Society.

USN middleware manages sensor networks and collects sensing value on the behalf of sensor information consumers. Consumers use USN middleware to collect various sensing information and control sensor networks. Conversely, various sensor networks use USN middleware to provide sensing information. Generally, the cost of establishing and maintaining sensor network is comparatively expensive. Therefore the collected sensing value must be provided to the authorized parties. Without any security considerations, valuable sensing data can flow to unauthorized consumers. In this paper, we analyze possible security threats using USN middleware. From this analysis, we can define security mechanisms to cope with and to use USN infrastructure securely.

2 What Is the USN Middleware?

There has been much research on sensor network middleware. *MiLAN*[1] operates sensor networks in a tightly-coupled-with-application way. By doing so, it satisfies user QoS and manages sensor network effectively. *Cougar*[2] considers the sensor networks as a big Database and runs query planner for in-network processing within sensor network. It optimizes whole sensor network resource usages, network traffic and sensor network lifetime. There are more sensor network middleware researches such as *SINA*[3], *DSWare* [4], etc. But until now, the heterogeneous characteristics of sensor networks are ignored or are considered as out-of-scope. ETRI USN middleware handles multiple different sensor networks and provides sensing data to multiple applications. Fig. 1 shows the concept of ETRI USN middleware [8].

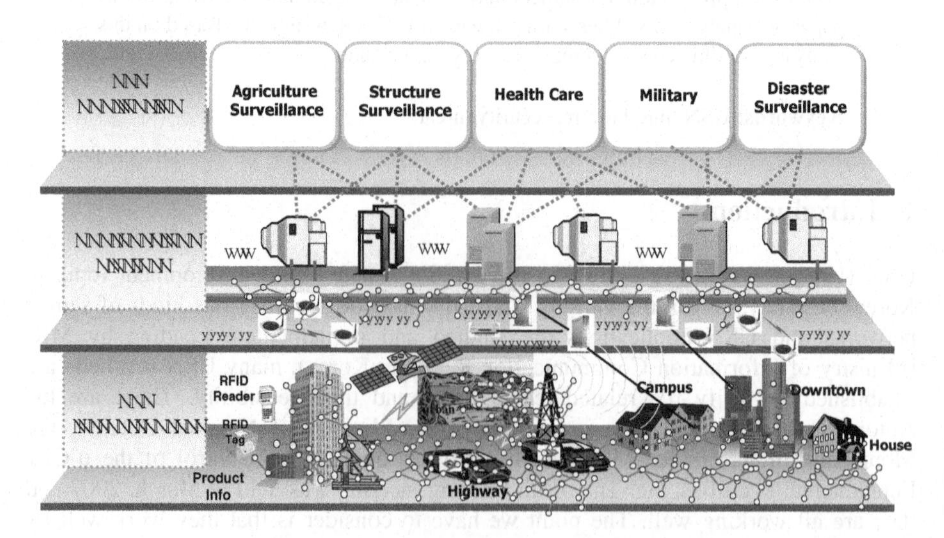

Fig. 1. ETRI USN Middleware Concept

The USN middleware is a service platform that enables integrated USN application to be developed easily and to control over the whole ubiquitous environment. USN middleware integrates various kinds of information providers such as sensor networks, RFID readers, and Mobile RFID readers, etc. It will not only collect sensing data, but also provide intelligent services such as sensing data query optimization, sensing data aggregation, sensing data mining, autonomous monitoring and recovery, etc. **Fig. 2** describes ETRI USN Middleware architecture.

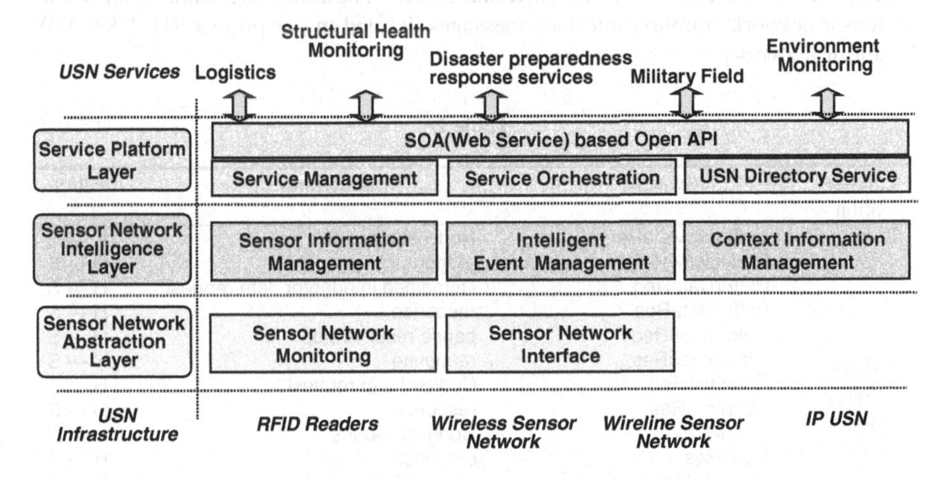

Fig. 2. USN Middleware Architecture

USN middleware comprises of the Service Platform Layer, the Sensor Network Intelligence Layer, and the Sensor Network Abstraction Layer.

- **Service Platform Layer: The** Service Platform Layer comprises of a Open API component, a Service Management component, a Service Orchestration component, and a USN Directory Service component. This layer provides an application interface for various services. Especially, the USN Directory Service contains various metadata to find appropriate resources such as components or sensor network like that.
- **Sensor Network Intelligence Layer: The** Sensor Network Intelligent Layer is composed of a Sensor Information Management component, an Intelligent Event Management component, and a Context Information Management component. This layer provides intelligent sensing data processing, intelligent event processing, and context information processing.
- **Sensor Network Abstraction Layer: The** Sensor Network Abstraction Layer comprises of a Sensor Network Monitoring component, a Sensor Network Interface component, an RFID Reader Adaptor component, and a Sink Node Adaptor component. This layer provides wireless information infrastructure abstraction and sensor network monitoring.

We'd implemented prototype of USN middleware in 2006. USN middleware prototype was implemented using SPRING framework for lightweight and platform independent implementation. The communication protocol between Sensor Network Abstraction Layer (Sensor Network Interface) and sensor networks is implemented on TCP/IP. We defined sensor network common interface messages and communication protocol. It was submitted to TTA (Telecommunication Technology Association) as Korea domestic standard. The communication between application and USN middleware is implemented using RMI and SOAP. The following Table 1. shows list of sensor network common interface messages defined in our project. (H: USN MW, S: sensor network)

Table 1.

classification	message	function	Direction
Control (Sync Mode)	NodeListMetaReq	Nodelist request	H → S
	NodeListMetaRes	response	H ← S
	BufferDataReq	Remained in adaptor data request	H → S
	BufferDataRes	response	H ← S
	CmdActionReq	pause/resume/stop	H → S
	CmdActionRes	response	H ← S
	ConnInfoReq	Connect info request	H → S
	ConnInfoRes	response	H ← S
	ConnReq	Connect request	H ← S
	ConnRes	response	H → S
	DisConnReq	disconnect request	H ← S
	DIsConnRes	response	H → S
	ChannelCheck	Check if alive	H → S
	ChannelConfirm	response	H ← S
Sensing (Async Mode)	InstantCmd	instant cmd	H → S
	InstantWithCondCmd	Instant cmd with condition	H → S
	ContinuousCmd	Continuous cmd	H → S
	ContinuousWithCondCmd	Continuous cmd with condition	H → S
	RunActuatorCmd	Run actuator	H → S
	RunActuatorRpt	response	H ← S
	StatusCheckCmd	Check status	H → S
	StatusCheckRpt	Status report	H ← S
	SensingValueRpt	Sensing value report	H ← S
	FinishRpt	Sensing finish report	H ← S
	ErrorRpt	Error report	H ← S
	SNUpdateRpt	Sensor network update report	H ← S

3 Security Analysis over USN Middleware

Before starting analyzing security threats in USN environment, we need to check how USN middleware is used first. We classify the users of USN middleware into two groups. One is USN application or administrator (consumer). Simply, we call it as an application. Application uses USN middleware to control USN infrastructure and acquire sensing information (raw sensing information or processed sensing information). The other group of users is USN infrastructure such as wireless/wired sensor network, RFID, mobile RFID, and IP-USN, etc. Simply, we call them as a sensor network.

A How Applications Use USN Middleware?

Applications use USN middleware to control sensor networks and collect sensing information from the sensor networks connected to the USN middleware. Applications send queries to USN middleware to acquire raw sensing value and/or processed information[1]. The USN middleware interprets application requests and sends requests to various sensor networks in each sensor network comprehensible ways. Multiple applications can share the sensing information through USN middleware. USN middleware can provide raw sensing value from sensor networks. In addition, USN middleware integrates several raw sensing values from different sensor networks and even more provide processed information from several sensing values and legacy data. Furthermore, USN middleware can derive processed information from raw sensing data, historical data and legacy data using mining technology, context-aware technology, and event processing technology.

Application can control sensor networks which are connected to USN middleware. Application may activate/deactivate some kinds of actuators, change sensor network topology, or even change application running on sensor node dynamically.

Usually, sensor network is powered by battery. And the devices such as sensor node, sink node, gateway are not cheap yet. Therefore, applications have to manage sensor network in a cost-effective way.

B How Sensor Networks Use USN Middleware?

Sensor networks use USN middleware to provide sensing values to the applications. Sensor network provides its sensing value as response to the request or without explicit request. Usually, sensor network is used for environmental surveillance. For example, Sensor Web[7] led by NASA JPL(Jet Propulsion Laboratory), has been used to implement a global surveillance program to study volcanos. Sensor network senses environmental parameters such as temperature, humidity, pressure, etc. The way of sensing is usually periodic with some specific interval and lifetime. Often it responds just one time on receiving the request from application. ETRI USN middleware classify the queries into 4 groups. They are Instant Query, Continuous Query, Instant Query with Condition and Continuous Query with Condition. "Instant Query" means sensor network responds only one time at receiving the request from application. "Continuous Query" means the query such as "get temperature every 30 minutes during 30 days." "Instant Query with Condition" means a kind of Instant Query restricting the response. For example, "get temperature if sensed temperature is over 30°C." "Continuous Query with Condition" means a kind of Continuous Query restricting the response. For example, "get temperature every 30 minutes during 30 days if sensed temperature is over 30°C."

---

[1] Information which are integrated and derived from raw sensing data from a(or multiple) sensor network(s).

From a USN middleware viewpoint, sensor networks are information providers. Sensing information flowing into USN middleware is flowing into several applications. Therefore, the genuineness of sensing information is very crucial to USN middleware and USN application.

C Security Threats by Applications

Application uses USN middleware to access sensor networks. As mentioned above, sensor networks are usually powered by batteries. It is the critical obstacle in USN industry. And the devices such as sensor nodes, sink nodes, gateways are not cheap until now. Therefore cost-effect usage of sensor network is very importance factor. Followings are possible security threats over USN middleware caused by applications.

- Authorized Sensing Information Acquisition
  Unauthorized user can acquire various sensing data from multiple sensor networks through USN middleware. In this case, sensor networks may work more for unauthorized users than authorized users.
- Sensor Network Disruption
  Application can control sensor network through USN middleware. By using USN middleware, application can invoke some kinds of actuators and change sensor networks configuration. By doing so, sensor networks may be configured in a wrong way. Sometimes, actuator such as alarm may cause some problems.
- Derive DoS attack
  If unauthorized user sends sensing information request to the specific sensor network continuously, then that sensor network is going to be losing all batteries. Ultimately, this sensor network is supposed to deny all requests from authorized users. This kind of DoS attack is very easily happened without any authentication and authorization mechanism support.

  The computing capability and storage available are very different among sensor networks. Some sensor network can't run multiple operations at the same time. In this case, request arrived late may be ignored or have to be queued. If unauthorized application requests sensing data and authorized application requests the same type of sensing data from the same sensor network later, then, authorized request may be ignored or queued until unauthorized request processing is over. Queued request may be discarded. This kind of DoS attack is also crucial to sensor network. That is because, sensor networks usually are powered by batteries, so they have lifetime.

D Security Threats by Sensor Networks

Sensor networks use USN middleware to provide sensing value. If unidentified malicious sensor networks flow sensing information into USN middleware, then all applications which use that information are going to be corrupted. Followings are such possible security threats to be happened by malicious sensor networks

- Wrong Sensing Value Injection

  In the USN middleware architecture, every application gets information from USN middleware. Those kinds of information are either raw sensing value or derived/integrated information from those raw ones. One of characteristics of using USN middleware is information-sharing among several applications through USN middleware, conditioned that they want sensing values from the same sensor networks. Fig. 3 depicts the wrong sensing value injection scenario. In this scenario, both of App #1 and App#2 use temperature values of SN#1. If SN#1 flow wrong values into the USN middleware, then App#1 and App#2 are disrupted together. It may cause activation of fire-alarm wrongly, or activate air conditioner at very low temperature.

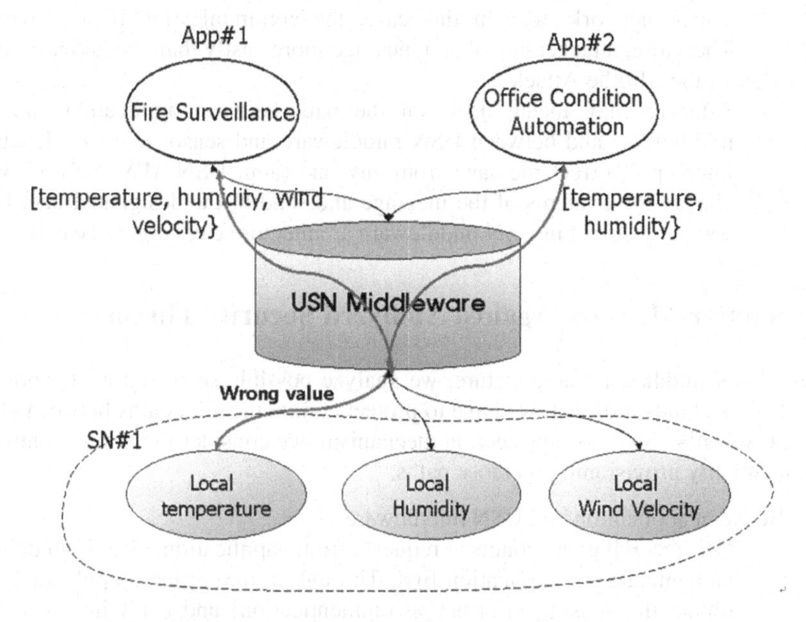

Fig. 3. Wrong Sensing Value Injection Scenario

E Threat Caused by External Object

There are two open paths may be intruded by an attacker. One is between application and USN middleware and the other is between USN middleware and sensor network. Especially, the communication path between USN middleware and sensor network is wireless. That is easy to be attacked. Followings are possible attacks to be happened by exploiting two open paths.

- Eavesdropping

  If attacker places between application and USN middleware, then it can obtain sensing information freely, without any charge. That is applied to the path between USN middleware and sensor network, too.

- Replay Attack

  If attacker locates between application and USN middleware, he can obtain commands from application. After he acquires some specific command from an application, he resends the same command later again. Then USN middleware resends the same command to the same sensor network. In this way, attacker may exhaust the power of the sensor network and it cause DoS from that sensor network.

  Or, attacker obtains sensing value from USN middleware. Now, he intercepts command to be sent later, and then he reuses the sensing value obtained from the previous command and doesn't send the command to USN middleware. In this case, an application can't detect real situation. Therefore it can't deal with emergent events appropriately.

  Above cases are applied to the interface between USN middleware and sensor network, too. In this case, the communication path is wireless. Therefore, attacker can obtain message more easily than the above cases.
- Man-in-the-Middle Attack

  Attacker may locate itself on the path between application and USN middleware and between USN middleware and sensor network. If attacker intercepts correct message from any one (app, USN MW, SN), he would change the contents of the message and send to the designated one. In this way, any object in USN middleware architecture can be failed easily.

4 Counter-Measure Against Analyzed Security Threats

In the USN middleware architecture, we analyze possible security threats above. To handle those kinds of threat, we have to protect communication paths between objects (app, USN MW, SN). As a protection mechanism, we consider the access control and confidentiality provisioning on those paths.

- Between application and USN middleware

  On receiving a connect request from application, USN middleware authenticate the application first. Through it, unregistered application can't obtain the sensing information [authentication] and can't invoke harmful operations to sensor network [authorization].

  In USN middleware project, we consider ID/password-based authentication and Certificate-based authentication. As for authorization, we classify authorization for administrator and authorization for normal application. In case of administrator, all kinds of control functions over USN middleware functionalities and sensor network infrastructure (such as security mechanism, sensor node add/delete, sensor network reset, etc) are permitted but not services provided by USN middleware. For normal application, they have to register first with requesting certain services. After registrations, they can use only those services through USN middleware.

  If secure channel (encrypted channel) is used between two objects and timestamp is used within the messages, then disclosure of sensing information, replay attack, DoS attack, and eavesdropping can be blocked.

In our project, application and USN middleware communicate with SOAP/HHTP. Therefore we consider XML-Encryption and TLS for secure channel.

- Between USN middleware and sensor network.

  When sensor network requests to connect USN middleware, USN middleware authenticate sensor network. Then sensor network, which is not trust-worthy, can't inject wrong value to USN middleware.

  In USN middleware project, we consider ID/password-based authentication and Certificate-based authentication. According to the computing and memory capability of sensor network, it can be configured either ID/password-based authentication, or Certificate-based authentication. We don't consider any authorization for sensor network.

  If secure channel is used between two objects and timestamp is used within the messages, then replay attack and eavesdropping can be blocked. For secure channel, we consider TLS.

Beyond this kind of counter-measure, we can think dynamic application download for sensor network through USN middleware. It means for the purpose of controlling sensor network securely, USN middleware downloads validated application on each sensor node. But this needs dynamic application download capability for sensor network.

If it comes to USN middleware, sensor network security is out-of-scope. The security of sensor network itself has to be check and validated by itself. It can think various wireless and wire line security between sensor nodes.

5 Conclusions

We are living in ubiquitous computing society already. We can communicate anywhere, anytime, anyone whenever we wish. There are mobile phone, laptop computers, and various kinds of Kiosk, displayers, and so forth. We can acquire any information anytime by using mobile internet, and send any information through various kinds of gadgets or computer. But to provide intelligent service to the user, the system needs the environmental information surrounding human being, that is context information, and need to access legacy system to get necessary facility information.

USN middleware is the system to provide useful environmental information around the world, such as local temperature, local humidity, and local wind velocity, etc. By using that information, many kinds of automated environmental service provisioning can be realized such as automatic fire surveillance system, automatic ocean environmental supervision system, and u-silvercare system, etc. But until now, the maintenance cost of various sensor networks is not that cheap. Therefore, sensor network needs to be operated effectively to be lasting for a long time.

To run various kinds of USN applications and sensor networks effectively, USN middleware is the one indispensable. USN middleware provides developmental and operational environment to develop USN applications easily and to manage sensor networks securely and cost effectively. But without consideration on security mechanism, whole USN computing society may be failed easily.

To proliferate USN industry, cost reduction of devices related with USN and consideration on security have to be studied and developed at the same time.

References

[1] Murphy, A., Heinzelman, W.: MiLAN: Middleware Linking Applications and Networks. Thesis, Rochester, NY (January 2003)
[2] Yao, Y., Gehrke, J.: The Cougar Approach to In-Network Query Processing in Sensor Networks. Thesis, Cornell University (2002)
[3] Shen, C.-C., Srisathapornphat, C., Jaikaeo, C.: Sensor Information Networking Architecture and Applications. IEEE Personel Communication Magazine, University of Delaware (2001)
[4] Li, S., Lin, Y., Son, S.H., Stankovic, J.A., Wei, Y.: Event Detection Services Using Data Service Middleware in Distributed Sensor Networks. Thesis, University of Virginia (2003)
[5] IEEE Standard for a Smart Transducer Interface for Sensors and Actuators-Network Capable Application Processor(NCAP) Information Model (June 1999)
[6] The Application Level Events(ALE) Specification, EPCglobal Ratified Spec. (September 2005)
[7] NASA Sensor Web, http://sensorwebs.jpl.nasa.gov/
[8] Kim, M., Lee, Y.J.: Design of Middleware for Pervasive Wireless Infrastructure. In: International Ubiquitous Conference (2006)

Security and Handover Designs for VoWLAN System*

Mi-Yeon Kim[1], Misook Lim[1], Jin-Soo Sohn[1], and Dong Hoon Lee[2]

[1] Service Development Department, Infra Laboratory, Korea Telecom
{miyeon,misook,jssohn}@kt.co.kr
[2] Center for Information Security Technologies (CIST), Korea University
donghlee@korea.ac.kr

Abstract. There is a growing interest for VoWLAN (Voice over Wireless LAN) services in the advent of network convergence and user mobility. As we design a VoWLAN system capable of handling both intra- and inter-handover real-time data, we should likewise consider its security architecture as compared to WLAN (Wireless LAN) systems. In this paper, we propose a method for VoWLAN that would provide handover and security standards designed to enhance and speed up the authentication process while handover.

Keywords: VoWLAN, Security, Handover.

1 Introduction

VoIP (Voice over Internet Protocol) is one of the fastest growing Internet services because of relatively low cost compared to the PSTN (Public Switched Telephone Networks). Concurrently, driven by huge demands for mobility, the WLAN (Wireless Local Area Network) market has grown up quickly. Thanks to the convergence of these two telecommunications trends, VoWLAN (Voice over WLAN) is expected to become an important Internet application in a few years [1].

The IEEE 802.11 standard is one of the most widely adopted WLAN standards. When the 802.11 was adopted in wireless commercial products, the biggest issue was about security. The security vulnerability of the 802.11 was principally the reason why a more-secured WLAN standard was introduced as an intermediate measure to take the place of 802.11, hence the birth of "IEEE 802.11i."

VoWLAN is a voice service based on WLAN, and that is why it should be noted that security is one of the most important functions in a VoWLAN system. Moreover, real-time voice packet for handover is also one of the most important functions needed to implement a VoWLAN system because—as far as mobility is concerned—voice communications takes precedence over data communications.

There are two types of handover. The first type is called "intra-handover." This happens when a user moves from one AP (Access Point) to another AP in the same subnet. In this particular category, handover delays are always a given. Handover

* "This research was supported by the MIC(Ministry of Information and Communication), Korea, under the ITRC(Information Technology Research Center) support program supervised by the IITA(Institute of Information Technology Advancement)" (IITA-2006-(C1090-0603-0025)).

S. Ata and C.S. Hong (Eds.): APNOMS 2007, LNCS 4773, pp. 447–456, 2007.

delays occur because of some re-authentication procedures for newly-associated APs. The IEEE 802.1x authentication protocol accepted by 802.11i consists of multi-step procedures in order to enhance the security level. However, the multi-step re-authentication causes the latency on handover. The second type is called "inter-handover." This takes place when a user moves over to different subnets. In this case, we should consider the method to control IP address transparently due to the fact that WLAN operates on L1 (Layer 1) and L2 security policies among the OSI 7(Open System Interconnection 7)-layer frames. The WLAN Access Point cannot support the inter-handover operated on L3.

In this paper, we propose the design method of VoWLAN to support handover as well as security practically. To support fast intra-handover, we present the method of separating voice traffic from data traffic using multi-VLAN technology. This method supports the fast intra-handover in the network layer and the security in the application layer for voice traffic. On the other hand, this method guarantees the security level of data traffic in network layer. To support the inter-handover, we present the method of extending the broadcasting area of voice traffic using VLAN tagging.

The remainder of this paper is organized as follows: In Section 2, we present several related items about WLAN security and VoIP security. In Section 3, we propose the practical design method solving the delay of the intra-handover as well as security support. In Section 4, we propose the extended design method to support inter-handover. Section 5 concludes this paper with a summary of the main idea and outline of future work.

2 Related Works

The VoWLAN service combines WLAN technology of network level and VoIP technology of application level. Generally, WLAN has been deployed for data communication. However, VoWLAN should be designed to support voice communication as well as data communication, effectively. Just recently, the IEEE 802.11 standard has been amended to support real-time data like voice. However, it is still a common practice that research on WLAN is focused on wireless data communications on the contrary that the VoIP is centered on wired voice communications. In this section, we present the related works about WLAN security and VoIP security. These works have been studied separately but we propose how each security mechanism should be applied together in the VoWLAN system.

2.1 WLAN Security

The IEEE 802.11 standard defined two basic security mechanisms [2]. The first mechanism is about entity authentication which includes open system and shared key authentication. The second is data encryption called Wired Equivalent Privacy (WEP). However, both have proved to be vulnerable to security attacks, and that is why IEEE developed a new security standard of WLAN known as IEEE 802.11i [3][4]. The 802.11i cooperates with IEEE 802.1X [5] to remove the vulnerability issues of entity authentication of 802.11. It also defined key management issues and enhanced encryption algorithm to provide stronger privacy.

IEEE 802.1X Standard. The IEEE 802.11 standard specified two types of authentication. The first type is the "open system" which means authentication method is not necessary. In other words, WLAN AP unconditionally permits or allows free access to all MNs (Mobile Nodes). The second type is the "shared key authentication." With this method, WLAN administrator distributes a static key to all APs and all users in its service scope. The users make use of the static pre-shared key to authenticate whenever they want to connect with the AP. The shard key authentication causes the following problems: the revealment possibility of the pre-shared key and the management difficulty of a pre-shared key for each AP.

Therefore, IEEE 802.11i combines IEEE 802.1X to address these problems. The IEEE 802.1X standard defines three main components: supplicant, authenticator and AS (Authentication Server). A supplicant means a MN equipped with NIC (Network Interface Card). An authenticator generally means the WLAN AP. On the other hand, an AS commonly represents a RADIUS (Remote Authentication Dial-In User Service) server. The AS manages users' information and MNs' information to authenticate users and MNs.

The 802.1X authenticates entities using unique information for users and MNs, and all of the authentication information is centrally managed in the AS. As a result, IEEE 802.1X solves the problems of shared key authentication of IEEE 802.11.

The 802.1X combines EAP (Extensible Authentication Protocol) in order to support various authentication protocols such as MD5 (Message Digest 5), TLS (Transport Layer Security), TTLS (Tunneled TLS), PEAP (Protected Extensible Authentication Protocol) and so on. The EAP-MD5 does not support data encryption because it does not generate a secure key after authentication. The EAP-TLS requires public key certificates of users and AS to make mutual authentication. This protocol is prone to some technical problems which often increase the cost and complexity of the system because of the maintenance of PKI (Public Key Infrastructure) to manage certificates. Hence, considering security and management, EAP-TTLS and PEAP are commonly used. Both protocols use the identifier and password of users instead of users' certificates in order to authenticate the users. Only when authenticating the AS, both protocols use the AS certificates.

Figure 1 presents the message flow when the authentication using EAP-TTLS is executed successfully. The EAP-TTLS negotiation comprises two phases: (1) The TLS handshake phase and, (2) the TLS tunnel phase [6]. In Phase 1, TLS is used to authenticate the Authentication Server and the supplicant, and optionally, the supplicant to the AS. Phase 1 results in the activation of a cipher suite, allowing Phase 2 to proceed securely. In phase 2, many of these functions are securely performed through the tunnel generated between a supplicant and an AS. These might include user authentication, negotiation of data communication security capabilities, key distribution and others.

The entity authentication, itself, may be EAP, or it may be a legacy protocol such as PAP (Password Authentication Protocol), CHAP (Challenge Handshake Authentication Protocol), and so on. Figure 1 shows the example of entity authentication using EAP-MD5. In Figure 1, EAP-REQ (EAP-request) and EAP-RES

(EAP-response) are generated by the AS and the supplicant, respectively. The authenticator only passes through the EAP information.

IEEE 802.1X authentication process is invoked after users move to new AP from original AP as well as after the users associate a new AP. As shown in Figure 1, IEEE 802.1X authentication consists of multi-round message exchange which causes the handover delay when we communicate with real-time data like voice.

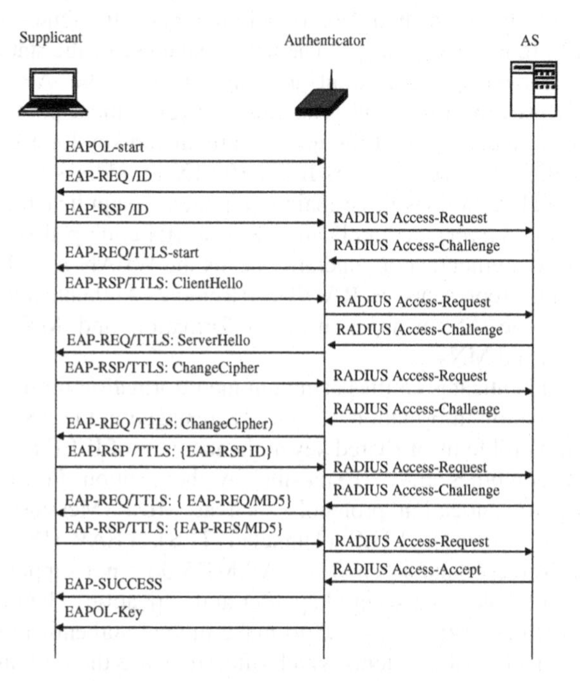

Fig. 1. Authentication using EAP-TTLS

IEEE 802.11i Standard. Weaknesses in WEP encryption mechanism are as follows [7]: The first is that key management is not specified. Hence, all users have one static pre-shared key called the "WEP key" with an encrypted data using the same WEP key for every each session. The second is the Initialize Vector (IV). The size of IV used in WEP at 24 bits is too small. Even if WEP chooses IV randomly, there is approximately 50% change of reuse after less than 5000 packets. If the cipher stream for a given IV is found, an attacker can decrypt packets with the same IV. The third is ICV (Integrity Check Value) algorithm. The WEP's ICV is based on CRC32 (Cyclic Redundancy Check 32) algorithm. It is the linear function of the message; hence, an attacker can modify an encrypted message and easily fix the ICV so that the modified message appears authentic.

On the other hand, the IEEE 802.11i proposes key management mechanisms and improved encryption algorithms. The IEEE 802.11i is based on IEEE 802.1X for entity authentication and IEEE 802.1X provides the base for key management. After succession of 802.1X authentication, the same PMK (Pairwise Master Key) is generated in the MN and an AS independently. The AS transmits the PMK to the AP and then the session key is derived by PMK in the MN, and the AP identically. After which, all data transported between the MN and the AP are encrypted using the session key.

The IEEE 802.11i defines better cryptographic protocols such as the TKIP (Temporal Key Integrity) protocol and CCMP (Counter-mode with Cipher block chaining Message authentication code Protocol). The TKIP is based on RC4 encryption algorithm which is the same as that of the WEP. However, the TKIP amends the vulnerability of WEP. The TKIP extends the size of IV to 48 bits. Secondly, the TKIP utilizes 802.1X to refresh the TK (Temporal Key). The TKIP combines the TK with the MAC (Media Access Control) address of MN and then adds a relatively large 16-bit IV to encrypt data. This method ensures that each MN uses a different encryption key. In addition, the TKIP uses sequence numbers to protect replay attacks. The TKIP also uses hash functions to generate MIC (Message Integrity Code) and prevents attackers from regenerating the ICV.

The CCMP, on the other hand, uses AES (Advanced Encryption Algorithm) with the CCM mode instead of RC4 to provide stronger encryption base.

The TKIP is considered as a short-term solution. But the important aspect of TKIP is that it only needs software upgrade using existing IEEE 802.11 AP. The CCMP is considered as a long-term solution since it is necessary to upgrade the hardware chipset of AP. Hence, it provides stronger encryption capability.

IEEE 802.11i defines pre-authentication and cache of information related with PMK as the primitive function for mobility. The pre-authentication is the function that MNs are authenticated to multi-APs including adjacent APs as well as associated APs well in advance. As the result of pre-authentication, adjacent APs can cache PMKs ahead of real association. This function makes preparations for handover in the near future. However, it is necessary to improve the 802.11i in order to support real-time handover functions for the application such as VoIP, etc. [8].

2.2 VoIP Security

The SIP (Session Initiation Protocol) [9], defined by the IETF (Internet Engineering Task Force), is a leading standard in VoIP infrastructure. The SIP architecture provides two main phases: (1) The call-setup phase and, (2) the communication phase. In the "call-setup" phase, the SIP Server establishes new sessions between SIP UAs (User Agents).

In the "communication" phase, UAs communicate to each other using real-time data. Therefore, SIP security should be considered on the two phases. In the call-setup phase, the SIP security mechanism is to protect the SIP message. Except for the difference in character sets—much of SIP message and header field syntax are identical to HTTP/1.1 specification. Hence, all security mechanisms available for HTTP can be applied to SIP sessions [10]. The use of MIME (Multi-purpose Internet Mail Extension) containers within SIP message suggests the potential use of email security mechanisms like PGP (Pretty Good Privacy), S/MIME (Secure/MIME) [11]. And similar to "http", URI (Uniform Resource Identifier) can build up a secure transport layer tunnel using TLS. Moreover, IPSec (IP Security) protocol can be used as a general purpose mechanism to encrypt all IP-based packets on the network layer.

But, version 2 of SIP deprecated the use of HTTP basic authentication and PGP in favor of S/MIME.

In the communication phase, SIP security mechanism protects real-time data transported utilizing RTP (Real-time Transport Protocol). The RTP is based on UDP (User Datagram Protocol) so that it is difficult to apply TLS based on TCP (Transmission Control Protocol). On using IPSec, the serious drawback is the large overhead incurred by the ESP (Encapsulation Security Payload) i.e., IPSec is inefficient to apply RTP. Therefore, SRTP (Secure RTP) [12] is standardized to secure RTP packets on the application layer. The SRTP can provide confidentiality, message authentication, and replay protection to the RTP traffic and the control traffic for RTP.

Before, the SIP security used to be studied on network level. However, only recently, the research for SIP security has been studied on application level because of the drawback of IPSec and TLS. The S/MIME has been studied as the security mechanism for SIP session, and the SRTP has been studied as the security mechanism for RTP.

3 The Design of Security Considering Intra-handover

When WLAN started to be used commercially, the most frustrating issue is security. In a wired line network, physical security exists potentially with controlling the entrance and for purposes of "eavesdropping," an attacker should do the task by tapping the wire. However, in a wireless network, anyone is able to intercept radio waves given the right frequency because the radio waves emanate outside the building.

The IEEE 802.11 standard organization developed IEEE 802.11i to address and give solutions to the vulnerability of WLAN. However, 802.11i is not sufficient to cover VoIP services which transfer real-time data. One of the reasons is the handover delay caused by multi-round process for re-authentication. On using voice communications, the delay of handover results in the packet loss and cascade voice distortions or session interruptions.

In this section we propose the design method to provide fast intra-handover as well as security in the VoWLAN system. The main point of our proposal is to separate the traffic route of time-sensitive voice and time-insensitive data respectively. And then each traffic route is configured in different security policies.

For data traffic on the WLAN level most mechanisms for security are applied. Both the 802.1X for user authentication and the 802.11i for data encryption are configured on the WLAN level. Loss of packet caused by 802.1X multi-round process less affects time-insensitive data.

On the other hand, for voice traffic, security mechanism is divided between on WLAN layer and on VoIP application layer. On the WLAN level, only MN authentication using MN's MAC address is applied. User authentication and voice packet encryption are performed on the VoIP application level. The user is authenticated using the challenge/response RADIUS authentication protocol. SIP session messages and real-time voice packets are encrypted using S/MIME and SRTP, respectively. By using this mechanism, re-authentication time for WLAN AP is reduced on intra-handover. In other words, minimal security is provided on the network layer, and the additional security mechanism is applied on the application layer.

In order to implement this mechanism, WLAN AP should support multi-VLAN (Virtual Local Area Network) functions and multi-SSID (Service Set Identifier) functions. Utilizing multi-VLAN and multi-SSID, Each AP is configured in several VLANs and each VLAN is named with a unique SSID. Besides, it is necessary that each VLAN is configured with an independent security policy.

Practically speaking, the VLAN for voice packet is generated on AP—named unique SSID, for example "voice"—and configured with MAC authentication as a security policy. Likewise, VLAN for data packet is generated—named unique SSID, for example "data"—and configured with 802.1X and 802.11i.

Figure 2 shows the system architecture to implement the voice VLAN and data VLAN. On this system environment, if a user wants to transfer data, the user executes the connection client for WLAN, chooses "data" SSID and connects the data VLAN. After connecting to the AP, the user inputs a user identifier and password and then requests for 802.1X authentication. If the request for authentication is successful, encryption key is derived on the user's MN and AP and encrypted data communication is performed.

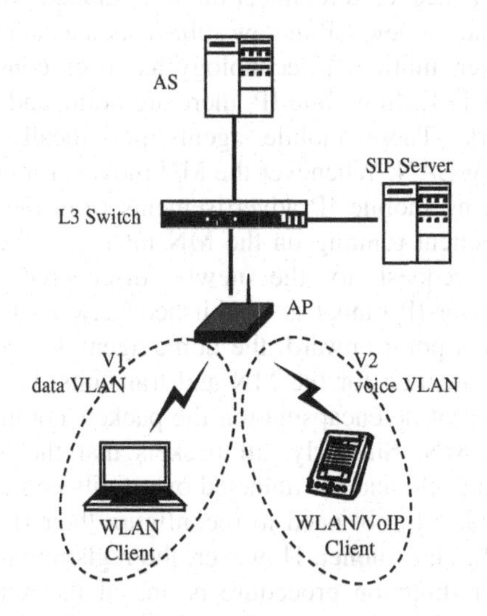

Fig. 2. VLANs for data and voice on WLAN AP

If a user wants to communicate with voice, the user executes the connection client for WLAN, chooses "voice" SSID and connects the voice VLAN. Because the voice VLAN authenticates a MN to use the MAC address, the user is unaware of it. After associating AP, the user executes VoIP client, inputs identifier and password and connects the SIP server. The SIP session is securely initiated using S/MIME protocol by SIP server and then a call requestor also securely communicates with a call responder using SRTP.

On the system architecture, a user may confuse the VLAN. The user chooses the voice VLAN and may execute e-mail client. The user may also choose the data VLAN and may execute VoIP client. To defend this case, the L3 switch ought to control the routing to prevent voice packets from transferring to the data VLAN and vice versa.

4 The Design Extension Considering Inter-handover

The IEEE 802.11 specifies only the physical layer and data link layer of the OSI 7 Layer. Therefore, WLAN AP cannot support the inter-handover performed on the network layer. In general, the network in one building consists of several subnets so that we should consider not only intra-handover caused by mobility in the same subnet but also inter-handover caused by mobility over the different subnets.

In this section, we propose the design method considered inter-handover with extending the design described in the previous section.

When a user moves from one subnet to another subnet, the IP address of the MN should be changed. However, because WLAN AP is L2 bridge equipment, it cannot treat IP address on the network layer. Therefore, in order to support inter-handover, the additional method is needed to connect the new subnet. After connecting the new subnet, the MN associates a new AP in new subnet such as intra-handover.

On the network layer, mobile IP technology has been considerably researched to support inter-handover [13]. In mobile IP, there are home and foreign agents running on the wired network. These mobile agents periodically broadcast mobile IP advertisements on the WLAN. Whenever the MN moves from home subnet to foreign subnet, it starts receiving mobile IP advertisements from the corresponding foreign agent. The mobile IP client running on the MN intercepts these advertisements and sends a registration request to the newly discovered foreign agent. After authentication, an IP-over-IP tunnel is established between the home agent and the foreign agent. From this point onward, the home agent acts as a proxy for the MN, intercepts all packets intended for the MN and transmits them over the tunnel. The foreign agent takes care of de-encapsulation the packets coming from the tunnel and forwards them to the MN. Similarly, all packets that the MN transmits are first received by the foreign agent and are tunneled over to its home agent. With mobile IP technology, mobile nodes do not need to reconfigure their IP addresses while move form home subnet to foreign subnet. However, the registration process to the foreign agent and the packet redirection procedure is one of the reasons for the handover latency. The scheme to solve the problem has been studied but the mobile IP is still far from commercial use. Besides, there is a drawback in this particular scheme since the mobile IP needs to install additional client on the MN. The installation of the mobile IP client decreases the performance of the MN because the MN—such as a small WLAN phones (shaped like mobile phones) and PDAs (Personal Digital Assistants)—has a very limited CPU-capability. As a result, the current Mobile IP technology is not suited for time-sensitive voice applications.

As a practice mechanism for inter-handover, we propose the method using VLAN in the same solution as intra-roaming. The main idea of this proposal is extending the broadcasting area of voice VLAN using 802.1Q VLAN tagging [14] in order that voice packets can be broadcasted to new subnets whenever the user moves from the original subnet to a new subnet.

Figure 3 is the conceptual system architecture applied to this mechanism. The broadcasting area of data VLAN is limited within one subnet. The voice VLAN is configured to extend broadcasting area over several subnets using VLAN tagging.

In this mechanism, whenever there is an increase in the number of users, the broadcast traffic load likewise increases. This situation may decrease the performance depending on the traffic overhead. But, this mechanism practically supports inter-handover functions.

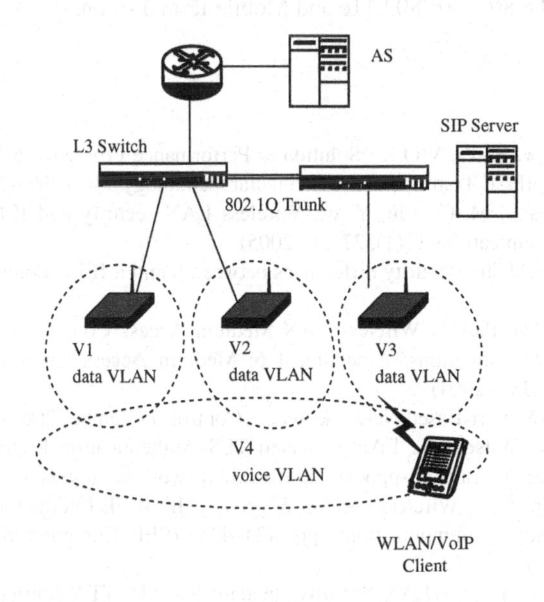

Fig. 3. Voice VLAN using 802.1Q tagging

5 Conclusions

At present, IEEE 802.11 and IETF are showing increasing interest in standardization related to real-time communications like voice. The IEEE 802.11r is working on standardizing fast handover functionalities and the IEEE 802.11e is working on QoS (Quality of Service). On the other hand, the IETF has already specified the mobile IP standard and has made considerable efforts to improve it.

Due to the increasing interest in VoIP and the growing mobile market, research and development in VoWLAN service is now more focused. However, the technology of the VoWLAN has yet to reach the practical phase as far as the security and handover are concerned.

In this paper, we propose the design method of VoWLAN in order to support the intra- and inter-handover functionalities while maintaining the security.

For this scheme, we make effective use of VLAN technology. To solve the delay of intra-handover, we separate the voice VLAN and data VLAN and configure an independent security policy considering the characteristic of each VLAN. Moreover, to support inter-handover, we extend the broadcasting area of voice VLAN using 802.1Q tagging.

The VoWLAN design which we are proposed in this paper may provide some inconveniences as the user may have to distinguish the voice VLAN and data VLAN. However, until the technology on fast handover, the QoS and mobile IP are eventually implemented and improved, the mechanism to apply a VoWLAN system is a practical necessity considering the trade-offs between security and convenience.

In the future, we plan to improve the technology of VoWLAN as we further do some studies on the 802.11r, 802.11e and Mobile IP and so on.

References

1. Wang, W., Liew, S., Li, V.O.K.: Solution to Performance Problems in VoIP Over 802.11 Wireless LAN. IEEE Transactions on Vehicular technology 54(1), 366–384 (2005)
2. Chen, J.-C., Jiang, M.-C., Liu, Y.-w.: Wireless LAN security and IEEE 802.11i. IEEE Wireless Communications 12(1), 27–36 (2005)
3. Brown, B.: 802.11: the security differences between b and i. IEEE Potentials 22(4), 23–27 (2003)
4. IEEE std 802.11i, Part11: Wireless LAN Medium Access Control (MAC) and Physical Layer (PHY) Specifications; Amendment 6: Medium Access Control (MAC) Security Enhancements (July 2004)
5. IEEE std 802.1X. Port-Based Network Access Control (December 2004)
6. Funk, P., Blake-Wilson, S.: EAP Tunneled TLS Authentication Protocol (EAP-TTLS), IETF Internet draft, draft-ietf-pppext-eap-ttls-05.txt (work in progress)
7. Borse, M., Shinde, H.: Wireless security & privacy. In: IEEE International Conference on Personal Wireless Communications, pp. 424–428. IEEE Computer Society Press, Los Alamitos (2005)
8. Kang, Yu-Sung, et al.: WLAN Security standard 802.11i. TTA Journal No 99, 123–129 (2005)
9. Rosenberg, J., et al.: SIP: Session Initiation Protocol. IETF RFC 3261, (June 2002)
10. Steffen, A., et al.: SIP Security, DFN-Arbeitstagung ber Kommunikationsnetze, pp. 397–412 (2004)
11. Ramsdell B.: S/MIME Version 3 Message Specification. IETF RFC 2633 (1999)
12. Baugher M., et al.: The Secure Real-time Transport Protocol (SRTP), IETF RFC 3711 (March 2004)
13. Sharma, S., Zhu, N., Chiueh, T.-c.: Low-latency mobile IP handoff for infrastructure-mode wireless LANs. IEEE Journal on Selected Areas in Communications 22(4), 643–652 (2004)
14. IEEE Std 802.1Q, Virtual bridged local area networks (May 2003)

An Effective Secure Routing for False Data Injection Attack in Wireless Sensor Network

Zhengjian Zhu, Qingping Tan, and Peidong Zhu

School of Computer Science, National University of Defense Technology
Changsha, Hunan, China 410073
nowaterfire@yahoo.com, 13973123266@hnmcc.com,
zpd136@sina.com

Abstract. The authenticity of the data collected by the sink is pivotal to a lot of WSN applications. But as the monitoring environment and objects are more and more complex, traditional secure protocols are not fit for the false data injection attack. In this paper, EASY, an effective secure routing for false data injection attack is presented. EASY can effectively resolve the contradiction between security requirement and additional load and guarantee to get the required security with the satiable overload.

1 Introduction

The research of wireless sensor network (WSN) currently focuses on the gathering, transmitting and processing of simple environment data [1]. However, as the monitoring environment and objects are becoming more and more complex, simple scalar data can't satisfy the requirements of applications. It is urgent to introduce information-abundant media data, such as image, audio and video, to the WSN based environment monitoring, so that we can perform fine-grained and precise monitoring [2]. Therefore stream data have become the main data processing objects of wireless multimedia sensor network (WMSN) instead of scalar data.

Secure routing is the important security guarantee of the WMSN applications. For stream data and scalar data, their network architecture and sensor capability are very different. Furthermore, their data content, size and correlation are greatly different. So the traditional secure routing protocols for scalar data in WSN cannot be applied to the stream data in WMSN.

This paper proposes an effective secure routing for false data injection attack (EASY) in WMSN. EASY is applicable to the real-time and energy-constrained applications.

The rest of this paper is organized as follows. Section 2 discusses related work. Section 3 shows EASY in detail. Section 4 gives complete security analysis. Finally, Section 5 gives concluding remarks and directions in future work.

S. Ata and C.S. Hong (Eds.): APNOMS 2007, LNCS 4773, pp. 457–465, 2007.

2 Related Work

Routing and energy efficiency are the main focus of the general routing protocols, however, the security issue is not considered [3] 4[], so the general routing protocols are vulnerable to various attacks.

The multi-path routing mechanism proposed by Ganesan etc. [5] can resist the selective forwarding attack. The main idea is to build multiple paths from the source to the destination, and transmit the copies of one packet on multiple paths. So long as there is no malicious node performing selective forwarding on one single path, the packet can be delivered to the destination.

INSENS proposed by Deng [6] limits the destruction of a malicious node to some extent. And it can continue providing routing service while not excluding the malicious node.

SEF proposed by Ye [7] focuses on identifying false data. When an event happens, SEF elects one center node CoS from all the sensors which detect the event. CoS gathers all the detecting data and generate an integrated report. Then CoS broadcasts the report to all the sensors which detect the event. If the detecting node agrees with the report, then it generates a message authentication code (MAC) and transmits it to CoS. After CoS gathers enough MACs, it notifies the sink of the event. However, the event reports about which CoS haven't gathered enough MACs will be abandoned.

Wang etc. [8] proposed a scheme to detect whether a node is faulty or malicious with the collaboration of neighbor nodes. In the proposed scheme, when a node suspects that one of its neighbors is faulty, it sends out messages to request opinions on the behavior of this suspected node from other neighbors of the suspect. After collecting the results, the node analyzes the results to diagnose whether the suspect has a fault.

Zhu etc. proposed an interleaved hop-by-hop authentication (IHOP) scheme in [9]. IHOP guarantees that the base station will detect any injected false data packets when no more than a certain number t of nodes are compromised.

Although stream data and scalar data have much common ground, they still have the following differences [10].

- The WSN for scalar data is always a flat, homogenous architecture. However the WMSN for stream data is generally heterogeneous architecture,
- The storage and runtime memory size of WMSN sensor has scaled from several KB to MB magnitude. And the CPU processing capability has increased from 6MHZ of Mica series to hundreds MHZ [11].
- The amount of data in WMSN is more in magnitude than that in WSN.
- In scalar data transmission, the source node sends packets to the sink periodically. However in stream data transmission, the packets derived from the source node will congest the whole routing path, and the source are continuously generating new packets.

Because of the above differences, if the existing research is applied to stream data in WMSN, the following problems will appear:

●The energy consuming is too large

The existing work mainly focuses on protecting every data packet. However the stream data amount is very huge. Obvious it is not practical for energy constrained sensors in WMSN.

●The additional delay worsens QoS

When the secure mechanisms such as SEF are validating data, in order to carry out voting they need transmit each packet repeatedly between event detecting nodes and data gathering node. Obviously this will lead to large delay between neighboring packets. So the delay will greatly worsen the application QoS.

To avoid these problems, EASY adopts random detection mechanism to reduce the packets to be validated. Then the whole network overload is decreased.

3 Easy Protocol

The goal of EASY is to try to avoid the false data injection attack, and at the same time increases little computation and low communication overload.

3.1 Assumptions and Main Idea

The main idea of EASY is as follows: EASY randomly samples and validates some packets of the data stream, so that the amount of the packets to be validated is largely reduced. Therefore, the additional network overload induced by security is finally decreased.

Before we describe EASY in detail, we firstly give the following assumptions about the application background.

1. The in-network key management is based on the random key pre-distribution protocol for heterogeneous networks [12] and the basic random key distribution [13].
2. The router nodes and cluster heads have much less probability to be captured than general nodes. In random capture mode, the probability can be regarded as zero.
3. There exists authenticity validation mechanism for the multiple nodes monitoring the same event.
4. The packets are numbered and transmitted according to the order with which they are generated. And they reach the sink sequentially.

3.2 Easy Description

The EASY protocol has four phases: node initialization, network initialization, data processing and malicious nodes excluding phase.

Node Initialization Phase

In EASY, there are four types of nodes: the sink, routers, cluster heads and general nodes. The sink gathers and processes the data from the whole network. Routers forward the packets from cluster heads to the sink. Cluster heads aggregate the intra-cluster data and validate their authenticity. A general node can not communicate with the sink directly. It communicates with the sink through cluster heads and routers. All the routers and cluster heads form the data transmission backbone of the network.

During the EASY node initialization phase, the communication keys are distributed to all sensor nodes. The main process is as follows:

1. Before deployment, two large key pools PR and PO are built at the sink.
2. The keys of key ring at routers are selected in the PR, and managed by basic random key distribution protocol.
3. Cluster heads have two key rings. The keys of one ring are selected from PR and the keys of another ring are selected from PO. The keys from PR are managed by basic random key distribution. However the keys from PO are managed by heterogeneous random key distribution.
4. The keys of key ring at general nodes are selected from PO, and managed by basic random key distribution.
5. We generate ID-based identity key K_i [14] shared with the sink for each sensor.

Network Initialization Phase

After node initialization phase, all sensor nodes are uniformly randomly deployed at the monitoring area. The secure communication channels among general nodes are built through heterogeneous random key distribution protocol, so does the secure communication channels between general nodes and cluster heads. The basic random key distribution protocol is used to build secure communication channels among routers and cluster heads.

After building network topology, each sensor generates an ID-based neighbor identity key NK_i shared with its direct neighbors. Then cluster head gathers the intra-cluster topology information, and builds up intra-cluster group key G_i [15]. Then it generates an ID-based intra-cluster identity key CK_i.

Data Processing Phase

Before the source node transmits a packet, it signs the packet with group key Gi and identity key Ki. If a packet reaches the cluster head through multi-hop, all the intermediate forwarding nodes should insert their own ID and signature with CKi. Therefore, the cluster head knows which nodes participate in packet forwarding and lookup the malicious nodes in the future. When the cluster head receives a packet, firstly it validates the source node itself with group key Gi. The packets from illegal nodes are discarded directly. Then the cluster head randomly samples some legal packets and validates their content authenticity. The other packets which are not sampled are forwarded directly to the sink by the cluster heads and routers.

To validate the sampled packet, the cluster head broadcasts it to notify the nodes around the source. After the node receives the notification, if it thinks the packet content is authentic, it adds its signature with intra-cluster identity key CKi into the packet, and then sends it back to the cluster head. If the cluster head gather enough signatures for the sampled packet during some period, it argues that the packet content is authentic. Then it inserts its own signature into the packet and sends it to the sink through the routers. When the cluster head is the direct neighbor of the sink, it can communicate directly with the sink. If the cluster head can not acquire enough signatures for the sampled packet, it argues that the sampled packet is a false one and stops forwarding the packet. At the same time it sends a report about it to the sink and starts to exclude the malicious node.

Malicious Nodes Excluding Phase

The steps to exclude malicious nodes are as follows. Firstly, EASY finds the set of suspicious nodes. Secondly, it discovers malicious nodes from the set of suspicious nodes. Finally, it avoids malicious nodes to transmit data streams.

1. Find the suspicious nodes set

In EASY, when the packet is transmitted inside the cluster, the intermediate forwarding node will insert its own ID into the packet. And the ID is signed with intra-cluster identity key CK_i and neighbor identity key NK_i. At the same time if it recognizes the last-hop node, it signs the last-hop node ID with CK_i. If the malicious node hasn't signed the packet with its own neighbor identity key NK_i, the packet generated by it will be discarded at the next hop node (named by A). Furthermore, node A will report to the cluster head that there exists a malicious node among its neighbors. Since the cluster head knows the intra-cluster topology information, it can argue that the neighbors of node A are the suspicious nodes. Another case is that the malicious node signed the packet with its own neighbor identity key NK_i. Therefore the cluster head can argue that the set of suspicious nodes consists of all the nodes on the path.

2. Locate the malicious nodes

After finding the set of suspicious nodes, the key step is to judge whether the suspicious nodes are malicious. In EASY, we adopt the overhearing-based mechanism. Since the cluster head knows exactly the intra-cluster topology information, it can use the approach in [9]. That is, for each suspicious node, the cluster head asks its neighbors to vote. The cluster head will make decision according to all the voting.

3. Avoid the malicious nodes

In EASY, it needs two steps to avoid the malicious node.

1) Broadcast the malicious node IDs to all cluster members

The cluster head can broadcast the malicious node ID through μTESLA protocol [16]. Furthermore, the corresponding intra-cluster identity key CK_i and neighbor identity key NK_i are revoked.

2) Choose a new path

Firstly, the cluster head gets rid of the malicious node from its intra-cluster topology information. Then it finds one new path between the last-hop and next-hop of the malicious node. Finally it notifies them of the new path. The subsequent packets will be transmitted along the new path.

4 Security Analysis

In this section we first analyze the miss ratio of EASY in theory. Then we compare EASY with SEF.

4.1 Analysis for Miss Ratio in Theory

Threat Model: There are malicious nodes on the intra-cluster routing path. The probability of which the packets are tampered is α. The probability of which the cluster head validates the packet authenticity is β. If the packets to validate are false data, it is assured that the cluster head will discover the false data. For simplicity, we suppose that the time for which the cluster head calculates the intra-cluster new routing path can be ignored, and no new malicious nodes exist on the new path.

Definition 1: Miss ratio. When the packets transmission ends, we define the miss ratio, termed μ, as the expectation of the percentage of the false packets in the total packets received by the sink.

We use the following notion and terms for simplicity:

$A(n)$	event that the cluster head hasn't found false packets among the first n packets when malicious nodes exist on the intra-cluster routing path
$B(n)$	event that the cluster head hasn't found false packets among the first $n-1$ packets and finds the *nth* packet is false when malicious nodes exist on the intra-cluster routing path
C	event that malicious nodes don't exist on the intra-cluster routing path
$D(n)$	event that the cluster head hasn't found false packets among the first n packets
N	the number of the packets of the data stream
$P(X)$	the probability of which the event X happens

Theorem 1. If there are malicious nodes on the intra-cluster routing and the time for which the data stream is transmitted is long enough, it is assured that EASY will find the false packets.

Proof

$$P(A(n)) = [(1-\alpha)+\alpha(1-\beta)]^n \tag{1}$$

Equation (1) is the monotone decreasing function of n. Then $P(A(n))$ will descend to zero with the increasing of transmission time. So the probability of which the cluster head discovers the false packets will ascend to 1 with the increasing of transmission time. That is, the cluster head always discovers the false packets, if the time is enough. So theorem 1 holds.

● The formula for Miss ratio, μ

When the cluster head finds the false packet, the subsequent packets will be transmitted to the sink along a new path where there are no malicious nodes. Thus, μ can be calculated by the following equation:

$$\mu = \begin{cases} \alpha & \text{if } \beta = 0 \\ 0 & \text{if } \alpha = 0 \text{ or } \beta = 1 \\ \sum_{n=1}^{N} \left(\frac{n-1}{N} * (1-\beta)^{(n-1)} * \beta \right) & \text{if } \alpha = 1 \text{ or } 0<\beta<1 \\ \sum_{n=\left(\frac{1}{\alpha}\right)}^{N} \left(\frac{n*\alpha-1}{N} * \alpha * \beta * ((1-\alpha)+\alpha*(1-\beta))^{(n-1)} \right) & \text{if } 0<\alpha<1 \text{ or } 0<\beta<1 \end{cases}$$

4.2 Simulation

To compare EASY with SEF, we define the following terms.

Definition 2: trueness ratio. When the packets transmission ends, we define the trueness rate, termed ω, as the expectation of the percentage of the true packets in the total packets received by the sink.

Definition 3: security communication overload. We define the security communication overload, termed ψ, as the number of the extra packets which be sent for security.

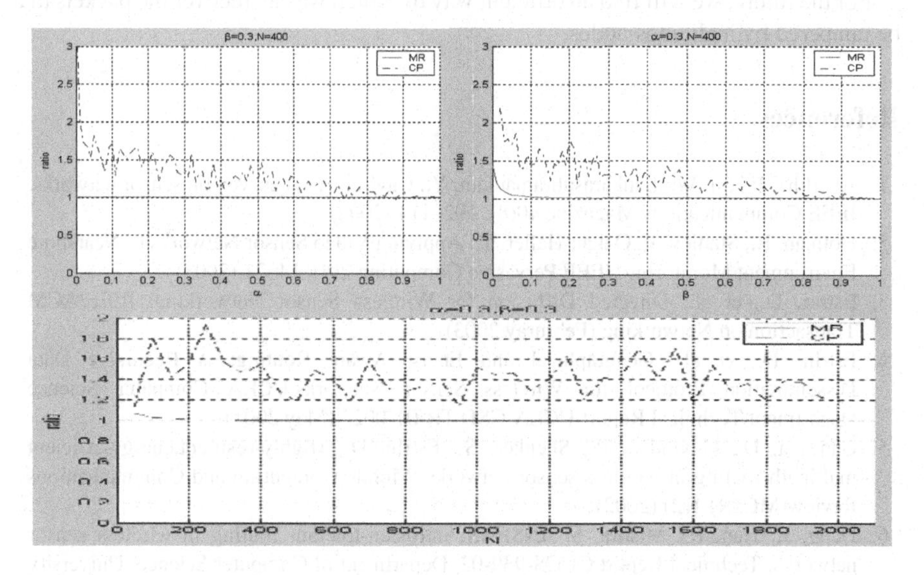

Fig. 1. The comparison between SEF and EASY

Because SEF won't exclude malicious nodes, we can't directly compare SEF with EASY. In simulation, we join the malicious nodes excluding phase with SEF. After finding the false packet, SEF will find the new path as EASY does.

In simulation, we compare SEF and EASY in ω and ψ. In figure 1, MR represents the ω of SEF to the ω of EASY and CP represents the ratio of the ψ of SEF to the ψ of EASY. We can make the following conclusions:

- In most conditions, the ω of EASY approximates the ω of SEF.
- The ψ of EASY is lower than the ψ of SEF. In some conditions, the ψ of SEF is 1.5 times the ψ of EASY.
- If α and β are fixed, CP isn't always increasing along with the increasing of N. The reason is that CP mainly relates to the time when cluster heads first find the false packet. If N is higher than a certain value, the time only relates to α and β.

5 Conclusion

By reducing the number of the packets to be validated, EASY lowers the overload which security protection generates. This paper proves that if there are malicious nodes on the intra-cluster routing path, EASY increases the probability of finding the existing malicious nodes with the increasing of detection time. At the same time, we give the formulae for miss ratio. By the simulation, we prove that in most condition EASY can work as well as SEF does with lower overload. Aiming at the attack of false packet injection, EASY can effectively resolve the contradiction between security requirement and additional overload in stream-data applications.

For the future, we will find an efficient way by which we can recover the packets that be tampered by malicious nodes.

References

1. Akyildiz, F., Su, W., Sankarasubramaniam, Y., Cayirci, E.: A survey on sensor networks. IEEE Communications Magazine 40(8), 102–114 (2002)
2. Holman, R., Stanley, J., Ozkan-Haller, T.: Applying Video Sensor Networks to Nearshore Environment Monitoring. IEEE Pervasive Computing 2(4), 14–21 (2003)
3. Estrin, D., et al.: Directed Diffusion for Wireless Sensor Networking. IEEE/ACM Transaction on Networking (Februray 2003)
4. Estrin, D., et al.: Geographical and Energy-Aware Routing: A Recursive Data Dissemination Protocol for Wireless Sensor Networks.UCLA Computer Science Department Technical Report,UCLA-CSD TR-01-0023 (May 2001)
5. Ganesan, D., Govindan, R., Shenker, S., Estrin, D.: Highly-resilient,energy-efficient multipath routing in wireless sensor networks. Mobile Computing and Communications Review(MC2R) 1(2) (2002)
6. Deng, J., Han, R., Mishra, S.: INSENS: intrusion-tolerant routing in wireless sensor networks. Technical Report CUCS-939-02, Department of Computer Science, University of Colorado (2002)

7. Ye, F., et al.: Statistical En-Route Filtering of Injected False Datasensor Networks. In: Proc. IEEE INFOCOM, Hong Kong (2004)
8. Wang, G., et al.: On supporting Distributed Collaboration Sensor Networks. In: Proc. MILCOM (2003)
9. Zhu, S., et al.: An Interleaved Hop-by-Hop Authentication Scheme for Filtering of Injected False Data in Sensor Networks. In: Proc. IEEE Symp. Security and Privacy, Oakland, CA, pp. 259–271 (May 2004)
10. Akyildiz, K.R.: A survey on wireless multimedia sensor networks. Computer Networks 51(4), 921–960 (2007)
11. Crossbow wireless sensor platform. http://www.xbow.com/Products/Wireless Sensor Networks.htm
12. Traynor, P., Choi, H., Cao, G., Zhu, S., La Porta, T.: Establishing Pair-wise Keys in Heterogeneous Sensor Networks. In: IEEE Infocom 2006, IEEE Computer Society Press, Los Alamitos (2006)
13. Eschenauer, L., Gligor, V.D.: A key-management scheme for distributed sensor networks. In: Proceedings of the 9th ACM conference on Computer and communications security, pp. 41–47. ACM Press, New York (2002)
14. Merkle, R.: Secure communication over insecure channels. Communications of the ACM 21(4), 294–299 (1978)
15. Chadha, A., Lius, Y., Das, S.: Group key distribution via local collaboration in wireless sensor networks. In: Proceedings of IEEE SECON 2005, Santa Clara, CA (September 2005)
16. Perrig, A., Sivalingam, R., et al.: SPINS: Security protocols for sensor networks. Wireless Networks 8(5), 521–534 (2002)

On a Low Security Overhead Mechanism for Secure Multi-path Routing Protocol in Wireless Mesh Network*

Muhammad Shoaib Siddiqui, Syed Obaid Amin, and Choong Seon Hong

Department of Computer Engineering, Kyung Hee University,
Sochen-ri, Giheung-eup, Yongin-si, Gyeonggi-do, 446-701, South Korea
{shoaib,obaid}@networking.khu.ac.kr, cshong@khu.ac.kr

Abstract. Secure multi-path routing is a critical issue in security management of WMNs due to its multi-hop nature as each node takes part in routing mechanism making it prone to routing attacks. Security management mechanisms are armed with features such as asymmetric cryptography which are costly in term of computations, transmissions and time delays. In this paper, we propose a security management mechanism for multi-path routing which efficiently uses the characteristics of WMNs, mutual authentication and secrete key cryptography to provide secure multi-path route management. Our management scheme takes less overhead than the available secure multi-path routing mechanisms. Simulation analyses and the performance of the mechanism are presented in support of the proposal.

Keywords: Security management, Secure multi-path routing, Wireless mesh networks, Security overhead, Public key cryptography.

1 Introduction

Wireless Mesh Network [1] is an emerging new technology which is being adopted as the wireless internetworking solution for the near future. Characteristics of WMN such as rapid deployment and self configuration make WMN suitable for transient on-demand network deployment scenarios such as disaster recovery, hard-to-wire buildings, conventional networks and friendly terrains. The form of mesh networks that are of most commercial interest are often called hybrid mesh networks [2], shown in Fig. 1. In hybrid mesh networks, the end users such as PDAs and laptops make up mesh client networks and mesh router nodes are part of the network infrastructure [2]. Here, the network consists of two types of links: short range wireless links (shown in Fig. 1 as dotted lines) among client mesh nodes and mesh relay links (shown in Fig. 1 as dashed lines) between router nodes to form the packet transport backbone.

WMN has been a field of active research in recent years. However, most of the research has been focused around various protocols for multi hop routing leaving the area of network and security management mostly unexplored. In this paper, we provide a management mechanism for hybrid wireless mesh networks, which reduces the security overhead in the network and in turns, increases the overall efficiency of

* "This paper was supported by ITRC and MIC".

S. Ata and C.S. Hong (Eds.): APNOMS 2007, LNCS 4773, pp. 466–475, 2007.

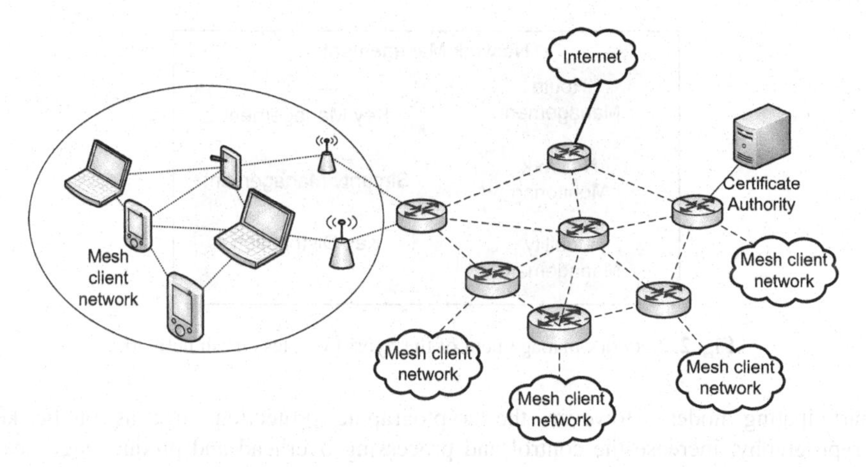

Fig. 1. A Hybrid Wireless Mesh Network

the secure routing protocol. In section 2, we discuss the various aspects of network management in WMNs with emphasis on security management. In section 3, we provide introduction to related approaches in the secure multi-path routing field. In section 4, we discuss the proposed management scheme. In section 5, we present the simulations and analytical comparison of our proposal with related work. In section 6, we conclude our proposal and discuss the future work.

2 Network Management

Network management refers to the maintenance and administration of large-scale computer and telecommunication networks at the top level. In general, network management is a service that employs a variety of tools, applications, and devices to assist human network managers in monitoring, maintaining and securing networks. The fundamental network management concepts in wireless mobile network are mobility management, route management, network monitoring and security management as shown in Fig. 2.

There are two ways of managing secure communication in WMNs: (1) Using the multiple paths [3] available in between the nodes. (2) Using the cryptographic key management to secure the communication in between two nodes. In first approach all the multiple paths between two nodes need to be node-disjoint (a node cannot participate in more than one path between two end nodes). If there are k multiple paths available then the adversary requires compromising at least k nodes – and more particularly at least one node in each path – in order to control the communication [4]. This approach is cost effective as it does not include any computation or transmission overhead and hardly inject delay in the network. But it does not ensure a certain level of security as there are not always multiple paths in between two end nodes and it is difficult to identify a compromised path.

Multi-path routing protocols need to be properly enhanced with cryptographic means which will guarantee the integrity of a routing path and the authenticity of the

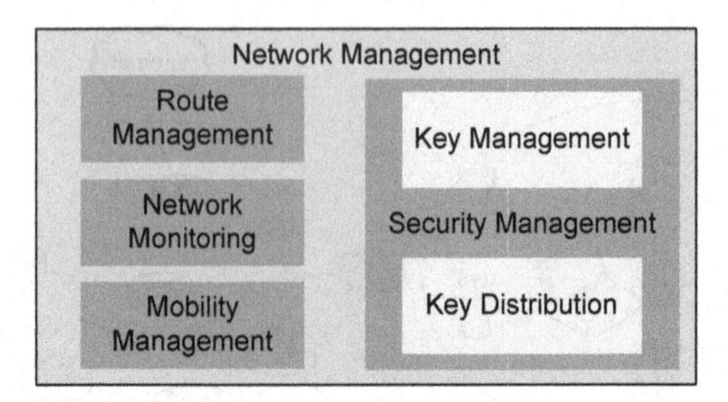

Fig. 2. Network management of the hybrid wireless mesh network

participating nodes. However, the cryptographic protection such as public key cryptography, increase the control and processing overhead and produce significant delay thus diminishing the efficiency of the secure multi-path routing protocol.

3 Related Works

Multi-path routing protocols [3] were initially designed for providing reliability [5] and QoS in the ad hoc networks. However, their nature of attack resilience was quickly identified as a significant security feature. Indeed, with single path routing protocols, it is easy for an adversary to launch routing attacks. A compromised node controlled by the adversary may participate in route discovery between end nodes without being noticed. Hence, the adversary can control the routing mechanism and disrupt the services at any instance.

Secure multi-path routing protocols are more resilient to routing attacks than typical routing protocols [6]. Although a lot of work is being done in the field of routing protocols in WMNs but little effort is put up for a security management in routing protocols. However, there are some protocols which are good enough to be implemented in WMNs and provide a secure multi-path route management such as [8], [7], [9] and [10].

A secure multi-path routing protocol called Secure Routing Protocol (SRP) [7] by Papadimitratos and Haas was initially developed considering the general security of ad hoc networks. Another approach was provided by Burmester and Van Le [8], which is based on the Ford-Fulkerson maximum flow algorithm. Kotzanikolaou et al presented Secure Multi-path Routing (SecMR) [10] protocol to reduce the cost of node authentication. SecMR works in two phases: mutual authentication and route discovery phase. At the end of route discovery, the end nodes use a symmetric key in order to verify the integrity of the discovered paths. SecMR provide multiple paths along with routing security and is better than the other two protocols. However, due to the use of digital signature in periodic mutual authentication phase, the computation cost and control overhead incurred render this scheme inefficient.

Michael Weeks and Gulshan Altan have provided a secure and efficient version of Dynamic Source Routing (DSR) in [9]. However, their security mechanism uses a shared network key, which is a single point of failure (if compromised), in the network. There scheme also provide secured communication using public key cryptography, which again results in high computational cost and delay.

4 The Proposed Mechanism

Although wireless mesh networks are self organizing but they are also scalable and as the number of nodes increase in the network the size of the network makes network management essential. Network management helps in detecting abnormalities in the network and may help in other issues such as routing, guaranteeing QoS and providing security. Currently, to the best of our knowledge very little research has been done on the network management issues in wireless mesh networks. We provide a mechanism which makes network management simple and efficient.

4.1 Assumptions

Wireless mesh network has a hierarchical structure with mesh router making a routing infrastructure and mobile wireless clients making up ad hoc networks at the second level of the network. Each ad hoc network of wireless mesh clients has one or more routers from the router infrastructure in the ad hoc region. Our mechanism assumes that these router nodes are powerful enough to provide management functionality to the wireless mesh network. The routers which are connected to the mesh client nodes are named as boundary routers or manager routers. The mesh client networks are also termed as ad hoc regions/components. Nodes in the client mesh are also termed as client nodes (as shown in Fig. 1).

4.2 Mechanism

By associating each mesh client network with one router of infrastructure mesh, the management of the whole wireless mesh network would become simple. Each mesh client network can be managed by a boundary router. Boundary router is responsible of provide addresses, routing assistance, mobility management, power management and network monitoring to the mesh client networks. Security mechanism can also be enhanced by centralizing the mesh client network.

Route management is the job done by the routing protocol, while our mechanism provides security as an add-on to the existing routing protocol. Manager router in each mesh client provides the key management and distribution responsibilities. Manager router manages each mesh client network such as providing addresses, assisting routing and providing security. We also assume that there is a Certification Authority (CA) [11] in the wireless mesh network, which is a trusted third party that can authenticate the digital certificates of the nodes. Every node is provided with a pair of public and private key during the deployment phase.

With the implementation of this scheme, each mesh client network is now centrally managed by the manager router of that region. But the over all mesh network is still distributed. Each manager router communicates with other routers, collaborates and manages the whole wireless mesh network. We discuss the addressing, routing assistance, mobility management, routing assistance, network monitoring and security assistance by this mechanism.

4.3 Addressing and Mobility Management

Addresses for mobile clients are allocated dynamically by the router of that region. This address defines the location of the mobile client i.e. in which ad hoc region the mobile node is present. As the WMN clients are mobile, they may change position from one ad hoc region to the other. Our mechanism uses the techniques of Mobile IP [12] to provide addresses to client nodes. Similar to mobile IP, a client node has two addresses; one to identify it in its home ad hoc region and the other one is for the other ad hoc regions. Whenever a node enters the network for the first time, an address is assigned to it by the manager router. This router in the home (ad hoc) region of the client node serves the purpose of 'Home Agent'. When a client node changes it location and goes into another region, it is provided a second address from the router of that region. The client node informs its 'home agent' and its 'foreign agent' about this new address and location [12], so that a packet directed to the client node is redirected to its new address.

Hence, mobility of each client node can be easily managed. Locality information of each node is maintained by the manager routers. Whenever a node moves from one region to another region, the manager router of the new region provides new address to the node and the node remains connected to the network. The home agent directs the communicating node to the mobile nodes' new location.

4.4 Routing Assistance

Our mechanism also helps the routing mechanism. As the border router manages the addresses and monitors the network, it can help in routing decisions. The manager router can find optimal paths between two nodes, detect link losses and find alternate paths within the client mesh network. Network monitoring may keep a topological view which can also help in routing. Localization can help in geographic routing protocols by helping in decisions such as which neighbor node to forward the data to reach the destination node. The manager router can also work as a gateway between the static router infrastructure and the mobile client mesh network.

4.5 Network Monitoring

Due to dynamic nature of mesh clients, monitoring the network topology is a desired feature for WMNs. We can designate the responsibility of network monitoring of a single ad hoc region to a single manager router. Then all the client mesh networks can be monitored in a centralized way. These routers collaborate to perform the task of monitoring for the whole WMN in a distributed environment.

4.6 Security Management

Security is the most critical concern of every network. These days resource consuming public key cryptography is used to provide security which is not feasible for the client nodes. Our architecture presents an efficient way of reducing the security overhead.

Whenever a new node comes into a mesh client network, its request for an address is sent to the manager router of that region. The router provides the address to this client node along with its public key and starts the process of mutual authentication with the node. The public key of the router node assures the authenticity and integrity of the following messages as all those messages are encrypted by the private key of the router node.

The client node and the router node encrypt the messages by their private keys before sending them to each other. This process authenticates both the nodes. For the authenticity of each other, the router node or the client node can contact the CA to verify the digital signature of each other. During this time of mutual authentication both nodes share a secret key using authenticated Diffie-Hellman [13] algorithm (shown in Algorithm 1) so that in the future they are not required to use public key cryptography. In the same way all the nodes within a mesh client network has a secret key shared by the manager router of that region. The algorithm is stated in the next sub-section.

The second phase is the key deployment phase among the client nodes. The router node distributes the keys calculated through a hash chain to all the client nodes for intercommunication. These are the secret keys which would be used by the client nodes to provide secure multi-path routing in the wireless mesh network.

4.7 Example

Let there be a wireless mesh network as shown in Fig. 3. The circle represents nodes and the dashed line shows the communication links. The cloud represents mesh infrastructure connected to several mesh clients. One such mesh client network is shown consisting of nodes **A, B, C, D**, and **R. R** is the router node managing the mesh client network while the other nodes are mesh client nodes. There is a **CA** connected to the mesh infrastructure somewhere in the wireless mesh network.

A new node **E** comes into the mesh client (shown in Fig. 3 as a grey node). First it sends an address request (such as DHCP request) in the network. The router node **R** provides the address to node **E**. After that they start the process of sharing a key using authenticated Diffie-Hellman.

At first, node **E** select two prime numbers g and p and a secrete integer a (e.g. $a=6$, $p=23$ and $g=5$) and calculate X and encrypt it with its own private key, make a digital signature and send it to node **R** along with p and g.

$$X = g^a \mod p \;=\; 5^6 \mod 23 \;=\; 8$$

R receives p, g and encrypted X and decrypts the message to get the value of X, using the public key of **E**. This authenticates the sender is **E**. **R** select an integer value b (e.g. $b=15$) and calculate Y, encrypt it with its own private key, make a digital signature and send it to **E**.

$$Y = g^b \mod p \;=\; 5^{15} \mod 23 \;=\; 19$$

E receives the encrypted Y and decrypts it using the public key of **R**. It then calculates the value of K.

$$K = [g^b \mod p]^a \mod p \;=\; Y^a \mod p = 19^6 \mod 23 = 2$$

Algorithm 1. The algorithm for authenticated Diffie-Hellman [13] for sharing a secret key between the router node R and the client node E is as follows:

Step 1. ***R** & **E** each possess a public/private key pair and a certificate for the public key.*
Step 2. ***R** & **E** agree to use a prime number p and g.*
Step 3. ***E** chooses a secret integer a, then sends **R** (g^a mod p) together with its signature and public key certificate.*
Step 4. ***R** chooses a secret integer b, then sends **E** (g^b mod p) together with its signature and public key certificate.*
Step 5. ***E** computes $K = (g^b \bmod p)^a \bmod p$*
Step 6. ***R** computes $K = (g^a \bmod p)^b \bmod p$*
Step 7. *Shared Secret key is K; **E**'s private key is 'a' and **R**'s private key is 'b'.*

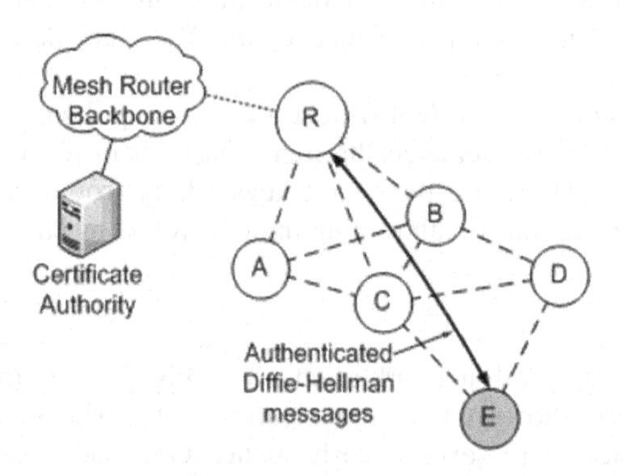

Fig. 3. Mutual authentication at the entrance of the node E in mesh client network

Similarly, **R** can calculate the value of *K*.

$$K = [g^a \bmod p]^b \bmod p \; = X^b \bmod p = 8^{15} \bmod 23 = 2$$

5 Simulation and Analysis

We compared our security mechanism with the SRP [7], secure multi-path routing protocol of Burmester and Van Le [8] and SecMR [10] routing protocols. We perform the simulation of each of these security schemes. The proposed scheme is implemented with ad hoc on demand multi-path distance vector (AOMDV) [14] which is a multi-path derivative of AODV.

We have compared the routing overhead of these schemes and also the amount of energy consumed by these scheme at each node. We performed the simulation in NS-2 [15]. The network model was consisted of 49 client nodes placed randomly within an area of 1000 x 1000 m^2. There are 16 mobile router nodes deployed in a grid environment to make up the mesh infrastructure. This scenario constructed 10 different mobile client networks. Each node has a propagation range of 150 meters with channel capacity 2 Mbps. The speed of mobile nodes is set to be 0 or 20 m/s. The size of the data payload is 512. Each run of simulation is executed of 900 seconds of simulation time. The medium access control protocol used is IEEE 802.11 DCF. The traffic used is constant bit rate (CBR).

5.1 Simulation Analysis

From Fig. 4 and Fig. 5, we observe that SRP works better than other schemes as it has less overhead and also consumes very little amount of energy. However, SRP does not provide optimal security; the intermediate nodes are not authenticated and the messages integrity is ensured by secret key cryptography. All this factors sum up to make SRP not feasible for wireless mesh networks.

The high routing overhead of scheme in [8] is due to the fact that it attaches the neighborhood information along with digital signatures with the route request and forward it towards the destination node. This information is increased at every node so the message size increases drastically and produces a huge amount of overhead. Although [8] is good for security and provides mutual authentication between the intermediate nodes as well as the end nodes but its overhead is very high; lot of energy is required at the client nodes and a share of bandwidth is wasted, plus delay in finding the route is also high.

SecMR protocol seems to be better than other schemes as it has less routing overhead and energy consumption than [8] and it also provides secure messaging. In SecMR, each node mutually authenticates its neighbor node at a periodic interval and public key cryptography is used to ensure security of the messages. Although the routing phase is separated from this authentication phase but this authentication is required after a constant interval, hence a considerable amount of energy is wasted in these periodic mutual authentications.

Our security mechanism does not require this periodic authentication, instead it uses public key cryptography only once and secret keys are used for further communication. This secret key deployment is not periodic and done after the mutual authentication by using public key cryptography. This reduces the energy consumption at each node and the routing overhead is also less than the other schemes.

5.2 Security Analysis

Our mechanism is secure enough that if a node is compromised then the whole network does not get affected by it. As all nodes communicates with each other with separate secret keys so, if a node is compromised and tries to adverse the network it is not possible for the node to be much hostile to the rest of the network. If there is a compromised node in the network, then there are two possibilities of an adversary node being in the network. In case 1, a node outside the network tries to attack the routing mechanism. Case 2 is the scenario in which the node entering the network is already a compromised node or the node is compromised during its participation in the network (such as due to the lack of physical protection etc).

In the first case, the messages by the compromised node would not be accepted by the other nodes as it cannot be authenticated by them. So the adverse messages would be dropped by the nodes as they cannot verify the adverse node as a member node.

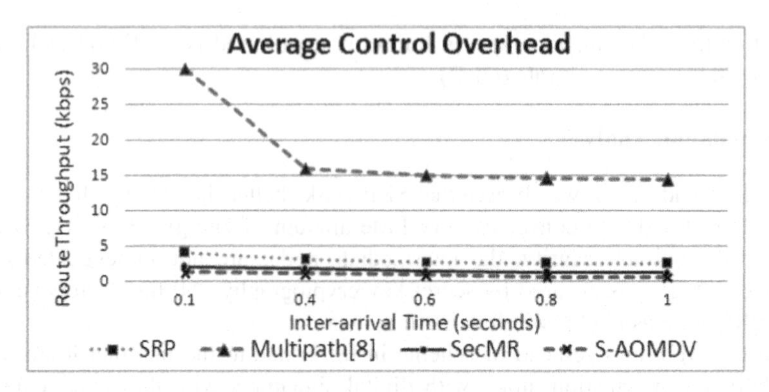

Fig. 4. Comparison of routing overhead of each protocol with function of time interval

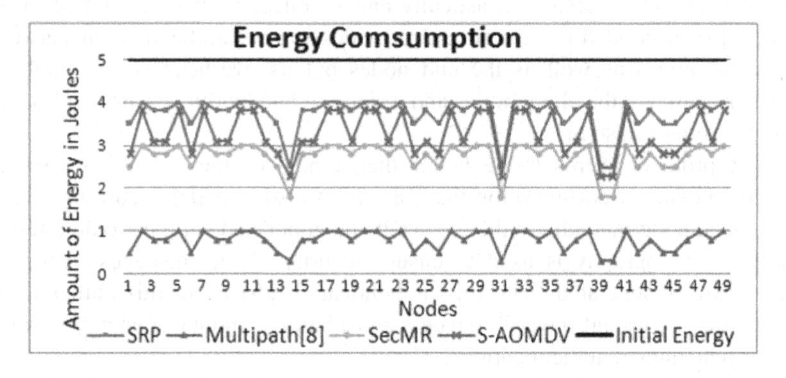

Fig. 5. Amount of energy left in joules at each node after the 900 s simulation

The second case can be harmful for the network as other nodes can verify the compromised node as a decent node. This node can communicate with its neighbor nodes and can inject false information in the network. But this compromised node cannot listen to other nodes' communications and cannot affect them. So if a node is compromised in the network all the other nodes are safe from this node and can communicate with other nodes securely. As our mechanism is for a multi-path routing protocol, hence, the messages are secure from the adversary as there are several paths to evade the compromised nodes. Even if the adversary have *'n'* compromised nodes with every compromised node is in a different path then with *'m'* paths in between two nodes, adversary require $n \geq m$.

6 Conclusion

In this paper, we have presented a security management mechanism for multi-path routing protocols in wireless mesh network. This scheme provides an efficient network management scheme, which enhances the life-time of the network as less energy is consumed in the network. Our security management scheme also sufficiently decreases the control overhead of a secure routing protocol. Currently, we are working on this security mechanism to implement it in our multi-path routing protocol, which promises to provide better performance than AOMDV which we used for our simulations. This scheme is the basic efficiency factor in our secure multi-path routing protocol for wireless mesh networks.

References

1. Bruno, R., Conti, M., Gregori, E.: Mesh networks: commodity multihop ad hoc networks. IEEE Communications Magazine 43(3), 123–131 (2005)
2. Akyildiz, I.F., Wang, X., Wang, W.: Wireless Mesh Network: A Survey. Computer Networks and ISDN Systems 47(4) (2005)
3. Garcia-Luna-Aceves, J.J., Mosko, M.: Multipath Routing in Wireless Mesh Networks. In: WiMesh 2005, Santa Clara, CA, September 26, 2005, IEEE Computer Society Press, Los Alamitos (2005)
4. Gupta, R., Chi, E., Walrand, J.: Different Algorithms for Normal and Protection Paths. Journal of Network and Systems Management archive 13(1), 13–33 (2005)
5. Ganjali, Y., Keshavarzian, A.: Load Balancing in Ad hoc Networks: Single-path Routing vs. Multi-path Routing. In: proceedings of IEEE Annual Conference on Computer Communications (INFOCOM), 1120–1125 (March 2004)
6. Papadimitratos, P., Haas, Z.J.: Secure Routing for Mobile Ad Hoc Networks. Mobile Computing and Communications Review 6(4) (2002)
7. Papadimitratos, P., Haas, Z.: Secure routing for mobile ad hoc networks. In: Proceedings of the SCS Communication Networks and Distributed Systems Modeling and Simulation Conference (CNDS), TX, San Antonio (January 2002)
8. Burmester, M., van Le, T.: Secure multipath communication in mobile ad hoc networks. In: ITCC 2004, IEEE, Las Vegas (2004)
9. Weeks, M., Altun, G.: Efficient, Secure, Dynamic Source Routing for Ad-hoc Networks. Journal of Network and Systems Management 14(4), 559–581 (2006)
10. Kotzanikolaou, P., Mavropodi, R., Douligeris, C.: Secure multipath routing for mobile ad hoc networks. In: Proceedings of the WONSS05 Conference, St. Moritz, Switzerland, January 19-21 2005, pp. 89–96. IEEE, Los Alamitos (2005)
11. Raghani, S., Toshniwal, D., Joshi, R.: Dynamic Support for Distributed Certification Authority in Mobile Ad Hoc Networks. In: ICHIT 2006, vol. 1, pp. 424–432 (November 2006)
12. Perkins, C.E.: Mobile networking through Mobile IP. IEEE Internet Computing 2(1), 58–69 (1998)
13. Diffie, W., van Oorschot, P., Wiener, M.: Authentication and authenticated key exchange. Designs, Codes and Cryptography 2(2), 107–125 (1992)
14. Marina, M.K., Das, S.R.: On-demand multipath distance vector routing in ad hoc networks. In: the proceedings of Ninth International Conference on Network Protocols, November 11-14, 2001, pp. 14–23 (2001)
15. UCB/LBNL/VINT Network Simulator - ns 2, http://www.isi.edu/nsnamjns

Performance Evaluation of a Mobile Agent Based Framework for Security Event Management in IP Networks

Ching-hang Fong, Gerard Parr, and Philip Morrow

School of Computing and Information Engineering, Faculty of Engineering, University of Ulster, Coleraine, Co. Londonderry, BT52 1SA, United Kingdom
{fong-c,gp.parr,pj.morrow}@ulster.ac.uk

Abstract. Conventional network management protocols are insufficient especially in dealing with frequent security attacks. Our research has developed a Mobile Agent Security framework for Autonomic network Management (MASAM) which intends to address the above limitations. In this paper, we aim to evaluate the performance of the MASAM framework when dealing with security attack event management. The evaluation focuses on the traffic cost comparison between the new framework and SNMPv3. Event management traffic models are proposed and utilised to facilitate this evaluation. In order to validate the derived formulas, a corresponding set of simulation experiments have been conducted and the results are analysed. Positive evaluation results have been obtained from three focus points: the entire network, the manager and the managed device. We conclude that the MASAM framework reduces the amount of management traffic generated to respond to security attacks and also scales better than SNMP as a function of network size.

Keywords: Security Management, Mobile Agent, SNMP, Performance Evaluation.

1 Introduction

Internet Protocol (IP) networks have been growing dramatically in size and functionality and are evolving into a global service communication infrastructure [1]. This growth requires management of the network at enormous scale with increasing sophistication. Traditional network management paradigms and conventional management protocols are insufficient due to lack of performance, scalability and functionality. The dominant SNMP standard utilises the client-server paradigm with a polling-based approach to achieve the Management Functional Areas (MFAs) defined by the ISO FCAPS model [9]. Due to the limitations of its design, the MFAs are neither completely achieved nor implemented in a sophisticated manner. Furthermore, it can be exposed to security threats from several network attacks [2]. A new security model [3-4] to the SNMPv3 is under construction by the ISMS charter of the IETF. However, it is only an integration scheme for authentication as a supplement to the USM with no large security enhancement is being offered. The rest of this paper is organised as follows. Section 2 briefly reviews the MASAM framework. Section 3

S. Ata and C.S. Hong (Eds.): APNOMS 2007, LNCS 4773, pp. 476–486, 2007.

discusses the methodology employed for this study. Section 4 proposes mathematical traffic models for facilitating our performance evaluation, and section 5 presents the results and analysis. Finally, we provide an outline of further work.

2 Background

MA-based Network Management (MANM) can be seen as an attractive approach in that management tasks can be partially moved from the manager to the managed devices (MDs), therefore performing certain micro-management operations locally on the MDs. With the concept of management by delegation, it offers easier management of wireless and ad-hoc formed networks with heterogeneous architectures [5]. In practice, MANM is also a prominent proposal that does not require abandonment of SNMP-embedded devices in this transitional time, but is also capable of developing its own management standard. Unfortunately, three major obstacles (Security, Standardisation and Implementation Difficulties) have been identified that are preventing the large scale employment of MANM systems over real-world networks. We believe that the first two issues cannot be addressed separately because of their interrelationship and the third obstacle cannot easily to be addressed unless standardisation has occurred and technology has migrated.

In our approach we intend to deal with these obstacles by introducing a MA-based Security framework for Autonomic network Management, the MASAM framework. After evaluation and optimisation, the ultimate goal of our research is to propose a new MANM standard. The framework aims to guide design and implementation of secure MANM systems. Two main contributions of the framework are provision of: 1) advanced network management functionality especially in enhancing security management, such as policy-based onside decision making and MA-based instruction detection, capability in order to address the largest weakness of SNMP; and 2) security of the entire MA-based Network Management Infrastructure (NMI) as the baseline protection guarantees effectiveness of the management functions. A modular-based design approach is employed in that we firstly designed an overview and logical architecture of the framework. This architecture maps the abstract components of the framework and defines their relationships, provides functions, and constraints. This is the same approach that was taken with the development of the SNMP standard. The MASAM framework is relatively complex in its architectural view [6], but it offers flexibility in design of the advanced management functions.

3 Methodology

In this evaluation study, we chose mathematical modelling for two reasons: 1) it is a formal quantitative method and 2) it can possibly eliminate many unpredictable variables. We focus on the traffic cost in terms of bandwidth consumption implications only, while the discussion of other parameters, such as delay, is outside the scope of this paper. Network Management Traffic (NMT) generally can be classified into three groups. 1) Event Management Traffic (EMT), specific traffic generated for detecting and recovering from network management events, such as network security attacks presented in table 1. 2) General Management Traffic (GMT), generated by periodic management operations aimed at two purposes: a) the manager wishes to keep acceptable levels of accuracy of the management information regarding the state of each MD, and b) the MD wishes to notify the manager regularly regarding its health. The final group is 3) Component Initialisation Traffic (CIT), generated by the initialisation processes of the management components. For the purpose of this paper, we consider GMT and CIT to be outside of the scope of the evaluation for several reasons. Many publications, such as [10], have already presented results showing that the MANM paradigm can significantly reduce traffic in the GMT group, but no publications can be found regarding a study of EMT. The reason for not considering the CIT is that it is only a one-shot operation while others are continuously consuming network bandwidth. The rest of this paper aims to examine the EMT traffic cost implications associated with the MASAM and SNMP approaches when dealing with security event management, and will concentrate on three different views, the network, the network manager and the MD.

4 Modelling

With the MASAM framework, the management functions can be achieved through three basic methods: 1) management by targeted or roaming agents, MAs can be generated by the manager and migrated to their desired MDs in order to perform management operations; 2) onside decision making, the policy agent residing on each MD is capable of performing management operations based on its policy rules without notifying the manager in advance; and 3) message exchange, management information or commands can be enveloped into Agent Communication Language (ACL) messages and exchanged between MANM entities, such as MAs, agent platforms or ACL-enabled interfaces. With the MASAM framework, the management traffic will be generated from two sources: MA migrations and ACL message exchanges. A sample function formulating the total traffic cost (λ_{MASAM}) for managing a general event is the following:

$$\lambda_{MASAM} = \sum_{a=1}^{A} \lambda_{MASAM_a}^{Agt} + \sum_{n=1}^{N} \lambda_{MASAM_n}^{ERA} \tag{1.1}$$

where $\lambda_{MASAM_a}^{Agt}$ denotes the total traffic consumed by the management MA a including migrations and message exchanges and $\lambda_{MASAM_n}^{ERA}$ denotes the total traffic generated by the Event Reporting Agent (ERA) residing on the MD n. A is the number of management MAs created for this event, while N is the number of MDs involved in this event. Regarding the total traffic generated by a management MA, it can be formulated as following:

$$\lambda_{MASAM}^{Agt} = \sum_{v=1}^{V} \left(S_v + \sum_{x=1}^{X_v} L_{v,x} \right) \tag{1.2}$$

where S_v represents the current size of the agent when migrating to node v. When a MA is visiting a node, it may report to the manager with partial management results by using externally sent ACL messages. Thus, in this formula, $L_{v,x}$ represents the size of the message x sent when visiting the node v while X_v is the total number of messages sent, and V is the number of nodes visited by this MA. Regarding the total traffic generated by an ERA, it can be simply formulated as follows:

$$\lambda_{MASAM}^{ERA} = \sum_{r=1}^{R} L_r \tag{1.3}$$

where L_r represents the size of the ACL message (event report) r sent by the ERA and R is the number of messages sent. Substituting (1.2) and (1.3) into (1.1) gives, the generic EMT model for the MASAM approach, as follows:

$$\lambda_{MASAM} = \sum_{a=1}^{A}\left[\sum_{v=1}^{V_a}\left(S_{a,v} + \sum_{x=1}^{X_{a,v}} L_{a,v,x}\right)\right] + \sum_{n=1}^{N}\sum_{r=1}^{R_n} L_{n,r} \tag{1.4}$$

With the SNMP approach, the management traffic will be generated from the exchanges of SNMP messages including requests, responses and unsolicited traps. A formula representing the total traffic cost λ_{snmp} for managing a general event is:

$$\lambda_{SNMP} = \lambda_{SNMP}^{Mgr} + \sum_{n=1}^{N} \lambda_{SNMP_n}^{Dev} \tag{2.1}$$

where λ_{SNMP}^{Mgr} denotes the total traffic generated by the manager and $\lambda_{SNMP_n}^{Dev}$ denotes the total traffic generated by the MD n while N is the number of MDs involved. Regarding the total traffic generated by the manger, it can be formulated as:

$$\lambda_{SNMP}^{Mgr} = \sum_{n=1}^{N}\sum_{g=1}^{G_n} M_{n,g}^{Mgr} \tag{2.2}$$

where $M_{n,g}^{Mgr}$ represents the size of the SNMP message g sent by the manager (request-based or response for InfoRequest) for the MD n, while G_n is the number of messages sent. On the MD side, the traffic generated can be formulated as follows:

$$\lambda_{SNMP}^{Dev} = \sum_{g=1}^{G} M_g^{Dev} + \sum_{j=1}^{J} T_j \tag{2.3}$$

where M_g^{Dev} represents the size of the SNMP message g (response-based or InfoRequest) sent by this MD, T_j represents the size of the trap message j, and J is the number of trap messages sent. We let M be the total size of a pair of request and response messages, i.e. $M_g = M_g^{Mgr} + M_g^{Dev}$. Substituting (2.2) and (2.3) into (2.1) gives, the generic EMT model for the SNMP approach, as follows:

$$\lambda_{SNMP} = \sum_{n=1}^{N}\left(\sum_{g=1}^{G_n} M_{n,g} + \sum_{j=1}^{J_n} T_{n,j}\right) \tag{2.4}$$

5 Evaluation

The proposed EMT models are generic. However, in order to facilitate our evaluation, the models need to be modified into an applicable form. The strategies include generalising the variables where their actual values are very similar or reformulating part of the model where the actual value may be unpredictable. In formula (1.4), we assume that the size of each ACL message is the same, no matter if it is sent by the management MAs or the ERAs. We also assume that, for the same kind of AECs, the number of messages sent by each management MA and each ERA are the same respectively, i.e. $\sum_{x=1}^{X_{a,v}} L_{a,v,x} \cong \sum_{x=1}^{X} L$ and $\sum_{n=1}^{N}\sum_{r=1}^{R_n} L_{n,r} \cong \sum_{n=1}^{N}\sum_{r=1}^{R} L$. In formula (1.2), which represents the total network bandwidth consumed by every migration of a MA, we reformulate it as $\sum_{v=1}^{V} S_v \cong \sum_{v=0}^{V-1} (S_0 + vI)$, where S_0 denotes the initial size of this MA and I is the difference in size after a node visit, i.e. $S_1 = S_0$, $S_2 = S_0 + I, \ldots, S_v = S_0 + (V-1)I$. The value of I can be positive, in the case of a information collection MA that collected data will be stored in the MA; or negative, in case of a command-based agent where the MA size will be slightly reduced after preferred a management task that removing a task definition in its data storage. We reformulate equation (1.4) as follows:

$$\lambda_{MASAM} = \sum_{a=1}^{A}\left[\sum_{v=0}^{V-1}\left(S_0 + vI + \sum_{x=1}^{X} L\right)\right] + \sum_{n=1}^{N}\sum_{r=1}^{R} L \tag{3.1}$$

Similarly, for (2.4), we assume that the size of each SNMP message pair and trap are the same respectively and that the total number of message pairs and traps sent by each MD are also the same respectively. The modified form of (2.4) is presented as:

$$\lambda_{SNMP} = \sum_{n=1}^{N}\left(\sum_{g=1}^{G} M + \sum_{j=1}^{J} T\right) \tag{3.2}$$

5.1 Traffic Evaluation and Simulation Focus on the Entire Network

The evaluation of the proposed EMT models is based on four security Attack Event Cases (AECs) as indicated in Table 1. An AEC involves a possible security attack (with three definite stages: attack, detection and recovery) where an intruder utilises security threats to attack the MDs ("multiple targets" based scenarios). Note that the traffic generated in the "attack stage" of the AECs will be ignored due to the focus of this study being the traffic generated for event detection and recovery while the attack traffic is mainly created by intruders. While [6] gives a detailed discussion on each AEC, Figure 1 illustrates the traffic flow difference between the MASAM and SNMP.

A set of prototype-based experiments have been conducted in order to estimate a reasonable value for each listing variable of the formulas. Based on the results obtained, we let the size of a pair of SNMP request-response messages (M), a trap message (T), an ACL message (L), and the initial size of a management MA (S_0) be 496, 211, 283 and 2132 bytes respectively. With the MASAM approach, we assume

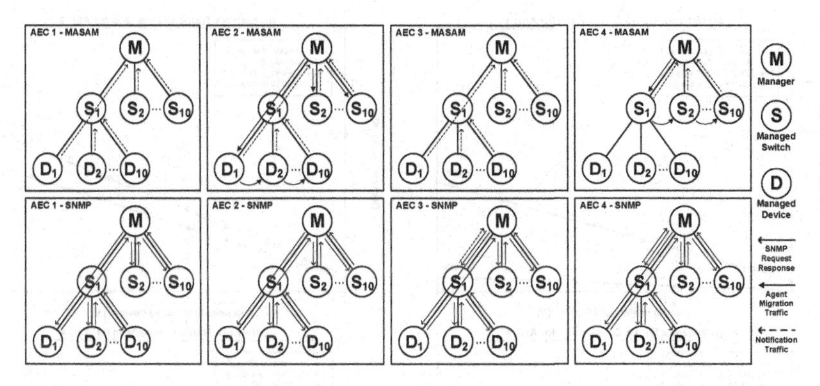

Fig. 1. Network Topology (Adopted for the OPNET Simulations) Shows Traffic Flow Differences between MASAM and SNMP for Each AEC with Multi-Targets Scenarios

that each created management MA will visit 25 MDs (a network segment consists of 24 workstations that are connected by using a managed switch), i.e. $v = 25$ and $A = N + 25$. Due to the MAs employed by every AEC being management command-based MAs, the size of the data difference (I) of a MA should be negative. Based on the same set of results obtained, we estimate the value of I be -20 bytes. Substituting the estimated values onto the formulas (3.1) and (3.2), Table 1 presents the calculated results for each AEC in terms of number of MDs (N) involved.

Table 1. Calculated Results Based on Formulas Derived from the EMT Models and Normalised Weight of Each AEC Based on Related Attacks/Misuses Presented in [7]

AECs	λ_{MASAM}	λ_{SNMP}	Normalised Weight
1. Network Footprinting Attack	283N	2121N	12%
2. Virus Infection Attack	2458N	3113N	44%
3. Message/Agent Reply Attack	283N	1699N	27%
4. Man-in-the-Middle Attack	2175N	4464N	17%

Figure 2 shows the traffic cost comparisons for each AEC, where the number of MDs involved ranges from 10 to 100 (10 MDs for each step). From the figure, the results show the MASAM approach generates less traffic with every AEC, especially in AEC 1 and 4, therefore higher performance could be obtained because of traffic reduction. In addition, a corresponding set of simulation experiments have been conducted by using OPNET 12.0 PL3 with the "custom application" facility. Figure 1 provides an abstract view of the network topology employed for the simulations. The testbed mainly consists of one network manager and 100 MDs, which are connected through Ethernet 100BaseT cables and switches running TCP/IP. The focused statistics for the simulations are the traffic generated by the custom application without any background traffic provided. The simulation results (shown on the same set of graphics with dotted lines) are very similar and their tendencies also closely conform to the calculated results obtained from the traffic models. Thus, the results

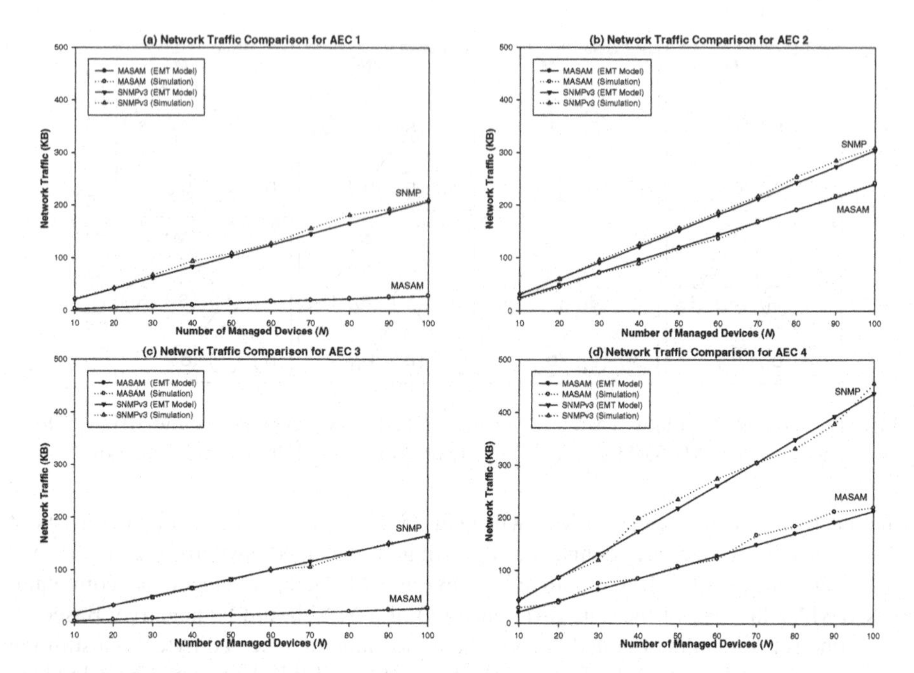

Fig. 2. Traffic Cost Comparisons for four AECs based on Calculated Results Based on the EMT Models and Simulation Results Obtained from the OPNET Experiments

support the accuracy of our proposed EMT models. The simulation results fluctuated because of queue delay or packet lost experienced in the simulations. Finally, we can predict that the MASAM always performs better than the SNMPv3 in every AEC.

In order to obtain more realistic results, we normalised the weight of each AEC (results in table 1) based on the frequencies of related types of attacks/misuses detected in 2006, according to the latest survey report on computer crime [7]. For further evaluation of the EMT, the frequency f of events (how many seconds passed before a new event occurred) should also be considered. We derived two equations based on formulas (3.1) and (3.2), the calculated results from table 1 and the normalised AEC weights and the event frequency, as $\lambda_{MASAM} = 2458N^{*} f$ and $\lambda_{SNMP} = 3113N^{*} f$. Adopting these equations, figure 3 presents the traffic cost comparison between the MASAM (on the left) and SNMP (on the right) where the number of MDs range from 1000 to 9000 and the frequencies of AECs range from 30sec to 2sec.

The figure shows that the total management traffic generated by the MASAM is close to half that of SNMP. This finding implies that, when the frequency of events is growing, the MASAM approach can provide a higher chance of survival to the entire network by reducing management PDUs and their exposure to a security attack. This is especially true if the scale of the managed networks is large and has high latency. Another implication is that the MASAM approach offers a higher opportunity to manage today's wireless network due to less bandwidth consumption offered even if the event frequency is higher.

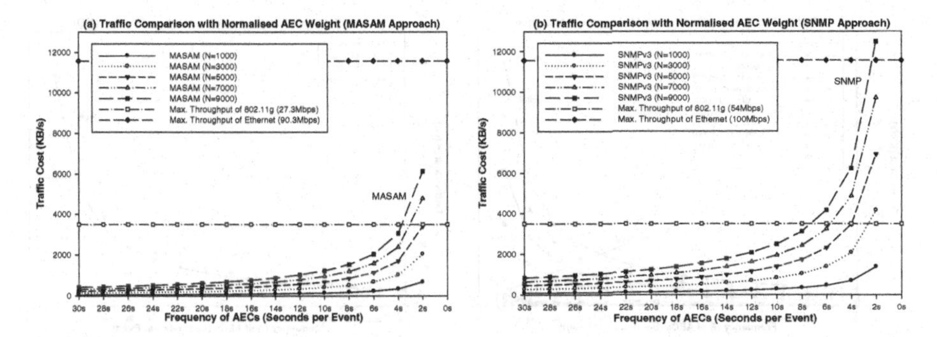

Fig. 3. Traffic Cost Comparisons based on the Calculated Result from the EMT Models (with Normalised AEC Weight) Against the Frequency of AECs

5.2 Traffic Evaluation Focus on the Manager and Managed Device

By reusing part of formula (3.1), we formulate the total traffic generated by the manager for managing a general event with the MASAM approach as follows:

$$\lambda_{MASAM}^{Mgr} = \sum_{a=1}^{A} S_0 \tag{4.1}$$

With the SNMP approach, we simply reuse formula (2.2) and then generalise it based on our assumptions mentioned above. Figure 4 shows the comparison focusing on the traffic cost generated by the manager, where the number of MDs involved ranges from 1000 to 9000 and the frequency of AECs ranges from 30sec to 2sec. The MASAM approach can reduce the large amount of traffic cost generated by the manager even if the frequency of events tends towards being extremely high. When the size of managed network grows, the MASAM approach can effectively protect the manager from rigorous resource consumption and traffic congestion on its network interface. We can therefore also argue that the performance of the manager in security event management will be significantly improved when employing the MASAM.

$$\lambda_{MASAM}^{Dev} = \sum_{a=1}^{A} \left(\sum_{v=1}^{V-1} \frac{S_0 + vI}{V} + \sum_{x=1}^{X} L \right) + \sum_{r=1}^{R} L \tag{4.2}$$

Considering the traffic generated by a MD, with the MASAM approach, it can be counted from the agent migrations from this node to another and the messages sent by the arriving MA or the ERA. By reusing part of formula (3.1), we formulate the total traffic generated by a MD for managing a general event as (4.2), while, for the SNMP approach, we simply reuse formula (2.3). Figure 5 shows the traffic cost generated by a MD, where the event frequencies range from 30sec to 2sec. Although some literatures, such as [8], argued that the MANM is just a "fair" approach that ships processes and traffic from the manager to the MDs side, the figure shows interesting results that the MASAM approach can even slightly reduce a certain amount of traffic generated by each MD when dealing with security event management. Finally, we can argue that the MASAM framework can benefit not only the network manager but also the MD side in terms of traffic generated.

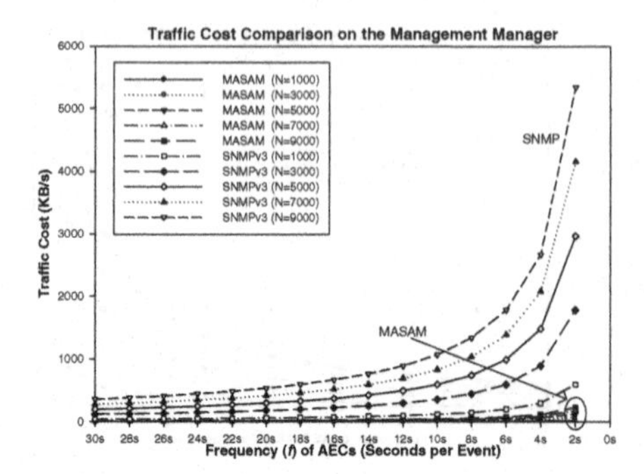

Fig. 4. Traffic Cost Comparison between MASAM and SNMP on the Manager Side

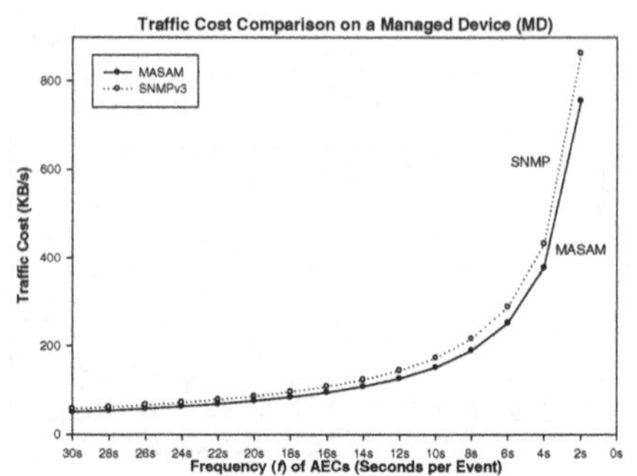

Fig. 5. Traffic Cost Comparison between MASAM and SNMP on a Managed Device

5.3 Scalability Implications Focus on the Number of Managers

The above evaluation results indicate that the number of MDs (N) involved and the frequency of events (f) will be significant factors, as a linear function (for MDs) and an exponential function (for frequency) respectively, that affect the volume of management traffic of the entire network (λ), the manager (λ^{Mgr}) and the MDs (λ^{Dev}). On the other hand, for managing a large scale network, the network should be split into several management domains and managed by different network management nodes. Thus, the strategy for management domain deployment is crucial and the most important parameter is the number of managers to be employed. Based on our EMT models, when N and f are fixed, the formulas show that a change of the number of managers employed will not affect the total traffic cost on the entire network nor the traffic generated by each MD if the management domains are not overlapped. With the MASAM approach, the traffic load (packets sent and received) on a manager (ξ^{Mgr}_{MASAM}) can be formulated as (5.1). Due to the polling-based paradigm of SNMP, in which all management traffic is either sent from or received by the manager (see figure 1), the formula representing the load on the manager with the SNMP (ξ^{Mgr}_{SNMP}) is the same as the total traffic cost of the entire network, i.e. $\xi^{Mgr}_{SNMP} = \lambda_{SNMP}$.

Considering a network with 100,000 MDs, figure 6 shows the total traffic load on each manager for managing this network, where the number of network managers employed ranges from 100 to 10 and the frequency of events is 15sec. The results show that the MASAM approach can effectively reduce the load on each manager employed even when only 10 managers employed, where each manager will manage up to 10,000 MDs. Therefore, we can argue that the MASAM framework also provides better scalability in terms of number of managers employed. Furthermore, when managing such kind of enterprise networks globally interconnected through the Internet, the evaluation results also imply that the MASAM approach has a higher chance of survival. This is because the MASAM framework can significantly reduce the management traffic generated and the number of management domains to be divided. Thus, it simplifies the design of management systems and may also reduce the number of security flaws caused by design complexity or human mistakes.

$$\xi_{MASAM}^{Mgr} = \sum_{a=1}^{A} S_0 + \sum_{n=1}^{N} \left(\sum_{x=1}^{X} L + \sum_{r=1}^{R} L \right) \tag{5.1}$$

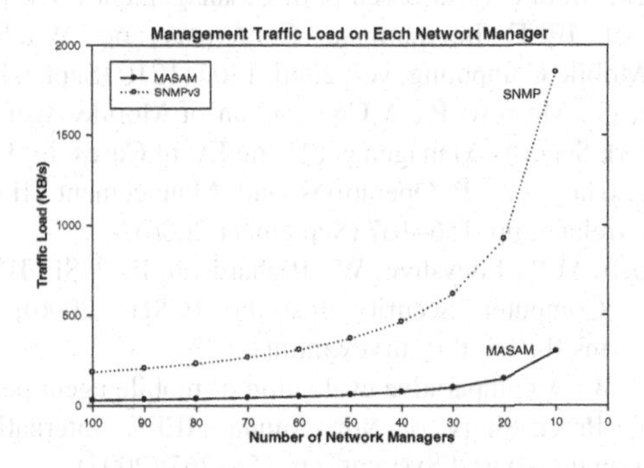

Fig. 6. Traffic Load Comparison on Each Manager when Managing 100,000 MDs

6 Conclusion

In this paper, we discussed our proposed event management traffic models. The derived formulas are validated by the results obtained from a corresponding set of OPNET simulation experiments. The formulas were then utilised to evaluate the performance of the MASAM framework based on the four AECs. The evaluation focused on the traffic implications of how the MASAM and SNMP approaches deal with network security event management. Positive evaluation results on both performance (focused on the entire network, the manager and the MDs) and scalability (focussed on the number of MDs, frequency of events and the number of network managers) have been obtained and we conclude that the MASAM framework is an effective approach for achieving high performance network event management and also improves scalability when compared to the SNMPv3.

Future work will involve further experiments and evaluations focussing on delay and internal process overheads which are another set of crucial parameters for performance and scalability. A new set of prototyping-based experiments and detailed simulation of the MASAM framework will be developed and deployed in order to obtain additional results. After the MASAM has been fully evaluated and optimised, the ultimate step of this research is to convert our philosophy into a new network management standard based on MA technology.

References

1. Cheng, Y., Farha, R., Kim, M.S., Leon-Garcia, A., Hong, J.W.-K.: A generic architecture for autonomic service and network management. Journal of Computer Communications 29, 3691–3709 (2006)
2. Chatzimisios, P.: Security issues and vulnerabilities of the SNMP protocol. In: Proceeding of IEEE Intl. Conference on Electrical and Electronics Engineering, pp. 74–77 (June 2004)

3. Harrington, D., Schoenwaelder, J.: Transport Subsystem for the Simple Network Management Protocol (SNMP). Internet-Draft, IETF (March 2007)
4. Harrington, D.: Secure Shell Transport Model for SNMP. Internet-Draft, IETF (May 2007)
5. Zhou, Y., Xiao, D.: Mobile agent-based policy management for wireless sensor networks. In: Proceeding of IEEE International Conference on Wireless Communications, Networking and Mobile Computing, vol. 2, pp. 1207–1210 (September 2005)
6. Fong, C.-h., Parr, G., Morrow, P.: A Comparison of Mobile Agent and SNMP Message Passing for Network Security Management Using Event Cases. In: Proceeding of 6th IEEE International Workshop on IP Operations and Management, IFIP/IEEE MANWEEK Conference 2006, Ireland, pp. 156–167 (September 2006)
7. Gordon, L.A., Loeb, M.P., Lucyshyn, W., Richardson, R.: CSI/FBI Computer Crime and Security Survey. Computer Security Institute (CSI) (2006), Available online at http://gocsi.com/forms/fbi/csi_fbi_survey.jhtml
8. Tang, L., Pagurek, B.: A comparative evaluation of mobile agent performance for network management. In: Proceedings of 9th Annual IEEE International Conference on Engineering of Computer-Based Systems, pp. 258–267 (2002)
9. Raman, L.: OSI systems and network management. IEEE Communications Magazine 36(3), 46–53 (1998)
10. To, H., Krishnaswamy, S., Srinivasan, B.: Mobile agents for network management: when and when not. In: Proceedings of ACM Symposium on Applied Computing, pp. 47–53 (2005)

Design and Implementation of User-Oriented Handoff Framework with VoIP Service

Hsu-Yang Kung[1], Chuan-Ching Sue[2], and Chi-Yu Hsu[2]

[1] Department of Management Information Systems
National Pingtung University of Science and Technology, 912 Pingtung, Taiwan
kung@mail.npust.edu.tw
[2] Department of Computer Science and Information Engineering
Natinoal Cheng Kung University, 701 Tainan, Taiwan
{suecc,p7895113}@mail.ncku.edu.tw

Abstract. This work design and implement a User-oriented Handoff Control Framework (UHCF) comprising Wireless Sensor Networks, SIP and VoIP. A two-stage strategy is proposed to support user-oriented handoff. The strategy is to detect and collect user's location information, and determine whether user-oriented is triggered. If a user-oriented handoff is triggered, then a progressive VoIP session is switched smoothly from the old to the new host with the user's mobility. We also implement User-oriented Handoff Control Framework (UHCF) to a SIP-based VoIP system. Implementation results show that the UHCF with SIP-based VoIP service has deployed successfully by wireless sensor networks, and the system can efficiently assist user-oriented handoff with VoIP service.

Keywords: User-oriented, Handoff protocol, Voice over IP, Session Initiation Protocol.

1 Introduction

Host-oriented handoff has been widely discussed in past decade [1, 2, 3]. In this paper, we proposed a User-oriented Handoff Control Framework (UHCF). The major difference between two kinds of handoff is the handoff starter. As the name applied, Host-oriented handoff is triggered by host, accordingly User-oriented handoff is triggered by users. The detail of User-oriented handoff is defined as follows. When the user moves to another terminal, the session in use is switched to a new terminal device with the user. The distinguishing characteristic of the user-oriented handoff is that the session switches with the position of mobile users on anytime and anywhere. Fig. 1 depicts user-oriented handoff situation. In the figure, three User Agents (UAs) represent different terminal hosts, supporting lower-layer mobility services such as Mobile IP, in which $User_A$ is moving between the coverage areas of these UAs. While $User_A$ is moving to UA_3, the user-oriented handoff is triggered automatically. Furthermore, the primitive conversation in UA_1 is smoothly switched to UA_3.

S. Ata and C.S. Hong (Eds.): APNOMS 2007, LNCS 4773, pp. 487–491, 2007.

This study attempts to solve the following challenges of user-oriented handoff.

- The user-oriented handoff mechanism is determined by the locations of the user and the host. If a user-oriented handoff is triggered, then a progressive VoIP session is switched smoothly from the old to the new host with the user's mobility. Hence, detecting and collecting user location, and making handoff decisions. Additionally, since handoff procedure consumes significant system and network resources, avoiding redundant handoffs is also important.
- To implement a user-oriented system with SIP-based VoIP service. The system can support re-establishing the session quickly at both ends, minimizing the packet loss rate and preventing packets from falling out of sync are also significant issues.

Section 2 introduces the proposed User-oriented Handoff Control Framework (UHCF). Section 3 presents system implementation with UHCF and results. Section 4 draws conclusions.

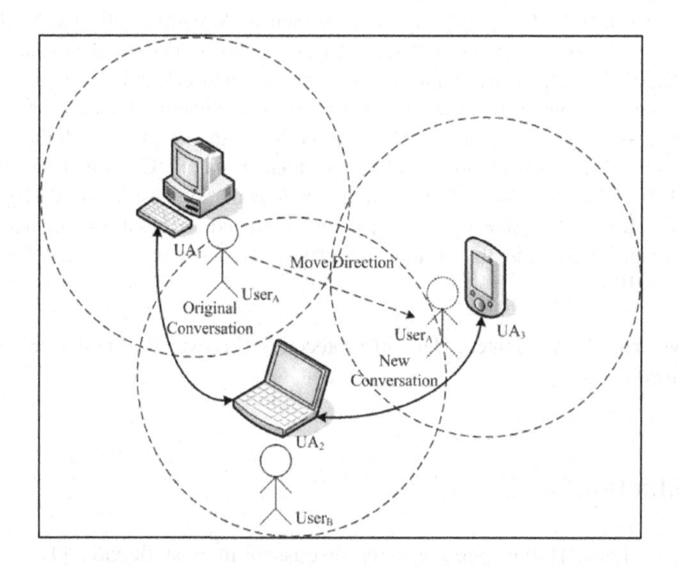

Fig. 1. The illustration of user-oriented handoff situation

2 The Overview of Proposed User-Oriented Handoff Control Framework

The UHCF is performed in two schemes, (i) User-Oriented Handoff Detection (UOHD) and (ii) Fast Session Re-establishment (FSR). Table 1 defines UOHD terminologies in this scheme. In UOHD, three handoff detection modes based on power constraint and scalability challenges are proposed, such as (i) User Agent Assistant Mode (UA^2M), (ii) Mobility Agent Assistant Mode (MA^2M) and (iii) Hybrid Assistant Model (HAM). The UA using UA^2M can periodically emit

"Detection" messages to detect and collect user's location, and send the measured data to MA and LS, respectively. The UA then asks the MA about a user's location of surrounding UAs. According to query results, the UA determines whether user-oriented handoff is triggered. For this mode, the detection and collection of user's location data and handoff decisions are conducted by each UA since this way can efficiently decrease network and MA loading. Nonetheless, UAs must be constantly active, and consumes substantial energy. Therefore, UA^2M is not appropriate for power constraint appliances such as PDAs and smartphones. In MA^2M, the MA initiates the detection and collection of user locations in MA^2M. Hence, MA will periodically send a message to UA commencing the handoff detection procedure. After completing measurement, MA returns and updates the measured user's location data to LS, and determines whether user-oriented handoff criteria are met. If the criteria are conformed, then MA transmits a trigger message to UAs to notify them that the user-oriented handoff procedure can be performed. The MA makes the handoff decision in this mode, and is therefore overloaded due to the large number of UAs in the network. The MA^2M control procedure is given in detail as follows. HAM combines UA^2M and MA^2M, and thus overcomes the problems of power constraint and system overload. The purpose of Fast Session Re-establishment (FSR) scheme is to achieve a user-oriented handoff framework with low handoff delay for the VoIP service. Refs [4] and [5] demonstrate that re-INVITE lengthen the voice interrupt time and require manual operation, e.g., user's URI input. To efficiently decrease the voice interrupt time, this work presents two fast re-establishment methods, namely (i) RS-based 3PCC and (ii) REFER with RACK, for intra-domain as well as inter-domain handoff, respectively.

Table 1. Terminology definitions of UOHD

Terminology	Description
User	A person using VoIP service and moving among different hosts
User Agent (UA)	A program installed in handheld devices and personal computers, and used to originate, receive, and terminate calls.
Mobility Agent (MA)	An agent that communicates between SIP components and non-SIP components in an SIP network, and to help UA to perform the handoff procedure.
Location Server (LS)	A server storing every user's URI, preferences and IP address.

3 System Design and Implementation

This study implements a SIP softphone as shown in Fig. 2. The softphone has origination, termination, and call transfer capabilities. In addition, user also can configure the used audio device and adjust audio volume. Fig. 2(a) shows an interface when you make a call to other ends, in which user can choose call type such as

PC-to-PC or PC-to-Phone, and input destination URI that can be user name, IP address. While a call is incoming, an alarm window will pop as shown in Fig. 2(b), in which the user can choose accept, reject, or forward the incoming call. As Fig. 2(c) shows, if the user is to accept the call, the conversation session will be connected. Location detection and collection of the user adopts MTS 410 kits. Fig. 2(d) demonstrates Listener Sensor (LiS) and Beacon Sensor (BS). LiS is connected to the laptop by RS-232 line, it is to monitor beacon message sent from BS.

4 Conclusions

This proposed a User-oriented Handoff Control Framework (UHCF) which provides a handoff framework for SIP-based VoIP service based on the user perspective. UHCF comprises a two-stage strategy to support seamless user-oriented handoff. The first stage is User-Oriented Handoff Detection (UOHD), in which three handoff modes were considered based on system performance as well as power consumption level. The second stage is Fast Session Re-establishment (FSR). In addition, we implement User-oriented Handoff Control Framework (UHCF) with VoIP service. Implementation results show that the proposed User-oriented Handoff Control Framework (UHCF) with SIP-based VoIP service has deployed successfully by wireless sensor networks, and deployed system can efficiently support VoIP handoff based on the user's location.

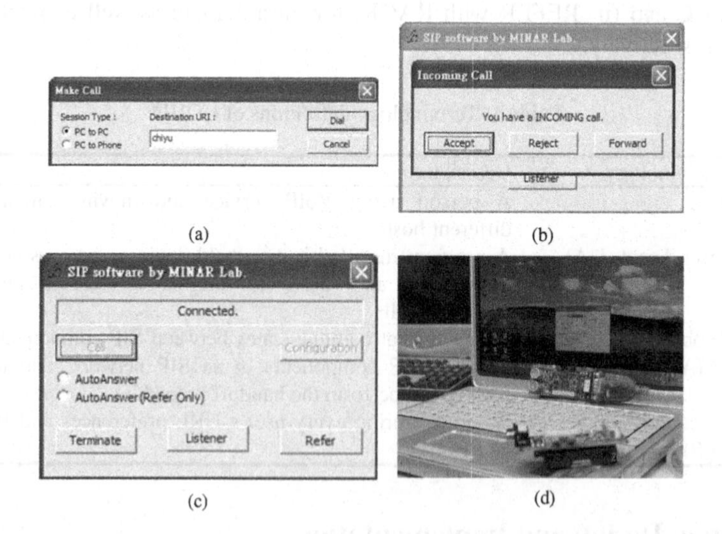

Fig. 2. (a) Make a call, (b) Incoming call, (c) The call is successfully connected, (d) An demonstration of LiS and BS.

Acknowledgments. This research was supported by the National Science Council, Taiwan, R.O.C., under Grant NSC 96-2221-E-020-034-MY2.

References

1. Andrew, T.C., Javier, G.C.: IP Micro-Mobility Protocols. Mobile Computing and Communication Review, 45–53 (2000)
2. Kwon, T.T., Gerla, M., Das, S.: Mobility Management for VoIP Service: Mobile vs. SIP. IEEE Wireless Communications, 66–75 (2002)
3. Lin, Y.B., Chlamtac, I.: Wireless and Mobile Network Architectures, pp. 15–33. Wiley, Chichester (2001)
4. Schulzrinne, H., Wedlund, E.: Application-layer mobility using SIP. ACM SIGMOBILE Mobile Computing and Communications Review 4, 47–57 (2000)
5. Sinnreich, H., Johnston, A.B.: Internet Communications Using SIP. Wiley, Chichester (2001)

A Study on Low-Cost RFID System Management with Mutual Authentication Scheme in Ubiquitous

Soo-Young Kang and Im-Yeong Lee

Division of Computer, Soonchunhyang University, #646, Eupnae-ri, Shinchang-myun, Asan-si, chungchungnam-Do, 336-745, Korea
{bbang814,imylee}@sch.ac.kr

Abstract. The RFID system is a core technology used in building a ubiquitous environment, and is considered an alternative to bar-code identification. The RFID system has become very popular, with various strengths such as fast recognition speed and non-touch detection. However, there are some problems remaining, as the low-cost tag can operate through queries, leading to information exposure and privacy encroachment. Various approaches have been used to increase the security of the system, but the low-cost tag, which has about 5K~10K gates, can only allocate 250~3K gates to security. Therefore, the current study provides a reciprocal authentication solution that can be used with low-cost RFID systems, by splitting 64 bit keys and minimizing calculations. Existing systems divided a 96 bit key into 4 parts. However, the proposed system reduces the key to 32 bits, and reduces communications from 7 down to 5. To increase security, one additional random number is added to the two existing numbers. The previous system only provided XOR calculations, however in the proposed system an additional hash function was added. The added procedure does not increase effectiveness in terms of the XOR calculation, but provides more security to the RFID system, for better use over remote distances.

Keywords: RFID, Mutual Authentication.

1 Introduction

With the advent of the ubiquitous environment, the use of RFID is increasing. RFID stands for Radio Frequency IDentification; it is a type of identification technology that will replace barcodes. An active tag has its own power source and a passive tag receives power from the reader. The price of an active tag is high due to its remote identification and high capabilities, hence the low-cost passive tag is generally used. However, the low-cost passive tag is easily activated by queries from the reader, which exposes it to privacy encroachment problems caused by user and product information, stored in the tag. To solve this problem, various researches, such as using hashed IDs instead of proper IDs, using random numbers, or using hash functions, have been carried out. However, while a hashed ID value can provide ID integrity, it can be traced because it is a fixed value. Also, random numbers carry the problem of attaching the random number generator(R.N.G.). Although the hash function method is the safest, since only 250~3K out of the 5K~10K gate can be allocated for security,

S. Ata and C.S. Hong (Eds.): APNOMS 2007, LNCS 4773, pp. 492–502, 2007.

a lighter approach is necessary. Therefore, methods to produce random numbers from the reader or database, to reduce hash function numbers, and to use simple algorithms for lightness are under research. Also, authorized subjects share an authentication value for mutual authentication, and check legitimacy through key values.

Therefore, this study uses random numbers and safely shared authentication values to verify the authenticity of each object. Also, the proposed method uses simple algorithms(XOR, OR) to be realistically used in low-cost RFID systems. The extended method uses hash functions to provide integrity, which could not be provided in the proposed method.

The new procedure is introduced based on existing procedures. Chapter 2 deals with security risks and requirements related with RFID systems. Chapter 3 studies previous RFID security models and chapter 4 illustrates the design of the proposed system. Chapter 5 analyzes solutions for requirements discussed in chapter 2, and chapter provides conclusions and directions for future research.

2 Security Threats and Requirements

A RFID system consists of a tag, reader, and database. User or item information is sent to the database after the tag receives power from the reader. After authorization, the database compares the tag information and stored information, and sends authorization data to the tag. For this procedure to be processed properly, fabrication of data transfer must be prevented. However, the communication channel between the tag and reader is wireless, thus exposed to third parties. Therefore, this chapter deals with the security risks associated with the RFID system and security requirements to solve these issues.

2.1 Security Threats

RFID is vulnerable to various forms of attack because it uses radio frequencies. Therefore, we must look at various forms of attack in RFID usage and solutions to counter these problems. The following describes various security threats associated with the RFID system.

▪ Man-in-the-Middle Attack : An illegal third party acquires the data being sent from the tag and reader, then copies and modulates the data, which might result in the illegal user obtaining authentication. (Ed-meaning changed, confirm as intended)

▪ Traffic Analysis : This is a process in which data, acquired from eavesdropping, is compiled to guess the next session's value or acquiring essential values. (Ed-please clarify this, its not clear what you are referring to) The data transfer traffic of each session must not be uniform, and random variable values should be used to prevent third party users from analyzing the data. (Ed-original slightly unclear, confirm as intended)

▪ Replay Attack : This form of attack re-sends data acquired through eavesdropping to induce or acquire essential values. When identical or essential values are exposed by the tag during the re-sending process, tracking problems may arise and privacy may be encroached upon. Therefore changing values should be used in communication to prevent replay attack.

▪ Tracking Attack : One of the most serious privacy problems of RFID is that if a set value is exposed during each session, the privacy of the user may be encroached upon even if it is not possible to locate his exact location. To counter this form of attack, changing random numbers or time stamps should be used.

2.2 Requirements

To solve the aforementioned security threats, the following requirements should be provided.

▪ Authentication : All components of the system should go through an authentication process. RFID is comprised of a tag, reader, and database. Each part should provide authentication to each other. The tag should send secret values, which have been previously agreed upon, to each component, to become authorized.

▪ Anonymity : Even if data is acquired from a tag, it should not be trackable to a tag. If identification values are set, anonymity cannot be guaranteed. Therefore essential values should not be exposed and different values should be used to send data.

▪ Integrity : Since data is sent through a radio channel, illegal users can copy and modify data. To solve this problem, data should be encrypted for integrity. Passive tags have little power and calculation skills. This allows them to use formulas to prevent data copying and modification.

▪ Confidentiality : Values used in the security protocol should not be exposed and only authorized users may share them. All components should share a secret value to authenticate each other. In order to provide confidentiality, a third party should not be able to access the secret value, and the security protocol should be protected based on this confidentiality.

▪ Efficiency : Although efficiency is not included in the security requirements, passive tags require hash formulas or XOR calculation. Passive tags require applicable security measures, and therefore require efficiency.

The security threats noted above require solutions. These solutions will provide adequate privacy protection.

3 Related Works

Many studies are being conducted in the RFID field on securing privacy. Hash formulas are usually used to increase security, but LMAP, M2AP, and EMAP procedures provide security using simple operations. By dividing the 96 bit key into 4 parts to be used as separate keys, and by refreshing the key value and message each session, the user can be protected from location tracking and re-transmission attacks. The current chapter deals with existing procedures.

3.1 LMAP(Lightweight Mutual Authentication Protocol)

This method can provide security, using only 300 gates. After the 96 bit key is divided into 4, 4 messages are produced, of which the reader sends A, B, C messages to the tag. To provide authentication, the tag replies with a D message, in accordance with messages A, B, and C[4]. However, there are risks of data forgery and fabrication during transfer. If the final session is concluded irregularly, the IDS and key values are not refreshed, thereby exposing identical values, which make location tracking possible.

- Message generation process

  $A=IDS_{tag(i)} \oplus K1_{tag(i)} \oplus n1$ $B=(IDS_{tag(i)} \vee K2_{tag(i)})+n1$

  $C=IDS_{tag(i)}+K3_{tag(i)}+n2$ $D=(IDS_{tag(i)}+ID_{tag(i)})) \oplus n1 \oplus n2$

- Key renewal process

  $IDS_{tag(i+1)}=(IDS_{tag(i)}+(n2 \oplus K4_{tag(i)})) \oplus ID_{tag(i)}$

  $K1_{tag(i+1)}=K1_{tag(i)} \oplus n2 \oplus (K3_{tag(i)}+ID_{tag(i)})$

  $K2_{tag(i+1)}=K2_{tag(i)} \oplus n2 \oplus (K4_{tag(i)}+ID_{tag(i)})$

  $K3_{tag(i+1)}=(K3_{tag(i)} \oplus n1)+(K1_{tag(i)} \oplus ID_{tag(i)})$

  $K4_{tag(i+1)}=(K4_{tag(i)} \oplus n1)+(K2_{tag(i)} \oplus ID_{tag(i)})$

3.2 M2AP(Minimalist Mutual Authentication Protocol)

This method uses the same 300 gates as the LMAP method but has an additional E value to add more security in database authentication than the LMAP method[5]. It is efficient in that it uses some calculations for authentication, but cannot provide integrity since it does not use hash formulas or encryption algorithms. If the final session is concluded irregularly, the IDS and key values are not refreshed, thereby exposing identical values, which make location tracking possible.

- Message generation process

  $A=IDS_{tag(i)} \oplus K1_{tag(i)} \oplus n1$ $B=(IDS_{tag(i)} \wedge K2_{tag(i)}) \vee n1$

  $C=IDS_{tag(i)}+K3_{tag(i)}+n2$ $D=(IDS_{tag(i)} \vee K4_{tag(i)})) \wedge n2$

  $E=(IDS_{tag(i)}+ID_{tag(i)})) \oplus n1$

- Key renewal process

  $IDS_{tag(i+1)}=(IDS_{tag(i)}+(n2 \oplus n1)) \oplus ID_{tag(i)}$

  $K1_{tag(i+1)}=K1_{tag(i)} \oplus n2 \oplus (K3_{tag(i)}+ID_{tag(i)})$

  $K2_{tag(i+1)}=K2_{tag(i)} \oplus n2 \oplus (K4_{tag(i)}+ID_{tag(i)})$

  $K3_{tag(i+1)}=(K3_{tag(i)} \oplus n1)+(K1_{tag(i)} \oplus ID_{tag(i)})$

  $K4_{tag(i+1)}=(K4_{tag(i)} \oplus n1)+(K2_{tag(i)} \oplus ID_{tag(i)})$

3.3 EMAP(Efficient Mutual Authentication Protocol)

This method is a more efficient method than the LMAP and M2AP methods, in that it provides security with 150 gates[3]. Here, the 4 keys, which have been divided from message E, produce the XOR algorithm sigma value (K1⊕K2⊕K3⊕K4) and provide a more accurate way of authentication. The 96 bit ID can be divided into two, resulting in the use of a 1~48 bit ID and a 49~96 bit ID, which results in the use of 2 identification values. By inputting the key value into the formula, the safety of the system is enhanced. This method is more effective than the two previously mentioned systems, and can provide security within close ranges. However it can be exposed to 3rd party tapping, message fabrication or forgery over long ranges.

- Message generation process

$A=IDS_{tag(i)} \oplus K1_{tag(i)} \oplus n1$ $B=(IDS_{tag(i)} \wedge K2_{tag(i)}) \vee n1$

$C=IDS_{tag(i)}+K3_{tag(i)}+n2$ $D=(IDS_{tag(i)} \vee K4_{tag(i)})) \wedge n2$

$E=(IDS_{tag(i)} \wedge n1 \vee n2) \oplus ID_{tag(i)} \oplus_{i=1}^{4} Ki_{tag(i)}$

- Key renewal process

$IDS_{tag(i+1)}=(IDS_{tag(i)}+(n2 \oplus n1)) \oplus ID_{tag(i)}$

$K1_{tag(i+1)}=K1_{tag(i)} \oplus n2 \oplus (K3_{tag(i)}+ID_{tag(i)})$

$K2_{tag(i+1)}=K2_{tag(i)} \oplus n2 \oplus (K4_{tag(i)}+ID_{tag(i)})$

$K3_{tag(i+1)}=(K3_{tag(i)} \oplus n1)+(K1_{tag(i)} \oplus ID_{tag(i)})$

$K4_{tag(i+1)}=(K4_{tag(i)} \oplus n1)+(K2_{tag(i)} \oplus ID_{tag(i)})$

4 Proposed Protocol

Previous methods used 4 keys, which are sections of the 96 bit key ID, for authentication, and the reader produced 2 random numbers for security. This method divides the 64 bit ID into 2 keys, thereby increasing the keys used by 8 bits, lowering the overall key size by 32 bits, and increasing efficiency by producing only 1 random number. The existing algorithms (XOR, OR, AND, ADD) in previous systems can provide security within short ranges. However since there are risks of tapping, message forgery and fabrication, security must be enhanced. The proposed method uses key partition and light algorithms for security, which are similarly used in previous methods, but reduces the size of the key and number of random numbers produced for efficiency. However, since there are risks of forgery and fabrication associated with light algorithms, unilateral hash functions were added for integrity. This chapter provides methods for security, while satisfying the above requirements.

4.1 Parameters

The parameters for the proposed protocols.

- Ki : i-th key value(i=1, 2)
- ID : Tag's unique ID
- $metaID$: Hashing ID H(ID)
- r : Random number
- V_1 : First various value $K1 \vee r$
- V_2 : Second various value $K2 \vee r$
- Kr_1 : First XOR operate value with key value and random number, $K1 \oplus r$
- Kr_2 : Second XOR operate value with $metaID$ and V_1, $metaID \oplus V_1$
- Kr_3 : Third XOR operate value with tag's ID and V_2, $ID \oplus V_2$
- H_1 : First hash value with $metaID$ and V_1, $H(metaID \| V_1)$
- H_2 : Second hash value with ID and V_2, $H(ID \| V_2)$
- $H(\,)$: One-way hash function
- $\oplus$: Exclusive OR function

4.2 Proposed Protocol Summary

The proposed method uses key partition and random numbers, as used in LMAP, M^2AP, EMAP methods, but reduces the 96 key to a 64 bit key and the number of random numbers from 2 to 1 for added efficiency. Existing methods may take a long time over a total of 7 passes, where the proposed one goes to 5 passes to save time. While efficiency and time were enhanced, the actual length of the key used is increased from 24 bits to 32 bits, which enhances security. Hash functions are used to provide security in long range communications, to prevent active attackers from encroaching upon the system and provide integrity. This chapter discusses ways to prevent safety threats and satisfy requirements.

4.3 Proposed Protocol

The proposed method can be used on the low-cost RFID system, provide security through XOR and OR algorithms, and divides the 64 bit key into two. The system is 32 bits smaller than previous systems and uses a key that is 8 bits larger, which results in increased storage space efficiency and enhanced security.

Step 1. To initiate the communications session, the reader produces the random number r from R.N.G and divides the 64 bit key into two. The first key K1 of the 1~32 bits uses an XOR algorithm with random number r, and the result Kr_1 is sent to the tag.
Step 2. The tag receives Kr_1 from the reader and uses an XOR algorithm with the tag's K1 to acquire r. The tag uses an OR algorithm with the acquired r and K1 to produce V_1, and sends Kr_2, a value produced from the XOR algorithm between metaID and V_1, to the reader.
Step 3. The reader receives Kr_2 from the tag, connects the initial r with Kr_2, then sends them to the database.

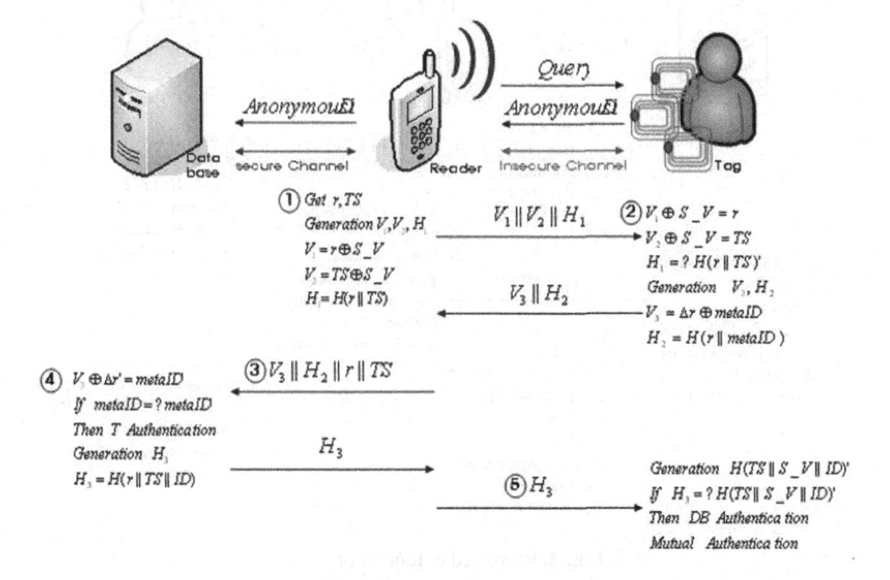

Fig. 1. Proposed protocol

Step 4. The database conducts an OR algorithm on the received r and K1 to produce V_1', and conducts an XOR algorithm with the received Kr_2 and V_1' to produce the metaID. If the received metaID' is identical to the metaID, which the database contains, the database authenticates the tag, and acquires an ID for the metaID. V_2 is acquired through the OR algorithm between K2 and r, and the acquired ID is run through an XOR algorithm to produce Kr_3, which is then transmitted to the reader.
Step 5. The tag receives the Kr_3, then produces V_2' by running an OR algorithm with K2 and r, then uses the XOR algorithm with Kr_3 to acquire the ID'. If the acquired ID' is identical with the tag's ID', the database is authenticated, and reciprocal authentication is provided.

4.4 Proposed Extended Protocol

The proposed method uses XOR and OR algorithms, and is therefore safe from passive attacks. However, it cannot provide security against active attacks, in which messages are forged or fabricated during message transfer. Therefore, the proposed extended method uses hash functions to prevent breaches from active attacks.

Step 1. To initiate the communications session, the reader produces a random number r from R.N.G and divides the 64 bit. The first key K1 of the 1~32 bit uses the XOR algorithm with random number r, and the result Kr_1 is sent to the tag.
Step 2. The tag receives Kr_1 from the reader and uses the XOR algorithm with the tag's K1 to acquire r. The tag uses the OR algorithm with the acquired r and K1 to produce V_1, and sends Kr_2, a value produced using the XOR algorithm between metaID and V_1, to the reader.

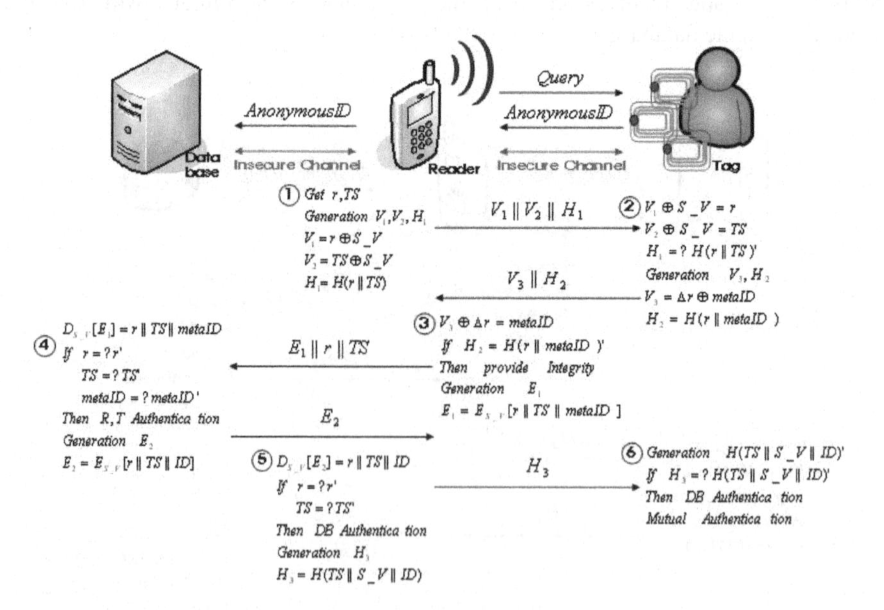

Fig. 2. Proposed extended protocol

Step 3. The reader receives Kr_2 from the tag, and connects the initial r with Kr_2, and sends them to the database.
Step 4. The tag uses the XOR algorithm between the tag's K1 and the received Kr_1 to produce r. By conducting an OR algorithm between the tag's K1 and the acquired r, V_1 is produced. Then this is run through an XOR algorithm with the tags metaID to produce Kr_2. For data integrity, the metaID and V_1 are hashed to produce H_1. Then the Kr_2 and H_1 are sent to the reader.
Step 5. The database runs an OR algorithm on the received r and K1 to produce V_1'. Then the metaID' is acquired by running an XOR algorithm with the transferred Kr_2 and V_1'. If the database has a metaID that is identical with metaID', an ID for the metaID is acquired. The metaID' and V_1' are hashed to produce H_1'. If this is identical with the transferred H_1, integrity is authenticated. To produce an authentication result, the database runs an OR algorithm on K2 and r to produce V_2, and the H_2, which was hashed between the ID and V_2, is sent to the reader.

5 Analysis

The proposed method analyzes whether the security requirements, discussed in chapter 2, can be satisfied. The LMAP, M2AP, EMAP methods, which were analyzed in 2006 as low-cost RFID systems, are compared with the proposed method. Although the proposed extended method is not as efficient as previous methods, it can be used to enhance the security of the RFID system, once the tag algorithm capabilities are enhanced. This chapter compares existing methods with the proposed method.

5.1 Security Threat Measures

The proposed method can solve the security threat related problems mentioned in chapter 2. There are 4 kinds of security threats, eavesdropping, man-in-the-middle-attack, replay attack, and tracking attack. The proposed method can overcome these problems.

- Eavesdropping : A third party could tap into the data transfer, since it uses a wireless channel, but important values cannot be acquired because of the use of XOR and OR algorithms.
- Man-In-The-Middle-Attack : The proposed extended method uses hash functions to provide integrity. However, the proposed method only uses XOR and OR algorithms, which make it susceptible to 3rd party data forgery and fabrication.
- Replay Attack : The system is safe from re-transfer attacks because it uses changing random numbers during communication.
- Tracking Attack : A new random number is produced for each session, and the random number is run through an OR algorithm with the key value, to produce V_1 and V_2. Since V_1 and V_2 change each session, the system is safe from location tracking.

Table 1. Analysis according to security threats and requirements

	LMAP	M2AP	EMAP	**Proposed protocol**	**Proposed extended protocol**
Eavesdropping	*	*	*	O	O
	IDS exposure	IDS exposure	IDS exposure	Random Number use	Random Number use
MITMA	X	X	X	X	O
	Data forgery possibility	Data forgery possibility	Data forgery possibility	Data forgery possibility	Data forgery impossibilit y
Replay Attack	O	O	O	O	O
	Update key	Update key	Update key	Random Number use	Random Number use
Tracking Attack	O	O	O	O	O
	Update key	Update key	Update key	Random Number use	Random Number use
Mutual Authentication	*	*	*	*	O
	When data forge impossibil ity	When data forge impossibilit y	When data forge impossibilit y	When data forge impossibilit y	Data forgery impossibilit y
Anonymity	*	*	*	O	O
	IDS exposure	IDS exposure	IDS exposure	Random Number use	Random Number use
Confidentiality	*	*	*	O	O
	IDS exposure	IDS exposure	IDS exposure	K1, K2, ID, metaID	K1, K2, ID, metaID
Integrity	X	X	X	X	O
	Data forgery possibility	Data forgery possibility	Data forgery possibility	Data forgery possibility	Hash function use
Efficiency	O	O	*	O	X
	$\oplus, \vee, \wedge, +$	$\oplus, \vee, \wedge, +$	$\oplus, \vee, \wedge, +, f()$	$\oplus, \vee$	$\oplus, \vee$ H( )

5.2 Requirements Measures

This chapter analyzes and provides solutions to the security requirements mentioned in chapter 2. The current chapter analyzes how the solutions are satisfied. The calculation tools of the components and frequency of usage of these tools are analyzed to see if they are efficient.

▪ Authentication : The proposed method is safe from passive attacks, but may be vulnerable to active attacks. If the attacker forges or fabricates the data, a legitimate user may be rejected. However, the proposed extended method utilizes hash function values to prevent forgery and fabrication, and provides mutual authentication.

▪ Anonymity : The tag's ID and key value are random numbers, anonymity is achieved even if an illegal user acquires these values.

▪ Integrity : The proposed method is used to respond to passive attacks and data forgery and fabrication. However, the proposed extended method uses hash functions to check the forgery and fabrication of transferred data and provide integrity.

▪ Confidentiality : The tag's ID and 64 bit key value are only shared between legitimate subjects, and confidentiality is provided through light algorithms and hash functions.

▪ Efficiency : The method uses unilateral hash functions rather than complicated encryption algorithms, considering the limited calculation and storage capabilities of the passive tag. These hash functions must have a small number of calculations, and random numbers are produced by the R.N.G, which is attached to the reader.

6 Conclusions

The RFID system, a core technology of the ubiquitous environment, can be easily activated with a reader signal. This exposes it to data encroachment and privacy invasions. Therefore, the current method divides the 64 bit key into two parts and uses random numbers and count values to provide security. Since the proposed method does not expose the key and identification values through XOR and OR algorithms, it is safe from passive attacks. However, since it is not safe from active attacks, in which data is forged and fabricated, the proposed extended method uses hash values to counter such attacks. The XOR algorithm results are appended to the hash value during transfer, which prevents forgery or fabrication. However, it might be difficult to apply it to current low-cost RFID systems. Low-cost RFID systems have a 5K~10K gate and only a 250~3K gate can be allocated to security. Therefore, the hash function must be light, and further study should be conducted to find ways to provide sufficient security. By finding a compromise between lightness and security, privacy invasion, the biggest problem associated with the RFID system, can be solved, and a solid foundation for popular adoption of RFID systems can be established.

References

1. Henrici, D., MÄuller, P.: Hash-based enhancement of location privacy for radio-frequency identification devices using varying identifiers. In: IEEE International Workshop on Pervasive Computing and Communication Security PerSec, pp. 149–153 (March 2004)
2. Vajda, I., Butty´an, L.: Lightweight authentication protocols for low-cost RFID tags. In: 2nd Workshop on Security in Ubiquitous Computing (2003)
3. Feldhofer, M., Dominikus, S., Wolkerstorfer, J.: Strong Authentication for RFID Systems Using the AES Algorithm. In: Joye, M., Quisquater, J.-J. (eds.) CHES 2004. LNCS, vol. 3156, Springer, Heidelberg (2004)
4. Ohkubo, M., Suzuki, K., Kinoshita, S.: Cryptographic approach to 'privacy-friendly' tags. In: RFID Privacy Workshop (November 2003)
5. Peris-Lopez, P., Hernandez-Castro, J., Estevez-Tapiador, J., Ribagorda, A.: EMAP: An Efficient Mutual Authentication Protocol for Low-cost RFID Tags. In: OTM Federated Conferences and Workshop: IS Workshop (2006)

6. Peris-Lopez, P., Hernandez-Castro, J., Estevez-Tapiador, J., Ribagorda, A.: LMAP: A Real Lightweight Mutual Authentication Protocol for Low-cost RFID tags. In: Workshop on RFID Security (2006)
7. Peris-Lopez, P., Hernandez-Castro, J., Estevez-Tapiador, J., Ribagorda, A.: M2AP: A Minimalist Mutual-Authentication Protocol for Low-cost RFID Tags. In: Ma, J., Jin, H., Yang, L.T., Tsai, J.J.-P. (eds.) UIC 2006. LNCS, vol. 4159, Springer, Heidelberg (2006)
8. Sarma, S., Weis, S., Engels, D.: Radio-frequency identification: security risks and challenges, Cryptobytes, RSA Laboratories, pp. 2–9, Spring (2003)
9. Sarma, S.E., Weis, S.A., Engels, D.W.: RFID Systems and Security and Privacy Implications. In: Kaliski Jr., B.S., Koç, Ç.K., Paar, C. (eds.) CHES 2002. LNCS, vol. 2523, pp. 454–470. Springer, Heidelberg (2003)
10. Weis, S., Sarma, S., Rivest, R., Engels, D.: Security and privacy aspects of low-cost radio frequency identification systems. In: International Conference on Security in Pervasive Computing SPC, pp. 454–469 (March 2003)
11. Weis, S.: Security and privacy in radio-frequency identification devices, Masters Thesis, MIT (May 2003)
12. Lee, S.-M., Hwang, Y.J., Lee, D.H., Lim, J.I.: Efficient authentication for low-cost RFID systems. In: Gervasi, O., Gavrilova, M., Kumar, V., Laganà, A., Lee, H.P., Mun, Y., Taniar, D., Tan, C.J.K. (eds.) ICCSA 2005. LNCS, vol. 3483, pp. 619–627. Springer, Heidelberg (2005)
13. Dimitriou, T.: A lightweight RFID protocol to protect against traceability and cloning attacks. In: Conference on Security and Privacy for Emerging Areas in Communication Networks SecureComm (September 2005)
14. Weis, S.: Security Parallels Between People and Pervasive Devices. In: Workshop on Pervasive Computing and Communications Security - PerSec (March 2005)

Security Management in Wireless Sensor Networks with a Public Key Based Scheme*

Al-Sakib Khan Pathan, Jae Hyun Ryu, Md. Mokammel Haque, and Choong Seon Hong

Department of Computer Engineering, Kyung Hee University
{spathan,jhryu,malinhaque}@networking.khu.ac.kr, cshong@khu.ac.kr

Abstract. This paper proposes an efficient approach for managing the security in WSN. Our approach uses the notion of public key cryptography in which two different keys are used for encryption and decryption of data. Our analysis and performance evaluations show that, our approach is viable with the specifications of today's Berkley/Crossbow MICA2dot motes.

1 Introduction

Among several public key (PK) schemes proposed for wireless sensor networks (WSNs), Elliptic Curve Cryptography (ECC) based algorithms have a proven and acceptable performance for low-powered sensor nodes [1]. However, the use of certificates in such a scheme consumes a huge amount of bandwidth and power. Considering the constrained resources of sensors, here we propose an efficient PK-based security scheme for WSN. Our analysis and simulation results show that our scheme demonstrates good performance for the current generation sensor nodes.

2 Network Assumptions and Preliminaries

The BS is a trusted entity and cannot be compromised in any way. The sensors deployed in the network have resources like modern era sensors (e.g., MICA2 motes [2]). Once the sensors are deployed over the target area, they remain relatively static.

The pseudoinverse matrix or generalised inverse matrix [3] has a very nice property that could be used for cryptographic operations. It is well known that, a nonsingular matrix over any field has a unique inverse. For a general matrix of dimension $k \times n$, there might exist more than one generalized inverse. This is denoted by, $M(k,n) = \{A\text{: } A \text{ is a } k \times n \text{ matrix}\}$. Let, $A \in M(k,n)$. If there exists a matrix $B \in M(n,k)$ such that, $ABA = A$ and $BAB = B$, then each of A and B is called a generalized inverse matrix (or pseudoinverse matrix) of the other.

* This paper was supported by ITRC and MIC.

S. Ata and C.S. Hong (Eds.): APNOMS 2007, LNCS 4773, pp. 503–506, 2007.

3 Our Proposed Scheme

The first part of our scheme is the key handshaking process and the second part is used for confidential and authenticated data transmissions between two nodes.

Key Handshaking between any Node and Base Station: Let n_i be a node in the network and S be the base station (BS). To derive a shared secret key between the node n_i and the BS, the following operations are performed:

1. Node n_i randomly generates a matrix X with dimension $m \times n$ and its psedoinverse matrix, X_g. These matrices are kept secret in the node.

2. n_i calculates $X_g X$ and sends it to the base station S.

3. In turn, S randomly generates another matrix Y (dimension $n \times k$), and finds out its pseudoinverse matrix Y_g. These matrices are also kept secret in the BS.

4. S calculates $X_g XY$ and $X_g XYY_g$. Then it sends the resultant to n_i.

5. Upon receiving the products of matrices from S, n_i calculates, $XX_g XYY_g = XYY_g$ and sends it back to the base station.

6. Now, both n_i and base station S can compute the common secret key. n_i gets it by calculating $X(X_g XY) = XY$ and the base station gets it by calculating $(XYY_g)Y = XY$. Both of these outcomes are the same matrix (dimension $m \times k$).

Our mechanism ensures that, the individually calculated keys are same and this common key is used for encrypting the messages in the network. The derived common key could be used for node to BS or BS to node secure communications.

Encryption and Decryption of Data for Node-to-Node Communications. The main module in secure node to node communications is a central key generator (CKG) which is located at the base station. The CKG helps any node in the network to decrypt the received encrypted messages from other nodes. If a node n_i wants to send message securely to another node n_j, it uses the key that it has derived using the key handshaking process. Say for example, the encrypted message sent from n_i to n_j is $E_{XY}(M)$. Here, *M* is the message sent from the sender to the receiver.

E_{XY} means the message is encrypted with the key XY which is actually the shared secret key between the base station and the sender n_i. Upon receiving the encrypted message, n_j places its own identity and the identity of the sender to the CKG. In turn, CKG generates a decryption key and transmits it to n_j encrypting it with the secret shared key that it has with n_j. As the CKG in the base station has prior knowledge about the shared secret keys of both the nodes, it uses that knowledge to generate the decryption key. Now, n_j first decrypts the encrypted message with its shared key, finds out the decryption key, and uses that key to decrypt the message sent from node n_i.

4 Performance Evaluation and Conclusions

To analyze the performance of our security scheme, we considered the specifications of MICA2dot [2] mote platform. In the key handshaking process, we have used linear matrix operations, more specifically matrix multiplication. The complexity of matrix multiplication is very low; hence it could be performed very quickly. In our shared secret key derivation scheme, total number of bits passed is, $n^2 + n(n+k) + mn = n(2n+k+m)$ bits. All the calculations here are linear and can be performed very easily. In the first part, for key handshaking we use public channel for message transmissions. However, capturing the messages like $X_g X$, $X_g XY$, $X_g XYY_g$ and XYY_g could not be helpful to construct the locally computed secret shared key XY. A potential attack could arise in the key handshaking process between a node and the BS, if there exists any sort of identification problem of the participating entities during communications. But, this threat is completely eliminated in our case because; (a) the base station is a trusted entity and could not be compromised in any way and (b) the ids of the communicating nodes are checked by the BS before further communications. In the second part, when the receiver node requests for the corresponding decryption key, the key is not sent as a plain message, instead it is encrypted with the shared secret key of the receiver. So, in no way, any adversary can get the decryption keys for a particular sender-receiver pair.

We compared our shared key derivation scheme with Diffie-Hellman's scheme [4]. We found that, to achieve a security level (complexity) of $2^{49.3}$ in D-H scheme, a key size of 200 bits is needed. On the other hand, to achieve almost the same level security (1/probability) of 2^{48}, 48 bit key is required in our scheme. Likewise, to get security of $2^{59.3}$, $2^{67.4}$, and $2^{74.4}$ in D-H scheme, 300, 400, and 500 bit keys are required respectively. On the other hand, to get security of 2^{60}, 2^{70}, and 2^{75} in our approach, 75, 84, and 105 bit keys are required respectively. In this analysis, the sizes of p in bits for D-H scheme are 200, 300, 400, and 500 respectively. For our approach, (m,n,k) are (4,8,12), (5,9,15), (6,11,14), and (7,12,15) correspondingly. Figure 1(a) shows the level of security to be achieved with required size of the keys in our approach and in D-H scheme. Here in all of these cases, our approach needs keys with much less size than that of D-H scheme. In the figure, along the x-axis we have shown the values of the exponent of 2 to plot the security level for various key sizes. That is, in the figure, a value say, 48 along the x-axis indicates 2^{48}.

The amount of traffic carrying the key related information is also dependent on the size of key. Figure 1(b) plots the key sizes versus the amount of traffic (considering only key related information) needed to pass through the open public channel in our

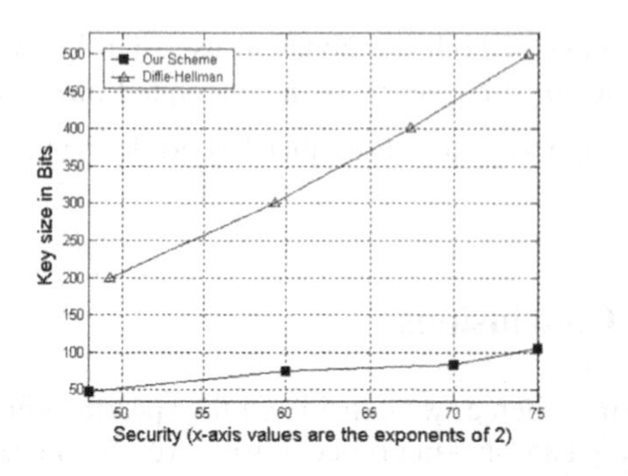

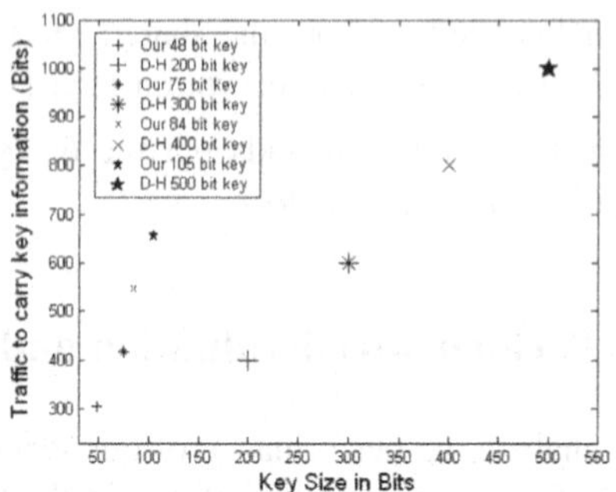

Fig. 1. (**a**) Required sizes of the key to provide almost same level of security in our key handshaking scheme and D-H scheme (**b**) Key size versus traffic to carry key information

entire scheme and Diffie-Hellman scheme. The data plotted here are based on the fact that, same (or almost same) level of security is to be ensured for both of the schemes.

MICA2dot motes [2] are equipped with 8-bit ATmega128L microcontrollers with 4 MHz clock speed, 128 kB program memory and Chipcon CC1000 low-power wireless transceiver with 433-916 MHz frequency band. According to our calculations, the cost of transmission of one byte is 59.2 μJ while the reception operation takes about half of the transmission cost (28.6 μJ). The power to transmit 1 bit is equivalent to roughly 2090 clock cycles of execution of the microcontroller. We considered a packet size of 41 bytes (payload of 32 bytes, header 9 bytes). With an 8 byte preamble (source and destination address, packet length, packet ID, CRC and a control byte) for each packet we found that, to transmit one packet $49\times 59.2 = 2.9008$ mJ ≈ 2.9 mJ energy is required. Accordingly, the energy cost for receiving the same packet is $49\times 28.6 = 1.4014$ mJ ≈ 1.4 mJ. Considering the same packet size for all the network operations, to set up a shared secret key with the base station each node needs (two transmissions and one reception) $((2\times2.9)+1.4) = 7.2$ mJ of energy. For node to node communication, the sender needs one transmission (2.9 mJ) and the receiver needs two receptions and one transmission $(((2\times1.4)+2.9) = 5.7$ mJ). As a whole, the entire scheme could be well-afforded by the energy resources of the current generation sensor nodes. Our scheme is also highly scalable, as any number of new sensors could be added to an existing wireless sensor network whenever needed.

References

1. Malan, D.J., Welsh, M., Smith, M.D.: A Public-Key Infrastructure for Key Distribution In TinyOS Based on Elliptic Curve Cryptography. In: Proc. IEEE SECON, Santa Clara, California, pp. 71–80 (2004)
2. Xbow Sensor Networks, Available at: http://www.xbow.com/
3. Boullion, T.L., Odell, P.L.: Generalized inverse matrices. Wiley-Int. New York (1971)
4. Rhee, M.Y.: Internet Security: Cryptographic Principles, Algorithms and Protocols. Wiley, Chichester (2003)

Scheduling Management in Wireless Mesh Networks*

Nguyen H. Tran and Choong Seon Hong

Department of Computer Engineering, Kyung Hee University
Giheung, Yongin, Gyeonggi, 449-701 Korea
nguyenth@networking.khu.ac.kr, cshong@khu.ac.kr

Abstract. We propose a greedy algorithm to investigate the problem of how to schedule of a set of feasible transmissions under physical interference model. We also consider the fairness in scheduling to prevent some border nodes from starvation. We evaluate our algorithm through extensive simulation and the results show that our algorithm can achieve better aggregate throughput and fairness performance than 802.11 standard.

1 Introduction

Recently, several works have focused on many typical problems of Wireless Mesh Networks (WMNs) like channel assignment, routing, scheduling [1], [2]. In literature, there are two main interference models in literature: *protocol* and *physical* interference models. The behavior of protocol interference model is similar the characteristic of CSMA/CA. We see that the characteristic of physical model is suitable with spatial reuse TDMA access scheme. Moreover, since the majority of traffic is transferred to and from management nodes, traffic flows will likely aggregate at the mesh routers close to the management nodes, which connect to the Internet. There is probably the starvation of the mesh client of border mesh routers. So, fairness must also be considered significantly. In this paper, we propose a heuristic scheduling algorithm using STDMA access scheme under the physical interference model to reach the objective of throughput improvement with fairness in WMNs. Simulation results show that the performance of our algorithm is significantly better than 802.11 CSMA/CA both in throughput improvement and fairness.

2 System Models

We consider the backbone of WMN modeled by a *network graph* $G(V,E)$, where V is the set of nodes (mesh routers) and E is the set of links. Each node is equipped with one or more wireless interface cards, referred to as radios in this paper. We assume there are K orthogonal channels available in the network without any inter-channel interference. We assume the packet length is normalized in order to be

* This research was supported by the MIC, Korea, under the ITRC support program supervised by the IITA (IITA-2006-(C1090-0602-0002)).

S. Ata and C.S. Hong (Eds.): APNOMS 2007, LNCS 4773, pp. 507–510, 2007.

transmittable in a unit time slot t. We denote $Q_e(t)$ the number of packets waiting to be transmitted on link e by the end of time slot t, also known as queue length of e.

Physical Interference Mode: Denoting $RSS_j^{\,i}$ as the signal strength of node j received when node i transmits to node j, and $ISS_j^{\,k}$ as interfered signal strength received by j from another node k which is also transmitting, packets along the link (i, j) are correctly received if and only if:

$$\frac{RSS_j^{\,i}}{N+\sum_{k\in V_c} ISS_j^{\,k}} \geq \alpha \tag{1}$$

where N is the white noise, V_c is the subset of nodes in V that are transmitting concurrently, and the threshold α is the constant.

Interference graph: In an interference graph [3], a node v' represents for the edge e in network graph and the directed edge between two nodes has a weight. The weight value $w_{e_2}^{e_1}$ represents for the interference contributed by e_1 to e_2 is:

$$w_{e_2}^{e_1} = \frac{\max(ISS_j^{v}, ISS_j^{u})}{\frac{RSS_j^{i}}{\alpha} - N} \tag{2}$$

We find the conditions to determine whether a certain set of concurrent transmissions on the same channel is feasible. 1) A necessary condition: The set $E_M = \{e_1,\dots,e_k\} \subseteq E$ is feasible only if none of its edges is incident with each other on the same node. 2) A sufficient condition: Every receiver of all links in E_M must have $SINR \geq \alpha$. So, we can state the following corollary:

COROLLARY 1. *A set $E_M \subseteq E$ of concurrent transmission on the same channel in a given network graph $G(V,E)$ is feasible if every vertex of the corresponding interference graph $G'(V',E')$ satisfies:*

$$\sum_{v_k'\in V'-\{v'\}} w_e^{e_k} \leq 1 \tag{3}$$

Proof: From Eq. (1) and (2), we can easily derive the result.

3 Scheduling Algorithm

In this section, we present a greedy algorithm to construct a feasible schedule for a set of transmissions by investigating WMN in a subgraph for fairness characteristic. When setting schedule for a subgraph in each period, the number of high priority links has been reduced, so the border links can transmit with higher probability. Consequently, we decide to choose Minimum Spanning Tree (MST) as the subgraph of the network graph $G(V,E)$ in our algorithm because MST has all characteristics appropriate for the purpose of our algorithm. First, MST is a spanning subgraph that contains all vertices of $G(V, E)$ so it gives an equal chance for all links incident with all nodes. Second, MST of a graph defines the cheapest subset of edges that keeps the graph in one connected component. Finally, it can be computed quickly and easily, e.g. Kruskal's minimum spanning tree algorithm [5] can have the running time $O(|E| \log |V|)$. It's an important factor to reduce time complexity of our algorithm. The fair scheduling algorithm is as follows.

1. Construct MST from network graph

forall $k = 1..K$ orthogonal channels in the MST

2. Order the set of links on the same channel k according to the decrease of queue lengths.
3. Find the maximal feasible set E_M^k. Beginning with the highest queue length link, transform next ordering links into vertices of the interference graph until there is a link making the interference graph unsatisfied with corollary 1.
4. Schedule each link in E_M^k from slot 0 to slot $Q_e(t)$.

endfor

Finally, we have aperiodic time slotted schedules in which the set of feasible transmission satisfies corollary 1 in every slot. The length of a period depends on the link which has the maximum queue length in set E_M^k, $T = \max_{e \in E_M^k} Q_e(t)$ with $k = 1..K$. And the algorithm schedules each edge e of E_M^k in $Q_e(t)$ time slots.

4 Performance Evaluation and Conclusion

We evaluate the performance of our scheduling algorithm by comparing with the algorithm of Alicherry, et al. [2], which uses CSMA/CA whose behavior is similar to protocol interference model. We have implemented our algorithm in ns-2 (ver2.28). The simulations are carried out for a $800 \times 800\, m^2$ area in which 50 nodes are placed randomly. We use the default transmission rates 11 Mbps to reflect realistic 802.11b data rates. We also use constant bit rate (CBR) over UDP and use Adhoc On-demand Distance Vector (AODV) as the base routing protocol. We choose Kruskal's algorithm [5] to construct the MST from the network for our algorithms.

Throughput Improvements Evaluation: We vary the number of orthogonal channels available from 1 to 8 and the number of radios is from 1 to 4 respectively. From Figure 1, we see that our algorithm can exploit effectively the increasing number of channels with different number of radios. For example, as the number of channels goes from 1 to 8, the network throughput goes from 1.3 Mbps to 4.6 Mbps, from 2.9 Mbps to 11.7 Mbps, from 5.8 Mbps to 16.86 Mbps and from 6.75 Mbps to 18.9 Mbps in case of 1, 2, 3 and 4 radios respectively. Compared with 802.11, we can see the average increase of our algorithm is respectively 45%, 36%, 30% and 25%.

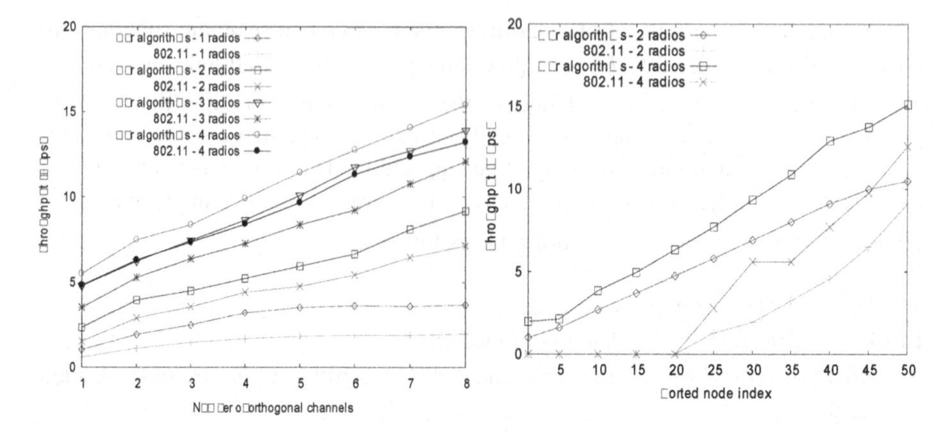

Fig. 1. Throughput Improvement Evaluation

Fig. 2. Fairness Evaluation

Fairness Evaluation: To evaluate the fairness of our algorithm and Alicherry's algorithm using 802.11, we compare the aggregate throughput of nodes starting from the border of network towards the nodes which are near the management node. Therefore, the nodes are sorted with the order of increasing queue length. We also vary number of radios (2 and 4 radios) to show their effects on fairness evaluation. We choose the fixed number of orthogonal channels in the network $K = 8$. From Figure 2, it can be observed that the border nodes throughput of our algorithm is higher than that of 802.11. The number of nodes which are starved in case of 802.11 is significant (nearly 20 nodes). With our algorithm, the fairness has been improved much when the border nodes still can transmit the data.

References

1. Ben Salem, N., Hubaux, J.P.: A Fair Scheduling for Wireless Mesh Networks. In: WiMesh, IEEE Computer Society Press, Los Alamitos (2005)
2. Alicherry, M., Bathia, R., Li, L.: Joint Channel Assignment and Routing for Throughput Optimization in Multi-Radio Wireless Mesh Networks. In: Proc. ACM Mobicom, pp. 58–72 (2005)
3. Jain, K., Padhye, J., Padmanabhan, V., Qiu, L.: Impact of Interference on Multi-Hop Wireless Network Performance. In: Proc. ACM Mobicom, pp. 66–80 (2003)
4. Nelson, R., Kleinrock, L.: Spatial-TDMA: A Collison-free Multihop Channel Access Protocol. IEEE Trans. on Communication 33, 934–944 (1985)
5. Cormen, T.H., Leiserson, C.E., Riverest, R.L., Stein, C.: Introduction to Algorithms, 2nd edn. MIT Press, Cambridge (2001)

Evolution of Broadband Network Management System Using an AOP

EunYoung Cho[1], Ho-Jin Choi[2], Jongmoon Baik[2], In-Young Ko[2], and Kwangjoon Kim[1]

[1] Electronics and Telecommunications Research Institute, Daejeon, Korea
{eycho,kjk}@etri.re.kr
[2] Information and Communication University, Daejeon, Korea
{hjchoi,jbaik,iko}@icu.ac.kr

Abstract. In broadband networks, dynamic service change affects the management system because of the variability in time to market, transmission quality and transport system. Therefore, reusability and maintainability become the important quality factors in Network Management System (NMS). An aspect-oriented software development method is a feasible solution for the evolvability of NMS. In this paper, we propose a method to generate aspects in a standard management information model using Aspect Conversion and Metric Calculation (ACMC) algorithm. We applied it ITU-T M.3100 and evaluated it via the ratio of reduced redundancy in point of crosscutting concerns.

1 Introduction

In broadband convergence network, various services with Quality of Service (QoS) are required for both customers and service providers. To guarantee reusability and maintainability, there are a number of approaches like Component-Based Software Development (CBSD), Software Product Line (SPL), design pattern. However, in this way, as the systems evolve, crosscutting concerns would be found to be present in legacy object-oriented NMS. This also makes a system difficult to maintain [2]. Therefore, in the presence of not only the new requirements but also *evolvability* it is necessary to have to meet time-to-market and stability issues. In this paper, we try to present an applicability and benefits of the AOP on the request for change matters as applicable to the NMS.

2 Aspects in NMS

In broadband network, quality of service, service level agreement and 99.999% availability are critical requirements because it is connected to many tributary systems sensitively. Recently, the convergence of circuit and packet transmission is provided in the backbone network for multimedia services. In this environment, NMS takes advantage of the reusability from modularity in AOP [1][3][4]. Figure 1 shows aspects category in each hierarchical layer and network domains.

S. Ata and C.S. Hong (Eds.): APNOMS 2007, LNCS 4773, pp. 511–514, 2007.

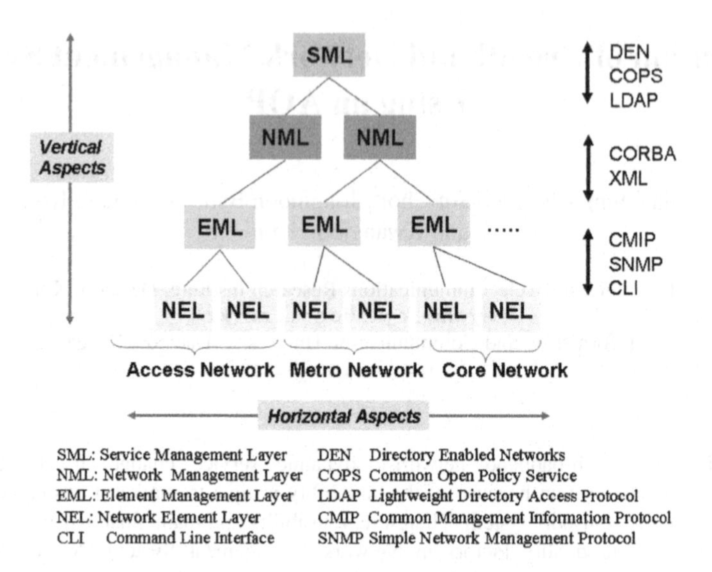

Fig. 1. Aspects category in Telecommunications Management Network

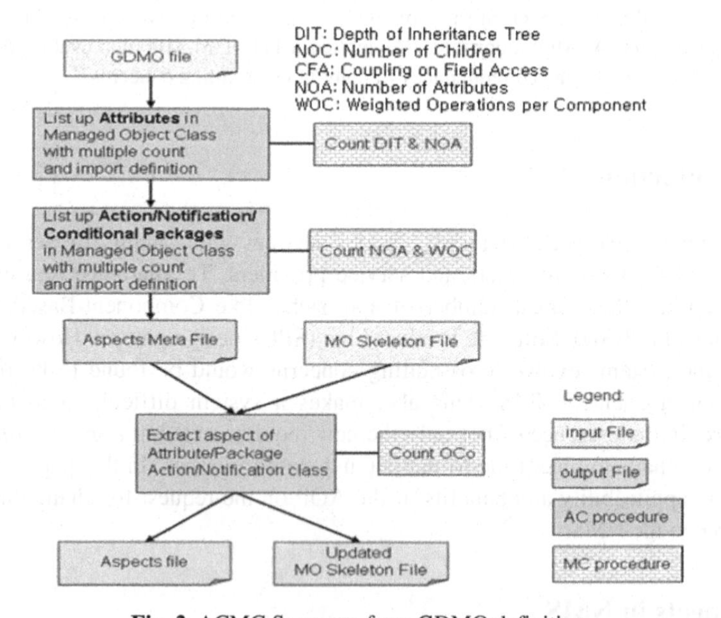

Fig. 2. ACMC Sequence from GDMO definition

3 Aspect Conversion and Metric Calculation

Besides logging, transaction, and security concerns, redundant attributes and operations are defined through the standard Management Information Base (MIB).

Moreover, current management systems operate in the hierarchical layer architecture and various vertical domains. Therefore, aspects are identified as functional aspects within a certain management system and the relationship among network management systems. Figure 2 shows the sequence of ACMC algorithm. The lists of candidate aspects based on key words are generated aspects meta-file. During the generation of aspects, all metrics suite are calculated as an option. At the last step, aspects codes are generated from legacy object-oriented files. If the user wants aspects selectively, it is possible to edit and choose aspects meta-file.

In ITU-T M.3100, there are 36 MOs with 91 conditional packages. After reducing the redundancy, an effectiveness of 48.4% is obtained.

MO	Cond. Pkg	DIT	NOC	CFA	NOA	WOC
networkR1	1	2	0..4*l*	2	2+(1)	2*n*
circuitPack	-	3	0..1*l*	1	2	3*n*
equipmentR2	18	2	k	0	3+(5)	2*n*
equipmentHolder	19	3	2*l*	0	4	4*n*+(7*n*)
managedElementR1	13	2	10*l*+*m*	10+(3)	5+(7)	4*n*+ (2*n*)
managedElementComple x	1	1	0..1*l*	1	2	3*n*+(2*n*)
softwareR1	12	2	0..1*l*	1	1+(8)	2*n*+(17*n*)
CTPBidirectional	4	3	0	(4)	3	5*n*+(19*n*)
TTPBidirectional	15	7	0..1	2	2	4*n*+(12*n*)
Total (9)	83	25	-	17+(7)	24+(21)	29*n*+(59*n*)
Average	9.22	2.78	-	2.1(0.2)	2.7(2.3)	3.2*n*(6.6*n*)

k: number of slot at runtime, real number of instances
l: number of children in a specific managed object inherited from managedElement
m: multiple children dependent on transmission capacity
n: : fixed input parameters generated from GDMO compiler
( ): optional or maximum value in case of including conditional packages

Fig. 3. Estimation of reduced ratio in ITU-T M.3100

4 Conclusion

This paper shows an application of aspect-oriented software development method in NMS through the application of an algorithm of automatic aspect conversion and metric calculation. It provides for the elimination of redundancy in standard object-oriented management information model and analysis for optimization of the degree of AOP via ACMC algorithm. We contribute to following points: the efficiency increase with the automatic conversion of the aspect-oriented implementation in the object-oriented management information model, the aspect optimization analysis by the automatically generated metric values, the code redundancy reduction, the improvement of reusability and maintainability.

References

1. Gandhirajan, A.: Modularizing and Code Reuse Using AOP, http://www.developer.com/design/article.php/3649681
2. Morris, S.B.: Reducing Upgrade Risk with Aspect Oriented Programming (2005), http://www.onjava.com/pub/a/onjava/2005/03/16/aop-mgmt.html
3. Raz, S.: A Set of Tools to Solve NMS Scaling Using Aspects (2004), http://www.cs.tau.ac.il/ amiramy/SoftwareSeminar/NMSAspect.ppt
4. Cavalli, V., et al.: Report on the availability and characteristics of equipment for next-generation networks, IST-2001-34925 (April 2003), http://www.serenate.org/publications/d9-serenate.pdf

Standby Power Control Architecture in Context-Aware Home Networks

Joon Heo[1], Ji Hyuk Heo[1], Choong Seon Hong[1], Seok Bong Kang[2], and Sang Soo Jeon[3]

[1] Department of Computer Engineering, Kyung Hee University, 1 Seocheon, Giheung, Yongin, Gyeonggi, 449-701 South Korea
{heojoon,jihyuk,cshong}@khu.ac.kr

[2] I'ware Inc., Ltd., Kangnam Internet Business Center #508 San 6-2 Gukal, Giheung, Yongin, Gyeonggi, 449-702 South Korea
sbyhkang@naver.com

[3] VITZRO SYS Co., Ltd., 233-3, 1Dong, Sungsu-2, Sungdong, Seoul, 133-826 South Korea
paul@vitzrosys.com

Abstract. Standby power is the energy consumed by appliances when they are not performing their main functions or when they are switched off. In this paper, we propose a Host-Agent based standby power control architecture in context-aware home network. The proposed architecture uses the IEEE 802.15.4 based ZigBee protocol between Host and Agent for communication and security. We have made an experiment on according to various context-aware scenarios using the implemented prototype devices.

1 Introduction

Products that have power switches consume certain amounts of standby power. Particularly, products such as consumer electronics, office equipments and white goods, where the 'standing by' time exceeds the operation time, have high standby power consumption. By 2020 standby power consumption is projected to be 1/4 of the total household energy consumption, and the main cause of such an increase can be attributed to the home network system [1]. The IEEE 802.15.4 standard defines transmission and reception on the physical radio channel (PHY), and the channel access, PAN (personal area network) maintenance, and reliable data transport (MAC) [2]. ZigBee defines the topology management, MAC management, routing, discovery protocol, security management and includes the 802.15.4 portions. ZigBee is a new low rate wireless network standard designed for automation and controlling network. The standard is aiming to be a low-cost, low-power solution for systems consisting of unsupervised groups of devices in houses, factories and offices [3]. In this paper, we propose standby power control architecture in context-aware home network. This paper is organized as follows: Section 2 describes the Host-Agent based system architecture, context modeling and context management. Implementation results of proposed mechanism are presented in section 3. Finally, we give some concluding remarks and future works in section 4.

S. Ata and C.S. Hong (Eds.): APNOMS 2007, LNCS 4773, pp. 515–518, 2007.

2 Proposed Standby Power Control Architecture

Proposed standby power reduction architecture is the Host-Agent based. Where Agent acquires the current local context information using the various embedded sensor and sends this information to the Host. The Host compares this context information from Agent with current database and sends standby power control message to Agents. This architecture uses the IEEE 802.15.4 based ZigBee communication protocol between Host and Agent for context information and control message transmission. Also, in order to enable natural and meaningful interactions between the context-aware home and its occupants, the home has to be aware of its occupants' context, their desires, whereabouts, activities, needs, emotions and situations. Such context will help the home to adapt or customize the interaction with its occupants. By context, we refer to the circumstances or situations in which a computing task takes place [4]. Context modeling is about providing a high level abstraction of context information. The diversity of contextual information and its use in diverse domains lead to different ways for modeling context. We define the use of dynamic sensing information (user, environment and system). Contextual information is modeled as main-value pairs called sensing information. Figure 1 shows an example of context modeling for standby power reduction.

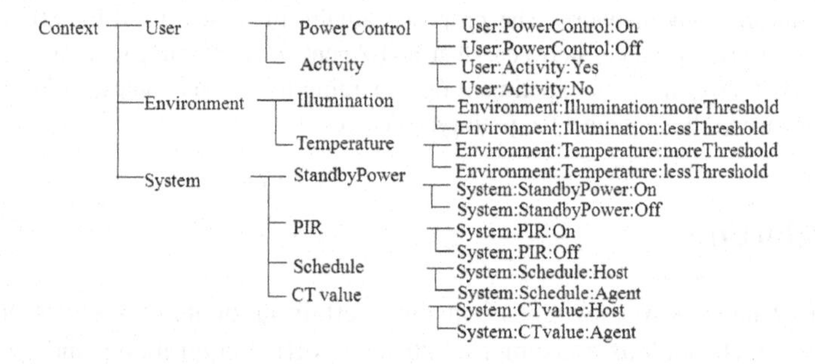

Fig. 1. Main-value based context modeling

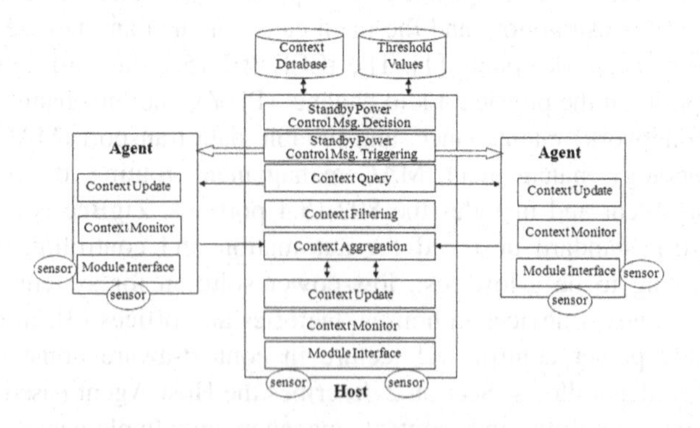

Fig. 2. Context-aware management architecture

Also, we have designed context management architecture for standby power control. A high-level view of the main components of this architecture is shown in Figure 2. Context gathering sensors of Agents acquire and send the sensed context information to Host. Host also acquires context itself using the embedded sensors. Context should be monitored and updated as schedules.

3 Implementation and Test Results

As we explained before, the proposed architecture of this paper uses IEEE 802.15.4 based ZigBee communication protocol. Therefore, we have implemented network and security functions at prototype devices according to ZigBee specification [3]. The proposed control architecture has been organized based on tree topology. Requirement functions such as routing, address allocation, encryption/decryption and message authentication have been tested. Implemented security module can support ZigBee security specification such as CCM* algorithm, MAC/NWK layer security, key establishment, message encryption/decryption and message integrity [4].

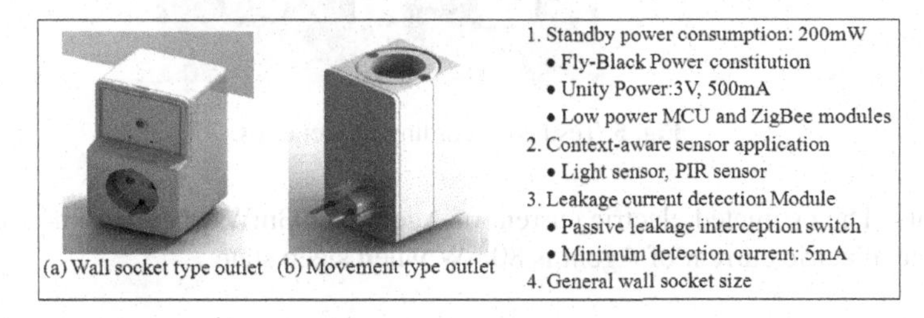

(a) Wall socket type outlet
(b) Movement type outlet
1. Standby power consumption: 200mW
• Fly-Black Power constitution
• Unity Power:3V, 500mA
• Low power MCU and ZigBee modules
2. Context-aware sensor application
• Light sensor, PIR sensor
3. Leakage current detection Module
• Passive leakage interception switch
• Minimum detection current: 5mA
4. General wall socket size

Fig. 3. Agent Prototype and Characteristics

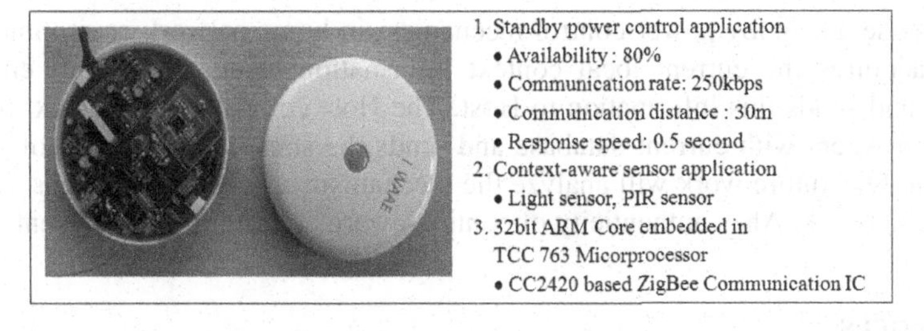

1. Standby power control application
• Availability: 80%
• Communication rate: 250kbps
• Communication distance : 30m
• Response speed: 0.5 second
2. Context-aware sensor application
• Light sensor, PIR sensor
3. 32bit ARM Core embedded in
TCC 763 Micorprocessor
• CC2420 based ZigBee Communication IC

Fig. 4. Host Prototype and Characteristics

Real features of implemented prototype device and characteristics are shown in Figure 3(Agent) and Figure 4(Host). Agent has implemented 3 types (wall socket type, movement type, switch type). Each type can be used according to control scenarios. To prove the necessity and the efficiency of the proposed control mechanism, we have made an experiment on control test according to various context-aware scenarios using the implemented prototype devices. Table 1 explains the test scenarios and Figure 5 shows the testbed. Similar standby power value has been estimated at all

Table 1. Test Scenarios according to context situation

Scenario	Contex-aware	Controlled Agent
1	indoor light sensing value of Agents	Agent 1 (socket type outlet)
2	PIR sensing value of Host	Agent 2 (socket type outlet)
3	light sensing value of desk lamp	Agent 3 (movement type outlet)
4	CT value of electronic device	Agent 4 (movement type outlet)
5	switch touching of user	Agent 5 (button type switch)

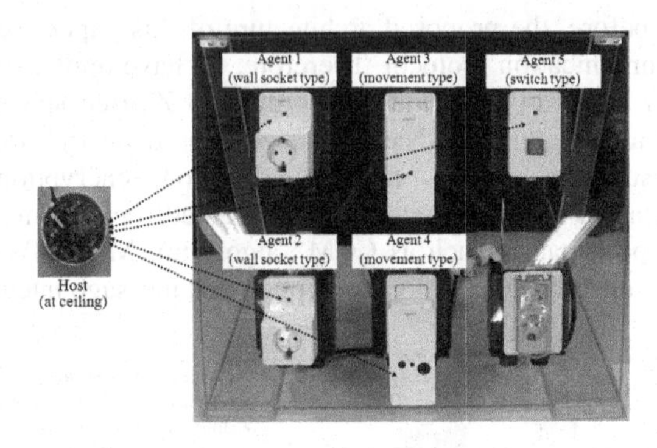

Fig. 5. Testbed according to scenarios

of Agents. The estimated electric current of Agent is 203mW when working state; the other side, electric current of Agent is 80mW when sleep state.

4 Conclusion

We propose a standby power control mechanism in home network environment. The Agent acquires the current local context information using the various embedded sensors and sends this information to Host. The Host compares this context information from Agent with current database and sends the standby power control message to Agent. Our future work will analyze the mechanism according to various scenarios in home network. Also, authenticity of context-aware algorithm should be enhanced.

References

1. Kim, Y.R.: KOREA 1W Policy. In: Standby Power Conference 2006 (November 2006)
2. Wireless Medium Access Control and Physical Layer Specification for Low-Rate Wireless Personal Area Networks. IEEE Standard, 802.15.4-2003 (May 2003)
3. ZigBee Document 053474r13 ZigBee Specification (December 2006)
4. Heo, J., Hong, C.S., Kang, S.B., Jeon, S.S.: Wireless Home Network Control Mechanism for Standby Power Reduction. In: Proceedings of the International Conference on Wireless Information Networks and Systems, pp. 70–75 (July 2007)

End-to-End Soft QoS Scheme in Heterogeneous Networks

Young Min Seo, Yeong Min Jang, and Sang Bum Kang

School of Electrical Engineering, Kookmin University
861-1 Chongnung-dong Songbuk-Gu, Seoul 136-702, Korea
{zzeromin,yjang,daggadae}@kookmin.ac.kr

Abstract. We propose an end-to-end Soft QoS scheme designed to provide effective multimedia data traffic. Using a greedy algorithm, this scheme can satisfy the conditions required by the multimedia data traffic with QoS by processing first those data that have a higher order of priority. In addition, it is a suitable method for the system reference model as it can effectively support a next-generation mobile communication service for the upcoming Converged Network. The proposed end-to-end Soft QoS scheme satisfies the QoS conditions required by the Converged Network where WCDMA, HSDPA, and WiBro coexist.

Keywords: Management of Heterogeneous Networks, Soft QoS.

1 Introduction

The IP Multimedia Subsystem (IMS) is not only a structure that offers mobile communication service subscribers easy access to the Session Initiation Protocol (SIP)-based packet service, but it also supports simultaneous use of real-time services including voice and video contents delivery as well as non-real-time services such as data or text message transmission. It should be noted, however, that once NGN is established, an end-to-end QoS scheme is necessary in order to offer transmission of high-quality and ultra-high-speed multimedia data using IMS. Until now, it has been relatively easy to implement a QoS policy among the same networks in a policy-based network management scheme by applying a DiffServ policy in the current 3GPP core network. But, as the network advances toward becoming a wire and wireless integrated network or a heterogeneous network, problems could occur where the QoS policy would not be set uniformly due to the specialization of each network. In a heterogeneous network, a different QoS policy has to be applied to each network during handover; thus, the same DiffServ policy cannot be used.

In this paper, we propose an end-to-end Soft QoS scheme using a greedy algorithm in order to satisfy the QoS of end-to-end multimedia data traffic.

In Section 2 of the paper, we introduce the Soft QoS scheme and the proposed end-to-end QoS. In Section 3, we discuss an analysis of the performance of the scheme proposed in Section 2. In Section 4, we conclude the paper.

S. Ata and C.S. Hong (Eds.): APNOMS 2007, LNCS 4773, pp. 519–522, 2007.

2 The IMS-Based E2E Resource Allocation Scheme Using a Greedy Algorithm

2.1 The Greedy Algorithm

The greedy algorithm is a method for selecting items among a series of data items which best fits under the currently given standard without regard to past or future selections.

We introduce, in this paper, a Soft QoS scheme using the Knapsack Algorithm [1]. Aiming at transmitting the maximum value without exceeding W, the total capacity of the network, the scheme lists and selects data in the order of profit per bandwidth, where, $D = \{\text{connection}_1, \text{connection}_2, ..., \text{connection}_n\}$, w_i is bandwidth, p_i is profit, and W is the maximum capacity.

The algorithm calculates a subset A of D so that it satisfies $\sum_{connection_i \in A} \left(\frac{p_i}{w_i}\right)$ and produce the maximum $\sum_{connection_i \in A} w_i \leq W$.
The allotted capacity to be calculated is:

$$C_{allocation} = max \sum_{connection_i \in A} (\frac{p_i}{w_i}), \tag{1}$$

where the conditions are $\sum_{connection_i \in A} w_i \leq W$.

1. A new connection request is made to the Access Network.

2. The algorithm finds the bottleneck spot, C_{bot}, by comparing the bandwidths of the two access networks and one CN using the following equation,

$$C_{bot} = min(C_{AN\#1}, C_{CN}, C_{AN\#2}). \tag{2}$$

3. If the C_{bot} is found to be small after comparing the bandwidth of the currently used network C_{used} and the requested bandwidth C_{req}, then multiply $1 - \xi$ to C_{req} so as to reduce the minimum requested capacity of the new connection's requested resource, and then to update the value of C_{req}

$$C_{bot} < C_{used} + C_{req} \tag{3}$$

$$C_{req,updated} = C_{req} \times (1 - \xi). \tag{4}$$

4. If the C_{bot} is still small after comparing again the updated C_{req} value and the bandwidth of the currently used network, then apply Soft QoS. Otherwise, the scheme receives the request

$$C_{bot} < C_{used} + C_{uploaded}. \tag{5}$$

5. The Soft QoS algorithm using the Knapsack Algorithm is as follows:

$$C_{allocation} = max \sum_{connection_i \in A} (\frac{p_i}{w_i}). \tag{6}$$

6. Once the requested resource is permitted by comparing C_{bot} and the re-alloted bandwidth after applying Soft QoS, the connection is granted

$$C_{bot} < C_{allocation} + C_{req}. \tag{7}$$

3 Numerical Results

To compare the performance of the DiffServ scheme, the Hard QoS scheme, and the Soft QoS scheme, we constructed a virtual network based on the IMS. We analyzed the relationship between the traffic flowing into the network and the usage ratio of the bandwidth, the change in the number of connections to the network, and in the Critical Bandwidth Ratio, ξ.

Now, let us discuss the results of the numerical analysis using the proposed Soft QoS scheme which was applied to the IMS network. We assumed the value of ξ based on two incidents of real-time video data traffic and network access (See Fig.1).

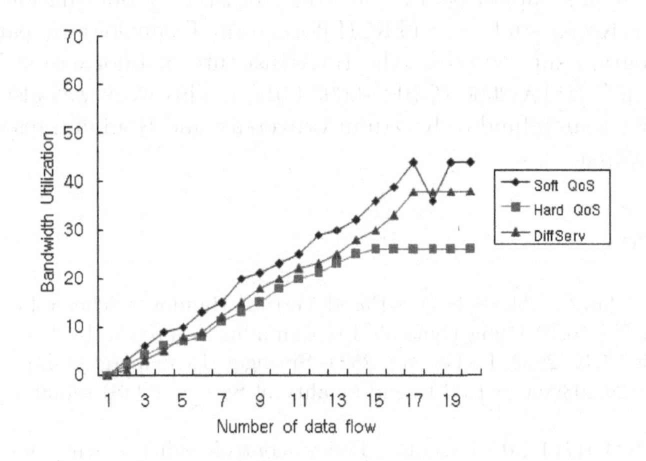

Fig. 1. The bandwidth utilization vs. number of data flow

The bottleneck points mean the conditions where the current bandwidth capacity has reached up to its maximum and will not receive any new connection. In the case of Soft QoS, the usage ratio goes down momentarily and comes back again at the end. This change was due to the fact that the currently active connection allotted its resource to he next connection.

We analyzed the relationship between the traffic flowing into the network, the usage ratio of the bandwidth, the change in the number of connections to the network, and in the Critical Bandwidth Ratio (ξ). The results showed that the smaller the value of ξ and the number of connections, the higher the bandwidth usage ratio became.

4 Conclusion

We proposed an end-to-end QoS scheme in order to provide high-quality multimedia data traffic in an environment where it may be difficult to guarantee QoS. The existing 3GPP core network suggested PBNM using the DiffServ policy.

Although it may be useful as a QoS policy in a same-kind network, it was difficult to apply in a heterogeneous network environment due to the specialization of each network. To solve this problem, in this study, we implemented a Soft QoS scheme where we could efficiently manage the network's resources by using the functional relationship between the overall bandwidth allotment capacity of each network and the level of user satisfaction. Based on a greedy algorithm, this scheme processed first data that had a higher priority to satisfy the required QoS conditions for the multimedia data.

Acknowledgment

This research was supported by the MIC(Ministry of Information and Communication), Korea, under the ITRC(Information Technology Research Center) support program supervised by the IITA(Institute of Information Technology Advancement)" (IITA-2006-(C1090-0603-0019)). This work was also supported by the 2007 research fund of Kookmin University and Kookmin research center UICRC in Korea.

References

1. Kim, S.H., Jang, Y.M.: Soft QoS-Based Vertical Handover Scheme for WLAN and WCDMA Networks Using Dynamic Programming Approach. In: Lee, J.-Y., Kang, C.-H. (eds.) CIC 2002. LNCS, vol. 2524, Springer, Heidelberg (2003)
2. 3GPP TS 29.208v6.6.1: End-to-end Quality of Service (QoS) signalling flows (R6) 2006-03
3. 3GPP TS23.917v1.2.0: Dynamic Policy control enhancements for end-to-end QoS(R6) 2004-01
4. 3GPP TS23.228v7.6.0: IP Multimedia Subsystem(IMS); stage 2 (R7) 2006-12

A Multi-objective Genetic Algorithmic Approach for QoS-Based Energy-Efficient Sensor Routing Protocol

Navrati Saxena[1], Abhishek Roy[2], and Jitae Shin[1]

[1] School of Info. & Comm. Eng., Sungkyunkwan University, Suwon, Korea
[2] WiBro System Lab,Samsung Electronics, Suwon, South Korea
{navrati,jtshin}@ece.skku.ac.kr, abhishek.roy@samsung.com

Abstract. As the trend of small, connected, computing devices weaved, with almost any type of daily useable object, increases drastically, the deployment of Wireless Multimedia Sensor Networks (WMSNs) becomes a necessity. Sensor networks stages to deliver multimedia content, forcing Quality of Service (QoS) to become an important issue on the already low-powered, energy constraint sensor nodes. Such a problem is proved to be NP-complete. We, in this paper, propose an energy-efficient, QoS-aware sensor routing protocol for such WMSNs. Our protocol is based on *multi-objective genetic algorithm* (MOGA) in particular. The protocol determines application-specific, near-optimal sensory-routes, by optimizing multiple parameters - QoS, and energy consumption. Simulation results demonstrate that the proposed protocol is capable of providing lower delay and lower energy consumption in comparison to other existing QoS-routing protocols for wireless sensor networks.

1 Introduction

The main component of wireless (multimedia) sensor network (W(M)SN), are the sensor nodes, which are small in size, capable of self-organizing, sensing, processing data and communicating with other typically over a RF channel [1]. Energy conservation is a primary concern and a well-accepted challenge in WSNs. With the convergence of wireless technologies and the Internet, real-time audio-visual transmissions are required. Such multimedia transmissions over WSN require strict quality-of-service (QoS) guarantee on bandwidth and delay. QoS routing offers significant new challenges over the already energy-constrained sensor networks. Although QoS based routing has been an active research area, QoS based routing for WSN has received relatively less attention. A state of the art in algorithms, protocols and hardware is surveyed in [1] for WSN and multimedia applications over it. The energy-aware QoS-routing protocol in [2] finds a least cost energy efficient path, while meeting certain end-to-end delay constraints. Another QoS-routing protocol SPEED [4] depends on state-less geographical non-deterministic forwarding to provide soft real-time end-to-end QoS guarantee. A QoS supporting scheme is developed in [7] for dynamic traffic conditions

S. Ata and C.S. Hong (Eds.): APNOMS 2007, LNCS 4773, pp. 523–526, 2007.

by controlling data generating rates at individual clusters in a cluster based WSN. Researches for considering QoS-routing with multiple parameters mostly rely on approximation algorithms and heuristics. Multi Objective Genetic Algorithm (MOGA) is a very efficient and useful tool for solving these intractable problems. The efficiency of MOGA has been demonstrated in [6] by obtaining QoS-based multicast routes in computationally feasible time. This motivates us to develop a QoS-aware energy-efficient routing protocol for WMSNs.

We claim that determining optimal routes satisfying multiple QoS parameters simultaneously, in an energy-aware WSN is a NP-complete problem. This claim can be intuitively justified by the fact that, the problem of satisfying multiple QoS parameters in WSN is equivalent to constrained Steiner Tree problem – a well-known NP-complete problem [5]. In this paper we propose a MOGA [8] based strategy that determines a set of near-optimal routes, satisfying application-specific QoS-parameters, in WMSNs. The protocol optimizes QoS parameters — namely, end-to-end delay, bandwidth requirement, while maintaining the energy constraints — without combining them into a single scalar objective function.

2 Routing Algorithm and Protocol Details

Genetic algorithms are guided random search and optimization heuristics, based on the basic principles of natural evolution: *survival of the fittest* and *inheritance* [3]. Many real world problems require simultaneous optimization of multiple objectives. There may not exist a single best solution with respect to all the objectives under consideration. Instead, there might exist a set of solutions superior to the rest in the entire search space. Such a solution set is termed as *Pareto-optimal* [8]. We can view the WSN as a graph $G = (V, E)$, a specific number of sensory source nodes $v_{s_1}, v_{s_2}, .., v_{s_n}$ and a single destination-sink v_d. There could exists lots of routes from each of the sensory sources to the sink forming routing trees. But not all the paths or routes in these trees necessarily meet the desired QoS requirements. The underlying idea behind our algorithm is not to combine the QoS and energy-based objective functions on an ad hoc basis to form a single scalar objective function, but to tackle the problem from the perspective of multi-objective optimization. MOGA is a perfect tool to deal with such problems. Our algorithm takes as input a network graph. It then finds a pool of possible routing paths from v_d to each of $v_{s_1}, v_{s_2}, ..., v_{s_n}$, using a depth first search (DFS) algorithm. Thus an initial set of routing trees is constructed for our proposed MOGA. Each of these routing trees is mapped to a string consisting of the sequence of nodes along the path from v_d to each of the destinations $v_{s_1}, v_{s_2}, ..., v_{s_n}$. The fitness value of each of these routes is computed with the help of a fitness function. Randomly chosen chromosomes undergo the genetic operations of crossover and mutation. As MOGA-based routing scheme executes, at every iteration the genetic operations dynamically update the chromosomes (routes) and try to improve the corresponding probabilities of the three QoS parameters. This entire process is repeated until the difference of fitness values between the current Pareto-optimal set and the previous one is less than a chosen precision, ϵ.

3 Simulation Experiments and Results

Simulation experiments are conducted with 802.11 MAC with $n = 500$ sensor nodes, and the average values of bandwidth and end-to-end delay requirement are 75 Kbps and 60 msec, respectively. The performance of our algorithm is compared with SPEED [4] and cluster-based QoS-routing [7] in WSN. Figure 1 explains that end-to-end delay for wireless video traffic is lowest for our protocol. Figure 2 delineates the comparative performance in terms of energy consumption of sensor nodes proving that the average energy consumption of our protocol is minimum.

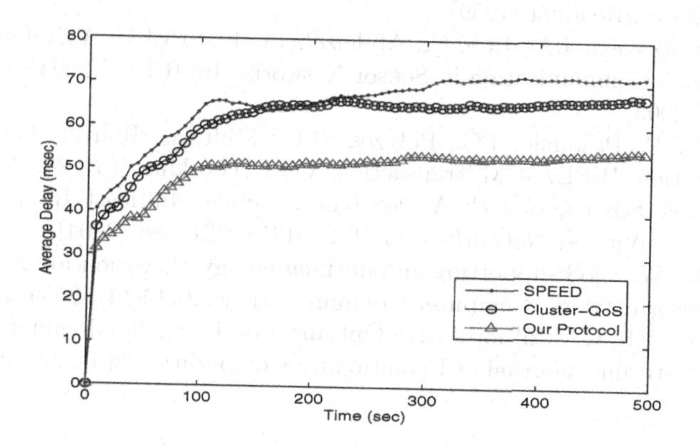

Fig. 1. Comparative Delay of Different Sensor-QoS-Routings

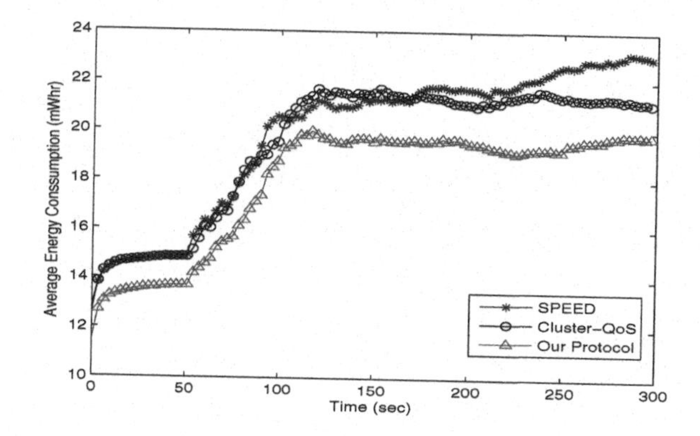

Fig. 2. Dynamics of Energy Consumption

Acknowledgements

This work was supported by the Korea Research Foundation Grant funded by the Korean Government (MOEHRD)(KRF-2006-331-D00358).

References

1. Akyildiz, I.F., Melodia, T., Chowdhury, K.R.: A survey on wireless multimedia sensor networks. The International Journal of Computer and Telecommunications Networking 51(4), 921–960 (2007)
2. Gao, Q., Blow, K.J., Holding, D.J., Marshall, I., Peng, X.H.: Radio Range Adjustment for Energy Efficient Wireless Sensor Networks. Ad-Hoc Networks 4(1), 75–82 (2006)
3. Goldberg, D.E.: Genetic Algorithms: Search, Optimization and Machine Learning. Addison-Wesley, Reading (1989)
4. Hea, T., Stankovica, J.A., Lub, C., Abdelzahera, T.: SPEED: A Stateless Protocol for Real-Time Communication in Sensor Networks. In: ICDCS 2003, Rhode Island, USA (May 2003)
5. Kompella, V.P., Pasquale, J.C., Polyzos, G.C.: Multicast Routing for Multimedia Communication. IEEE/ACM Transactions On Networking 1(3), 286–292 (1993)
6. Roy, A., Das, S.K.: QM^2RP; A QoS-based Mobile Multicast Routing Protocol. ACM/Kluwer Wireless Networks (WINET) 10(3), 271–286 (2004)
7. Tang, S., Li, W.: QoS supporting and optimal energy allocation for a cluster based wireless sensor network. Computer Communications 29(13-14), 2569–2577 (2006)
8. Srinivas, N., Deb, K.: Multiobjective Optimization Using Nondominated Sorting in Genetic Algorithms. Journal of Evolutionary Computation 2(3), 221–248 (1995)

A Density Based Clustering for Node Management in Wireless Sensor Network*

Md. Obaidur Rahman, Byung Goo Choi, Md. Mostafa Monowar, and Choong Seon Hong

Department of Computer Engineering, Kyung Hee University, South Korea.
{rupam,bgchoi,monowar}@networking.khu.ac.kr, cshong@khu.ac.kr

Abstract. This paper represents a new clustering approach for wireless sensor network. It is a decentralized algorithm having the topology control information in each sensor node. A post leader selection algorithm is acted upon each of the clusters just after their formation. Experimental validation shows that the proposed scheme is an efficient approach for sensor node management.

1 Introduction

Sub grouping of network by clustering promotes efficient use of network resources like, battery power or energy consumption, processing, routing etc. Spontaneously load balancing among several parts of the network also increase the network life time. In the past, clustering of the network has been studied both theoretically and in perspective of ad-hoc networks [1] [2] [3]. But in recent days decomposition or grouping issues of sensors have become the prominent research field for wireless sensor network. In this paper, we have gone through the issues and characteristics of sensor node density in terms of standard deviations. We proposed a hierarchical architecture of sensor network with cluster formation and cluster head selection algorithm using various parameter metrics related to sensor node density.

2 Network Model and Assumptions

In our proposed network model, deployment density variation of sensor nodes is indicated by the edge or link lengths standard deviations. Each node connected to its neighbor nodes via wireless link. The average link length gives a good assumption of the inter node distance within a cluster. Proposed algorithm identifies two types of links: i) intra-cluster link and ii) inter-cluster link. Network discontinuity can be identified using the inter-cluster link. Always inter-cluster links are larger than the intra-cluster links and based upon this criterion we define the clusters.

* "This research was supported by the MIC (Ministry of Information and Communication), Korea, under the ITRC (Information Technology Research Center) support program supervised by the IITA (Institute of Information Technology Advancement)" (IITA-2006-(C1090-0602-0002)).

S. Ata and C.S. Hong (Eds.): APNOMS 2007, LNCS 4773, pp. 527–530, 2007.

3 Cluster Formation

Variability of different parameters makes the identification of inter-cluster links easier. Definitions of different parameters are given below:

Definition 1. The mean link length of a sensor node is denoted as $\bar{L} = \frac{\sum_{j=1}^{p} l_j}{p}$. Here p is total number of links to a sensor and l_j is the length of each link to that sensor.

Definition 2. The standard deviation of link lengths of a particular sensor node is denoted by σ and calculated as, $\sigma = \sqrt{\frac{1}{p}\sum_{j=1}^{p}(\bar{L}-l_j)^2}$.

Definition 3. The global mean of standard deviations of the network is denote by μ and can be defined as, $\mu = \frac{\sum_{i=1}^{N}\sigma_i}{N}$. N is the total number of nodes and σ_i is the link lengths standard deviations of all nodes, where $i = 1, 2, \ldots, N$.

The mean is an average value of the length distribution, so we can come to an end with the following considerations.

Consideration 1: Let $\bar{L}$ be the mean link length of a sensor node and μ be global mean of standard deviations. Now $(\bar{L}-\mu)$ represents a very small value compare to the average link length distribution of a sensor. Thus it is used as a maximum threshold for selecting short distanced link for clustering.

Consideration 2: In contrast to *consideration 1*, $(\bar{L}+\mu)$ represents a large value than the average of link length distribution of a sensor. So, we can use this as a minimum threshold for selecting the long distanced link and these links can separate the clusters from each other.

Sensor nodes of the whole network perform the proposed clustering algorithm individually to be a member of a cluster. Figure 1 and 2 presents the scenario of clustering and the clustering algorithm respectively.

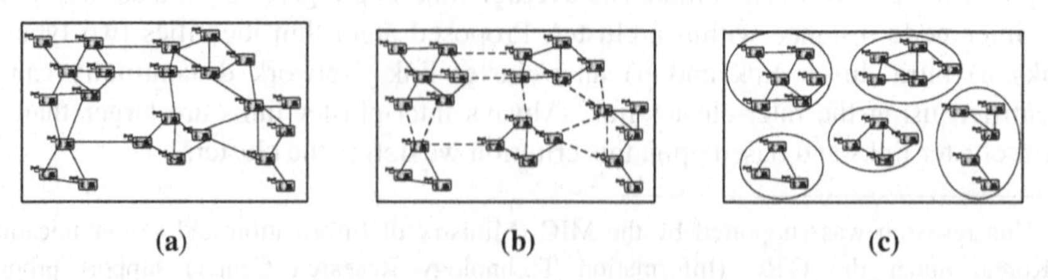

Fig. 1. Clustering Algorithm Implementation Scenario; a) Sensors with links before Clustering, b) Identification of inter-cluster links and intra-cluster links, c) Clustered Network

```
Algorithm Cluster Formation
Inputs: Number of Nodes N
Output: Clusters

1. Calculate mean link length L̄;
2. Calculate standard deviation of link lengths σ;
3. Broadcast σ value into the network;
4. Calculate global mean of standard deviations μ;
5. for each links j = 1,2,...p
       if (l_j < (L̄ − μ))
           form group with corresponding link's node;
       else if (l_j ≥ (L̄ + μ))
           leave group of corresponding link's node;
```

Fig. 2. Clustering Algorithm

4 Leader Selection Algorithm

After cluster formation, cluster head has been selected using the information of: Degree and Residual energy of a sensor node according to the following phases:

- Each node advertises the neighbor discovery packet in its transmission range.
- Each node calculates its degree having acknowledgement of neighbor discovery packet.
- Each node multicast a control message having the information of its degree and residual energy to other nodes of same cluster.
- The node with highest degree and having a minimum residual energy but more than threshold will be selected as cluster head.
  - If a node has highest degree but residual energy is less than threshold, then node degree will be calculated again among the nodes excluding that node.
  - If node degree is same for more than one node then the node with higher residual energy will be elected as cluster head.

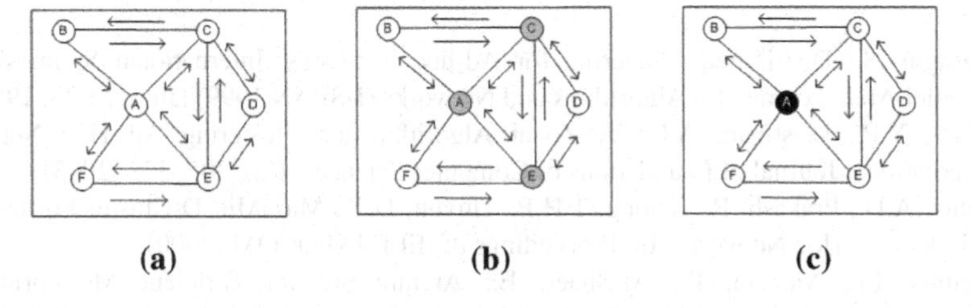

Fig. 3. Implementation Scenario for Leader Selection Algorithm

Implementation scenario is presented in figure-3. It shows three nodes: A, C and E has the same number of degree 4 (four). Node A has the residual energy higher than node C and E and obviously which is more than or equal to the threshold energy level. Thus node A declares it as a cluster head in figure 3 (c).

5 Experimental Validations and Conclusions

The effectiveness of our proposed clustering method is validated through simulation. In the simulation environment we deployed a set of 100 sensors randomly in an area of 1000×1000 m^2 area. Environmental setup continued with the energy constraint. We use the energy consumption model used in [4] and consider the *threshold energy level* for our leader selection algorithm as follows:

$$Threshold\ Energy\ Level \geq E_{tx} + E_r + E_{sensing}$$

Here, E_{tx}, E_r, $E_{sensing}$ are the energy required per sensor node for transmitting, receiving and sensing r bits data respectively.

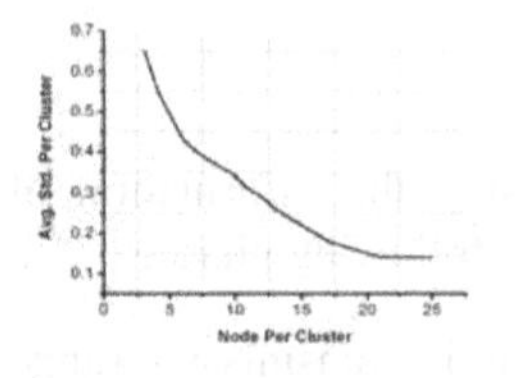

(a) Average Standard Deviation Per Cluster vs □ode Per Cluster

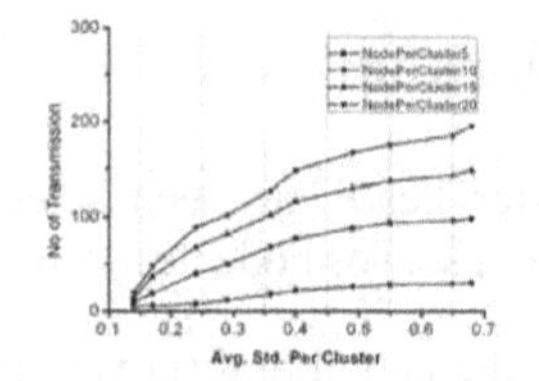

(b) □o□of Transmission vs Average Standard Deviation Per Cluster

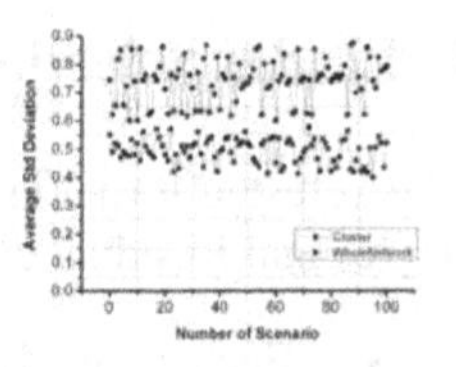

(c) Average Standard Deviation vs □o□of Scenario

Fig. 4. Experimental Results

Proposed leader selection algorithm can establish an efficient hierarchical routing to the sink using cluster heads. Experimental results of this paper proved that the proposed method of clustering could be implemented in a larger extent of wireless sensor network. Though the sensor node management is a critical issue for wireless sensor network, still we hope our promote progression of this paper will contribute enormously.

References

1. Basagni, S.: Distributed Clustering for Ad-hoc Networks, International Symposium of Parallel Architectures. In: Algorithms and Networks (I-SPAN 1999) (June 23-25, 1999)
2. Chen, Y.P., Liestman, A.L.: A Zonal Algorithm for Clustering Ad Hoc Networks. International Journal of Foundations of Computer Science 14(2), 305–322 (2003)
3. Amis, A.D., Prakash, R., Vuong, T.H.P., Huynh, D.T.: Max-Min D-Cluster Formation in Wireless Ad Hoc Networks. In: Proceedings of IEEE INFOCOM (2000)
4. Younis, M., Munshi, P., Al-Shaer, E.: Architecture for Efficient Monitoring and Management of Sensor Networks. In: Marshall, A., Agoulmine, N. (eds.) MMNS 2003. LNCS, vol. 2839, pp. 488–502. Springer, Heidelberg (2003)

Multimedia Service Management for Home Networks with End to End Quality of Service

Ralf Seepold, Javier Martínez Fernández, and Natividad Martínez Madrid

Departamento de Ingeniería Telemática,
Universidad Carlos III de Madrid, Leganés (Madrid), Spain
{ralf.seepold,javier.martinez,natividad.martinez}@uc3m.es

Abstract. This article describes an architecture providing capabilities to integrate multimedia services into a hardware-independent platform, enabling services via dynamic remote configuration on demand, and finally, establishing end-to-end QoS connections between devices located in different home networks. Based on existing standards like OSGi, UPnP and SIP, a multimedia service management is developed that maintains the current standards but proposes capability extensions in order to cover more complex demands beyond a local (private) network.

1 Introduction

This article proposes device integration, service registration and multimedia service delivery with Quality of Service (QoS) between devices located in (different) home networks. The first step is to integrate devices into a home network infrastructure. The OSGi platform has been selected, since it allows hot-plug of services and it supports a remote configuration [1]. A quite elegant way of device interfacing is proposed by the UPnP (Universal Plug and Play) Forum [2].With the help of the UPnP standard, new devices can be quickly detected, services can be used and devices are capable to start direct communication for passing multimedia data from one device to another, without any mediator. Within this group a specific architecture has been proposed that supports multimedia applications (UPnP AV architecture [3]). Additionally, the Session Initiation Protocol (SIP) [4] has been selected to provide capabilities to establish an IP session. The UPnP Forum has also defined a specification for a QoS architecture for local networks [5]. This UPnP QoS framework consists of three types of services: *QosManager, QosPolicyHolder* and *QosDevice*. A control point requests to the *QosManager* the provision of QoS for a certain traffic specification. The manager establishes the traffic policy with the *QosPolicyHolder* and calculates the intermediate points in the flow path from origin to destination. The manager checks the status of the *QosDevices* on the path and performs admission control on their behalf. Only prioritized traffic is considered in the current version. The work integrates the use of UPnP AV and QoS, extending them to cover end-to-end communications considering prioritized and parameterized traffic flows with QoS manager based on SIP and in the IP Multimedia Subsystem (IMS).

S. Ata and C.S. Hong (Eds.): APNOMS 2007, LNCS 4773, pp. 531–534, 2007.

2 Multimedia Service Management on an OSGi Platform

The objective of the multimedia sub-system is to define an architecture that is integrating the devices and the multimedia services (Fig. 1). The *AV Subsystem UPnP* Provides an UPnP Control Point and an UPnP Media Server in the Residential Gateway (RGW). In addition, the RGW will host modules like proxies for multimedia components that are going to model the access to external multimedia resources.

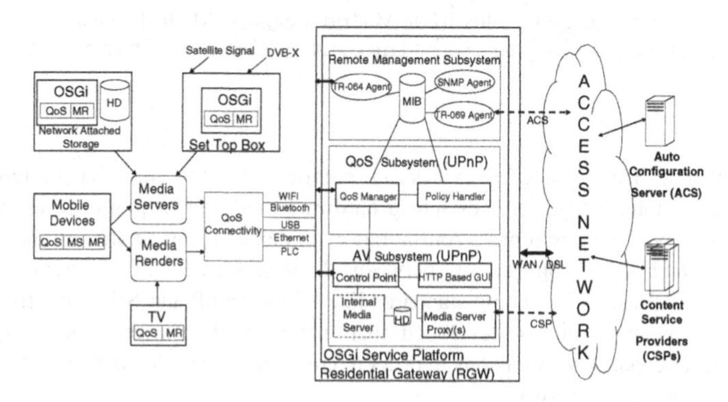

Fig. 1. Service architecture

Two modules are defined in *The QoS Subsystem*: the QoS Manager as a central QoS management unit and the Policy Holder, the unit that stores rules for the QoS. Policies are required since the resources in the network are limited. Taking into account the definitions included in UPnP QoS version 2 only priority based QoS is possible, so it is necessary to extend the QoS architecture in order to provide an end-to-end QoS that is really required.

The *Remote Management Subsystem* take care of the configuration of the RGW and it stores also the current configuration. Since the implementation is based on OSGi services, different agents are available which allow remote configuration of the system without re-initialisation of the device. Since the definition of a connection with QoS may change the RGW router configuration, a communication link is provided between QoS Manager, Policy Handler and a Management Information Base (MIB.)

3 Extension of the UPnP AV/QoS Architecture to Provide End to End Quality of Service

The model proposed in this paper addresses discovery and interaction of multimedia devices among two different home networks. The system uses the SIP protocol with two goals. Firstly, it allows mobile location of the remote services on the gateway and negotiation of the multimedia contents and capacities on the remote network. And secondly, it uses SIP/IMS signaling on the access network to manage QoS in the access networks. The functional architecture of the system can be seen in Fig. 2.

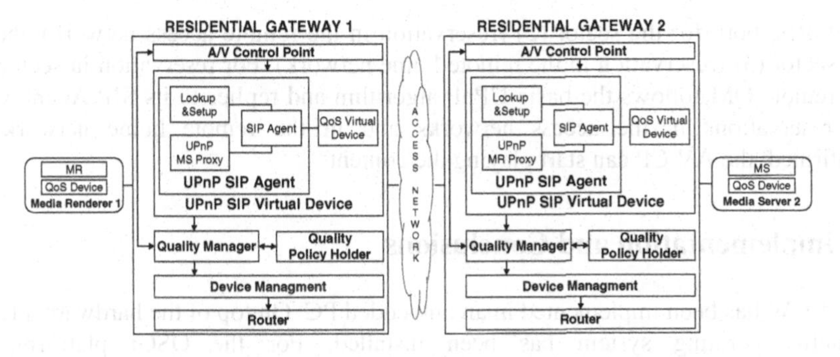

Fig. 2. Functional architecture of the end to end QoS system

The main new element is the *UPnP SIP Virtual Device*, which is actually an OSGi service. It contains an *UPnP SIP Agent* able to perform the remote lockup of devices and setup of contents and protocol, an UPnP Proxy representing the remote Media Server (respectively the remote Media Renderer) and a SIP Agent implementing the SIP protocol stack. It also contains a *QosVirtualDevice*, which is a particular implementation of the UPnP *QosDevice* service that is able to embed the end-to-end QoS management. The *Quality Management* element is an extended version of an UPnP *QosManager* taking into account also parameterized traffic, and finally, the *Device Management* that is the implementation of the UPnP QosDevice service of the gateway itself, which configures the router.

Currently, the standard UPnP protocol only works on local networks. To allow remote discovery, the procedure has been tunneled into a SIP communication between the audiovisual control points of both gateways through their UPnP SIP agents. The local AV CP requests a list of multimedia devices from the remote home network through the lookup service of the local UPnP SIP Virtual Device, and finally, it is able to select one remote Media Server. The content, the transfer protocol and format negotiation with the remote Media Server needs to go also through the UPnP SIP Virtual Devices. With the help of the UPnP Media Server Proxy and the UPnP Media Renderer Proxy this functionality is implemented.

Once the content and format negotiation with the remote device is completed, the local AV CP configures the proper parameters in the local renderer, and it can proceed with the configuration of the QoS for the traffic.

To provide end-to-end quality of service for the multimedia traffic from the media server in the remote network to the media renderer in the local network, the complete path is divided in three sectors: (1) between the local renderer and the local gateway, (2) between both gateways, (3) between the remote gateway and the remote device.

The local AV CP requests from the QosManager (QM) the reservation of QoS for the traffic. The QM detects that the communication is remote and estimates the distribution of the traffic parameters for the three sectors. The QM handles locally the QoS reservation for sector (1) and delegates to the UPnP SIP Virtual Device the reservation for sectors (2) and (3) via the QoS Virtual Device service of the UPnP SIP Virtual Device. This service performs a remote QoS reservation through the SIP Agent. The local SIP Agent sends a INVITE request with an SDP (Session Description Protocol) offer/response part containing the necessary characteristics of the traffic both for the sector (2) (reservation in the remote access network) and for the sector (3) (reservation in the remote home network). For reservation in sector (3), the remote QM follows the basic UPnP algorithm and replies to its SIP Agent. Once all reservations in the access networks and in the remote home network are confirmed, the AV CP can start playing the content.

4 Implementation and Conclusions

The RGW has been implemented in an embedded PC. On top of the hardware a Linux Ubuntu operating system has been installed. For the OSGi platform, the implementation Oscar 1.5 has been used [6], an UPnP Control Point [7], a Media Renderer [8] and Media Server [9] have been installed. Since these services are offered as OSGi services, both the control point and the server have been converted into OSGi bundles. The server and the UPnP control point have been installed in the RGW, a notebook has been used as a renderer and the Nokia 770 has been used as remote control. It is possible to select multimedia data, to list possible media renderers and to start the multimedia data stream, i.e. a direct connection between media server and renderer has been established. After integrating UPnP AV capabilities in the middleware platform, devices can connect to services independent from the UPnP capabilities of the services provider. In a second step, the proposed extension of the UPnP QoS capabilities and the development of SIP compliant modules in the gateway have been implemented. Finally, it has been demonstrated that is possible to provide QoS connections beyond the border of a home network.

Acknowledgments. The work has been partly funded by the projects PLANETS (Spanish Ministry of Industry; FIT-330220-2005-111), MUSE (IST-Program; MUSE-IST 026442) and MARTES "Model-based Approach to Real-Time Embedded Systems development" (ITEA 04006, FIT-340000-2006-166).

References

1. Open Service Gateway Initiative (OSGi) Alliance (2007), URL http://www.osgi.org
2. Universal Plug and Play (UPnP) Forum, UPnP AV Architecture:1 (June 2002), URL http://www.upnp.org/specs/av/UPnP-av-AVArchitecture-v1-20020622.pdf
3. Universal Plug and Play (UPnP) Forum, MediaServer V2.0 and MediaRenderer V2.0 (March 2006), URL http://www.upnp.org/specs/av/
4. SIP, Session Initialization Protocol, Internet Engineering Task Force (IETF), Network Working Group RFC: 3261 (2002), URL: tools.ietf.org/html/rfc3261
5. Universal Plug and Play (UPnP) Forum, Quality of Service V2.0 (October 2006), URL http://www.upnp.org/specs/qos/
6. Oscar, OSGi Implementation (2007), URL http://forge.objectweb.org/projects/oscar/
7. Cidero Software Solutions for the Digital Home (2007), URL http://www.cidero.com
8. UPnP Media Render implementation for Linux (2007), URL http://soggie.soti.org/gmediarender/
9. Reference Implementation of UPnP AV, Satoshi Konno, Tokyo, Japan (June 2007), URL http://www.cybergarage.org/net/cmgate/java/index.html

An OSGI-Based Model for Remote Management of Residential Gateways

Mario Ibáñez[1], Natividad Martínez Madrid[1], Ralf Seepold[1], Willem van Willigenburg[2], and Harold Balemans[2]

[1] Universidad Carlos III de Madrid, Avda. de la Universidad 30, 28911 Leganés (Madrid), Spain
{mario.ibanez,natividad.martinez,ralf.seepold}@uc3m.es
[2] Bell Labs Europe, Alcatel-Lucent, Larenseweg 50, 1221 CN Hilversum, The Netherlands
{willigenburg}@alcatel-lucent.com

Abstract. Devices connected to the home network demand in many cases services which are located outside the home network. For this reason one of the most important elements in the home network is the Residential Gateway. It will manage all subscribed services, work as central point to control all devices of the home network and provide an integration platform for various internal and external services. This paper proposes a solution based on OSGI and TR-069 to support auto-configuration of a residential gateway located in home network.

1 Introduction

A continuously growing number of services are available to be used in our homes. Examples are applications like Video on Demand (VoD) or Voice over IP (VoIP). The distribution of those services inside the Home Network (HN) has to maintain the same quality of service (QoS) that is provided in the Access Network (AN). Furthermore, different devices can be connected via different technologies and protocols to the HN. The device managing all services of our home is called Residential Gateway (RGW). Due to the mentioned heterogeneous environment, the RGW needs to be as open as possible in order to adapt to the scenario. The scenario is completed when adding a set of external servers to the heterogeneous home scenario which also provide different services to the home.

The working scenario is shown in Fig. 1, the RGW is placed between the HN and the AN, and connects the home to the Internet. The RGW offers a platform used by service providers to install the demanded services and by the end user to install own applications. The scenario has two different parts: On the left hand side is the HN where the user can connect different devices using different protocols and technologies. This variety of elements will cause a complex handling of service delivery to the device of the end user. However, this task will not be covered in this paper and the proposed solution for auto-configuration will concentrate on the AN.

On the right hand side, the RGW is connected to AN. This network provides access to the different services that the user can contract. For example VoD, SIP proxies for

S. Ata and C.S. Hong (Eds.): APNOMS 2007, LNCS 4773, pp. 535–538, 2007.

establishing connections or an e-care server that controls some home devices. One possibility to provide access to all these services is to configure manually each service in the RGW. That means each time a customer contracts a service from the provider, the provider must access to the RGW configuration to modify it.

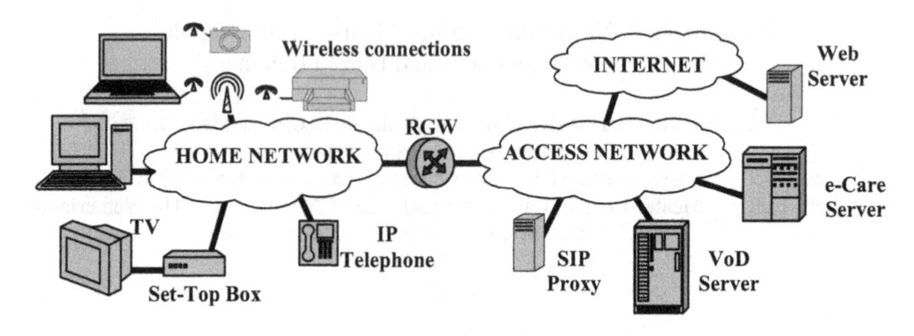

Fig. 1. Network environment

The requirements and solutions for remote management of RGWs will be described in section 2. Section 3 details the architecture for remote auto-configuration. Finally, the results and conclusions are presented.

2 Relevant Technologies

The RGW has been split into two main elements: the service platform (OSGI) and the underlying router (Click). The TR-069 protocol is used for the communication between the RGW service platform and the ACS.

OSGI has been selected as the platform for the services in the RGW [1]. It provides a Java-based open, common architecture for network delivery of managed services. Services are added through software components (*bundles*). The OSGI specification does not pre-define a remote management standard.

TR-069 [2] is a protocol used for communication between a CPE (Customer Premises Equipment) and ACS (Auto-Configuration Server) that encompasses secure auto-configuration as well as other CPE management functions within a common framework. The data model for TR-069 is defined in TR-098 [3].

Click [4] is software architecture for building flexible and configurable routers. Click consists of different modules which have different functionalities. These modules are called elements, and these elements can be used for packet classification, queuing, scheduling and interfacing with network devices.

3 Detailed Architecture

In order to implement a solution that supports the TR-069 protocol and that allows dynamic reconfiguration of a running system without rebooting, the main communication will take place between the ACS and the OSGI service platform. A solution can be provided in case a connection between the RGW and the ACS is available. The functional blocks of the solution are presented in Fig. 2.. Bundles that allow the communication with ACS and the configuration of router are explained next.

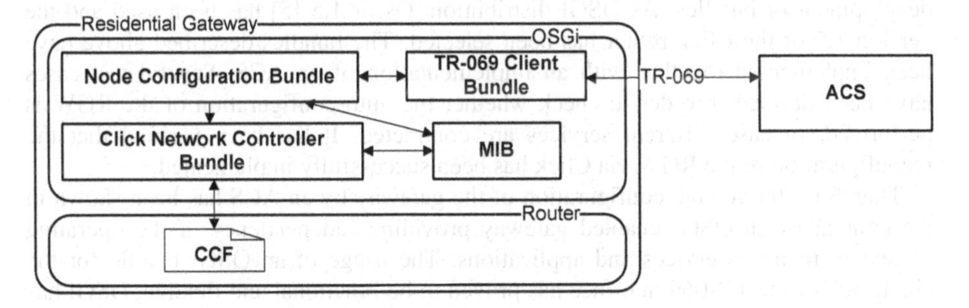

Fig. 2. Functional blocks for automatic management

Auto Configuration Server (ACS): The server is placed in the access network and is in charge of sending configuration parameters to the RGW in case a new service is contracted at the service provider. It runs a TR-069 server that is capable to communicate with the TR-069 client of the RGW.

TR-069 Client Bundle: This bundle is in charge of communication with the ACS using the TR-069 protocol. It manages the different messages that arrive from the ACS and processes them to configure the behaviour of the RGW, which is carried out by another bundle that is called Node Configuration Bundle (NCB).

Node Configuration Bundle (NCB): This bundle provides services for the TR-069 Client Bundle for configuring the RGW. It will provide these services when other basic bundles are available; these bundles are the CNC bundle and the MIB bundle. It communication with CNC only allows the reboot of the machine and with the MIB allows the management of the configuration parameters of the RGW.

Management Information Base (MIB): The MIB represents the entire set of Configuration Parameters specified for use with a certain management protocol. As in most management protocols, in TR-069 the MIB is built up from hierarchical objects. For example, the root object of a RGW is the InternetGatewayDevice, which contains other objects that describe the functionality of the device. The objects presented in the MIB determine the range of configuration possibilities.

Click Network Controller Bundle (CNCB): This bundle is in charge of reconfigure the router. It subscribes to selected objects in the MIB and waits for configuration changes notifications. The reconfiguration is made writing a file that is constantly checked by Click, the Click Configuration File (CCF). Before writing the CCF, the CNCB has to translate between the data formats of TR-069 and Click.

Click Configuration File (CCF): This is file of Click where all the information of configuration is stored. Click uses a proprietary language to store the configuration.

4 Results and Conclusion

The prototype developed has been tested over a Lex Light embedded PC with a motherboard CV860A and a CPU VIA C3 533MHz (i386 compatible). The PC is running under Linux Ubuntu 6.06. J2SE 1.5 has been selected for the execution and development of bundles. As OSGI distribution, Oscar 1.5 [5] has been used and the version 1.5 of the Click router has been selected. The bundles described above have been implemented together with an implementation of an ACS. Several test cases have been defined in order to check whether the auto-configuration of the RGW is performed, in case different services are contracted. It has been verified that the reconfiguration of the RGW via Click has been successfully implemented.

Therefore, the remote configuration of the gateway by an ACS has been shown in the context of an OSGI enabled gateway providing independency of the operating system to manage services and applications. The usage of an OSGI bundle for the client side of the TR-069 interface has proven to be functional and flexible. OSGI has been also used to provide an easily maintainable interface from the TR-069 client towards the software boxing the actual implementation of the gateway control. It can be concluded that a proof of concept has been given for a TR-069 remote configured gateway while using the OSGI flexible bundle mechanism for the implementation the TR-069 client side of the protocol.

Acknowledgment

The work has been partly supported by project MUSE (IST, MUSE-IST 026442) [6] and project PLANETS (Spanish Ministry of Industry; FIT-330220-2005-111) [7].

References

1. Open Service Gateway Initiative (OSGI) Alliance (July 2007), http://www.osgi.org
2. DSL Forum, TR-069: CPE WAN Management Protocol (May 2004), http://www.dslforum.org/aboutdsl/Technical_Reports/TR-069.pdf
3. DSL Forum, TR-098: DSLHomeTM Internet Gateway Device Version 1.1 Data Model for TR-069 (September 2004) http://www.dslforum.org/techwork/tr/ TR-098.pdf
4. The Click router (July 2007), http://www.pdos.lcs.mit.edu/click/
5. Oscar (July 2007), http://forge.objectweb.org/projects/oscar/
6. Multi Service Access Everywhere (MUSE) European Project, MUSE-IST 026442 (July 2007), http://ist-muse.org/
7. Platforms for Networked Service Delivery (PLANETS), MEDEA+ project A-121, financed by the Spanish Ministry of Industry under project Num. FIT-330220-2005-111 (July 2007), URL: www.medeaplus.org

Design and Implementation of TPEG Based RFID Application Service

HyunGon Kim

Dept. of Information Security, Mokpo National University,
Muan-Gun, Jeonnam, 534-729, Korea
hyungon@mokpo.ac.kr

Abstract. This paper presents the design of the TPEG-RFID event container and the results of a trial implementation. With regard to the designed TPEG-RFID event container, an example of its coding rules providing for RFID based bus-line information is presented. Implementation details of DMB data server and user terminal are also designed as software functional blocks. The presented implementation is an attempt to verify the feasibility of the designed TPEG-RFID application service system. In particular, the RFID based bus-line information service is specified. The test results show that users could utilize the current location based RFID application services with the designed system.

1 Introduction

DMB is a digital transmission standard based on the digital audio broadcasting (DAB) Eureka-147 standard[1]. It is a multiplexed system capable of delivering audio, video, IP and other data files. It is expected that DMB will be fully converged with other services such as wireless broadband (WiBro), radio frequency identification (RFID), and telematics. From this point of view, an integrated service model was introduced for supporting RFID application services on the T-DMB as an initial draft[2]. The model is called TPEG-RFID application service and is based on the T-DMB transport protocol expert group (TPEG) technology. It could allow users to support an identical RFID application service delivery on the user's T-DMB terminal. For the application service in the draft, however, TPEG-RFID application message details are not defined specifically. The main objective of this work is to present the design of TPEG-RFID application message details and conduct a trial implementation. The implementation is an attempt to verify the feasibility of the designed system. With regard to the designed TPEG-RFID message details, the structure of TPEG-RFID event container is designed and an example of its coding rules providing for RFID based bus-line information is presented. Implementation aspects of the DMB data server and user's terminal are designed as software functional blocks.

2 Design of the TPEG-RFID Event Container

For realization of the TPEG-RFID application service[2], the RFID Event Container schematically is designed. The elements are used to describe, with the

S. Ata and C.S. Hong (Eds.): APNOMS 2007, LNCS 4773, pp. 539–542, 2007.

end-user in mind, the serving RFID service information based on the T-DMB channels. The RFID Event Container consists of six elements, that is, Classification, RFID Tag ID, Content Server's URL, content provider (CP) Information, Advertisement, and Reserved field, as shown in Fig. 1. The TPEG-RFID service is classified into 7 categories: RFID-based (RB) Movie Information, RB Advertisement and Marketing, and so on. Each category may in turn include sub-categories.

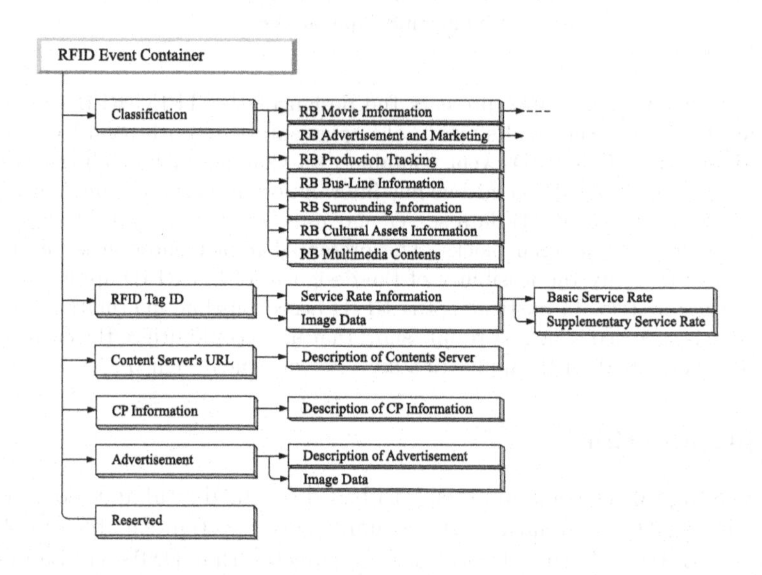

Fig. 1. Structure of TPEG-RFID event container

Code	Word	Entry in related TPEG table(subtype), table name)
0	Unknown	(subtype=0), no TPEG table available
1	RB Movie Information	TPEG table rfid02(...), Movie_Information_Class
2	RB Advertisement and Marketing	TPEG table rfid03(...), Advertisement_Marketing_Class
3	RB Production Tracking	TPEG table rfid04(...), Production_Tracking_Class
4	RB Bus-Line Information	TPEG table rfid05(...), Bus-Line_Information_class
5	RB Surrounding Information	TPEG table rfid06(...), Surrounding_Information_Class
6	RB Cultural Assets Information	TPEG table rfid07(...), Cultural_Assets_Information_Class
7	RB Multimedia Contents	TPEG table rfid08(...), Multimedia_Contents_Class

Code	Word	Desc.
0	Unknown	
1	Bus route MAP	
2	Transfer information	Transfer by bus,,,,
3	Bus arrival time	
4	Bus current location	
5	Traffic information	
6	Bus ticket buying	
7	Station surrounding infor.	
8	Bus operator information	

* RB = RFID-Based

Fig. 2. TPEG-RFID reference table and Bus-Line information class

As a sample, Fig. 2 is used to describe classified bus line information. The information allows road user to obtain RFID service related to bus information, such as route MAP, transfer information, bus arrival time, etc. via a user terminal without RFID reader capability.

3 Implementation of TPEG-RFID Application

Regarding real deployment for the TPEG-RFID application, we designed the implementation architecture as shown in Fig. 3. The figure presents logical software architecture of physical entities in terms of software blocks and their respective functionalities.

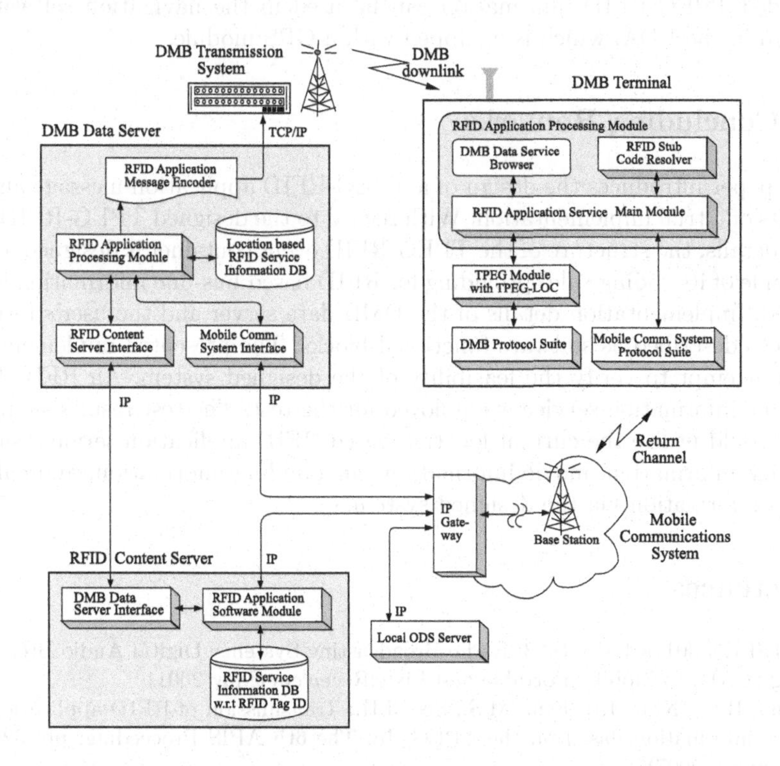

Fig. 3. Implementation architecture of TPEG-RFID application

We implemented the DMB data server and the terminal to compose and encode TPEG-RFID information. The DMB receiver with a TPEG-RFID decoder platform then parsed the information using the proposed application message, as shown in Fig. 4. The encoded TPEG stream is inserted into the ensemble multiplexer through the TCP/IP connection and is transmitted via a transparent data channel (TDC) with stream mode. The data rate is set to 64kbps and the average code rate is about 1/2. In the receiving side, a TPEG-RFID decoder has been implemented in a PDA that is connected to the DMB receiver through the universal asynchronous receiver/transmitter (UART). Thus, the

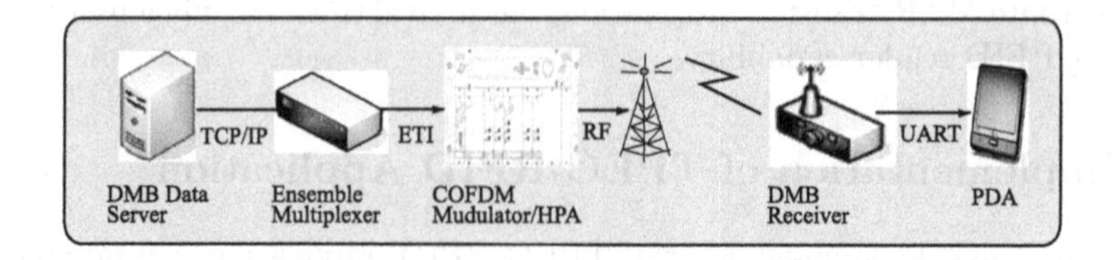

Fig. 4. Test-bed for TPEG-RFID application service

decoded TPEG-RFID information can be used in the navigation software installed in the PDA, which is equipped with a GPS module.

4 Concluding Remarks

This paper introduces the design of a TPEG-RFID application message and the results of a trial implementation. With regard to the designed TPEG-RFID message details, the structure of the TPEG-RFID event container is designed and an example of its coding rules providing for RFID based bus-line information is presented. Implementation details of the DMB data server and the user's terminal are also described as software functional blocks. The presented implementation is an attempt to verify the feasibility of the designed system. An RFID based bus-line information service is employed for the test. The test results show that users could utilize the current location based RFID application services such as bus-line information, movie information, surrounding information, and cultural assets information via the designed system.

References

1. ETSI EN 301 401 Ver.1.3.3: Radio Broadcasting Systems; Digital Audio Broadcasting (DAB) to Mobile, Portable and FixedReceivers (May 2001)
2. Kim, H.G., Kim, J.B, Kim, M.S., Seo, J.H.: Transmission of RFID application service information based on the TPEG. In: The 6th APIS Proceeding, pp. 328–331 (January 2007)
3. EBU B/TPEG: Transport Protocol Experts Group (TPEG) TPEG specifications - Part 1: Introduction, Numbering and Versions, TPEG-INV/002, draft (October 2002)
4. `http://www.tpeg.org`

Energy-Efficient Distance Based Clustering Routing Scheme for Long-Term Lifetime of Multi-hop Wireless Sensor Networks*

Young-Ju Han, Jung-Ho Eom, Seon-Ho Park, and Tai-Myoung Chung

Dep. of Computer Engineering, Sungkyunkwan Univ.
300 Cheoncheon-dong, Jangan-gu,Suwon-si, Gyeonggi-do, 440-746, Korea
{yjhan,jheom,shpark}@imtl.skku.ac.kr and tmchung@ece.skku.ac.kr

Abstract. To balance energy consumption among cluster-heads in multi-hop wireless sensor networks, we propose an energy-efficient distance based probabilistic clustering scheme. Our scheme considers a distance from the sink(base station) to each node as well as the residual energy of each node as the criterion of cluster-head election. Through simulation experiments, we show that our scheme is more effective than LEACH and EEUC in prolonging the lifespan of multi-hop wireless sensor networks.

1 Introduction and Motivation

Clustering scheme enabling the efficient utilization of the limited energy resources of the deployed sensor nodes can effectively prolong the lifetime of wireless sensor networks[1,2]. Wireless sensor networks based on multi-hop communication has the hop spot problem that the CH(cluster-head) closest to the sink is burdened with a heavy relay traffic load and die first[4]. Most of clustering schemes utilize mainly CH frequency or residual energy of each node as criterion of CH election and don't consider the distance from the sink to the node[1,3]. So, those schemes occur unbalanced energy consumption among CHs.

In this paper, we propose an energy-efficient distance based probabilistic clustering scheme in multi-hop wireless sensor networks. We consider the distance from the sink to each node as well as the residual energy of each node as the criterion of CH election for load-balancing of CHs. Our scheme provides fully distributed manner by utilizing local information and good energy-efficiency by load-balanced clustering scheme.

* This research was supported by the MIC(Ministry of Information and Communication), Korea, under the ITRC(Information Technology Research Center) support program supervised by the IITA(Institute of Information Technology Advancement) (IITA-2006-C1090-0603-0028)

S. Ata and C.S. Hong (Eds.): APNOMS 2007, LNCS 4773, pp. 543–546, 2007.

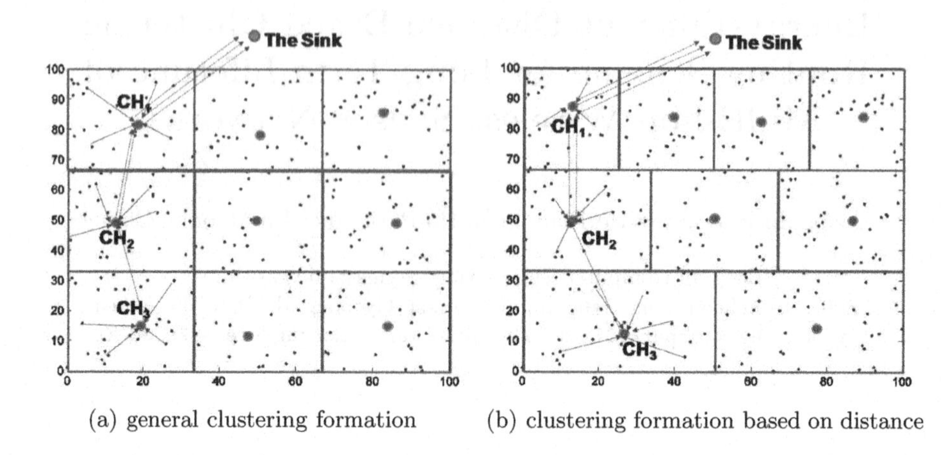

(a) general clustering formation (b) clustering formation based on distance

Fig. 1. The basic idea of the proposed scheme

2 The Proposed Scheme

2.1 The Concept of the Proposed Scheme

The objective of the proposed scheme is to balance energy consumption among the CHs for prolonging network lifetime of multi-hop wireless sensor networks. For doing so, we consider the unbalanced energy consumption of CHs which is resulted from unbalanced cost between inter-cluster and intra-cluster communication.

The basic idea behind the proposed scheme is that the closer the CH is to the sink, the smaller cluster size it could have. Let us assume clustering network which is composed with 6 clusters. As shown in Fig.1-(a), general clustering scheme partitions the sensor nodes into clusters of equal size. In Fig.1-(a), inter-cluster communication cost of CH_1, $E_{inter}(CH_1)$, is higher than $E_{inter}(CH_2)$ or $E_{inter}(CH_3)$ because CH_1 is relay point between the sink and the other CHs(CH_2 and CH_3), but intra-cluster communication cost of CH_1, $E_{intra}(CH_1)$, is the same as $E_{intra}(CH_2)$ or $E_{intra}(CH_3)$. So, the comparison of energy consumption among the CHs is the same as $E(CH_3) < E(CH_2) < E(CH_1)$. To solve this problem, we partitions the sensor nodes into clusters of unequal size as shown in Fig.1-(b). In Fig.1-(b), the intra-cluster communication cost among CHs is the same as $E_{intra}(CH_3) > E_{intra}(CH_2) > E_{intra}(CH_1)$, the energy consumption among CHs will be balanced as $E(CH_1) \cong E(CH_2) \cong E(CH_3)$.

2.2 The Operation of the Proposed Scheme

The operation of the proposed scheme is broken up into rounds like LEACH[1]. Each round consists of cluster-setup phase and data-transmission phase.

Cluster-setup phase

In the proposed scheme, the operation of cluster-setup phase is the same as LEACH except the difference of probability formula, Eq. 1, of becoming a CH.

$$P_i = 3P_{opt} \times \frac{d_{max} - d(S_i, SINK)}{d_{max} - d_{min}} \times \frac{E_{res}(i)}{E_{init}(i)}, \quad 0 <= P_{opt} <= \frac{1}{3} \tag{1}$$

where P_{opt} is an optimal value of a probability of probabilistic clustering scheme, S_i is i^{th} sensor node, d_{max} represents the distance of the farthest sensor node from the base station and d_{min} represents the distance of the closest sensor node, $d(x, y)$ is the distance between node x and y, $E_{init}(i)$ and $E_{res}(i)$ represent the initial energy and the residual energy of ith node respectively. Equation 1 shows that the farther the sensor node is to the sink, the lower probability of becoming a CH it could have. In other words, they can have more members than the CHs closer to sink since the few CHs are elected in the area farther to sink. Besides, among the nodes having the same distance from the sink, the more the sensor node has the residual energy, the higher probability of becoming a CH it could have.

Let us examine this phase in detail. At the beginning of this phase, sensor node chooses a random number between 0 and 1. If the random number is less than its P_i, the sensor node becomes the CH. And then each CH broadcasts the advertisement message containing its residual energy, $E_{res}(i)$. When the plain node receives the advertisement message from the CHs, it chooses the CH_j with more cost $C(i, j)$ among the CHs and joins to the CH_j as the member. The formula of $C(i, j)$ is $\frac{E_{res}(CH_j)}{d(S_i,CH_j)^2}$ and means the relative amount of the CH's residual energy on the basis of the distance from itself to the CH.

Data-transmission Phase

In a intra-cluster communication, the operation of data transmission is the same as LEACH[1]. Next, let us consider an inter-cluster communication. Each CH must find relaying node through following steps. Firstly, if it is located at the closest distance to sink among all CHs within transmission range, it communication directly with sink. Otherwise, its relay node is determined as CH having the minimum value of $\frac{E_{res}(CH_j)}{(d(CH_i,CH_j)^2+d(CH_j,SINK)^2)} < \frac{E_{res}(CH_i)}{d(CH_i,SINK)^2}$ for all CHs within transmission range.

3 Performance Evaluation

To evaluate the performance of the proposed scheme in the aspect of energy-efficiency, we implemented the proposed scheme, LEACH and EEUC with MATLAB. Simulated sensor network is composed of 400 sensor nodes distributed uniformly on a square area of size 100 × 100. For the simulation, we used the value of parameters of system, network and radio model in [1,3]. For simplicity, we assume that simulation environments are a congestion-free MAC layer and

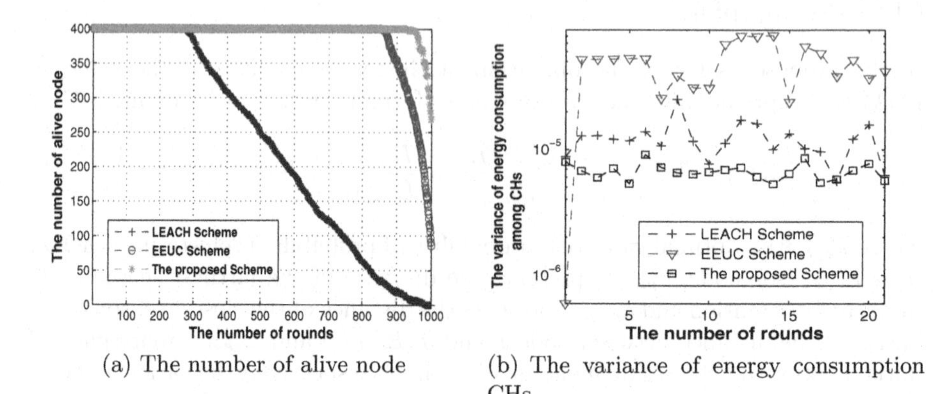

(a) The number of alive node

(b) The variance of energy consumption CHs

Fig. 2. Simulation results

an error-free communication. Firstly, to compare the network lifetime of each scheme, we measure the number of round when the first node dies. Figure 2-(a) shows the network lifetime of each scheme. In Fig.2-(a), the proposed scheme is more energy-efficient than other schemes. Secondly, we measure the variance of energy consumption among CHs of each scheme. The lower the variance is, the better energy consumption among CHs is balanced. As shown in Fig.2-(b), the proposed scheme is more load-balanced than EEUC and LEACH.

4 Conclusion

To prolong the lifetime of wireless sensor networks, an energy-efficient routing protocol is necessary. We introduced an energy-efficient distance based probabilistic clustering scheme which can prolong the lifetime of multi-hop wireless sensor networks through load-balancing of energy consumption among CHs. As shown in the simulation result, the proposed scheme is more energy-efficient and load-balanced than LEACH and EEUC.

References

1. Heinzelman, W.R., et al.: An application-Specific Protocol Architecture for Wireless Microsensor Networks. IEEE Transactions on Wireless Communications 1(4) (2002)
2. Karl, H., Willing, A.: Protocols and Architectures for Wireless sensor Networks. John Wiley & Suns, Chichester (2005)
3. Li, C., et al.: An Energy-Efficient Unequal Clustering Mechanism for Wireless Sensor Networks. In: MASSC2005. Proceedings of IEEE Mobile Adhoc and Sensor Systems Conference2005, IEEE Computer Society Press, Los Alamitos (2005)
4. Soro, S., et al.: Prolonging the Lifetime of Wireless Sensor Networks via Unequal Clustering. In: Proceedings of IEEE PDPS 2005, IEEE Computer Society Press, Los Alamitos (2005)

Single Sign on System Architecture Based on SAML in Web Service Environment Using ENUM System

Jiwon Choi and Keecheon Kim

Department of Computer Science & Engineering, Konkuk University,
1 Hwayang-dong, Kwangjin-gu, Seoul, 143-701, Korea
{jackeroo,kckim}@konkuk.ac.kr
http://mbc.konkuk.ac.kr

Abstract. Recently, with development of Internet telephone, we can say that the ENUM could be the most suitable Internet identification system standard in the BcN(NGN). When embody ENUM, there are problems on security for personal information and convergence with existing domain. And it is need that ENUM service provides convenient function to users. In this paper, we design the system that offers Single Sign On without division of user terminal using SAML when it requires URL with a telephone number to ENUM DNS.

1 Introduction

As superhigh Internet is spread, it is smartening on discussions about BcN (Broadband convergence Network). To combine PSTN to BcN, it needs study about a development plan of Internet address structure and identification number that is suitable for providing voice, data and multimedia service for BcN. For development of the shape that combines all communication method, it must communicate each other through single identification structure between terminals that have different characteristic and communication system. It is ENUM (tElephone Number Mapping) as a standardization of Internet identification structure that is suitable for BcN environment. But because general users don't feel inconvenience with previous Internet identification structure, for revitalization of the ENUM new technology has to introduce for advantage of communication manufacturer and general user.

This paper research DNS security method that is used first for ENUM security and designs a system that can offer Single Sign On at the same time that run safe authentication using SAML.

2 SAML Based Single Sign on System Architecture in Web Service Environment Using ENUM System

Figure 1 is schematic diagram for single sign on system between terminal users and service providers using ENUM DNS. If terminal users want to use this service, they first connect to ENUM DNS by their terminals. ENUM DNS offers form that requests a domain name with telephone number and user input s corresponding telephone number. Then ENUM DNS requests to user certification. In case user uses a cellular

S. Ata and C.S. Hong (Eds.): APNOMS 2007, LNCS 4773, pp. 547–550, 2007.

phone, own number can be an identifier. In case user uses the other devices like laptop or PC, user who has private telephone number can use the number. If user doesn't have a telephone number he can use new identifier instead of telephone number. Figure 1 shows three entities.

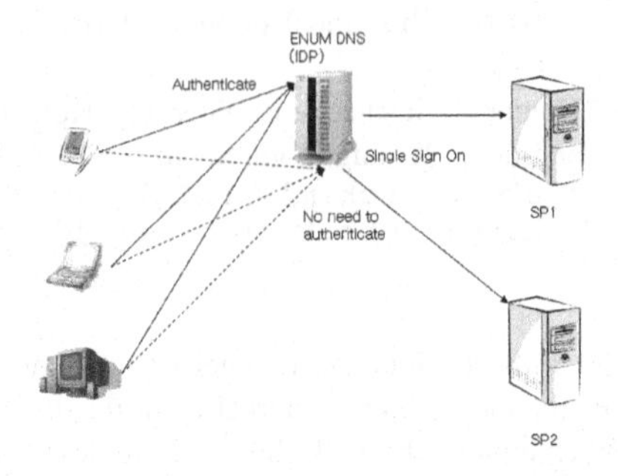

Fig. 1. Single Sign On System Architecture using ENUM DNS

Certification order that use above three entities is as following. User terminal queries URL to ENUM DNS with SP1's telephone number. ENUM DNS requests certification to user terminal. User inputs ID (one's telephone number) and password. ENUM DNS sends a certificate that certified on the basis of user information to SP1. SP1 confirms whether user certification is valid on the basis of received certificate. If it is valid, SP1 provides service to user terminal.

Then when user terminal uses SP2's service, certification order is as following. User terminal queries URL to ENUM DNS with SP2's telephone number. SP2 requests user information to ENUM DNS. ENUM DNS examines certificate with session value from SP2, it recognizes that user had logged in to SP1 before. ENUM DNS informs that user is valid authenticator to SP2. After SP2 receives valid certification response from ENUM DNS, SP2 provides service to user terminal.

Figure 2 shows that it pulls assertion which contains certification information for special one from the system that requires certification information using SAML

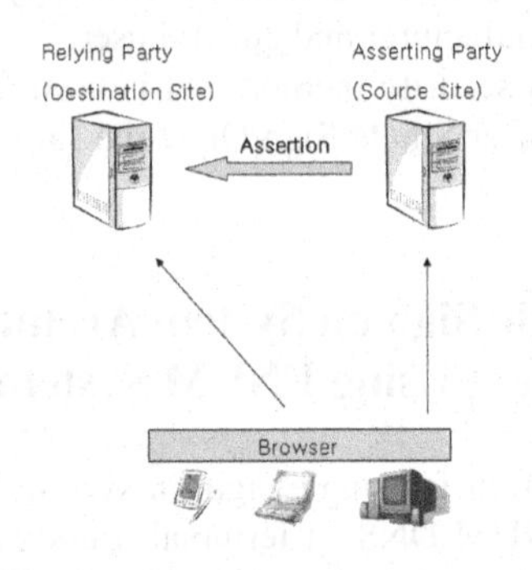

Fig. 2. SAML Assertion model

artifact. Relying party's SSO process is as following. User has certificated session from asserting party (ENUM DNS). When the user approaches in resource in relying party, artifact is string that was encoding in base-64. Relying party has to judge user's identification and capacity. And it sends SAML requirement to asserting party, then it asks how local cite can prove the user. Assertions are return as SAML response. Relying party can decide about certification and certification with received assertions.

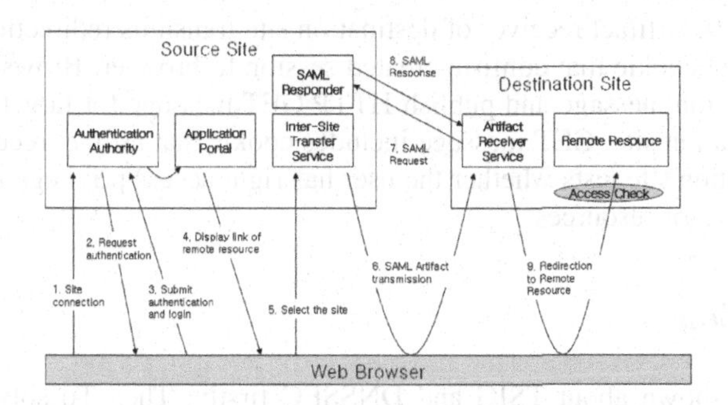

Fig. 3. Message flow from web browser to source site and destination site

Figure 3 shows a message flow that query with a destination telephone number in web browser on user terminal. ENUM DNS includes component that is called ITS (Inter-Site Transfer Service). This component offers function for SAML process such as artifact and redirect creation. Process is as following.

a) Step 1: User requests target address to ENUM DNS.
b) Step 2: ENUM DNS checks access and confirms whether the user has a sess ion or not and then decides authentication process.
c) Step 3: When user needs authentication, user submits authentication informa tion such as identification and password.
d) Step 4: If authentication process is complete, session for the user is created a nd suitable message is displayed.
e) Step 5: The user selects the menu of displayed screen. It means that the user tries to approach resource or application in destination web site. This generat es HTTP request that sends to ITS in ENUM DNS. This request includes U RL of resources in destination site.
f) Step 6: ITS creates assertion and artifact about the user. Artifact includes ref erence (AssertionHandle) about source ID of SAML responder that responds to SAML request and asserting. ITS sends HTTP redirection message that i ncludes URL of artifact service in destination site, target URL and artifact to web browser. Browser that receives redirect message publishes HTTP get m essage and <artifact> is value that is coded that is by base 64. This message i s sent to server that hosts target URL.
g) Step 7: Artifact receiver that receives HTTP message in destination site dra ws source ID. This time, mapping of source ID and SAML responder is poss ible because it has a relationship beforehand. Artifact receiver knows that it must communicate with SAML responder that corresponds in abstracted sou

rce ID. Artifact receiver in destination site transmits SAML request includin g artifact that offer in ENUM DNS's ITS to ENUM DNS's SAML responder.

h) Step 8: SAML responder of ENUM DNS transmits SAML response messag e including assertion that create at 6 steps before. In most implementation, s ession for the user's destination site is created when it receives valid assertion.

i) Step 9: Artifact receiver of destination site transmits redirection message inc luding cookie that confirm created session to browser. Browser processes re direction message and publish HTTP GET message for target resources of d estination site. GET message includes cookie that artifact receiver sends. De stination site tests whether the user has right access privilege before it allow access for resources.

3 Conclusion

This paper is shown about TSIG and DNSSEC firstly. Then To solve the problem which can use like general DNS security and provide user convenience, we make a architecture of Single Sign On system using ENUM DNS. In this architecture, user can connect by personal number. After user login in ENUM DNS, no need to get authentication.

References

1. Laura Taylor.: Understanding Single Sign-On (2002) http://www.intranetjournal.com/ articles/ 00205/e_05_28_02a.html
2. Assertions and Protocol for the OASIS Security Assertion Markup Language (SAML) V2.0 http://docs. oasis-open.org/security/saml/v2.0/saml-core-2.0-os.pdf
3. RFC 3281. An Internet Attribute Certificate Profile for Authorization. http://www.ietf org/rfc/rfc3281.txt
4. Web Services Security: SOAP Message Security 1.0 (WS-Security 2004) (2004) http://docs.oasis-open.org/wss/2004/01/oasis-200401-wss-soap-message-security-1.0.pdf
5. e-authentication Guidance for Federal Agencies, Joshua B. Bolten (Director, Executive Office of the President, Office of Management and Budget), (December 2003) http://www.white house.gov/omb/memoranda/fy04/m04-04.pdf
6. NIST Special Publication 800-63 Electronic authentication Guideline, Recommendations of the National Institute of Standards and Technology Version 1.01, William E. Burr, Donna F. Dodson, W. Timothy Polk (September 2004), http://csrc.nist.gov/publications/ nistpubs/800-63/SP800-63v6_3_3.pdf
7. IBERTY ALLIANCE PROJECT White Paper: Liberty Alliance and WS Federation?Comparison Overview, http://www.projectliberty.org
8. Web Services Security: SAML Token Profile, http://docs.oasis-open.org/wss/oasis-wss-saml-token-profile-1.0.pdf
9. Web Services Security Rights Expression Language (REL) Token Profile, http://docs.oasis-open.org /wss/oasis-wss-rel-token-profile-1.0.pdf

Providing Seamless Services with Satellite and Terrestrial Network in Mobile Two Way Satellite Environments

NamKyung Lee, HoKyom Kim, DaeIk Chang, and HoJin Lee

Electronics and Telecommunications Research Institute
YuSeong-gu GaJeong-dong, DaeJeon 449-701 Korea
nklee@etri.re.kr

1 Introduction

Satellite communication environment can be divided into LOS and NLOS environments. While, terminal can get satellite signal in LOS environment, NLOS needs additional function to get the signal such as cooperation between gap filler or terrestrial network (WLAN or WIMAX). In this paper, the two way satellite communication system is based on DVB-RCS specification and terrestrial networks are based on WLAN(IEEE 802.11) or WIMAX(IEEE 802.16). This paper adopt proper algorithms to handover among satellite and wireless network for providing seamless service when moving object enter NLOS environment in collaboration area and estimate the performance of the method by simulations.

2 Related Work

2.1 DVB-RCS

DVB-RCS system is based on satellite environment and it has two channels which are established between the service provider and the user. One is broadcast channel and the other is interaction channel.

2.2 Classifications of Moving Environment

The two way satellite system can move and embark on vehicle, ship and aircraft. To provide Internet access service with high-speed moving object, we need to consider satellite network area(LOS), wireless network area(N-LOS), and satellite-wireless network collaboration area(N-LOS), respectably.

3 Requirements for Collaborations

In case that satellite network can't expect any LOS and there is only any wireless network, client gets seamless service through the cooperation with terrestrial wireless network. To provide seamless service to any client of two way satellite communication system with is enabled to the moving object, it is need that network mobility, handover from satellite to wireless network, and handover from wireless network to

S. Ata and C.S. Hong (Eds.): APNOMS 2007, LNCS 4773, pp. 551–554, 2007.

satellite network. For these handovers, we should consider all physical characteristics and handover algorithms together.

4 Collaboration of Satellite and Terrestrial Network

ETRI developing the two way satellite (DVB-RCS compatible) system. Figure 1 shows block diagram of the collaboration system. The system is based on DVB-RCS specifications and divided into two parts – hub station and group terminal (RCST : Return Channel Satellite Terminal). The hub station is composed of satellite network gateway subsystem(SNG) and Satellite Interface Subsystem(SIS). The terminal group is composed of Active Phased array Antenna subsystem (APA), Gap Filler Subsystem (GFS), Satellite Mobile Terminal (SMT) and Wireless Network Access Subsystem (WINAS).

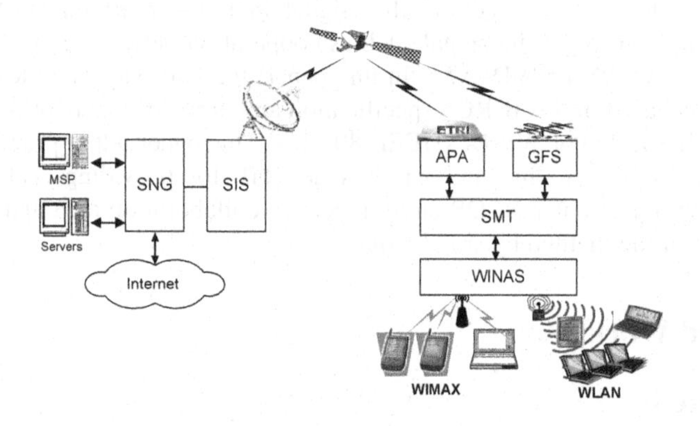

Fig. 1. Block diagram of collaboration system

In the hub station side, the SNG connects DVB-RCS system to terrestrial network and passes data from Multimedia Service Provider(MSP), Internet Service Provider(ISP) or broadcasting TV to the SIS for transmission to terminal and the SIS sends data to satellite or receive data from it. Each MSP which is located in hub station and ISP which are connected to internet use forward link for transmission data from hub station to group terminal. The SIS converts data for MPEG-2 TS and transmits them to group terminal. An end-user who is located in terminal station use return link for transmission data to hub station. The SMT converts data for ATM cell and transmits them to hub station.

In the terminal side, the APA provides send/receive interface in mobile terminal with using active phased array antenna. The GFS provides send/receive interface in tunnel area and the SMT works as does hub station. The WINAS does send/receive traffic from/to AP in rail station area (N-LOS).

Below is a basic transmission specification for high-speed moving object's Internet satellite-wireless cooperation system.

- Satellite Network
  -Forward Link : DVB-S, max. 80*Mbps*
  -Return Link : DVB-RCS, max. 10*Mbps*
- Terrestrial Network (WLAN or WIMAX)
- Gap filler Network (-ISM band or other wireless network specification)

5 Simulations

Most important thing to consider the Handover between satellite network and wireless LAN network is RTT difference between these networks. Using satellite network, end-to-end RTT is more than 600ms; on the other hand, WLAN has less than 100ms. These will reduce the performance of IP Handover. For this simulation, we also considered channel model and transport layer model.

1) Handover from satellite to terrestrial network
We simulated the TCP Throughput in handover from satellite to WLAN. The train entered the N-LOS area after 18seconds and compare the throughput between normal handover and fast handover. Figure 2. shows little throughput reduction over general IP handover. However, it is a handover among different types; it is not possible to have physical layer switching delay. Thus, if the delay time is larger than used link's router buffer, it occurs data loss.

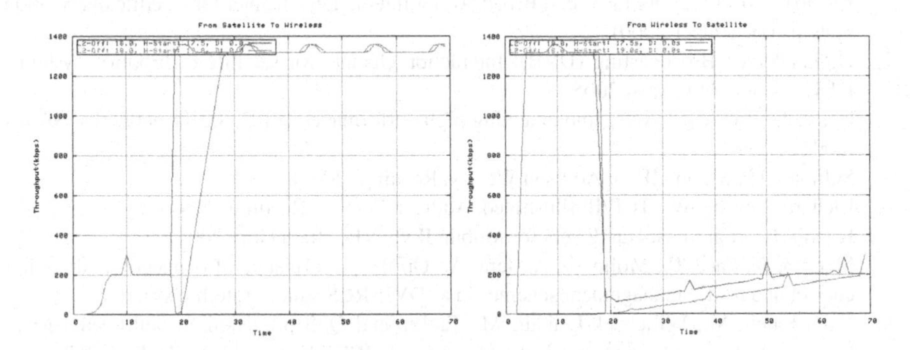

Fig. 2. Handover from 'Satellite to WLAN' and 'from WLAN to Satellite' (switching delay: 300ms)

2) Handover from terrestrial to satellite network
We simulated the TCP Throughput of system handovers from WLAN to satellite network. The train has 18seconds of link loss with PAR, and these graphs show the results of comparing fast handover with 500ms ahead to normal IP handover. There is TCP Time out even though we set the physical layer switching delay as 0 sec in handover from WLAN to satellite network. The reason why it has this time out is that transmission delay was increased or RTT was increased due to reducing of bandwidth. Because RTT value is used for retransmission timer, there will be more chance to have TCP congestion control in handover quicker network to slower network. In

the case of congestion control, we can minimize the throughput reduction by performing handover a little bit earlier.

6 Conclusion

This paper shows characteristics of two way satellite internet system. And we classified the NLOS environment into tunnel area and WLAN cooperation area. In addition that, we considered several factors and collaboration methods with satellite and wireless network to provide seamless service in each environment, Adopting brief Fast IP Handover algorithms, we simulated satellite to WLAN and WLAN to satellite handover. With these simulations, we put several variations on time difference, switching delay, starting time of brief fast handover protocol to get various throughputs. In conclusion, Fast Handover has less throughput reduction (at minimum 500kbps) over IP Handover. And Fast Handover with predicted mobility has minimum loss in the case of train that commute with fixed direction and performs handover among different networks. However, in case of wireless to satellite communication, rapid RTT increase should have TCP congestion control, it is better to have an additional TCP timeout-proof method besides Fast Handover.

References

1. EN 301 790 v1.2.2 Digital Video Broadcasting Interaction Channel for satellite distribution systems (December 2000)
2. Digital Video Broadcasting (DVB); Interaction channel for satellite distribution systems, ETSI EN 301 790 (April 2005)
3. Geier, J.: Wireless LANs; Implementing high performance IEEE 802.11 networks. SAMS (2002)
4. Soliman, H.: Mobile IPv6. Addison Wesley, Reading (2004)
5. Richard Stevens, W.: TCP/IP Illustrated. Addison Wesley, Reading (1995)
6. Koodli, R., et al.: Fast Handovers for Mobile IPv6, RFC 4068 (July 2005)
7. Marano, S., Pace, P., Molinaro, A., Iera, A.: On the performance of connection admission control and traffic management schemes in a 'DVB-RCS suited' satellite system
8. Naghshineh, M., Willebeek-Lemair, M.: End-to-end QoS provisioning multimedia wireless/mobile networks using an adaptive framework. IEEE Comm. Mag., 72–81 (1997)

Evaluation of Processing Load in the Network with DACS Scheme

Kazuya Odagiri[1,3], Rihito Yaegashi[2], Masaharu Tadauchi[3], and Naohiro Ishii[1]

[1] Aichi Institute of Technology, 1-38-1 Higashiyamadouri Chikusa-ku
Nagoya-city Aichi,Japan
odagiri@toyota-ti.ac.jp
ishii@aitech.ac.jp
[2] Shibaura Institute of Technology, 3-7-5 Toyosu Koutou-ku
Tokyo 135-8548
rihito@sic.shibaura-it.ac.jp
[3] Toyota Technological Institute, 2-12-1 Hisakata Tenpaku-ku
Nagoya-city Aichi,Japan
tadauchi @toyota-ti.ac.jp

Abstract. The study of DACS Scheme, which manages a whole network system by communication control on a client computer, has been advanced. The problem for applying DACS Scheme to practical network is that, processing load is heavy at the time of controlling communication. In this paper, we propose models for managing a whole network system; Course management type model (COMTM) and Client management type model (CLMTM). Each processing load which occurs at the time of controlling communication in these two models is compared. It is shown that, DACS Scheme which is CLMTM is advantageous in comparison with COMTM from the point of processing load, which occurs at the time of control communication.

Keywords: DACS Scheme, processing load, communication control.

1 Introduction

As the network management scheme for the efficient network management in the university network, DACS (destination addressing control system) Scheme has been proposed[1] [2] . However, when DACS Scheme is introduced into practical network, processing load for communication control by a user unit is heavy in comparison with communication control in existing network scheme. As methods of communication control by a user unit, there are a method of controlling the communication from the outside of local area network by SSL-VPN [3], and another method of controlling the communication between local area network and outside network such as Opengate[4][5]. In these methods, communication control by a user unit is performed partially. The study for the purpose of managing a whole network system by communication control by a user unit, isn't found besides DACS Scheme. Therefore, in this paper, we propose two models; Course management type model (COMTM) and Client management type model (CLMTM). Processing load which occurs at the time of

S. Ata and C.S. Hong (Eds.): APNOMS 2007, LNCS 4773, pp. 555–558, 2007.

controlling communication in these two models are compared. It is shown that, DACS Scheme which is CLMTM is advantageous in comparison with COMTM from the point of processing load, which occurs at the time of control communication.

2 Existing DACS Scheme

First, summary of DACS Scheme is explained. Fig.1 shows the basic principle of the network services by DACS Scheme. At the timing of the (a) or (b) as shown in the following, DACS rules are distributed from DACS Server to DACS Client.

(a) At the time of user's logging in the client
(b) At the time of a delivery indication from the system administrator

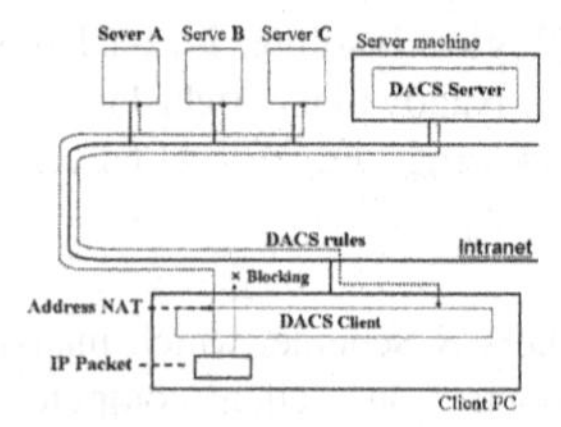

Fig. 1. Basic Principle of DACS Scheme

According to distributed DACS rules, DACS Client performs (1) or (2) operation as shown in the following. Then, communication control of the client is performed for every login user.

(1) Destination information on IP Packet, which is sent from application program, is changed.
(2) Packet from the client, which is sent from the application program to the outside of the client, is blocked.

3 Comparison and Calculation of Two Models

The model of managing a whole network system with communication control by a user unit on the network not introducing DACS Scheme is called COMTM. The model of managing a whole network system with communication control by a user unit on the network introducing DACS Scheme, is called CLMTM.

From here, processing load for communication control is calculated and compared. When the communication between a client software and a network service in these two models is performed, processing load as follows occurs.

(1) Processing by the network service on the server for the request from the client software
(2) Processing for communication control by a user unit
(3) Processing by the client software on the client

When the communication situation of two models is all the same, processing of (1) and (3) in two models is all the same. In Fig.2, processing load in COMTM is described. In Fig.2 and Fig.3, each network service is shown by the description of "NS1,NS2・・". Each client is shown by the description of "CL1,CL2・・". Processing load is shown by the description of "p1,p2・・".

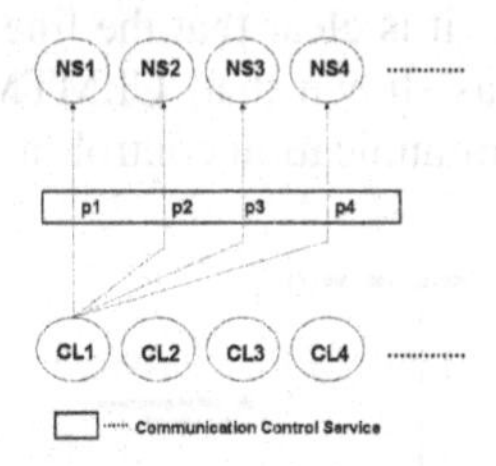

Fig. 2. Processing load in COMTM

Here, the point of (P1) is examined, and processing load for communication control is calculated. In Fig.2, processing load occurring in Communication Control Service by the communication between CL1 and each server is described. When the number of network services is assumed x, and the number of clients is assumed y, processing load occurring in Communication Control Service of COMTM is described as follows.

$$y \times (p1 + p2 + p3 + p4 + \cdots + px) = y \sum_{n=1}^{x} Pn \quad \cdots\cdots\cdots \quad \text{(a)}$$

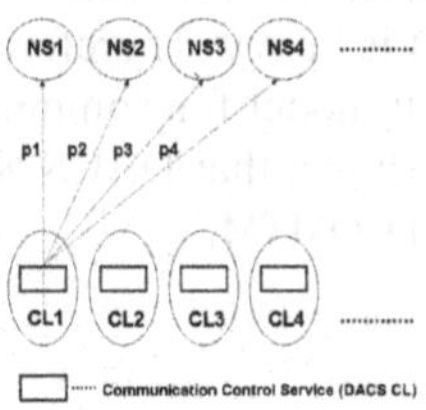

Fig. 3. Processing load in CLMTM

Next, processing load occurring in Communication Control Service of CLMTM is described in Fig.3. When the number of network services is assumed x, processing load by Communication Control Service in CLMTM is described as follows.

$$(p1 + p2 + p3 + p4 + \cdots + px) = \sum_{n=1}^{x} Pn \quad \cdots\cdots\cdots\cdots \quad \text{(b)}$$

Based on these calculation results, in both models, processing load for communication control is compared and examined. When $\sum_{n=1}^{x} Pn$ is assumed L and the upper limit of processing load in COMTM and CLMTM is assumed M(=kN) (k:fixed number) and N, the ratio of processing load spent against that upper limit (processing load rate) is shown as follows.

COMTM : $y \times L/M \times 100 = y/k \times L/N \times 100$ (%)
CLMTM : $L/N \times 100$ (%)

When the number of clients connected to network is added one by one, processing load rate in COMTM continues increasing to 100 of the upper limit as shown in (A) of Fig.4. As shown in (B) of Fig.4, processing load rate in CLMTM doesn't change as L/N (%) regardless of the number of clients. Then the point of (P2) is examined. When a client is used normally, it is clear that the line of (B) doesn't arrive at 100(%) of the upper limit value. It was shown that, CLMTM was more advantageous at a point of processing load for communication control in comparison with COMTM.

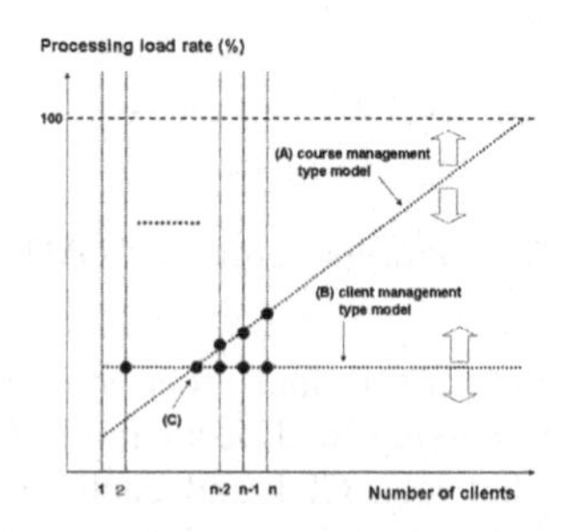

Fig. 4. Comparison of processing load

4 Conclusion

When DACS Scheme is introduced into practical network, processing load for communication control by a user unit is heavy. Therefore, two models were shown. Then, processing load occurring in each model for communication control is compared. As the result of comparison, it was shown that DACS Scheme, which was CLMTM, was advantageous in comparison with COMTM.

References

1. Odagiri, K., Yaegashi, R., Tadauchi, M., Ishii, N.: Efficient Network?Management System with DACS Scheme: Management with communication control. International Journal of Computer Science and Network Security 6(1), 30–36 (2006)
2. Odagiri, K., Yaegashi, R., Tadauchi, M., Ishii, N.: Secure DACS Scheme, Journal of Network and Computer Applications, Elsevier (to appear)
3. Shiraishi, Y., Fukuta, Y., Morii, M.: Port randomized VPN by mobile codes. In: CCNC, pp. 671–673 (2004)
4. Watanabe, Y., Watanabe, K., Hirofumi, E., Tadaki, S.: A User Authentication Gateway System with Simple User Interface, Low Administration Cost and Wide Applicability. IPSJ Journal 42(12), 2802–2809 (2001)
5. Tadaki, S., Hirofumi, E., Watanabe, K., Watanabe, Y.: Implementation and Operation of Large Scale Network for User' Mobile Computer by Opengate. IPSJ Journal 46(4), 922–929 (2005)

Empirical Testing Activities for NeOSS Maintenance

Dae-Woo Kim, Hyun-Min Lim, and Sang-Kon Lee

Network Technology Lab, Korea Telecom
463-1, Jeonmin-dong, Yuseong-ku, Daejeon 305-811, Korea
{daewoo,hmlim,sklee}@ kt.co.kr

Abstract. This paper describes the testing activities for the maintenance of the NeOSS (New Operations Support System) of Korea Telecom. To ensure the successful maintenance of the NeOSS without any effect on the existing functions and performance, we performed various tests related to functionality, efficiency and others before the added and modified parts were applied to the NeOSS. In this paper, we show the maintenance process, the various tests related to it, the test phases, and the test environment for controlling the quality of the NeOSS maintenance.

Keywords: Operations support system, Maintenance, Test phases, Test-bed, Testing activities.

1 Introduction

Generally, software maintenance activities can be classified into corrective, adaptive, perfective and preventive maintenance [1]. After a software is developed and released, the correction and modification items for it are continuously and the requests related to them are generally reduced as time passes [2].

However, in the case of the NeOSS as a telecommunication operations support system, the new and continuous requirements for emerging services and the evolution of the network do not reduce the additional functions for the system. Also, the new requirements, including the improvement of existing functions and the correction of errors, are requested by the users and operators of the system. In addition, because the NeOSS was developed in the MS (Microsoft) .Net framework, whenever MS patches are released, the impact of the MS patches needs to be checked through functionality and performance testing in each occasion. That is, the maintenance activities of the NeOSS focus on corrective, perfective and preventive maintenance. Also, by continuously adding new emerging services to the NeOSS, the system gradually becomes larger.

Therefore, in order to successfully ensure the maintenance of the NeOSS, we need to check if the maintenance activities affect the existing system through verification and validation processes. If the newly added functions and error corrections create problems in the NeOSS, it will adversely affect KT's business.

In this paper, we dealt with the testing activities for the NeOSS maintenance to ensure the success of maintenance as a case study.

S. Ata and C.S. Hong (Eds.): APNOMS 2007, LNCS 4773, pp. 559–562, 2007.

2 Testing Activities for Maintenance

The NeOSS development period was three years. After the first version was developed, we performed the test and evaluation activities to continuously improve the quality of the NeOSS for two years. Functional testing was performed for 52 times.

We maintained the number of test cases until the 23rd test and then we decreased the number of test cases. In general, this is because it became increasingly laborious and time-consuming to go back and check every possible effect every time a small change is made. But this sharply reduced the testing coverage for the system functions and led to hidden problems in finding defective processes.

In addition to the above, we need to take the following into consideration in NeOSS maintenance:

- Continuous checking of the performance impact of the additional functions, modification of existing functions and correction of errors, and additional new services.
- Continuous checking of the impact of MS product patches for the MS.Net , MS SQL, Windows 2003 Server and others on functionality and performance.

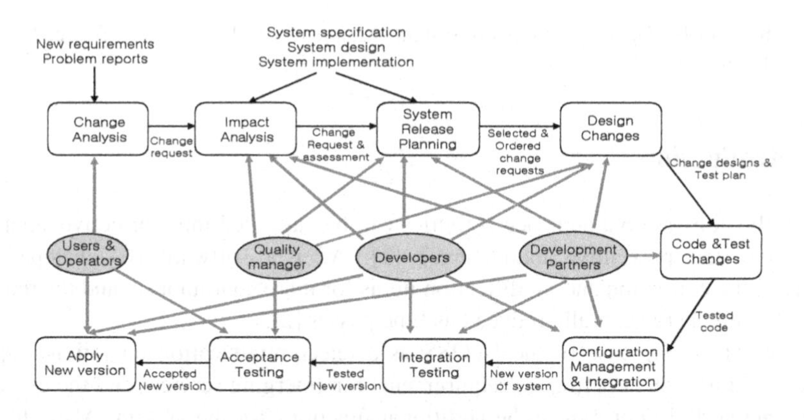

Fig. 1. The maintenance process for the NeOSS

The testing activities to ensure the quality management of NeOSS maintenance are as follows: Figure 1 shows the maintenance process for the NeOSS, which consists of the activities and the stakeholders related to its maintenance. It refers to the maintenance reference model [3] [4]. In Figure 1, the activities from Change Analysis to System Release Planning are related to the handling of new requirements such as error correction, the new functional requirements, and the new service requirements, which come from the users, operators and business departments.

Figure 2 shows the testing phases for maintenance, which consist of three phases. The first phase is where the developers and development partners test the new or modified functions through a unit test and a integration test. In the second phase, the new version of the NeOSS is tested by the quality department, which is comprised of experts in NeOSS operations. In addition, they perform the efficiency test for the added and modified functions to check if their performance affects the performance of the NeOSS. And the regression test is performed on the main functions of the NeOSS to check if the existing functions of the NeOSS are affected by the added and modified functions with an automatic testing tool. In the third phase, the representative of the users and operators finally performs an acceptance testing on the added and modified functions of the new NeOSS version.

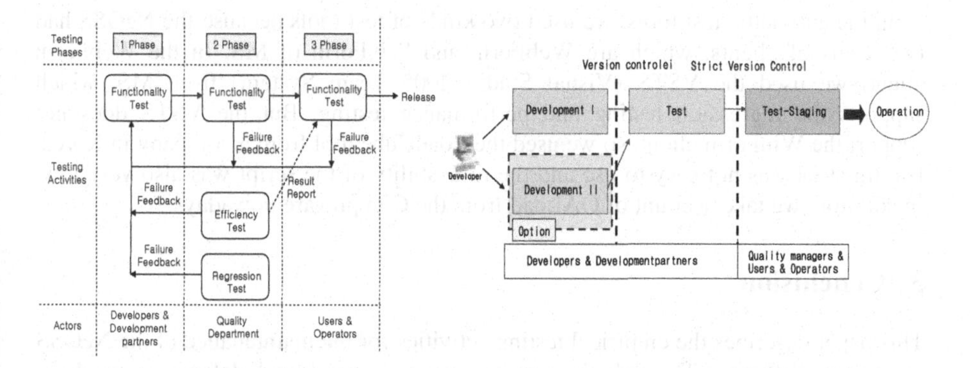

Fig. 2. The testing phases for maintenance

Fig. 3. The test-bed

Figure 3 shows the test-bed for the maintenance [5]. As shown in Figure 1, the developers and development partners develop the added and modified functions and test them through a unit test in the Development I environment. They perform the integration test for the NeOSS with the added and modified functions in the Test environment. After that, the new version of the NeOSS is controlled by the quality manager and applied to the test-staging environment, where the quality managers perform functional testing, efficiency testing and regression testing on it, and then lastly, the representative of the users and operators test it. In particular, the Test-Staging environment is usually the same version as that in the operation environment. The reason why the new version of the NeOSS is first applied to the Test-Staging before it is applied to the operation environment is to find the hidden problems that can take place. The Development II environment is for the next version of the NeOSS, which has a big difference from the operational version of the NeOSS in order not to affect the maintenance version. This environment is optional and only made when they are needed.

The tests for maintenance of NeOSS include the functionality, efficiency and regression, and the MS patch testing. The functionality testing is for the functions of the new requirements from the users and operators, the added functions for the new service, and the correction of error. The efficiency testing is performed to check if the added functions, for the new services in particular, affect the performance of the existing NeOSS. The load of the efficiency testing is about 8% of the load in the operating environment because the scale of the Test-staging is about 8% of the operating environment. And this testing includes performance test for the response time and processing time, load test for the maximum number of simultaneous users, stress test for the overloading, and stability test for stability under peak load. The regression testing is to check if the added and modified functions and the MS patches have any impact on the existing functions of the NeOSS. This test is performed through automatic tools to reduce the manpower and test time. The MS patches related to the NeOSS are for the MS.Net framework for the development of the system, MS-SQL, MS BizTalk and Windows 2003 Server. Whenever a new version of the MS patches is released, its reliability and stability for the NeOSS must be checked before it is applied to the NeOSS. The test cases for the MS patches consist of the main functions of the NeOSS.

In the automatic test tools, we used two kinds of test tools because the NeOSS had two kinds of clients, which are WebForm and WinForm of MS. In the WebForm client, we used the VSTS (Visual Studio 2005 Team System) from MS, which supports the regression testing and performance testing. But the VSTS does not support the WinForm client, so we used the LoadCube tool from a company in Korea. But this tool was not easy to use and the re-usability of the script was also very low. In addition, we take account of QALoad from the Compuware nowadays.

3 Conclusion

This paper describes the empirical testing activities for the maintenance of the NeOSS as a large software. Through the testing activities, we found defects as much as possible and could ensure the stability and reliability of the NeOSS in the system maintenance before the error corrections, the new requirements of the users and operators, the new services from the business departments, and the application of MS patches are applied to the NeOSS. Also, we performed regression testing to increase the coverage of the tests for the NeOSS using the testing tools.

In the near future, we need to study the following areas:

1. Expanding the regression testing to the business flow to further increase the coverage of the tests for the NeOSS
2. The testing method for the abnormal DTC (Distributed Transaction Coordinator) increased in operating environment because the added and modified functions sometimes abnormally affect the NeOSS
3. The analysis for the types of defects and the added and modified functions. The result of the analysis would be applied to the maintenance process to improve it.

References

1. Takang, A.A., Grubb, P.A.: Software Maintenance: Concepts and Practice, Department of Computer Science, University of Hull, International Thomson Computer Press (1996)
2. Burch, E., Kung, H.-J.: Modeling Software Maintenance Requests: A Case Study. In: ICSM'97. Proceedings of the 13th International Conference on Software Maintenance (1997)
3. Layzell, P.J., Macaulay, L.: An Investigation into Software Maintenance Perception and Practices. Journal of. Software Maintenance and Practice 6(3), 105–120 (1994)
4. Pfleeger, S.L., Bohner, S.A.: A Framework for Software Maintenance Metrics. In: Proceedings of Software Maintenance 1990, pp. 320–327 (November 1990)
5. Kim, D.-W., Lim, H.-M., Lee, S.-K.: Test-bed for Verification and validation Activities in Developing an Operations Support System. In: SERP'06. Proceedings of the International Conference on Software Engineering Research & Practice, vol. 1, pp. 45–51 (2006)

A Study on Service Problem Management and Resource Trouble Management on a Telecommunication Network

Byeong-Yun Chang, Hyeongsoo Kim, Seongjun Ko, and Daniel Wonkyu Hong

KT Network Technology Lab
463-1, Jeonmin-dong, Yuseong-gu,
Daejeon, South Korea
{bychang,essence,sun1go,wkhong}@kt.co.kr

Abstract. We study service problem management and resource trouble management on a telecommunication network with eTOM (Enhnanced Telecom Operations Map). As results of this research, we derive primary necessary functions for process elements of eTOM and compare them with KT NMS functions. In addition, we propose an improvement direction of KT network operations management.

Keywords: eTOM, network operations management, service problem management, resource trouble management.

1 Introduction

Nowadays, the telecommunication industry and markets in the world are changing very fast. In the view of markets and technology changes, closed and partial environments are changing to open environments, the circuit and voice network based on PSTN to IP and data network, wireline to wireless. In addition, in the view of service the environment based on service providers is changing to the environment based on customers like service level agreement, quality of service. [1]

In the customer-based business environment, we can consider price, quality and the variety of services as the most important factors. Among these factors this paper considers network operations management related to the quality of telecommunication network.

As result of this research, we give primary necessary functions for telecommunication network problems operations management and compare them with KT NMS functions. These results are derived with the deep experience of network operations management along with studying through eTOM. We also propose an improvement direction of KT network operations management.

In the next section we analyze the network problems operations management in KT. Then in the section 3 we derive primary necessary functions and compare them with KT NMS. In the section 4 we propose an improvement direction of KT network operations management. In the final section we summarize our results in this research.

S. Ata and C.S. Hong (Eds.): APNOMS 2007, LNCS 4773, pp. 563–566, 2007.

2 Network Problems Operations Management in KT

In this section we analyze the current network problem operations management process of KT. In KT, There are two routes to detect network problems. One is from VOC (voice of customer) which is from customers. The other is from Network Management Systems (NMS). For VOC, NeOSS SA which is a system in KT takes the problem claims. Then it orders Netis to create trouble tickets (TT). After Netis creates TT, Network Service Center (NSC) deals with problems. For the other, NSC monitors alarm event. When alarm is created from NMS, then NSC deals with that alarm event. Figure 1 describes network problems operations management in KT

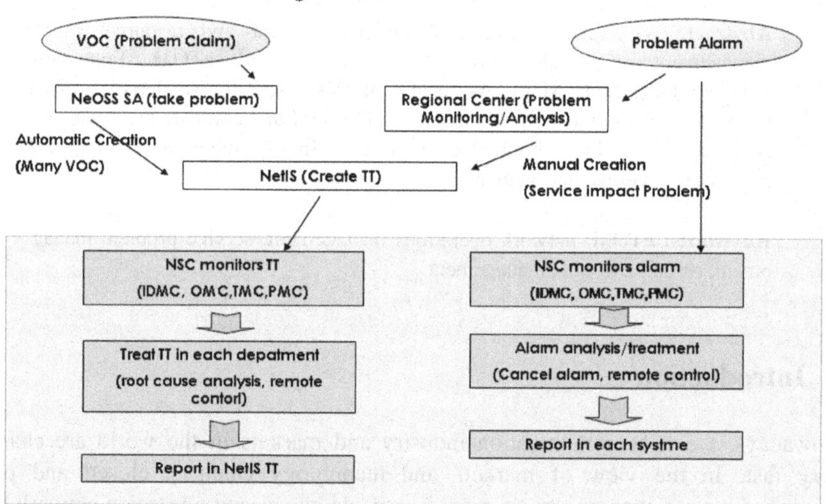

Fig. 1. Network Problems Management Process in KT

3 eTOM and Primary Necessary Functions for Network Problems Operations Management

In this section we review eTOM and derive primary necessary functions for process elements of eTOM. eTOM is an element of NGOSS (Next Generation Operations Systems and Software) [2]. It defines all major business processes in service providers. It is also recognized as a national standard in ITU-T M.3050. It gives the framework and common language of business process which are used in the telecommunication industry. In addition, it could be used to classify the telecommunication business process or service providers and system integrator can communicate with business process framework[3,4].

In eTOM, we focus on service problem management and resource trouble management (SP&RT) in Assurance process of Level 2 process of eTOM. To derive primary necessary functions for network problems operations management, we first understand and draw process flow diagrams of process elements in SP & RT management processes based on deep experience of network operations management. After analyzing roles and responsibilities of the process elements in SP&RT, we derive these primary necessary functions as in Table 1. In Table 1, we also compare KT NMS with these functions. In Table 1 'A' means that KT NMS needs automatic function. '0' means that KT NMS has these functions.

Table 1. Primary necessary functions for SP&RT Management Process Elements

Resource Trouble Management

Process Elements	Primary Necessary Functions	KT NMS
Survey & Analyze Resource Trouble	Alarm correlation analysis	A
	Alarm filtering	0
	Alarm record management	0
	Real time alarm status monitoring	0
	Service impact alarm management	A
	Inform service impact alarm and change	A
	Real time alarm table	0
	Inform urgent problem list (SMS)	0
	Hearing real time alarm	0
Localize Resource Trouble	Diagnose and test resource trouble	0
	Aanaize root cause	A
Correct & Resolve Resource Trouble	Inform control task list	0
	Remote troubeleshooting	0
Track & Manage Resource Trouble	Monitor current RTR status	0
	See/modify/add/change RTR status	0
	See RTR in emergency	0
	Allocate problem shooting task	0
Report Resource Trouble	Manage and inform RTR status	0
	Create various RTR statistics and report	0
	Provide report	0
Close Resource Trouble	RTR CLOSE	0
Create Resource Trouble Report	Create RTR	0
	Modify/cancel RTR	0
	Estimate recovery time	0

Service Problem Management

Process Elements	Primary Necessary Functions	KT NMS
Create Service Trouble Report	Create SPR	0
	Estimate recovery time	0
Diagnose Service Problem	Diagnose and test service trouble	0
	Analyze root cause	A
Correct & Resolve Service Problem	Remote troubleshooting	0
	Automatic recovery	A
Track & Manage Service Problem	Monitor current SPR status	0
	See/modify/add/change SPR status	0
	Allocate problem shooting task	0
	See SPR in emergence	0
Report Service Problem	Monitor current SPR status	0
	Provide current SPR status	0
	Inform completion of SPR	0
	Create various SPR status and report	0
	Infrom SPR change	0
	Provide report	0
Close Service Trouble Report	SPR close	0
Survey & Analyze Service Problem	Alarm correlation analysis	A
	Alarm record management	0
	Real time alarm status monitoring	0
	Customer impact alarm management	A
	Inform customer impact alarm to CRM process	0
	Service problem impact analysis	A
	Infrom urgent problem list (SMS)	0

4 Improvement Directions of KT Network Problems Operations Management

In the process of analyzing eTOM, obtaining primary necessary functions for network problems operations management and comparing KT NMS with these functions in section 3 we establish improvement directions of KT network problems operations management.

First KT network problems operations management processes need to improve service problems alarm route structure and enhance the functions of RM&O layer in eTOM. To improve service problems route structure we need to give information of service problems caused by resource troubles to NeOSS-SA which is in SM&O layer in KT. Enhancement of the functions of RM&O indicates that RM&O layer in KT need to have functions to support functions of SM&O layer and E2E network management.

The Figure 2 represents the network problem operations management architecture in KT. In Figure 2, RTR and RPDR indicates resource trouble report and resource performance degradation report, respectively.

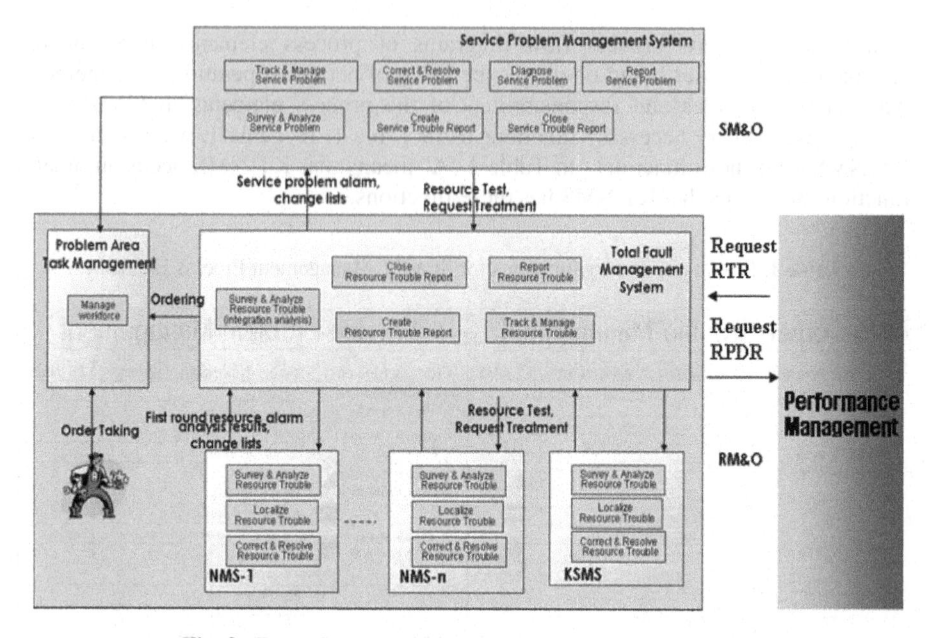

Fig. 2. Future KT network problems operations management

5 Conclusions

In this paper we study service problem management and resource trouble management in a telecommunication network. We derive primary necessary functions for telecommunication network problems operations management and compare them with KT NMS. In addition, we also introduce KT network problems operations management architecture with intuition while we study service problem management and resource trouble management in a telecommunication network. These results could be used as a reference for process improvement in telecommunication operations management.

References

1. Choi, M.J., Kim, D.H., Hong, J., Choi, H.S.: Analysis of NGOSS Implementation Example. KNOM Review 6(2), 21–32 (2004)
2. Reilly, J.P., Creaner, M.J.: NGOSS Distilled. Cambridge University Press, Cambridge (2005)
3. TM Forum: Enhanced Telecom Operations Map (eTOM). GB 921, Release 6.0 (2005)
4. TM Forum: Enhanced Telecom Operations Map (eTOM). GB 921D, Release 6.0 (2005)

Distributed and Scalable Event Correlation Based on Causality Graph

Nan Guo, Tianhan Gao, Bin Zhang, and Hong Zhao

Information Science and Engineering College, Northeastern University, China 110004
{guonan,gaotianhan,zhangbin,zhaoh}@ise.neu.edu.cn

Abstract. The traditional centralized event correlations are less of simplicity, scalability and robustness. DSEC is presented to solve this problem. It is based on the Divide-and-Conquer strategy that makes parallel local correlation at multiple agents and combines local causes on the global level. It regards a management task as an event probe, thus brings high-level and rich-semantic information to events and achieves the simplicity and scalability of causality graph. Meanwhile, it limits event explosion at each agent and improves robustness to noise.

Keywords: distributed network management, event correlation, causality graph, divide-and-conquer.

1 Introduction

To solve the problem of event explosion, a number of event correlation systems have been proposed and implemented [1-9], which vary widely in the underlying representation structures used to model the relationships between events and consequently the algorithms used in their correlation engines. However, there are three problems remained: (1) less of simplicity. The events are traditionally structured of low-level and less semantic information, which increase the complexity of correlation; (2) less of scalability. Event probes are mostly statically preinstalled at agents; both events and their relations can not be updated dynamically; (3) less of robustness. The centralized event correlations are inevitably sensitive to the noise (such as loss, delay, etc.) in an event stream.

To solve these problems, Distributed and Scalable Event Correlation (DSEC) is presented in context of the distributed network management in which management tasks are assigned to agents to monitor event locally.

2 DSEC Principle

The premise of DSEC is regarding a management task as an event probe. Thus, events comprise high-level and rich-semantic information; meanwhile, the event correlation is scalable as the management tasks can be updated on-the-fly.

S. Ata and C.S. Hong (Eds.): APNOMS 2007, LNCS 4773, pp. 567–570, 2007.

The principle of DSEC is Divide-and-Conquer strategy, which indicates it is often easier to solve several small instances of a problem than one large one. DSEC divides event correlation into two phases: Local Correlation and Global Correlation.

- Local Correlation: making correlation of the events emitted locally to identify a local cause.
- Global Correlation: collecting local causes from a given domain and making correlation of them in the view of network topology.

Either Local Correlation or Global Correlation is processed by the same algorithms but the different causality graphs. We define a complete causality graph as composed by the all events which are the outputs of all the management tasks of a network management system. Based on divide-and-conquer, DSEC cuts the complete causality graph into some particular local causality graphs corresponding to the assignment of management tasks and sends each local causality graph to the corresponding agent. As management tasks are extended to an agent to implement more complicated function, the local causality graph is easy to be scaled; it will be expanded as tasks extended as well as shrunk as tasks retracted.

3 DSEC Algorithms

Because a management task is regarded as an event probe, each event in a causality graph can be physically detected by the corresponding management task. In the run-time of correlation, events (instances) can be strictly classified in three categories:

- Reported Event: indicates the observable event; it is the output parameter of the corresponding management task.
- Inferred Event: indicates the indirectly observable event which has occurred; it may be either lost or failed to emit because of disabled host (such as management task or physical resource).
- Unrelated Event: indicates the events not affected by the occurrent problem.

DSEC includes a series of five algorithms as following: (The effects of event e is described as ImmEff(e) = {e' | e → e'}. G_{GC} is indicated the global causality graph of all the events and their causal relations of a given network management system. G_{LC} is indicated the local causality graph of an agent.)

```
Extracting Algorithm
  Add E to V(G_LC);
  /* E is the set of the events attached with the
  management tasks INITIALIZED to a given agent */
  If ∃e((e ∈ E) ∧ (ImmEff(e) ⊆ E))
    Add {e→ImmEff(e)} to E(G_LC);
```

The Extracting Algorithm is to decide which edges, presenting cause-effect relations between the events attached with the management tasks initialized to a given agent, of a global causality graph will be extracted to construct a local causality graph.

```
Expanding Algorithm
  Add Ee to V(GLC);
  /* Ee is the set of the events attached with the
  management tasks EXTENDED to a given agent */
  If ∃e((e ∈ V(GLC)) ∧ (ImmEff(e) ⊆ V(GLC))
    Add {e→ImmEff(e)} to E(GLC);
```

The Expanding Algorithm is to decide which edges, presenting cause-effect relations between the events attached with the management tasks extended to a given agent and the previous events of $V(G_{LC})$, of a global causality graph will be extended to a local causality graph.

```
Shrinking Algorithm
  Delete {e→ImmEff(e)|e ∈ Er} from E(GLC);
  /* Er is the set of the events attached with the
  management tasks RETRACTED from a given agent */
  Delete {e'→ImmEff(e')|(e'→e) ∈ E(GLC) ∧ (e ∈ Er)}
  from E(GLC);
  /* delete the relationships of each event of Er
  respectively in the case of being a cause or an effect
  */
  Delete Er from V(GLC);
```

The Shrinking Algorithm is to decide which edges, presenting cause-effect relations between the retracted events attached with the management tasks retracted from a given agent and the remained events of $V(G_{LC})$, of the local causality graph will be deleted.

```
Local Correlation Algorithm
  If ∃e((e∉S) ∧ (ImmEff(e) ⊆ S))
  /* S is the set of the reported events emitted within
  a given agent */
    Add e to S;
  LocalCause(S) = e|((e∈ S) ∧ (¬∃e'((e' ∈ S) ∧
  (e'→e) ∈ E(GLC)));
```

The Local Correlation Algorithm is to deduce local cause in the view of a given local causality graph for a set of reported events.

```
Global Correlation Algorithm
  If ∃e((e∉R) ∧ (ImmEff(e) ⊆ R))
  /* R is a set of the local causes collected from a
  given domain of agents */
    Add e to R;
  RootCause(R) = e|((e ∈ R) ∧ (¬∃e'((e' ∈ R) ∧
  (e'→e) ∈ E(GGC)));
```

The Global Correlation Algorithm is to make final reasoning of local causes in the view of network topology (presented in global causality graph).

Specially note that, DSEC is able to deal with the event relationship of not only logical conjunction but also disjunction. Concerning the isolating nodes of any local causality graph when extracted, expanded, and shrunk, they will be deleted because they are not related to any other events. Besides, to event e affected by more than one causes, i.e. $d^-(e) > 1$, it needs to be sent together with local cause to management station to enjoy global correlation in the case that its causes are distributed to the different agents.

4 Conclusion and Future Works

DSEC is proposed as an event correlation mechanism working effectively in the distributed network management. It decomposes the complexity of event correlation into multiple correlation engines working on local and global network fault management levels, and improves robustness to noise. We are implementing the prototype of DSEC in the distributed network management test bed of Northeastern Network Center of CERNET (China Education and Research Network). The experimental data by now proves that DSEC can figure out the problem with the maximum probability in a given simulated event stream. Our focus is on enhancing the power of DSEC from three aspects: dealing with "NOT" relation, establishing temporal relationships and associating a probability with a relationship.

References

1. Masum, H., Binay, S., Ramesh, V.: A Conceptual Framework for Network Management Event Correlation and Filtering Systems. In: The Sixth IFIP/IEEE International Symposium on Integrated Network Management, Boston, USA, pp. 233–246. IEEE Computer Society Press, Los Alamitos (1999)
2. Minaxi, G., Mani, S.: Preprocessor Algorithm for Network Management Codebook. In: The first Workshop on Intrusion Detection and Monitoring, California, USA (1999)
3. Subramanian, M.: Network Management Principles and Practice. Pearson Education Press, USA (2001)
4. Rajeev, G.: Layered Model for Supporting Fault Isolation and Recovery. In: IEEE/IFIP Network Operations and Management Symposium, Hawaii, USA, IEEE Computer Society Press, Los Alamitos (2000)
5. Boris, G.: Integrated Event Management: Event Correlation Using Dependency Graphs. In: The Tenth IFIP/IEEE International Workshop on Distributed Systems: Operations & Management, Zurich, Switzerland, IEEE Computer Society Press, Los Alamitos (1999)
6. Tiffany, M.: A Survey of Event Correlation Techniques and Related Topics (May 2002), http://www.tiffman.com/netman/netman.html
7. Yemini, S., Kliger, S., Mozes, E.: High Speed and Robust Event Correlation. IEEE Communications 34(5), 82–90 (1996)
8. Lo, C.-C., Chen, S.-H., Lin, B.-Y.: Coding-Based Schemes for Fault Identification in Communication Networks. International Journal of Network Management 10(3), 157–164 (2000)
9. Mouayad, A., Bruce, B., Evens, M., et al.: A Framework for Event Correlation in Communication Systems. In: Proceedings of the 4th IFIP/IEEE International Conference on Management of Multimedia Networks and Services: Management of Multimedia on the Internet, pp. 271–284. IEEE Computer Society Press, Los Alamitos (2001)

Detection and Identification of Neptune Attacks and Flash Crowds

The Quyen Le[1], Marat Zhanikeev[2], and Yoshiaki Tanaka[1,3]

[1] Global Information and Telecommunication Institute, Waseda University
1-3-10 Nishi-Waseda, Shinjuku-ku, Tokyo, 169-0051 Japan
[2] School of International Liberal Studies, Waseda University
1-21-1 Nishi-Waseda, Shinjuku-ku, Tokyo, 169-0051 Japan
[3] Research Institute for Science and Engineering, Waseda University
17 Kikuicho, Shinjuku-ku, Tokyo, 162-0044 Japan
quyenlt@fuji.waseda.jp, maratishe@aoni.waseda.jp, ytanaka@waseda.jp

Abstract. Neptune attack and Flash Crowd are two typical threats to web servers. These two anomalies have many identical features that make them difficult to distinguish. In this paper, we propose a statistical packet-based method to detect Neptune attacks and Flash Crowds and more importantly, by performing separate analysis by source address aggregation, we also propose additional efficient means to differentiate these two similar anomalies.

1 Introduction

Nowadays, web services have become an indispensable part of the Internet society. Sometimes, Flash Crowds may occur when the number of legitimate requests to a server increases much faster than in normal condition. Besides, a server may become a target for attackers who are likely to try Neptune or TCP SYN attack to disable the service. J. Jung et al. [1] also proposed a research on Flash Crowds and DoS attacks which requires detailed analysis of Web logs of HTTP requests. We propose a packet-based detection method for Flash Crowds and Neptune attacks based on some differences in their nature to distinguish them. Traffic traces containing Neptune attack (created by attack tool) and Flash Crowd (provided by the Internet Traffic Archive [2]) are used to verify the proposed method.

2 Detection Schemes

In this section, we present a statistical detection scheme based on the fact that the number of requests (or connections) to a specific destination grows much faster than in normal condition when either Neptune attack or Flash Crowd occur. Therefore, we count the number of TCP SYN and FIN packets in the traffic to detect these anomalies. If all connections are normal, the subtraction of total number of SYN and FIN packets will be zero. But if Neptune attack or critical Flash Crowd occurs, we will see this difference growing to an unusually

S. Ata and C.S. Hong (Eds.): APNOMS 2007, LNCS 4773, pp. 571–574, 2007.

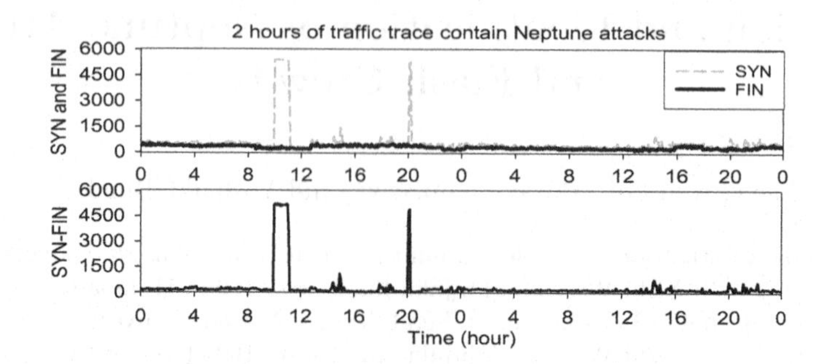

Fig. 1. Counting SYN and FIN packets

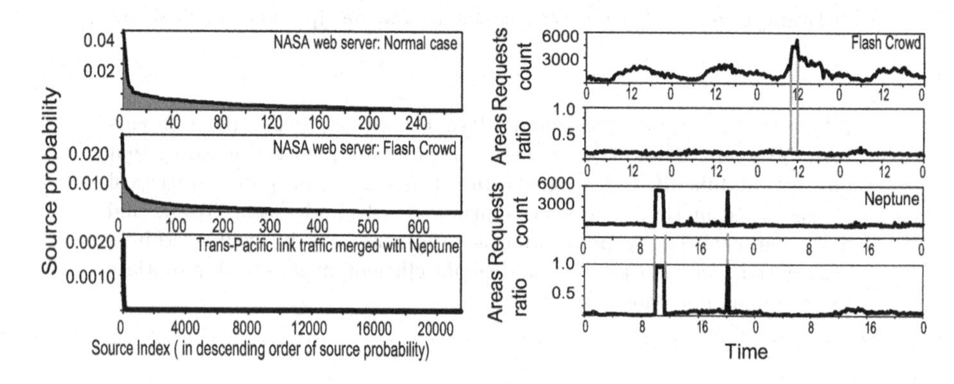

Fig. 2. Probability distribution of source addresses

large number because there are no FIN packets offered for any connection. Fig. 2 shows the charts of the total number of SYN, FIN packets and SYN-FIN counted every 1 minute from the traffic containing Neptune attacks.

3 Distinguishing Neptune Attacks and Flash Crowds

First, we calculate the probability that a source will be found among server requests. The results are displayed in the left chart of Fig. 3 in the order of decreasing probability of source address occurrence.

In case of normal access and Flash Crowd, the smoothness of the curve tells us that many users access the web server more than once even in a short time. User community includes enthusiastic users that explore many webpages (hight probbility) and curious or rare access users (low probability). This chart indicates a typical behaviour of legitimate web service user community. We noticed that, because of the sufficient number of regular users (higher source probability) in the user community, the areas covered by the plot line (dark area in the graphs) are clearly visible. However, in case of Neptune attack, we can see that the distribution now is clearly different when the curve has a sharp breakpoint close to the (0, 0) root point. There is a small set of legitimate source IPs in

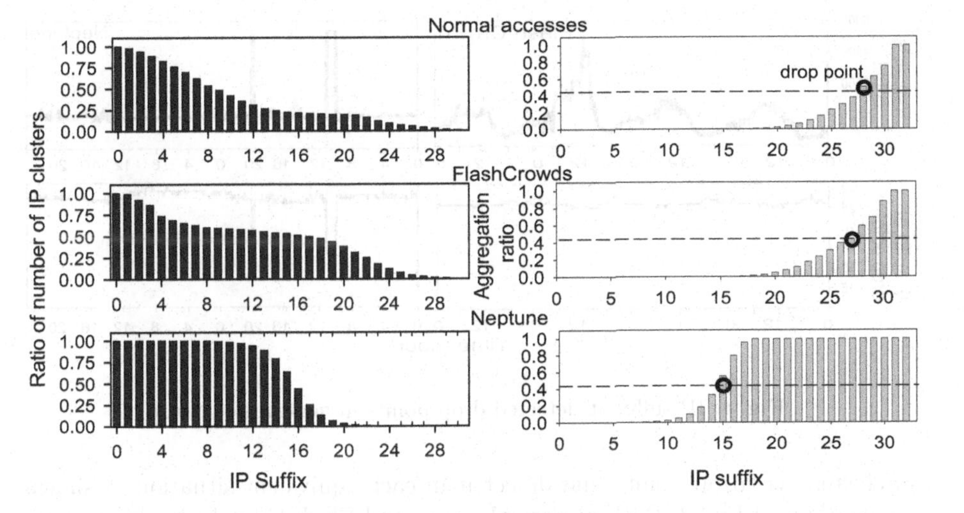

Fig. 3. Statistical analysis of source address aggregation

the lowest index area that has high probability. The rest with low probability are sources spoofed by the Neptune attack. Therefore, the area covered by the plot line is very small and almost invisible. This means that when Neptune attack occurs, the plot line area suddenly become very small and this change is significant enough to distinguish Neptune attacks from Flash Crowds. Therefore, we calculate the ratio of the graph area and plot line area (regardless of the number of distinct sources) with notice that when Neptune attacks occur, this value will become very large. The right charts of Fig. 3 shows the results with distinguishable signals for Flash Crowd and Neptune.

In the second analysis, we count the number of distinct source networks (clusters) while decreasing the number of network bits (IP prefix), or increase the number of host bits (IP suffix) in IP address to see how server sources are aggregated. We calculate the decreasing ratio of the number of clusters from IP suffix /0 to IP suffix /31. The results for the cases of normal access, Flash Crowd and Neptune attacks are displayed on the left of Fig. 3. In normal accesses and Flash Crowd, the number of clusters decreases gradually as the IP suffix increases because after each step of IP suffix, there is a sufficient number of closely located IP addresses that are gathered into same clusters. This is the nature of the web server user community distribution. However, in Neptune attack, the chart retains high values (close to 1) and then quickly drops at around IP suffix 13 to 16. The reason for this is that all the source IPs in Neptune attack are randomized so they scarcely share the same networks at small IP suffixes. The number of clusters gets closer to the maximum number of cluster allowed ($2^{IPprefix}$) as IP suffix grows. According to our experiments, when the aggregation ratio (number of clusters divided by maximum number of cluster allowed) reaches 0.4, the number of clusters will decrease as fast as the maximum number of clusters allowed (2 times smaller for every step of IP suffix). We refer to this drop of number

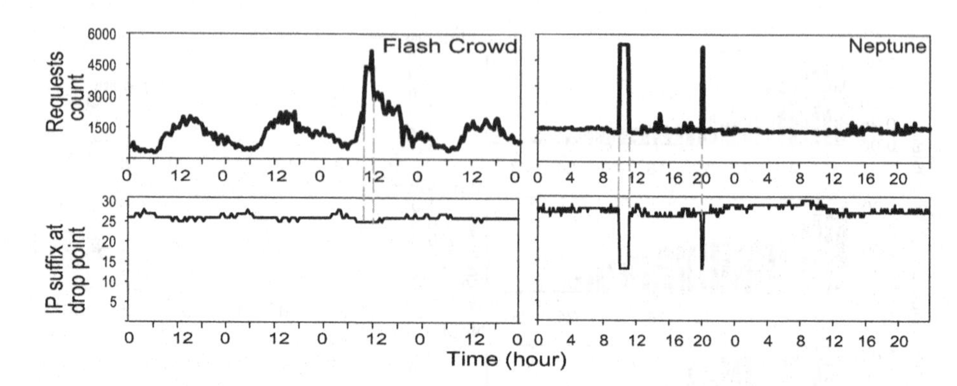

Fig. 4. IP suffix at detected drop points in aggregation ratio

of clusters as "drop point" and detect it in each equivalent situation as shown on the right of Fig. 3. Both in normal access and Flash Crowd, the drop points are detected at IP suffixes around 27 and 28 as a result of source aggregation of the same web user community. But when Neptune attack occurs, the drop point moves much further to the left and is detected at IP suffix 15.

Therefore, statistically, we can detect and distinguish Neptune attacks from Flash Crowds by detecting the IP suffix of drop point in aggregation ratio. If the IP suffix of the drop point suddenly decreases, it obviously signals the occurrence of Neptune attack. Fig. 4 shows the chart for IP suffix of drop point along with traffic time. We can see that Neptune attack can be easily detected and correctly distinguished from Flash Crowd.

4 Conclusion

In this paper, we presented a method for detecting and identifying Neptune attacks and Flash Crowds. The packet-based detection method can be used with high speed network connections and provides accurate results. Then we applied further analysis using the unique features of these anomalies to distinguish them effectively. These analysis methods provided results with clear differences between these two kinds of anomalies regardless of whether a Flash Crowd was non-congested or critical.

References

1. Jung, J., Krishnamurthy, B., Rabinovich, M.: Flash Crowds and Denial of Service Attacks: Characterization and Implications for CDNs and Web Sites. In: Proc. Int'l World Wide Web Conference, pp. 252–262. ACM Press, New York (2002)
2. ClarkNet and NASA Web server logs are available on the Internet Traffic Archive, `http://ita.ee.lbl.gov/html/traces.html`

Deploying Application Services Using Service Delivery Platform (SDP)

Jae Hyoung Cho, Bi-Feng Yu, and Jae-Oh Lee*

Information Telecommunication Lab,
Dept. of Electrical and Electronics Engineering,
Korea University of Technology and Education, Korea
{tlsdl2,yubifeng,jolee}@kut.ac.kr

Abstract. Nowadays in communication and network businesses, architectures and systems are independently constructed with existing service or new service deployment without considering common functions and services between them. Consequently, the study of standardized Service Delivery Platform (SDP) structure and its related interfaces is needed to provide the methodology of existing or new services in an efficient way. Therefore, in this paper, we intend to view various solutions about SDP structure with the proposed interfaces which are able to support them, and then recommend various application services which SDP can offer to end users.

1 Introduction

Currently, some services development and deployment methodologies provided by Service Providers are made up of the form of separated vertical service structures. And these may cause complication, intensiveness of resource and cost increasing of the management. Therefore, we are in face of a platform that can develop, integrate and preserve the horizontal common services to eliminate these complications. This anticipant platform enables the IP Multimedia Subsystem (IMS) structure to provide the control of service consistently and integrate wire and wireless network to meet those demands. The IMS calls for the creation of SDP that can develop new service based on existing service of the IMS with valued additional services [1].

Service providers care about not only the appropriate cost of the new innovative services but also the integration of the existing services. They demand their environment which can deliver these services in the IMS. Therefore, SDP satisfactorily provides open valued services for service providers to exploit the various types of user devices. Nowadays there isn't a common standard structure about SDP. So, this paper carries out the research about various interfaces of SDP structure. In section 2 we present the structure for SDP in detail and then in section 3 make its usefulness through one network emulator and presence service. Finally in section 4, we make some conclusions.

* Corresponding author.

S. Ata and C.S. Hong (Eds.): APNOMS 2007, LNCS 4773, pp. 575–578, 2007.

2 Composite Services and Interfaces

Fig. 1 represents the SDP composition elements and interfaces in detail. The IMS services are able to access the SDP through SIP AS or OSA-SCS (through Parlay API). Also, the intra-bus that composes SDP is defined by SOA/Web Service (WS). And SIP interface is defined for the inter-bus of each SDP. To access the IMS network, a module is existed as various protocol gates in a middleware structure. The developer uses the middleware to provide high-level API about network protocol. The middleware carries out the necessary protocol conversion between high-level API command and protocol. Supported APIs can be defined as private one or standard ones (e.g., Parlay API, SS7-family SIP, MM7, LIP, SMPP, PAP, SMTP, POP3).

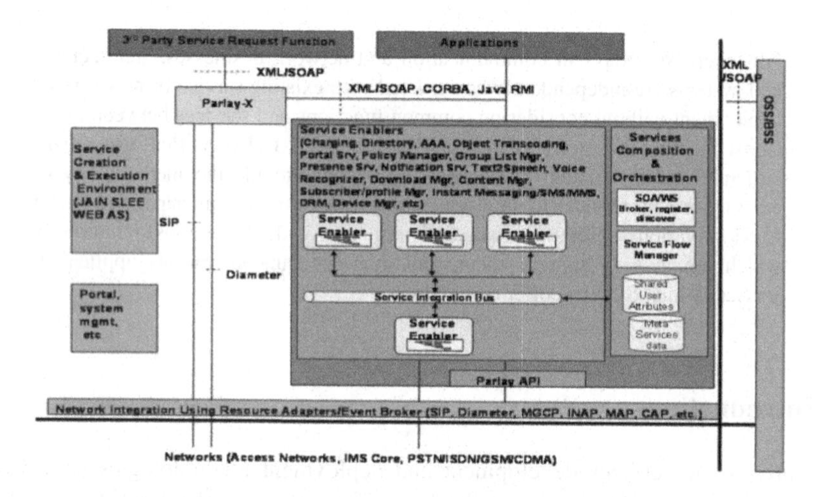

Fig. 1. Composition Elements and Interfaces in Detail

Service Logic Execution Environment (SLEE), dealing with scalability, distribution, configuration of service, is a real-time environment for service providers. The candidate technologies are based on J2EE/.NET, SIP Servlet, JAIN SLEE and Web Service. A common service enabler supports a basis structure within the service domain. A service developer is able to control and prepare application service that uses the service capabilities provided by common service enablers. Accordingly, service providers don't have to re-create these service functions as stove-pipes. These service enablers include contents management, devices management, policies control, logging, subscribers/profiles management and so on. Service composition and orchestration is used to create service in a fast way. There is a modeling tool that might compose a desirable service. The communication among services is accomplished as Service Integration Bus (SIB) that is based on Business Process Engineering Language (BPEL). The SIB uses SOA/WS technology as an interface between one service enabler and other service enablers. The Service Creation & Execution Environment (SCEE) uses a GUI-based tool for making a real-time service bundling. Portal System is used as a third-party management tool. And resource adapter as a SDP service component takes a role of access gateway to IMS network.

3 Deployment of Application Services

The simulation we use here is Application Test Suite (ATS) Network Emulator component [8]. Suppose a group of teenagers of our university planning to meet together next month in another university to attend an International Telecommunication Conference. Jacky is intended to tell his professor about this, and then reply to Lucy the manager of another university academic conference.

Firstly Jacky simply selects the person-to-talk group number from his phone books (e.g. to Professor Lee, 20070417), then just presses the push-to-talk button. The call signal transforms to SDP Parlay/OSA GW, and then decided into Professor Lee according to the phone number. Also Jacky can see his professor's online status that the professor is now with a busy line. So Jacky chooses the SMS service to inform his professor (see Fig.2). The professor who is now driving on a highway realizes this SMS message, later during his free time he also presses his push-to-talk button to answer to Jacky. After that Jacky makes a replying phone call to Lucy's computer. Lucy as she is now busy with typing some document on her computer realizes this call (according to her pre-configuration, the SDP center holds on a waiting line for Jacky and Lucy until Lucy hands up this session), she speedup finishing her work then makes the video session with Jacky about their concerning detail. Push-to-Talk over Cellular (PoC) makes point-to-point (private) and point-to-multipoint (group) connections among members of a group and combines the fix and mobile network. This Push-to-talk can be extended to include other media such as text, pictures and real-time video. SDP is able to provide that connecting SMS to PoC enabler which creates new Emulator services. Connection of SMS and PoC is performed by orchestration and interaction. SDP is able to create that common enabler interacts with access enablers (e.g., IMS enabler). SDP abstracting a function of network and defining standardized interface is possible to be used whether third-party service providers have a professional knowledge or not.

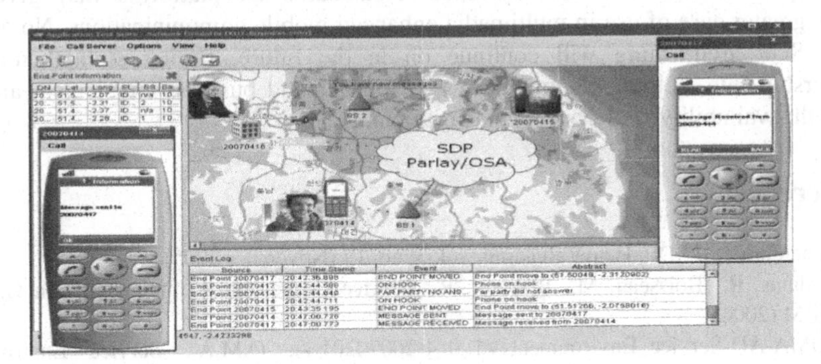

Fig. 2. Simulation of PoC

As another example, Fig.3 describes a simple presence-based application service using SDP. SDP is able to control the location of mobile user and present the state of devices through Parlay interface [9]. Let's suppose that Elice wants to find a suitable boiler-repairman. She notifies the message to Presence-based Application Service

(PAS), via SMS Messaging Gateway (SMSG), using her cell phone to find the closest boiler-repairman from her. When this request message reaches PAS, Profile enabler collects the list of candidate boiler-repairmen. By using this list, Availability enabler can infer their availability status. Location enabler interacts with Parlay gateway to obtain and track the locations of registered boiler-repairman and her. Feedback enabler makes a decision of their ranking. The ranking represents their priority. The boiler-repairman with the highest priority might receive her request. If he is busy or has any other problems in doing her request, PAS recommends that she can contact with a boiler-repairman who has the next priority. Orchestration and interaction can be accomplished by performing the coordination among various service enablers using SDP which can deploy application services in a dynamic and uniform execution environment.

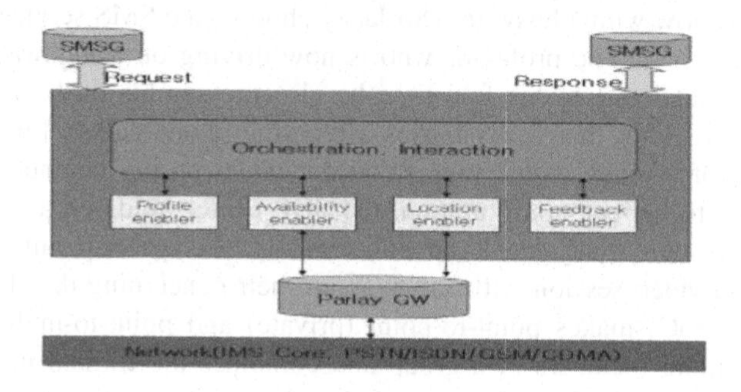

Fig. 3. A Simple Presence-based Application Service Using SDP

4 Conclusions

Now we have presented the main function and beautiful perspective of SDP. In Application Server (AS) Layer, SDP architecture holds the promise of seamless converged voice/data services by kinds of standards and platforms. SDP gives the users greater ease of use in multimedia enhanced mobile communications. No doubly IMS SDP architecture will continue on in the future, further strengthening its leadership role, helping customers in their on demand business enablement and the growth, availability, and systems management.

References

1. The Future of SDP, Heavy Reading, vol. 3, no. 19 (November 2005)
2. Adams, R., Boisseau, M., Bowater, R.: Service Orchestration Implementation Options, IBM (2004)
3. OMA-AD-Service_Environment-V1_0_4-20070201-A, O.M.A.: Service Environment (February 2007)
4. ETSI ES 202 391-1 v1.1.1, Open Service Access (OSA): Parlay X Web Services, Part1: Common (2005)
5. Open API Solutions, Application Test Suite & Parlay/OSA framework simulations, http://www.openapisolutions.com
6. Chakravorty, D., et al.: BusinessFinder: Harnessing Presence to Enable Live Yellow Pages for Small, Medium and Micro Mobile Business

A Study of Recovering from Communication Failure Caused by Route Hijacking

Toshimitsu Ooshima[1], Mitsuho Tahara[1], Ritsu Kusaba[1], Souhei Majima[1], Satoshi Tajima[2], Yoshinori Kawamura[2], and Ryousuke Narita[2]

[1] NTT Network Service Systems Laboratories, NTT Corporation, 9-11 Midori-Cho, 3Chome Musashinoshi, Tokyo, 180-8585 Japan
[2] NTT Communications Corporation 1-6 Uchisaiwai-cho 1-chome Chiyodaku, Tokyo, 100-8019 Japan
{ooshima.toshimitsu,tahara.mitsuho,kusaba.ritsu, majima.souhei}@lab.ntt.co.jp

Abstract. In this paper, we discussed methods for recovering from communication failures due to route hijacking. Looking ahead we intend to pursue further work on systemization, authentication, and evaluation.

Keywords: BGP, route hijacking, communication recovery.

1 Introduction

The Internet consists of over 20,000 ASs, which exchange routing information with each other using BGP [1]. Since BGP does not have any function for verifying the authenticity of routing information, if erroneous routing information is advertised, this false routing information is transmitted across the whole Internet. The phenomenon is known as route hijacking. In order to solve this problem, improved protocols have been proposed, including soBGP [2] and sBGP [3]. Despite these proposals, it would be difficult for the more than 20,000 ASs to simultaneously convert from the BGP protocol that they are all using to one of these improved protocols. Therefore, we set out not only to improve the BGP protocol, but also, with an eye to countering the continued threat of route hijacking, to carry out R&D to develop a technique capable of rapidly detecting route hijacking, of recovering from the resulting communication failures, and of preventing route hijacking through the construction of a highly reliable routing information database. In this paper, our discussion focuses on methods for recovering from communication failures resulting from route hijacking.

2 Recovery from Communication Failure Caused by Route Hijacking

When an victim AS suffers a communication failure caused by route hijack, a recovery from the communication failure needs to be implemented at the victim AS, at least temporarily. Recovery methods like this can be categorized as follows: (1) Recovery methods applied only to the victim AS, and (2) Cooperative recovery methods, in which requests are made to all the ASs that are relaying the hijack route.

S. Ata and C.S. Hong (Eds.): APNOMS 2007, LNCS 4773, pp. 579–582, 2007.

(1)Recovery Methods Applied Only to the Victim AS

Recovery methods that are applied only to the victim AS make use of the same mechanism as the route hijacking itself, by advertising from the victim AS on the hijack route ("hijacking back"). The main issues of this method are (1-1) advertising routes, and (1-2) advertising location. Here we will discuss each of these points.

(1-1) Advertising Routes

Proposal A-1 is a method for advertising a route having a prefix longer than the hijack route. The reason is that when there is a multiple number of prefixes containing destination addresses, the routers conduct forwarding based on the longest match rule. Note, however, when these routes are filtered at an AS having a policy to filter routes of long prefix, it is possible that advertised routes are not adequately transmitted over the Internet. Proposal A-2 is a method for advertising the same prefixes as the hijack routes. Accordingly, the chance of routes being filtered due to prefix length is lower than in the case of proposal A-1. However, since hijack routes and the advertised routes have the same prefix, the range of communication recovery becomes the range where the advertised route is selected as the "best path". So communication can be recovered over only about half of the entire Internet. A side-effect of this is that hijack routes can become invisible, depending on the places where hijack routes are monitored, making it impossible to recognize whether hijacking is occurring or not. Proposal A-3 is a combination of the proposal A-1 and A-2. Thus, the range of communication recovery is wider than that of either A-1 or A-2 applied alone, but the side-effect of proposal A-2 remains. Proposal A-4 is a two-stage method. Firstly, proposal A-1 is implemented and then the communication recovery situation is assessed. If it is found that the recovery range does not expand, proposal A-3 is additionally applied. This method is subject to the same side-effect of proposal A-2, but in this case it is possible to minimize the side-effect.Considering the above proposals as a whole, the most favorable proposal is A-4.

(1-2)Advertising Locations

In Fig. 1 we show an example of the internal structure of a victim AS (AS65050) that implements a recovery procedure, and we discuss the locations (devices) within an AS that implement route advertising. When the above route is advertised from the victim AS, there are two methods that can be applied. Proposal a): A method implemented at the router advertising hijacked routes; and Proposal b): A method implemented at a dedicated advertising device. The advertised routing information is a combination like proposal A-3 or A-4, and the NEXTHOP attribute of these must be NH-A. Thus, in proposal a), when advertised from the router, routing information must be advertised from RT-A. So, it is necessary to set route advertising to RT-A, and also necessary to establish an algorithm for searching RT-A, which stores a value of 10.0.0.0/16, from an IP address management system, using NH-A—the NEXTHOP attribute of 10.0.0.0/16—as a key. There are other challenges, such as coordination with an account management system for obtaining accounts to enable login to RT-A, and dealing with setting commands for advertising route. Accordingly, the practicability of systemization and versatility of applicability to many ASs, are both low. Since the routing information that is desired for advertising is clearly defined, proposal b) utilizes a dedicated advertising device rather than a router, which is limited to advertise. The operation of a dedicated advertising device is explained in

Fig. 1. The device is connected to the router, receives routing information, and stores the routing table. Here, when a NOC operator places a request for communication recovery of the hijack route to the device, the prefix at the device is a combination like proposal A-4; and for the NEXTHOP attribute, NH-A is searched from the routing table inside the device, using the hijack route as a key. Next, routing information is formed from these prefixes and NEXTHOP attributes, and routes are advertised to connected routers. Like this, the practicability of systemization is high, and versatility of applicability to a diversity of ASs is also high. From the above, we can conclude that proposal b) is desirable in terms of practicability and versatility.

Summary of Recovery Methods at a Victim AS

From the results of the above investigation, as a recovery method that is applied only at the victim AS, it is desirable to use a two-stage method, utilizing "longer" and "exact" routes, and to use dedicated advertising devices for advertising route.

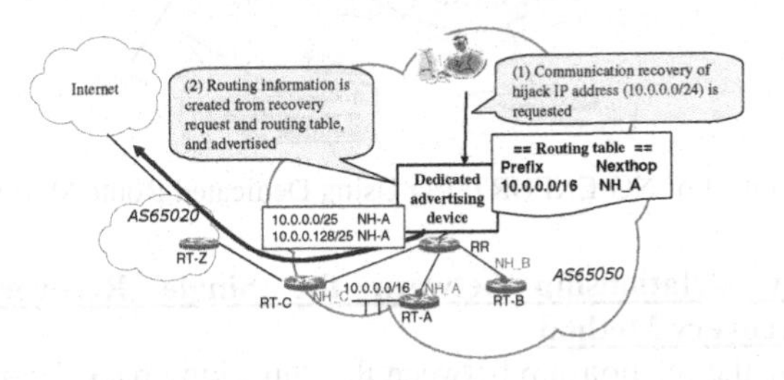

Fig. 1. Route Advertising Using Dedicated Advertising Device

(2) Coordinated Recovery Methods

Here we discuss coordinated recovery methods that implement recovery by placement of requests from the victim AS to the ASs that relay the hijack route . There are three proposals for coordinated recovery methods. Response C is a method for filtering hijack routes at the border routers of an AS. When a hijack route is filtered, it does not flow inside that AS, so communication is recovered. Transmission of hijack routes to other ASs is also stopped. Like this, this kind of response is highly effective, but it is difficult to establish an algorithm for filter settings of routers. So, the practicability of systemization is poor. Response D is a method to assign a community tag, "NO-EXPORT," to a hijack route. For the routing information to which "NO-EXPORT" is assigned, no routes are advertised between the border routers connected to other ASs and other ASs. Thus, hijack routes are not transmitted to other ASs, and there is a high possibility of communication recovery at other ASs. On the other hand, since hijack routes transmit within ASs to which "NO-EXPORT" is assigned, communication is not restored. Thus, this method is less effective than response C. This response can be further categorized, in terms of where the response is implemented, into response D-a), which is a response at the router, and response D-b) which is a response at a dedicated route modification device. Response D-a is a method for assigning "NO-EXPORT" at a router, so it is necessary to develop an algorithm for assigning, and to set up a filter. Therefore, it is difficult to achieve systemization like Responce C. Response D-(b) is a method for using a dedicated route modification device. When a route with "NO-EXPORT" assigned is selected as the best path, RR advertises this routing information to each router connected via iBGP. Thus, routes with "NO-EXPORT" assigned are also advertised to RT-C, so the hijack routes advertised up to now to a neighbor AS will be stopped. Thus, hijack routes disappear from neighbor ASs and communication is restored. Response E is a method that the advertised routes are also the same as hijacking route, and routing information is created as routes of the victim AS and advertised. However, such a practice would be considered unethical Internet behavior, so it would be inappropriate to use this method. From the above, it is clearly desirable to assign a "NO-EXPORT" tag using a dedicated route modification device, by means of responce D-b).

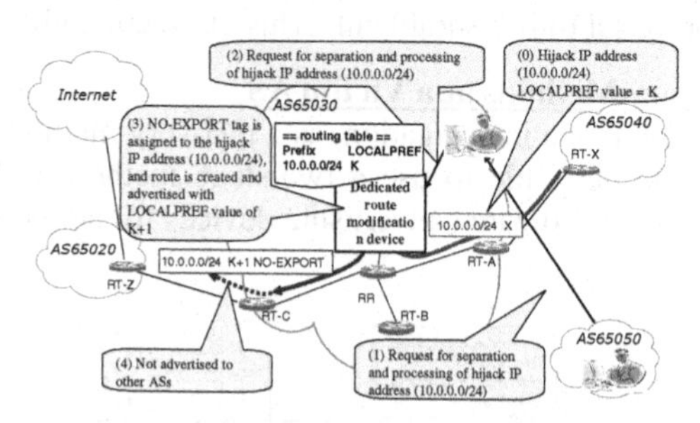

Fig. 2. Assignment of NO-EXPORT Tag Using Dedicated Route Modification Device

Complementary Relationship between the Single Recovery Method and Cooperative Recovery Method

Here we describe the relationship between the "hijacking back," corresponding to the single recovery method response A), and the assignment of "NO-EXPORT," as in the cooperative recovery method response D-b). Basically, communication is restored by response A. However, a victim AS readvertises routes that are "hijack back," if the upstream AS has a policy to "filter routes having long prefixes," or to "filter routes not registered in IRR (Internet Routing Registry)," then these advertised routes will be filtered, resulting in the possibility that the routes that were hijack back and readvertised by a victim AS may not be transmitted over the Internet. In such a situation, the cooperative recovery method is more effective.

3 Conclusion

In this paper, we discussed methods for recovering from communication failures due to route hijacking. Looking ahead we intend to pursue further work on systemization, authentication, and evaluation.

This paper is the output of a research project for the Ministry of Internal Affairs and Communications (Japan), "Research and Development Relating to the Detection, Recovery, and Prevention of Route Hijacking."

References

1. Rekhter, Li: A Border Gateway Protocol 4 (BGP-4), RFC1771
2. White, Architecture and Deployment Considerations for Secure Origin BGP (soBGP), draft-white-sobgp- architecture-02.txt
3. Lynn, et al.: Secure BGP (S-BGP), draft-clynn-s-bgp- protocol-01.txt

Multi-agent Learning and Control System Using Ants Colony for Packet Scheduling in Routers

Malika Bourenane[1], Djilali Benhamamouch[1], and Abdelhamid Mellouk[2]

[1] Dept of Computer Science, University of Es-Senia, Oran, Algeria
[2] LISSI/SCTIC, University of Paris XII-Val de Marne France
{Malika Bourenane,mb_regina}@yahoo.fr,
{Djilali Benhamamouch,d_benhamamouch}@yahoo.fr,
{Abdelhamid Mellouk,mellouk}@univ-paris12.fr

Abstract. This paper describes a novel method of achieving packet scheduling in several routers of network, in order to optimize the end to end delay. We use a multi-agent system to model this problem, where each agent of this system tries to optimize the local scheduling and through a communication with each other, attempts to make global coordination in order to optimize the total scheduling. The communication between agents is done by mobile agents like ants colony. A pheromone-Q learning approach is presented in this paper, which consists to applying the standard Q-learning technique adapted to our architecture with a synthetic pheromone that acts as a communication medium speeding up the learning process of cooperating agents.

Keyword: Multi-agent system, reinforcement learning, mobile agent, ant colony, packet scheduling.

1 Introduction

Data, voice, and video services that are currently carried on multiple service-specific networks, will be carried on one single flexible and ubiquitous converged IP network. However, current IP technology is still mainly best effort and cannot provide guaranteed distinct Quality of Service (QoS) required by QoS-sensitive services. Over-provisioning is the currently applied strategy for QoS in IP networks, and it serves as a reasonable solution for the core networks where technologies have made bandwidth abundant and relatively low-cost. However, efficient transport of kind of applications requires new network capabilities such as packet scheduling mechanisms to prioritize multimedia traffic [2]. A good scheduling discipline should allow the network to treat heterogeneous applications differently in accordance with their QoS requirements and in a dynamically changing environment it should be adaptive to the new traffic conditions. Reinforcement learning (RL) [4] provides a natural support for the development and improvement of a policy that can adapt by interacting with the environment, to the conditions (load and resources availability) of the network. We use a multi-agent system (MAS) to model this problem, where each agent tries to optimize the local scheduling and through a communication with each other, attempts to make global coordination in order to optimize the total scheduling. A mobile agent

S. Ata and C.S. Hong (Eds.): APNOMS 2007, LNCS 4773, pp. 583–586, 2007.

called ant [5] located at each router can keep the first agent informed of the conditions at the next router. For example, the buffer may be bursting or the bandwidth is not sufficient, so to prevent the packet from being dropped, the first agent could improve its learning policy to service an other queue until the second agent indicates the possibility to receive the packet indicated beforehand. In this paper, we adopt the framework of Markov decision processes [1] applied to multi-agent domain. While centralized MAS may be considered as a huge MDP, we work with decentralized systems where each agent is independent from the other as far as decision and learning are concerned.

2 Proposed Approach

We modify somewhat the definition of a multi-agent MDP to define a decentralized multi-agent MDP.

Definition: The MDP for each agent i is defined as $MDP^i = (S, A^i, P^i, r^i)$, where S is a finite set of system states, A^i is a finite set of actions of agent i, $P^i: S \times A^i \rightarrow \Delta(S)$ is the state transition function where $\Delta(S)$ is the set of probability distributions over the set S and $r^i: S \times A^i \times S \rightarrow đ$ is the individual reward such that $r^i(s, a^i, s')$ is the reward obtained by agent i when it performs in state s, an action a^i and move to state s'.

Again, we need a local policy π^i for each agent i:

$$\pi^i : S \rightarrow A^i \tag{1}$$

The expected discounted value function of agent i is the following:

$$V^i(s, \pi^i) = E(\pi^i)[\, r^i \mid s_0 = s] \tag{2}$$

We consider also Q^i as local Q-function, defined for each state-action pair as:

$$Q^i\pi^i (s, a^i, s') = r^i(s, a^i, s') + \gamma \Sigma\, P(s' \mid s, a^i)\, V^i(s', \pi^i) \tag{3}$$

In MAS it will not be possible for the agent to have complete view of the system state. Hence, we improve our decentralized multi-agent MDPs by the concept of mobile agent like ant to make the current state of each agent known throughout the system. In this way, scheduling is made based on local decisions and instantaneous information (policies, etc) communicated periodically by mobile agents, so they can enhance each others state or action value functions. The Phero-Q technique combines Q-learning [4] with a synthetic pheromone [5] introducing a belief factor into the update equation. The formula bellow describes the belief factor [3]:

$$B(s, a) = \frac{\Sigma_{s \in Na}\, \Phi(s)}{\Sigma_{\sigma \in Na}\, \Phi_{max}(\sigma)} \tag{4}$$

where $\Phi(s)$ is a scalar value that integrates the basic dynamic nature of the pheromone, namely aggregation, evaporation and diffusion. The Q-learning update equation modified with synthetic pheromone and applied to our model is given by:

$$Q^i(s, a^i) \leftarrow Q^i(s, a^i) + \alpha\, \{R + \gamma \max_{a^{i'}} [Q^i(s', a^{i'}) + \xi\, B(s', a^{i'})] - Q^i(s, a^i)\} \tag{5}$$

$$R = \max \sum_{i=1}^{n} r^i(s, a^i, s') \quad (6)$$

and the parameter ξ is a sigmoid function of time periods such that $\xi \geq 0$. Also, the optimal policy π^{i*} for each agent i can be obtained as:

$$V^i(s^1, \ldots, s^n, \pi^{i,*}) = \max Q^{i,*}(s^1, \ldots, s^n, a^i) \quad (7)$$

The optimal action value function Q*of the multi-agent MDP is the sum of the optimal action value functions $Q^{i,*}$:

$$Q^*(s^1, \ldots, s^n, a^1, \ldots, a^n) = \sum_{i=1}^{n} Q^{i,*}(s^1, \ldots, s^n, a^i) \quad (8)$$

An optimal policy can be directly derived from the optimal action value Q*(s, a) by:

$$\pi^*(s^1, \ldots, s^n) = \arg\max_{(a^1, \ldots, a^n)} Q^*(s^1, \ldots, s^n, a^1, \ldots, a^n) \quad (9)$$

3 Learning Algorithm

We present in this section a Decentralized Multi-Agent Learning (DEMAL) algorithm that uses Q-learning. To converge, Q-learning requires the knowledge of the actual state. In our case this information is gathered and distributed by mobile agents.

Algorithm DEMAL

Repeat
Initialize $s = (s^1, \ldots, s^n)$
  Repeat
    For each agent i
    Choose a^i using Boltzman formula (see [6])
    Take action a^i, observe reward r^i and state s'
    $Q^i(s, a^i) \leftarrow Q^i(s, a^i) + \alpha \{R + \gamma \max_{a^{i\prime}} [Q^i(s', a^{i\prime}) + \xi B(s', a^{i\prime})] - Q^i(s, a^i)\}$

    $s \leftarrow s'$
  Until s is terminal
Until algorithm converges

4 Evaluation

The objective of the simulation is to compare the global decision time of the multi-agent system when it acts alone and when it acts with mobile agents. This simulation is carried out for two cases. In the first case we consider that the output link capacity of all routers is sufficient. We assume, in the second case, that the output link is not sufficiently large in some routers. We observe (Fig 1) in the two scenarios that the global decision time is better with presence of the mobile agents.

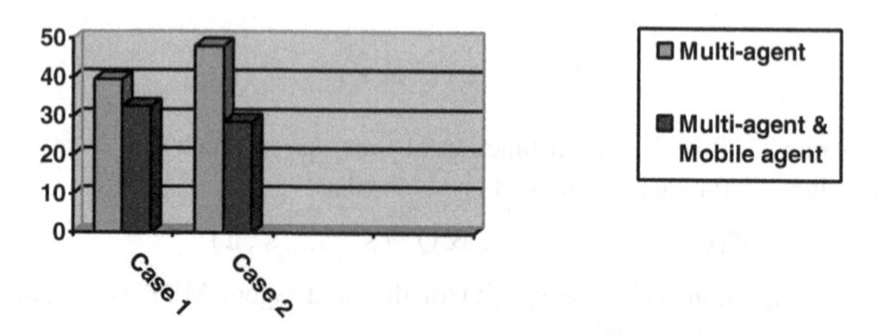

Fig. 1. Global decision time of the multi-agent system

5 Conclusion

This paper addresses the problem of performing a decision making task using a set of agents through the problem of scheduling optimization between several routers. We formulated this problem as decentralized multi-agent MDPs, improved by ant-like mobile agent on the level of each router to guarantee a global view of the system's state. Our simulation shows that the proposed approach leads to better results than when the multi-agent system acts alone. As a future work, we will evaluate how adaptive our approach is, by performing it when the network environment changes. We will also compare it with others models.

References

1. Valckenaers, P.H., Kollingbaum, M., Van Brussel, H.: Multi-Agent Coordination and Control Using Stigmergy. Computers in Industry 53, 75–96 (2004)
2. Nichols, K., Blake, S., Baker, F., Black, D.: Definition of the differentiated services field (DS field)in the IPv4 and IPv6 headers, RFC 2474 (1998)
3. Monekosso, N., Remagnino, P.: The analysis and performance evaluation of the pheromone- Q-learning algorithm. Expert Systems 21(2), 80–91 (2004)
4. Sutton, R.S., Barto, A.G.: Reinforcement Learning: An introduction. Mit press, Cambridge (1998)
5. Nouyan, S., Ghizzioli, R., Birattari, M., Dorigo, M.: An insect-based algorithm for the dynamic task allocation problem. Kunstliche Intelligenz 4/05, 25–31 (2005)
6. Kapetanakis, S., Kudenko, D.: Reinforcement learning of coordination in cooperative multi-agent systems. In: AAAI/IAAI, pp. 326–331 (2002)

A Framework for an Integrated Network Management System Base on Enhanced Telecom Operation Map (eTOM)

A.R. Yari and S.H. Hashemi Fesharaki

Management and Economics Study Department, Iran Telecom Research Center, Karegar Ave., Tehran, Iran
a_yari@itrc.ac.ir

Abstract. This paper provides a summary of development process for the conceptual design of an integrated network management system (NMS) platform for incumbent with a huge legacy system. Based on proven international standards such as Enhanced Telecom Operation Map (eTOM) [1] and New Generation Operations System and Software (NGOSS) [2] framework an approach has been develop to address the current situation of incumbent as well as to make provision for NGN. In other world, the solution has been developed in a way that a relatively smooth transformation/ development of the existing fragmented management landscape towards an integrated NMS is possible. A service centric concept has been deployed for our conceptual design.

Keyword: Network Management system, eTOM, NGOSS.

1 Introduction

As a result of deregulation, network service providers that were once monopolies are now subject to intense competition. Deregulation has been accompanied by developments in mainstream computing (notably the advent of Internet technologies) that have led to an explosive increase in the amount of data traffic carried by telecom networks.

This trend has continued as demand has grown for seamless communications anywhere at any time between a variety of devices. In addition to sophisticated telephony services with attractive pricing, telephone subscribers now expect other services such as voice mail, Internet access, multi-media, email, wireless access, and IP telephony.

The many telecom network management systems in existence today are designed to meet a common set of challenges. Some of these challenges arise from the competitiveness of the telecom market. Others arise from the technical complexity of the telecom network management domain. This complexity comes from the need to manage performance, alarms, and states in networks consisting of many millions of network elements of many different types. To meet the challenges of today's telecom

S. Ata and C.S. Hong (Eds.): APNOMS 2007, LNCS 4773, pp. 587–590, 2007.

market, telecom network management systems must be highly reliable, offer high performance, and be easy to operate.

Current attempts to meet the challenges of telecom network management are based on approaches such as: Telecommunications Management Network (TMN) [5], Transaction Language 1 (TL1), Simple Network Management Protocol (SNMP), Common Object Request Broker Architecture (CORBA).

After this in section 2, the service group model has been explained. Then in section 3, the management model has been proposed in three levels. Finally a short summary of paper has come.

2 Service Group Model

Service groups need to be defined to create logical domains. These logical domains group services that are either hosted on the same and/or adjacent OSI layers (e.g. transmission services such as SDH and IP transport) or form a group with distinct functional boundaries (e.g. mobile services). The following four main groups have been considered and service groups have be defined accordingly [3]:

Transmission Services: WDM and DWDM, SDH transport (optical fibre and micro wave), PDH transport, Satellite transport, IP/MPLS transport and Network time services for transmission

Fixed Line Services: International gateways services, analog and digital switching, soft switching via NGN, value added services (VAS), access services (e.g. two wire copper loop, xDSL), network time services for switching

Data Services: The service group "Data Services" comprises the following services, VPN services, internet services (bulk for resellers), remote access and dial up services, hosting services, E-mail services, DNS services, security, content filtering and caching services

Mobile Services: Mobile voice services, mobile VAS, GPRS, network time services for mobile switching

It should be noted that the classification is not free of ambiguity and henceforth infrastructure needed to provide these services could be attributed to another group as well. For the conceptual design however, this ambiguity is of no concern.

Customer groups also have to be defined to enable company to define develop (specialized) products based on the service groups outlined in this section. Examples for customer groups are: residential customers, business customers, wholesale etc. The detailed definition of customer groups is not considered in this paper.

3 Management Model

The proposed management model itself is eTOM based and focuses around the three core areas of fulfilment assurance and billing [1]. The proposed model is based on the following three levels:

3.1 High Level Design

In the high level design, the management plane interfaces with the entire stack from customer equipment via access, aggregation, transport and control to the application plane in one side and with the business management plane on the other side.

Except for the core components fulfilment assurance and billing as defined in eTOM, two more components the common data model/management and the common integration and process automation can be identified. These two components are essential for the deployment of a common platform or interoperable platforms for incumbents.

3.2 Management Model Decomposition in Level 1

Applying these requirements a generic level 1 model has been developed that contains all major functional blocks. As a new element a mediation/abstraction layer between the network elements/element managers and the proposed solution has been introduced. This layer is responsible for the mediation between the functional components using the common data model of the solution and the multiple sets of proprietary data supported by the network elements/element managers.

3.3 Level 2 Decomposition of Management Model

A further decomposition into level 2 functions has been performed. The level 2 architecture has shown the logical interfaces/relationships between the modules as well as where data will be held.

So far the management model can be applied to any of the service groups. Process automation and an end to end service view would be possible within the boundaries of a service group, but not across service group boundaries. This can be achieved by introducing interfaces for data/process information exchange between 2 or more management platforms.

In this level each module contains a brief description of its key functions as well as associated databases. The logical interfaces are also considered between the modules; core interfaces to an external umbrella management system. It should be noted that the architecture of the solution is essentially the same for all service groups, differences will only occur at the specific, technology dependent implementation.

A central Master Service Control System would follow the same architecture, however lower layer functionalities such as element and resource management, resource and usage data collection would be kept at the service group level of the solution.

4 Summary

A framework design for an integrated NMS solution has been presented consisting of a component model, a service group and a management model leading to a functional system architecture. The creation of a common data model and a common integration and process framework is proposed.

The following advantages can be seen if this framework is being adopted by telecom companies

- Consistency throughout all NMS instances at company
- End to end service view supported by cross domain management
- Common data structures, hence easier comparability of key quality and key performance indicators
- Better support for cross domain and end to end service level implementation
- Pre-requisite for automated cross-domain service management

References

1. Enhanced Telecom Operation Map. GB921. TeleManagement Forum (January 2004)
2. The NGOSS Technology-Neutral Architecture, NGOSS Release 4.0. TMF053. TeleManagement Forum (January 2004)
3. Information technology – Open distributed processing – Reference Model: Overview. ISO/IEC 10746-1 (1998), http://www.iso.org/
4. Fleck, J.J.: Overview of the Structure of the NGOSS Architecture. Hewlett-Packard Company (May 2003)
5. ITU-T TMN Recommendation M.3400 (TMN Management Functions, ITU-T, 4/97), M.3010 (Principles for a telecommunication management network, ITU-T), M.3200 (TMN Management Services, ITU-T, 1996) and Related Recommendations

High Performance Session State Management Scheme for Stateful Packet Inspection

Seungyong Yoon, Byoungkoo Kim, Jintae Oh, and Jongsoo Jang

Electronics and Telecommunications Research Institute, 305-350,
161 Gajeong-dong, Yuseong-gu, Daejeon, Republic of Korea
{syyoon,bkkim05,showme,jsjang}@etri.re.kr

Abstract. This paper relates to a method for performing Stateful Packet Inspection(SPI) in real time using a session table management scheme that allows more efficient generation of session state information. SPI is an important technique to reduce false positive alerts in network intrusion detection system(NIDS). As the number of session increases, this technique requires a higher processing speed, thereby causing performance problems. However, existing software-based solutions cannot perform real-time packet inspection ensuring the wire speed. To guarantee both performance and functionality with respect to statefulness, we designed and implemented SPI-based intrusion detection module in a FPGA to help alleviating a bottleneck in network intrusion detection systems in this paper.

Keywords: Stateful Packet Inspection, High Performance Intrusion Detection.

1 Introduction

One of the major problems and limiting factors with Network Intrusion Detection System(NIDS) is the high false positive alert rate. In order to reduce these false positive alerts, a lot of methods and techniques are proposed. Stateful Packet Inspection(SPI) is one of these solutions. SPI was originally developed for Firewall[1][2], but it became a very important factor in NIDS. Stateless NIDSs generate tremendous false positive alerts while stick or snot attempts to attack[3][4]. Most existing NIDSs have SPI module which is supported statefulness but they don't satisfy high-performance in gigabit internet environment. It is so difficult that we manage a lot of session state information with limited hardware resource and satisfy performance of high-speed internet. In other words, the rapid evolution of recent network technologies to gigabit network environments require existing SPI module to have more improved functions and performance. SPI basically requires a session table which stores source and destination IP addresses and port numbers. It is necessary to perform real-time packet inspection by checking, for each input packet, whether or not a corresponding entry is present in the session table. Real-time packet processing at wire speed should not cause any packet delay or loss even when the number of managed sessions is increased to more than one million.

Previously developed software-based solutions cannot meet these requirements. One software-based technique has attempted to use a distributed system. However, as

S. Ata and C.S. Hong (Eds.): APNOMS 2007, LNCS 4773, pp. 591–594, 2007.

the number of session increases, this technique requires a higher processing speed, thereby causing performance problems. Thus, software-based solutions cannot perform real-time packet inspection ensuring the wire speed. To guarantee both performance and functionality with respect to statefulness, we designed and implemented SPI-based intrusion detection module in a FPGA to help alleviating a bottleneck in network intrusion detection systems in this paper. The performance of SPI-based intrusion detection system mainly depends on the performance of processing session table[5] and pattern matching[6]. In this paper, we focused on session state management scheme and omitted pattern matching method. Our work related to pattern matching method is described by Byoungkoo Kim at al. [7] in detail.

2 Session State Management Scheme

Fig.1 shows the SPI-based intrusion detection module. Legitimate TCP sessions are established through 3-way handshake and terminated through 4-way handshake. State manager has session table and tracks these session state. If input packet doesn't exist in session entries, this packet will drop or forward to Intrusion Detection Engine(IDE) with additional state information according to security policies.

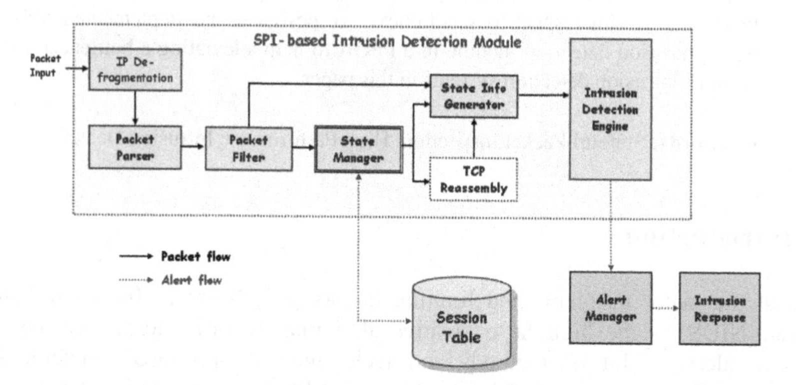

Fig. 1. SPI-based Intrusion Detection Module

Fig. 2 is a basic architecture of session state manager for stateful packet inspection. Session state manager includes a hash key generator, a session table, a session detection module, a session management module, and a state info generation module. The session table stores session entries that are indexed and managed by the hash key generator. 4-tuple information including a source IP address, a destination IP address, a source port, and a destination port is input, as information used to hash a newly received packet, to the hash key generator. Hash key generator has a dual hash structure with two different hash functions Hash1(x) and Hash2(x). The hash functions Hash1(x) and Hash2(x) are well-known functions that are used to hash packets. One hash function Hash1(x) is used to generate indices that point to hash sets permitting hash collisions in order to achieve faster session table search. The other hash function Hash2(x) is used to generate hash addresses that are used to identify session entries in a hash set pointed by the hash function Hash1(x). Session table may be designed and implemented using two or more SRAM devices, if necessary. For efficient session table management, the session table has an N-way set associative session table structure in which each hash set in the session table can include N session entries. The session table shown in Fig. 2 is a 32-way set associative session table that is constructed using two 72-Megabit SRAMs with each session entry having a length of 36 bits.

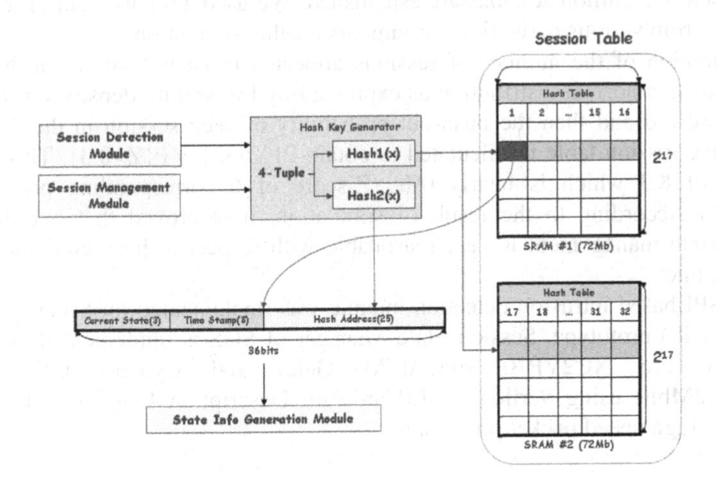

Fig. 2. Basic Architecture of Session State Manager

Each session entry stored in the session table includes current state, time stamp, and hash address parts. The current state part includes current connection state information of a corresponding session, the time stamp part is used to determine which session entry is to be deleted when the session table is full, and the hash address part is used to identify each session entry in the same hash set. According to current state, State Info Generation Module generates the state information of the packet. Then, inspection of the packet is performed based on the generated state information

3 Simulation and Implementation

Our session state management scheme in SPI-based intrusion detection system is affected by two major factors, hash collision rate and push-out rate. There is every probability of hash collision occurrence because hash function for faster session table search is used. The wrong state information is generated if the hash collision is occurred. Therefore, the SPI-based intrusion detection module generates the false positive alert. The hash collision rate is determined by the Hash2(x). Theoretically, the probability of hash collision is $1/2^{20}$ if the session table is full. As the number of session entries increase gradually, the session table is filled with new session. Also, there is every probability of push-out occurrence because the size of hash set has limitation(32-way set). When the session table is full, the probability that each session is brought into a push-out state is very important in a session table management scheme because wrong session state information is generated if any existing session, which has not yet been terminated, is replaced with a new session. In this case, since the SPI-based intrusion detection module generates the false negative alert, the push-out rate can be said to be the factor which is important than the hash collision rate. In order to ensure that push-out rate is reasonable in our design, we made a simulation for distribution of the number of sessions allocated to each hash set in the session table when one million sessions are established. We used a separate set of traffic data collected from various network environments for this simulation.

Distribution of the number of sessions allocated to each hash set in the session table follows a normal distribution as expressed by Probability density function. This is standardized and then the push-out probability of each session in the 32-way set associative session table is calculated to obtain $P\{X>32\} = P\{Z>8.3\}$. This indicates Z-score of 8.3 which is nearly 0%. (Z-score of 6 corresponding probability is 0.0003%) According to the result of simulation, it is proved that our design for session state management is very reasonable with respect to hash collision rate and push-out rate.

Our SPI-based intrusion detection module was implemented on Security Gateway System(SGS) prototype. Session State Manager of SGS is implemented on a Xilinx Vertex-II Pro XC2VP70 FPGA(7M Gate) and Cypress CY7C1470V33 SRAM(72Mbit) using verilog HDL(Hardware Description Language) that is best suited for high-speed packet processing.

4 Conclusion

In this paper, we proposed session state management scheme which can perform stateful packet inspection in real time by performing session table processing that allows more efficient generation of state information. And we designed and implemented SPI-based intrusion detection module in a FPGA to help alleviating a bottleneck in network intrusion detection systems.

References

1. Firewall-1 Product, http://www.checkpoint.com
2. Spitzner, L.: Understanding the FW-1 State Table, http://www.spitzner.net/fwtable.html
3. Caswell, B., Beale, J., Foster, J.C., Faircloth, J.: Snort 2.0 Intrusion Detection(Syngress Publishing, February 2003)
4. Snort Preprocessor Stream4, http://www.snort.org
5. Li, X., Ji, Z.-Z., Hu, M.-Z.: Stateful Inspection Firewall Session Table Processing. In: ITCC'05. Proc. Of the International Conference on Information Technology: Coding and Computing, vol. 2, pp. 615–620 (April 2005)
6. Sergei, et al.: SNORTRAN: An Optimizing Compiler for Snort Rules, Fidelis Security Systems, Inc. (2002)
7. Kim, B., Heo, Y., Oh, J.: High-Performance Intrusion Detection in FPGA-based Reconfiguring Hardware. In: Proceeding of APNOMS (2005)

A Parallel Architecture for IGP Weights Optimization

Visa Holopainen and Mika Ilvesmäki

TKK Helsinki University of Technology, Networking Laboratory, P.O. Box 3000, 02015 TKK, Finland
{visa.holopainen,mika.ilvesmaki}@netlab.tkk.fi
http://www.netlab.tkk.fi

Abstract. Currently ISPs use tools, such as Interior Gateway Protocol Weight Optimizer (IGP-WO), to tune link weights of routing protocols so that offered traffic load is balanced evenly for network's links. IGP-WO uses a sequential algorithm that prevents efficient use of parallel computing resources. We introduce a parallel link weight optimization architecture and compare its performance to IGP-WO. We find that our system achieves good load balancing much faster than IGP-WO.

Keywords: Traffic Engineering, Weight Optimization, Load Balancing.

1 Introduction

A near-optimal intra-domain load balancing can be achieved by altering the link weights of the network's routing protocol (OSPF or IS-IS) [1,2]. Open-source Traffic Engineering software [3] provides a weight optimizer called IGP-WO [4]. We compared the performance of IGP-WO (in terms of execution time and the quality of load balancing) to our proposed system.

2 The Architecture

The IGP link weight optimization process can be conceptually divided into three parts:

1. Delivery of topology and traffic demand information to Traffic Engineering Unit (TEU)
2. Determination of the near-optimal link weights for load balancing based on traffic demand and topology information (link weight optimization)
3. Transferring of the link weights to routers

We implemented a prototype network that contains all of these functions, however, in this work we focus on the second.

In our architecture, the master node of the TEU periodically checks if traffic measurement files and the topology file have arrived and if weight re-configuration is needed. Whenever it decides that weights need to be altered in the network,

S. Ata and C.S. Hong (Eds.): APNOMS 2007, LNCS 4773, pp. 595–598, 2007.

it uses a novel mapping system to pre-select 20 candidate weight sets, and then distributes the sets to 20 slave nodes, along with the current topology and demand matrix. Each of the slave nodes then uses simulation to test the quality of load balancing the given weight set actually produces with the current demand matrix and topology (maximal link load). The slaves then inform the master about the quality of load balancing they obtained, after which the master can select the best weight set (the one that produced smallest maximal link load). Overview of our architecture is presented in Figure 1.

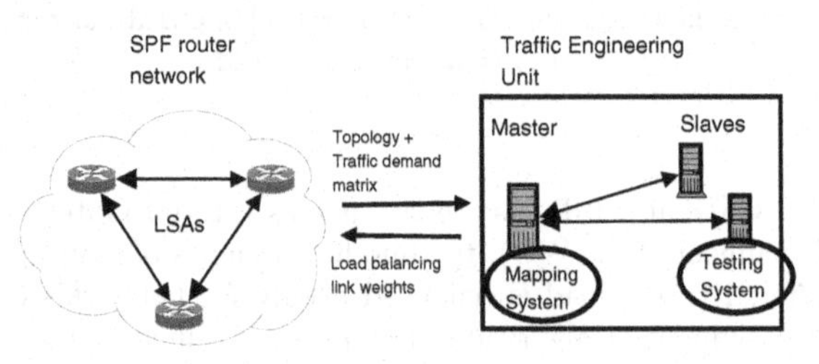

Fig. 1. Overview

The main logic of the master node of TEU is presented in Figure 2.

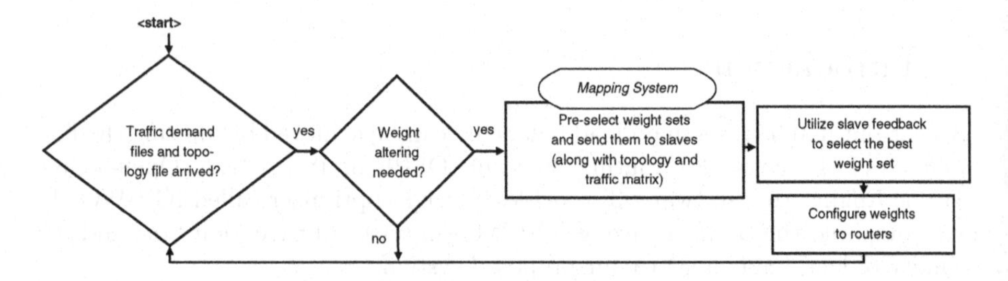

Fig. 2. Main logic of the master node

The slave node is logically very simple. It sleeps until the master node sends topology, demand matrix, and weight set files to it. When the slave receives these files, it simulates the quality of load balancing that is achieved with the given weight set and returns the result to master.

Overview of the mapping system located at the master node is presented in Figure 3. The *off-line* process of the mapping system creates a database of link load vectors that are obtained by "flooding" randomly generated demand matrices to the network using hop count weights and the current topology. The process uses IGP-WO to calculate near-optimal link weights with respect to each of the vectors (demand matrices). These near-optimal weights are also moved to the database.

Whenever the *on-line* process of the mapping system decides that link weights need to be altered in the network, it finds 20 most similar demand vectors in terms of their *1-norm distance* from the database compared with the current

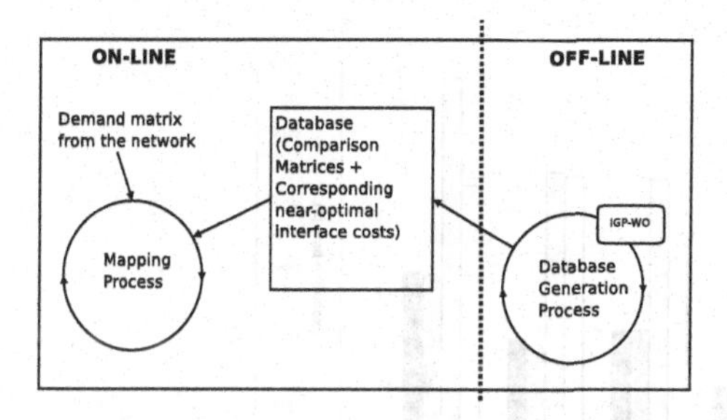

Fig. 3. Overview of the mapping system

traffic demand vector (for links) of the network. We define similarity of two matrices (say **A** and **B**) as the opposite number of difference D of **A** and **B**, which can be calculated as follows:

1. Convert matrices **A** and **B** to link load percentage vectors **a** and **b** by flooding them into the network using hop count weights.
2. Calculate the 1-norm distance between load percentage vectors **a** and **b**: $D = \sum_{i=1}^{n} |a_i - b_i|$, where n is the amount of links in the network, and i represents a link.

3 Simulation Results

We simulated the performance of our system on one PC (2.0 GHz CPU, 516 MB RAM), which contained the master and all of the slaves. We tested the quality of load balancing obtained by 1) 50 iterations of IGP-WO, 2) random weight pre-selection, and 3) the mapping system with a database of 200 matrices. Figure 4 presents the *maximal link loads* obtained with 1) hop count link weights (default), 2) random weight pre-selection, 3) IGP-WO and 4) mapping system, and the *times* it took to obtain the presented maximal link loads with help of IGP-WO and the mapping system (20 slaves located on the same PC), as well as an estimate of the running time with 20 separate slave PCs.

Our topologies contained 12, 36, 60, 84, and 108 nodes, and 15, 53, 95, 133, and 167 links respectfully. Topology 1 was the Abilene backbone [8], and topologies 2-5 were generated by combining multiple Abilene-topologies with multiple links at different locations. All results are averages over 50 test matrices, that were generated randomly from a uniform distribution with parameters $0, a$, with the exception that a certain, relatively small percentage of the demands were generated from a uniform distribution $0, b$, where $b \gg a$.

It can be seen that the random pre-selection cannot find good weight for load balancing in larger topologies. However, the mapping system is able to balance the load comparably to IGP-WO and in a much shorter time even when no parallelism is utilized.

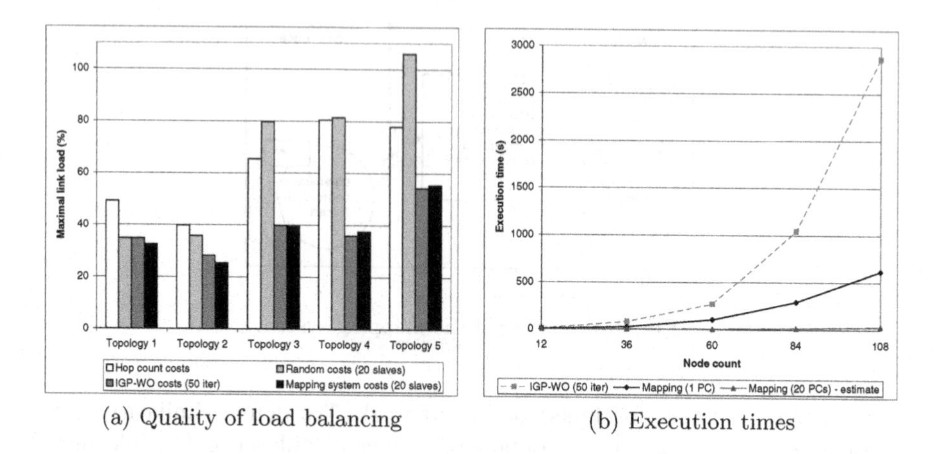

(a) Quality of load balancing

(b) Execution times

Fig. 4. Simulation results

4 Conclusions

We presented an architecture that can be used to find near-optimal IGP link weights for load balancing and compared it to an open-source optimization tool (IGP-WO). Quality of load balancing obtained by the architecture is comparable to that obtained by IGP-WO, while the time reduction is significant even in a simulation scenario where all slave nodes are located on the same PC and no parallelism is utilized.

References

1. Fortz, B., Thorup, M.: Increasing Internet Capacity Using Local Search. Optimization and Applications 29(1) (October 2004)
2. Fortz, B., Thorup, M., Rexford, J.: Traffic Engineering With Traditional IP Routing Protocols. IEEE Communications Magazine 40(1) (2002)
3. TOTEM Project, Web site (2006), `http://totem.run.montefiore.ulg.ac.be/`
4. IGP-WO, Web site (2006), `http://www.poms.ucl.ac.be/totem/`
5. Cariden MATE Framework, Web site (2006), `http://www.cariden.com/products/`
6. Opnet SP Guru, Web site (2006), `http://www.opnet.com/products/spguru/`
7. Evans, J., Gous, A.: Practical Strategies for IP Traffic Engineering and Enhancing Core network Availability, RIPE 48 EOF Tutorial (May 2004)
8. Abilene-network, Web site (2007), `http://abilene.internet2.edu/`

Internet Management Network

Jilong Wang, Miaohui Zhang, and Jiahai Yang

Network Research Center, Tsinghua University, Beijing 100084, China
wjl@cernet.edu.cn

Abstract. Currently, network management systems on Internet are lots of islands floating around. They seldom get connected with each other, especially among those of different networks. Under such situation, overlapping implementation of network management functions in the same network domain commonly exists, and end-to-end management can't be smoothly implemented across different network domain. To solve these problems, a novel model of network management architecture is proposed, by name, Internet Management Network (IMN). IMN is an overlay network of NMSes. The structure of IMN is almost the same like Internet itself. A prototype system based on SNMP is also introduced with an interesting application, multi-domain search engine of network management information.

Keywords: Network Management, Architecture.

1 Introduction

Currently, network management systems on Internet are lots of islands floating around. They seldom get connected with each other, especially among those of different networks. The isolation among NMSes prevents end-to-end management across different network domains. For example, Inter-domain troubleshooting, security events detection, QoS and traffic engineering, etc. In a single NMS, in spite of centralized or distributed system, different components are always tight coupled, scalability is very weak, and cost of maintenance and update is very high.

In this paper, we propose a novel model of network management architecture, by name, Internet Management Network (IMN). IMN is an overlay network of NMSes. In IMN world, NMSes can get connected with each other and a NMS is just like a LAN composed of loose coupled unit systems. IMN is a network world while traditional SNMP based NMS is a system world.

Given that almost all current NMSes are based on SNMP, which is the industrial standard of network management, we implemented IMN based on SNMP by adding an overlay networking mechanism to SNMP, which make it possible for IMN to be accepted by industry community.

2 Architecture Design

In the beginning we need get a picture of management domain. Just like Internet is composed of numerous ASes, IMN is made up of management domains (or domains).

S. Ata and C.S. Hong (Eds.): APNOMS 2007, LNCS 4773, pp. 599–602, 2007.

Management domain can inherit the structure of AS in Internet. Usually an AS can be taken as a management domain. The NMSes in a domain belongs to the same organization and trust each other.

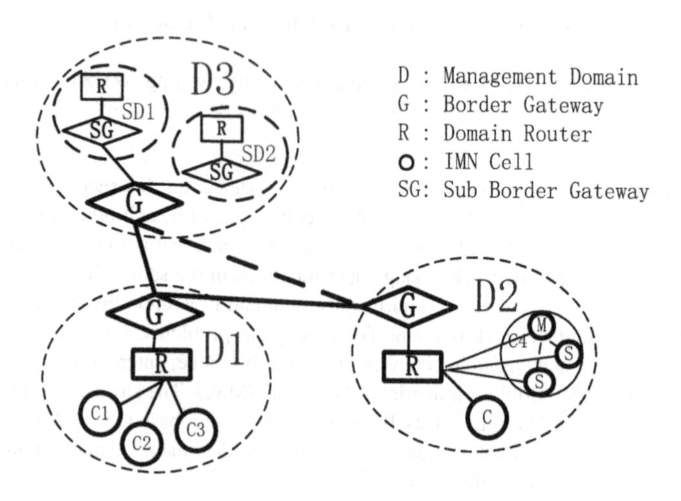

Fig. 1. IMN Architecture

IMN has 3 type of networking entity, IMN Cell (IC), Domain Router (DR) and Border Gateway (BG).

1) IC is a specific network management system. A NMS becomes an IC by implementing IMN MIB and IMN networking functions. IMN MIB plays an important role in IMN, for it prescribes a unified way to store information for local network management and an interface for interaction between NMSes. IC must register its basic system information and meta data of its resource to DR, which can publish its existence to the IMN. Once joining IMN, an IC may receive lots of requests for interaction, and it usually provides service at different levels to different ICs. IC is also able to find needed resource in the IMN.

2) DR serves for intradomain networking. Assuming that each management domain or sub management domain has one DR, It performs three functions in working. Firstly a DR manages IC system information and monitors the running of ICs in local domain. Secondly it generates and exchanges IMN routing information within domain. Thirdly it provides search service with system information and routing information. Search service produces a NMS list as an answer to given query for resource.

3) BG serves for interdomain networking. BG plays a key role in connecting different management domains into an overlay network. Every management domain has a top-level BG, while a sub management domain has a sub BG. Like BGP routers in an AS, BG manages exchanges of routing information between domains. All top-level BGs are organized in Peer-to-Peer model, which constitutes the backbone of IMN. Neighbor BGs in trust relationship exchange route information periodically. BGs can be seen as bridges that connects ICs in different management domains.

ICs are the basic components of IMN. There are four types of relationship between ICs.

1) Master/Slave relationship: It is the most intimate relationship between ICs. There are lots of integrated NMSes that have a master system and some sub-systems in large scale domains. Sub systems of an integrated NMS communicate and interact in their private manner. This weakens system's extensibility and openness. Master/Slave relationship is proposed to resolve this problem. An integrated NMS can be divided into some ICs that maintain their intrinsic function. These ICs can reconstruct the former integrated NMS. Information exchange and interaction between master and slaves is advised to go on the basis of IMN MIB. Therefore, the integrated NMS is composed of standard ICs, in which system's coupling is decreased. Components of an integrated NMS are easily put in or taken off, while every component can be reused by other systems. As Fig.1 shown, C4 in D2 is an integrated system, which has a master system and two slave systems.

2) Groupmate relationship: ICs in the same management domain have such relationship. Interaction between ICs in the same domain is of key importance to accomplish more powerful management or to avoid overlap function. ICs in the same domain are usually owned by the same organization. So the interaction request from groupmates is conducted with high priority. For example, as Fig.1 shown, three ICs in D1 have groupmate relationship.

3) Peer relationship: NMSes in different domain that trust each other is in peer relationship. Peer relationship is established by putting each other's BG as peer BG. BGs in peer relationship change route items periodically. ICs can easily find and visit resource in need from peer domain. As Fig.1 shown, D1 and D3 have peer relationship.

4) Strangers: Strangers refers to ICs in different domains that haven't peer relationship. An IC can't acquire resources in a strange IC, though the latter has the needed resources through IMN addressing mechanism.

There are four types of information communicated on IMN network.

1) Network Management Information. Every IC has two MIBs, information MIB and operation MIB. The former is used to store data for local function and to be accessed by another NMS. The latter helps to make located operation invoked by other NMSes.

2) System Information. Information expresses the basic function of an IC and provides the key to access it. It contains {IC IP, community, MIB tree list}. IP is used to identify an IC. Community is the key to access SNMP MIB of an IC to get needed information. MIB tree list shows what information and public operation an IC has, and what function it supports.

3) Control Information: Keep-alive and updated information between IMN entities. An IC sends keep-alive message to register periodically at runtime. BGs also send keep-alive message to their neighbors in P2P network. When ICs or BGs change their basic information they will release updated message.

4) Route Information: The meta data that describes the information or public operation an IC has in detail. Route information is generated in DR. DR collects announcement from ICs announce items such as {information, IC Information sub MIB tree OID, IP section}. DR generates route items from announcements by

expressing abstractly. Format of information route item is {information, BG IP, IC IP, Information MIB tree OID, IP Section}, while format of operation route item is {operation, BG IP, IC IP, Operation MIB tree OID}.

3 Routing

Routing in IMN is divided into intradomain routing and interdomain routing. It's easy to implement intradomain routing because a DR has the route information of the whole domain. It's easy to implement interdomain routing among peers because a BG has the route information of its peer domains. The question is how to locate resource in a stranger domain?

Chord [3] is a scalable P2P lookup service, which presents a valuable routing method in a P2P network. Given a key, it maps the key onto a node. Data location can be easily implemented by associating a key with each data item, and storing the key/data item pair at the node to which the key maps. In IMN, BGs can be seen as nodes of a P2P network. Every BG has an IP, and all IMN route items are based on IP. We can choose IP as the key, while route item as data. So it's very suitable for IMN to utilize Chord mechanism to implement routing among stranger domains.

4 Conclusion and Future Work

The next generation of the network management [4] should have the scalability corresponding to the Internet. Therefore, we propose a networked architecture for future Internet management. Currently, we are developing a search engine of network management information based on IMN architecture. It will give a choice to share information among ISPs in a loose-coupled way.

Reference

1. RFC 3165, Definitions of Managed Objects for the Delegation of Management Scripts, (August 2001)
2. RFC 2925, Definitions of Managed Objects for Remote Ping, Traceroute, and Lookup Operations Script MIB (September 2000)
3. Stoica, I., Morris, R., Karger, D., Kaashoek, M F., Balakrishnan, H.: Chord A Scalable Peer-to-peer Lookup Service for Internet Applications. In: Proceeding of ACM SIGCOMM 2001, San Diego, California, USA, ACM Press, New York (2001)
4. Jilong, W., Chunsheng, N., Jianping, W.: Next Generation Internet. In: ISPAN (2004)

A Hybrid Staggered Broadcasting Protocol for Popular Video Service

Yong-Hwan Shin, Seong-Min Joe, and Sung-Kwon Park

Hanyang University, 17 Haengdang-dong, Seongdong-gu,
133-791 Seoul, Korea
twolf3@hanyang.ac.kr, smjoe@korea.com, sp2996@hanyang.ac.kr

Abstract. In a popular video broadcasting method, two major technical issues are focused on both to reduce the viewer's waiting time maintaining a given bandwidth allocation and the client's buffer requirements. In this paper, we propose a pagoda staggered broadcasting (PSB) method which has a simple structure and substantially improves broadcasting efficiency in given bandwidth. The proposed method performs significantly better than the existing broadcasting methods in the viewer's waiting time and the buffer requirement with simplicity. The numerical results demonstrate that the viewer's waiting time of the PSB method is nearly equal to the existing pagoda broadcasting methods. However the complexity of this method decreases up to 60% if dividing coefficient h is 1 by adjusting the short front part of a video.

Keywords: broadcasting, video-on-demand, pagoda staggered broadcasting, simple structure, video-on-demand efficiency, dividing coefficient.

1 Introduction

With the growth of broadband networking technology, video-on-demand (VoD) services have become possible. VoD system generally denotes a platform to provide real-time multimedia delivery services. VoD allows clients to select any given video from a large on-line video library and watch it through their set-top box (STB) at any time without waiting via communication networks. For providing VoD service, Many VoD broadcasting protocols have been proposed, such as fast broadcasting 1, harmonic broadcasting 2, staircase broadcasting 3, pagoda 4. A drawback of these various broadcasting methods is that a client using these methods can not avoid managing many segments of a video, managing a frequency of channel hoping and joining many channels at the same time. Unavoidably these increase the complexity of VoD system design.

In order to overcome these problems mentioned above, this paper proposes a new VoD method entitled the pagoda staggered broadcasting (PSB) method. In the PSB method, the viewer's waiting time is nearly equal or less than that of previous methods. In addition, the complexity of this method is significantly decreased and the buffer requirement can be adjusted as dividing coefficient h. These merits make it possible to easily implement VoD service and more practical than previous proposed VoD methods.

S. Ata and C.S. Hong (Eds.): APNOMS 2007, LNCS 4773, pp. 603–607, 2007.

2 Pagoda Staggered Broadcasting (PSB) Method

The pagoda method has bandwidth efficiency in a video delivery and the staggered method has a simple system structure. The PSB method contains all these features. This method divides a video into a short front part and a long rear part. The short front part of a video is then broadcasted in the pagoda method and the long rear part of a video is broadcasted in the staggered method. This framework leads to decreasing not only viewer's waiting time and the buffer requirement of a client, but also the complexity of the VoD method.

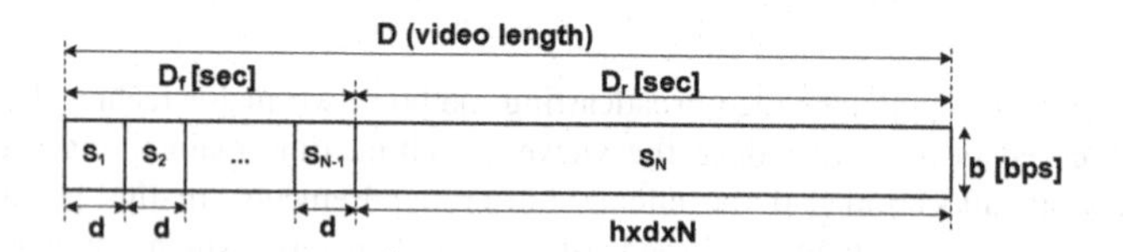

Fig. 1. The basic partition operating of the PSB method for a video when h is a video dividing coefficient, N is the number of segments and d is the length of each segment in short front part

Referring to Fig. 1, on the server side, suppose there is a video with length $D[\text{sec}]$, where $D_f[\text{sec}]$ is the front data length of the video and $D_r[\text{sec}]$ is the rear data length of the video. The consumption rate of the video is $b[bps]$. Thus, the size of the video is $V[bit] = D \times b$. Suppose the bandwidth, which can be allocated for the video, is $B[bps]$, $B = \beta \times b$, $\beta \geq 1$. On the server's respect, the PSB method involves the following steps.

1. The video length D is divided into a short length of front part D_f and a long length of rear part D_r. Thus, the size of D_f is $V_f[bit] = D_f \times b$, and the size of D_r is $V_r[bit] = D_r \times b$. The each part relation is given as
$$D_r = h(D_f + d),\ h \geq 1. \tag{1}$$
where, h is the video dividing coefficient, d is the length of each segment in D_f. Here, h is only a positive integer and is equal to the number of the assigned channels in D_r part.
2. The bandwidth B is equally divided into k logical channels, where $k = \lfloor B/b \rfloor = \lfloor \beta \rfloor$. Assume the number of the assigned channels in D_f as m ($m \geq 1$) and the number of the assigned channels in D_r as n ($n = h$, $n \geq 1$). Thus, the number of the logical channels is $k = m + n$. Let $\{C_0^{D_f}, ..., C_{m-1}^{D_f}\}$ represent the m channels and $\{C_0^{Dr}, ..., C_{n-1}^{Dr}\}$ represent the n channels, where The notation C_p^q represents the p th channel with q's part.

3. A video length D is divided into N segments, where N is the number of total segments and the video length of the front data D_f also is uniformly divided into N_f segments, where N_f is the number of the front data segments. Since the video length of the rear data D_r is one segment, the total number of segments is $N = N_f + 1$, where

$$N_f = \begin{cases} 4(5^{(m/2)-1}) - 1, \ (m = even) \\ 2(5^{\lfloor m/2 \rfloor}) - 1, \ (m = odd) \end{cases}, \quad (2)$$

referring [8]. Suppose S_i is the i th segment of the video. All the segments concatenated in the ascending order constitute the whole video.

4. Within the continuous data segments on $C_0^{D_f}$, broadcast segment S_1 periodically. For $C_i^{D_f}$ channels, $i = 1, ..., \lfloor (m-1)/2 \rfloor$, periodically broadcast on channel $C_{2i-1}^{D_f}$ the segments $\{S_w, S_{2w}, S_{w+1}, S_{2w+2}, ..., S_{\frac{3w}{2}-1}, S_{3w-2}, S_w, S_{2w+1}, S_{w+1}, S_{2w+3}, ..., S_{\frac{3w}{2}-1}, S_{3w-1}\}$ and on channel $C_{2i}^{D_f}$ the segments $\{S_{\frac{3w}{2}}, S_{3w}, S_{4w}, ..., S_{2w-1}, S_{4w-2}, S_{5w-2}, S_{\frac{3w}{2}}, S_{3w+1}, S_{4w+1}, ..., S_{2w-1}, S_{4w-1}, S_{5w-1}\}$, where $w = 2(5^{i-1})$.

5. If m is even, then periodically broadcast on the last channel C_{m-1} the segments $\{S_w, S_{w+1}, ..., S_{2w-1}\}$, where $w = 2(5^{i-1})$.

6. On $C_j^{D_r}$, where $j = 0, ..., n-1$, the D_r is broadcasted using the n channels in the staggered method, with the time of staggered channel interval T_s[sec], $T_s = D_f + d$. Therefore, the number of the allocated staggered channels is given by $n = D_r / T_s$, and also the number of the assigned channel in D_f is obtained as $m = k - h$.

At the client side, suppose there is plentiful memory space to save the data segments of the playing video. For watching a video, the following steps are involved.

1. Get the channels' assignment and the data segment size of a movie from the server. Then, wait until the earliest start time of new channel $C_0^{D_f}$.
2. Begin to download the first data segment S_1 at the earliest new start time on $C_0^{D_f}$ and then to download other data segments from $C_1^{D_f}, ..., C_{m-1}^{D_f}$ concurrently. If the play point of the first data segment is equal to the start of a staggered point, the last data segment S_N simultaneously records the ongoing staggered channel's segments. Otherwise, the last data segment S_N buffering waits till the start of the staggered point, only if buffering is necessary to the client.
3. Right after downloading the data segments, we can start to consume the video with its consumption rate in the order of $S_1, S_2, ..., S_N$.

4. Stop downloading from the channel $C_i^{D_f}$, $i=0,...,m-1$, when the client has received all data segments of the front data part from that channel. Stop downloading from channel C_j^{Dr}, $j=0,...,n-1$, when the client has completely received the S_N data segment from that channel.

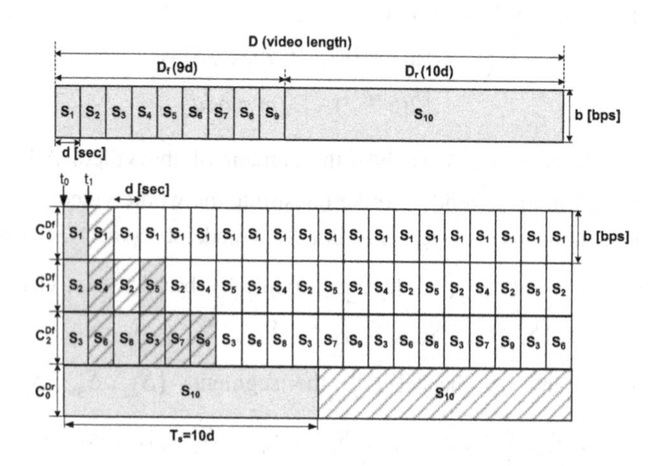

Fig. 2. Example of the PSB method when k=4, β=4, h=1, m=3, and n=1 channels

Fig. 2 shows the operation of the PSB method for a video with k=4, β=4, h=1, m=3 and n=1 channels. This picture describes the server side broadcasting operation. The video length D is divided into D_f and D_r by the coefficient $h=1$, $D_r=(D_f+d)$. The D_f is partitioned into nine segments and these segments are broadcasted over three channels ($C_0^{D_f}$, $C_1^{D_f}$, and $C_2^{D_f}$). The D_r is also broadcasted using one channel with the staggered manner. The time of staggered interval T_s is $10d$, where d is the segment length in D_f. In this example, segment length d is $D/19$, which is also the maximum viewer's waiting time. In the same case, for the pagoda method the maximum viewer's waiting time is $D/19$, for the fast and staircase method $D/15$ and for the harmonic method $D/30$. For the more, the number of segments of the video is obtained as 10 for the PSB method, 19 for the PB method, 15 for the fast broadcasting method, 85 for the staircase method, 465 for the harmonic method, respectively.

References

1. Juhn, L.-S., Tseng, L.-M.: Fast data broadcasting and receiving scheme for popular video service. IEEE Transactions on Broadcasting 44, 100–105 (1998)
2. Juhn, L.S., Tseng, L.M.: Harmonic broadcasting for video-on-demand service. IEEE Trans. Broadcasting 43(3), 268–271 (1997)

3. Juhn, L.S., Tseng, L.M.: Staircase data broadcasting and receiving scheme for hot video service. IEEE Trans. Consumer Electron. 43(4), 1110–1117 (1997)
4. Paris, J.F., Carter, S.W., Long, D.D.E.: A hybrid broadcasting protocol for video on demand. In: Proc. 1999 Multimedia Computing and Networking Conf, pp. 317–326 (1999)
5. Lee, S.W., Seo, K.J., Park, S.K.: Improving Channel Efficiency for Popular Video Service Using Dynamic Channel Broadcasting. IEICE Transaction on Communications E87-B(10), 3068–3075 (2004)
6. Lee, S.W., Park, S.K.: Hybrid Video-on-Demand System Using Dynamic Channel Allocation Architecture. IEICE Transaction on Communications E88-B(7), 3036–3045 (2005)

Efficient Congestion Control Based on Awareness of Multistage Resources (CC-AMR)*

Jijun Cao, Xiangquan Shi, Chunqing Wu, Jinshu Su, and Zhaowei Meng

School of Computer, National University of Defense Technology, Changsha 410073, Hunan, China
caojijun0@sina.com.cn

Abstract. Congestion control is an important part of IP QoS. The traditional congestion control algorithms judge the network congestion status and make drop decisions mainly according to the congestion information of local buffer resources. This causes the bandwidth-wasting problem when data flow congested. The paper analyzes the problem theoretically and proposes a new congestion control algorithm (CC-AMR) based on awareness of the congestion status of multistage resources. CC-AMR accounts the congestion status of resources in remote forward engines and their ports synthetically so that more reasonable congestion control decisions can be made. The CC-AMR has been implemented and tested on a core router which is implemented basing on network processor. Results show that this algorithm can enhance the total throughput of router effectively during periods of congestion.

1 Introduction

The typical router congestion control algorithms are Tail-Drops, RED, WFQ, and so on. Tail-Drops algorithm used buffer to buffer packets in FIFO mode. While in AQM [1] algorithms, packets are dropped (or marked) before the buffer overflow occurs. The most representative AQM algorithm is RED [2]. Aiming at overcoming some shortcomings of the RED, many researchers have put forward many improved schemes and algorithms [3, 4]. FQ algorithm maintains many independent queues for each data flow. WFQ algorithm is an extension of FQ algorithm [5]. It allocates the resources weighted fairly for each queue according to the bandwidth requirement.

In essence, all the algorithms mentioned above are to manage the shared buffer resources in routers. They make drop decisions just according to the congestion information of local buffer, that is to say it does not take the congestion status of multistage resources into account synthetically. This will likely cause the bandwidth-wasting problem. This paper proposes a new scheme and algorithm for congestion control, which is based on awareness of multistage resources.

---

* This research was supported by the National Basic Research Program of China (973) under Grant No. 2003CB314802 and the National Natural Science Foundation of China under Grant No. 90604006.

S. Ata and C.S. Hong (Eds.): APNOMS 2007, LNCS 4773, pp. 608–611, 2007.

2 Problem Definition

2.1 Multistage Independent Congestion Control

The current popular routers are mostly adopting switch-fabric and multiprocessor based architecture at large. In the router, there are four stages of queues so four stages of congestion control schemes (scheme-1 to scheme-4) are performed in router correspondingly. They are as follows:

(1) Congestion control scheme-1: When packets arrive at the ingress of NP from network, the congestion control scheme-1 will manage *WFWQ* (Wait ForWarding Queue) resources.

(2) Congestion control scheme-2: When the packets that have been processed in ingress of NP arrive at the queue for target NP, the congestion control scheme-2 will manage *WSQ* (Wait Switching Queue) resources.

(3) Congestion control scheme-3: When the packets that have been switched to the egress of NP arrive at the egress of NP through the switch fabric, the congestion control scheme-3 will manage the *WOPQ* (Wait Output Processing Queue) resources.

(4) Congestion control scheme-4: When the packets that wait for output arrive at the target port queues, the congestion control scheme-4 will manage the *WOQ* (Wait Output Queue) resources.

2.2 Analysis of Bandwidth-Wasting Problem

We model the multistage independent congestion control as follows (see Fig. 1). In this simple model, we assume that the two stages of queue resources in the system are *Qm* and *Qn* (include *Qn1* and *Qn2*), respectively. And as a result, this system has two stages of congestion control schemes, which are *G(m)* and *G(n)*. Both *G(m)* and *G(n)* are using RED algorithm. The *s1* and *s2* are respectively the data sources of Flow1 and Flow2. The sending bandwidth of Flow1 and Flow2 are respectively T1 and T2. The *d1* and *d2* are the destinations of Flow1 and Flow2, respectively. The receiving bandwidth of Flow1 and Flow2 are respectively *R1* and *R2*. Suppose the maximal supported bandwidth of *Qm* and *Qn* are respectively *Bm* and *Bn*.

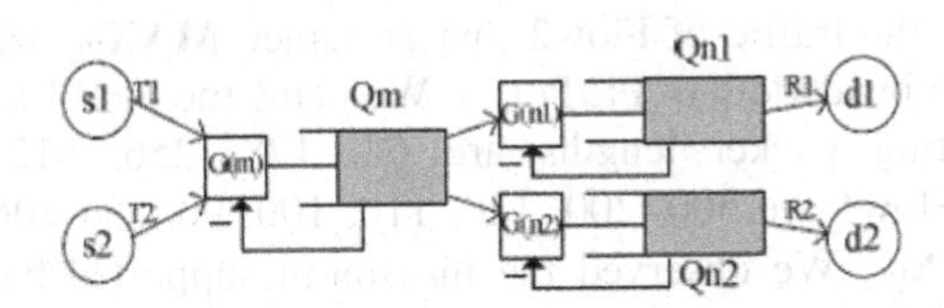

Fig. 1. The model for analysis of bandwidth-wasting problem

In the model, let us suppose $T1>Bn$ and $T2\leq Bn$, so the system is congested. The *R2* and the sum of *R1* and *R2* are summarized in Table 1 in this condition.

We can conclude from Table 1 that (1) when both *Qm* and *Qn1* are congested, with the increase of *T1*, *R2* will decrease and the *R1* will hold the line; (2) when *Qm* is non-congested and *Qn1* is congested, both *R1* and *R2* will hold the line. So in the

Table 1. The bandwidth of receiving data flow in system

Status of *Qm*	Status of *Qn1*	*R2*	*R1+R2*
congested	congested	$Bm{\cdot}T2/(T1+T2)$	$Bm{\cdot}T2/(T1+T2)+Bn$
congested	non-congested	$Bm{\cdot}T2/(T1+T2)$	Bm
non-congested	congested	$T2$	$T2+Bn$
non-congested	non-congested	$T2$	$T2+Bm{\cdot}T1/(T1+T2)$

model, when both *Qm* and *Qn1* are congested, with the increase of *T1*, the *R2* and *R1+R2* will decrease. We can see the congested Flow1 wastes the bandwidth. The bandwidth-wasting problem caused by congested data flow always occurs in router system when many data flows share multistage resources. The method to solve this problem is to be aware of congestion and do drop decision earlier.

3 CC-AMR Scheme and Algorithm

To overcome the shortcomings of multistage independent Congestion Control, we put forward the new scheme. Compared with the congestion control scheme-2 of multistage independent congestion control, this scheme uses the congestion information not only of *WSQ* but also of *OPQ* and *WOQ* to make drop decisions.

We define 128 transmit probability tables to represent different congestion grades. Every table has 64 entries, and each entry is corresponding to the drop probability of different packet types. The transmit probability table is indexed by *QQQTCC*, of which *QQQ,T* and *CC* respectively represent the type of QoS in DSCP (DiffServ Code Point), the congestion information of *WOPQ*, and the drop priority in DSCP.

4 Implementations and Experimental Evaluation

We implemented the CC-AMR scheme and algorithm in our router whose main components are PowerNP 4GS3s. To evaluate the performance, the AX/4000 instrument generates two data flows (Flow1 and Flow2). The data flows outflow the router at FE/Port 1 and FE/Port 3. The traffic of Flow1 exceeds the receiving ability of FE/Port 1 while the traffic of Flow2 that is named $\mathbf{MAX_{PL}}$ exactly equals to the maximal processing bandwidth of FE/Port 3. We name the Flow1 *background flow*.

Firstly, the testing packet lengths are 64, 128, 256, 512 and 1500 bytes respectively. The Flow1 are 500, 200, 120, 110, 100, 90, and 80Mbps respectively. The Flow2 is $\mathbf{MAX_{PL}}$. We observed the maximum supported bandwidth of Flow2 when the Flow1 changes. We compared the CC-AMR with the unmodified SARED at different lengths of testing packets.

We also recorded the total supported bandwidth of Flow1 and Flow2. Fig.2 shows the compared results. When Flow1 is 200Mbps and Flow2 is $\mathbf{MAX_{PL}}$, We observed the total supported bandwidth of Flow1 and Fow2 at various length of testing packets. From the five observing points, the average supported bandwidth of SARED is 71.12% of that of CC-AMR, that is to say, the average supported bandwidth is increased by CC-AMR algorithm for about 40.60% than SARED.

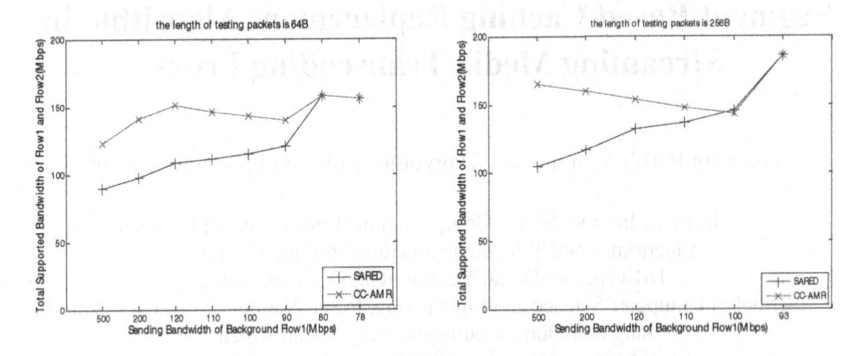

Fig. 2. Total supported bandwidth of Flow1 and Flow2 for different background Flow1

We choose forwarding delay as the criterion to evaluate the overhead of CC-AMR algorithm. The experimental results show that the delays of SARED and CC-AMR are very approximate to each other.

5 Conclusions and Future Works

The CC-AMR algorithm accounts the congestion status of multistage resources synthetically. Utilizing the characteristics of our router that is based on IBM NP4GS3 we implemented the CC-AMR algorithm. The testing result shows this algorithm enhances the throughput of routers by large margin with low overhead.

References

1. Braden, et al.: Recommendations on Queue Management and Congestion Avoidance in the Internet. RFC 2309 (April 1994)
2. Floyd, S., Van Jacobson: Random Early Detection Gateways for Congestion Control. IEEE/ACM Transactions on Networking 1(4), 397–413 (1993)
3. Cisco System. Distributed weighted random early detection (2005), available http://cco.cisco.com
4. Ott, T.J., Lakshman, T.V., Wong, L.H.: SRED:Stabilized RED. In: Proceedings of IEEE INFOCOM, IEEE Computer Society Press, Los Alamitos (1999)
5. Demers, A., Keshav, S., Shenker, S.: Analysis and Simulation of a Fair Queuing Algorithm. In: Proceedings of ACM SIGCOMM, Austin, TX, pp. 1–12. ACM Press, New York (1989)

Segment Based Caching Replacement Algorithm in Streaming Media Transcoding Proxy

Yoohyun Park[1], Yongju Lee[1] Hagyoung Kim[1], and Kyongsok Kim[2]

[1] Dept. of Internet Server Group, Digital Home Research Division,
Electronics and Telecommunications Research Institute,
161 Gajeong-Dong, Yuseong-Gu, Daejeon, Korea
[2] School of Computer Science and Engineering, Pusan National University, San-30,
Jangjeon-Dong, Geumjeong-Gu, Busan, Korea
{bakyh,yongju,h0kim}etri.re.kr,
gimgs@asadal.cs.pusan.ac.kr

Abstract. To support various devices for multimedia services, it needs to some adaptation technologies. A transcoding is one of major fields in adaptive multimedia delivery. A transcoding proxy is usually necessary to provide adapting multimedia object to various mobile devices by not only transcoding multimedia objects to meet different needs on demand, but also caching them for later use. In this paper, we propose a caching algorithm in transcoding proxy that can reduce the startup time, improve caching ratio with segment-based proxy management. Our simulation results demonstrate that the proposed algorithms outperform the competing algorithms.

Keywords: Transcoding; Proxy Caching; Multimedia Caching, Segment based caching.

1 Introduction

Recent technology advances in mobile networking have guided in a new era of personal communication. Users can access the Internet via various mobile devices such as PDAs, cellular phones and laptop computers. To support these devices for multimedia services, it is need to some adaptation technologies. A transcoding is one of major fields in adaptive multimedia delivery.

Transcoding is defined as a transformation that is used to convert a multimedia object from one form to another. From the aspect of the place where transcoding is performed, the transcoding can be classified into server-based, client-based and proxy-based. For many reasons, proxy-based transcoding technology will be better to transcoding the multimedia object[1].

Proxy caching has been widely used to cache static objects on the Internet so that subsequence requests to the same objects can be served directly from the proxy without contacting the original content server. However, the proliferation of multimedia content makes caching challenging due to the typical large size and the low latency and continuous streaming demands of media objects[2].

S. Ata and C.S. Hong (Eds.): APNOMS 2007, LNCS 4773, pp. 612–615, 2007.

To solve the problems caused by large size media objects, researchers have developed a number of segment-based proxy caching strategies that cache partial segments of media objects instead of entire media objects[2]. And currently, most cache architectures are designed for end users with similar network links and device profiles. So, for the various clients, it needs transcoding mechanism and management various versions of same content.

So, in this paper, we propose the considering the aggregate effect of caching multiple versions of the same content with segment-based proxy caching.

2 Segment-Based Cache Replacement Algorithm for Streaming Media in Transcoding Proxy

The *Segment base Profit Function*(SPF) generates a value that is used for find a victim object at replacement procedure. The SPF is composed of two factors – delay time saving and reduction in the number of bytes transmitted. Initially, the individual profit function is derived to estimate the profit from caching a single version of a video object(definition 1). Then, the aggregate profit function is derived to estimate the profit when caching multiple versions of a video object simultaneously(definition 2). Finally, depending on the aggregate profit function, the complete profit function is formulated for estimating the profit from caching a version of a video object when other versions of the object are already cached(definition 3). We use the symbols $o_{n,v}$ for the version v of content n, $r_{n,v}$ for access count of $o_{n,v}$, $TRL(v, v')$ for the transcoding length from version v to v', $R_{n,v}$ for the bitrate of $o_{n,v}$, $d_{n,v0}$ for the fetch delay of $o_{n,v0}$, $w_n(v,v')$ for the transcoding weight value from version v to v', and $S_{n,v}$ for the cached size of $o_{n,v}$.

Definition 1 : SPF($o_{n,v}$) is a function used to calculate the individual profit for caching $S_{n,v}$ size of $o_{n,v}$, when no other version of object n is cached. $SPF(o_{n,v}) =$

$$\frac{\sum_{(v,v')\in E(G_n)} r_{n,v'} \times TRL(v,v') \times R_{n,v'} \times (d_{n,v_0} + w_n(v_0,v') - w_n(v,v'))}{S_{n,v}}$$

Definition 2 : SPF($o_{n,m1}, o_{n,m2}, \ldots, o_{n,mk}$) is a function used to evaluate the aggregate profit for simultaneously caching $o_{n,m1}, o_{n,m2}, \ldots, o_{n,mk}$, where G_n' is the subgraph obtained by the MATC procedure1.

$$SPF(o_{n,m1}, o_{n,m2}, \ldots, o_{n,mk}) = \sum_{v\in V(G_n')} \frac{\sum_{(v,v')\in E(G_n')} r_{n,v'} \times TRL(v,v') \times R_{n,v'} \times (d_{n,v_0} + w_n(v_0,v') - w_n(v,v'))}{S_{n,v}}$$

Definition 3 : SPF($o_{n,m} | o_{n,m1}, o_{n,m2}, \ldots, o_{n,mk}$) is a function used to calculate the individual profit for caching $S_{n,v}$ size of $o_{n,v}$, when $o_{n,m1}, o_{n,m2}, \ldots, o_{n,mk}$ are already cached, where m $\neq$ m1, m2, …, mk.

$$SPF(o_{n,m} \mid o_{n,m1}, o_{n,m2},\ldots,o_{n,mk})$$

$$=SPF(o_{n,m}, o_{n,m1}, o_{n,m2},\ldots,o_{n,mk,}) - SPF(o_{n,m1}, o_{n,m2},\ldots,o_{n,mk,})$$

Based on the segment based profit function(SPF), cache-replacement algorithm is designed. The replacement is happened when the new segment is needed at transcoding or fetching a content from original server. In other transcoding algorithms such as FVO[5], TVO[5], TeC[6], and transcoding graph based methods, they reserve a space area with total size of a version or average duration size in advance at saving the first parts of the content. And when the proxy space has no more space, the proxy selects some contents for victim and fully dumps them. They are simple methods, but those methods have more chances to miss requested data. In our proposed segment based algorithm, it reserves a segment at new data caching by a small size unit and when replacement needs, it dumps the victim content by segment size unit. So, more different data can remain in transcoding proxy than other methods.

3 Simulation Results

We implemented a simulator for performance analysis. In the client model, client devices can be partitioned into 5 classes (15%, 20%, 30%, 20%, 15%). That is, each data item can be transcoded to 5 different versions by the transcoding proxy to satisfy the users' requirements. The bitrate of the 5 versions of each media object are assumed to be 512, 256, 128, 64 and 32 kbps. The transcoding delay for the first sufficient segments of version from i to j is determined to $(j - i) * 500$ ms. The popularity of the video object follow a Zipf distribution with a skew factor α of 0.47. And we use 500 CBR video clips, whose lenghs are uniformly distributed (30 sec ~ 12 min). The simulation lasts 400 simulation hours with 100,000 accesses that follow a Poisson distribution. The delays for fetching the first several segments of video objects from the original server are exponentially distributed (μ= 1.5) and the ratio of the access duration to the total duration of a video sequence in a partial viewing environment is random distributed. The cache capacity is assumed to be (0.3 ~ 0.9) * (Σ 128 kbps bit-rate object size).

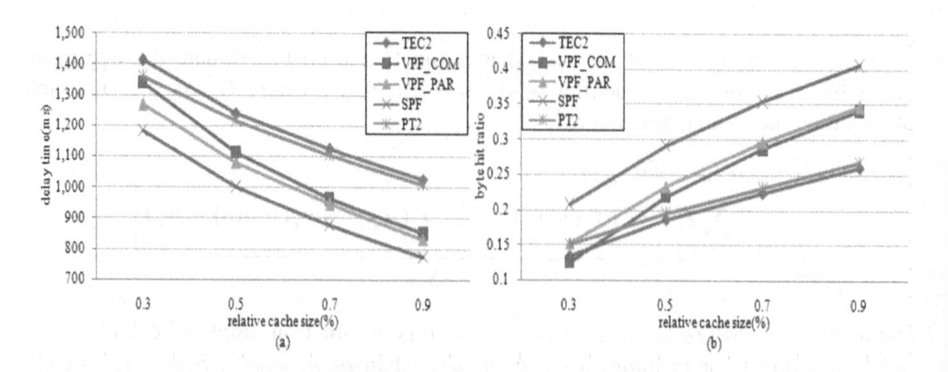

Fig. 1. Startup delay time (a) and byte-hit ratio (b) under various cache capacities

In our simulation, the variation of the performance of the proposed cache-replacement algorithm with cache capacity is investigated. Fig. 1 plot the results of the evaluations of SPF and other policies such as TeC2[6], PT2[4], VPF-complete[3] and VPF-partial[3]. It shows the startup delay time and byte-hit ratio as a function of relative cache size. Byte hit ratio is defined as the number of bytes served from the cache to clients over the number of bytes requested from the clients. Specially, TeC system has worst performance in delay time, because it is based on LRU algorithm ignore the size and the popularity of the objects as well as the delay.

4 Conclusions

Recent technology advances in mobile networking have guided in a new era of personal communication. Users can access the Internet via various mobile devices. To support these devices for multimedia services, it is need to some adaptation technologies. A transcoding is one of major fields in adaptive multimedia delivery. A transcoding proxy is usually necessary to provide adapting multimedia object to various mobile devices by not only transcoding multimedia objects to meet different needs on demand, but also caching them for later use. In this paper, we proposed the segment-based caching algorithm for efficiently caching multiple versions of the same multimedia object. We proposed a segment based profit function(SPF) using aggregate profit values in streaming media transcoding proxy. The simulation results demonstrate that the proposed algorithm outperforms the competing algorithms in delay time, byte hit ratio.

References

1. Chang, C.-Y., Chen, M.-S.: On Exploring Aggregate Effect for Efficient Cache Replacement in Transcoding Proxies. IEEE Transactions on Parallel and Distributed Systems 14(6) (2003)
2. Chen, S., Shen, B., Wee, S., Zhang, X.: Segment-Based Streaming Media Proxy: Modeling and Optimization. IEEE Transactions on Multimedia 8(2) (2006)
3. Kao, C.-F., Lee, C.-N.: Aggregate Profit-Based Caching Replacement Algorithms for Streaming Media Transcoding Proxy Systems. IEEE Transactions on Multimedia 9(2) (2007)
4. Lee, Y., Bak, Y., Min, O., Kim, H., Lee, C.: The PT-2 Caching Algorithm in the Transcoding Proxy Cluster to Facilitate Adaptive Content Delivery. In: International Workshop on Multimedia Content Analysis and Mining 2007(MCAM 2001), WeiHai, China (2007)
5. Tang, X., Zhang, F., Chanson, S.T.: Streaming Media Caching Algorithms for Transcoding Proxies. In: ICPP 2002 (2002)
6. Shen, B., Lee, S.-J., Basu, S.: Caching Strategies in Transcoding-Enabled Proxy Systems for Streaming Media Distribution Networks. IEEE Transactions on Multimedia 6(2) (2004)

Author Index

Lecture Notes in Computer Science

Sublibrary 5: Computer Communication Networks and Telecommunications

For information about Vols. 1–4396
please contact your bookseller or Springer

Vol. 4773: S. Ata, C.S. Hong (Eds.), Managing Next Generation Networks and Services. XIX, 619 pages. 2007.

Vol. 4745: E. Gaudin, E. Najm, R. Reed (Eds.), SDL 2007: Design for Dependable Systems. XII, 289 pages. 2007.

Vol. 4725: D. Hutchison, R.H. Katz (Eds.), Self-Organizing Systems. XI, 295 pages. 2007.

Vol. 4712: Y. Koucheryavy, J. Harju, A. Sayenko (Eds.), Next Generation Teletraffic and Wired/Wireless Advanced Networking. XV, 482 pages. 2007.

Vol. 4686: E. Kranakis, J. Opatrny (Eds.), Ad-Hoc, Mobile, and Wireless Networks. X, 285 pages. 2007.

Vol. 4685: D.J. Veit, J. Altmann (Eds.), Grid Economics and Business Models. XII, 201 pages. 2007.

Vol. 4581: A. Petrenko, M. Veanes, J. Tretmans, W. Grieskamp (Eds.), Testing of Software and Communicating Systems. XII, 379 pages. 2007.

Vol. 4572: F. Stajano, C. Meadows, S. Capkun, T. Moore (Eds.), Security and Privacy in Ad-hoc and Sensor Networks. X, 247 pages. 2007.

Vol. 4549: J. Aspnes, C. Scheideler, A. Arora, S. Madden (Eds.), Distributed Computing in Sensor Systems. XIII, 417 pages. 2007.

Vol. 4543: A.K. Bandara, M. Burgess (Eds.), Inter-Domain Management. XII, 237 pages. 2007.

Vol. 4534: I. Tomkos, F. Neri, J. Solé Pareta, X. Masip Bruin, S. Sánchez Lopez (Eds.), Optical Network Design and Modeling. XI, 460 pages. 2007.

Vol. 4517: F. Boavida, E. Monteiro, S. Mascolo, Y. Koucheryavy (Eds.), Wired/Wireless Internet Communications. XIV, 382 pages. 2007.

Vol. 4516: L. Mason, T. Drwiega, J. Yan (Eds.), Managing Traffic Performance in Converged Networks. XXIII, 1191 pages. 2007.

Vol. 4503: E. Airoldi, D.M. Blei, S.E. Fienberg, A. Goldenberg, E.P. Xing, A.X. Zheng (Eds.), Statistical Network Analysis: Models, Issues, and New Directions. VIII, 197 pages. 2007.

Vol. 4479: I.F. Akyildiz, R. Sivakumar, E. Ekici, J.C.d. Oliveira, J. McNair (Eds.), NETWORKING 2007. Ad Hoc and Sensor Networks, Wireless Networks, Next Generation Internet. XXVII, 1252 pages. 2007.

Vol. 4465: T. Chahed, B. Tuffin (Eds.), Network Control and Optimization. XIII, 305 pages. 2007.

Vol. 4427: S. Uhlig, K. Papagiannaki, O. Bonaventure (Eds.), Passive and Active Network Measurement. XI, 274 pages. 2007.

Vol. 4396: J. García-Vidal, L. Cerdà-Alabern (Eds.), Wireless Systems and Mobility in Next Generation Internet. IX, 271 pages. 2007.

Vol. 4373: K.G. Langendoen, T. Voigt (Eds.), Wireless Sensor Networks. XIII, 358 pages. 2007.

Vol. 4357: L. Buttyán, V.D. Gligor, D. Westhoff (Eds.), Security and Privacy in Ad-Hoc and Sensor Networks. X, 193 pages. 2006.

Vol. 4347: J. López (Ed.), Critical Information Infrastructures Security. X, 286 pages. 2006.

Vol. 4325: J. Cao, I. Stojmenovic, X. Jia, S.K. Das (Eds.), Mobile Ad-hoc and Sensor Networks. XIX, 887 pages. 2006.

Vol. 4320: R. Gotzhein, R. Reed (Eds.), System Analysis and Modeling: Language Profiles. X, 229 pages. 2006.

Vol. 4311: K. Cho, P. Jacquet (Eds.), Technologies for Advanced Heterogeneous Networks II. XI, 253 pages. 2006.

Vol. 4272: P. Havinga, M. Lijding, N. Meratnia, M. Wegdam (Eds.), Smart Sensing and Context. XI, 267 pages. 2006.

Vol. 4269: R. State, S. van der Meer, D. O'Sullivan, T. Pfeifer (Eds.), Large Scale Management of Distributed Systems. XIII, 282 pages. 2006.

Vol. 4268: G. Parr, D. Malone, M. Ó Foghlú (Eds.), Autonomic Principles of IP Operations and Management. XIII, 237 pages. 2006.

Vol. 4267: A. Helmy, B. Jennings, L. Murphy, T. Pfeifer (Eds.), Autonomic Management of Mobile Multimedia Services. XIII, 257 pages. 2006.

Vol. 4240: S.E. Nikoletseas, J.D.P. Rolim (Eds.), Algorithmic Aspects of Wireless Sensor Networks. X, 217 pages. 2006.

Vol. 4238: Y.-T. Kim, M. Takano (Eds.), Management of Convergence Networks and Services. XVIII, 605 pages. 2006.

Vol. 4235: T. Erlebach (Ed.), Combinatorial and Algorithmic Aspects of Networking. VIII, 135 pages. 2006.

Vol. 4217: P. Cuenca, L. Orozco-Barbosa (Eds.), Personal Wireless Communications. XV, 532 pages. 2006.

Vol. 4195: D. Gaiti, G. Pujolle, E.S. Al-Shaer, K. Calvert, S. Dobson, G. Leduc, O. Martikainen (Eds.), Autonomic Networking. IX, 316 pages. 2006.

Vol. 4124: H. de Meer, J.P.G. Sterbenz (Eds.), Self-Organizing Systems. XIV, 261 pages. 2006.

Vol. 4104: T. Kunz, S.S. Ravi (Eds.), Ad-Hoc, Mobile, and Wireless Networks. XII, 474 pages. 2006.

Vol. 4074: M. Burmester, A. Yasinsac (Eds.), Secure Mobile Ad-hoc Networks and Sensors. X, 193 pages. 2006.

Vol. 4033: B. Stiller, P. Reichl, B. Tuffin (Eds.), Performability Has its Price. X, 103 pages. 2006.

Vol. 4026: P.B. Gibbons, T. Abdelzaher, J. Aspnes, R. Rao (Eds.), Distributed Computing in Sensor Systems. XIV, 566 pages. 2006.

Vol. 4003: Y. Koucheryavy, J. Harju, V.B. Iversen (Eds.), Next Generation Teletraffic and Wired/Wireless Advanced Networking. XVI, 582 pages. 2006.

Vol. 3996: A. Keller, J.-P. Martin-Flatin (Eds.), Self-Managed Networks, Systems, and Services. X, 185 pages. 2006.

Vol. 3976: F. Boavida, T. Plagemann, B. Stiller, C. Westphal, E. Monteiro (Eds.), NETWORKING 2006. Networking Technologies, Services, and Protocols; Performance of Computer and Communication Networks; Mobile and Wireless Communications Systems. XXVI, 1276 pages. 2006.

Vol. 3970: T. Braun, G. Carle, S. Fahmy, Y. Koucheryavy (Eds.), Wired/Wireless Internet Communications. XIV, 350 pages. 2006.

Vol. 3964: M.Ü. Uyar, A.Y. Duale, M.A. Fecko (Eds.), Testing of Communicating Systems. XI, 373 pages. 2006.

Vol. 3961: I. Chong, K. Kawahara (Eds.), Information Networking. XV, 998 pages. 2006.

Vol. 3912: G.J. Minden, K.L. Calvert, M. Solarski, M. Yamamoto (Eds.), Active Networks. VIII, 217 pages. 2007.

Vol. 3883: M. Cesana, L. Fratta (Eds.), Wireless Systems and Network Architectures in Next Generation Internet. IX, 281 pages. 2006.

Vol. 3868: K. Römer, H. Karl, F. Mattern (Eds.), Wireless Sensor Networks. XI, 342 pages. 2006.

Vol. 3854: I. Stavrakakis, M. Smirnov (Eds.), Autonomic Communication. XIII, 303 pages. 2006.

Vol. 3813: R. Molva, G. Tsudik, D. Westhoff (Eds.), Security and Privacy in Ad-hoc and Sensor Networks. VIII, 219 pages. 2005.

Vol. 3462: R. Boutaba, K.C. Almeroth, R. Puigjaner, S. Shen, J.P. Black (Eds.), NETWORKING 2005. XXX, 1483 pages. 2005.